D1534721

SMOOTH MUSCLE

SMOOTH MUSCLE

Edited by

Edith Bülbring
University of Oxford

Alison F. Brading
University of Oxford

Allan W. Jones
University of Pennsylvania

Tadao Tomita
Kyushu University

BALTIMORE
The Williams & Wilkins Company
1970

MOUNT UNION COLLEGE
LIBRARY

146089

591.8
B933

© E. Bülbring, A. F. Brading, A. W. Jones, T. Tomita 1970

First published 1970

Distributed in the U.S.A. by
The Williams & Wilkins Co.

First published in Great Britain by
Edward Arnold (Publishers) Ltd.

Printed in Great Britain by
William Clowes and Sons, Limited, London and Beccles

PREFACE

P-13-71- Publ. $29.75

The chapters in this book have all been written by people who, at one stage in their career, have worked under Professor Edith Bülbring in the Department of Pharmacology, Oxford. As assistant editors, and authors of chapters, we have the pleasant task of expressing the appreciation and gratitude of the authors for the opportunity of working under her guidance. Our understanding and interest in smooth muscle behaviour has been extended by her enormous experience and enthusiasm for the subject. The present addresses of the authors will show how far afield this influence has spread. There are few groups now engaged in smooth muscle research that have not had some connection with the Oxford laboratory. During the 24th International Congress of Physiological Sciences in Washington, D.C., 28th August, 1968, a 'Smooth Muscle' luncheon was held at which Professor Bülbring was honoured as a distinguished pioneer in the physiology of smooth muscle. A message of greetings, signed by 49 investigators, was sent expressing gratitude for her contributions.

In addition to paying tribute to an outstanding figure in smooth muscle research, a second objective of this book is to present a comprehensive account of the current activity in this field. The range of topics reflects the scope of Professor Bülbring's interests. Although much progress has been made in recent years, the diversity of approaches and theories currently applied underlines the opportunities available for future development. It is therefore hoped that this volume will serve as a useful review for workers in the field as well as a stimulus for future investigators.

ALISON F. BRADING
ALLAN W. JONES
TADAO TOMITA

ACKNOWLEDGEMENTS

We would like to thank Mrs Carole Trinder for her help with the bibliography and Dr T. B. Bolton for his help with the index. We are also indebted to Dr T. B. Bolton, Mrs Nesta Dean and Dr H. Ohashi for helping with the correction of proofs.

CONTRIBUTORS

EDITH BÜLBRING, University Department of Pharmacology, South Parks Road, Oxford, England.

Y. ABE, Department of Physiology, Faculty of Medicine, Kyushu University, Fukuoka, Japan.

J. AXELSSON, Department of Physiology, University of Iceland, Reykjavik, Iceland.

G. V. R. BORN, Department of Pharmacology, Royal College of Surgeons, Lincoln's Inn Fields, London, England.

ALISON F. BRADING, University Department of Pharmacology, South Parks Road, Oxford, England.

G. BURNSTOCK, Department of Zoology, University of Melbourne, Parkville, Victoria, Australia.

G. CAMPBELL, Department of Zoology, University of Melbourne, Parkville, Victoria, Australia.

R. CASTEELS, Institute of Physiology, University of Louvain, Dekenstraat, Louvain, Belgium.

A. CREMA, Pharmacological Institute, University of Pisa, Via Roma, Pisa, Italy.

M. D. GERSHON, Department of Anatomy, Cornell University Medical College, York Avenue, New York, U.S.A.

K. GOLENHOFEN, Institute of Physiology, University of Marburg, Deutschhausstrasse, Marburg, W. Germany.

P. J. GOODFORD, Laboratory of Biophysics and Biochemistry, The Wellcome Research Laboratories, Langley Court, Beckenham, Kent, England.

MOLLIE E. HOLMAN, Department of Physiology, Monash University, Clayton, Victoria, Australia.

A. W. JONES, Bockus Research Institute, University of Pennsylvania, 19th & Lombard Street, Philadelphia, U.S.A.

S. R. KOTTEGODA, Department of Pharmacology, Faculty of Medicine, University of Ceylon, Kynsey Road, Colombo, Ceylon.

H. KURIYAMA, Department of Physiology, Faculty of Dentistry, Kyushu University, Fukuoka, Japan.

C. Y. LEE, Department of Pharmacology, College of Medicine, National Taiwan University, Taipei, Taiwan, Republic of China.

H. LÜLLMANN, Department of Pharmacology, Christian-Albrechts University, Hospitalstrasse, Kiel, W. Germany.

J. SETEKLEIV, Institute of Pharmacology, University of Oslo, Blindern, Oslo, Norway.

R. N. SPEDEN, Department of Physiology, University of Tasmania, Hobart, Tasmania, Australia.

T. TOMITA, Department of Physiology, Faculty of Medicine, Kyushu University, Fukuoka, Japan.

CONTENTS

INTRODUCTION

EDITH BÜLBRING

The decision to embark on an investigation of the properties of smooth muscle arose for me, a pharmacologist, from the wish to understand the mode of action of drugs. For many years I had applied drugs to isolated smooth muscle preparations, observing contraction or relaxation and measuring the responses in biological assays.

'The simplest method (Magnus's method) of studying the action of drugs on plain muscle is to remove the muscle from the body and hang it up in salt solution, so that the lower end is fixed and the top end is attached to a lever which writes on a drum' (Gaddum, 1940). Many fundamental discoveries have been made with this method and many more will be made in the future. However, this method is 'simple' only in the hands of the experienced worker who is familiar with the rules, which have been established empirically over many years, concerning the precautions which have to be taken to obtain the most suitable conditions.

It is important to choose the piece of muscle for a particular purpose from a particular part of the organ. The strip has to be cut to the most suitable dimensions. The temperature and the load has to be adjusted according to the type of response to be studied. A special salt solution is preferable among a large number of 'modified' Ringer solutions which differ in ionic composition, pH and osmolarity. Such modifications influence spontaneous activity, they diminish or augment the muscle tone, and they may increase the sensitivity to the drug under investigation.

All these precautions having been taken, the muscle response to a given dose of a drug still depends on the pre-history within the experiment, e.g. on the size of the preceding dose, on the time of exposure and on the time interval between doses. Reproducible responses are obtained only by careful control of dosage, exposure time and spacing of drug applications.

More disturbing than the inconsistency of the size of a smooth muscle response was, to my mind, the fact that the direction, i.e. the sign of the response of the same muscle to the same dose of a drug could reverse in the course of an experiment though conditions were, apparently, identical. An interpretation of such confusing behaviour on the basis of mechanical records alone was bound to remain speculative.

During the year 1949/50, I worked for some time in the Department of Biophysics, Johns Hopkins University, Baltimore. I measured oxygen uptake of sympathetic nerves and ganglia and became interested in the relationship between tissue metabolism and function. I had already been interested in automatic rhythmic activity of various tissues and was now fascinated by the rhythmic discharge from neurones set up by various stimuli, especially the chemical excitation of nerves and ganglia, in which the frequency of discharge was proportional to the concentration of the chemical applied. The fundamental cellular mechanisms initiating the rhythmic discharge and, once established, modulating the frequency appeared to be determined by intrinsic characteristics of the tissue. This was the time when intracellular electrical recording techniques from nerve and striated muscle were introduced. The new lines of approach seemed to me also applicable to smooth muscle. When I returned to England I had made up my mind to try an electrophysiological study of smooth muscle in order to arrive at a better understanding of the mechanism of the action of drugs.

While I was abroad, 'a discussion on the action of local hormones' had been held (Burn, 1950) in which G. L. Brown expressed the need for the supplementation of pharmacological techniques by an attack on processes of excitation and inhibition in smooth muscle. He concluded his remarks by saying: 'I am only too fully aware of the difficulties confronting the electrophysiologist when dealing with cells of the size and complexity of those surrounding hollow organs, but the paucity of even the most elementary information holds out the promise of rich rewards to the experimenter with a sufficient temerity to begin such an investigation.' If I had this temerity it must have been due to my total lack of expertise. The real beginning was only made when young colleagues, who were or became experts in different fields and who then developed their special line of approach further, joined in the work.

The common aim was to obtain information on the electrical properties of smooth muscle, on the ionic basis for the resting and active membrane potential and on the processes which control excitability. At the back of my own mind there was always the guiding motive: to establish the electrophysiological basis for an investigation of the mode of action of drugs.

In the first place it was necessary to measure the membrane potential with microelectrodes. It took about five years before the values of the membrane potential, recorded with intracellular electrodes, could be accepted with some confidence. It took several more years before the method was perfected to such a degree that spike potentials with an overshoot were regularly observed. However, while we persisted in our efforts to obtain quantitative results, we became familiar with some of the properties of smooth muscle and began to understand its behaviour.

The taenia coli is a typical example of smooth muscle with autorhythmicity. In conditions suitable for intracellular recording, when short lengths of tissue are held under some tension to minimize movement, spikes are discharged almost continuously. Each individual spike is followed by a small increment of tension. The time course of the mechanical change is slow. Hence, it depends on the interval between spikes whether increments summate or not, i.e. whether the tension builds up or declines. 'Tone' is the result of rhythmic spike discharge. The rate of discharge fluctuates with spontaneous fluctuations of the membrane potential. Each slow wave of depolarization is associated with an acceleration of spike discharge and increased tone, each intervening repolarization with slowing of spike discharge and lower tone. In many ways the behaviour of the taenia resembles that of a discharging sensory organ. For example, if the muscle is stretched the membrane is depolarized, this leads to firing of impulses and development of tension. A similar sequence of events is observed in response to stimulating drugs, and it can be produced by applying depolarizing current. The changes are, however, only superimposed on the intrinsic spontaneous activity which depends on the load.

The basic cause for automatic myogenic rhythmic activity is not known. Bozler (1948) suggested that cell metabolism fluctuates spontaneously and leads to fluctuations in membrane potential which are associated with mechanical changes, contraction being produced when a wave of depolarization leads to the discharge of spikes. The so-called 'minute rhythm' (see Chapter 10) has also been considered an oscillation which is mainly connected with basic processes of cell metabolism, while the rhythm of the spike discharge, i.e. the spike frequency, is governed by processes in the membrane. The manifest rhythm of membrane activity can be resolved into a main component (the basic minute rhythm) with superimposed whole-number multiplications. The spike frequency can be altered, for example by stretch or by cooling, while the basic oscillation remains quite stable. Metabolic inhibitors, however, abolish the minute rhythm while spike discharge continues, often at an increased rate.

The nature of the inherent oscillations of cell metabolism and the processes by which these oscillations cause changes in membrane potential have remained obscure. I began research on smooth muscle by measuring the oxygen consumption of the taenia coli. Since this muscle is continuously active, the resting rate of oxygen consumption cannot be determined. Moreover, the oxygen uptake is influenced, firstly, by the muscle length and, secondly, by the muscle tone; i.e. by 'the resistance of the muscle to extension' (Lovatt Evans, 1926). In isotonic conditions the two factors operate in opposite directions so that the effect of shortening on oxygen uptake may mask the effect of the work done in overcoming the load. In isometric conditions, though the muscle length is kept constant, the tension nevertheless fluctuates and oxygen uptake fluctuates in parallel. But it is not known how much of the total oxygen uptake is due to 'basic processes of cell metabolism', how much to processes influencing membrane polarization, how much to processes involved in the initiation of, and recovery from, electrical activity and last, not least, how much to processes which activate the contractile apparatus.

The finding that the 'minute rhythm' can be selectively suppressed by metabolic inhibitors is strong evidence in favour of a metabolic basis. But it remains an open question how cell metabolism is linked to the processes at the membrane which determine the membrane potential and excitability. The influence must be strong. It is easier to drive a muscle by electrical stimulation, or mechanically by rhythmic stretch, if the applied rhythm is close to the spontaneous rhythm. On the other hand, the inter-action may not be only in one direction. Bozler (1948) mentions the possibility that metabolism may fluctuate 'as a result of previous activity, or distension, or the action of drugs'. The accentuation of the oscillatory behaviour after *any* disturbance, be it excitatory or inhibitory, is a common observation. Since a disturbance, e.g. by drugs, probably causes changes in the membrane permeability to certain ions, the oscillations during the recovery period may well be caused by the influence of a changed ionic composition inside the cell on the rates of metabolic processes which are connected with the control of membrane polarization and the threshold of excitation.

An interpretation of electrophysiological observations on the basis of the distribution of ions has been, and still is, one of the greatest difficulties. In the early stages there was no electrophysiological method available for measuring the membrane conductance. Nevertheless, the observations on the effects of changing the external ionic environment left no doubt that the membrane potential was mainly a potassium potential. It was also influenced by altering the external Cl concentration, while Na had relatively little influence (see Chapter 12). On the other hand, in the tracer experiments, the extremely fast exchange of Na, which seemed, at first, to have a large cellular component, was puzzling. There was some doubt concerning the size of the extracellular space. Many years have been spent in the effort to dispel this uncertainty and, although no definite solution has been found, this has led to the introduction of new methods. They include the use of ethanesulphonate and sorbitol instead of inulin and, more recently, the introduction of an extracellular marker, ^{60}Co EDTA, which has the advantage that it can be used simultaneously with other tracers so that the extracellular space can be measured in each individual efflux determination. One explanation for the extremely fast Na-exchange seems to be that the amount of intracellular Na ions has been over-estimated. Confidence is growing, but, even so, no agreement has been reached between different workers concerning the exact distribution of ions in extracellular and intracellular compartments.

It is, therefore, interesting to consider the different approaches to these problems and to compare the different moods, optimistic or despondent, in which this aspect is discussed in this book. An interpretation of the relation between membrane potential and ion distribution on the basis of the classical membrane theory (see Chapter 2) can be weighed against an interpretation based on the association-induction hypothesis (see Chapter 4). An attempt at reconciliation— or perhaps at overcoming the language barrier—considers ion fluxes on the basis of several compartments, including the competition between cations for membrane sites with fixed anionic charges (see Chapter 3).

The studies of osmotic behaviour (see Chapter 6) have revealed fundamental differences between striated and smooth muscle in the mechanisms which control cell volume and ion distribution. These differences may possibly be explained, in part, by morphological differences. The surface area of smooth muscle cells in relation to cell volume is much larger than in striated muscle and it is probably still further increased by the presence of many plasmalemmal vesicles, seen in the electron-microscope, apparently mostly connected with the outside through narrow openings (see Chapter 1).

These vesicles, at present an attractive structure for much speculation, may yet, as proper methods become available, provide the morphological basis for the interpretation of ion distribution and ion exchange. Tracer flux measurements can be brought into line with electrophysiological observations (Brading, unpublished work) if most of the rapidly exchanging sodium is considered to be extracellular. This would be possible if there were an extracellular compartment, not available to extracellular markers, with an ion exchange limited by the rate of diffusion into the extracellular space. Investigations of the changes in fine structure produced by experimental conditions known to cause changes of electrophysiological properties and of ion exchange are needed.

The information derived from studies with the electron microscope has been immensely valuable and has advanced our knowledge considerably. In the first place, the question whether smooth muscle is a syncytium has been answered: electron microscopy (see Chapter 1) has shown beyond doubt that there is no protoplasmic continuity between smooth muscle cells. However, there are special regions in which the opposite membranes of adjacent cells are fused. There is strong evidence that these areas have a low electrical resistance, allowing current to spread from cell to cell, so that the tissue, consisting of many separate cells, can behave like a syncytium.

It has taken years of patient work before it became possible to determine the membrane parameters (see Chapter 7). At first, it was surprising how difficult it was to evoke a spike by applying depolarizing current to a single cell through a microelectrode. For example, in the taenia coli a spike could hardly ever be triggered in spite of the fact that the same cell discharged spikes spontaneously and also fired in response to external field stimulation. Moreover, in the presence of low resistance cell-to-cell connections, one would have expected electrotonic spread from one active cell to its neighbours. But when, occasionally, a spike was triggered by intracellular stimulation, no electrotonic potential could be detected in a neighbouring cell. This problem has been solved by a careful analysis in which the electrotonic potentials evoked and recorded with the same intracellular microelectrode, were compared with those evoked by external stimulation and recorded intracellularly. It has become clear that, if a stimulus is applied to one single cell, the current is shunted by the large membrane area of the surrounding cells which are all inter-connected in three dimensions. On the other hand, if the stimulus is applied externally, by large electrodes, many cells are exposed to a one-dimensional current flow and excitation spreads along functional bundles of inter-connected muscle fibres. Consequently, the whole tissue has cable properties, and electrical membrane parameters can be calculated using the cable equations. The cable properties had been predicted by Bozler (1938) who reached his conclusion on the basis of mechanical and external electrical records. They have thus been largely substantiated and important quantitative data have been added.

On the other hand, the evidence obtained with electronmicroscopy and with intracellular electrical recording has shown that the classification of smooth muscles into multi-unit and unitary types (Bozler, 1941) requires some modification. A distinction can be made as far as innervation is concerned, although all intermediates probably exist between one extreme, in which all muscle fibres are innervated, and the other extreme, in which most cells are not innervated. However, the passive properties (membrane parameters) of the muscle fibres themselves appear to be very similar in all (see Chapters 7 and 13). Cell-to-cell connexions, 'nexuses', seem to be present in all smooth muscles, although the number of interconnections differs with tissues and species. In the guinea pig the current spread is almost the same along a multi-unit tissue (e.g. the vas deferens) as it is along a unitary tissue (e.g. taenia or uterus).

There remains, nevertheless, a difference in the physiological mechanism of excitation and synchronization. In the multi-unit muscle, activity is initiated and synchronized by nervous activity; the release of transmitter by nerve impulses causes excitatory junction potentials which trigger well synchronized muscle spikes (see Chapter 8). In the unitary muscle, however, activity is initiated spontaneously. The focal unit for activity is not a single cell but a group of cells. Automaticity is maintained by many cells and the activity is synchronized by cell-to-cell conduction. Such muscular synchronization is particularly efficient in the longitudinal intestinal muscle but it is never quite perfect—it becomes perfect only by nervous activation. In hollow organs, the sensitivity of the muscle to stretch is probably important in preparing it for the arrival of the nerve impulse. The depolarization caused by stretch brings the membrane potential closer to the firing threshold. This not only improves synchronization of spontaneous discharge but also increases the probability of a synchronous response to the nerve impulse. This mechanism must be evident to anybody who has watched the sequence of events leading up to the reflex contraction when the lumen of an isolated loop of intestine is gradually distended. The spontaneous tone of the longitudinal muscle and the size of the pendular contractions progressively increase until, suddenly, as the reflex is triggered, the spontaneous rhythm gives way to a single powerful contraction.

It is possible to study, with the electronmicroscope, the relation between the nerve terminals and the muscle fibres, and to measure the distance between the nerve and muscle membrane. On the basis of such measurements, combined with the determination of transmitter content, and of its release and uptake, the effectiveness of the transmitter has been estimated (see Chapters 1, 8 and 20). When the innervation is sparse and when the separation of nerve and muscle membrane is wide, the transmitter can reach distant muscle cells only by diffusion through the extracellular space and the interaction between muscle cells is an important factor. This is, to a large extent, the pattern of adrenergic innervation in some blood vessels (see Chapter 20) and in much of the gastro-intestinal tract, though there are considerable variations between different species. When the innervation is dense and if nerve and muscle membranes make close contacts, the transmitter action can be recorded as a junction potential in each individual smooth muscle cell (see Chapter 8). This is the pattern of the adrenergic innervation of the vas deferens. It is also the pattern of the intrinsic (non-adrenergic) inhibitory innervation in the gastro-intestinal tract.

The analysis of the nervous pathways involved in the peristaltic reflex in the intestine (see Chapters 17, 18 and 19) might seem at first remote from a study of smooth muscle. It is, however, essential for our understanding of the mechanisms which integrate and coordinate smooth muscle activity for the function of the whole organ. It has also led to new discoveries.

Hitherto unknown types of neurones, both excitatory and inhibitory, have been shown to be present in the intrinsic nervous system of the gastro-intestinal tract. When activated by electrical stimulation, the effects can be demonstrated on the motility of the stomach or the intestine, and also electrophysiologically by recording the potential changes in single smooth muscle cells. The responses to the unknown transmitters are clearly distinguishable from those produced by cholinergic or adrenergic nervous activity on the basis of pharmacological and electrophysiological criteria. It is remarkable how much the results obtained with the 'simple' method of recording the muscle contractions on a smoked drum, and those obtained by intracellular recording from single muscle cells, have contributed and supplemented each other in the discovery of new components in the intrinsic nervous system (see Chapters 8, 15, 16, 17, 18 and 19). Although the nature of the new transmitters has not yet been identified, advances are being made in disentangling the complex nervous pathways in the intrinsic ganglionic plexuses.

The role of 5-hydroxytryptamine as a possible transmitter in the interaction between neurones within the intrinsic nerve plexus is only one of the intriguing problems. 5-HT is a strong stimulant of sensory receptors. Its release from the argentaffin cells, in proportion to the intraluminal pressure, in close proximity to muscosal sensory nerve terminals, facilitates the peristaltic reflex activation. In addition, 5-HT has ganglion-stimulating and smooth muscle stimulating properties though the two receptors are different. Interesting new developments are taking place in the identification of the membrane components and binding sites which bring about the specific reaction (see Chapter 14).

Much work is in progress concerning the connection of receptor sites with special processes at the cell membrane which control membrane conductance and which are influenced by pharmacologically active substances, for instance the work on catecholamines (see Chapter 12) and the possible connection between adrenergic receptors and the binding of calcium at the membrane.

The function of Ca at the smooth muscle cell membrane has, so far, been investigated mainly on the taenia coli. The function is twofold. In the first place, there is strong evidence that the action potential is due to the entry of Ca and that the main source providing this Ca may be Ca bound at the membrane itself (Bülbring & Tomita, 1969d). Secondly, membrane bound Ca controls the permeability to other ions, and, in the taenia, the dependence of the potassium permeability on the amount of Ca at the membrane seems to be significant.

Furthermore, if it is accepted that Ca is required for the activation of the contractile proteins (see Chapter 9), then the reabsorption of Ca from the cytoplasm into stores and its recurrent release from binding sites must entail a rapid turnover. The distribution of Ca and Ca-exchange (see Chapter 5) is notoriously difficult to estimate and has, so far, largely evaded interpretation. The reason for this is probably that Ca-exchange involves exceedingly small amounts of Ca (see Chapter 3) and continuous transition from a bound to a free state, and vice versa, as well as continuous movement between compartments which cannot, with the use of our present tracer methods, be allotted to the intra- or the extracellular phase.

The lability of Ca-binding at the cell membrane may be an important factor in the mechanisms underlying the 'pacemaker' or 'generator' potential and may provide a key for the occurrence of automaticity. Periodic changes in the rate of Ca-accumulation on the one hand, and of Ca-removal on the other, may be linked with changes in metabolic energy supply. Investigations of the mode of action of catecholamines have lent support to such a hypothesis (see Chapter 12). The metabolic effects, e.g. the increase in the rate of $3',5'$-AMP formation and synthesis of energy-rich phosphate compounds are well established. Both the α- and β-effects can best be explained by an action on processes in the membrane controlling calcium transport. It has also been suggested that the two receptors are two different sites for hormone-enzyme interaction on the adenylcyclase molecule. However, the enzyme systems involved have not been identified and here is a field wide open for a combined biochemical, physiological and pharmacological investigation.

This question brings us back to the point of departure. We may now ask ourselves: what have we learnt and what are the most important open questions? There is no doubt that we have learnt a great deal about the electrical properties of the smooth muscle membrane and about the ionic basis for the membrane potential, though the information cannot as yet be as precise as in other excitable tissues in which the cells are larger and their arrangement less complex. Nevertheless, in contrast to twenty years ago, electrophysiological methods have now been developed for determining the membrane parameters and for observing changes in membrane conductance which may be produced by drugs. Thus we are now in a position where an analysis of the mode of action of drugs can be attempted.

One example is the study of the mechanisms underlying the large increase in membrane potential which occurs in the uterus during pregnancy, or under treatment with oestrogen and progesterone. It was found that this change is not associated with an increase in the intracellular K concentration but with an increase in the K permeability of the membrane. Electrophysiological data agree with the evidence of the measurements of K flux which is much larger in the progesterone dominated uterus than under the influence of oestrogen alone (see Chapters 4, 11 and 13).

It is known that the response of the uterus to drugs changes with the hormonal condition. This may be due to a change in the ratio of the membrane permeabilities to different ions. This ratio may, in fact, be specific for each smooth muscle type. A comparative study along these lines might be profitable.

Another example concerns the mode of action of catecholamines. The electrophysiological observations of the changes in membrane conductance produced by adrenaline agree with the changes in membrane permeability observed by measuring ion flux with radio-active tracers (see Chapters 11 and 12). However, while a correlation between physiological and biochemical effects has been demonstrated, a causal connection is not proven. Unless we discover the nature of the processes which connect cell metabolism to events at the membrane and understand the origin of autorhythmicity we will not be able to understand fully the mechanisms of excitation and inhibition.

An important aspect for investigation is, of course, the innervation of smooth muscle. Our conception of the autonomic nervous system is expanding (see Chapters 15 and 16). New methods for tracing nervous pathways and identifying transmitters are available. Moreover, the time is near when it will be possible to investigate the reflex activity of the intrinsic nervous system of visceral organs by recording electrical activity not only from smooth muscle cells but also from ganglion cells, and thus to study functional units.

In writing this introduction, I have not intended to give the reader a summary of the contents of this book. Instead, I have indulged in developing my own thoughts, referring to individual chapters to reassure the reader about the existence of experimental facts. Actually, this book is the result of a concerted effort by twenty authors to put on record how far twenty different lines of approach in the study of smooth muscle have advanced our knowledge. It is not yet possible to assemble the multitude of observations and construct a complete picture of the whole subject. Although much has been observed, little is known. Yet the foundations have been laid and we can look forward to future experiments with confidence.

1

STRUCTURE OF SMOOTH MUSCLE
AND ITS INNERVATION

G. BURNSTOCK

INTRODUCTION

The first descriptions of the structure of smooth muscle were published in the mid-nineteenth century (Schwann, 1847; Henle, 1841; Kölliker, 1849). Since this time a tremendous volume of work has been carried out on the light microscopy of smooth muscle, particularly silver and methylene blue studies of the nature of the innervation apparatus (Hillarp, 1960; Grigor'eva, 1962; Botár, 1966).

The first electronmicroscope studies of visceral smooth muscle were published in the mid-fifties (Csapo, 1955; Bergman, 1958; Gansler, 1956; Häggqvist, 1956; Mark, 1956) but perhaps the most important early paper in the field was by Caesar, Edwards & Ruska (1957), who included the first description of the fine structure of the relationships of autonomic nerves to smooth muscle cells.

Electronmicroscope studies of vascular smooth muscle appeared later than those on visceral smooth muscle (Fernandez-Moran, 1958; Moore & Ruska, 1957; Pallie & Pease, 1958; Parker, 1958; Karrer, 1959; Fawcett, 1959) and the nervous supply of the medial musculature began to be resolved at the electronmicroscope level only after the publications of the results of applying the fluorescent histochemical method for the localization of monoamines (Falck, 1962) to various blood vessels (Appenzeller, 1964; Brettschneider, 1964; Barajas, 1964; Lever, Graham, Irvine & Chick, 1965; Simpson & Devine, 1966; Rhodin, 1967; Graham, Lever & Spriggs, 1968; Devine & Simpson, 1968; Verity & Bevan, 1968).

Much of the early work on smooth muscle concerned the nature of inter-muscle fibre relationships (Mark, 1956; Bergman, 1958; Thaemert, 1959; Prosser, Burnstock & Kahn, 1960) but it was not until new methods of fixation were developed that the existence of 'tight junctions' between smooth muscle cells was discovered (Dewey & Barr, 1962). Studies of the contractile apparatus (Gansler, 1961; Choi, 1962; Needham & Shoenberg, 1964; Conti, Haenni, Laszt & Rouiller, 1964; Lane, 1965; Panner & Honig, 1967; Kelly & Rice, 1968) have in general lagged behind those made on other muscle systems.

As in other fields, fine structural studies in more recent work have been undertaken in close relation to physiological problems. In the smooth muscle field these are centered in two main areas, namely, the nature of the autonomic neuro-effector junction (Richardson, 1962; Merrillees, Burnstock & Holman, 1963; Lane & Rhodin, 1964a; Evans & Evans, 1964; Taxi, 1965; Hökfelt, 1966a; Rogers & Burnstock, 1966b; Bennett & Merrillees, 1966; Bennett & Rogers, 1967) and the nature and function of intra-axonal vesicles (de Robertis & Pellegrino de Iraldi, 1961; Lever & Esterhuizen, 1961; Grillo & Palay, 1962; Richardson, 1964, 1966; Burnstock & Merrillees, 1964; Taxi, 1965; Van Orden, Bloom, Barrnett & Giarman, 1966; Van Orden, Bensch & Giarman, 1967; Burnstock & Robinson, 1967; Hökfelt, 1958).

Three important extensions of electronmicroscopy are currently being developed in relation to these and other problems. Firstly, the exacting task of making serial or semi-serial sections has been undertaken (Taxi, 1965; Thaemert, 1966; Merrillees, 1968) so that quantitative correlations can be made with electrophysiological data. Secondly, the localization of various substances in innervated smooth muscles has been made by the application of electronmicroscopy to histochemical methods developed previously for light microscopy. These include the fine structural localization of acetylcholinesterase (Burnstock & Robinson, 1967; Robinson & Bell, 1967; Esterhuizen et al, 1968; Graham et al, 1968; Robinson, 1969), catecholamines (Bloom & Barrnett, 1966b), and ATPase (Rostgaard & Barrnett, 1964; Lane, 1967). Electronmicroscopic autoradiography following injection of tritium-labelled noradrenaline has also been carried out on autonomically-innervated tissues (Wolfe, Potter, Richardson & Axelrod, 1962; Taxi & Droz, 1966 Esterhuizen et al, 1968; Devine & Simpson, 1968). The third field which is beginning to receive more attention is that of the development of smooth muscle and its innervation (Yamamoto, 1961; Shestopalova, 1964; Leeson & Leeson, 1965a, b; Yamauchi & Burnstock, 1969a, b).

The future of this field is bright. There is little doubt that many of the outstanding problems concerning smooth muscle that faced the early light microscopists, and more recently posed by the physiologists and biochemists, can be resolved by the localization of specific chemicals at the cellular level. One can look to a clarification of the diversity of the structure and innervation of smooth muscles in different systems with widely different physiological roles.

SMOOTH MUSCLE

Size, shape and arrangement of muscle cells

Muscle coats

Light microscope studies have shown that in most hollow organs there is an outer longitudinal muscle coat and an inner circular coat. In some hollow organs this arrangement is reversed, e.g. 'possum bladder' (Burnstock & Campbell, 1963). There is often a further inner longitudinal coat, for example in mammalian ureter and rat vas deferens. In the developing mouse vas deferens, the circular muscle coat is present at birth, but the longitudinal muscle coat does not appear until 5 days after birth (Yamauchi & Burnstock, 1969a).

In blood vessels, the muscles are usually confined to the media and are arranged in a spiral or helical fashion with the dominant orientation being circular (Strong, 1938; Pease & Paule, 1960; Keech, 1960). The angle between the helical turn and the long axis is about 30° in mouse femoral arteries (Rhodin, 1962), 72° in arterioles (60 µ diameter) (Rhodin, 1967) and with decreasing vessel diameter, the angle increases until in small arterioles the muscle cells are arranged in a truly circular fashion.

Muscle bundles

In the muscle coats of most visceral tissues, the muscle cells are arranged in branching bundles or fasciae of irregular cross-section, surrounded by connective tissue sheaths (see Fig. 1 and Csapo, 1955, 1959, 1962; Prosser et al., 1960; Merrillees et al., 1963; Evans & Evans, 1964; Bennett & Rogers, 1967). The muscle cells in the media of many arteries have also been shown to be arranged in bundles (Boucek, Takashita & Fojaco, 1963).

From physiological studies, Burnstock & Prosser (1960b) concluded that effector muscle bundles were about 100 µ in diameter; when the diameter of strips of guinea-pig taenia coli or cat intestinal muscle were reduced to below 100 µ, there was no propagation of action potentials. This has been confirmed in further experimental studies (see Bennett & Burnstock, 1968). Stimulation of individual muscle cells with an intracellular electrode failed to initiate a propagating action potential (or contraction) in the guinea-pig taenia coli (Kuriyama & Tomita, 1965; Tomita, 1966a) or in the vas deferens (Hashimoto, Holman & Tille, 1966; Bennett, 1967b); the minimum diameter of the tip of an external micro-electrode for producing propagated action potentials in intestinal muscle was about 100 µ (Nagai & Prosser, 1963a). Muscle bundles of approximately this size have been demonstrated microscopically in most tissues (Prosser et al., 1960; Csapo, 1962; Nagasawa & Mito, 1967; Bennett & Rogers, 1967), although there is considerable variation in size and shape (Fig. 1).

The arrangement of muscle bundles to each other within muscle coats has not yet been examined in any detail, i.e. with serial sections. This has become

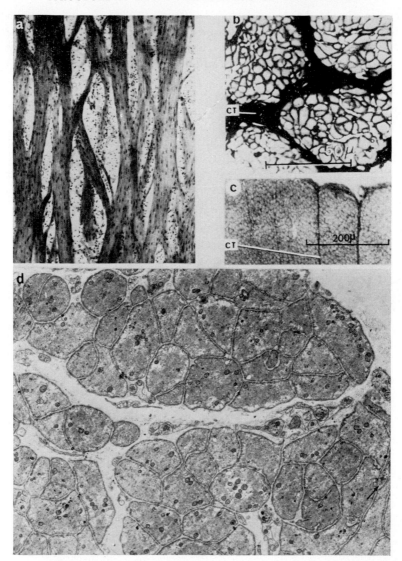

FIG. 1 *a* Structure of the longitudinal muscle layer of the parturient rabbit uterus. Light microscopy, ×41. (From Csapo, 1962) *b* & *c* Photomicrographs of cross-sections of smooth muscles. b, dog retractor penis; c, cat circular intestinal muscle. (From Prosser, Burnstock & Kahn, 1960) *d* Low power electronmicroscope section showing muscle bundle arrangement. Cat nictitating membrane. ×4200. (From Taxi, 1965).

an important problem, since it has been established that muscle bundles rather than individual muscle cells are the effector units in smooth muscle systems (Bennett & Burnstock, 1968). Merrillees (1968) in a study with serial sections of the guinea-pig vas deferens concluded, after following a small bundle which quickly lost its identity, that neighbouring bundles frequently anastomose and that the relative positions of neighbouring muscle fibres within a bundle change because of losses and gains of fibres within the group.

Cell dimensions

For many years it was assumed that smooth muscle cells were spindle-shaped with an approximately centrally-placed nucleus and with long tapering ends.

However, serial section sampling from electron microscope studies of the guinea-pig vas deferens (Merrillees, 1968) and frog and mouse intestine (Taxi, 1965; Thaemert, 1966) have made it possible to analyse the shape of smooth muscle cells more accurately, and several surprising features have emerged:

1. The nucleus is not always centrally placed but varies up to 50 μ from the centre point of the fibre.
2. Cells appear to vary in size, but this may be due to different degrees of contraction of different cells during fixation.
3. Cells are not arranged in a regular transverse or diagonal array.
4. Cells are not really fusiform or double cones, but show extremely uneven contours along their length. In the vas deferens, few fibres had a regular poly-hedral profile in cross section and the shapes changed considerably from level to level, transverse sections were polyhedral, triangular or even extremely flattened elipses; many cells were ribbon or rod-like.
5. The length of muscle cells in the longitudinal coat of the guinea-pig vas deferens varied between 380 and 465 μ which compared with a mean of 200 μ when the cells were measured individually after separation by nitric acid and glycerine treatment (Burnstock, unpublished observations). In general, the largest cells are found in pregnant uterus (up to 10 μ × 600 μ) (Csapo, 1962) and in urodel amphibians (up to 12 μ × 1200 μ) (Burnstock & McLean, un-published observations). The smallest cells are found in arterioles (approx. 2 μ × 15–20 μ) (see Table 1).
6. Volumes of single fibres ranged from 2300 to 3500 μ^3 and surface areas from 4000 to 5400 μ^2 in the vas deferens (Merrillees, 1968).

Smooth muscle cells of arteries have irregularly-shaped branched processes (Keech, 1960; Pease & Paule, 1960; Prosser et al., 1960). Muscle cells at an early stage of development are also characterized by numerous fine cytoplasmic processes which are closely apposed to neighbouring cells (Yamamoto, 1961; Leeson & Leeson, 1965a, b; Yamauchi & Burnstock, 1969a). Cells of the mus-cularis mucosae of the mouse intestine have many processes (Lane & Rhodin, 1964b), unlike those of the pig oesophagus (Prosser et al., 1960).

The iris dilator muscle consists of very narrow cells with unusually complicated outlines, closely packed together with interlocking processes (Richardson, 1964). The myoepithelial character of the dilator pupillae, which was debated for so

TABLE 1. Size of smooth muscle cells

Tissue	Animal	Technique	Length (μ)	Diameter (μ) (in nuclear region)	Reference
Intestine	mouse	E.M.	*400	—	Taxi, 1965
	cat	N.A.M.	120	6	Prosser et al., 1960
longitudinal coat (jejunum)	mouse	E.M.	150 (relaxed)	2–3·5	Lane, 1965
			70 (contracted)	6	
longitudinal coat	mouse	E.M.	100	2	Rhodin, 1962
circular coat		E.M.	—	5	
	axolotl	N.A.M.	1200	12	Burnstock & McLean (unpublished data)
taenia coli	guinea-pig	N.A.M.	150	6	Prosser et al., 1960
		E.M.	70–90	3–4	Yamauchi, 1964
		E.M.	*200 (relaxed)	2–4	Bennett & Rogers, 1967
muscularis mucosae (oesophagus)	pig	N.A.M.	220	6	Prosser et al., 1960
ciliary muscle	cat	N.A.M.	100	—	Taxi, 1965
nictitating membrane	cat	E.M.	350–400	—	Taxi, 1965
retractor penis	dog	N.A.M.	90	6	Prosser et al., 1960
vas deferens	guinea-pig	E.M.	*450 (slightly contracted)	—	Merrillees, 1968
vascular smooth muscle					
small arteries	mouse	E.M.	60	1·5–2·5	Rhodin, 1962
arteriole	rabbit	E.M.	30–40	5	Rhodin, 1967
renal vein	pig	N.A.M.	110	3	Prosser et al., 1960

* Asterisks are placed against figures measured from serial electronmicroscope sampling, since they are likely to be the most reliable.
     (E.M. = Electron Microscopy)   (N.A.M. = Nitric Acid Maceration)

many years (Lowenstein & Lowenfeld, 1962) is now well established (Richardson, 1964). The outer surface of smooth muscle cells in small mesenteric veins of the rat tend to be crenulated (Devine & Simpson, 1967).

Electronmicroscope studies have been made of the changes in form of smooth muscle cells of the longitudinal coat of the mouse jejunum (Lane, 1965) and of the guinea-pig taenia coli (Nagasawa & Suzuki, 1967) in different degrees of contraction. The relaxed cell is long and narrow with smooth cytoplasmic and nuclear contours. As contraction progresses, the cell becomes ellipsoid and its borders exhibit invaginations at the points of myofilament attachment to the plasma membrane (see p. 33) alternating with micropinocytotic vesicle-containing projections of the intervening membranes. The nucleus of the contracting cell is shortened and widened, with convolution of its limiting membrane (see Fig. 4a). Iris sphincter pupillae muscles contracted down by pilocarpine, showed extreme folding of the muscle borders (Richardson, 1964).

There is a rapid increase in the length of smooth muscle cells of the mouse vas deferens between 1 and 20 days after birth; this is followed by a marked increase in cell diameter during the next 5 days (Yamauchi & Burnstock, 1969a). Fully developed smooth muscle is present in the mouse vas deferens between 1 and 3 months after birth (Yamauchi & Burnstock, 1969a) in contrast to rat ureter muscle cells which are fully developed between 1 and 5 days after birth (Leeson & Leeson, 1965a).

Arrangement of muscle cells within bundles

Since the muscle bundle is the effector unit (Burnstock & Prosser, 1960b; Tomita, 1966a; Bennett & Burnstock, 1968), the exact relationship of the muscle cells to each other within a bundle is of great functional significance (see Chapters 7 and 8).

It was assumed for a long time that the thick middle portion of the cell which contains the nucleus, lies adjacent to the long tapering ends of surrounding cells. This appears from serial sectioning (Bennett & Merrillees, 1966) to be approximately true for the guinea-pig vas deferens, but in many blood vessels the branched muscle cells appear to meet in an end-to-end fashion (Rhodin, 1962). Interdigitations between cells are common (Caesar et al., 1957) especially at their ends; these are prominent in pregnant uterus (Kameya, 1964).

The separation of neighbouring muscle cells is generally between 500 and 800 Å in most organs (Prosser et al., 1960; Merrillees et al., 1963; Merrillees, 1968). Basement membrane material and sometimes scattered collagen filaments fill the narrow spaces between muscle cells within bundles.

Various models of cell interrelationships have been put forward in relation to theoretical studies of the equivalent circuit of the effector bundle (Burnstock & Prosser, 1960b; Prosser, 1962; Nagai & Prosser, 1963a; Barr, 1963; Vayo, 1965; Bennett & Merrillees, 1966; Tomita, 1966b; Bennett and Rogers, 1967; Abe & Tomita, 1968). The situation in the guinea-pig vas deferens as revealed by serial sampling (Merrillees, 1968) appears to be that in any one plane, a muscle cell is surrounded by 6 others but there is an irregular longitudinal splicing of neighbouring cells, such that each muscle cell is surrounded by about 12 others over its whole length. As each fibre in the vas deferens was followed from end to end, it became displaced by 2 to 3 muscle diameters relative to its original position amongst its neighbours. This interweaving is important in relation to the innervation of individual cells, i.e. muscle fibres which otherwise would have no close approach to a particular nerve trunk, insinuate themselves between their neighbours to lie near the trunk. In the taenia coli also, Bennett & Rogers

(1967) concluded from the relative disposition of nuclei in a series of sections, that muscle cells have a staggered arrangement with adjacent cells overlapping, such that each cell is surrounded by 10 to 12 others.

Extracellular space

A variety of materials, including collagen, blood vessels, nerves and Schwann cells, macrophages, fibroblasts, mucopolysaccharides and elastic tissue may be seen in the extracellular space (Caesar *et al.*, 1957).

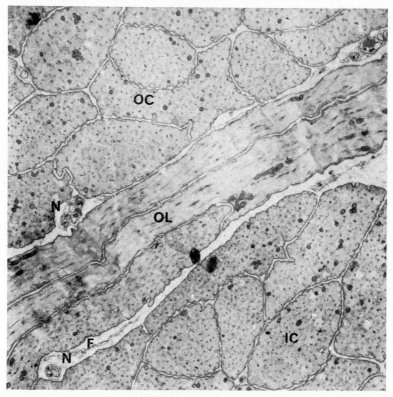

FIG. 2 Vas deferens musculature showing inner circular (IC), outer longitudinal (OL) and circular (OC) layers. The musculature is four-layered in this area and two-layered in others. The muscle cells are closely packed and demonstrate abundant peg-and-socket intercellular connections. The space between muscle layers is small, ranging between 3 and 10 μ, and contains collagen, small nerve bundles (N), and occasional fibroblastic processes (F). × 4500. (From Lane & Rhodin, 1964a)

Although the extracellular space is bounded by distinct membranes (Fig. 2), the problem is further complicated by the existence of abundant 'micropino- cytotic' or 'plasmalemmal' vesicles (Fig. 11b) some of which appear to be connected with an intracellular endoplasmic tubular system (Fig. 4b).

Both these structures would considerably increase the effective extracellular space.

Values for extracellular space (see Table 2) have been obtained from electron-micrographs of various smooth muscles, by measuring the relative areas occupied by muscle and space with fine transparent grids superimposed over large prints (Prosser *et al.*, 1960; Rhodin, 1962), or by weighing the pieces of tracing paper not occupied by muscle (Yamauchi & Burnstock, 1969*a*). It has been argued that this is an unreliable method of measuring extracellular space because of the unknown degree of swelling or shrinking of the tissue that takes place during

TABLE 2. Extracellular space

Tissue	Percentage extracellular space		Reference
	Electron-micro-scope measurements	Measurement by uptake of solutes (Inulin, unless otherwise stated)	
Visceral smooth muscle			
guinea-pig taenia coli	12		Prosser *et al.*, 1960
guinea-pig taenia coli		30–39	Goodford & Hermansen, 1961; Born, 1962; Nagasawa, 1963
guinea-pig taenia coli		40 (Sorbitol-^{14}C)	Goodford & Leach, 1966
cat intestine	9		Prosser *et al.*, 1960
mouse intestine	13		Lane & Rhodin, 1964*a*
rat colon		38 (Thiosulphate)	Laszt, 1960
toad stomach		29 (I^{131}-tagged serum albumen)	Burnstock, Dewhurst & Simon, 1963
mouse vas deferens	8		Lane & Rhodin, 1964*a*
mouse vas deferens	12 (adult) 50 (neonatal)		Yamauchi & Burnstock, 1969*a*
rat uterus		40–46	Daniel, 1963*b*; Melton, 1962
rabbit uterus		32–36	Kao, 1961; Horvath, 1954
		43–57 (SO_4)	Daniel & Daniel, 1957
cat uterus		31–40	Daniel & Daniel, 1957
		46–63 (SO_4)	
Vascular smooth muscle			
pig carotid artery	39		Prosser *et al.*, 1960
mouse femoral artery	30		Rhodin, 1962
rat aorta		35	Hagemeijer, Rorive & Schoffeniels, 1965
rabbit aorta		62	Bevan, 1960
cow carotid artery		39 (Thiosulphate)	Laszt, 1960
dog carotid artery		25	Headings, Rondell & Bohr, 1960

fixation. However, examination of the appearance of the fine structure in a particular electronmicrograph can give a reliable indication of the preservation of the tissue. Furthermore, if a comparative study is made of various tissues all fixed under the same conditions of preparation and stretch, the figures are consistent and reliable.

The values for extracellular space obtained with the electronmicroscope measurements are consistently lower than those calculated from the uptake of solutes (see Table 2). This is mainly because the electronmicroscope method offers a unique opportunity to select for measurement those areas of a preparation which consist solely of functional muscle bundles. In contrast, the large pieces of muscle used with the uptake technique include such tissues as connective tissue sheaths, blood vessels, fibroblasts, interstitial cells, nerve trunks and glandular tissue. However, the higher values obtained with the latter method, while unrepresentative of the true value of extracellular space within muscle effector bundles, are of course valid when used as a basis for calculations of intracellular ion concentrations when large pieces of muscle tissue are used for the experiments.

Comparison of the extracellular space of various smooth muscles prepared under similar conditions was carried out by Prosser et al., (1960). It was concluded that muscle fibres are more closely packed in fast conducting than in slow or non-conducting muscles. During pregnancy there is an increase in extracellular space between uterine muscle cells (Jaeger, 1962, 1963).

Intermuscle fibre relationships

Introduction

For most of their surface, the cells of the vas deferens have been shown to be separated from their neighbours by a basement membrane-filled gap of 500 to 800 Å (Merrillees, 1968). Similar separations were reported in a variety of other visceral smooth muscles by Prosser et al., (1960) but separations of 2000 to 3000Å were characteristic of large arteries. Serial sectioning has also shown that muscle cells often begin and end against the sides of their neighbours, sometimes in a gutter or even a tunnel (Merrillees, 1968). Rarely does the end of one fibre interdigitate with the beginning of the next.

Protoplasmic continuity has been claimed to exist between smooth muscle cells from a few early studies with the light microscope (Barfurth, 1891; Aunap, 1936) and also with the electronmicroscope before the techniques of tissue preservation and resolution were refined (Häggqvist, 1956; Mark, 1956; Thaemert, 1959). However, later electronmicroscope studies with better techniques have not confirmed these observations.

A variety of specialized interfibre relationships have been described, ranging from 'areas of close contact', to 'bridges', 'intrusions', 'desmosomes', 'intermediate junctions' and 'nexuses' (see Burnstock & Merrillees, 1964). One factor should be borne in mind in discussing these structures, namely that different methods of preparation, fixation and staining of the tissue for electronmicroscopy make the different relationships more or less prominent. While this is partly a matter of the best technical conditions for displaying different structures, it also indicates a certain lability in the nature of the intercellular relationships. For example, even such a specialized area as the fusion of membranes at a nexus (or 'tight junction') has been shown to be labile (Dewey & Barr, 1964) although this has been contested recently (Cobb & Bennett, 1969). The nexus was never observed in dog intestinal muscle except when the tissue was prevented from shortening and incubated long enough with physiological saline to recover responsiveness prior to fixation in permanganate. Fixation by perfusion *in situ* or immediate fixation of small excised strips (which leads to marked contraction) was unsatisfactory for demonstration of nexuses. Osmium tetroxide fixative provided no evidence of membrane fusion in dog intestinal smooth muscle, although bridges with membrane separation of 100 Å were demonstrated (Dewey & Barr, 1964). However, Yamauchi and Burnstock (1969a) were able to demonstrate tight junctions between some smooth muscle cells of the mouse vas deferens if fixed with osmium tetroxide or glutaraldehyde, but not with potassium permanganate. Trelstad, Revel and Hay (1966) have claimed that the best technique for demonstration of tight junctions is paraformaldehyde and glutaraldehyde fixation followed by uranyl acetate stain.

Intercellular 'bridges'

Many workers have reported the presence of areas of close contact, termed 'bridges' by Bergman (1958), between muscle cells in different tissues, where the membranes of opposing cells are intact, with a separation of about 100 to 200 Å (Richardson, 1958; Prosser *et al.*, 1960; Rhodin, 1962; Merrillees *et al.*, 1963; Bennett & Rogers, 1967). The area of contact of cells at 'bridges' is about 0·2 to 0·4 μ. Bergman (1958) concluded that there were probably several bridges between adjacent cells in the rat ureter, while Rhodin (1962) claimed that there were at least 150 points of contact between one smooth intestinal muscle cell and its neighbours, which he calculated to represent a surface area of about 20 μ². This represents about 5% of the total cell surface membrane-to-membrane contact with other cells. The points of contact at bridges did not show any structural specialization similar to cardiac intercalated discs.

'Intrusions' (Richardson, 1962, 1964; Merrillees *et al.*, 1963; Nagasawa & Suzuki, 1967), 'protrusions' (Thaemert, 1963; Yamauchi, 1964), 'projections' (Bennett & Rogers, 1967), or 'peg and socket' structures (Lane & Rhodin, 1964a; Verity & Bevan, 1966) are common in a variety of tissues, where an evagination of one cell fits into an invagination of the adjacent cell (Fig. 2 and Fig. 3a). The area of close contact involved is about 0·6 μ and the membranes are about 100 Å apart in mouse vas deferens and are numerous such that the percentage of their total cell area is 3 to 5% (Lane & Rhodin, 1964a). The cytoplasm beneath the membranes at these points does not usually contain organelles.

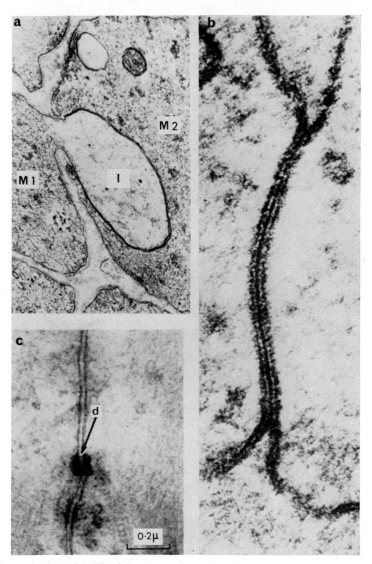

FIG. 3   *a* An intrusion (I) of a process of muscle cell (M1) into a neighbour (M2). It contains no recognizably organized myofilaments. The plasma membranes of both cells still confine the intrusion, and are dense and closely applied. × 27,000. (From Merrillees, Burnstock and Holman, 1963)   *b* Circular muscle of mouse intestine. Note the fusion of the inner lamellae of the unit membranes of neighbouring muscle cells to form a nexus. × 280,000. (From Taxi, 1965)   *c* Chick amnion. 10 days' incubation. Longitudinal section of two smooth muscle cells showing area of close apposition of plasma membranes, with no intervening basement membrane. An attachment plaque is shown at (*d*). (From Evans and Evans, 1964)

The intrusion is usually attached to the parent cell by a very narrow stalk (Merrillees *et al.*, 1963; Merrillees, 1968). It has been suggested that intrusions are more common after asphyxia and maltreatment of the tissue (Merrillees, 1968).

'*Tight junctions*' or nexuses and '*gap junctions*'

Regions of apposition of smooth muscle cells in intestine have been reported where there is fusion of plasma membranes, such that a 5 rather than a 7-layer system of membranes results (Intestine: Dewey & Barr, 1962, 1964; Lane & Rhodin, 1964*a*, *b*; Oosaki & Ishii, 1964; Bennett & Rogers, 1967; Uterus from oestrogen-treated rat: Bergman, 1968. Constrictor pupillae and chick amnion: Evans & Evans, 1964; Taxi, 1965. Ureter: Taxi, 1965. Mouse vas deferens: Yamauchi & Burnstock, 1969*a*. Vascular smooth muscle: Verity & Bevan, 1966; Cliff, 1967; Rhodin, 1967; Foussman, 1967 — personal communication). This type of connection in smooth muscle has been termed a 'nexus' (Dewey & Barr, 1962) and is comparable to the 'tight junctions' or *zonula occludens* (Farquhar & Palade, 1963) found in other systems (Sjöstrand, Andersson-Cedergren & Dewey, 1958; Karrer, 1960; Robertson, 1961, 1963; Robertson, Bodenheimer & Stage, 1963; Bennett, Aljure, Nakajima & Pappas, 1963). Furshpan (1964) has shown that electrical coupling can take place across tight junctions, so that the structure is of great importance in relation to the mechanism of intermuscle fibre conduction (see Chapter 7). The chick amnion (Fig. 3*c*) was selected for study of intermuscle fibre relationships by Evans & Evans (1964) because it is nerve-free, but nevertheless capable of propagating a wave of contraction (Evans, Schild & Thesleff, 1958; Cuthbert, 1963*b*).

Nexal regions of two types have been described in different systems:
1. Simple abutment of adjacent cells.
2. Projections of one cell into another (intrusions).

Simple abutments are more frequent in dog intestinal muscle, while projections are dominant in cat and guinea-pig intestine (Dewey & Barr, 1964). Dewey & Barr, (1964) suggest that the simple abutment type of nexus may act as a mechanical tie between contracted cells, whereas projections are more likely to be involved in electrical coupling. When contraction was suppressed (with procaine, atropine, low temperatures, nitroglycerine) projection-type nexuses were seen more often.

The overall thickness of the 5-layered nexus is remarkably constant, about 140 Å (Dewey & Barr, 1964). Occasionally, the fusion of plasma membranes at the contact zone is incomplete, with alternating zones of tight junctions and areas of close (about 125 Å) apposition only (Oosaki & Ishii, 1964). Interlocking muscle processes of a similar nature with intermittent tight junctions have also been observed in invertebrates (Rosenbluth, 1965*a*). The variations in tightness at such junctions may indicate the labile nature of the tight junctions in relation to different cellular conditions as reported by Farquhar & Palade (1963) for contact regions between epithelial cells. Recently, high resolution electronmicrographs of regions of close muscle apposition suggest that, in some tissues at least, the nexus does not represent true fusion of membranes, but rather 'gap junctions' with separation of apposing membranes, by about 20–30Å. (Revel et al., 1967; Uehara & Burnstock, 1969).

Taxi (1965) showed, from serial section studies of mouse intestinal smooth muscle, that there were about 2000 points of close connection (bridges with 60 Å separation of membranes with OsO_4-fixation, or nexuses, after permanganate fixation) on muscle cells 350 to 400 μ long (see Fig. 3b). He calculated that these formed about 6% of the total surface area of the cell. Areas of close apposition between cells of the mouse duodenum, about 1·5 μ in diameter, represent about 3 to 5% of the total cell membrane area according to Lane & Rhodin (1964a). The cell membrane at these points appeared to be fused, forming nexuses, but it was difficult to be certain since permanganate fixation was not used. In terminal arterioles (50 μ diameter), the lateral membrane-to-membrane contacts between smooth muscle cells were more frequent and were established for longer distances than in arterioles with several smooth muscle layers (Rhodin, 1967). No special organelles appeared to be located beneath the nexus regions, but glycogen granules, ribonucleoprotein, endoplasmic reticulum, vesicles, motochondria and myofilaments were all sometimes present (Lane & Rhodin, 1964a).

Zonula occludens (tight junctions) and Zonula adherens (intermediate junctions) (Farquhar & Palade, 1963) occurred profusely between the processes of the rabbit iris dilator muscle cells (Richardson, 1964). On the other hand, no 'tight junctions' have been observed between smooth muscle cells in the nictitating membrane or vas deferens by most workers (Richardson, 1962, 1964; Merrillees et al., 1963; Lane & Rhodin, 1964b; Evans & Evans, 1964; Taxi, 1965). However, some have been described in both mouse (Yamauchi & Burnstock, 1969a) and guinea-pig (Cobb, personal communication) vas deferens. It seems likely that there are considerable differences in the type, distribution and number of tight junctions in different tissues.

Desmosomes, 'attachment plaques' and intermediate junctions

Desmosomes or *maculae adherens* (Farquhar & Palade, 1963) consisting of small, heavily-staining electron-dense areas on adjacent muscle membranes, are common in the chick amnion (Fig. 3c), although cytoplasmic tonofibrils have not been clearly seen running into the dense zones (Evans & Evans, 1964). Similar attachment plaques have been seen between cat constrictor pupillae muscles (Evans & Evans, 1964). A 'desmosome-like' structure has also been reported between both pregnant and non-pregnant uterine smooth muscles (Kameya, 1964) and in guinea-pig urinary bladder (Nagasawa & Suzuki, 1967).

In the rabbit iris dilator muscle, processes of adjacent cells interlock and there are special 'contact zones' or 'junctional complexes' (*fascia adherens*) where the cell membranes are specialized (Richardson, 1964). However, although 'tight junctions' are abundant, desmosomes are rare in this tissue.

Desmosomes probably perform an important mechanical role in maintaining the integrity of the tissues through considerable 3-dimensional changes.

Intercellular junctions, where a narrow intercellular space (30 to 200 Å) is occupied by a homogeneous and amorphous material of moderate density and associated with an increased density of the cytoplasmic opaque components of strictly parallel plasma membranes, have been described in rat intestine (Oosaki & Ishii, 1964) and in rabbit colon (Nagasawa & Susuki, 1967). This type of junction should be distinguished from the desmosome and has been described as an intermediate junction (*zonula adherens*) in other tissues (Sjöstrand & Elfvin, 1962; Farquhar & Palade, 1963; Elfvin, 1963).

Summary
1. For most of their surface, visceral smooth muscle cells are separated from their neighbours within effector bundles by a basement membrane-filled gap of 500 to 800 Å.
2. There are areas of close contact between most smooth muscle cells within effector bundles. These often take the form of 'bridges' or 'intrusions'. About 3 to 5% of the total surface area of a muscle cell has specialized regions of contact with neighbouring cells.
3. A method of fixation can usually be found for a particular tissue, where a close apposition ('gap junction') or fusion of apposing membranes ('tight junction' or 'nexus') can be demonstrated. These structures probably provide a morphological basis for electrical coupling and the propagation of electrical activity through smooth muscle bundles. However, it is possible that this type of intermuscle fibre relationship is a labile formation, its integrity depending on both the physiological and fixation conditions.
4. 'Desmosomes', or desmosome-like structures consisting of small, heavily staining electron-dense areas on adjacent muscle membranes, are common in some tissues, and probably represent specialized areas concerned with maintaining the mechanical integrity of the tissue during extension and contraction.

Cell organelles
Cell membrane and plasmalemmal (micropinocytotic) vesicles

The plasma membrane is clearly defined. It is about 70 to 110 Å thick and is composed, as in most other tissues, of three layers, 2 outer darkly osmiophilic layers (25 Å thick) separated by a dense layer (about 30 Å thick). The earlier estimates (Mark, 1956; Bergman, 1958) of plasma-membrane thickness were greater (up to 250 Å in rat ureter), but this can probably be explained in terms of technical difficulties.

A basement membrane of variable thickness in different visceral tissues (50 to 250 Å) is separated from the plasma membrane by a clear interspace (60 to 130 Å) (Caesar *et al.*, 1957; Oosaki & Ishii, 1964; Yamauchi, 1964; Lane & Rhodin, 1964*a*). In the chick amnion there are large areas where the membranes of neighbouring cells are separated by a narrow zone where basement membrane is absent, as shown in Fig. 3c. (Evans & Evans, 1964). The basement membrane about smooth muscle cells in arteries and arterioles is often well developed and of the order of 800 Å (Yamamoto, 1962; Rhodin, 1967). Elastic fibres are sometimes directly attached to the plasma membrane of arterial muscle cells (Yamamoto, 1962).

The plasma membrane of smooth muscle shows several features not characteristic of other muscles. There are numerous bottle-shaped *caveolae intracellulares* (700 × 2500 Å) (Caesar *et al.*, 1957; Prosser *et al.*, 1960; Yamauchi, 1964; Simpson & Devine, 1966; Nagasawa & Suzuki, 1967) (see Fig. 12b) which are comparable to plasmalemmal vesicles (Bruns & Palade, 1968) or micropinocytotic vesicles seen in endothelial or epithelial cells involved in resorptive processes (Bennett, 1956; Clark, 1959; Palay & Karlin, 1959). Some workers have said that caveolae are less abundant (Moore & Ruska, 1957; Rhodin, 1962) and others that they are more abundant, in vascular than in visceral smooth muscles (Pease & Molinari, 1960; Keech, 1960; Karrer, 1961; Verity & Bevan, 1966). The explanation for this controversy may lie in the age of the tissue examined; old vascular smooth muscle has many more caveolae than young tissue (Burnstock, unpublished observations). In rabbit pulmonary artery, about 45% of the plasmalemma surface was estimated to be occupied by vesiculation (Verity & Bevan, 1966). In a quantitative study of guinea-pig coronary and splanchnic arterioles, many more plasmalemmal vesicles were seen on the outer (towards adventitia) and inter-faces of muscle cells than on the inner (towards lumen) surfaces (Lever, Ahmed & Irvine, 1965; Irvine, Lever & Ahmed, 1965). Caveolae are extremely rare in smooth muscle cells of the mouse vas deferens up to 8 days after birth (Yamauchi & Burnstock, 1969a). The number of caveolae in uterine cells increases during pregnancy (Kameya, 1964).

The appearance of vesicles on both sides of two opposing cell membranes has been described in rabbit colon and guinea-pig bladder by Nagasawa & Suzuki (1967) and designated as a special structure called 'paired pinocytosis'. Dark areas (see pp. 31–33) alternate with regions of 'pinocytotic activity' (Caesar *et al.*, 1957) in most smooth muscles and often appear to be arranged in longitudinal rows down the muscle length (Prosser *et al.*, 1960). Plasmalemmal vesicles are particularly abundant at the ends of muscle cells in the rabbit colon (Nagasawa & Suzuki, 1967).

The sarcoplasmic area beneath membrane areas with abundant vesicles usually contains subsurface cysterna and/or vacuoles (0·1 to 0·3 μ diameter) (Prosser *et al.*, 1960; Yamauchi, 1964). Burnstock & Merrillees (1964) observed (Fig. 4b) occasional continuity between micropinocytotic vesicles and the endoplasmic reticulum in smooth muscle cells of the rat vas deferens, comparable to the relationship in endothelial cells (Moore & Ruska, 1957). Gansler (1961) claimed that vesicles increased in size and number after contraction of smooth muscle cells of the guinea-pig colon. The cell membrane of muscle cells in chick amnion

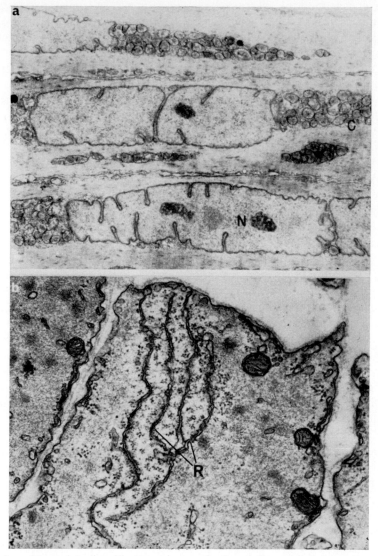

FIG. 4 *a* Mouse intestine. Electron micrograph of partially contracted smooth muscle cells in longitudinal section. The cell borders are slightly ruffled. The cell is roughly the shape of a double cone. The nucleus (N) is cylindrical but shows convolution of its membrane. The organelles at the nuclear pole form a broad short column (C). These mitochondria and others scattered among the myofilament bundles are not oriented in the long axis of the cell. × 5600. (From Lane, 1965) *b* Rat vas deferens, outer longitudinal muscle coat, oblique section. After a fixative containing potassium dichromate and calcium chloride, the endoplasmic reticulum is well defined. Many vesicles and short segments of tubules are seen under the plasma membrane. Large arrays of endoplasmic reticulum (R) are commonly seen in rat muscle, often closely associated with the plasma membrane. It may be that the reticulum is continuous with vesicles invaginating the plasma membrane. × 24,000. (From Burnstock and Merrillees, 1964)

frequently shows foldings forming cisterns which extend into the cytoplasm (Evans & Evans, 1964). Similarly the surfaces of the smooth muscle cells facing the adventitia in terminal arterioles often have a series of grooves containing basement membrane material (Rhodin, 1967).

The function of the plasmalemmal or micropinocytotic vesicles in smooth muscle is still not known. However, their association with other sarcoplasmic features such as mitochondria and endoplasmic reticulum suggests that they may be involved in active transport processes. It has been suggested that they may compensate for the sparse supply of blood capillaries in this tissue (Caesar et al., 1957). Other suggestions are that they may form part of a system to facilitate the speed of excitation by decreasing electrical resistance (Yamauchi, 1964), or that they are associated with 'relaxing' factor (Nagasawa & Suzuki, 1967).

In an electron microscope study of the localization of nucleoside phosphatases in the rat intestine, a high level of ATPase activity was found in the pinocytotic vesicles (Rostgaard & Barrnett, 1964; Lane, 1967).

Since the number of plasmalemmal vesicles varies from one specimen of the mouse vas deferens to another, even at the same postnatal age, according to whether fixation is with osmium tetroxide, glutaraldehyde or potassium permanganate, it appears that they are labile formations easily affected by fixation conditions (Yamauchi & Burnstock, 1969a).

Nucleus and centrosome

The nucleus is an ellipsoidal body. It is relatively large and, unlike the cytoplasm, remains about the same size from a very early stage in development (Yamauchi & Burnstock, 1969a). From serial sections, the nucleus in the muscle cells of guinea-pig vas deferens was shown to be 11 to 13 μ long (Merrillees, 1968). It has been described as up to 25 μ in length in relaxed intestinal and vascular smooth muscle (Rhodin, 1962) and was claimed to be 15 μ in relaxed mouse jejunum longitudinal muscle cells (Lane, 1965).

The nucleus in smooth muscle cells is characterized by spiral indentations which disappear on stretching and become more marked after contraction (see Fig. 4a). The diameter of the stretched nucleus can be as little as 1 μ and on contraction as much as 3 μ (Lane, 1965).

The nuclear membrane is a double structure with a total thickness of about 270 Å (Caesar et al., 1957) and is clearly perforated by nuclear pores (Bergman, 1958; Yamamoto, 1960; Evans & Evans, 1964; Yamauchi & Burnstock, 1969a). These pores are bridged by a thin diaphragm (Rhodin, 1962). A perinuclear space of about 130 Å separates the outer (80 Å) and inner (80 Å) components of the membrane.

The nucleoplasm contains one or two nucleoli which are particularly abundant in dioestrous uterine smooth muscle cells (Burnstock & Laparta — unpublished results) and prominent, mainly peripheral aggregations of chromatin.

A centrosome has been observed a few times in smooth muscle cells (Caesar, Edwards & Ruska, 1957; Rogers & Burnstock, 1966b) and has an identical structure to that seen in other cells (de Harven & Bernhard, 1956). It is often found to be set into a small depression of the nucleus and appears as a rounded body with a dense central vesicle encircled by 9 others (Caesar et al., 1957).

The nucleo-cytoplasmic ratio estimated at the sectional level in cells of mouse vas deferens showed a rapid reduction in neonatal stages followed by a slower, but consistent decrease during subsequent development even up to 6 months after birth (Yamauchi & Burnstock, 1969a).

Endoplasmic reticulum

The endoplasmic reticulum is not a well-developed system in most smooth muscles, especially when compared with striated muscles (Mark, 1956; Caesar et al., 1957; Peachey & Porter, 1959; Prosser et al., 1960; Gansler, 1961) but appears to be more prominent in fast contracting smooth muscles such as rat vas deferens (Burnstock & Merrillees, 1964) (see Fig. 4b).

A rough-surfaced reticulum is mostly found at cell margins near areas with pinocytotic vesicles and in the perinuclear regions (Caesar et al., 1957). Rough-surfaced reticulum is particularly prominent in developing smooth muscle (Yamamoto, 1961; Shestopalova, 1964; Leeson & Leeson, 1965a, b; Yamauchi & Burnstock, 1969a) and in rat uterus during oestrus or after treatment with oestrogen (Laguens & Lagrutta, 1964; Ross & Klebanoff, 1967). It has been suggested that since oestrogens have been shown to promote the synthesis of uterine RNA, collagen and noncollagenous protein, smooth muscle cells of the oestrous uterus may participate in the synthesis of connective tissue proteins (Ross & Klebanoff, 1967). It has also been suggested that muscle cells in the developing rat aorta have two functions, one being contractile and the other being secretory within sharply-defined cytoplasmic regions of each cell (Cliff, 1966).

A longitudinal tubular system in relation to myofilaments has been described (Yamauchi, 1964; Lane, 1967) (see p. 33) and the presence of transversely-orientated tubules has been claimed in some smooth muscles (Rogers, 1964).

Mitochondria, lysosomes and Golgi complex

The mitochondria may be spherical but are usually cigar-shaped structures about 0·2 to 0·25 μ in diameter and 0·5 to 0·8 μ long (Caesar et al., 1957; Prosser et al., 1960). As in other tissues, they are bound by a double membrane and contain prominent cristae. These cristae sometimes, but not always, run parallel to the long axis (e.g. intestine). 'Giant' mitochondria, more than 1 μ in minor diameter are sometimes seen in developing muscle cells of the mouse vas deferens (Yamauchi & Burnstock, 1969a). During pregnancy there is an increase in numbers of mitochondria in uterine muscle cells (Mark, 1956; Yamamoto, 1961; Jaeger, 1963; Kameya, 1964). Mitochondria are most abundant near the ends of the nucleus where they often extend out in long chains (see Fig. 4*a*) and also close to the cell membrane behind regions with many micropinocytotic vesicles (Prosser *et al.*, 1960).

Dense granules are sometimes seen within smooth muscle mitochondria (Merrillees *et al.*, 1963; Rogers & Burnstock, 1966*b*). They appear to increase with age in cells of the mouse vas deferens (Yamauchi & Burnstock, 1969*a*). It has been suggested that similar granules in the mitochondria of rat liver are concerned with Ca transport (Greenawalt, Rossi & Lehninger, 1964; Peachey, 1964; Greenawalt & Carafoli, 1966).

Few structures resembling lysosomes (De Duve, 1959) have been described in smooth muscles (Rhodin, 1962).

The perinuclear sarcoplasm is characterized by rough-surfaced endoplasmic reticulum with dispersed ribosomes (or RNP particles), many mitochondria, smooth surfaced vacuoles, and a Golgi complex (Caesar *et al.*, 1957; Prosser *et al.*, 1960; Merrillees *et al.*, 1963; Evans & Evans, 1964; Lane, 1965). The relatively small Golgi complex is composed of a series of vacuoles in close relation to several parallel membranes and smaller vesicles (Caesar *et al.*, 1957) as described for other tissues (Dalton & Felix, 1956). Smooth muscle cells show well-developed Golgi apparatus and ribosomes in early development (Yamamoto, 1961; Leeson & Leeson, 1965*a, b*; Shestopalova, 1964; Yamauchi & Burnstock, 1969*a*).

Cell inclusions
1. Lipid granules have been seen in both visceral and vascular smooth muscle (Rhodin, 1962). These are usually spherical, about 0·3 μ diameter, densely and homogeneously osmiophilic.
2. Glycogen granules, sometimes in the form of rosettes are common in smooth muscle and are particularly prominent in pregnant uterine muscle (Kameya, 1964; Nemetschek-Gansler, 1967) and in the smooth muscle of amphibia (Boyd, Burnstock and Rogers, 1964; Rogers & Burnstock, 1966*a, b*; Burnstock & Robinson, unpublished data) and of reptiles (Burnstock & Merrillees, 1964). The abundance of glycogen, characteristic of embryonic cardiac muscle cells (Yamauchi, 1965), was never observed at any stage in postnatal development of smooth muscle cells of the mouse vas deferens (Yamauchi & Burnstock, 1969*a*).
3. 'Juxtaglomerular granules' have been identified in arteriolar smooth muscle cells in the sheep kidney (Simpson & Devine, 1966). These are conspicuous, homogeneously staining, membrane-bound granules, round or oval in shape up to approximately 1 μ in diameter. Muscle cells containing these granules tend to occupy a peripheral position in the arteriole.
4. Multivesicular bodies have been described in smooth muscle cells of chick amnion (Evans & Evans, 1964).
5. Rounded membranous structures of unknown significance were described in the cells of the guinea-pig vas deferens (Merrillees *et al.*, 1963). These consisted

2*

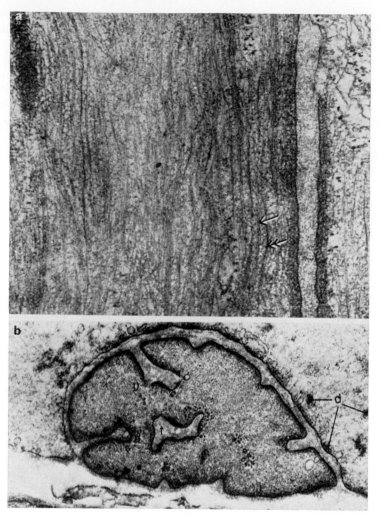

Fig. 5   *a* High-power view of myofilaments from fresh chicken gizzard embedded in Araldite. This micrograph shows light and dark filaments. The dark filaments appear thick in the centre (100 to 140 Å wide) and tapered at both ends (50 Å wide) but at present it is not possible to decide whether this represents a true picture of their shape, or whether it is a consequence of the filaments passing out of the plane of these very thin sections. The light filaments are much more numerous, thinner and of uniform diameter (50 Å). × 63,000. (From Needham & Shoenberg, 1964)   *b* Guinea-pig vas deferens. A transverse section through the tapering end of a smooth muscle fibre lying in a hollow in another fibre. The filamentous material is becoming uniformly compacted at the expense of other organelles, including the scattered densities (d in the neighbouring fibre). A similar dense material, normally found in patches against the inside surface of the plasma membrane, now forms an almost continuous peripheral zone. Deep grooves and tunnels (t), filled with basement membrane material, frequently penetrate the tips of the fibres, parallel to the long axis. × 28,000. (From Merrillees, Burnstock & Holman, 1963)

of concentric flattened cysternae, each cysterna being separated from the next by a narrow zone containing a row of granules, somewhat smaller than ribosomes. It was suggested that these bodies might be complex formations of ergastoplasm.

Fine structure of contractile apparatus

Myofilaments

A number of authors using osmium fixative methods have described tightly packed myofilaments lying approximately parallel to the long axes of the muscle cell and occupying the entire cell area except the terminal perinuclear zones (Gansler, 1956; Mark, 1956; Caesar et al., 1957; Shoenberg, 1958; Pease & Molinari, 1960; Karrer, 1961; Conti et al., 1964; Lane, 1965; Panner & Honig, 1967). These filaments often appear to be arranged in pairs (Prosser et al., 1960) and are about 30 to 80 Å in diameter in most smooth muscles, i.e. approximately the same size as *actin filaments* in striated muscle. There is some variation in diameter reported in different tissues (Table 3). The thin actin filaments could not be followed for distances longer than 2 µ in most preparations whether thick or thin sections were studied (Rhodin, 1962; Lane, 1965). Thus, it has been suggested that either actin myofilaments have a maximum length of about 1 to 2 µ, or that they are longer, but twisted around each other. The ends of muscle cells contain the most dense aggregation of filaments (Merrillees et al., 1963) (Fig. 5b).

Several authors have distinguished filaments of two sizes (Fig. 5a) in smooth muscles (Choi, 1963; Rhodin, 1963; Needham & Shoenberg, 1964; Evans & Evans, 1964; Lane, 1965; Kelly & Rice, 1968; Yamauchi & Burnstock, 1969a). However the status of the thicker filaments remains in considerable doubt (see p. 27). In mouse intestinal muscle, thin filaments (20 to 30 Å diameter) were numerous and formed distinct bundles of 5 or more (Lane, 1965). He claimed to distinguish thicker filaments (50 to 60 Å diameter) which were fewer in number and rarely seen in discrete bundles. In chicken gizzard, Choi (1962) noted the presence of thick filaments 50 to 70 Å in diameter interspersed with finer filaments less than 30 Å diameter. In both these studies, the tissue was fixed in osmium tetroxide without prior glycerol extraction or glutaraldehyde fixation. This may account for the diameters of the two filament types being different from those measured after glutaraldehyde fixation, as described by Needham & Shoenberg (1964) for guinea-pig taenia coli, rabbit uterus and chicken gizzard (100 to 140 Å for thick and 50 Å, for thin filaments), by Yamauchi & Burnstock (1969a) for the mouse vas deferens (70 to 140 Å and 30 Å) and by Conti et al., (1964) for cow carotid artery (90 to 130 Å and 60 to 90 Å). In a study of sections of glycerinated chicken gizzard smooth muscle, Kelly & Rice (1968) claimed to be able to demonstrate a well-ordered arrangement of thick filaments surrounded by 7 or 8 thin filaments, providing the pH on fixation was less than 6·6.

TABLE 3. Diameter of myofilaments found in smooth muscle

Tissue	Animal	Thin filaments (Å in diam)	Thick filaments (Å in diam)	Fixation method	Reference
Uterus	Rabbit	—	150	OsO_4	Csapo, 1955
Uterus	Rat Human	—	100–200	OsO_4	Mark, 1956
Urinary bladder	Mouse	—	100–200	OsO_4	Caesar et al., 1957
Uterus	Rabbit (pregnant)	–80	—50	OsO_4	Shoenberg, 1958
Stomach	Frog	20–40	—	OsO_4	Gansler, 1960
Pial vessels	Cat Monkey	30	—	OsO_4	Pease & Molinari, 1960
Uterus Colon	Rat Guinea-pig	40–60	—	OsO_4	Gansler, 1961
Aorta	Mouse	40	—	OsO_4	Karrer, 1961
Uterus	Rabbit	40–100	—	OsO_4	Yamamoto, 1961
Gizzard	Chicken	30	50–70	OsO_4	Choi, 1962
Circular intestinal muscle	Mouse	50–80	—	OsO_4	Rhodin, 1962
Femoral artery	Mouse	50–80	—	OsO_4	Rhodin, 1962
Vas deferens	Mouse	40	100	OsO_4	Rhodin, 1963
Carotid artery Relaxed	Cow	91	125		Conti, Haenni, Laszt, Rouiller, 1964
Tonic contraction (KCl)		63	91	Glutaraldehyde	
Phasic contraction (electrical stimulation)		—	128		
Uterus	Rabbit	—	100	OsO_4	Kameya, 1964
Gizzard	Chicken	50	100–140	Glutaraldehyde	Needham & Shoenberg, 1964
Intestine	Mouse	20–30	50–60	OsO_4	Lane, 1965
Gizzard	Turkey	20–60	—	Glutaraldehyde or OsO_4	Panner & Honig, 1967
Vas deferens	Mouse	30	70–140	Glutaraldehyde	Yamauchi & Burnstock, 1969a
Gizzard (glycerinated fibres)	Chicken	50–70	110–210	Glutaraldehyde (pH 5·8–6·6)	Kelly & Rice, 1968

Weinstein and Ralph (1951) using fragmented material of chicken and turtle intestine treated with nickel, observed a filament periodicity of 320 to 740 Å. This has not been shown in other tissues fixed with osmium although, after staining Epon sections with lead hydroxide and/or uranyl acetate, 'beaded' filaments were seen with a periodicity of 100 Å (Rhodin, 1962). Regions of myofilaments in cow carotid artery muscle showing low electron density after glutaraldehyde fixation have also been described (Conti *et al.*, 1964). In glycerin-ated muscle fixed in glutaraldehyde, postfixed in OsO_4, and doubly stained, thin filaments from chicken gizzard smooth muscle have been shown to exhibit substructure (Panner & Honig, 1967). Many filaments appeared to be composed of globular subunits about 50 Å in diameter, which were sometimes arranged in a fashion consistent with a two-dimensional projection of a helix; this is further evidence for the identity of thin filaments with actin.

The myofilaments in the iris dilator muscle are a particularly prominent feature (Richardson, 1964), but no special study has been made of them yet in this tissue. The arrangement of myofilaments in the smooth muscle cells of the nictitating membrane is also unusual. Numerous lightly-staining myofilaments with a minority of thicker darkly staining myofilaments form a sleeve around the nucleus and perinuclear sarcoplasm (Fig. 6), from which processes extend to the cell membrane (Evans & Evans, 1964).

As pregnancy proceeds there is an increased number of myofilaments in uterine muscle cells (Mark, 1956; Yamamoto, 1961), but very few were seen in castrated uterus (Kameya, 1964). Uterine muscle cells after oophorectomy and those seen after ovulation and during pregnancy contain many more filaments than do those stimulated by oestrogen (Nemetschek-Gansler, 1967).

During the development of smooth muscle cells of the mouse vas deferens, fine myofilaments (about 30 Å diameter) were seen one day after birth, whereas coarse myofilaments (70 to 140 Å diameter) were first observed at 6 days (see Fig. 5a). The coarse myofilaments increased in thickness and number up to 6 months. A transient appearance of ribosomal rosettes also occurred between 6 and 12 days postnatal, suggesting involvement in myosin synthesis (Yamauchi & Burnstock, 1969a). In the rabbit uterus, myofilaments did not appear in smooth muscle cells until 12 days after birth (Yamamoto, 1961).

The properties of actin isolated and purified from uterine smooth muscle (Carsten, 1965) resemble closely those of actin prepared from skeletal and car-diac muscle. Skeletal muscle myosin will react normally with purified prepara-tions of uterine actin to give a highly viscous actomyosin (Needham & Williams, 1963b). Similarly, purified uterine myosin will react with skeletal muscle actin.

Tropomyosin B or soluble tropomyosin has also been found in smooth muscle (Sheng & Tsao, 1955; Jaisle, 1960; Needham & Williams, 1963a), and is present in much higher proportion to actomyosin than in skeletal muscle, i.e. 1:2–3 in smooth muscle (Needham & Williams, 1963b) compared to 1:5–6 in skeletal muscle (Perry & Corsi, 1958). Both troponin and α-actinin have been isolated recently from smooth muscle (Ebashi, 1969); it is suggested that α-actinin is localised in the dense bodies.

Location of myosin

In most muscles (striated or invertebrate non-striated) there are distinct filament types, thin filaments containing actin and thicker filaments containing

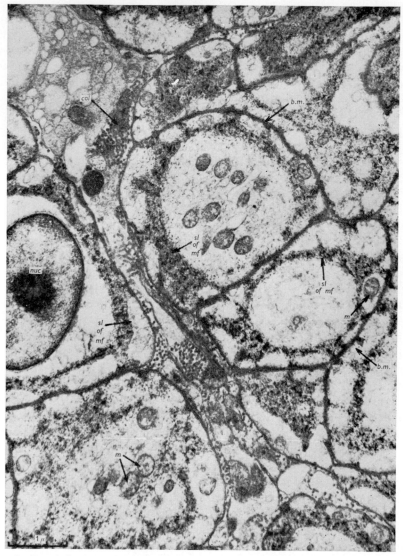

FIG. 6. Cat nictitating membrane. Transverse section through smooth muscle showing annular disposition of myofilaments. col = collagen fibres; b.m. = basement membrane; nuc = nucleus; m = mitochondria; sl. of mf. = sleeve of lightly staining myofilaments with a minority of thicker, darkly staining myofilaments around the nucleus and perinuclear sarcoplasm. (From Evans and Evans, 1964)

myosin (Hanson & Lowy, 1964b). F-actin filaments have been clearly identified in smooth muscle, but there is some debate about the presence of myosin filaments. Hanson & Lowy (1963, 1964b) using homogenates of the guinea-pig taenia coli and Shoenberg, Rüegg, Needham, Schirmer & Nemetchek-Gansler

(1966) using bovine artery were able to identify actin, but not myosin filaments. Furthermore, Elliot (1964, 1967) could only demonstrate actin patterns in high and low angle x-ray diffraction. However, it is well known that the solubility properties of smooth muscle myosin differ from those of skeletal muscle (Hasselbach & Ledermair, 1958; Needham & Williams, 1959, 1963a, b, c; Laszt & Hamoir, 1961; Filo, Rüegg & Bohr, 1963; Mallin, 1965) and this may account for some of the difficulties. Furthermore, smooth muscle contains very much less actomyosin than striated muscle (6 to 10 mg/g wet wt in smooth muscle compared to about 70 mg/g wet wt in skeletal muscle (Needham & Williams, 1963a)) although purified uterine actomyosin has been shown to contain actin and myosin in a ratio of about 1:4, as in skeletal actomyosin (Needham & Williams, 1963b).

As mentioned above, apart from the thin filaments, thicker and more darkly staining filaments have been described in sections of various smooth muscles especially if prepared with glutaraldehyde (Needham & Shoenberg, 1964; Yamauchi & Burnstock, 1969a). These are uneven in length and irregular in distribution. They may not be true filaments, but rather random superimposition of thin filaments, which would be quite possible in the relatively thick sections used. In any case, there appears to be an insufficient number of these filaments to account for all the myosin in the cell. The sparseness and variation in diameter of myosin filaments could be explained if the major part of the myosin were in the form of soluble 'tonoactomyosin' in various stages of aggregation (Laszt & Hamoir, 1961; Needham & Shoenberg, 1964; Elliot, 1964) or, to put it another way, in relaxed muscle the myosin may exist largely in a colloidally-dispersed phase (Shoenberg et al., 1966). Another possibility is that the thick filaments represent aggregates of myosin, or actin filaments upon which myosin had been deposited during fixation. Dark, thick filaments were prominent in fresh fixed material but were not seen in glycerinated muscle or in muscle washed in solutions of low ionic strength before fixation, after the cell membrane had been damaged (Needham & Shoenberg, 1964). These authors suggest that this could be explained in terms of the extraction of tonomyosin in the damaged preparations and aggregates of tonomyosin by fixation in fresh material. In a more recent study (Shoenberg et al., 1966), it was concluded that 'in relaxed vertebrate smooth muscle the contractile protein consists of actin in filament form and disaggregated myosin, or myosin aggregated to such a small degree that the aggregates are not visible by negative staining'.

Panner and Honig (1967) denied the existence of thick myosin filaments in smooth muscle and made the interesting suggestion that myosin exists in a relatively disaggregated form as single molecules or dimers either free or attached to actin. In glycerated sections, lateral processes about 30Å wide and resembling heavy meromyosin from skeletal muscle were demonstrated at intervals along the actin filaments. On the basis of these results, Panner & Honig (1967) put forward a model of the contraction mechanism of smooth muscle in which shortening occurs by interdigitation and sliding of actin filaments to which myosin dimers are attached as small functional units.

Homogenates of smooth muscle — synthetic myofilaments

Negatively-stained filaments of actin (60 to 80 Å), but not myosin, were found in electronmicrographs taken from extracts of actomyosin prepared from homogenized smooth muscles of the guinea-pig taenia coli and pregnant rat uterus (Hanson & Lowy, 1963, 1964*b*; Shoenberg *et al.*, 1966), cow artery (Shoenberg *et al.*, 1966) and turkey gizzard (Panner & Honig, 1967). These were indistinguishable from the F-actin filaments found in other muscles. This result was true whether fresh muscles or glycerol-stained muscles treated with EDTA and ATP-Mg were used. However, it was necessary to homogenize the muscle for relatively long periods, because of the high amounts of connective tissue present, and because, under these conditions, actin filaments have been shown to be preserved, but myosin filaments to disintegrate in other muscle systems.

However, if the method, used so successfully by Huxley (1963) for isolating the contractile proteins of striated muscle, is applied under modified conditions to smooth muscle, myosin, as well as actin filaments can be demonstrated (Hanson & Lowy, 1964*b*; Shoenberg, 1965; Kelly & Rice, 1968). Smooth muscle from the taenia coli and rat uterus was disintegrated mechanically in solutions of low ionic strength, and the resulting suspension was treated with ethylenediamine-tetraacetate (EDTA), magnesium chloride and ATP to separate the actin and myosin. Even then, unless the actomyosin was first extracted with 0·6 M KCl, ultracentrifuged at 0°C in the presence of ATP and magnesium and the supernatent solution of myosin dialysed against 0·1 M KCl (pH 6·8), the myosin filaments could not be demonstrated in negative-staining electronmicroscopy. Shoenberg (1965) also demonstrated myosin as well as actin filaments from homogenized taenia coli, but only in the presence of sigma collagenase. The myosin filaments were more numerous and more clearly defined after standing overnight, ageing or lowering the pH.

Myosin filaments obtained in this manner were from 0·3 to 0·7 μ long and about 350 Å in diameter (Fig. 7*b*). They often had thickened endings and some-times irregularly-distributed protuberances like those described by Huxley (1963) in synthetic myosin filaments from skeletal muscle. Smooth muscle myosin filaments were somewhat shorter and thicker than those from skeletal muscle. Only occasionally were these filaments seen partly attached to actin filaments (Needham & Shoenberg, 1967).

It may be that the failure so far to demonstrate more than a few, if any, myosin filaments in intact smooth muscle might not be because of their absence, but rather because of the problem of finding a technique capable of demonstrating them under the unusual conditions of myosin activity characteristic of this tissue. On the other hand, it may be that few, if any, aggregates of myosin into thick filaments occur normally in smooth muscle, and that the demonstration of myosin filaments, comparable to those seen in skeletal muscle from homogenates, is merely a demonstration that under special conditions, naturally-occurring smooth muscle myosin molecules and dimers can be induced to

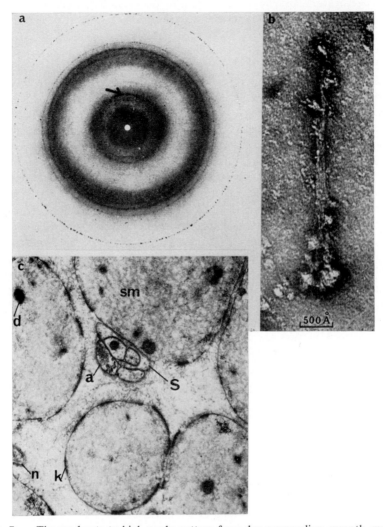

FIG. 7 *a* The moderate-to-high angle pattern from dry mammalian smooth muscle (the taenia coli muscle of the guinea-pig). For calibration purposes the meridonial actin reflexion at 9·14 Å is marked with an arrow. (From Elliot, 1964) *b* Electron micrograph of synthetic myosin filament prepared from uterus muscle, negatively stained with uranyl acetate. (From Hanson & Lowy, 1964*b*) *c* Toad large intestine, circular muscle coat. Note the dense bodies (d) and dense areas (k) which are a prominent feature of the smooth muscle cells. Small nerve fascicle composed of axons (a) and Schwann cell processes (S) lying in close apposition with smooth muscle cell (sm). Note Schwann cell separates axons from muscle cell. (n), single naked axon (no Schwann cell sheath). × 17,500. (From Rogers & Burnstock, 1966*b*)

aggregate to form filaments. A study of the influence of ATP and calcium on the aggregation and filament formation of smooth muscle myosin at low ionic strength may help to clarify the situation.

X-ray diffraction studies

Studies have been made of diffraction patterns from smooth muscle of the dog retractor penis and the guinea-pig taenia coli (Bear, 1945; Elliot, 1964, 1967; Huxley, 1966).

The best patterns were obtained from guinea-pig taenia coli which had been slightly stretched and then dried, but it has also been examined wet, after fixation with formaldehyde and phosphotungstic acid stain, after glycerol extraction and in the living relaxed state (Fig. 7a).

At high angles, the taenia did not give the typical-pattern characteristic of all the other muscles investigated. The equatorial diffraction in the 10 Å region resembles the diffraction from dried F-actin films (Astbury, Perry, Reed & Spark, 1947) rather than the horseshoe-shaped reflexion typical of muscles containing myosin and paramyosin (Cohen & Holmes, 1963). At moderate angles a typical actin pattern is observed, while at low angles, collagen, but neither myosin nor paramyosin patterns is present. Both high and low angle diffraction evidence suggests that the myosin in smooth muscle is not organized into distinct orientated filaments.

Huxley (1966) confirmed these observations, but suggested that, since the myosin pattern is very easily disturbed, this evidence was not conclusive. However in a more recent study of taenia coli under a wide variety of preparative conditions (dried muscle; living muscle without stimulation; glycerol extracted muscle; muscle fixed in formaldehyde, glutaraldehyde and osmium, with or without phosphotungstic acid staining), no low or high-angle reflections from myosin or its homologues have been observed (Elliot, 1967).

Myofilaments in relaxed and contracted muscle

Comparison of electronmicrographs taken of relaxed and contracted taenia coli showed that the filaments did not increase in diameter (Shoenberg, 1962). A few filament counts were made by Shoenberg (1962) of the number and diameter of end-on filaments in relaxed and contracted muscle at the nuclear level. The diameter of the filaments did not change, but the total number of filaments was much greater in contracted than relaxed cells. This was taken to indicate that some kind of sliding mechanism might be operating, but it is difficult to visualize this unless the myosin is predominantly present in the form of filaments.

In contracted mouse intestinal muscle cells there was an increase in density of the myofilament bundles, a straighter and more parallel course of the myofilaments within a bundle, and an increase in the number and length of dense bodies in their interstices (Lane, 1965). Strong contraction also resulted in forces which were exerted on the cell surface and which deformed it. The changes in position and orientation of cell organelles during contraction were probably a reflection of the forces involved in the reorientation of myofilaments. For example, in the relaxed cell, the central column of mitochondria and RNP particles extending from the nuclear poles was long and narrow; as the cell contracted these organelles tended to form a mass near the poles of the nucleus and in the perinuclear zone. The mitochondria in these regions were usually rotated through 90° and came to lie perpendicular to the long axis of the cell. The mitochondria interspersed amongst the myofilaments however, became more clearly aligned longitudinally, parallel with the myofilaments.

Some studies were also made of the changes occurring during phasic and tonic contraction of cow carotid smooth muscle (Conti *et al.*, 1964). Two completely different contractile mechanisms were postulated. During tonic contraction produced by raising the extracellular potassium concentration, the filaments were arranged in a network, with both longitudinally and laterally orientated myofilaments. The stability and density of this network increased as tonus was increased and the diameter of the filaments was reduced. These changes were related to a reduction in hydration, associated with the entry of K^+ ions, leading to a reorientation and stabilization of the polypeptide chains of the contractile proteins. During phasic contraction produced by electrical stimulation, the ultrastructure of the myofilaments showed that they became longitudinally orientated, parallel with the main axis of the cell. A network of filaments was never seen, nor were regions corresponding to the points of attachment of lateral filaments.

'Light' and 'dark' cells

Some workers have demonstrated cells which vary considerably in the electron density of the cytoplasm in the same electronmicroscope section, especially in conditions of great shortening to approximately one-third resting length (Gansler, 1960, 1961; Merrillees, personal communication; Conti *et al.*, 1964). In general, the number of 'light' cells is smaller than the number of 'dark' cells. Uterine smooth muscles, under the influence of progesterone, after ovulation and during pregnancy show a predominance of dark cells (Nemetschek-Gansler, 1967). Whether this difference represents cells in different states of contraction is not known. However, this observation has been used as the basis of a theory that the mechanism of smooth muscle contraction depends on synaeresis, i.e. changes in state of actomyosin (Gansler, 1961; Conti *et al.*, 1964). Dark cells are believed to be highly contracted, having lost water and decreased in volume. During tonic contraction of the cow carotid artery produced by potassium, light and dark cells were revealed, but during phasic contraction produced by electrical stimulation all the cells were of equal density (Conti *et al.*, 1964).

Dark (or dense) bodies and dark (or dense) areas

Darkly-staining areas, measuring about 4000 Å by 700 Å are found in the cytoplasm of most smooth muscle cells (Figs. 2, 4, 7). In some tissues they seem to be randomly dispersed (Mark, 1956; Shoenberg, 1958; Charles, 1960; Harman, O'Hegarty & Byrnes, 1962; Rhodin, 1962; Lane, 1965; Rogers & Burnstock, 1966b; Simpson & Devine, 1966); in others they appear to be arranged in a distinct evenly spaced pattern and it has been suggested that they may form a kind of skeletal lattice (Prosser et al., 1960). Dark bodies are not present in the extremities of cells where the myofilaments are closely packed (Merrillees et al., 1963) (Fig. 5b).

Myofilaments often appear to converge upon and enter these dark bodies (Mark, 1956; Prosser et al., 1960). Lane (1965) claimed that in mouse intestinal muscle the dense bodies are the only places where both thick and thin filaments converge. In some electronmicrographs, dark bodies appeared to represent an aggregation of thin filaments (Yamauchi, 1964). In rabbit arteriole smooth muscle cells the dense bodies form long bar-like structures regularly distributed and projecting inwards from the cell surface (Rhodin, 1967).

According to Prosser et al. (1960), treatment with alkaline KCl, lipase, lecithinase, RNA-ase, hyaluronidase, lysozyme, ethanol-ether-chloroform mixture and 50% glycerol failed to remove the dark bodies. Papain and trypsin removed the myofilaments first and, after prolonged treatment, the dark bodies as well, suggesting that they are formed of protein, but not of contractile protein, although the latter could be a component contributing to them. In ultra-thin sections, fine canaliculi appear to form part of the structure of dark bodies (Rogers, 1964).

The dark bodies are longer (up to $0.8\ \mu$), but not wider in contracted than in relaxed cells (Lane, 1965). They are several times more numerous per unit area in contracted cells than in relaxed cells. Rhodin (1962) has claimed that the dark bodies, while present in contracted mouse intestinal muscle, disappear completely in 'totally relaxed muscle', but most workers have not seen this correlation. On the basis of this result, he suggested that 'these structures are the structural evidence for a smooth muscle contraction, although the dark bodies may not be actively engaged in the contraction process'.

Several possible functions have been suggested for 'dark bodies'. Pease & Molinari (1960) referred to them as 'attachment devices' for the myofilaments. Prosser et al. (1960) suggested further that they might form a well-organized lattice for the muscle filaments. The possibility that they might correspond in function to the Z-membranes in skeletal muscle has also been mentioned (Needham & Shoenberg, 1964, 1967; Rogers, 1964). Support for this view is the demonstration of dense bodies with actin filaments emerging from both ends, in homogenates of smooth muscle (Panner & Honig, 1967), since these resemble isolated Z discs from skeletal muscle (Huxley, 1963). Furthermore, in sections, short, dark-staining lateral filaments 15 to 25 Å in diameter link adjacent actin filaments with dense bodies, again suggesting homology with skeletal Z-disc filaments (Panner & Honig, 1967).

Dark bodies are said to increase heavily in the cells of the pregnant uterus (Gansler, 1961; Kameya, 1964). In the developing mouse vas deferens, dense bodies were first observed in the cytoplasm 6 days after birth, coincident with the appearance of coarse, thick myofilaments and continued to increase in number with increasing age even up to 6 months (Yamauchi & Burnstock, 1969*a*). Dark bodies were seldom seen in uterine muscle cells of the rabbit even 14 to 16 days after birth (Yamamoto, 1961).

Dark areas or patches have been observed on cell membranes of many smooth muscles (Caesar *et al.*, 1957; Prosser *et al.*, 1960; Pease & Paule, 1960; Feeney & Hogan, 1961; Pease & Molinari, 1960; Harman *et al.*, 1962; Lane, 1965). They are particularly prominent in the toad rectum (Rogers & Burnstock, 1966*b*, see Fig. 7*c*), and rabbit arterioles (Rhodin, 1967). They usually alternate with areas rich in plasmalemmal vesicles. This is particularly obvious in contracted muscle (Lane, 1965), since the muscle is invaginated in these regions alternating with evaginated pinocytotic regions. Myofilaments appear to be attached to these dark areas (Pease & Molinari, 1960; Lane, 1965; Rosenbluth, 1965*b*; Rogers & Burnstock, 1966*b*; Panner & Honig, 1967). It has been suggested by Rosenbluth (1965*b*) that filaments are inserted at an angle of about 10° in the dark areas, which he argues would lead to greater development of tension at less velocity than in striated muscle.

Sarcotubular system

The sarcotubular system is very poorly developed in smooth muscle when compared to the elaborately organized structure found in skeletal and cardiac muscles (Porter & Franzini-Armstrong, 1965). The existence of a longitudinal smooth sarcotubular system has been described in intestinal muscle (Yamauchi, 1964; Lane, 1965, 1967; Rogers, 1964). Microtubules 200 to 300 Å in diameter are scattered throughout the cytoplasm together with a system of sarcotubules 0·5 to 1 μ in diameter coursing in the longitudinal axis of the cell for up to 30 μ (Lane, 1965).

In a study of smooth muscle cells in various systems, including guinea-pig ureter and taenia coli, Rogers (1964) claimed to be able to identify transversely-orientated tubules. These tubules appeared to be continuous with the longitudinal tubules.

ATPase activity has been localized in the longitudinal tubular system (agranular endoplasmic reticulum) of smooth muscle cells of the mouse bladder (Lane, 1967). By analogy with striated muscle, where ATPase is also deposited in the longitudinal tubules of the sarcotubular system it seems likely that the longitudinal tubular system in smooth muscle is involved in the process of excitation-contraction coupling.

The existence of a poorly-developed granular endoplasmic reticulum presumably involved in protein synthesis in smooth muscle has been discussed previously.

Summary

1. Actin is present in tightly-packed filaments, 30 to 80 Å in diameter, lying largely parallel to the longitudinal axis of the cell.

2. Myosin is probably present mainly in dispersed form perhaps as single molecules, dimers or possibly tetramers attached at intervals along actin filaments. However, under certain conditions, thick filaments, 70 to 200 Å in diameter, have been seen or perhaps have been induced to form from the disaggregated myosin molecules normally present. Actin, but no myosin patterns have been seen in x-ray diffraction studies of smooth muscle. Synthetic thick myosin filaments can be formed under certain conditions from homogenized smooth muscle.

3. The myofilaments are attached to the plasma membrane at numerous areas (dark areas) which alternate with micropinocytotic vesicle-bearing regions over the cell periphery. The free ends of the bundles of myofilaments interact with other bundles at the dense bodies, which resemble Z-discs from skeletal muscle. Dense bodies and dense areas are filament attachment points and denote units analogous to sarcomeres. The whole system is a meshwork, orientated so that contraction results in forces being applied to the entire periphery of the cell. However, since the majority of the fibres lie in the longitudinal axis, the main force is extended from end to end. Further, the ends of the cells contain more filaments so that the greatest force is applied to the plasma membrane in these regions. In strongly-contracted muscle, the cell is changed from an elongated cylindrical shape to an oblate spheroid and the nucleus becomes shorter and wrinkled. Moreover, the distance between the dark areas of myofilament attachment in the membrane is reduced, causing protrusion of the vesicle-bearing regions of the membrane.

4. The proportion of actin to myosin in smooth muscle actomyosin is 1 to 4, comparable to that of skeletal muscle actomyosin, but the total amount of actomyosin present is 6 to 10 mg/g wet wt which is considerably lower than that found in skeletal muscle (about 70 mg/g wet wt). The proportion of tropomyosin B to actomyosin in smooth muscle is 1 to 2–3 compared to 1 to 5–6 in skeletal muscle. Both troponin and α-actinin are present in smooth muscle.

5. The diameter of myofilaments does not change in contracted smooth muscle suggesting that a sliding filament mechanism of contraction is operative.

6. A longitudinal sarcotubular system is present. This shows high ATPase activity, indicating involvement in excitation-contraction coupling.

INNERVATION OF SMOOTH MUSCLE

Light microscopy of intramural plexuses

Introduction

The literature describing the morphology of autonomic innervation of smooth muscle as seen from silver and methylene blue staining is vast (see Jabonero,

1959, 1965; Hillarp, 1959, 1960; Richardson, 1960; Taxi, 1965; Botár, 1966). However, although these methods allow detailed observation of the organization of intramural plexuses and of their gross connections with extrinsic nerves, very little can be learned about the relationship of single fine autonomic nerve fibres and individual smooth muscle cells. Nevertheless, the existence of a fine-meshed plexus of nonmyelinated fibres in relation to smooth muscle was established and called variously a 'Schwann plasmodium' (Lawrentjew, 1926), 'sympathetic ground plexus' (Boeke, 1949), 'terminal reticulum' (Stöhr, 1954) and 'autonomic ground plexus' (Hillarp, 1959). It is remarkable that, despite the technical limitations, Hillarp's general concept of the nature of this plexus was in many ways correct and has been confirmed and extended by the more recent work with various cellular techniques.

Myenteric plexus

As an example of the complex organization of an intramural plexus, a brief description of the innervation of the gut will follow:

1. Primary, secondary and tertiary plexuses defined according to the size and position of the nerve bundles can be distinguished in Auerbach's plexus (Fig. 8*a*) lying between circular and longitudinal muscle coats (Auerbach, 1864; Richardson, 1958, 1960; Dupont & Sprinz, 1964). The primary plexus consists of large nerve bundles forming a wide meshwork containing ganglion cells. The interstices of this network are traversed by thinner secondary nerve bundles and by a close tertiary plexus of continuous fine nerve strands between 3 and 10 μ in diameter.

2. Extrinsic nerves penetrate the longitudinal muscle coat from the mesentery to fuse with nerve bundles in the primary plexus.

3. Fine bundles of nerves from these plexuses pass inwards, mainly together with blood vessels to Meissner's plexus and the mucosa (Meissner, 1857).

4. An 'interstitial cell network' (Cajal, 1893) interweaves amongst the tertiary nerve bundles and sometimes forms close relationships with muscles (Fig. 8*b*). These are not nerve or Schwann cell elements as was originally suggested (Lawrentjew, 1926; Schabadasch, 1934; Leeuwe, 1937; Boeke, 1949; Meyling, 1953; Stöhr, 1954; Jabonero, 1960; Taxi, 1965) but have been shown with the electron microscope to be composed primarily of fibroblast cells (Richardson, 1958; Yamauchi, 1964; Taxi, 1965; Rogers & Burnstock, 1966*a*). However their function is still unknown.

5. Nerve bundles from the tertiary plexus run into the muscle coats at intervals and then run parallel to the main axes of the muscle fibres, branching into smaller bundles and finally terminating in restricted areas as single fibres, free of Schwann cell covering (Richardson, 1960; Bennett & Rogers, 1967). The tertiary plexus, together with the nerve bundles which run from it into the muscle coats, is essentially equivalent to the autonomic ground plexus as defined by Hillarp (1959).

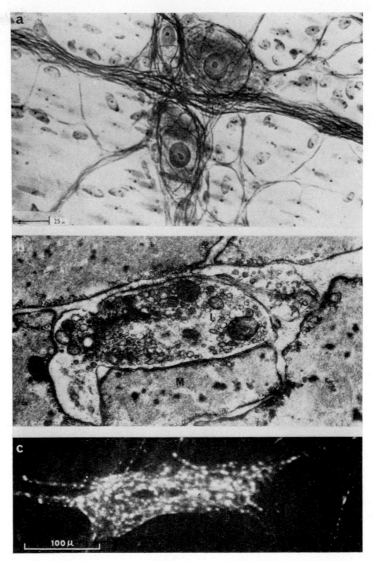

F<small>IG</small>. 8   *a* Cat small intestine. A small cluster of ganglion cells associated with a secondary bundle of Auerbach's plexus. The considerable number of fine neurites traversing the surfaces of the perikaryons makes the density of a larger ganglion too great for photomicrography. The cat intestine is too thick for satisfactory immersion fixation of the tertiary bundles which appear distorted in this preparation. (From Richardson, 1960)   *b* Longitudinal section through small intestine (*B. marinus*). A process of an interstitial cell (I) in close apposition with a smooth muscle cell (M). Palade fixation. × 14,000. (From Rogers & Burnstock, 1966a)   *c* Stripped stretch preparation of the wall of guinea-pig colon, showing green fluorescent, varicose, noradrenergic nerve fibres in Auerbach's plexus. These fibres are arranged about non-fluorescent ganglion cell bodies at the node. One hour formaldehyde treatment for demonstration of tissue catecholamines by the Falck-Hillarp technique. (Read and Burnstock, unpublished)

6. Ganglion cells of at least 3 main types lie mainly at the nodes of the myenteric plexus (Dogiel, 1898; Hill, 1927; Gunn, 1951). There are over 5 million cells in Auerbach's plexus in the small intestine of the cat and 2 to 3 times as many in Meissner's plexus (Sauer & Rumble, 1946; Leaming & Cauna, 1961). There is about a 4-fold increase in the number of nerve cells between new-born and adult rat intestine (Gabella, 1967).

7. The existence of noradrenergic terminals about ganglion cells in both Auerbach's and Meissner's plexuses has been demonstrated recently (see Fig. 8c) with the fluorescent histochemical method (Norberg, 1964; Jacobowitz, 1965; Hollands & Vanov, 1965; Gabella & Costa, 1967; Campbell & Burnstock, 1968; Read & Burnstock, 1968). This has been correlated with the presence of numerous granular vesicles in some presynaptic terminals on ganglion cells (Honjin, Takahashi, Shimasaki & Maruyama, 1965) and of silver grains seen with electronmicroscopic-autoradiography after infusion of tritium labelled noradrenaline (Taxi & Droz, 1966).

8. At least 3 different functional nerve types have been demonstrated physiologically in the gut, i.e. cholinergic, noradrenergic and non-adrenergic inhibitory nerves (Burnstock, Campbell & Rand, 1966, see also Chapter 15). It seems likely that these nerves can be correlated morphologically with three types of vesicles found in different axon profiles, i.e. predominantly agranular vesicles in cholinergic nerves; predominantly small granular vesicles in noradrenergic nerves and predominantly large granular vesicles in non-adrenergic nerves (Burnstock & Robinson, 1967).

Fluorescent histochemistry of sympathetic nerves

The highly sensitive fluorescent histochemical method for localizing monoamines (Falck, 1962; Falck *et al.*, 1962; Corrodi & Hillarp, 1963, 1964; Dahlström & Fuxe, 1964a; Eränkö, 1967; Jonsson, 1967) has been applied to the question of the nature of the innervation of smooth muscle and thus allowed the problem to be carried further at the light microscope level (see Norberg & Hamberger, 1964; Malmfors, 1965a; McLean & Burnstock, 1966, 1967a, b, c). This technique has recently been further developed to allow semi-quantitative measurements to be made (Van Orden, Vugman & Giarman, 1965; Gillis, Schneider, Van Orden & Giarman, 1966; Ritzén, 1966, 1967).

The following features of the relation of the terminal regions of nerves to smooth muscle cells have become established (see Figs. 9, 10):

1. Sympathetic nerve fibres run long distances through the smooth muscle effector system before terminating (Malmfors & Sachs, 1965). Dahlström & Häggendal (1966b) calculated that the terminal system in each neurone was about 10 cm long.

2. The nerve fibres within the muscle effector system have a varicose structure, with 10 to 30 varicosities per 100 μ (Fig. 9a). Serial sampling electron microscopy has shown that the diameter of the nerve at a varicosity can be up to 2 μ, while in intervaricose regions it is 0·1 to 0·2 μ (Merrillees, 1968).

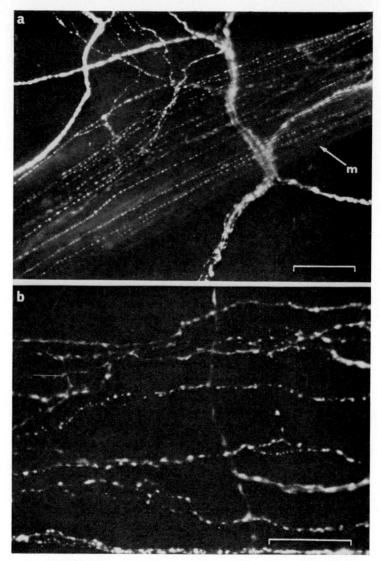

FIG. 9   *a* Band of smooth muscle (m) in the lung of the lizard (*Trachysaurus rugosus*) which is innervated by a number of fine fluorescent varicose nerve fibres containing noradrenaline. These nerves run along its length in a manner suggestive of a functional innervation. Whole mount, incubated in formaldehyde vapour for 1 hr. Calibration 100μ. (From McLean & Burnstock, 1967c)   *b* Whole mount preparation of the sheep mesenteric vein at the level of the inner surface of the adventitia, showing innervation of the medial muscle coat by an autonomic ground plexus consisting of bundles of fine varicose nerves containing noradrenaline. Incubated in formaldehyde vapour for 1 hr. Calibration 50μ. (McLean & Burnstock, unpublished)

3. The transmitter concentration in varicosities is high and there is evidence that it can be released during the passage of an impulse down the nerve, i.e. there are 'en passage', as well as terminal synapses, so that a single nerve can influence many muscle cells (see Bennett & Burnstock, 1968; Malmfors 1965a).

4. Evidence has been presented that noradrenaline (NA) is synthesized in the preterminal varicosities as well as in the perikaryon (Malmfors, 1965a; Austin, Livett & Chubb, 1967).

5. There is considerable variation in the density of the sympathetic ground plexus in different tissues. For example, it is very dense in the vas deferens, nictitating membrane, dog retractor penis, and rat iris (Falck, 1962; Norberg & Hamberger, 1964; Malmfors, 1965a) but sparse in the uterus, ureter and intestine (Norberg, 1964; Read & Burnstock, 1969b; Nakanishi, McLean, Wood & Burnstock, 1969; Campbell & Burnstock, 1968). The length of the preterminal varicose regions of the nerve and the number of close (200 Å) neuromuscular junctions per muscle cell also vary between tissues. For example, in the adult mouse vas deferens, there may be up to six close junctions on a single muscle cell (Yamauchi & Burnstock, 1969b), whereas in arteries, although the density of varicose nerve fibres is high, they are mostly confined in large bundles in the adventitia (Fig. 9b).

6. Combined biochemical and fluorescent histochemical studies (Dahlström & Häggendal, 1966a; Dahlström, Häggendal & Hökfelt, 1966; Ritzén, 1967) have revealed the following data: the amount of NA in an adrenergic nerve cell body of the cat superior cervical ganglion is about 100 μg/g wet wt; the fluorescent varicosities in peripheral adrenergic nerves have been calculated to have an average catecholamine (CA) content of about 5×10^{-9} μg or $1-4 \times 10^3$ μg/g wet wt; the total NA content of cell bodies is about 50 ng compared to 32 μg in the preterminal and terminal varicose portions (or, to put it another way, the ratio of NA content between cell bodies and terminals is approximately 1 : 300); the NA is transported down sympathetic nerves at a rate of about 5 mm/hr (Dahlström, 1965; Dahlström & Haggendal, 1966a; Livett & Geffen, 1967).

There is some confusion in the current literature concerning the terminology used to describe the varicose regions of sympathetic nerves. Fig. 10, based on both fluorescent histochemical studies and on serial electronmicroscope sampling (Merrillees, 1968) is an attempt to clarify the situation. 'Terminal varicosity' is restricted to the last varicosity of the nerve, i.e. the ending. 'Preterminal varicosities' refer to all varicosities proximal to the terminal varicosity, but three functional regions are distinguished: a, refers to varicosities in the nerve before it enters the effector organ. These varicosities are usually completely enveloped by Schwann cell processes and are unlikely to be sources of transmitter release: b, refers to the bulk of the preterminal varicosities, which lie within the effector organ and are usually only partially-enclosed in Schwann cell. Many of these varicosities represent 'en passage' synapses: c, refers to the last few preterminal varicosities before the nerve terminates. These are usually completely naked of

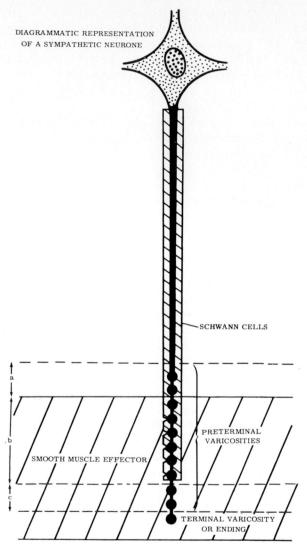

DIAGRAMMATIC REPRESENTATION
OF A SYMPATHETIC NEURONE

SCHWANN CELLS

a

b

SMOOTH MUSCLE EFFECTOR

PRETERMINAL
VARICOSITIES

c

TERMINAL VARICOSITY
OR ENDING

Fig. 10. Diagrammatic representation of a sympathetic neurone. (For explanation see p. 39.)

Schwann cell, lie in close apposition with muscle cells and represent functional synapses. In some systems examined with both silver staining and fluorescent histochemical methods, the diameter of the varicosities appears to become progressively smaller as the nerve ending is approached (Willis & McLean, personal communication). In the terminology introduced by Hillarp (1959), the term 'terminal varicosities' refers to regions *b* and *c* as well as to the terminal or end varicosity.

Fine structure of autonomic nerve-smooth muscle junctions

Introduction

The electronmicroscopy of the innervation of smooth muscle was described for the first time by Caesar et al. (1957), on the mouse urinary bladder. They established the existence of neuromuscular synapses with close apposition (70 to 200 Å) of nerve and muscle membranes. This relationship has since been confirmed in a variety of different tissues (Richardson, 1962, 1964; Merrillees et al., 1963; Thaemert, 1963; Lane & Rhodin, 1964a; Evans & Evans, 1964; Taxi, 1965; Nagasawa & Mito, 1967). The idea that transmitter could be effective even when released from nerves separated by more than 200 Å from muscle was suggested by Gansler (1960), and 'en passage' release of transmitter was proposed by Richardson (1962) and Merrillees et al. (1963). Finally, the detailed relationship of nerves and muscles has been examined with serial electronmicroscope section sampling (Taxi, 1965; Thaemert, 1966; Bennett & Rogers, 1967; Merrillees, 1968), and the varicose nature of the nerves established. Vesicles and mitochondria have been shown to be confined to varicosities while intervaricosities contain only neurotubules and/or neurofilaments. Schwann cell processes have been shown to leave varicose regions of the nerve partially naked where they pass close to muscle cells, and completely naked just prior to and at termination.

The density and pattern of the innervation apparatus varies widely in different smooth muscle systems. This is perhaps not surprising when one recognizes the wide variety of physiological roles that different organs containing smooth muscles perform. For example, the vas deferens and ciliary muscles exhibit fast co-ordinated responses, and this is reflected by specific innervation of every muscle fibre. In contrast, spontaneously contracting smooth muscles such as those found in the intestine and uterus, are geared primarily for slow graded responses, and the innervation involves diffuse release of tramsmitter from bundles of nerves of different types and a high degree of intermuscle fibre spread of excitation.

Vas deferens

Hypogastric nerves divide into numerous fine branches as they enter the vas deferens and consist predominantly of non-myelinated fibres (0·2 to 1·75 μ diameter) embedded in Schwann cells. The axons contain neurofilaments, mitochondria and, even within these large bundles, sometimes contain vesicles (Merrillees et al., 1963). Occasionally, one or two myelinated fibres (2 to 3 μ diameter) are also seen in these bundles (Merrillees et al., 1963; Nagasawa & Mito, 1967). The Schwann cells and axons are surrounded by collagen filaments and a delicate perineurium.

When the nerve branches pass into the muscle coats, they split up into smaller bundles of 2 to 8 axons (Fig. 11a). The varicose regions (about 0·5 to 1 μ diameter) of these axons are packed with vesicles and mitochondria and in these

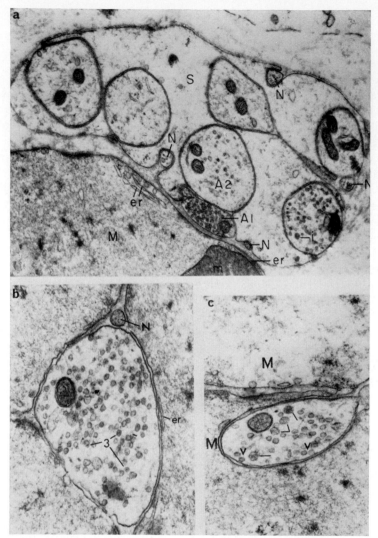

FIG. 11   *a* Guinea-pig vas deferens. A medium-sized intramuscular bundle of axons within a single Schwann cell (S). There is no perineurial sheath. The lower axons contain 'synaptic' vesicles. The axons A1 and A2 have the appearance of an axon-axon synapse because of the large number of vesicles in A1, but the significance of the crowded vesicles may be related more to the proximity (800 Å) of this uncovered axon (A1) to the muscle cell (M). Most of the axons in bundles of this size have few vesicles in the plane of section, but they resemble the vesicle-containing axons of the larger trunks in that they have very few large neurofilaments. The small profiles (N), less than 0.25μ in diameter, probably represent sections through intervaricose regions of nerves. $\times 15,000$. (From Merrillees, Burnstock & Holman, 1963)   *b* Guinea-pig vas deferens. An unusual case in which a large naked axon is in close apposition (200 Å) with three muscle fibres at once. The large size of this axon may be due to a varicosity. The small profile (N) probably represents a section through an intervaricosity. There is a small portion of a subsurface cisterna of the endoplasmic reticulum at er. $\times 26,000$. (From Merrillees, Burnstock & Holman, 1963)   *c* Guinea-pig vas deferens. An axon almost buried in the surface of a muscle fibre (an unusual condition). The axon contains vesicles and a mitochondrion. Four large granular vesicles are indicated by the lines. The muscle cell boundaries are confused because the membranes are cut obliquely. The two vesicles (v) are lying against the plasma membrane of the axon, but are not opening through it. $\times 26,000$. (From Merrillees, Burnstock & Holman, 1963)

regions the Schwann sheath is often incomplete, leaving them naked; the inter-varicose regions (about 0·1 to 0·2 μ diameter) contain neurotubules and/or neurofibrils only. In the mouse vas deferens, there is a ratio of one tertiary axon bundle to every 3 to 4 muscle cells and each bundle contains 1 to 5 axons (Lane & Rhodin, 1964a).

Single varicose nerve fibres leave the bundles, run separately, lose their Schwann sheath and make close apposition (200 Å) at varicosities and often end in shallow depressions in muscle cells (Fig. 11b, c). (Richardson, 1962; Merrillees et al., 1963; Thaemert, 1963; Lane & Rhodin, 1964a; Clementi, 1965; Taxi, 1965; Hökfelt, 1966a, b; Taxi & Droz, 1966; Nagasawa & Mito, 1967; Yamauchi & Burnstock, 1969b; Merrillees, 1968). Sometimes they form several 'en passage' close contacts with the same or different muscle cells before they end (Merrillees et al., 1963; Merrillees, 1968). The separation of nerve and muscle membranes at close neuromuscular junctions in the vas deferens is about 200 Å in guinea-pig (Merrillees et al., 1963), 180 to 250 Å in rat (Richardson, 1962) and 70 to 80 Å in mouse (Lane & Rhodin, 1964a). After osmium tetroxide or glutaraldehyde fixation, tight junctions have been observed occasionally between axons and smooth muscle cells in the mouse vas deferens (Yamauchi & Burnstock, 1969b) and the possible significance of this finding in terms of electrical coupling is being investigated.

There is little evidence of post-synaptic specialization of muscle membranes at close neuro-muscular junctions. The membrane is usually neither thickened nor increased in density and micropinocytotic vesicles are often but not always present (Richardson, 1962; Merrillees et al., 1963; Lane & Rhodin, 1964a; Nagasawa & Mito, 1967). Basement membrane material does not extend into the junctional cleft, at least not in significant amounts, to form a layer compar-able with the intermediate electron-dense layer of the skeletal motor end plate. (Palade, 1954; Robertson, 1955; Birks, Huxley & Katz, 1960). An elongated sac of endoplasmic reticulum (or sub-membranous sac) which lies close to the muscle membrane has been described in rat and mouse vas deferens (Richardson, 1962; Merrillees et al., 1963; Lane & Rhodin, 1964a) and more rarely in guinea-pig vas deferens (see Fig. 11 a and b) (Merrillees et al., 1963). The functional significance of this structure, assuming that it is not a fixation artefact, is unknown. Pinocytotic vesicles in the postsynaptic muscle region in the mouse vas deferens sometimes contain a dense body similar to those seen in the granular vesicles contained in the nerve terminal (Lane & Rhodin, 1964a). A series of electronmicrographs suggestive of transfer of intra-axonal granules across the post-junctional muscle membrane in trout heart has been published by Yamauchi & Burnstock (1968).

Quantitative measurements of the number of close (200 Å) neuromuscular contacts relative to muscle cell profiles seen in one plane of section in the vas deferens (see Table 4) suggests the occurrence of individual innervation of all the smooth muscle cells in the vas deferens of adult rat (Richardson, 1962;

TABLE 4. Innervation density in different tissues

Tissue	Animal	Number of axons within 200 Å or less of a muscle cell per 1000 cells in section (montage)	Number of 'close' neuro-muscular contacts per 1000 cells	Reference
Vas deferens	guinea-pig	4–18	127 (within 500 Å)	Merrillees *et al.*, 1963
Vas deferens	rat	*48	—	Taxi, 1965
Vas deferens	mouse	*7	—	
Nictitating membrane	cat	*2·5	—	
Ciliary muscle	cat	—	41 (within 120 Å)	
Intestine (circular muscle)	toad			
small		—	74 (within 1000 Å)	Rogers &
large		—	25 (within 1000 Å)	Burnstock, 1966b
Vas deferens	developing mouse			
1 day		10	10 (within 200 Å)	Yamauchi &
10 day		270	300 (within 200 Å)	Burnstock,
6 months		†390	†500 (within 200 Å)	1969b
(adult)			840 (within 500 Å)	

* A separation of within 150 Å was selected which may account for the low figures compared to those claimed by Yamauchi & Burnstock (1969b).

† The discrepancy here is because some terminal axons form close neuromuscular contacts with more than one muscle cell.

N.B. Very few, if any, contacts with a neuromuscular interval of less than 800 Å have been seen in vascular, uretal, uterine or intestinal longitudinal muscles.

Taxi, 1965) and mouse (Lane & Rhodin, 1964a; Yamauchi & Burnstock, 1969b), but not of the guinea-pig (Merrillees *et al.*, 1963). This conclusion for the guinea-pig has been confirmed in a recent examination with a serial sampling method; less than half of the cells were shown to have close (200 Å) neuromuscular contacts (Merrillees, 1968).

This latter painstaking study has also added the following information about the innervation of the guinea-pig vas deferens. The ground plexus consists of repeatedly branching and anastomosing small nerve trunks; individual axons branch within these trunks. The terminal nerve branches pass in both directions. The length of Schwann cell processes has been estimated to be several thousands of microns. All axons are more or less encased in Schwann cell cytoplasm until close to their terminations, when they invariably become naked. Bare terminations are from 10 to 60 μ long, and usually end after a variable number of varicosities at a terminal varicosity in a shallow depression in the surface of a muscle fibre. The great majority of nerve trunks contain only 2 or 3 axons. A few contain up to 6 or 7, but all of these appear to be equally capable of 'en passage' innervation. Larger trunks are probably not an effective part of the ground plexus, because the axons rarely contain varicosities with vesicles and are rarely closer than 2000 Å from a muscle fibre. The density of 'innervation' of muscle fibres (including 'en passage' episodes within 1000 Å or less separation) is extremely variable. Some fibres have no nerves within this distance, others have as many as 50 to 75 approaches when measured from 4·5 μ interval serial sampling. Some muscle fibres have more than one termination. Close (200 Å) contacts tend to be in the nuclear region.

Axons within bundles in the vas deferens rarely, if ever, show close apposition to each other (Richardson, 1962; Merrillees *et al.*, 1963; Lane & Rhodin, 1964*a*), in contrast to the occurrence of numerous close (about 100 Å) axon-axon relationships within nerve bundles in intestine and blood vessels (Taxi, 1965; Rogers & Burnstock, 1966*a*, *b*; Simspon & Devine, 1966). Both thin neurofilaments (60 to 100 Å) and thick neurotubules (200 Å) have been seen within axons in the rat vas deferens (Richardson, 1962).

Lane & Rhodin (1964*a*) claimed that in mouse vas deferens about 1 axon in 200 contained a very large population of agranular vesicles, and that these axons may serve a receptor function and may interact with the more common endings containing granular vesicles in whose company they usually appear. However, in view of the recent evidence of a separate cholinergic nerve supply to the guinea-pig vas deferens (Bell, 1967*a*; Burnstock & Robinson, 1967, Robinson, 1969*b*), it may be that the profiles described in the mouse vas deferens containing predominantly agranular vesicles represent cholinergic motor axon profiles. Some examples of cholinergic and adrenergic nerve profiles closely apposed within a common Schwann cell process have also been described in the guinea-pig vas deferens (Thoenen, Tranzer, Hürlimann & Haefely, 1966).

Yamauchi & Burnstock (1969*b*) studied postnatal development of the innervation of smooth muscle in the circular layer of the mouse vas deferens (see Table 5). They concluded that the *density* of innervation, both in terms of the number of axons per 100 muscle cells seen in section and in the number of close (200 Å) neuromuscular junctions, continues to increase even up to 6 months of age; between 3 months postnatal, when the mouse becomes fully pubescent (Rugh, 1962), and 6 months there is about a 50% increase in the nerve supply. The noradrenergic nerves become fully developed at 15 days in the sense that the proportion of the various intra-axonal vesicle types (see p. 56) remains constant after this stage; however, functional transmission does not occur until after this time (Furness, McLean & Burnstock, 1969). About 12 to 17% of the axons contain exclusively agranular vesicles, suggesting that there is a constant cholinergic nerve component in the innervation of the mouse vas deferens from about 10 days onwards.

Intrinsic eye muscles

Ciliary muscle. There have been many studies of the innervation of ciliary

3+

TABLE 5. Development of innervation in the mouse vas-deferens (adapted
from Yamauchi & Burnstock 1969*b*)

Age of specimens	Number of axons per 100 muscle cells	Number of close (200 Å) neuro-muscular contacts per 100 muscle cells	Percentage of different vesicle types found in profiles through varicose regions of noradrenergic nerves		
			Small granular vesicles	Large granular vesicles	Agranular vesicles
1 day	15	1	37	10	53
5 days	63	17	51	8	41
10 days	80	30	65	9	27
15 days	71	12*	81	4	15
20 days	87	19*	81	4	14
25 days	99	25*	82	3	15
3 months	112	32	84	4	12
6 months	121	50	85	3	12

* These low figures can probably be explained in terms of the increase in the size of muscle cells at this stage (Yamauchi & Burnstock, 1969*a*).

muscle of the cat and man with the light microscope (Boeke, 1933; Clark, 1937; Jabonero, 1954; Génis-Galvez & Clements, 1957; Tichowa, 1961; Taxi, 1965), and a few with the electronmicroscope (de Lorenzo, 1959; Krapp, 1962; Ishikawa, 1962; Taxi, 1965).

The innervation of the ciliary muscle is dense (see Table 4), resembling that of the vas deferens, with many examples of close (200 Å) nerve-muscle contacts (Taxi, 1965). Many more axons are separated by 800 to 1000 Å from muscle fibres. Taxi (1965) calculated from the distribution of close neuromuscular junctions in sections that there were about 4 synapses per muscle fibre in cat ciliary muscle. There does not appear to be any post-junctional specialization of the muscle membrane or submembrane area at the neuro-muscular junctions.

Iris muscles. There have been a number of electronmicroscope studies of the muscles of the iris, since these muscles offer a good opportunity to study adrenergic nerves (in dilator muscle) and cholinergic nerves (in sphincter pupillae) within the same tissue (Tousimis & Fine, 1959, 1961; de Lorenzo, 1959; Krapp, 1962; Evans & Evans, 1964; Richardson, 1964; Nilsson, 1964; Hökfelt & Nilsson, 1965; Hökfelt, 1966*a*; Tranzer & Thoenen, 1967*a*). It is significant (see p. 60) that the nerve profiles in the iris dilator muscle contain predominantly small granular vesicles, while those supplying the sphincter pupillae contain predominantly agranular vesicles (see Fig. 14*a, b*) (Richardson, 1964; Hökfelt, 1966*a*; Tranzer & Thoenen, 1967*a*), As in the vas deferens, varicose fibres may influence a number of effector cells (Richardson, 1964) and there is a relatively high density of close (200 Å) neuro-muscular junctions (see Table 4). 'Tight junctions' between muscle cells have also been seen (Evans & Evans, 1964).

Areas with increased density on both pre- and post-synaptic membranes have been observed at close neuromuscular junctions in the rabbit iris sphincter pupillae (Richardson, 1964).

Nictitating membrane. There are several light microscope studies of the innervation of the nictitating membrane (de Kleijn & Socin, 1915; Rosenblueth & Bard, 1932; Lawrentjew & Borowskaja, 1936; Bullon & Stiefel, 1955; Génis-Galvez & Clements, 1957; Thompson, 1961; Jabonero, 1962) and some electronmicroscope studies (Evans & Evans, 1964; Taxi, 1965; Esterhuizen *et al*, 1967, 1968). The type of innervation resembles that of the vas deferens and ciliary muscle, with a predominance of single axons, free of Schwann cell processes in close apposition (about 200 Å) to the muscle cells, but the density of the innervation is lower and many muscle cells may not have close neuromuscular junctions (see Table 4). Two to 28 days after superior cervical ganglionectomy, very few axon profiles were seen in the denervated muscle (Esterhuizen *et al.*, 1968).

Intestinal smooth muscle

Most electronmicroscope studies of the innervation of intestinal smooth muscle have been confined to descriptions of the nerves supplying the longitudinal muscular coat (Richardson, 1958, 1960; Hager & Tafuri, 1959; Thaemert, 1959; Yamamoto, 1960; Taxi, 1961, 1965; Hampton, 1960; Brettschneider, 1962; Grillo & Palay, 1962; Lane & Rhodin, 1964a; Tafuri, 1964; Yamauchi, 1964; Taxi & Droz, 1966; Bennett & Rogers, 1967; Nagasawa & Mito, 1967).

It has been generally assumed that the innervation of the circular muscle coat follows the same pattern. However studies of the innervation of the circular muscle coat of the intestine of toad (Boyd *et al.*, 1964; Rogers & Burnstock, 1966a, b), frog and rat (Thaemert, 1963, 1966) and rabbit (Nagasawa & Mito, 1967) suggest that the number of close (200 Å) neuromuscular junctions in this coat, at least in some species, resembles more closely that of the vas deferens (Table 4). Hence the two coats will be treated separately.

Longitudinal muscle coat. Large nerve bundles (50 or more axons) emerge from Auerbach's plexus and pass to the muscle coats, where they split up into smaller bundles which run in parallel with the main axes of the muscle fibres (Richardson, 1958; Rogers & Burnstock, 1966b; Bennett & Rogers, 1967).

Close (90 to 150 Å) axon-axon contacts are common within large nerve bundles in Auerbach's plexus or running between the muscle cells (Taxi, 1965; Rogers & Burnstock, 1966a, b) and have also been seen between axons in blood vessels (Simpson & Devine, 1966). Whether an axon can modify the activity of neighbouring axons by chemical or electrotonic means is not known. This arrangement is in contrast to that seen in nerve trunks, where axons are all invaginated into the surface of Schwann cells (Hess, 1956; Gasser, 1958; Elfvin, 1958, 1961).

Most of the nerves supplying the muscles run in bundles (Richardson, 1958;

Hager & Tafuri, 1959; Thaemert, 1959, 1963; Yamamoto, 1960; Hampton, 1960; Prosser *et al.*, 1960; Gansler, 1961; Taxi, 1961, 1965; Grillo & Palay, 1962; Lane & Rhodin, 1964*a*; Yamauchi, 1964; Bennett & Rogers, 1967; Nagasawa & Mito, 1967). In the mouse duodenum there is approximately one tertiary axon bundle for every 11 or 12 muscle cells, each bundle containing 4 to 20 axons (Lane & Rhodin, 1964*a*). In the taenia coli, the largest bundles (20 to 70 axons) lie in connective tissue in the underlying muscle (Fig. 12*a*) while only small bundles (3 to 5 axons) are found towards the serosal surface. Only small nerve bundles (3 to 5 axons) actually enter the muscle bundles of the taenia (Bennett & Rogers, 1967). At intervals of about 1 mm along the taenia there is a marked decrease in the number of small nerve bundles. Since very few examples of single axons have been seen (Yamauchi, 1964) it seems likely that all the axons in a small nerve bundle terminate within a short distance of each other.

The fibres on the periphery of the bundles are varicose, contain synaptic vesicles and mitochondria and are free of Schwann sheath at intervals (Taxi, 1965; Bennett & Rogers, 1967). Thus they give the impression that they form functional junctions with muscle cells. The separation of nerve and muscle membranes even in these parts is rarely less than 800 Å, so that diffusion of transmitter over unusually long distances in comparison with other known synapses seems likely. The more central fibres in the bundles contain predominantly neurotubules and fibrils.

Close apposition (180 to 250 Å) of nerve and muscle is extremely rare (Richardson, 1958; Yamamoto, 1960; Brettschneider, 1962; Grillo & Palay, 1962; Rhodin, 1962; Thaemert, 1963, 1966; Lane & Rhodin, 1964*a*; Yamauchi, 1964). Serial sections show that some muscle fibres are never sufficiently close to nerve bundles to make direct action of diffused transmitter likely (Taxi, 1965). It has been suggested (Bennett & Rogers, 1967) that varicosities may have to be within 3000 Å of a smooth muscle cell in the taenia coli if they are to produce a detectable effect on the muscle membrane during transmission. Furthermore, in the intestine of small rodents (mouse, rat, guinea-pig) there are no nerve bundles running inside the longitudinal muscle coat (Taxi, 1965). The sparse innervation of these muscles makes it likely that intermuscle fibre spread of excitation plays an important role. This is supported by observations of an extensive system of intermuscle fibre junctions in this muscle coat (see p. 14).

As a result of semi-serial sectioning through nerve bundles in the frog intestine (Thaemert, 1966) and in the guinea-pig taenia coli (Bennett & Rogers, 1967; Bennett & Burnstock, 1968), it has been shown that axons change their position in the bundle as the nerve passes through the muscle coat, such that many different muscle cells can be influenced by transmitter released 'en passage'.

The relative size of the varicosities and intervaricosity regions of nerves in the intestine appears to be similar to that seen in the vas deferens; intervaricosity diameters in the gut are rarely less than 0·08 μ (Bennett & Rogers, 1967)

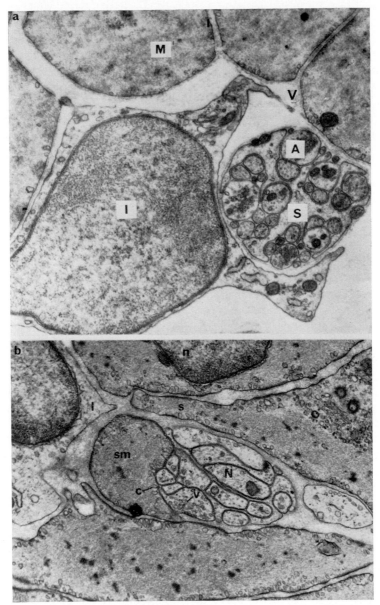

FIG. 12   *a* Transverse section through guinea-pig taenia coli. I, interstitial (connective tissue) cell; v, agranular vesicles; S, Schwann cell process enveloping numerous axons (A); M, smooth muscle cell. Palade fixation. × 13,000. (From Rogers & Burnstock 1966*a*)   *b* Toad small intestine. Multiaxonal junction between small nerve bundle (N) and tip of smooth muscle cell (sm). Note caveolae intracellulares (c) and agranular vesicles (v). I, interstitial cell (fibroblast); n, nucleus of smooth muscle cell; s, crest or spine-like process of muscle cell. × 29,000. (From Rogers & Burnstock, 1966*b*)

compared to 0·1 µ in the vas deferens (Merrillees *et al.*, 1963), while varicosity diameters are up to 1 µ in both organs.

Even in those regions where nerves closely approach muscle cells, there appears to be little post-junctional specialization of the membrane (Lane & Rhodin, 1964*a*; Taxi, 1965). However, areas with increased density of both pre- and post-synaptic membranes have been described in rat colon (Thaemert, 1963).

Circular muscle coat. The density of close neuro-muscular junctions in the inner, circular muscle coat is higher than in the outer, longitudinal muscle coat of rat intestine (Thaemert, 1963) and toad intestine (Rogers & Burnstock, 1966*b*). Indeed the figure of 25 to 78 close neuromuscular junctions in the toad intestine per 1000 muscle cells in one plane of section is closely comparable to that observed in the rat vas deferens (see Table 4) and suggests that every cell receives at least one close 'en passage' varicose synapse or ending.

There also appear to be some regional differences in the number of close (200 Å) neuromuscular junctions. For example, the density is greater in the large intestine than in the small intestine of the toad (Rogers & Burnstock, 1966*b*). Multiaxonal junctions (Fig. 12*b*) are characteristic in the circular muscle coat of the intestine (Brettschneider, 1962; Rogers & Burnstock, 1966*b*; Thaemert, 1966) but whether these represent polyaxonal or multi-terminal junctions has not yet been resolved. In the upper layers of the circular muscle coat of toad large intestine, the majority of axon-smooth muscle contacts occur towards the tip of the muscle cells and less than 3% in the vicinity of the nucleus.

The number of axons in the final ramifications of the ground plexus supplying the circular muscle coat is usually one, two or three with, or without, a Schwann cell sheath (Rogers & Burnstock, 1966*b*). These axons diverge from the smallest bundles (7 to 20 axons) directly, i.e. there is no gradual splitting off of smaller and smaller fascicules. Where the axons come into close relation to smooth muscle cells, the Schwann cell is either arranged in such a way that only a small opening is left facing the muscle cell, or it leaves almost the entire axon free. In the circular muscle layer of the rabbit colon, axons occur in groups, and no single axons are seen (Nagasawa & Mito, 1967).

Axons within bundles in toad large intestine contain agranular vesicles, large granular vesicles, clumps of very dense, irregularly shaped granules (probably glycogen), small oval mitochondria often containing dense granules, neurofilaments and sometimes delicate branching tubules (Rogers & Burnstock, 1966*b*).

It has not been possible yet to distinguish, at the electronmicroscope level, afferent from efferent nerve fibres in smooth muscle (Rogers & Burnstock, 1966*b*; Merrillees, 1968), and it has been pointed out that many afferent nerve processes show similar features to efferent fibres (De Lorenzo, 1958; De Robertis, 1958; Engström, 1958; Sjöstrand, 1958; Ross, 1959; Cauna & Ross, 1960; Smith & Sjöstrand, 1961; Thaemert, 1963). However some axon profiles have been observed to contain an unusually high proportion of mitochondria to vesicles (e.g. Fig. 9 in Rogers & Burnstock, 1966*b*), which may be sensory by analogy with the structure of sensory terminals in other systems (Pease & Quilliam, 1957; Merrillees, 1960; Cauna & Ross, 1960; Bleichmar & De Robertis, 1962).

Seminal vesicles, arrectores pilorum and retractor penis

The innervation pattern of the rat seminal vesicles appears to be closely comparable with that of the vas deferens (Nagasawa & Mito, 1967; Robinson, 1969*a*) as is that of the dog retractor penis (Yamauchi, Orlov, McLean & Burnstock, in preparation). There is brief mention of the electronmicroscopy of the nerves supplying the arrectores pilorum in cat (Richardson, 1962) and human (Charles, 1960) but no details are given.

Ureter

Light microscopists have not agreed about the nature of the innervation of the mammalian ureter. There appears to be little functional innervation of most of the length of the ureter of guinea-pig and cat, although there is possibly some innervation of the pacemaker region (Schabadasch, 1934). On the other hand, Hukuhara, Nanba & Fukuda (1964) have demonstrated functional sympathetic and parasympathetic innervation of dog ureter, and even shown the presence of some intramural ganglion cells.

The electronmicroscopy of the mammalian ureter has been neglected. There are some studies of the musculature but few descriptions of nerves (Bergman, 1958; Leeson & Leeson, 1965*b*). Some nerves have been observed in the rat ureter (Taxi, 1965; Nagasawa & Mito, 1967), although the most prominent feature of this tissue was the abundance of close junctions between muscle cells, which is consistent with the ease of intermuscle fibre propagation demonstrated electrophysiologically and the absence of junction potentials (Bozler, 1942; Burnstock & Prosser, 1960*b*; Bennett, Burnstock, Holman & Walker, 1962). In general, the arrangement of nerves in the ureter appears to be similar to that described earlier for the longitudinal muscle coat of the intestine, i.e. the nerves mostly run in bundles and close (200 Å) apposition between axons and muscles are rare. No 'interstitial cells' (as discussed for the gut, page 35) are present in the ureter (Taxi, 1965). Whenever close neuro-muscular junctions have been observed no specialization of the post-synaptic muscle membrane was apparent (Taxi, 1965).

Uterus

There have been a number of electronmicroscope studies of the uterus (Csapo, 1955; Gansler, 1956; Mark, 1956; Shoenberg, 1958; Siliotti & Grebella, 1959; Lauricella, D'Alessandro & Fiumara, 1960; Jaeger, 1962; Jaeger & Pohlmann, 1962; Kameya, 1964; Needham & Shoenberg, 1964; Laguens & Lagrutta, 1964; Ross & Klebanoff, 1967), but few observations of nerves (Caesar *et al.*, 1957;

Gansler, 1961; Yamamoto, 1961; Clementi, 1962). Varicose nerve fibres, mostly in bundles have been seen, but no close junctions. Thus, uterine innervation appears to resemble that of the longitudinal intestinal muscle.

Urinary bladder

The electronmicroscopy of the innervation of the mouse, rat, guinea-pig, possum and toad urinary bladder has been described (Caesar, *et al.*, 1957; Thaemert, 1963; Burnstock & Merrillees, 1964; Burnstock & Robinson, 1967; Robinson & Bell, 1967; Nagasawa & Mito, 1967). Häggqvist (1956); Choi (1963); Lane (1967) and Nagasawa & Suzuki (1967) described the fine structure of bladder muscle cells, but did not refer to their innervation.

The innervation of the urinary bladder appears to be comparable to that of the guinea-pig vas deferens, i.e. there are many single axons naked of Schwann cell processes in close (200 Å) apposition with muscle cells, as well as bundles of varicose axons running through the muscle coats. Caesar *et al.*, (1957) concluded that 'the number of smooth muscle axon connections we have seen so far in the urinary bladder, give a strong indication that each and every muscle cell shows a close relationship to the axon at a well defined locus'.

The axon profiles in the rat bladder appear to be more heavily endowed with mitochondria than those seen in the gut (Thaemert, 1963). This may indicate a higher proportion of sensory nerves.

Vascular smooth muscle

Most arterial smooth muscles, contrary to the prediction of Bozler (1948), appear to fit more closely the 'unitary' than the 'multiunit' type, at least as far as their innervation is concerned. However, there is a wide variation in form and innervation of vascular smooth muscles, perhaps as great as that seen between different visceral smooth muscles. For example, the umbilical artery and vein and the small vessels in the body of the uterus do not appear to be innervated, whereas arteries such as those found in the rabbit ear and mesentery have a dense nerve plexus. Some vessels, e.g. superior mesenteric vein and renal artery are spontaneously active and have an outer longitudinal muscle coat. The pulmonary vein has some striated musculature (Karrer, 1959).

One feature, which has emerged from the combination of morphological studies with silver and methylene blue staining (Grigor'eva, 1962), fluorescent histochemistry (Falck, 1962; McLean & Burnstock, 1966, 1967*a*, *b*, *c*) and electronmicroscopy (see p. 53), is that the nerves supplying most vascular smooth muscles are confined to the outside of the media. Known exceptions where many nerves are present within the media are sheep carotid artery (Keatinge, 1966*a*), renal vein and artery (Gannon, Iwayama & Burnstock, unpublished observations) and cutaneous veins (Ehinger, Falck & Sporrong, 1966).

Of more than 20 electronmicroscope studies of various blood vessels, there is not a single convincing demonstration of axon profiles within less than about

800 Å of a muscle cell (Pease & Paule, 1960; Pease & Molinari, 1960; Feeney & Hogan, 1961; Lever & Esterhuizen, 1961; Zelander, Ekholm & Edlund, 1962; Stahl, 1963; Thaemert, 1963; Wolfe & Potter, 1963; Appenzeller, 1964; Barajas, 1964; Brettschneider, 1964; Uchizono, 1964; Lever, Ahmed et al., 1965; Lever, Graham et al., 1965; Samarasinghe, 1965; Verity & Bevan, 1966; Sato, 1966; Simpson & Devine, 1966; Devine & Simpson, 1967; Rhodin, 1967). There are two reports to the contrary. Suwa (1962) claimed that many smooth muscle cells in the human aortic media were in close (180 Å) contact with nerves, axons often being seen in depressions or pockets in the muscle. However, the tissue preservation was not good. More recently, Rhodin (1967) has shown some oval profiles which he interprets to represent nerve endings within smooth muscle cells of small arterioles in rabbit with a membrane separation of only 45 Å. It has also been claimed (Rhodin, 1963) that the specialized sheath artery of the spleen is the only bood vessel whose muscle cells are richly provided with nerve endings.

The majority of nerve fibres supplying the media run in bundles of various sizes in the dense autonomic ground plexus lying in the adventitia. Arterioles of 50 μ diameter are surrounded by about 8 bundles of axons, while arterioles of 5 to 10 μ diameter are associated with about 2 axon bundles in the rat jejunum submucosa (Devine & Simpson, 1967). Up to 20 bundles of 3 to 8 axons have been found about small (150 μ diameter) mesenteric arteries of the rat (Devine & Simpson, 1967). These bundles run approximately in the same direction as the vessel and form part of the autonomic ground plexus in the sheep mesenteric vein (see Fig. 9b). There are between 2 and 6 axon bundles (containing 3 to 5 axons) around arterioles in the sheep renal cortex (Simpson & Devine, 1966). The basilar artery of the rat has been found to have 160 to 180 nerve fibres arranged in 16 bundles surrounding the media, the middle cerebral artery only about 8 bundles, and most intra-cerebral arterioles have none, in contrast to arterioles in other parts of the body (Samarasinghe, 1965).

As in other terminal portions of sympathetic nerves, varicosities (about 1 μ diameter) contain vesicles and mitochondria, while intervaricose regions (about 0·1 μ diameter) contain neurofilaments only. Varicosities nearly always occur at the boundary between 2 or more muscle cells (Simpson & Devine, 1966). Sometimes single varicose axons are seen in the plexus, naked of Schwann cell processes. However, axons usually appear to remain in bundles even when making relatively close (800 to 1200 Å) contact with muscle cells. In these situations, the side of the axon closest to the smooth muscle cell is usually free of Schwann cell. In these closest contacts, the basement membranes of the axon and muscle cells are fused together and the axon tends to follow the contour of the smooth muscle cell (Lever et al., 1965; Devine & Simpson, 1967). Contact at close (800 to 1200 Å) junctions is often maintained over distances of 5 μ or more (Simpson & Devine, 1966). Despite the width of the cleft between nerve and muscle membranes, it seems likely that these structures represent functional

3*

synapses in vascular smooth muscle, but whether some of the axons actually terminate on muscle cells cannot be resolved until a serial section study has been made. Some post-junctional specialization of muscle at close neuro-muscular contacts consisting of sub-membranous sacs of endoplasmic reticulum has been described in arterioles (Simpson & Devine, 1966; Devine & Simpson, 1967) similar to those seen in the vas deferens (Richardson, 1962).

In general, arterioles appear to have more close (800 to 1200 Å) neuro-muscular contacts than small arteries (Devine & Simpson, 1967), perhaps in relation to their exacting role in the regulation of peripheral resistance. The innervation of precapillary sphincters appears to be particularly dense (Rhodin, 1967). Devine & Simpson (1967) claimed that many smooth muscle cells in arterioles in rat jejunum have multiple innervation and that some axons probably innervate more than one muscle cell.

Several authors have speculated that, although the minimum neuromuscular contacts are 800 to 1200 Å in most vessels, transmitter released from nerves up to 3000 to 10,000 Å away from muscle cells is likely to be effective (Lever et al., 1965; Devine & Simpson, 1967; Bennett & Burnstock, 1968; Bell, 1969). Even if this is the case, the problem still remains as to how the cells on the inner (luminal) surface of the media are activated, especially in the larger arteries, where the muscle coat is over 500 μ thick. It has usually been assumed that there is effective diffusion of transmitter over these long distances from the nerves in the adventitial plexus. However it has recently been suggested that activation of many of the cells in the media is by electrotonic spread of current from 'key' cells directly affected by transmitter, which leads to the initiation and propagation of action potentials down muscle effector bundles (see Bennett & Burnstock, 1968; Burnstock, 1968). The recent descriptions of 'tight junctions' between smooth muscle cells in the media of several arteries (Verity & Bevan, 1966; Cliff, 1967; Rhodin, 1967; Foussman, 1967, personal communications) gives some support to this view.

The mitochondria within axons in periarterial plexuses are small and often elongated, up to 0·8 μ in length and about 1000 to 1500 Å in width (Simpson & Devine, 1966). Thick (150 to 200 Å) neurofilaments (Elfvin, 1958, 1961), but no thin neurofilaments, were found in axons about sheep arterioles (Simpson & Devine, 1966) in contrast to splenic nerve or vas deferens where both types have been described (Elfvin, 1961; Richardson, 1962). Dahl and Nelson (1964) claimed to see specialized sensory endings which were comparable to Meissner's corpuscles near medial smooth muscle cells in human intracranial arteries.

Few studies have been made of the electron microscopy of veins (Karrer & Cox, 1960; Prosser et al., 1960; Policard, Collet & Prégermain, 1960; Devine & Simpson, 1967). The nature of the innervation of some veins does not appear to differ radically from that of small arteries or arterioles, but the adrenergic nerve supply to veins is generally less dense than that of the corresponding arteries (Norberg & Hamberger, 1964). Other vessels such as the portal, anterior mesenteric and renal veins have a dense innervation.

Myoepithelial cells, glands and chromaffin cells

Close (200 Å) axon-gland cell junctions have now been described from electronmicroscope studies of a variety of glandular tissues, including salivary and lacrimal glands (Scott & Pease, 1959; Yamauchi & Burnstock, 1967) pancreatic acinar (Stahl, 1963), pineal gland (Milofsky, 1957; De Robertis & Pellegrino de Iraldi, 1961; Pellegrino de Iraldi & De Robertis, 1961, 1963; Wolfe, Potter, Richardson & Axelrod, 1962; De Robertis, 1963; Bondareff, 1965), endocrine gland and pancreatic islets (Bencosme, 1959; Stahl, 1963). Nerves with greater separation (> 400 Å) have been seen in submandibular (Scott & Pease, 1959; Fujita, Machino, Nakagami, Imai & Yamamoto, 1964), sweat (Yamada & Miyake, 1960; Munger, 1961) and thyroid (Brettschneider, 1963) glands.

The density of close junctions on the lacrymal gland has been shown to be particularly high (Yamauchi & Burnstock, 1967). A single fibre may contact several effector cells through more than one varicosity, (Scott & Pease, 1959; Coupland, 1965; Yamauchi & Burnstock, 1967). Polyaxonal innervation of single gland cells has also been claimed (Scott & Pease, 1959; Stahl, 1963; Coupland, 1965; Yamauchi & Burnstock, 1967).

The epithelial cells of most glands are enveloped by processes of myoepithelial cells, which are closely identified with smooth muscle cells by their fine structure. Close (200 Å) nerve-myoepithelial as well as nerve-epithelial cell junctions have been described (Yamauchi & Burnstock, 1967) but it will not be known until serial sections are made whether the nerves in close contact with epithelial cells represent nerves passing to myoepithelial cells and/or functional synapses with epithelial cells.

Since Fusari (1891) and Dogiel (1894) first described the innervation of the adrenal medulla, there has been much discussion by light microscopists about the nature of the innervation apparatus (see Hillarp, 1959). Electronmicroscope studies (Sjöstrand & Wetzstein, 1956; De Robertis & Ferreira, 1957; Eränkö & Hänninen, 1960; Coupland, 1962, 1965; Wood & Barrnett, 1963; Clementi, 1965; Elfvin, 1965) have shown that nerve fibres form typical synaptic-type endings on chromaffin cells. The presynaptic axon profiles are free of Schwann cell processes and contain abundant vesicles and mitochondria. The pre- and especially the post-synaptic membranes are often thickened (Coupland, 1962). Sometimes axon profiles appear in section to be 'inside' chromaffin cells. However there is complete integrity of axonal and chromaffin cell membranes, so that they really represent axon terminals embedded in pit-like invaginations of the chromaffin cell surface, and not true intracellular nerve endings as were often described by light microscopists (Fusari, 1891; Alpert, 1931). Some multiple synaptic-type endings have also been seen on single chromaffin cells, but appear to involve the same nerve fibre (Coupland, 1962). Unlike the autonomic neuro-muscular junction, the junction formed on adrenal medullary cells shows specialization; there are thickenings at the contacting membranes and envelopment of the nerve endings by glial cytoplasm. Thus, this junction resembles a ganglionic synapse rather than an autonomic neuroeffector junction.

Summary

1. There is a spectrum of type of innervation in different tissues: (a) the rat vas deferens, and cat ciliary muscle have many single axons, naked of Schwann cell processes, in close (200 Å) apposition with muscle cells, and with at least one such synapse supplying each cell; (b) heavy innervation by close (200 Å) neuro-muscular junctions, but probably not one to every cell, has been described in guinea-pig vas deferens, nictitating membrane, circular intestinal muscle constrictor pupillae and urinary bladder; (c) in the longitudinal muscle coat of the intestine, most vascular smooth muscles, uterus and ureter, close (200 Å) neuro-muscular junctions are rare and limited to discrete areas, while 'tight junctions' between muscle cells are prominent; (d) in chick amnion there are no nerves, but many 'tight junctions' between muscle cells.

2. The closest neuro-muscular contacts in rat, mouse and guinea-pig vas deferens, ciliary muscle, nictitating membrane, bladder, where there is fusion of nerve and muscle basement membranes, are 80 to 200 Å. This appears to constitute a true neuro-muscular synapse; the electrophysiological evidence (see Burnstock & Holman, 1963) supports this conclusion. However, it seems likely that transmitter released from nerves within about 1000 Å of a muscle fibre, at least in the guinea-pig vas deferens, is also effective.

In the longitudinal muscle coat of the intestine and in arteries, few neuro-muscular contacts of less than 800 Å have been observed, even though there is usually fusion of basement membranes in these regions. Thus the neuromuscular synaptic cleft might be greater in these tissues, and there is some evidence to suggest that transmitter released from nerves within about 3000 to 10,000 Å of muscle cells is also effective.

3. In general there appears to be little, if any, specialization of the post-synaptic muscle membrane at close neuromuscular junctions. However, thickening of both pre- and post-synaptic membranes and the presence of subsynaptic sacs of endoplasmic reticulum have been described in some smooth muscles.

4. No distinction has been found yet between sensory and motor neurones in smooth muscle.

Intra-axonal vesicles
Types and occurrence of vesicles

Three main types of synaptic vesicles (Fig. 13) have been observed in the pre-terminal and terminal varicosities of autonomic nerves in different systems (Lever & Esterhuizen, 1961; De Robertis & Pellegrino de Iraldi, 1961; Taxi, 1961, 1965; Grillo & Palay, 1962; Richardson, 1962; Burnstock & Merrillees, 1964; Lane & Rhodin, 1964*a*; Richardson, 1964, 1966; Tafuri, 1964; Bondareff, 1965; Bloom & Barrnett, 1966*b*; Grillo, 1966; Thoenen *et al.*, 1966; Van Orden

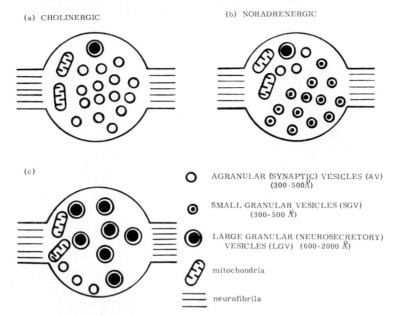

FIG. 13   Diagrammatic representation of the three major types of intra-axonal vesicle composition seen in different autonomic nerves. In each case a preterminal varicosity (approximately 1 to 2μ diameter) is drawn containing vesicles, bordered by intervaricose regions of the nerve (approximately 0·1 to 0·5μ diameter) containing neurofilaments. (a) Represents the vesicle composition of most parasympathetic cholinergic nerves with a predominance of agranular vesicles; (b) represents the vesicle composition of most sympathetic noradrenergic nerves. The actual proportions of the different vesicles found in these adult nerves are approximately as follows: 80% small granular vesicles; 15% agranular vesicles; 5% large granular vesicles. The proportion of small granular vesicles to agranular vesicles is increased after intra-axonal loading of noradrenaline; (c) represents the vesicle composition of many nerves seen in the vertebrate gastro-intestinal tract, the amphibian lung and bladder, some mammalian blood vessels, and in the molluscan viscera. The granular cores of the small, but not the large, granular vesicles are not present after treatment with reserpine or metaraminol.

et al., 1966; Burnstock & Robinson, 1967; Tranzer & Thoenen, 1967a, b; Van Orden, Bensch & Giarman, 1967; Hökfelt, 1968; Yamauchi & Burnstock, 1969b).

1. *Agranular or synaptic vesicles* (250 to 600 Å). These vesicles are found predominantly in all fibres of cat ciliary muscle, in some axon profiles in intestine, ureter, lung, bladder and sphincter pupillae (Fig. 14b). They are identical with those first described in the central nervous system (De Robertis & Bennett, 1955; Palay, 1956a) and at the motor end plate (Robertson, 1956). Some large granular vesicles are nearly always found within the same axon profiles as those containing predominantly agranular vesicles (Fig. 13a).

2. *Small granular vesicles* (250 to 600 Å). These vesicles have a dense central core of about 150 Å. They are predominant in all mammalian sympathetic nerves (Fig. 14c). Axon profiles containing predominantly small granular vesicles

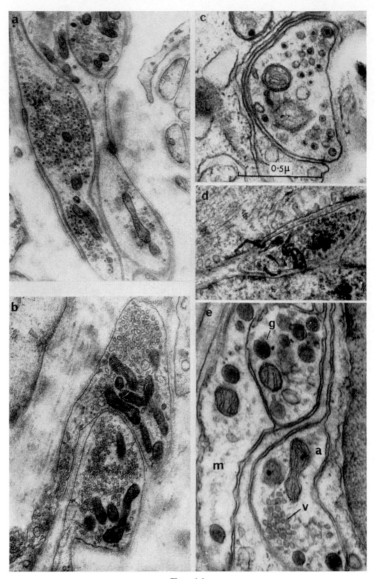

FIG. 14

are sometimes related closely to profiles containing predominantly agranular vesicles suggesting a possible interaction (Lane & Rhodin, 1964a; Thoenen *et al.*, 1966). Quantitative studies have shown that the proportions of vesicles of different types in sympathetic nerves in the mouse vas deferens (Yamauchi & Burnstock, 1969b) were as follows (Table 5): small granular vesicles 81 to 85%; large granular vesicles 3 to 4%; agranular vesicles 12 to 20%. These figures are comparable to those described for adult rat vas deferens (van Orden *et al.*, 1966; van Orden *et al.*, 1967).

3. *Large (neurosecretory) granular vesicles* (700 to 1600 Å). The central core (300 to 900 Å) in these vesicles varies considerably in size and density. A few large granular vesicles are often found within the same profiles which contain predominantly small granular vesicles or predominantly agranular vesicles (Fig. 13b). Large granular vesicles predominate in certain intramuscular bundles of nerves in the intestine (Hager & Tafuri, 1959; Thaemert, 1963; Lane & Rhodin, 1964a; Tafuri, 1964; Yamauchi, 1964; Taxi, 1965; Taxi & Droz, 1966; Bennett & Rogers, 1967; Nagasawa & Mito, 1967) and in axon profiles of some blood vessels (Verity & Bevan, 1966; Devine & Simpson, 1967). Large granular vesicles are also dominant in amphibian nerves supplying the lung, bladder and gastrointestinal tract (Fig. 14e) (Burnstock & Merrillees, 1964; Boyd, Burnstock & Rogers, 1964; Rogers & Burnstock, 1966a, b; Burnstock & Robinson, 1967). In transverse sections of nerve bundles in the toad intestine, 32% of the axons contained large granular vesicles only, while 35% contained both large granular and agranular vesicles; this distribution suggests that the

---

FIG. 14   *a* Rabbit iris. A small nerve bundle entering one of the larger spaces between two groups of *dilator muscle* processes. The Schwann sheath is becoming attenuated, leaving parts of the axonal surfaces covered only by basement membrane. The axons contain many small granular vesicles and some agranular vesicles which is typical of adrenergic sympathetic nerve fibres. × 33,000. (From Richardson, 1964)   *b* Rabbit iris. Two closely associated axons within a wide space separating two muscle fibres of the *sphincter pupillae*. The Schwann sheath has disappeared, but the axons are still invested with basement membrane. The axons appear to have bead-like enlargements which are packed with agranular vesicles and mitochondria. These are typical of cholinergic axons. × 33,000. (From Richardson, 1964)   *c* Rat vas deferens. The axon contains small granular vesicles and two mitochondria, one of which is cut tangentially through its external membrane. The interval between nerve and muscle membranes is maximal in this ending, being about 250 Å. The subsurface cistern between the muscle membrane and the muscle mitochondria can be traced almost to the surface of the muscle fibre above. Below the ending the cistern diverges from the muscle membrane to become continuous with an enlarged sac of endoplasmic reticulum. A row of pinocytotic vesicles associated with the muscle membrane is located within part of the synaptic area. (From Richardson, 1962)   *d* Rat vas deferens. Radioautographs obtained 30 min after injection of noradrenaline-³H. Silver grains are associated with an axon which contains granular vesicles and lies deep inside the musculature. × 27,600. (From Taxi & Droz, 1966)   *e* Toad small intestine. Portion of a small nerve bundle in the circular smooth muscle layer. Some axon (a) profiles are crowded with agranular vesicles (v) while others contain numerous large granular vesicles (g). m, mitochondrian. × 48,000. (From Rogers & Burnstock, 1966b)

majority of the axons contain both granular and agranular vesicles, although not always grouped together (Rogers & Burnstock, 1966b).

Large granular vesicles are the predominant vesicle type in the perikaryon of neurones (Taxi, 1961, 1965; Cravioto, 1962; Grillo & Palay, 1962; Pick, 1963; Grillo, 1966) which contain predominantly small granular vesicles in their preterminal varicosities and terminals. Multivesicular bodies are also sometimes found in the perikaryon of autonomic neurones (Grillo, 1966; Yamauchi & Burnstock, 1969c). Large granular vesicles have also been isolated from brain homogenate (Gray & Whittaker, 1962; Pellegrino de Iraldi, Duggan & De Robertis, 1963; De Robertis, Pellegrino de Iraldi, Arnaiz & Zieher, 1965; Whittaker, 1966; Wood, 1966).

Yamauchi (1964) presented evidence to suggest a mitochondrial origin of small granular vesicles, while he suggested that large granular vesicles originate from smooth-surfaced endoplasmic reticulum in the terminal axoplasm.

It should be pointed out that some methods of fixation for electronmicroscopy do not bring out the granules in some tissues (Merrillees et al., 1963; Richardson, 1966; Hökfelt, 1966a). In general, potassium permanganate or glutaraldehyde are the most reliable fixatives for revealing granular vesicles (Richardson, 1966).

Function of vesicles

The functions of the various types of intra-axonal vesicles have been discussed in several reviews (Burnstock & Merrillees, 1964; Grillo, 1966; Richardson, 1966; Burnstock & Robinson, 1967; Hökfelt, 1968).

Agranular vesicles. These are identical with those seen in the skeletal neuromuscular junction and at ganglionic synapses and by analogy have been equated with the storage of acetylcholine (Ach). Further support for this suggestion comes from comparative studies, where in preparations innervated predominantly by cholinergic nerves (e.g. lizard lung, sphincter pupillae, urinary bladder, ciliary muscle) agranular vesicles were overwhelmingly the dominant vesicle type in all axon profiles (Burnstock & Merrillees, 1964; Evans & Evans, 1964; Richardson, 1964; Hökfelt, 1966a) (see Fig. 14b).

The question has been asked whether the small proportion of agranular vesicles found in sympathetic axon profiles represent '*empty*' small granular vesicles which predominate in these profiles or whether they contain Ach. The latter possibility has been discussed in relation to the theory that a cholinergic link is involved in sympathetic transmission (Burn & Rand, 1959, 1965). It seems more likely that they represent empty granular vesicles in view of the results of experiments in which various acute and chronic drug treatments have increased intra-axonal stores of noradrenaline (NA) (Pellegrino de Iraldi & De Robertis, 1961, 1963; De Robertis, 1963; Clementi & Garbagnati, 1964; Pellegrino de Iraldi, Zieher & De Robertis, 1965; van Orden et al., 1966, 1967; Tranzer & Thoenen, 1967a). In these experiments, the relative number of a-granular vesicles in the profiles was decreased, suggesting that these vesicles are capable of taking up and storing NA. The suggestion has also been made that the agranular vesicles found in profiles containing predominantly small granular vesicles contain dopamine (Richardson, 1963).

Small granular vesicles. There is now very strong evidence that small granular vesicles are associated with stores of bound NA (see Burnstock & Robinson, 1967; van Orden *et al.*, 1967). They are seen predominantly in mammalian sympathetic nerves (see Burnstock & Merrillees, 1964; Taxi, 1965; Clementi, 1965; Simpson & Devine, 1966) which have been shown either by spectro-fluorometric assay or fluorescent histochemistry to contain high levels of nor-adrenaline in the varicose regions (Norberg & Hamberger, 1964). The increase in number of axons containing small granular vesicles in the developing mouse vas deferens is accompanied by a parallel increase in the number of noradrenergic nerves demonstrated histochemically (Yamauchi & Burnstock, 1969*b*).

Electronmicroscope-autoradiography of sympathetic nerves after injection of tritium-labelled NA into animals reveals silver grains in close relation to nerve profiles containing an abundance of small granular vesicles (Fig. 14*d*) (Wolfe *et al.*, 1962; Wolfe & Potter, 1963; Taxi & Droz, 1966; Esterhuizen, *et al.*, 1968).

Treatment with drugs which deplete NA stores in nerves (e.g. reserpine, guanethidine, α-methyl-m-tyrosine, metaraminol) removes the granules from small granular vesicles in sympathetic nerves (Pellegrino de Iraldi & De Robertis, 1961, 1963; Richardson, 1963; Clementi & Garbagnati, 1964; Shimizu & Ishii, 1964; Clementi, 1965; Bondareff & Gordon, 1966; Hökfelt, 1966*b*; Devine *et al.*, 1967; van Orden *et al.*, 1966, 1967). The number of granular vesicles and NA levels in the rat vas deferens were in good agreement up to 2 weeks after a single dose (5 mg/kg) of reserpine (Richardson, 1963). After NA depletion of the rat vas deferens to 20% of normal levels, with α-methyl-m-tyrosine, the percentage of small granular vesicles in varicosities was reduced from 53–64% to 6–11% (van Orden *et al.*, 1966). Subsequent reinfusion with NA, and its uptake into the nerve was associated with the reappearance of granules within these vesicles (Pellegrino de Iraldi, Zieher & De Robertis, 1965; Bondareff & Gordon, 1966; van Orden *et al.*, 1966).

The use of other drugs, such as monoamineoxidase (MAO) inhibitors, dopa and dopamine, which increase intraneuronal NA stores has not produced conclusive evidence of a parallel increase in small granular vesicles when given alone, but greater increases in small granular vesicles were seen with NA, dopamine or dopa given in addition to MAO inhibitors (Pellegrino de Iraldi & De Robertis, 1961, 1963; Clementi & Garbagnati, 1965; Clementi, Mantegazza & Botturi, 1966; van Orden *et al.*, 1967). A marked increase in small granular vesicles in sympathetic nerves was demonstrated after treatment of the animal with angiotensin (Panagiotis & Hungerford, 1966).

It should be pointed out that electronmicroscopy is basically a sampling method, so that some of the results of drug action must be interpreted with great care and are often inconclusive (Pellegrino de Iraldi & De Robertis, 1963; De Robertis, 1963; Clementi, 1965; Burnstock & Robinson, 1967). For example, De Robertis & Ferreira (1957) reported a depletion of vesicles present in the nerve terminals after long stimulation of the splanchnic nerve supplying the adrenal medulla. However, this has not been confirmed in other tissues, but it is very difficult to see quantitative changes of this kind on material where even the control distribution is extremely patchy and variable (Burnstock & Merrillees, 1964).

Further support for the relationship between small granular vesicles and noradrenaline has been obtained by electronmicroscopy and catecholamine (CA) assay of granules isolated by differential centrifugation of homogenized splenic sympathetic nerves (von Euler & Hillarp, 1956; von Euler, 1958; Schümann, 1958; Potter, Axelrod & Wolfe, 1962; Potter & Axelrod, 1962, 1963a, b; Gillis, 1964; Michaelson, Richardson, Snyder & Titus, 1964; Stjärne, 1964; von Euler & Swanbeck, 1964; Potter, Cooper, William & Wolfe, 1965; Potter, 1966), and rat salivary gland, vas deferens and heart (Wegmann & Kako, 1961; Potter, Axelrod & Wolfe, 1962; Potter & Axelrod, 1962, 1963b; Michaelson et al., 1964; Potter, 1966). The particulate fraction of rat heart which contains NA, contains many vesicles, 500 to 2000 Å in diameter, some of which have a dense core (Potter, 1966). Small and large granular vesicles and agranular vesicles have been isolated from autonomic nerves in dog heart, but small granular vesicles were not found in the nerves remaining in transplanted hearts where NA was absent, although the other vesicles remained (Potter, Cooper, Willman & Wolfe, 1965).

Finally NA has been localized histochemically at the electronmicroscope level in the cores of 80% of the vesicles seen in nerves in the rat vas deferens (Bloom & Barrnett, 1966b).

Large granular vesicles. The nature of the large granular vesicles is not yet clear. Furthermore, there is a wide range in size and density of granular cores. It is therefore possible that several types with different roles will be differentiated.

Large granular vesicles in mammalian nerves are unaffected by depletion of CA stores by reserpine treatment, e.g. in the vas deferens (Hökfelt, 1966b), the gut (Taxi, 1965) or in sympathetic nerves about small arteries (Devine, Robertson & Simpson 1967). However, more recent studies (Mayor & Kapeller, 1967; Hökfelt, 1968; Tranzer & Thoenen, 1968), suggest that the large granular vesicles found in sympathetic nerves are associated with the presence of monoamines.

The large granular vesicles seen in many axon profiles in the vertebrate alimentary tract (Taxi, 1965; Grillo, 1966; Rogers & Burnstock, 1966b) may be associated with the unknown inhibitory transmitter released from intrinsic enteric inhibitory neurones (Burnstock, Campbell, Bennett & Holman, 1963a, b, 1964; Burnstock et al., 1966; Bennett, Burnstock & Holman, 1966b).

The large granular vesicles found in amphibian sympathetic nerves may be associated with adrenaline (AD), since this is the dominant CA in the nerves of this vertebrate class (von Euler, 1963; Falck, Häggendal & Owman, 1963; Angelakos, Glassman, Millard & King, 1965; McLean & Burnstock, 1966, 1967*b*). However it is interesting that reserpine treatment does not appear to affect the large granular vesicles in amphibian nerves, although the store of AD is considerably reduced (Taxi, 1965; Robinson & Burnstock, unpublished results). A relation between large granules and AD has been established in the adrenal medulla (Blaschko & Welch, 1953; Hillarp, Lagerstedt & Nilson, 1953; Hagen & Barrnett, 1960). Large granular vesicles are also found in snail nerves which contain 5-hydroxytryptamine (Campbell & Burnstock, 1968). Sympathetic nerve endings, but not preterminal varicosities to the pineal body of the rat, contain 5HT as well as NA, which they apparently take up from the adjacent parenchymal cells (Bertler, Falck & Owman, 1964; Owman, 1964).

It is possible that in mammals the large granular vesicles represent sites of synthesis of transmitter, and small granular vesicles the sites of storage of CA. This view would be consistent with the fluorescent histochemical observations that NA synthesized in the perikaryon is passed down the nerve to the terminal varicose region (Dahlström, 1965), but that some noradrenaline may also be synthesized in pre-terminal and terminal varicosities (Malmfors, 1965*a*).

In a recent study of the fine structure of autonomic nerves from cats pre-treated with 5-hydroxy-dopamine (5-HDA), an amine acting as a 'false' sympathetic transmitter, a wide size range of large granular vesicles was shown to have accumulated in the sympathetic nerve terminals, although the NA content was severely reduced (Tranzer & Thoenen, 1967*b*). This experiment was taken as strong evidence for an association of dense material in these vesicles with 5-HDA. The agranular vesicles in cholinergic nerve endings were unaffected by the drug treatment.

Summary

1. Three main types of vesicles appear in the terminal regions of autonomic nerves: agranular or synaptic vesicles (250 to 600 Å); small granular vesicles (250 to 600 Å); large (neurosecretory) granular vesicles (700 to 1600 Å).

2. Agranular vesicles when found as the predominant vesicle type in autonomic nerves are associated with acetylcholine.

3. Small granular vesicles in mammalian autonomic nerves are associated with noradrenaline, i.e. with noradrenaline bound in the granular cores.

4. The proportions of the different vesicle types found in sympathetic noradrenergic nerves are approximately as follows: small granular vesicles 81 to 85%; large granular vesicles 3 to 4%; agranular vesicles 12 to 20%. Agranular vesicles found in sympathetic nerves appear to be 'empty' small granular vesicles rather than acetylcholine-containing vesicles.

5. The nature of large granular vesicles is not yet clear. There is evidence that those present in small numbers in sympathetic noradrenergic nerves contain monoamines. However their predominant appearance in various kinds of axons in tissues containing adrenaline or 5-hydroxy-tryptamine or unknown transmitter

in non-sympathetic myenteric nerves is, at this stage, no more than coincidental evidence of possible storage sites.

Electron microscope histochemistry at autonomic neuroeffectors

A recent and important technical advance, with considerable potential in the study of autonomic transmitters and associated substances, is the application of histochemical methods to sections prepared for electron microscopy (see Eränkö, 1967b).

Localization of cholinesterase

The basic light microscope methods (Koelle, 1963) have been modified for electronmicroscopy and have been applied to a variety of different tissues (Zacks & Blumberg, 1961; Barrnett, 1962, 1966; Torack & Barrnett, 1962; Karnovsky & Roots, 1964; Lewis & Shute, 1964, 1966; Miledi, 1964; Mori, Maeda & Shimizu, 1964; Davis & Koelle, 1965; Koelle & Foroglou-Kerameos, 1965; Smith & Treherne, 1965; Shimizu & Ishii, 1966; Shute & Lewis, 1966; Eränkö, Rechardt & Hänninen, 1967).

Several autonomically-innervated smooth muscle tissues have been examined, including the toad bladder, which is predominantly innervated by cholinergic nerves and the guinea-pig vas deferens which is innervated predominantly by adrenergic nerves (Burnstock & Robinson, 1967; Robinson & Bell, 1967; Robinson, 1969b). There are also descriptions of the localization of AChE in a few nerves in the cat nictitating membrane (Esterhuizen *et al.*, 1968), pancreatic arterioles (Graham, Lever & Spriggs, 1968) and uterine arteries (Bell, 1969). Although granules are not well preserved with formalin fixation, these early results suggest that there is a heavy deposition of AChE on the cell membranes of cholinergic nerves which, with longer incubation time, fills the cleft between the axon and the Schwann cell processes (Fig. 15a). Many of the noradrenergic nerves in the guinea-pig vas deferens appear to have a light staining for AChE on their membranes, but are easily distinguished from cholinergic nerves (Fig. 15b).

In those instances where an axon showing heavy AChE staining was within 800 Å of a muscle cell, in the guinea-pig vas deferens, uterine artery and toad bladder, the muscle membrane adjacent to the nerve also stained heavily for AChE, suggesting a functional cholinergic synapse. (Burnstock & Robinson, 1967; Robinson & Bell, 1967; Robinson 1969b; Bell, 1969).

The studies of the localization of AChE in the guinea-pig vas deferens show a small proportion of axons with heavy AChE staining on their membranes, which suggests that there is a small component of cholinergic fibres in the sympathetic nerves supplying this organ. (Burnstock & Robinson, 1967; Robinson, 1969b). This suggestion has also been made on the basis of evidence from pharmacology (Leaders, 1965), electrophysiology (Bell, 1967a), and the combination of fluorescent histochemistry and AChE staining using light microscopy (Jacobowitz & Koelle, 1963). The demonstration of some cholinergic nerves in the vas deferens may explain why many axon profiles in the guinea-pig (Merrillees *et al.*, 1963) and some in the rat (Richardson, 1962) and mouse (Lane &

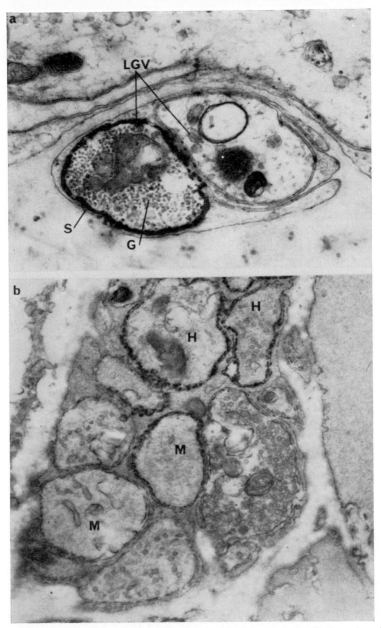

Fig. 15 *a* Toad sympathetic ganglion. Acetylcholinesterase incubation. A pair of axons passing close to (but not synapsing with) a sympathetic ganglion cell. One axon is ringed by a heavy deposit of stain (S) while the other has only a very small amount. Large granular vesicles (LGV) are present in both axons but may be difficult to see in the stained axon because of a high concentration of glycogen (G). Fixation — Formalin. Incubation — Karnovsky medium. × 24,000. (From Burnstock & Robinson, 1967) *b* Guinea-pig vas deferens. Acetylcholinesterase incubation. Portion of an axon bundle which demonstrates the range of cholinesterase activity found in axons in this tissue. About 5 to 10% of the axon profiles seen in the vas deferens show heavy (H) to moderate (M) staining. The remainder show little, if any, detectable enzyme, but this depends to some extent on the incubation time. Fixation—Formalin. Incubation—Karnovsky medium. × 24,000. (From Burnstock & Robinson, 1967)

Rhodin, 1964a, Yamauchi & Burnstock, 1969b) vas deferens contain predominantly agranular vesicles. There is also evidence (assay and histochemical) that the proportion of adrenergic to cholinergic fibres varies in the vas deferens of different mammalian species (Sjöstrand, 1965; Jacobowitz & Koelle, 1963).

Localization of catecholamines

Several techniques for fine-structural localization of NA and AD have been developed in recent years. The method introduced by Wood & Barrnett (1963) for the specific demonstration of NA in granules of the adrenal medulla used the selective oxidation of NA, but not AD, by potassium dichromate at pH 4·1, after glutaraldehyde fixation. For demonstration of adrenaline, the tissue was incubated with dichromate at pH 6·5, after glutaraldehyde fixation, and postfixed in osmium tetroxide. These techniques have substantiated (Wood & Barrnett, 1963, 1964; Coupland, Pyper & Hopwood, 1964; Tramezzani, Chiocchio & Wasserman, 1964; Coupland & Hopwood, 1966) the earlier claims that NA and AD occur in different cell types in the adrenal medulla (Hillarp & Hökfelt, 1953; Eränkö, 1955, 1956, 1960).

The same methods have been applied to nerve endings on chromaffin cells (Coupland, 1965), adrenal medulla (Coupland, Pyper & Hopwood, 1964; Bloom & Barrnett, 1966b), and more recently to autonomic nerves in the rat vas deferens (Bloom & Barrnett, 1966b). Nerve endings in the adrenal medulla contain agranular and large granular vesicles, but neither of these vesicle types stained for NA. In contrast, the small granular vesicles contained in most axons in the rat vas deferens stained for NA with intensity equal to that seen in the adrenal medullary cells. Following depletion of noradrenaline stores with either reserpine or α-methyl-m-tyrosine, few or no densely-staining granular vesicles were seen.

Methods for simultaneous localization of CA and ChE at the electron microscope level are just beginning (Lewis & Shute, 1967). The differential staining of NA and AD with glutaraldehyde-osmium tetroxide fixation was still retained in the rat adrenal medulla after the tissues had been subjected to the thiocholine technique, and chromaffin cells of two types could still be distinguished.

Diversification and classification of smooth muscle

Bozler (1941, 1948) divided vertebrate smooth muscle into two categories:
1. 'Unitary' muscles, which include gut, ureter and uterus. These muscles behave like single units and conduction is from muscle fibre to muscle fibre. They are usually spontaneously active.
2. 'Multiunit' muscles, which include nictitating membrane, iris sphincter, ciliary muscle, pilomotors, urinary bladder and most vascular smooth muscles. These muscles are normally activated by nerves and consist of numerous independent units. They are usually not spontaneously active.

This classification has been useful, but it is now clear that it cannot be regarded as rigid; many muscles (e.g. guinea-pig vas deferens, bladder) show features of both types (Burnstock & Holman, 1961, 1963; Ursillo, 1961). Furthermore, recent work indicates that many vascular smooth muscles do not belong to the multiunit type.

In view of the considerable advance in knowledge of smooth muscle systems since Bozler's classification, mainly as a result of microelectrode and electron-microscope studies, a new and more explicit classification was proposed recently (Burnstock, 1968) and a schematic representation of three different categories of smooth muscle, based largely on the nature of the autonomic neuromuscular junction is shown in Fig. 16. Models A and C approximate Bozler's miltiunit and unitary muscle types respectively, while model B lies between and is the best substantiated experimentally (see Bennett & Burnstock, 1968). It is likely that there is a range of muscles showing features intermediate between models B and C.

Model B, representing guinea-pig vas deferens, and probably also cat nictitating membrane and constrictor pupillae, urinary bladder, dog retractor penis, circular intestinal muscle and seminal vesicles, shows part of a muscle effector bundle with low resistance pathways between most of the cells. Some (20 to 50%) of the cells have one or more close (200 to 500 Å) neuro-muscular junctions. In the model these have been termed *directly-innervated or key cells* and are directly affected by transmitter released from the nerve. The remainder of the cells, called *indirectly-innervated or coupled cells*, exhibit junction potentials by electro-tonic coupling with directly-innervated cells. Some of the coupled cells may also be affected by release of transmitter from 'en passage' varicosities further than 500 Å and up to about 1000 Å away. All the cells are capable of carrying a propagated action potential when a sufficient area of cells is depolarized by nerve action.

Model A, representing mouse and rat vas deferens and probably also cat ciliary muscle shows individual innervation of every cell by one or more close (200 Å) neuromuscular junctions. These muscles are geared for fast simultaneous contraction. There are probably some low resistance pathways between cells allowing electrotonic coupling of junction potentials and decremental spread of spike activity, but the membrane properties are such that all-or-none action potentials do not propagate through the tissue (Furness & Burnstock, 1969).

Model C, representing longitudinal muscle coat (alimentary canal) and probably also uterus, ureter and most vascular smooth muscles is organized into effector bundles with highly developed intercellular couplings (Bennett & Burnstock, 1968). There are only a few key cells in limited areas of an effector bundle (in some tissues included in this group, e.g. vascular smooth muscle, minimum nerve-muscle membrane separation may be as great as 800 Å). Coupled cells in the vicinity (within about 1·4 mm, which is the space constant) of key cells exhibit junction potentials carried by electrotonic coupling, while a

MODELS OF AUTONOMIC INNERVATION OF SMOOTH MUSCLE .

A. <u>Every</u> cell with close n. m. j. ; some electrotonic coupling,
<u>no</u> propagated action potentials

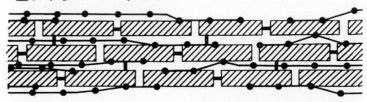

B. <u>Many</u> cells with close n. m. j. ; electrotonic coupling,
propagated action potentials

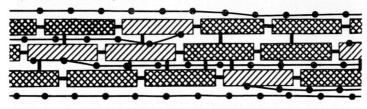

C. <u>Few</u> cells with close n. m. j. ; electrotonic coupling,
propagated action potentials

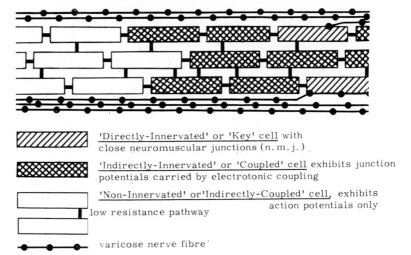

'Directly-Innervated' or 'Key' cell with
close neuromuscular junctions (n. m. j.)

'Indirectly-Innervated' or 'Coupled' cell exhibits junction
potentials carried by electrotonic coupling

'Non-Innervated' or 'Indirectly-Coupled' cell, exhibits
action potentials only
low resistance pathway

varicose nerve fibre

FIG. 16   Schematic representation of autonomic innervation of smooth muscle. For
explanation see text (page 67).

third group of cells, termed here *non-innervated or indirectly-coupled cells* are
activated by the passage of a propagated action potential set up in the region
of the key and coupled cells. Some coupled and indirectly-coupled cells may
also be effected in a minor way by transmitter released from 'en passage'

varicosities up to about 3000 to 10,000 Å away. These muscles, which are usually spontaneously active, are geared for complex, slow, graded, local tension changes. It is unlikely that the three cells described in this model represent rigid types with different properties; on the contrary it is probable that the same cell might act as a key, coupled or indirectly-coupled cell when different areas of muscle bundles are stimulated by nerves activated at different times during the complex physiological control of the organ in an intact animal.

It should be emphasized that these models are by way of provocative speculation, and need further experimental verification.

2

THE RELATION BETWEEN THE MEMBRANE POTENTIAL AND THE ION DISTRIBUTION IN SMOOTH MUSCLE CELLS

R. Casteels

INTRODUCTION

The study of the fundamental mechanism responsible for the membrane phenomena of smooth muscle cells requires a hypothesis which allows an interpretation of the experimental facts and which directs further investigation. There are not sufficient reasons as yet for assuming that this fundamental mechanism in smooth muscle may be different from the one proposed for other excitable tissues.

Several theories have been proposed to explain the potential difference across the cell membrane. According to some authors (Troshin, 1966; Ling, 1962) the membrane potential and ion distribution in excitable cells should be attributed to the selective adsorption of potassium ions to fixed negative charges within the cells. However the most widely-accepted theory is the membrane hypothesis which originated from the work of Bernstein (1912). Applying the physico-chemical concepts of Nernst and Ostwald, Bernstein put forward that the resting cell membrane was selectively permeable to potassium only and that this selectivity was lost during excitation, resulting in an indiscriminate penetration by other small ions such as sodium and chloride.

The fundamental idea of this theory of a membrane with selective permeability for ions has not been modified by recent investigations. Our present picture of the membrane function has, however, grown much more complex. First, it was found that excitation leads to a transient reversal of the membrane potential and not to a simple abolition, as proposed by Bernstein (Hodgkin & Huxley, 1939; 1945). It also became obvious that the membrane is not only permeable to potassium ions, but also to chloride ions (Boyle & Conway, 1941) and that there was also a slight permeability to sodium ions (Levi & Ussing, 1948; Keynes, 1954). This means that external sodium ions must diffuse into the cells under the combined influence of the concentration gradient and the membrane potential. The maintenance of a steady state of ion distribution in cells is only possible if there is some active mechanism capable of extruding sodium continuously across the surface membrane. Such a pumping system has been proposed first by Dean (1941) and has been fully investigated by Hodgkin and Keynes (1955a) in squid giant axon. This uphill movement of sodium from the cytoplasm to the outside depends on the presence of energy-yielding substrates, but the downhill movement in squid giant axon is not affected by inhibition of the metabolism. An active Na-extrusion, against an electrochemical gradient, also exists in other nerve fibres and skeletal muscle fibres (Caldwell, 1968) and in smooth muscle cells (Casteels, unpublished). It is only by a continuous active exchange of ions that these cells are able to maintain their steady state, i.e. a low intracellular sodium content and a high potassium content. However, by demonstrating the existence of a sodium pump one does not completely explain the relation between the ion distribution and membrane potential, because such a pump can affect the membrane potential in several ways. This pump could act as a direct source of current by transferring positive charges outward, and thus generate a potential difference across the cell membrane. The other diffusible ions like potassium and chloride would then be distributed between the intra- and the extracellular compartment until their concentrations are in equilibrium with the potential difference. Such a pump can make the membrane potential more negative than the potassium equilibrium potential in non-steady state conditions (Connelly, 1959; Kernan, 1962; Cross, Keynes & Rybová, 1965; Adrian & Slayman, 1966; Frumento 1965). The extrusion of sodium ions creates a potential difference across the membrane which attracts K^+-ions inwards. The coupling between sodium and potassium movement in this pumping system would be purely electrical. An alternative mechanism is an electrically-neutral pump, which links the outward transfer of each sodium ion to the inward transfer of a K^+-ion so that there is no separation of electrical charge across the membrane.

For most experimental observations an electroneutral pumping system is the most suitable working hypothesis because of the relation between uphill sodium and potassium movement, and because the membrane potential is less negative than the potassium equilibrium potential. However, one has to be aware of the fact that, even under steady state conditions, the coupling between the Na- and K-movement could be partially chemical and partially electrical. Such an intermediate coupling mechanism could be responsible for some discrepancies between measured membrane potentials and membrane potentials arising from the steady-state diffusion of univalent ions, as calculated by the Goldman equation.

It can be concluded that the unequal distribution of ions and the maintenance of the potential difference between the inside and outside of the cell always depends on a metabolic process, which expels sodium and accumulates potassium. In the presence of a neutral pump, the membrane potential results from the ionic gradients and the relative conductances for sodium, potassium and

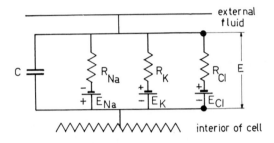

FIG. 1   Diagram of the excitable membrane. E_{Na}, E_K and E_{Cl} are the equilibrium potentials for sodium, potassium and chloride. E is the membrane potential; R_{Na}, R_K and R_{Cl} represent the resistance to the movement of the three ions; C is the membrane capacity. (After Hodgkin, 1958)

eventually for chloride. The distribution of the latter ion will only affect the membrane potential, if these ions are not distributed according to the existing membrane potential. If the pump works, in addition, partially electrogenically the membrane potential will be determined by some combination of diffusion potentials and the rate of transfer of electrical charges across the membrane.

The electroneutral system can be represented by an electric circuit diagram consisting of three electromotive forces (Hodgkin, 1958), being the equilibrium potentials for sodium, potassium and chloride and of three corresponding conducting channels represented by the resistance to the movement of the three ions. The channels are connected in parallel with the membrane capacity (Fig. 1). The actual value of the membrane potential in such a system must lie between the values of the equilibrium potentials for sodium, potassium and chloride ions, at a level which depends on the predominance of an ionic conductance.

A simple equation for such a system has been derived by Goldman (1943), assuming that the voltage gradient through the membrane is constant and that the ions only move under the influence of diffusion and the electric field. It is also assumed that the concentrations of ions immediately adjacent to the membrane are directly proportional to the concentrations in aqueous solution. Hodgkin & Katz (1949) obtained from these assumptions the following expression for the membrane potential:

$$E = -\frac{RT}{F} \ln \frac{P_K[K]_i + P_{Na}[Na]_i + P_{Cl}[Cl]_o}{P_K[K]_o + P_{Na}[Na]_o + P_{Cl}[Cl]_i}$$

where $(K)_i$, $(Na)_i$ and $(Cl)_i$ are the activities of the ions inside the cells, and where the same symbols with the subscript o are activities outside the cells. P_K, P_{Na} and P_{Cl} are permeability constants of the membrane (cm/sec) for the individual ions.

The equation

$$E = -\frac{RT}{F} \ln \frac{P_K[K]_i + P_{Na}[Na]_i}{P_K[K]_o + P_{Na}[Na]_o}$$

will represent the membrane potential for zero internal chloride in a chloride-free solution or in tissues where chloride ions are distributed passively (Hodgkin & Horowicz, 1959b; Mullins & Noda, 1963). Under these experimental conditions and if the only ion pump in the cell were an electroneutral cation-exchange pump, the equation would be valid, irrespective of the contour of the voltage profile through the membrane (Hodgkin & Horowicz, 1959b; Patlak, 1960; Barr, 1965).

For smooth muscle cells, the distribution of chloride ions is not passive (Casteels & Kuriyama, 1966) and accordingly the voltage profile must be determined to make an application of the Goldman equation possible. One of these possible restrictions is the constant field in the membrane. It is very likely that this constant-field assumption does not correspond perfectly to the charge distribution in complex biological membranes. It is thus not unexpected to find discrepancies between the actual observations and the predictions of the Goldman equation. These however are mostly not sufficient to invalidate the membrane concept and they can often be accounted for by a modification of the theory. Moreover this theory gives a starting point, and can be extended and adapted to new findings. In the present state of our knowledge of smooth muscle physiology there is not yet sufficient reason to replace a hypothesis which has been extremely fruitful in the study of other tissues.

MEMBRANE POTENTIAL AND ION CONTENT OF SMOOTH MUSCLE CELLS IN PHYSIOLOGICAL SOLUTION

Although the small cell diameter is a limiting factor for an accurate measurement of the membrane potential, several investigators have succeeded in obtaining reproducible values for the membrane potential in various smooth muscle tissues. Some values have been summarized in Table 1. It is obvious that an essential feature of the resting potential in smooth muscle cells is its relatively low value compared with that of skeletal and cardiac muscle. The interpretation of this resting potential by the membrane theory requires the knowledge of the intracellular ion concentrations and the permeability of the membrane for the various ions.

TABLE 1. Resting potential and action potential of some smooth muscle cells, measured with micro-electrodes

Tissue	Resting potential (mV)	Action potential (mV)	Authors
Taenia coli guinea-pig	51·5 ± 0·36 (105)	59·3 ± 0·27 (144)	Holman, 1958
Taenia coli guinea-pig	55 ± 0·5 (240)	62 ± 0·6 (240)	Bülbring & Kuriyama, 1963a
Colon rabbit	49·6 ± 0·6 (261)	33·5 ± 0·8 (261)	Gillespie, 1962a
Myometrium rat (ovariectomized)	35·2 ± 1·2 (25)	—	Marshall, 1959
Myometrium rat (oestrogen-dominated)	57·6 ± 0·49 (48)	65·3 ± 0·67 (48)	Marshall, 1959
Myometrium rat (progesterone-dominated)	63·8 ± 0·51 (82)	—	Marshall, 1959
Bladder rabbit	40	35	Ursillo, 1961
Vas deferens guinea-pig	57 ± 0·85 (50)	68 ± 1·08 (55)	Burnstock & Holman, 1961

The transfer of a smooth muscle tissue to an artificial solution is accompanied by a modification of its ion content (Goodford & Hermansen, 1961), because of the difference in ion content between the *in vivo* fluid and the so called physio-logical solution, and also because of the absence of colloid osmotic particles. In the physiological solution the tissue comes, however, to a steady state which makes a discussion of the relation between membrane potential and ion gradients possible. The present discussion is limited to the guinea-pig's taenia coli and vas deferens. The membrane potentials of these tissues (Table 1) and the total ion contents (Casteels & Kuriyama, 1966) have been determined after prolonged exposure to a modified Krebs solution. On the simplifying assumption that the tissues consist of an extracellular and an intracellular compartment in which the ions are dissolved in an aqueous phase, we can calculate the intracellular ion concentrations from the total ionic content, the extracellular space and the dry wt/wet wt ratio of the tissues. Such a calculation can only give approximate values for the intracellular ion concentrations because of possible binding of ions and because of the uncertainty concerning the extracellular space. Moreover the assumption of a single uniform extracellular and intracellular compartment cannot be completely justified. Inulin has been used extensively to determine the extracellular space in frog striated muscle and, in this tissue, the inulin space corresponds to the sucrose space and to the real extracellular space (Boyle,

Conway, Kane & O'Reilly, 1941; Edwards & Harris, 1957). However there is now very convincing evidence that the inulin space in smooth muscle cells does not correspond to the real extracellular space. Goodford & Leach (1966) have detected mucopolysaccharides in the extracellular space between the smooth muscle fibres of taenia coli. These substances prevent a uniform distribution of large molecules such as inulin over the whole extracellular space. The exclusion from the extracellular space increases in proportion to the molecular dimensions of the substance (Ogston & Phelps, 1961). Moreover the work of Phelps (1965) shows that inulin is a rather heterogeneous and labile substance. Therefore smaller molecules such as ethanesulphonate, sorbitol or mannitol, should be used to estimate the extracellular space. It is however essential that these substances should not be metabolized and should only slowly penetrate into the cells.

The values for ion content, extracellular space (measured with sorbitol-^{14}C) and the dry wt/wet wt ratios are summarized for taenia coli and vas deferens of the guinea-pig in Table 2. On the basis of a simple two-compartment model, the intracellular sodium, potassium and chloride concentration per litre cell water and the equilibrium potentials have been calculated (Table 3). The values

TABLE 2. Total ion content (m-mole/kg wet wt $\pm$ S.E.), extracellular space (ml/kg wet wt) and dry wt/wet wt ratio (%)

	Taenia coli of the guinea-pig	Vas deferens of the guinea-pig
K	$80 \cdot 7 \pm 0 \cdot 6$ (80)	$82 \cdot 5 \pm 1 \cdot 4$ (7)
Na	$56 \cdot 3 \pm 0 \cdot 8$ (80)	$56 \cdot 9 \pm 1 \cdot 5$ (7)
Cl	$72 \cdot 4 \pm 0 \cdot 6$ (80)	$71 \cdot 0 \pm 1 \cdot 3$ (7)
Sorbitol-^{14}C-space	343 ± 19 (12)	310 ± 20 (10)
dry wt/wet wt	$18 \pm 0 \cdot 4$ (22)	$17 \cdot 8 \pm 0 \cdot 5$ (10)

TABLE 3. Intracellular ion concentration calculated from the total ion content, dry wt/wet wt ratio and sorbital-^{14}C-space. The values for [Na]$_i$ and [Cl]$_i$ in brackets have been obtained by an extrapolation procedure after loading the tissue in radioactive solutions. This content is expressed in m-moles/l. cell water. The equilibrium potentials have been calculated by the Nernst equation

$$E = \frac{RT}{ZF} \ln \frac{[C]_o}{[C]_i}$$

	Taenia coli	Vas deferens
[K]$_i$	164	158
[Na]$_i$	19 (13·5)	28 (10·2)
[Cl]$_i$	55 (58)	57 (50)
E_K	-89	-88
E_{Na}	$+52$ ($+62$)	$+42$ ($+69$)
E_{Cl}	-24 (-22)	-23 (-26)

of these equilibrium potentials are similar to the values given by Casteels & Kuriyama (1965) for rat myometrium and by Casteels & Kuriyama (1966) and Casteels (1966) for taenia coli, using ^{35}S-ethanesulphonate for measuring the extracellular space.

It is not impossible that some ions are bound in smooth muscle tissue, and such binding would modify the calculated intracellular ion concentrations and the equilibrium potentials. Kao (1961) tried to determine the bound fraction of the ions in rabbit myometrium by extracting the tissues in isotonic sucrose solution. The amount of sodium, potassium and chloride left in the tissue after the extraction procedure was considered as the bound fraction. The bound potassium amounted to 29 to 8 mEq/kg tissue, for sodium between 14 and 4 mEq, and for chloride between 0·8 and 0·4 mEq, depending on the hormonal dominance. This procedure of determining the bound ions by extraction in sucrose can be criticized because the presence of negative macromolecular charges might lead to overestimation of any bound cations normally occurring. The procedure of Harris & Steinbach (1956) to determine the specific radio-activity counts min^{-1} μequiv^{-1} for Na or K in successive extracts from muscles soaked in the respective tracer solution will give clearer results, because the uniformity of the different fractions can be investigated as well. Some experiments have been performed along these lines on taenia coli and so far the results are not in favour of an important fraction of bound ions (Casteels, unpublished).

The quantity of intracellular sodium and chloride in the taenia coli has also been determined with radioactive isotopes. The tissues were exposed to a radioactive solution during a period which was sufficiently long to obtain a similar specific activity in the intra- and extracellular compartment (counts min^{-1} mM^{-1}). Thereafter, effluxes were done at 4°C, to obtain a clear separation between intra- and extracellular labelled ions. The procedure of doing the efflux at 4°C is necessary because the ratio of the amount of sodium or chloride in the cells to the amount in the extracellular space is unfavourable and because the rate of movement for these ions is rather high at 35°C (Harris, 1960). By extrapolation to zero time of the linear part of the efflux curve, plotted on a semilogarithmic scale, it is then possible to calculate accurately the amount of intracellular sodium or chloride. The amounts for taenia coli obtained by this method were 6·5 mEq sodium and 27·8 mEq chloride/kg wet wt, and for vas deferens 5·3 mEq sodium and 25·8 mEq chloride/kg wet wt. The intracellular chloride concentrations calculated from these values are similar to the values obtained by the analytical procedure, but the corresponding intracellular sodium concentrations are for both smooth muscle tissues lower than the analytical values (Table 3). The intracellular sodium concentrations, obtained by the extrapolation procedure will be considered as the correct values, because it is not impossible that some sodium is loosely bound to extracellular or intracellular structures.

The calculation of the equilibrium potentials from the intra- and extracellular ion concentrations implies the assumption that the activity coefficients in both compartments are similar. From the work of Hinke (1961) on the activity of sodium and potassium ions in the cytoplasm of squid giant axon and of Lev (1964) on the activity in the cytoplasm of frog sartorius muscle, it seems that the ratio of sodium activity/sodium concentration was always much smaller than the ratio of potassium activity/potassium concentration. It can be concluded either that the activity coefficient in the intracellular compartment for sodium ions is much lower than in the extracellular fluid or that an important fraction of sodium is bound in the cells. In contrast, the activity coefficient for potassium is only slightly lower in the cytoplasm than in the extracellular fluid. The activity of chloride ions has been studied by Keynes (1963) in squid giant axon and found to be in the same range as in the extracellular solution. The procedure of intracellular activity measurement or of measurement on isolated cytoplasm cannot be applied to smooth muscle cells. Therefore activity measurements for chloride ions have been performed on homogenized smooth muscle tissues (Casteels, 1965). These measurements gave no indication for a lowered activity coefficient of the chloride ions. All these studies on the activity of intracellular ions can certainly be criticized, but they give at least some indication that the equilibrium potentials for potassium and chloride can be calculated from the concentrations without introducing too much error. If the activity coefficient for intracellular sodium were much lower than in the external solution or if some sodium were bound, the sodium equilibrium potential would tend to a more positive value than the value given in Table 3.

A comparison between the membrane potentials and the equilibrium potentials indicates already that the membrane potential is determined largely by the potassium equilibrium potential and that the action potential could be explained on the basis of the sodium equilibrium potential. The chloride equilibrium potential is about 25 mV more positive than the resting potential and there is little reason to consider this high intracellular chloride concentration as an artefact, due to excessive binding of chloride or to a compartmentalization of this ion.

MEMBRANE PERMEABILITIES IN PHYSIOLOGICAL SOLUTION

Hitherto, the membrane permeabilities of smooth muscle cells have only been investigated with radioactive tracers. Recently Bülbring & Tomita (1968a, 1969a) have started to study membrane conductances with electrophysiological techniques. Only a comparison of direct conductance measurements with data from tracer experiments will allow a more precise interpretation of membrane phenomena in smooth muscle cells.

The flux experiments up till 1964 have been reviewed by Schatzmann (1964 a,b). The values for the sodium fluxes from the literature vary between 200 pM cm^{-2} sec^{-1} to 40 pM cm^{-2} sec^{-1} (Goodford & Hermansen, 1961; Durbin & Monson,

4+

1961), for the potassium flux between 2·3 and 6 pM cm^{-2} sec^{-1} (Born & Bülbring, 1956; Durbin & Monson, 1961) and the chloride flux amounted to 16 pM cm^{-2} sec^{-1} (Durbin & Monson, 1961).

The ratio of P_{Na}/P_K calculated from these flux values from the external sodium and potassium concentration varies, according to Schatzmann, between 0·23 and 3·0. Such high permeability ratios are difficult to reconcile with the predominant role of potassium ions in the generation of the membrane potential. Therefore it was found necessary to investigate this problem further.

Potassium fluxes

The study of potassium fluxes is easy because the rate constant is small, and because the intracellular concentration is so much higher than the extracellular concentration. The association of these two factors makes it possible to separate clearly the extracellular and the intracellular part of the efflux curve. If measured over a limited time (150 min) the efflux of K proceeds along a simple exponential curve, with a half-time of 68·2 ± 1·75 min ($n = 21$) for the taenia coli and of 111 ± 3 ($n = 10$) min for vas deferens. If the efflux is followed over a prolonged period, there is often a deviation from a single exponential function. According to Van Liew (1967) the simplest explanation for such deviations is the variability in the operation of individual cells of the tissue, so that, though each of the cells operates in a simple manner, the sum of their contributions appears complex.

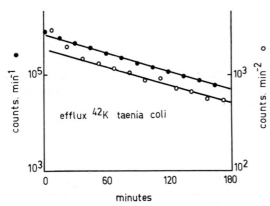

FIG. 2 Efflux of ^{42}K from taenia coli in counts/min (●) and counts/min^2 (○) plotted on a logarithmic scale against time in minutes.

In such a system there exists a distribution of simple exponential processes about an average value. Another possible reason for deviation from a simple exponential function would be a distribution of the radioactive ions over several compartments having different exchange rates. For potassium this is rather unlikely because the line expressing the decrease of the radioactive potassium remaining in the cells (counts min^{-1}) and the line expressing the decrease of the activity in the effluent (counts min^{-2}) are nearly parallel when plotted on a logarithmic scale (Fig. 2) suggesting that the ions are lost from a homogeneous compartment (Persoff, 1960).

The transmembrane flux of ions may be calculated by the equation of Keynes & Lewis (1951a)

$$\text{flux} = k\frac{V}{A}C_i$$

where k is the rate constant or the reciprocal of the time constant calculated from the half-time of efflux $t_{1/2}(k = 0.693/t_{1/2})$; V/A is the volume over area ratio which equals $1.5\,\mu$ in taenia coli (Goodford & Hermansen, 1961) and which is estimated at $1.75\,\mu$ in the vas deferens. The intracellular potassium concentration (C_i) is given in Table 3.

We obtain for the potassium efflux at 35°C in taenia coli 4.1×10^{-12} moles $\text{cm}^{-2}\,\text{sec}^{-1}$ and in vas deferens 2.7×10^{-12} moles $\text{cm}^{-2}\,\text{sec}^{-1}$.

The influx of potassium can also be calculated from the amount of ^{42}K, which penetrates into the cells contained in 1 mg of tissue, if we know the cell surface corresponding to 1 mg of tissue. For a density of the tissue of 1.05, for an extracellular space of 35% and for a mean cell diameter of $6\,\mu$, a surface has been calculated of $4.1\,\text{cm}^2$ for 1 mg of tissue. The calculated potassium influx for taenia coli in physiological solution was 3.9×10^{-12} moles $\text{cm}^{-2}\,\text{sec}^{-1}$.

Chloride fluxes

The chloride efflux (counts min^{-1}) proceeds sometimes along a simple exponential function, but often there is a deviation from linearity after 30 min. The mechanism is probably similar to the one described for the potassium efflux. The activity of the effluent (counts min^{-2}) can mostly be considered as a single exponential, which in some cases remains parallel to the decrease of the total intracellular activity (Fig. 3). The mean value for the half-time of chloride efflux is 11 ± 0.8 min (10) in taenia coli and 10.0 ± 0.7 min (6) in vas deferens.

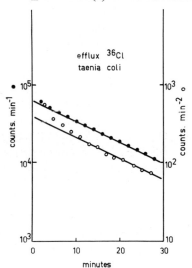

FIG. 3   Efflux of ^{36}Cl from taenia coli. Counts/min ($\bullet$) and counts/min^2 ($\bigcirc$) plotted on a logarithmic scale against time in minutes.

The efflux calculated by the equation of Keynes and Lewis amounts to $7\cdot5 \times 10^{-12}$ moles cm^{-2} sec^{-1} for taenia coli, and $11\cdot0 \times 10^{-12}$ moles cm^{-2} sec^{-1} for vas deferens.

An important conclusion that can be drawn from these chloride effluxes is that the largest part of the tissue chloride behaves as a single fraction and that consequently the high tissue chloride content does not have to be explained by compartmentalization of the chloride ions.

The chloride uptake by taenia coli cells in a normal solution at 35°C has also been determined and is 45×10^{-12} moles mg^{-1} sec^{-1} corresponding to 11×10^{-12} moles cm^{-2} sec^{-1}, using a surface over weight ratio of $4\cdot1$ cm^2/mg.

Sodium fluxes

Here we face an important and controversial problem in the study of membrane permeability of smooth muscle cells, because the sodium efflux proceeds rapidly and because there is often, in the semilogarithmic plot, a marked deviation from linearity which makes it difficult to determine accurately the half-time

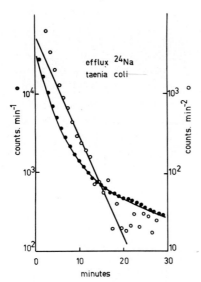

FIG. 4   Efflux of ^{24}Na from taenia coli. Counts/min (●) and counts/min^2 (○) plotted on a logarithmic scale against time in minutes.

of efflux (Fig. 4). The most likely explanation for this deviation from linearity is again the variation in efflux rates between the cells of the tissue and perhaps the presence of a small fraction of bound sodium. The efflux of sodium represents a process depending on metabolism, which should however equal the influx of sodium under steady conditions. However, only the influx of sodium corresponds to a passive movement of sodium ions into the cells, from which the sodium permeability of the membrane could be calculated. The use of the calculated intracellular sodium concentration, instead of the unknown intracellular sodium activity, could give an overestimation of the sodium efflux, and it is therefore interesting to compare the sodium efflux with the influx under the same experimental conditions.

The half-time for the sodium efflux from taenia coli can be calculated by two different procedures. The first procedure consists of measuring the half-time of the efflux rate (counts min^{-2}), which was linear over the main part of the efflux curve, if plotted on a semi-log scale against time (see Fig. 4). The second procedure has been described by Van Liew (1967) and is based on the assumption that there is a certain variability in the operation of sodium extrusion between the cells. Although it gave essentially the same results as the first procedure it has not been used in the present calculation because it is not impossible that the deviation between the curve representing the counts min^{-1} and the one representing the counts min^{-2} is due to the existence of a slower exchanging sodium fraction.

The mean value for the half-times of sodium efflux, obtained by the first procedure, was for taenia coli $2 \cdot 9 \pm 0 \cdot 22$ min (7) and for vas deferens $3 \pm 0 \cdot 15$ min. (5)

The efflux in moles $cm^{-2} sec^{-1}$, calculated from these half-times and from the intracellular sodium concentrations equals in taenia coli 11×10^{-12} moles $cm^{-2} sec^{-1}$, using an intracellular Na-concentration of 19 m-moles/l cell water (see Table 3), or $8 \cdot 1 \times 10^{-12}$ moles $cm^{-2} sec^{-1}$ using a value for the intracellular sodium of $13 \cdot 5$ m-moles/l cell water. In vas deferens the sodium efflux equals $18 \cdot 5 \times 10^{-12}$ moles $cm^{-2} sec^{-1}$ using a value for the intracellular sodium of 28 m-moles/l cell water and $6 \cdot 9 \times 10^{-12}$ moles $cm^{-2} sec^{-1}$ using a sodium concentration of $10 \cdot 2$ m-moles.

The influx of sodium has also been calculated for the taenia coli. The uptake is about $21 \cdot 7 \times 10^{-12}$ moles $mg^{-1} sec^{-1}$ corresponding to a transmembrane influx of $5 \cdot 3 \times 10^{-12}$ moles $cm^{-2} sec^{-1}$. For vas deferens an uptake of 35×10^{-12} moles $mg^{-1} sec^{-1}$ has been calculated and, using membrane surface over weight ratio of $3 \cdot 8$ cm^2 mg^{-1}, this uptake corresponds to a transmembrane influx of $9 \cdot 2 \times 10^{-12}$ moles $cm^{-2} sec^{-1}$.

The approximate values for some factors in these calculations could be the reason for the small discrepancy between influx and efflux. One may, however, certainly conclude that there is a rather good agreement between the lower values for the Na efflux (obtained by using the $[Na]_i$ found in the extrapolation experiments (cf. Table 3)) and the influx values. This suggests that the real sodium flux would have a value between 5 and 9 p-moles $cm^{-2} sec^{-1}$. Two factors are responsible for the lower value of the sodium efflux, compared with the results given in the literature; firstly the slightly longer half-time as calculated from the curve expressing counts min^{-2}, and secondly the lower value for the intracellular sodium concentration.

Knowing the intracellular ion concentrations (neglecting the activity coefficient), the flux values and the membrane potential, we can calculate the permeability coefficients, P_K, P_{Na} and P_{Cl} by the constant field assumptions of

Goldman (1943) and Hogdkin & Katz (1949). We have to assume, however, that the inward and outward movements of an ion species are independent of each other and that, in applying these equations to determine the membrane permeabilities to ions, the effluxes of potassium and chloride ions and the influx of sodium ions are due only to their own electrochemical forces. Because we are dealing with ions moving also under the influence of an electric field, the permeabilities (cm/sec) are determined by the ratio of flux/concentration, multiplied by a factor that represents the effect of the electric field in lowering or raising the chances of passage. This factor is according to Goldman

$$\frac{EF/RT}{1 - \exp(-EF/RT)}$$

The value of this factor for the outward potassium movement (against the electric field), at a membrane potential (E) of -55 mV and at 35°C, equals 0·3 and for the inward movement of sodium and the outward movement of chloride (both following the electric gradient) 2·46.

From the potassium efflux and $[K]_i$ in taenia coli we obtain

$$P_K = \frac{\text{efflux}}{K_i \times 0\cdot3} = \frac{4\cdot1 \times 10^{-12}}{164 \times 10^{-6} \times 0\cdot3} = 8\cdot3 \times 10^{-8} \text{ cm/sec}$$

From the sodium influx and the external sodium concentration we obtain

$$P_{Na} = \frac{\text{influx}}{Na_o \times 2\cdot46} = \frac{5\cdot3 \times 10^{-12}}{137 \times 10^{-6} \times 2\cdot46} = 1\cdot6 \times 10^{-8} \text{ cm/sec}$$

From the chloride efflux and the internal chloride concentration we obtain

$$P_{Cl} = \frac{7\cdot5 \times 10^{-12}}{55 \times 10^{-6} \times 2\cdot46} = 5\cdot45 \times 10^{-8} \text{ cm/sec}$$

These permeability constants of the smooth muscle membranes of taenia coli are markedly different from the ones given by Hodgkin & Horowicz (1959b) for frog skeletal muscle fibres. P_K and P_{Cl} are much smaller in smooth muscle ($8\cdot3 \times 10^{-8}$ versus $1\cdot4 \times 10^{-6}$ cm sec^{-1} and $5\cdot5 \times 10^{-8}$ versus 4×10^{-6} cm sec^{-1}). For the sodium permeability the difference between the two values is probably not significant. The low values of P_K and P_{Cl} are an explanation for the fact that the membrane potential in smooth muscle is less affected by potassium and chloride ions than in striated muscle.

From the permeability constants we can calculate the ratio of these permeability constants $P_{Na}/P_K = 0\cdot19$ and $P_{Cl}/P_K = 0\cdot65$.

The membrane potential calculated from these values by the Goldman equation

$$E = -\frac{RT}{F} \ln \frac{[K]_i + P_{Na}/P_K[Na]_i + P_{Cl}/P_K[Cl]_o}{[K]_o + P_{Na}/P_K[Na]_o + P_{Cl}/P_K[Cl]_i}$$

equals -35 mV for taenia coli cells.

This value is much lower than the actual value of the membrane potential and, accordingly, the question arises whether the assumption of independent ion movement is valid. A first possibility is that potassium ions move in single file along a chain of sites (Hodgkin & Keynes, 1955b). Under these conditions the potassium conductance calculated from the potassium efflux, would be n times lower than the real potassium conductance (g_K)

The equation

$$g_K = n \frac{P_K F^3 E K_o}{R^2 T^2} \cdot \frac{1}{1 - \exp(-EF/RT)}$$

would have to be used instead of

$$g_K = \frac{P_K F^3 E K_o}{R^2 T^2} \cdot \frac{1}{1 - \exp(-EF/RT)}$$

An alternative mechanism to bring the membrane potential to its actual higher value of -55 mV would be an electrogenic extrusion of sodium ions. Such a mechanism would be in agreement with the observation that the membrane quickly depolarizes after exposure to ouabain and that there is also an early repolarization after removal of ouabain (Casteels, 1966).

However, we can as yet not draw a firm conclusion as to the origin of the membrane potential in smooth muscle and we have to consider that it might be due to a combination of diffusion potentials and an electrogenic sodium pump.

EFFECTS OF MODIFICATIONS IN THE EXTERNAL SOLUTION ON ION CONTENT AND ION FLUXES OF SMOOTH MUSCLE CELLS

Modifications of the external solutions are an important tool in the study of membrane permeabilities, because it is possible to eliminate an ion species in the Goldman equation. For instance if chloride has been replaced by a nonpermeant anion, we can eliminate the [Cl] factor from the Goldman equation and we obtain

$$E = -\frac{RT}{F} \ln \frac{P_K[K]_i + P_{Na}[Na]_i}{P_K[K]_o + P_{Na}[Na]_o}$$

This equation would be easier to use for an estimation of the P_{Na} and P_K factors.

The present results will however indicate that this procedure cannot be applied without restriction to smooth muscle cells because a modification of the concentration of one ion in the extracellular solution can also change the permeability of the membrane to other ions.

We will consider the effects of changes in the external sodium, chloride, potassium, calcium and hydrogen ion concentration on the membrane permeabilities and the extracellular ion content. The results will be interpreted in relation to the membrane potentials, measured under the same experimental conditions and given in the literature.

Modification of the sodium concentration

Tris chloride, choline chloride and sucrose are the sodium substitutes that
have been used in the present experiments. The use of sucrose is, however,
limited because a reduction of the external sodium is accompanied by a reduction of the external chloride concentration and of the ionic strength. A substitution of chloride ions by sucrose will affect the membrane properties by
reducing the potassium permeability (Fig. 5) and conclusions drawn from such
experiments are not valid for physiological conditions (see below p. 91). An
advantage of using sucrose is however that the tissue potassium is maintained
at a steady value. Tris chloride and choline chloride (with atropine $1 \cdot 5 \times 10^{-6}$
g/ml) are not perfect sodium substitutes either. Choline penetrates slowly into
skeletal muscle fibres (Renkin, 1961) and into cardiac muscle fibres (Boulpaep,
1963). Also in taenia coli a progressive fall of the potassium content of the tissue

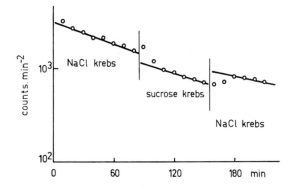

FIG. 5   Efflux of ^{42}K from vas
deferens in counts/min^2, plotted
on a logarithmic scale against
time. Replacing NaCl by sucrose
results in a pronounced decrease
of the rate of potassium efflux.

is observed (Casteels, unpublished observations). A similar effect on potassium
content of taenia coli has also been found for tris Cl, and this effect can be
related to the fact that at pH 7·40 a significant fraction (30%) of tris is un-
ionized and can thus easily penetrate into the cells. Moreover according to
Mahler (1961) tris could affect certain enzymatic reactions and this action could
be partly responsible for the decrease of the internal potassium and chloride
concentration.

The main purpose of using sodium-deficient solutions is to find out whether
part of the sodium flux in smooth muscle cells is due to exchange diffusion
(Ussing, 1949), whether there is some coupling between potassium uptake and
sodium extrusion, and to what extent sodium ions affect the membrane potential.

The problem of the exchange diffusion for sodium has been studied at 35°C and
20°C using sucrose, choline chloride or tris chloride. If exchange diffusion
exists, one would expect to find a decrease of the efflux of sodium, when the
external sodium concentration is reduced. Fig. 6 shows the efflux of ^{24}Na at
35°C, after loading the tissue in a ^{24}Na Krebs solution, in normal Krebs and in

a solution where NaCl has been replaced by sucrose. No statistically significant difference could be observed between the sodium efflux rates in these two different solutions. Similar experiments giving no evidence for Na-exchange diffusion, have also been performed at 20°C because the half-time of sodium efflux is larger (6 to 8 min) at this temperature and because the tissue still maintains a steady ion content (Table 4).

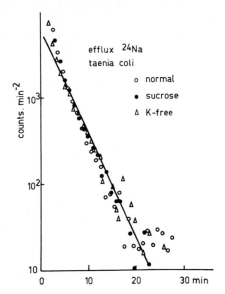

efflux ^{24}Na
taenia coli

o normal
● sucrose
△ K-free

FIG. 6 Efflux of ^{24}Na from taenia coli in counts/min^2 at 35°C in a normal Krebs solution (○), in a Na-free solution (NaCl replaced by sucrose) (●) and in a K-free solution (△). (The tissues were loaded in normal Krebs solution)

TABLE 4. Ion content (m-mole/kg wet wt ± S.E.) and extracellular space (ml/kg wet wt ± S.E.) of taenia coli in Krebs solution at 35°C and at 20°C after 2 hr and 4 hr

	Cl	Na	K	E.C.S.
Krebs 35°C	87 ± 2 (6)	59 ± 2·5 (6)	72 ± 1·5 (6)	390 ± 31 (7)
Krebs 20°C — 2 hr	85 ± 1·5 (7)	58 ± 1·5 (7)	75 ± 1·5 (7)	349 ± 23 (7)
Krebs 20°C — 4 hr	86 ± 2 (6)	60 ± 2·5 (6)	73 ± 2·5 (7)	375 ± 25 (7)

The exchange diffusion has also been investigated in sodium-rich and potassium-depleted tissues because the efflux of such tissues proceeds slowly in the absence of external K. No modification of the efflux rate was caused in these experiments by the presence or the absence of sodium in the efflux solution (Fig. 7a). Accordingly the results do not show that the sodium efflux is affected by sodium-deficient solutions and they do not suggest the existence of exchange diffusion for sodium ions in taenia coli. However for all experimental conditions one has to consider the possibility that the sodium leaking out of the tissue supplies a sufficient amount of sodium to allow exchange diffusion to go on in

4*

sodium-free solution. Moreover the values of P_{Na}, calculated from the sodium fluxes are not very different from the value calculated for frog striated muscle and are compatible with the influence of sodium ions on the membrane potential, observed in electrophysiological experiments. Sodium ions affect the membrane potential only to some extent, a sodium-free solution producing some hyper-polarization. According to the data of Kuriyama (1963a) the membrane potential is higher in a sodium-free solution with tris-Cl, than in one with sucrose. This

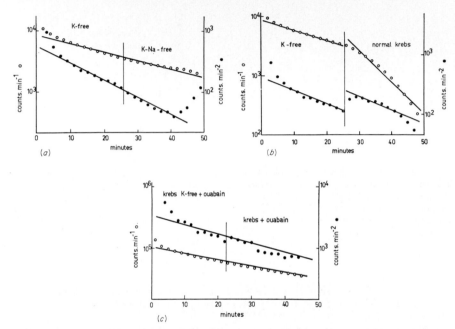

FIG. 7   Efflux of ^{24}Na from taenia coli, after prolonged exposure and loading in a K-free solution. The efflux is represented in counts/min (●) and counts/min^2 (○). (a) Efflux in K-free and K-Na-free solution; (b) efflux in K-free and normal solution; (c) efflux in K-free and normal solution containing ouabain.

difference could be due to the effect of replacing chloride by a non-permeant substance, because such a substitution reduced the potassium permeability (cf. p. 91).

The coupling between potassium influx and sodium efflux has been investigated by following the K-uptake by sodium-depleted tissues obtained by immersion in sodium-deficient solutions and by following the efflux of sodium in K-free solution. Reducing the external sodium concentration to one tenth of its normal value using tris as a Na-substitute does not affect the potassium uptake, but replacing all sodium by tris, causes a very important reduction of the potassium uptake (Fig. 8). Replacing sodium by choline or by tris also increases the passive loss of potassium from the cells (Fig. 9), probably because of the inability of the cells to extrude the penetrated choline or tris cations. The latter effect however is not directly related to the functioning of the Na-K exchange pump.

The efflux of sodium in a K-free solution at 35°C or 20°C proceeds at a rate which is not significantly different from the efflux rate in normal krebs solution,

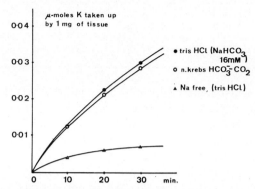

FIG. 8   Uptake of ^{42}K by taenia coli expressed in μ-moles taken up by 1 mg tissue, as a function of time in min. The uptake values are means of 10 experiments. The experimental conditions are: normal Krebs solution (○), a solution containing only 16 mM Na$^+$ (●) and sodium-free solution (▲).

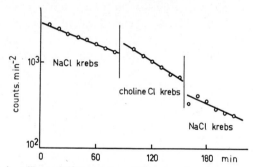

FIG. 9   Changes in efflux of ^{42}K produced by exposing the tissue to choline-Krebs solution. Rate of efflux (counts/min^2) plotted on a logarithmic scale against time in minutes.

if the tissue has been loaded in a normal radioactive Krebs solution (Fig. 6). Only in some experiments at 20°C was there a fall of the rate of sodium efflux, during exposure to K-free Krebs solution. The close relation between Na-extrusion and the external K-concentration could however be clearly demonstrated in potassium-depleted tissues. After 2 or 3 hr exposure to K-free solution (see Table 5), the tissues were loaded in a K-free ^{24}Na-Krebs solution. The efflux was followed in a K-free solution and after 25 min potassium was readmitted. Hereupon the ^{24}Na-efflux markedly increased and this increase could be inhibited completely by 10^{-5} M ouabain (Fig. 7b, c).

Probably two factors intervene here in demonstrating the effect of K-ions on the Na-efflux. Firstly the fact that the sodium efflux proceeds slower because

TABLE 5. Ion content (m-mole/kg wet wt $\pm$ S.E.) of taenia coli after immersion
for 1, 2 and 3 hr in a potassium-free Krebs solution

	K	Na	Cl	Extracellular space (ml/kg wet wt $\pm$ S.E.)
Control	80·7 $\pm$ 0·6 (80)	56·3 $\pm$ 0·8 (80)	72·4 $\pm$ 0·6 (80)	343 $\pm$ 19 (12)
1 hr K-free	45 $\pm$ 1·5 (20)	87 $\pm$ 2 (20)	60 $\pm$ 1 (20)	370 $\pm$ 15 (10)
2 hr K-free	12 $\pm$ 1 (11)	100 $\pm$ 3 (11)	61 $\pm$ 1·2 (11)	345 $\pm$ 20 (11)
3 hr K-free	7 $\pm$ 0·8 (11)	121 $\pm$ 2 (11)	62 $\pm$ 1·5 (11)	394 $\pm$ 21 (11)

of the high intracellular sodium concentration, and that this slower efflux allows
a longer observation period. The second and more important factor is that,
because of the very low intracellular potassium, a sufficiently high perimembrane
potassium concentration can no longer be maintained. The fact that a K-free
solution does not modify the Na-efflux, after loading in a normal Krebs solution
is probably due to this continuous supply of potassium by the cells.

The sodium efflux at 20°C is not affected by an increased external potassium
concentration.

Modification of the potassium concentration

A potassium-free solution decreases the potassium permeability of the mem-
brane in some tissues. This effect has been observed in *Carcinus* axon by Keynes
& Lewis (1951a), in Purkinje fibres by Carmeliet (1961a, b) and in frog ventricle by
Brady (quoted by Carmeliet, 1961a). In smooth muscle, however, a K-free solution

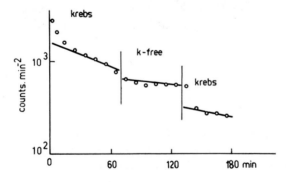

FIG. 10   Efflux of ^{42}K from
taenia coli at 35°C (counts/min^2).
Increase of the K-efflux on
changing to a potassium-free
solution. Decrease of the efflux
on return to normal solution.
Semi-log scale.

causes, after about 20 min, a progressive increase of the K-efflux (Fig. 10). This
increase is not observed if the external potassium is reduced only to one-fifth
or one-tenth of its normal value.

This increase of the K-efflux could explain the rapid fall of the intracellular
potassium concentration in K-free solution, as shown in Table 5, and also the
initial maintenance in a potassium-free solution of a normal or even slightly

increased membrane potential (Holman, 1958; Kuriyama, 1963a). A second effect of a potassium-free solution is a decrease of the intracellular chloride concentration to a value which would be compatible with a passive chloride distribution (25 m-moles/l cell water). Moreover the absence of K in the external solution causes a pronounced increase of the chloride efflux reducing the $t_{1/2}$ from 11 min to 4.8 ± 0.15 min (8) (Fig. 11). The efflux calculated from these data is about 9×10^{-12} moles cm^{-2} sec^{-1}, a value not very different from the value found in normal solution, in spite of the fall of [Cl]$_i$. This means that P_{Cl} has increased from its normal value of 5.5×10^{-8} cm/sec to a value of 14.5×10^{-8} cm/sec.

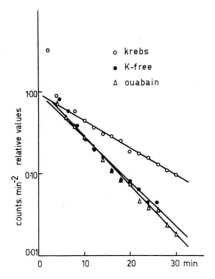

FIG. 11 Efflux of ^{36}Cl from taenia coli in counts/min^2 in normal Krebs solution (○), in K-free solution (●) and in a Krebs solution containing 10^{-6} g/ml ouabain (△).

This influence of potassium on the intracellular chloride is suggestive of a coupled inward transport of potassium and chloride ions. A decrease of the intracellular chloride concentration by exposure of the cells to ouabain (10^{-6} g/ml) (Casteels, 1966) could be explained by the same mechanism. However we do not know which factor is primarily responsible for all these changes in a K-free solution. The increase of the potassium efflux in a potassium-free solution can only be due to the change of the potassium concentration in the external solution. It seems rather unlikely that the change in external chloride would be responsible for the increase of the potassium loss, because replacing chloride by proprionate, resulting in the absence of internal chloride, decreases the K-efflux. The observation that the K-loss does not increase in a solution containing one-tenth of the normal K-concentration explains the delay between application of K-free solution and the increase of K-efflux. During this initial exposure time the potassium concentration at the membrane will fall below a critical concentration. One could speculate that this increased loss of potassium in K-free solution is due to the single-file movement of potassium ions. The potassium ions would leave the cells more easily when the number of potassium ions moving in the opposite direction is reduced.

The effect of a K-free solution on the chloride efflux also appears rather slowly. It cannot be explained completely by the fall in internal chloride, because a similar reduction of $[Cl]_i$ obtained by a reduction of $[Cl]_o$ is accompanied by a much smaller change of the rate of chloride efflux.

It is not impossible that the disturbance of the K-uptake due to the absence of K in the external solution, interferes with the inward movement of chloride. Under these conditions the decreased inward movement of chloride would not only affect the $[Cl]_i$, but could also facilitate the chloride efflux. Such a mechanism is suggested by the fact that the Cl efflux is as fast in the presence of 10^{-5} M ouabain as in K-free solution (Fig. 11).

An increase of the external potassium concentration (5·9 to 19·5 or 59 mM) increases the potassium efflux as observed in other tissues.

From the investigation of the effects of various ions on the resting potential and ion content (Casteels & Kuriyama, 1966) it is obvious that potassium ions are the main factor determining the membrane potential. A special feature of smooth muscle cells is that the tissue does not swell in an isotonic solution in which part of sodium has been replaced by potassium. In contrast, Boyle & Conway (1941) and Adrian (1956) found that frog skeletal muscle takes up an appreciable amount of water, because chloride ions penetrate into the cells

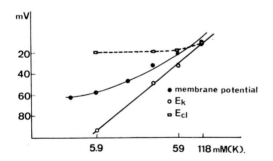

FIG. 12   The changes of the chloride equilibrium potential ($\square$), the membrane potential ($\bullet$) and the potassium equilibrium potential ($\bigcirc$) as a function of the external potassium concentration in solutions where NaCl + KCl = constant. Ordinate, the potential in mV, abscissa, the external K in mM on a logarithmic scale.

during depolarization and potassium ions accompany them to maintain electroneutrality. In this way the Donnan distribution of potassium and chloride is maintained in striated muscle, whereas in smooth muscle of taenia coli such a Donnan distribution of chloride does not exist. The depolarization of taenia coli cells in excess $[K]_o$ is not accompanied by an increase of $[Cl]_i$, as long as the chloride equilibrium potential is more positive than the membrane potential (Fig. 12). Only when $[K]_o$ is increased above 59 mM (corresponding to a depolarization to about -20 mV), will the passive distribution of chloride also come into play and will the internal chloride concentration increase (Casteels & Kuriyama, 1966).

Modification of the chloride concentration

The chloride distribution is a very interesting aspect of ion distribution in smooth muscle cells. The intracellular chloride concentration is too high to fit a passive distribution and, to explain this non-equilibrium distribution, one has to consider an active inward movement of chloride ions.

Replacement of chloride ions in the extracellular solution by other anions does not affect the intracellular Na or K-concentration. This has been observed for solutions in which chloride has been replaced by $C_2H_5SO_3^-$, I^-, NO_3^-, benzenesulphonate, acetylglycinate, propionate and pyroglutaminate (Casteels, unpublished observations).

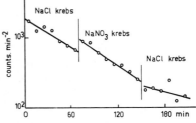

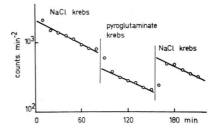

FIG. 13   Efflux of ^{42}K from taenia coli in counts/min². The rate of potassium efflux slightly increases in a nitrate Krebs solution and decreases in a pyroglutaminate Krebs solution.

This finding immediately raises the question of how the electroneutrality inside the cell is maintained since the intracellular chloride has disappeared. If permeant anions are used, one can assume that intracellular chloride is replaced by the new anion species, but for non-permeant anions this mechanism is excluded and, because (Na)ᵢ and (K)ᵢ are not decreased, we have to assume that some new anionic groups are formed by an unknown process.

The replacement of chloride by other anions has a pronounced effect on K-efflux. The efflux of potassium is slightly increased if Cl^- has been replaced by NO_3^-, a more permeant anion (Fig. 13). In contrast, K-efflux is reduced by a factor of 2 if chloride is replaced by a less permeant anion, independent of the nature of the anion. Ethanesulphonate, benzenesulphonate, propionate, acetylglycinate, pyroglutaminate and even sucrose (replacing NaCl) all have the same effect (Fig. 13). Moreover, replacement of chloride by non-permeant anions also reduces the influx of potassium ions, so as to maintain a steady state (see Fig. 16). The intracellular potassium concentration is probably an important factor for the maintenance of this steady state.

It is not clear by which mechanism anions affect the potassium efflux. A first possibility is that ion pairs are formed in the membrane between K and Cl (Shanes, 1958). This would mean that K and Cl can cross the membrane without contributing to the membrane conductance. But such a mechanism implies that replacement of chloride by other anions would not affect the membrane potential under steady state conditions because the potassium conductance would not be modified. However, the observations of Kuriyama (1963a) on the influence of anions on the membrane potential of taenia coli cannot be reconciled with such a hypothesis. This author found that the membrane potential is reduced if non-permeant anions are used and increased in the presence of NO_3^- or I^-. Therefore an effect of the anions on the potassium conductance seems more likely. The effect of sucrose on the resting potential of taenia coli cells is difficult to interpret, because at least four factors are changed: the sodium, potassium and chloride equilibrium potential (Brading, Bülbring & Tomita, 1969b) and the potassium permeability of the membrane.

The influence of anions on the K efflux could be explained by assuming that the permeant anions function as counter-ions for positive charges in the membrane pores, so as to facilitate the passage of potassium ions. To explain the effect of non-permeant anions on the K-influx one has to accept that under steady state conditions there exists a sensitive feed-back system which regulates the K-uptake as a function of the intracellular potassium concentration. It is not yet clear in which way the inward movement of potassium is coupled quantitatively to the sodium extrusion and the inward movement of permeant anions.

Modification of the calcium concentration

In squid giant axon, it has been demonstrated indirectly, by voltage clamp technique, that excess calcium brings about a decrease of P_{Na} and P_K and that a reduction of the external calcium causes an increase of both permeabilities (Frankenhaeuser & Hodgkin, 1957). A similar observation has been made by Jenerick (quoted by Shanes, 1958) on striated muscle fibres. The effect of calcium-deficient and calcium-excess solutions on the membrane potential of the taenia coli cells has been studied by Bülbring & Kuriyama (1963a). They found that excess calcium increases the membrane potential, while calcium deficiency decreases it. This observation might also be explained by a decrease of P_{Na} and of P_K in excess calcium and by an increase of P_{Na} and P_K in a calcium-deficient solution; the change of P_{Na} being predominant over the change of P_K.

However, recent observations by Bülbring & Tomita (1968a) indicate that the K- and Cl-conductances of taenia coli were increased by excess calcium, while they were decreased by lowering the external calcium although a reduction of $[Ca]_o$ to zero depolarized the membrane and increased the membrane conductance dramatically (see Chapter 7). This effect of calcium-free solution on the membrane potential of taenia coli cannot be explained by a change of the intracellular ion concentration, because exposure of tissues for 30 or 60 min to calcium-free Krebs solution did not modify the ion content (Casteels, unpublished observation).

Bülbring & Kuriyama (1963a) also investigated whether magnesium could substitute for calcium. It was found that an increase of the Mg^{2+}-concentration of six times normal (7·2 mM) scarcely altered the effect produced by the absence of calcium. In contrast, Niedergerke & Orkand (1966a) found that in frog heart muscle the effect on the resting potential of high magnesium is similar to that of strontium or calcium. The finding of Goodford (1966) that there is a striking loss of tissue potassium in a calcium-magnesium-free solution only indicates that Mg^{2+} can partly replace Ca^{2+} in preventing a complete breakdown of the membrane function. Bülbring & Tomita (1968b) found that Ba^{2+} and Mn^{2+} could also to some extent substitute for Ca^{2+} in controlling the membrane potential (see Chapter 7).

From the electrophysiological results one would expect to find a modification of the passive ions fluxes through the cell membrane after changing the external calcium concentration. However, a change of the external calcium concentration neither modifies the chloride efflux nor the potassium efflux in normal Krebs solution. If, on the other hand, the chloride has been replaced by benzene-sulphonate or propionate it is found that calcium-excess decreases the efflux of potassium and that calcium-deficiency causes an increase of the efflux (Fig. 14). Calcium does not seem to modify the sodium-efflux in normal solution. The action of calcium on the sodium uptake has not yet been studied.

It is impossible at the moment to interprete these results, and further investigation is necessary to elucidate the role of calcium in the control of the membrane permeability. It is also not clear in which way chloride ions can interfere with the effect of calcium on the potassium flux.

Modification of the pH

It is likely that ionic groups influence the selective permeability of the pathways for the different ions (Michaelis, 1926). The importance of these groups can be investigated by studying the influence of pH on the ion permeabilities.

The effect of various pH values (5·6, 7·4 and 8·9) on the chloride, potassium and sodium efflux and on the potassium uptake have been investigated, using a tris-maleate and acetylglycine buffer described by Hutter & Warner (1967a).

The chloride efflux is not affected by a change of pH between 5·6 and 8·9 in a normal Krebs solution. However, in a K-free solution, the chloride efflux rate increases by a factor of 2 at pH 8·9, while the efflux rate at pH 5·6 remains constant, compared with the efflux rate at pH 7·4 (Fig. 15). This effect of pH on the chloride fluxes is much less pronounced than the effect in frog skeletal

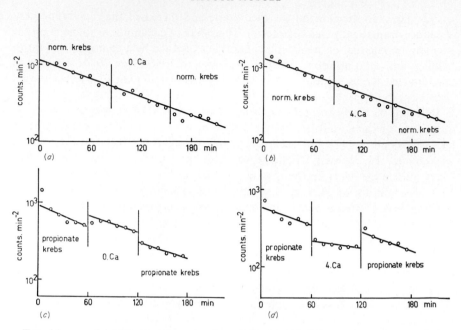

FIG. 14 *a* and *b* ^{42}K-efflux from taenia coli (counts/min^2) in normal Krebs-solution. Calcium-deficiency or calcium-excess does not affect the efflux rate. Semi-log scale. *c* and *d* efflux of ^{42}K from taenia coli in chloride-deficient solution (propionate or benzenesulphonate). Modification of the K-efflux by calcium-free and by calcium-excess solution. Semi-log scale.

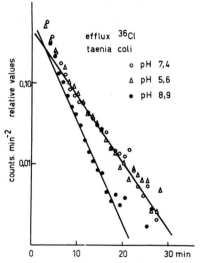

FIG. 15 Effect of pH on ^{36}Cl efflux (counts/min^2) in a K-free solution. A basic pH (●) increases the efflux rate by a factor of 2. An acid pH (△) has no effect on the chloride efflux.

muscle (Hutter & Warner, 1967*b*) and, moreover, it can only be observed in a potassium-free solution. Again there seems to be a relation between the chloride and the potassium movements, but the exact nature of this relation remains obscure.

The influx of potassium is decreased appreciably in a solution at pH 5·6, but the interpretation of these results is difficult because of the pronounced loss of intracellular potassium under these experimental conditions. The uptake of potassium at pH 8·9 is similar to the uptake at pH 7·4, using a tris-maleate buffer. But, as shown in Fig. 16, the type of buffer also affects the potassium

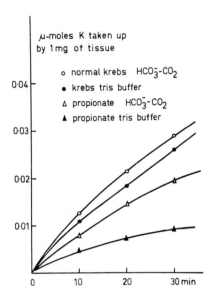

FIG. 16. Uptake of K by taenia coli, expressed in µ-moles taken up by 1 mg tissue as a function of time in min. The uptake values are means of 10 experiments. The experimental conditions are: normal Krebs solution (○), Krebs solution with tris buffer instead of bicarbonate buffer (●), solution with propionate instead of Cl^- and with bicarbonate buffer (△), and solution with propionate and with tris buffer (▲).

uptake. This uptake in a chloride or in a propionate Krebs solution proceeds faster in the presence of the HCO_3^-—CO_2 buffer than in the presence of a tris-maleate buffer. This observation points to a more complex role of bicarbonate ions than hitherto accepted.

The potassium efflux is reduced at pH 5·6 and slightly increased at pH 8·9. If however chloride is replaced by a non-permeant anion, the effect of pH 8·9 on K-efflux is no more significant and the effect of pH 5·6 is very much reduced or absent (Fig. 17). The sodium efflux is not affected by a basic pH, but is slowed down at pH 5·6 by a factor of about 1·5. The effect of a low pH on the sodium-efflux and the potassium uptake could be due to an inhibition of the mechanism for the sodium extrusion and potassium uptake. This action could also explain the decrease of $[K]_i$ in acid medium. But the potassium efflux which is considered as being passive, is also affected by an acid pH. For this reason an additional effect of a low pH on the potassium permeability is likely.

The variations of permeability with pH in artificial and natural membranes can be ascribed to the presence of fixed charges, whose ionization is modified with pH. Since the membrane takes on positive charges upon acidification, a rise in permeability for anions has been predicted and found in red blood cell membranes (Passow, 1961). But the results obtained in smooth muscle cells do not conform to this classical model: the cell membrane becomes more permeable

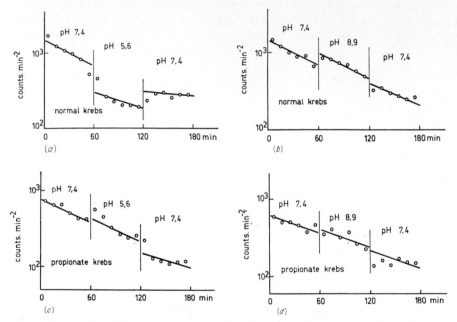

FIG. 17   *a* and *b* Effect of pH on ^{42}K efflux (counts/min^2) in normal Krebs solution at 35°C. Tris maleate-acetylglycine buffer. An acid pH decreases the efflux and an alkalinization increases the K-efflux. Semi-log scale.   *c* and *d* If chloride has been replaced by a non-permeant anion, the effect of pH 5·6 and 8·9 on the K-efflux has disappeared.

to chloride and potassium in alkaline medium. The increase of the effect of pH on the chloride flux by a K-free solution and the disappearance of this effect of pH on the K-efflux in chloride-free solution, suggest that the pH might exert its influence primarily on the chloride flux. It is however too early to propose a mechanism, which could explain these actions of hydrogen ions on ion movements.

ACTION POTENTIAL IN SMOOTH MUSCLE CELLS

The ionic basis of the electrical activity as described in the membrane theory (Hodgkin & Huxley, 1952d) presents some difficulty for its application to smooth muscle cells, although the positive value of the sodium equilibrium potential could certainly explain the depolarizing phase of the action potential by a sudden, shortlasting increase of P_{Na}. It is found that in most excitable tissues, the amplitude and the rate of rise of the action potential are a function of the external sodium concentration. In taenia coli however there is no change in spike amplitude if only one tenth of the normal sodium concentration is present and some authors (Axelsson, 1961b; Holman, 1958) even observed that spike discharge of taenia coli cells continues for some time in the total absence of sodium. The action of sodium ions on membrane activity has been further investigated by Bülbring & Kuriyama (1963a). They confirmed that spike height is not dependent on the external sodium concentration. Only the rate of rise and fall of the spikes were to some extent a function of the sodium concentration. However, excess calcium consistently increased the rate of rise and the amplitude of the action potential. The properties of the spike in taenia coli have recently been more fully investigated by Brading, Bülbring & Tomita (1969b) and are described in Chapter 7. Bennett (1967a) has investigated the effect of sodium and calcium on the action potential in the smooth muscle cells of the vas deferens. Reducing the external sodium from 180 to 25 mM did not alter significantly the threshold for firing or the overshoot of the action potential. An increase of external calcium up to 10 mM increased the rate of rise and the amplitude of the action potential, as in taenia coli.

In muscle fibres of crayfish the action potentials are also more or less independent of external sodium ions and action potentials dependent on calcium ions could be obtained in these fibres following prolonged exposure to tetraethylammonium ions. Propagated spikes could be elicited in $SrCl_2$ solutions and in these solutions the membrane resistance during the action potentials was found to decrease with increasing Sr concentration, while the resting resistance increased. According to Fatt & Ginsborg (1958) these action potentials would result from an increase of the membrane permeability to divalent ions.

Comparing the observations made on smooth muscle cells with the findings in crayfish muscle fibres, it does not seem unlikely that calcium ions contribute to the depolarizing current during the upstroke of the action potential and that these ions would carry the depolarizing current completely in a sodium-free solution. In relation to the spikes of smooth muscle in sodium-deficient solution, we have however to consider a mechanism proposed by Niedergerke & Orkand (1966b). According to these authors the overshoot could also be maintained in a sodium-deficient solution by a change of the P_{Na}/P_K ratio. Lowering the external sodium could increase the ratio P_{Na}/P_K and so the effect of the lowered $[Na]_o$ on the overshoot would be diminished. This hypothesis relates the magnitude of P_{Na}/P_K to the ratio of $[Ca]_o/[Na]_o^2$ in the external fluid and thus to the concentration of membrane calcium (Lüttgau & Niedergerke, 1958). The lack of effect of tetrodotoxin on the spikes of taenia coli and their abolition by manganese ions (Nonomura, Hotta & Ohashi, 1966) are arguments in favour of the hypothesis that the spikes in smooth muscle are mainly due to Ca-entry.

Another strong argument in favour of an important calcium influx during the action potential is the decrease of the membrane resistance in the vas deferens during the active response in the presence of a high external calcium concentration (Bennett, 1967a).

The structure of smooth muscle cells could also be suggestive for a calcium spike mechanism. The relatively small amount of endoplasmic reticulum is compatible with the hypothesis that calcium ions cross the membrane during the action potential and activate the contractile proteins. The amount of calcium necessary to reverse the potential of the membrane from -55 mV to $+7$ mV (Bülbring & Kuriyama, 1963a) can be calculated from the membrane capacity of 3 μF/cm^2 (Tomita, 1966b) and from the total cell surface/mg tissue. This surface/weight ratio corresponds to 4·1 cm^2/mg. A change of the potential from -55 mM to $+7$ mV for a mg of cells would require 0·76 $\times$ 10^{-6} Coulomb. Such an electric charge corresponds to 4 $\times$ 10^{-12} moles of Ca^{2+}. This means that an action potential in the cells of one mg of smooth muscle could be produced by a transfer of 4 picomoles of Ca^{2+}. A similar amount of calcium influx during activity has been calculated by Goodford (1967). The action potential could however be accompanied by an influx of a larger amount of calcium if the height of the overshoot would be reduced by an early increase of the potassium permeability. Inversely the amount of penetrating calcium would be lower if sodium participated in the generation of the upstroke of the spike. It should be stressed that a Ca influx of 4 $\times$ 10^{-12} moles mg^{-1} during the upstroke of the action potential could be sufficient to initiate contraction. We know from the work of Weber, Herz & Reiss (1964) and of Portzehl (1957) that the threshold concentration of calcium for contraction is in the range of 0·2 to 1·5 $\times$ 10^{-6} M. For taenia coli 1 mg of tissue corresponds to about 6 $\times$ 10^{-4} cm^3 of cell volume, and an influx of 4 $\times$ 10^{-12} moles would cause a change in the intracellular calcium concentration of about 4 $\times$ 10^{-12} moles/6 $\times$ 10^{-7} l = 7 $\times$ 10^{-6} M. This important change of the intracellular calcium by an influx of 4 $\times$ 10^{-12} moles calcium per mg tissue is due to the small cell diameter.

It should however be pointed out, that the elucidation of the relation between ion currents and action potential in smooth muscle cells might depend, as in other excitable tissues, on the voltage-clamp technique, as applied recently by Anderson & Moore (1968) on uterine smooth muscle.

CONCLUSION

The results presented in this paper can be explained by the membrane theory. This conclusion is not invalidated by the many observations which indicate a complex interaction in the membrane between the different ions. It is possible that these peculiarities are responsible for the special characteristics of smooth muscle cells. A further study of these peculiarities by a combined investigation with electrophysiological techniques and with tracers is here of paramount importance.

The central role of calcium for the membrane phenomena of smooth muscle cells has up till now only been demonstrated in the electrophysiological studies. The tracer studies in this field have been less conclusive, but it may be expected that the role of calcium will be elucidated in the coming years and that a more fundamental interpretation of the electrophysiological results will then be possible.

3

IONIC INTERACTIONS IN SMOOTH MUSCLE

P. J. GOODFORD

INTRODUCTION

When I started to work on ion contents and ion movements in smooth muscle, I had the naively extravagant hope that solutes might be distributed across the cell membrane of smooth muscle between the cytoplasm and the extracellular space, so that there were in effect only two options for each solute molecule: either it was freely dissolved in the cytoplasm or it was outside the cell. However, the briefest examination of an electron-micrograph shows that this interpretation is inherently improbable because of the multitude of intracellular structures, and analytical evidence soon accumulated to suggest that some substances were located in more specialized regions of the muscle (Schatzmann, 1961; Goodford & Hermansen, 1961; Weatherall, 1962b; Bauer, Goodford & Hüter, 1965; Stephenson, 1967). The concept of sequestration was therefore developed and is proving to be of increasing value in correlating the observed ionic distribution in a tissue with its electrophysiological properties. For example Bennett & Burnstock (1966) concluded, on the basis of *electrical* measurements, that 'smooth muscle cells obey the predictions of the Nernst equation for a potassium electrode', and Buck and Goodford (1966) made the tentative postulate after *analysing* another smooth muscle 'that KCl might be distributed across the quiescent cell membrane according to a Donnan equilibrium'. These compatible conclusions were reached independently but almost simultaneously, and the satisfactory agreement was only achieved because evidence was presented that some ions were sequestered in the tissue.

Buck & Goodford (1966) also explicitly pointed out that the amount of solute sequestered in a tissue might change during physiological stimulation or drug action, and an experimental basis for their prediction has recently been discovered in another type of cell, the toad oocyte. Dick & Lea (1967) have found that ouabain causes a release of sequestered sodium in these cells, and such an effect may account for some observations which cannot be explained easily by a single cytoplasmic compartment. The general concept must therefore be accepted that there may be more than two locations for a solute in a tissue, but the sequestration of slowly exchanging material in smooth muscle will not be emphasized in the present article because a brief discussion has recently been published (Goodford, 1968). On the contrary, more recent evidence will be reviewed which indicates that a small proportion of the smooth muscle electrolytes may be located at superficial sites in the tissue, and one may conclude if this evidence be accepted that there are no less than four regions where solutes can be found: (1) The extracellular space; (2) The superficial sites; (3) The other sequestering regions; (4) The cytoplasm. Such a breakdown into regions has been of great help in developing our present ideas, but progress in this direction becomes more and more difficult as more regions are conceived, and it must ultimately be acknowledged that the tissue is a continuum rather than a succession of infinitely sharp boundaries. Nevertheless the classical two-compartment model of an excitable tissue will be introduced in the present discussion, and some evidence will then be reviewed which is incompatible with this system but which may be explained by the presence of superficial sites at which cations can interact in smooth muscle. The hypothesis will then be advanced that these sites may influence the behaviour of the plasma membrane, and a more developed physical model of the ionic distribution in smooth muscle will be constructed. This will finally be used in combination with the new hypothesis to make further predictions for experimental testing.

THE CONCEPTS OF COMPARTMENTAL ANALYSIS

Model systems

If smooth muscle cells be treated as an array of long flexible close-packed cylinders, and light-microscopic observations are compatible with such an arrangement (Fig. 1 and McGill, 1909), the ratio of cell volume V to cell surface area A would depend on the absolute size of the cylinders, but the total volume of the cells would remain a constant fraction (0·9) of the whole tissue volume. This would leave 0·1 of the tissue to be outside the cells, which is well within the range of extracellular spaces (0·04 to 0·18) estimated by Prosser, Burnstock & Kahn (1960) from electronmicrographs of excitable smooth muscle, and the close-packed cylinders therefore have some properties in common with the tissue itself. They form a 'compartmental model' which is relatively simple if all

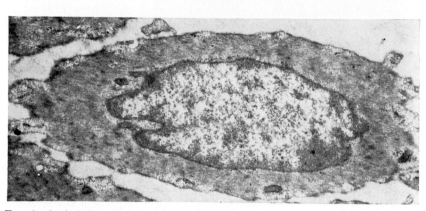

FIG. 1   (*top*) A line drawing of an early light-microscopic observation of a smooth muscle cell. (From McGill, 1909) (*bottom*) A recent electronmicrograph by Professor R. Johnson of a smooth muscle cell from the guinea-pig taenia coli, oblique section.

the cylinders are treated as identical compartments, but which can be deceptively complex if there is variation between cylinders, or if they are not close-packed, or if they are close-packed in bundles with spaces in between. This model was initially used in order to design experiments and, on the basis of the results obtained, more sophisticated models were then developed for use as further guides.

The two-compartment system

The distribution of materials in a system at every time may be described by defining:

i The distribution at a particular time $_0t$
ii The natural laws which govern changes of distribution

An initial simplifying assumption which has often been applied to the solute distribution in biological systems is that: *the rate at which material tends to leave a compartment is proportional to the 'amount' in that compartment*. This hypothesis, which is by no means universally true and does not have the status of a natural law, nevertheless affords a useful simplification. It contains the ill-defined concept of 'amount' which may be most accurately interpreted as *chemical activity* or, failing that, as *concentration* or, if constant compartmental volume be assumed, *weight*. Considering n constant volume compartments, each containing an 'amount'A, it leads to a series of equations:

$$\frac{d}{dt} A_1 = -\lambda_1 A_1 \quad \frac{d}{dt} A_2 = -\lambda_2 A_2 \quad \cdots \quad \frac{d}{dt} A_n = -\lambda_n A_n \qquad (1)$$

in which $(d/dt)A_n$ represents the rate of loss of material from each compartment n and λ_n is a rate constant. Integration gives:

$$\frac{A_1}{{_0}A_1} = e^{-\lambda_1 t} \quad \frac{A_2}{{_0}A_2} = e^{-\lambda_2 t} \quad \cdots \quad \frac{A_2}{{_0}A_n} = e^{-\lambda_n t} \qquad (2)$$

where ${_0}A_n$ is the 'amount' in each compartment at the initial time ${_0}t$, and A_n is the 'amount' at a later time t. It follows that:

$$\frac{A_1^{\alpha_1} A_2^{\alpha_2} \ldots A_n^{\alpha_n}}{{_0}A_1^{\alpha_1} {_0}A_2^{\alpha_2} \ldots {_0}A_n^{\alpha_n}} = e^{-(\alpha_1 \lambda_1 + \alpha_2 \lambda_2 + \cdots + \alpha_n \lambda_n)t} \qquad (3)$$

where $\alpha_1 \alpha_2 \ldots \alpha_n$ are arbitrary constants, and the simplifying assumption therefore yields well-known kinetic equations which may describe the behaviour of multicompartment systems.

The equations have been applied to the loss of material from the close-packed cylinder model of smooth muscle, but more complex equations are needed if there is both uptake and loss, or if there is interchange of material between compartments. A particularly interesting and simple case arises if all the cylinders have equal rate constants λ_q since the functions then show the same time dependence, and the cylinders may be treated as the equivalent of a *single cellular compartment*. The intervening spaces may be treated similarly if they are characterized by another common rate-constant λ_p, and in this case the system should show ideal two-compartment kinetics. The validity of such a model has been tested experimentally in smooth muscle (Bass, Hurwitz & Smith, 1964; Bauer et al., 1965; Born & Bülbring, 1956; Bozler, 1963; Bozler, Calvin & Watson, 1958; Buck & Goodford, 1966; Burnstock, Dewhurst & Simon, 1963; Chujyo & Holland, 1962; Durbin & Monson, 1961; Freeman-Narrod & Goodford, 1962; Goodford, 1964; Goodford, 1965; Goodford & Hermansen, 1961; Hurwitz, Battle & Weiss, 1962; Hurwitz, Tinsley & Battle, 1960; Schatzmann, 1961; Sperelakis, 1962) by equilibrating the tissue with a radioactive isotope and then studying the loss of tracer into an inactive medium. In such experiments it has been repeatedly observed that:

a. To a rough approximation the residual isotope R left in the tissue at time t after ${_0}t$ is:

$$R = {_0}A\, e^{-\lambda_p t} + {_0}B\, e^{-\lambda_q t} \qquad (4)$$

as predicted,

where $\lambda_p > \lambda_q$ since extracellular exchange is more rapid than cellular and $_0A$ and $_0B$ are the 'amounts' of isotope in the two compartments at $_0t$.

b. Accurate observations invariably show deviations from this double exponential behaviour, and these may be most readily detected by the method of Persoff (1960). Thus, after long periods of tracer washout $\lambda_p t$ becomes so large that $e^{-\lambda_p t}$ is virtually zero and equation (4) simplifies still further to

$$R = {}_0B\,e^{-\lambda_q t} \tag{5}$$

corresponding to the state when all extracellular tracer has escaped from the tissue. However, radioactivity is still leaking away from the cells at a rate:

$$\frac{dR}{dt} = -{}_0B\lambda_q\,e^{-\lambda_q t} \tag{6}$$

and taking logarithms of equations (5) and (6) and differentiating with respect to time leads to:

$$\frac{d}{dt}(\ln R) = -\lambda_q \tag{5'}$$

and

$$\frac{d}{dt}\left(\ln\left(\frac{-dR}{dt}\right)\right) = -\lambda_q \tag{6'}$$

These equations now show that the natural logarithm of R (the residual tracer left in the muscle) and the natural logarithm of $-dR/dt$ (the instantaneous rate of loss of tracer from the muscle) should eventually give *parallel* straight lines of slope $-\lambda_q$ when plotted against time, if the function finally becomes a single exponential, as it must, if the two-compartment model and the initial assumption are both valid.

In practice such parallel behaviour has rarely if ever been observed *after equilibration of smooth muscle with radioactive tracers* and Fig. 2 illustrates a typical experiment. One must conclude that the tissue does not show the ideal behaviour of a two-compartment system, and some causes for the observed discrepancies will now be considered.

Deviations from two-compartment behaviour

Deviations from ideality may be grouped into three main classes, and the first includes those *factors related to the bulk of the tissue, and to its inhomogeneity.* Some parts of the extracellular space are so filled with structural material (collagen, mucopolysaccharide, lipoprotein, etc.) that the uptake and loss of tracer are inhibited (Ogston & Phelps, 1961; Goodford & Leach, 1966) and the corresponding values of $_0A$ and λ_p reduced. Other parts have ready access to

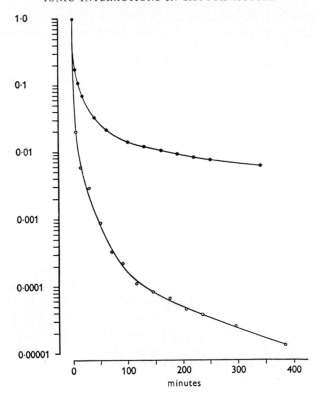

FIG. 2 Tracer washout observations on the smooth muscle of the guinea-pig taenia coli in Krebs solution at 35°C, after 210 min immersion in radioactive ^{45}Ca Krebs. Abscissa, time in min after transfer of the taenia coli to inactive solution. Logarithmic ordinate scale with a common origin for both curves. Open circles, the rate dR/dt of ^{45}Ca efflux (counts min^{-2}) as a fraction of the initial rate. Closed circles, the radioactivity R remaining in the muscle (counts min^{-1}) as a fraction of the initial amount. Note that exchange is not a single exponential process. (From Goodford, 1965)

sinuses or arteries or veins and therefore exchange more rapidly, so that there is a range of values of λ_p and a large or infinite number of differing exponential functions which, taken together, describe the extracellular exchange in terms of classical diffusion theory (Crank, 1956; Harris & Burn, 1949; Keynes, 1954; Goodford & Hermansen, 1961; Creese, Jenden & Steinborn, 1969). The values of $_0B$ and λ_q may also vary between cells (Goodford, 1968) and so may the cell surface area A and cell volume V (Creese, Neil & Stephenson, 1956) and, furthermore, tracer leaving one cell may be taken up by another before it has escaped from the tissue (Huxley, 1960; Weatherall, 1962a; Goodford, 1966). Thus there can be interactions between cells and, indeed, covariance would be expected between all the factors which have so far been considered (Keynes & Lewis, 1951a).

It is fortunate that the quantitative importance of these sources of variance can be reduced by the choice of appropriate experimental conditions. Thus the range of values of λ_p for potassium exchange is of a different order of magnitude to λ_q, and the cellular exchange may be studied without excessive extracellular interference. Furthermore, the variation of λ_q itself may be neglected if the observations are restricted to a time interval which is short (say under 25%) compared with the mean half-time $T_{1/2} = 0.69315/\lambda_q$ of cellular exchange, and it is experimentally quite feasible to study ^{42}K exchange in smooth muscle over such a period (Goodford, 1967). On the other hand the exchange of sodium in smooth muscle at body temperature is not so easily interpreted (Goodford & Hermansen, 1961; Freeman-Narrod & Goodford, 1962; Burnstock, Dewhurst & Simon, 1963; Buck & Goodford, 1966) because the values of λ_p and λ_q are so close that they tend to overlap, and the various components of the exchange processes cannot be distinguished with confidence from each other. Indeed a full description of the cell-sodium exchange kinetics in smooth muscle is still awaited.

The second major source of deviations from two-compartment theory is related to *the starting conditions of an isotope experiment*. Some regions will rapidly attain radioactive equilibrium when a tissue is exposed to tracer but others may take several hours, and during such a period the health and viability of the preparation may suffer. In general one cannot wait for equilibrium, and one does not know the extent or distribution of tracer exchange which has taken place during the loading phase before a tracer washout experiment begins. It would be rash to assume without proof that complete radioactive equilibrium had been established, and an element of uncertainty may therefore be present. Fortunately the bias which this could introduce may be eliminated by making observations in the tracer-uptake mode, when it is known that at the beginning of the experiment the tissue contained no isotope at all. Figure 3 shows an experiment of this type which has been interpreted on the two-compartment model with the uptake equation:

$$R = A(1 - e^{-\lambda_p t}) + B(1 - e^{-\lambda_q t}) \tag{7}$$

in which t is now the time after uptake began and A and B measure the final equilibrium uptake which would, in theory, be reached after an infinite period. It may be noted that an attempt has been made to restrict the observations to a short period (7 min in comparison with the mean half-time of 55 min) and that the earliest observations have been rejected. Thus at all times of observation one may assume that $e^{-\lambda_p t} \to 0$ and equation (7) simplifies to:

$$R = A + B(1 - e^{-\lambda_q t}) \tag{7'}$$

The final major source of deviations from two-compartment behaviour is *the over-simplified model of smooth muscle* which has been considered. Although the close-packed cylinders may be compatible with light microscopic observations (see Fig. 1), their crudeness is at once appreciated when electronmicroscopic sections (see Fig. 1) are examined. A host of subcellular structures are now exposed, each of which could have its own individual compartmental properties and kinetics. In some cases movements across the cell membrane might be rate-limiting, so that subcellular exchange would not be detected easily in the

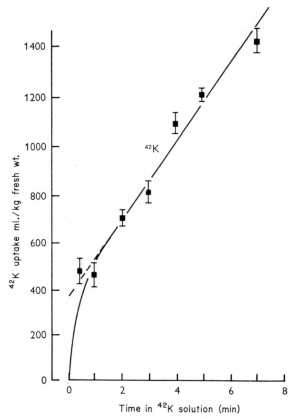

FIG. 3   Tracer ^{42}K uptake observations on the smooth muscle of the guinea-pig taenia coli in Krebs solution at 35°C, after 120 min preincubation in inactive solution. Abscissa, time in min after application of tracer. The ordinate is an inverted logarithmic scale showing the uptake of tracer potassium (see text and table 1). It does not differ appreciably from linearity at this early stage of exchange. (From Goodford 1967)

intact tissue; in other cases the subcellular exchange might be the slower and a slowly exchanging component has then, in fact, often been detected (Goodford & Hermansen, 1961; Weatherall, 1962*b*; Buck & Goodford, 1966; Goodford, 1965). One must therefore accept that there may be later slower phases of isotopic exchange, and consider how much bias these may introduce into a simple two-compartment analysis.

THE EXTRACELLULAR SPACE

The concept of extracellular space may be defined in histological, electrical, mechanical or analytical terms, and these spaces may have different sizes in the same smooth muscle (see Goodford 1968 for a summary of recent evidence). Thus the 'histological' space is that region outside the plasma membrane of the cells, which can be measured on electronmicrographs as typically 0·01 to 0·2 of the total tissue volume. It has been correlated with the electrical resistance of the tissue and with the relative refractory period and the conduction velocity of the action potential, and any of these parameters can be used to define an extracellular space which is relevant to 'electrical' observations (Prosser et al., 1960). The 'mechanically' measured extracellular space of a muscle may be determined by squeezing out the extracellular liquid and, in the heart at least, it corresponds closely to the analytical space (Fisher & Young, 1961). However, observations of tracer exchange and ionic distribution are themselves analytical measurements, and the analytical method of space determination is therefore most relevant. This gives values ranging from 0·2 to 0·5 of the total smooth muscle volume, and these usually exceed the histologically measured spaces. The analytical spaces are to be preferred for present purposes because they can be measured in the very same muscle samples under the various experimental conditions which are being simultaneously investigated.

The extracellular sorbitol space

The ideal material for analytical measurement would be a small molecule of comparable size to the ions being studied (e.g. Na^+, K^+, Ca^{2+}, Cl^-, HCO_3^-), but which would not enter the cells or only do so very slowly. It should be stable, cheap, easily measured and it should not be metabolized. Sorbitol has been used as a reasonable compromise between these desirable properties (Fisher & Lindsay, 1956; Fisher & Young, 1961; Goodford & Leach, 1966; Goodford, 1967). This molecule may be conveniently detected by labelling with radioactive ^{14}C, and Fig. 4 shows the uptake of $[^{14}C]$-sorbitol by smooth muscle samples after transfer from a Krebs solution containing inactive sorbitol to another containing radioactive sorbitol. One may observe that the uptake reaches an almost steady level within the first few minutes, after which more than an hour must elapse before any slower significant phase of exchange can be detected.

It has been shown (Goodford & Hermansen, 1961; Freeman-Narrod & Goodford, 1962) that the initial rate of uptake of solutes by smooth muscle is in accord with bulk diffusion kinetics in the extracellular space and that, in the absence of appreciable cellular exchange, this extracellular diffusion is initially complete within 2 or 3 min in tissues such as the taenia coli which are less than 0·6 mm thick. Thus the sorbitol uptake measured between 10 and 20 min may be taken as a convenient measure of the extracellular space of the taenia and it is, in fact, in adequate accord with the analytical spaces measured using other

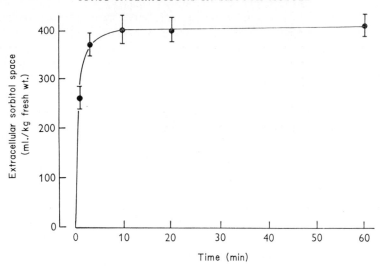

FIG. 4 Tracer [^{14}C]-sorbitol uptake observations on the smooth muscle of the guinea-pig taenia coli in Krebs solution at 35°C. Abscissa, time in min after application of tracer. Linear ordinate, uptake of tracer measuring the extracellular sorbitol space. Note that the uptake reaches an almost steady value within a few minutes. (From Goodford & Leach, 1966)

small ions and molecules (Goodford & Leach, 1966).

The 'extracellular potassium space'

Figure 3 shows the uptake of ^{42}K by the guinea-pig taenia coli. The results have been plotted on the usual semi-logarithmic axes but the whole picture has been inverted so that the uptake phenomenon is shown as a rising curve in contrast to the falling curves for efflux shown in figure 2. The logarithmic ordinate scale is appropriate to an exchange process (equation 7) that reaches equilibrium when $R = A + B$. In Fig. 3 the earliest stages of exchange are shown, and only a tenth of the ordinate is drawn in. The whole ordinate goes from 0 to 13,400 g solution/kg fr wt, this final value being calculated (Goodford, 1968) from the relation:

$$\frac{\text{Tissue potassium concentration}}{\text{Solution potassium concentration}} = \frac{80 \text{ m-mole/kg fr wt}}{5\cdot96 \text{ m-mole/kg solution}}$$

The straight line in the figure has been fitted to the observations by the method of least squares after logarithmic transformation, and corresponds to a single exponential process with rate-constant $\lambda_q = 0\cdot0126$ min^{-1}, and intercept $A = 362$ g/kg fr wt. A is called 'the uncorrected extracellular potassium space' on simple two-compartment theory, but it is a biased value due to the extracellular-intracellular interactions already discussed. A correction for these may be applied (Huxley, 1960; Goodford, 1966) giving $A' = 490$ g/kg fr wt for the 'corrected extracellular potassium space'.

5+

The values of $T_{1/2}$ λ_q A and A' so far calculated could still be distorted if some of the sequestered cell potassium were exchanging more slowly than the remainder, and the same parameters have therefore been recalculated in order to test the bias that could be introduced if 10, 20, 30, 40 or even 50% of the cell potassium were completely inexchangeable. The results are set out in Table 1 and it may be seen that the rate constant λ_q is indeed dependent upon the amount of inexchangeable potassium, but that the calculated extracellular potassium space is less sensitive. The sources of error in the calculation of A and A' in fact tend

TABLE 1

Inexchangeable K %	$T_{1/2}$ min	λ_q min^{-1}	A g/kg fr wt	A' g/kg fr wt
0	55	0·0126	362	490
10	49	0.0141	361	490
20	43	0·0161	360	490
30	37	0·0186	358	489
40	31	0·0221	356	489
50	25	0·0273	352	488

The effect of inexchangeable potassium in the guinea-pig taenia coli upon the calculated parameters which describe the uptake of ^{42}K shown in Fig. 3. It is assumed that all the exchangeable potassium exchanges according to a single exponential function. The intercept A at zero time, and the corrected intercept A', do not depend appreciably upon the amount of inexchangeable potassium, but the half-time $T_{1/2}$ and rate constant λ_q of the process are critically dependent.

to cancel each other giving a relatively unbiased estimate of the extracellular potassium space. Thus the value of 490 g/kg fr wt should give a reliable value for the extracellular potassium space and exceeds the sorbitol space of 410 g/kg fr wt. It is therefore necessary to account for this small but reproducible inequality.

IONIC INTERACTIONS

The 'excess rapidly-exchanging cation'

The 'extracellular potassium space', determined from ^{42}K kinetics, in fact represents a proportion of the total tissue potassium which exchanges more rapidly than the remainder. This rapidly-exchanging potassium can be calculated as: $(^{42}K\text{-space}) \times (K$ concentration in solution) so that in normal solution (Fig. 3): rapidly exchanging potassium $= (490 \text{ ml/kg}) \times (5\cdot96 \text{ m-mole/l})$
$$= 2\cdot92 \text{ m-mole/kg}$$

which may be compared with the amount of potassium dissolved in free solution in the extracellular [^{14}C]-sorbitol space at the same concentration as the potassium in the bathing solution:

potassium in [^{14}C]-sorbitol space $= (410 \text{ ml/kg}) \times (5\cdot96 \text{ m-mole/l})$
$$= 2\cdot44 \text{ m-mole/kg}$$

It may be seen that the conventional two-compartment intracellular-extracellular distribution does not completely account for the rapidly-exchanging potassium, but that there is a slight excess of 2·92–2·44 = 0·48 m-mole K/kg in reasonable agreement with an earlier estimate of 0·42 m-mole K/kg (Goodford, 1966).

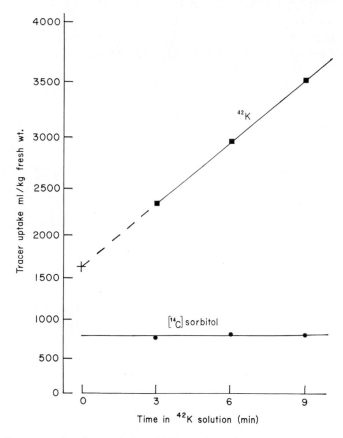

FIG. 5   Tracer uptake observations at 35°C on the smooth muscle of the guinea-pig taenia coli immersed in isotonic sucrose containing 1 mM sorbitol and 2·98 mM KCl. Abscissa, time in min after application of ^{42}K. Uptake time of [^{14}C]-sorbitol was always 10 min. Ordinate, (inverted logarithmic, see text), uptake of tracer. Note that the [^{14}C]-sorbitol space was much smaller than the ^{42}K intercept in this solution where 2·98 mM K$^+$ was the only cation.

Although the *excess* rapidly-exchanging potassium is small, and may not even be statistically significant in normal solution, much larger excesses have been observed in solutions of reduced ionic strength. When the sodium, calcium, and magnesium components of the solution were replaced with isotonic sucrose, and 2·98 mM K$^+$ was the only cation used to prepare the medium, the uncorrected extracellular ^{42}K-space rose to 1600 ml/kg fr wt (Fig. 5). Under these conditions the rapidly-exchanging potassium could not possibly be dissolved in the muscle at the same concentration as in the bathing fluid because the total weight of the tissue was only 1140 g/kg fr wt, and the average concentration of rapidly-exchanging cation would have to be somewhat raised even if it were distributed throughout the whole tissue volume. A plausible explanation for this increased concentration would be the presence of fixed anionic groups in the tissue. Indeed intracellular fixed charge is a well accepted rationale for the transmembrane Donnan distribution (Boyle & Conway, 1941) and the present results could be explained similarly if there were fixed negative groups outside the cells where the counter-cations could exchange quickly with externally-applied tracer ^{42}K.

The corrected potassium space calculated from Fig. 5 was 1830 ml/kg fr wt and corresponded to an *excess* rapidly-exchanging potassium of (1050 ml/kg) × (2·98 m-mole/l) = 3·1 m-mole/kg. This should be an estimate of the relevant fixed negative charge assuming, since potassium cations alone were used to

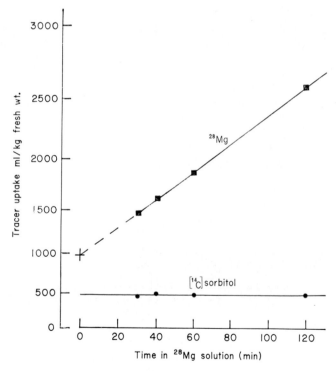

FIG. 6   Tracer uptake observations at 35°C on the taenia coli immersed in a modified Krebs solution containing only 0·1 mM Ca^{2+} and 1 mM K^{+}. Abscissa, time in min after application of ^{28}Mg. Uptake time of [^{14}C]-sorbitol was always 10 min. Logarithmic ordinate, uptake of tracer. Note that the [^{14}C]-sorbitol space was less than the ^{28}Mg intercept.

prepare the medium, that a potassium ion was the counter-cation at each anionic site. Estimates of the same order of magnitude have been observed previously (Goodford, 1966, 1967) in solutions containing 2·98 to 11·92 m-mole K/l and it may therefore be tentatively concluded that there is a region in the taenia coli in which metal ions can exchange rapidly, while being held as counter-cations to some 4 mEq/kg wet wt of fixed anionic sites.

It should be possible to detect '*excess* rapidly-exchanging cations' with other metals besides potassium, if metallic cations are indeed competing for superficial fixed negative charge in smooth muscle. Observations have been made on magnesium which is concentrated in intestinal smooth muscle (Potter & Sparrow, 1968; Sparrow & Simmonds, 1965; Sparrow, Mayrhofer & Simmonds, 1967), because its small ionic radius and double positive charge would give it a high deforming power (Fajans, 1923). It should therefore be well-favoured as a counter-cation, and recent results (Fig. 6 and Goodford and Sparrow, unpublished) are compatible with this prediction. The exchange of radioactive ^{28}Mg shows an initial rapid phase followed by a long slow and approximately exponential uptake and, as expected, the amount of rapidly-exchanging magnesium exceeded the amount in free solution in the extracellular space. Furthermore it was not necessary to remove all other cations to demonstrate this effect with magnesium, as was necessary with potassium, but reduction of calcium and potassium concentrations to 0·1 and 1 mM respectively allowed magnesium to occupy a substantial proportion of the available anionic sites. When, moreover, a higher calcium concentration was used in further experiments, the '*excess* rapidly-exchanging magnesium' was reduced and this would again be compatible with the counter-cation competition hypothesis.

The Nernst equation and the concept of regional electroneutrality

A convenient form of the Nernst equation is:

$$aA_1 = A_2\, e^{nFE/RT} \tag{8}$$

in which the constant 'a' is specific for each ionic species and incorporates any factors for:

  i any specific chemical affinities.
 ii activity coefficients.
iii any differences in the units of measurement in the two regions.

In one region, for instance, it might be most convenient to measure A_1 as a chemical activity, whereas in another region of constant volume it might be more suitable to measure A_2 as a weight. Equation 8 may therefore be rewritten in the form:

$$\left\{\frac{A_2}{aA_1}\right\}^{1/n} = e^{-FE/RT} \tag{8'}$$

in which the right-hand side is independent of the specific chemical nature of A.

It depends only on the electrical potential E at constant temperature, and (8′) may therefore be extended to give a useful series of equalities for all the ions which are taking part in the equilibrium distribution:

$$\frac{K_2}{kK_1} = \frac{Na_2}{naNa_1} = \left\{\frac{Ca_2}{caCa_1}\right\}^{1/2} = \left\{\frac{Mg_2}{mgMg_1}\right\}^{1/2}$$

$$= \frac{clCl_1}{Cl_2} = \frac{hco_3HCO_{3_1}}{HCO_{3_2}} = e^{-FE/RT} \quad (8'')$$

In any finite space the total number of positive and negative charges are normally almost equal, and this principle may be applied to the region of superficial fixed anionic charge in the taenia coli (Boyle & Conway, 1941; Buck & Goodford, 1966; Goodford, 1968). When K^+ and HCO_3^- are the only ions in the bathing solution the principle leads to

$$K_2 = HCO_{3_2} + R_2 + \epsilon \quad (9)$$

in which R is the 'amount' of fixed negative charge; and ϵ is the charge imbalance. ϵ is very small compared with K_2 or HCO_{3_2} or R_2.

One should specially note that equation (9) describes an electric charge distribution and that the terms are not chemical activities but concentrations or, assuming in the present case constant regional volume, are weights on the chemical equivalent scale.

It may now be seen that the estimate of superficial fixed anion in the taenia coli was made on the assumption that $K_2 = R_2$ in equation (9), and that the term in HCO_{3_2} was neglected. The experimental justification for this assumption will now be considered.

The potassium-bicarbonate interaction

Equation (8″) includes the equality:

$$\frac{K_2}{kK_1} = \frac{hco_3HCO_{3_1}}{HCO_{3_2}}$$

which may be combined with equation (9) to give:

$$R_2 = K_2 - k.hco_3K_1.HCO_{3_1}/K_2 \quad (10)$$

In this K_1 and HCO_{3_1} are the known concentrations in the bathing solution and ϵ is neglected. K_2 is the measured '*excess* rapidly-exchanging potassium', so that only R_2 and $k.hco_3$ are unknown. When, therefore, observations are made in two solutions, one of low $KHCO_3$ concentration and one of high, two simultaneous equations are obtained which can be solved for R and for the product $k.hco_3$ which has the units of:

$$k.hco_3 = \frac{K_2.HCO_{3_2}}{K_1.HCO_{3_1}} = \frac{(m\text{-mole/kg})^2}{(m\text{-mole/l})^2} = \frac{(litres)^2}{(kg \ wet \ wt)^2}$$

In this way it has been established that $k.hco_3$ is substantially less than $0\cdot1$ litre^2 kg^{-2}, and it may be seen by inspection of equation (10) that the approximation $R_2 = K_2$ is most accurate when $k.hco_3$ is in fact low.

Cation-cation interactions

Equation (9) may be extended when sodium ions as well as potassium are present and ϵ is neglected:

$$Na_2 + K_2 = HCO_{3_2} + R_2 \tag{9'}$$

and from equation (8″):

$$\frac{k}{na} = \frac{K_2Na_1}{K_1Na_2}$$

which is a pure number and therefore has no units. This ratio has been measured in a similar experiment to the last, comparing one solution containing K^+ ions alone with another containing both cations, and using the established smal value of $k.hco_3$ to calculate HCO_{3_2}. The resultant ratio k/na is not critically dependent upon the exact value of $k.hco_3$ so long as this is small, and it has been shown (Goodford, 1966) that k/na is of the order of 1 with, if anything, k slightly exceeding na. This means that *the rapidly-exchanging K^+ and Na^+ ions have comparable affinities for the superficial fixed anionic charge in the taenia coli*, although potassium might be rather better favoured as a counter-cation since the best estimates for k/na are $1\cdot8$ and $2\cdot2$ (Goodford, 1966).

IONS IN ASSOCIATION WITH CELL MEMBRANES

It would be as well at this point to review the discussion so far. The fundamental experimental observation under consideration is that the rapidly-exchanging cation in the taenia coli is too much to be simply dissolved in the extracellular space. Under some conditions the excess is so great that it could not even be dissolved in the whole tissue volume, without being concentrated above the concentration in the bathing solution. Several possible artifacts have been considered but do not account for the effect, and the proposal has therefore been made that the '*excess* rapidly-exchanging potassium' is a counter-cation at superficial anionic sites in the taenia coli. This hypothesis has been further tested by studying other ions, and ionic interactions have been described which are quantitatively compatible with theoretical equations derived from the principle of regional electroneutrality and with the Nernst equation. The calculated weights of superficial counter-ion are set out in the left-hand half of Table 2 for a kilogram of taenia coli immersed in normal solution, assuming plausible rounded values of the ratios k/na, etc.

The membrane counter-cation hypothesis

The possibility that these counter-ions are associated with the cell membrane

TABLE 2

	Extracellular concentration mM	Calculated extracellular counter-ion m-mole/kg wet wt	Calculated intracellular counter-ion m-mole/kg wet wt	Intracellular concentration mM
K^+	5·96	0·45	4·2	100
Na^+	137	5·2	0·7	35
Ca^{2+}	2·5	1·45	0	0
Mg^{2+}	1·25	0·72	0·09	0·5
Cl^-	150·46	6·0	1·08	15
R^-	0	4·0	4·0	121

The weights of counter-ion in 1 kg of taenia coli calculated from the cubic equation in K_2:

$$K_2^3(caCa_1 + mgMg_1) + K_2^2(naNa_1kK_1 + k^2K_1^2) - K_2(k^2K_1^2R_1) - clCl_1k^3K_1^3 = 0 \quad (11)$$

assuming that:

$$\frac{k}{na} = 2$$

$$\frac{k^2}{ca} = \frac{k^2}{mg} = 0.1 \text{ litre kg}^{-1}.$$

$$k.cl = 0.003 \text{ litre}^2 \text{ kg}^{-2}.$$

In this situation the calculated extracellular counter-cation approximates to the 'excess rapidly-exchanging cation' in normal solution. The intracellular counter-ion has been calculated similarly on the assumption that there is a similar set of binding sites in equilibrium with the intracellular solution.

must now be considered. A representation of the ionic balance across the membrane might then be obtained by reading across Table 2, the extracellular counter-ions being in rapid equilibrium with the external solution, and the intracellular counter-ions shown in the right-hand half of the Table in a similar equilibrium with the cytoplasm. The 2 counter-ion layers are, of course, not in equilibrium with each other but are, on the Davson-Danielli hypothesis, separated by the high-resistance lipid layer of the membrane.

Inspection shows that on this formulation of the ionic distribution, the tissue calcium is present as an extracellular counter-cation, and it may be significant that the role of calcium in excitable tissue physiology has been appreciated for over 80 years (Ringer, 1883). The actual presence of calcium in association with the cell membrane was specially considered by Shanes (1958), and in 1961 Schatzmann suggested that some of the rapidly-exchanging calcium in the taenia coli might be adsorbed at the membrane. The concept of membrane-bound calcium was used by Bülbring & Kuriyama (1963a, b) in order to interpret their observations of both the spontaneous electrical activity of the taenia coli and the inhibitory action of adrenaline. Bauer et al. (1965) then extended Schatzmann's observations of the tissue calcium distribution, and explicitly suggested that the second component of calcium exchange might be 'in equilibrium with the ionized extracellular calcium', and Goodford (1965) tentatively deduced that the amount of membrane calcium might be modified by competition with potassium ions. Since that time further evidence of cation competition has emerged (Goodford, 1966, 1967), and it is now necessary to consider some more quantitative aspects of the counter-cation observations in order to assess the quantitative feasibility of the hypothesis that they are associated with the cell membrane.

The area of cell membrane

It has already been pointed out that the ratio of cell volume V to cell surface area A should depend on the absolute size of the cells, if the close-packed cylinder model of smooth muscle be accepted, and a tissue such as intestinal smooth muscle with long thin cells would have a relatively high proportion of membrane and would therefore be particularly suitable for the direct measurement of membrane ions. The ratio V/A would be 1.5×10^{-4} cm if 6 μ were the average cell diameter (Freeman-Narrod & Goodford, 1962; Sparrow, unpublished observations), and a kilogram of cells would have a surface area of 6.6×10^6 cm². If the analytical measurement of extracellular space be chosen, a kilogram of tissue would contain some 500 ml of cells with a surface area of 3.3×10^6 cm², but there are reasons for regarding this as a low estimate. Firstly the close-packed cylinders should have a much larger cell volume which would be compatible with the histological observations of Prosser et al. (1960), and would lead to a calculated surface area of 6×10^6 cm². Secondly the surface of the cells is by no means smooth (Fig. 1) and any irregularity would further increase the membrane area and, in particular, large numbers of vesicles have been observed in the plasma membrane of smooth muscle. Rhodin (1962) has calculated that the vesicular pockets alone might increase the cell surface area of the longitudinal smooth muscle of the mouse small intestine by 25%, and we have recently estimated a still greater increase in the guinea-pig taenia coli (Goodford, Johnson, Krasucki & Daniel, 1967, and unpublished observations). In these experiments pieces of taenia were equilibrated for two hours in oxygenated Krebs solution at 35°C, and when spontaneous mechanical activity had been observed to make sure they were being maintained in a satisfactory condition they were fixed for electronmicroscopy by adding to the organ bath a warm oxygenated 50% glutaraldehyde solution at pH 7·3 giving a final glutaraldehyde concentration of 6%. The rhythmic activity of the muscle immediately stopped but oxygenation was maintained at 35°C for the next 10 min and during progressive cooling to 7°C over the next 30 min. After four hours' fixation under tension with continuous oxygenation at 7°C, small blocks of tissue were cut from each muscle and stored overnight at 4°C in phosphate buffer (pH 7·4). The tissues were then post-fixed for 1 hr in 1% osmium tetroxide buffered with veronal acetate; were rapidly dehydrated in an ascending series of methanol, and were embedded in Araldite. Thin sections were stained with either uranyl

5*

acetate or lead citrate and were observed at primary magnifications ranging from 10,000 to 40,000.

In this study muscles were fixed while still fully active at 35°C, and examination of many electron-micrographs confirmed that the vesicles were present predominantly in clumps with intervening smoother areas of cell surface. The proportion of the cell surface covered with such clumps was measured in both transverse and longitudinal sections selected at random, and this showed $54 \pm 2\%$ of the total surface of the cells to be covered with vesicles. When the plasma membrane of these vesiculated areas was measured at still higher magnifications by means of a curvimeter, it was found that their surface area was increased by a factor of $2 \cdot 34 \pm 0 \cdot 2$ times, compared with adjacent smooth surfaces, so that the vesicles increased the overall surface area of the cells by $54(2 \cdot 34 - 1)$ which is by more than 70%.

The observations which have just been summarized show that it is not easy to obtain an exact value for the area of plasma membrane in smooth muscle, and estimates range upwards from $3 \cdot 3 \times 10^6$ $cm^2 kg^{-1}$ tissue. Taking 5×10^6 $cm^2 kg^{-1}$ as a typical value, and assuming the thickness of the plasma membrane to be 75 Å, one may calculate that a kilogram of smooth muscle should contain some 4 g of membrane, and if this contained 4 mEq kg^{-1} of fixed anion uniformly distributed, the average concentration would be 1 equivalent/kg which is not an unreasonable order of magnitude. If the anion were spread smoothly over the outer surface of the cells there would be one electronic charge to every 22 $Å^2$ of membrane, and this is close to previous estimates (Goodford, 1967; Sparrow, unpublished observations) and may be compared with the known molecular cross-section (20 $Å^2$) of a condensed lipid film. Furthermore, in order to assess the significance of the present calculations, it is important to appreciate the size of error which could arise if the counter-cation hypothesis were seriously at fault. Typical conversion factors which have been used in the present calculations include Avogadro's number (6×10^{23} $mole^{-1}$) and the ratio $cm^2/Å^2$ (10^{16} $Å^2$), and when such vast factors are used the final result could err by many orders of magnitude if it were based on ill-founded assumptions. The present results, which are in agreement to within a factor of only 2 or 3, must therefore be regarded as compatible with the hypothesis that there may be fixed anionic groups on the surface of smooth muscle cells, at which counter-cations can exchange.

EXCITATION-CONTRACTION COUPLING

The detailed mechanism of excitation-contraction coupling in smooth muscle is still obscure, but it is reasonable to assume that it is of a similar nature to the mechanism in other muscle tissues, because the biochemical constituents of muscle contractile mechanisms have much in common. If this assumption is justified the calcium ion may be involved, and there is ample evidence that the total calcium content of intestinal smooth muscle is in the range of 1 to 4 m-mole/ kg wet wt when the extracellular concentration is 2·5 mM (Bauer *et al.*, 1965; Meigs & Ryan, 1912; Urakawa & Holland, 1964; Bozler, 1962; Sparrow & Simmonds, 1965). Roughly 1 m-mole Ca/kg wet wt would be freely dissolved in the extracellular space (Goodford, 1968) leaving up to 3 m-mole Ca/kg wet wt to be accounted for. *If this were uniformly distributed within the cells* the average intracellular concentration would be anything up to 5 mM Ca, and by analogy with skeletal muscle (Ashley, Caldwell, Lowe, Richards & Schirmer, 1965; Ashley, 1967) it would almost all be bound to the contractile mechanism which would itself be fully contracted. On the other hand *if the cell calcium were all bound extracellularly*, as the present formulation implies (Table 2), the contractile mechanism would be completely relaxed.

There is little or no evidence to suggest that the total smooth muscle calcium changes drastically when the muscle relaxes and contracts (Lüllmann, chapter 5), and hence there must be a migration of calcium from one region in the tissue to another if the cycle of contraction and relaxation is in fact associated with a shift of calcium towards and away from the contractile mechanism. The membrane sites now under consideration could accommodate all the muscle calcium during relaxation and yet this calcium could still be available to the contractile mechanism because the diameter of smooth muscle fibres is so small. Indeed, the membrane is at most 2 or 3 μ away from the contractile mechanism, and the rate of development of tension after a suitable excitatory stimulus in smooth muscle is compatible with simple diffusion over such a distance (Hill, 1928).

In skeletal muscle 0·1 to 0·5 m-mole Ca/kg wet wt is needed to initiate a maximal contraction, but the amount of actomyosin in smooth muscle is only about one tenth as great (Needham & Williams, 1963a), and if there is a correspondence between the weight of calcium needed to contract a given weight of actomyosin, (see Weber & Herz, 1963) this would suggest that 0·01 to 0·05 m-mole Ca/kg wet wt should suffice to contract smooth muscle maximally. Such a small amount could be accommodated on less than 1% of the postulated cell membrane sites and it is possible that there may actually be specialized membrane receptors which can take up this calcium from the contractile mechanism during relaxation. This has yet to be established, but it is reasonable to assume *a priori* that there may be different types of anionic site on the cell surface and that some of these may be completely unselective in their choice of counter-cation; some may have a preference for one ion but may not be completely selective, some may show different preferences according to their allosteric form and, finally, some may bind specifically to one cation alone. These last would not necessarily be detected by experiments designed to observe cation competition, and it should be born in mind that there could be many anionic sites in smooth muscle which have not been detected because they are tightly and specifically bound to one particular cation under all conditions. For the sake of simplicity one may now choose to assume that all sites are similar, but further evidence on this point should be sought.

The electrical capacity of the cell membrane of the taenia coli is 3 μF cm^{-2} (Tomita, 1966b), and if 5×10^6 cm^2/kg wet wt are discharged by an action potential of 60 mV (Holman, 1958), a total of 0·9 Coulombs/kg wet wt would be needed. A flow of 0·005 m-mole Ca/kg wet wt of divalent calcium ions could carry this charge across the membrane, and would thereby account for the inward current of the rising phase of the action potential. It would also increase

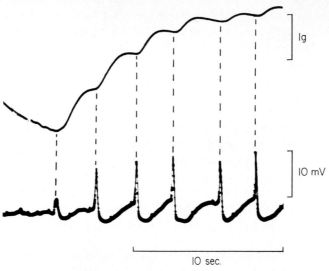

10 sec.

FIG. 7   Electrical and mechanical activity in the smooth muscle of the guinea-pig taenia coli. Upper tracing: tension, lower tracing: membrane potential. Six spontaneous action potentials were discharged after a quiescent period, and each was associated with an increment of tension (as shown by the dashed lines). These increments summated, and a longer run of more rapidly occurring action potentials would have been necessary for maximum tension development. (From Bülbring, 1957c)

the intracellular calcium content and thereby initiate a small contraction, although several such events would be needed to introduce 0·01 to 0·05 m-mole Ca/kg wet wt and so cause a maximal effect. On the basis of the present hypothesis one might therefore predict that several action potentials would be needed in order to initiate a maximal contraction in smooth muscle, and this is in fact compatible with well-established observations of the electrical and mechanical activity of the tissue (Fig. 7).

CONCLUSIONS

The electrical potential, measured at the tip of a glass capillary micro-electrode, changes suddenly when the electrode is driven slowly through a quiescent sample of smooth muscle, and it either registers the earth potential or a negative 'resting potential' of about -50 mV. It has become almost axiomatic to associate these readings with an extracellular and an intracellular region, and until recently the emphasis has therefore been placed on a two-compartment model of smooth muscle. Attempts to interpret the solute distribution have previously been made on this basis, but such a simple model has not been compatible with the observations on the tissue itself, and some of the problems which arose were discussed at the start of this chapter. The two-compartment model now seems to be unreasonably restrictive in view of the evidence, accumulated by electron-microscopic and ultracentrifugation and radioactive tracer exchange methods for specialized subcellular regions. *It may therefore be concluded that a multi-compartment model should give a better representation of smooth muscle.*

Fixed anionic charges are present in smooth muscle, and these charges may not all be equivalent. Some are apparently associated with slowly-exchanging counter-cations, and these may well correspond to the intracellular fixed anions proposed by Boyle & Conway in 1941. Others, however, have rapidly-exchanging counter-cations, and the possibility is considered whether they could be part of the cell plasma-membrane and to what extent its electrical properties might be interpreted in this way. On the author's philosophy absolutely sharp divisions occur very rarely in biological materials, and compartments and boundaries are more often figments of the scientist's imagination. It is this scientist's belief that the electromechanical properties of smooth muscle may eventually be associated with a fluctuating array of ionic charges and their associated atoms, and that compartmental boundaries such as the plasma membrane will be described as special parts of that overall array. Compartments, regions, boundaries and membranes should only be with us as short term expedients. Ars longa vita brevis.

4

APPLICATION OF THE 'ASSOCIATION-INDUCTION HYPOTHESIS' TO ION ACCUMULATION AND PERMEABILITY OF SMOOTH MUSCLE

A. W. JONES

INTRODUCTION

It has long been known that muscle cells contain relatively high concentrations of potassium and relatively low concentrations of sodium and chloride in comparison to the extracellular fluid. The processes controlling electrolyte permeability and accumulation in smooth muscle have been regarded as closely related to the electrical, mechanical and osmotic behaviour, yet only in recent years have systematic investigations been conducted into the ion-exchange properties of this tissue. These investigations have not, however, led to a uniform interpretation of the underlying mechanisms.

One area of development was the application of the membrane concept to the ion-exchange behaviour. The cell membrane was taken to function as a discrete resistance for the exchange of material between two well-mixed compartments. Various mechanisms have been postulated for the movement of ions through the smooth muscle membrane including; diffusion of independent ions, diffusion of ion pairs, coupled sodium-potassium exchange via an active transport mechanism, net outward transport of sodium in an electrogenic fashion, sodium–sodium exchange via a diffusion carrier or vesicular mechanism, surface adsorption of competing ions with subsequent inward diffusion. Discussion of such mechanisms have appeared in several reviews of smooth muscle behaviour (Kao, 1967; Goodford, 1968; Somlyo & Somlyo, 1968a) and in the present volume (Casteels, Chapter 2; Goodford, Chapter 3). An assumption common to the membrane models is that once the ions in question leave the membrane, they pass into a salt solution and undergo relatively little interaction with immobilized tissue elements such as proteins. Most of the electrolyte present in the intracellular and extracellular compartments would therefore turn over as a uniform or homogeneous pool irrespective of the intervening exchange mechanisms. The small amounts of ion undergoing transit through the membrane would constitute an important, but quantitatively small phase. The critical evaluation of the assumption of a uniform intracellular electrolyte pool will be undertaken in the present review and will serve as a basis for the consideration of an approach often overlooked in current work.

The adoption of membrane models to smooth muscle behaviour has not been universal. One of the earliest investigations of osmotic and permeability properties of smooth muscle led to the conclusion that frog stomach behaved more like a gel than a fibre surrounded by a semipermeable membrane (Meigs, 1912; Meigs & Ryan, 1912). Although much of the tissue sodium was readily mobilized, considerable amounts of fibre potassium and magnesium were postulated to be in a non-diffusible form. After a lapse of over 40 years the work of Meigs was confirmed and extended by Bozler and co-workers (Bozler, Calvin & Watson, 1958; Bozler & Lavine, 1958; Bozler, 1959, 1961a; 1962, 1965). Bulk properties of the smooth muscle cytoplasm rather than membrane properties were emphasized in the interpretation of permeability and osmotic characteristics. Cell volume was regarded as being controlled by elastic forces within a cross-linked system of macromolecules with differences in permeability resulting from variations in the size of channels within the matrix (Bozler, 1959, 1962, 1965). The suggestion was put forth in one of the earlier studies that intracellular potassium may be immobilized (Bozler et al., 1958). Later work did not develop this concept, however.

Models based on ion-protein-water interaction in the cell cytoplasm have been developed for ion permeability and accumulation behaviour of a wide variety of tissues. A review of the 'sorption theory of cell permeability' has recently been published in English (Troshin, 1966). In this approach the permeability of the cell is determined by solubility, adsorption, and chemical binding which take place in the cytoplasm as a whole. Evidence was put forth for the coacervate structure of the protoplasm, with water being organized by hydrophilic sub-stances which were capable of associations with other large molecules and ions.

A unified theory of cellular behaviour, based on the interaction of ions in a proteinaceous fixed-charge system, was independently formulated by Ling and termed the 'association-induction hypothesis' (Ling, 1962). This approach has been further developed in recent work to account for quantitative as-pects of ion permeability and selective accumulation (Ling, 1965a, b; Ling & Ochsenfeld, 1965, 1966; Ling, 1966a, b). The placing of this model on an analytical basis allows it to be tested in a rigorous manner and also offers an opportunity to take the work on ion permeability and accumulation in smooth muscle from a descriptive phase into one more closely based on biophysical principles.

The objectives of this review are to briefly summarize salient points of the 'association-induction hypothesis', relevant to smooth muscle research; and to examine ion permeability and accumulation properties in the light of this model. The reader may anticipate the author's conclusions that a model which con-siders both surface or membrane as well as bulk properties provides a basis for interpreting a wide variety of smooth muscle behaviour and for designing future experiments.

'ASSOCIATION-INDUCTION HYPOTHESIS'

Evidence has been reviewed by Ling (1962) which supports the premise that proteins, ions, and water exist in close association within the living cell. Later work has developed quantitative aspects of this approach and has provided further experimental support (Ling, 1965a, b; Ling, 1966a, b; Ling & Ochsen-feld, 1965, 1966; Ling, Ochsenfeld & Karreman, 1967). In this section concepts fundamental to this approach will be summarized and in some instances con-trasted to predictions of membrane models. The experimental evidence for these considerations are presented in the references noted above and will not be quoted in detail. One recent experimental development will, however, be dis-cussed in the last section because it represents a new approach which shows great promise in answering some fundamental questions concerning the physico-chemical state of ions and water in the living cell.

Selective accumulation of ions

A consideration of the interactions which can take place between ions, proteins and water led Ling to conclude that both short and long range forces must be taken into account (Ling, 1962). A theory of interactions based on proteins dissolved in dilute solutions was deemed to be inadequate for conditions in the living cell. For instance it was noted that the macroscopic dielectric constant for water would be inappropriate under conditions in which only one to three water molecules could separate a mobile charge from one fixed onto a macromolecule. The major source of the fixed charges were the β and γ carboxyl groups (negative sites) of aspartic and glutamic acid; and amino and guanidyl groups (positive sites). The net charge of the lattice was negative.

Although many of the fixed groups of opposite charge may interact forming salt linkages, the interactions with mobile ions were of particular significance for the accumulation properties. The associational energies for various cations were calculated on the basis of this approach (Chapter 4 in Ling, 1962). Relatively long-range forces (electrostatic effects) and short-range forces (London dispersion energy and Born repulsion energy) were taken into consideration for sites having varying polarizability. The interacting charges were treated in a linear model in which water molecules separated the groups (a three-dimensional model was beyond the computing facilities). It was noted that when three or less water molecules separated the charges the water would be 'frozen in' or dielectrically saturated. Because of the relatively close spacing of fixed charges (20 Å in frog sartorius) it was proposed that only a limited number of water molecules can be present. The computations indicated that the associational energies for various ions were not the same. The energies also changed with the degree of polarizability of the fixed sites resulting in different orders of selectivity. For instance, for one set of parameters K^+ association would exceed Na^+ association, while a shift in polarizability led to a reversal in the calculated selectivity. On theoretical grounds this rather simple model can account for the selective behaviour exhibited by many biological systems.

Selective accumulation of ions is therefore taken to result from the association of energetically-favoured ions at polarized sites on proteins. One important concept in the interpretation of interactions and control is that the sites are attached to a highly-polarizable resonating chain through which events at one functional group can induce changes in electron density at neighbouring groups (termed inductive effect). Such interactions can have a reinforcing effect, that is the association of an ion at one site can induce a shift in electrons which makes it energetically more favourable for the association of a similar ion at a neighbouring site and so on. Such interaction is called *auto-cooperative*. One having the opposite effect is termed *hetero-cooperative*. *Auto-cooperative* interaction provides a basis for the transition from one metastable equilibrium to another in an 'all-or-none' manner.

Recent work has led to the derivation and experimental testing of a general equation for *cooperative* adsorption of solute onto a limited number of fixed sites (Ling, 1965a, 1966a; Karreman, 1965). The adsorption isotherm took into consideration saturation behaviour, selective association of one ion over another, and the free energy of the nearest neighbour interaction designated $-\gamma/2$.

The *cooperative* adsorption isotherm for potassium interaction with sodium is:

$$K_{ad} = \frac{F_T^-}{2}\left[1 + \frac{\dfrac{[K]_{ex}}{[Na]_{ex}}K_{K,Na} - 1}{\left[\left(\dfrac{[K]_{ex}}{[Na]_{ex}}K_{K,Na} - 1\right)^2 + 4\dfrac{[K]_{ex}}{[Na]_{ex}}K_{K,Na}n^{-2}\right]^{1/2}}\right] \tag{1}$$

where

F_T^- = total fixed charge available, mEq/kg d.s.*
$K_{K,Na}$ = intrinsic equilibrium constant
n = an interaction parameter
$[Na]_{ex}$ = external sodium concentration, mEq/l
$[K]_{ex}$ = external potassium concentration, mEq/l

It was further noted (Ling, 1965a, 1966a; Karreman, 1965):

$$n = \exp\left(-\frac{\gamma}{2RT}\right) \tag{2}$$

where

$-\gamma/2$ = free energy of nearest neighbour interaction, kcal/mole
R = gas constant, kcal/mole/°C
T = absolute temperature, °K

When $n > 1$; $-\gamma/2 < 0$, it is energetically more favourable for two adjacent sites to adsorb the same solute, and is referred to as *auto-cooperative* adsorption. When $n < 1$; $-\gamma/2 > 0$ this process is referred to as *hetero-cooperative* adsorption. When $n = 1$; $-\gamma/2 = 0$ equation (1) reduces to the familiar Langmuir adsorption isotherm:

$$K_{ad} = F_T^- \frac{[K]_{ex}K_{K,Na}}{[K]_{ex}K_{K,Na} + [Na]_{ex}} \tag{3}$$

Another simplification of equation (1) takes place where: $K_{ad} \simeq F_T^-/2$. The following equation is then applicable (Karreman, 1965; Karreman and Jones, 1965; Ling, 1965a):

$$\log \frac{K_{ad}}{F_T^- - K_{ad}} = n \log \frac{[K]_{ex}}{[Na]_{ex}} + n \log K_{K,Na} \tag{4}$$

The slope of a log–log plot of the fraction sites occupied by K versus

---

* Three common references employed for normalizing electrolyte data in smooth muscle are: dry solid, d.s.; wet weight, wet wt (weight at end of *in vitro* incubation); and fresh weight, fr wt (weight after dissection, but before exposure to incubation media). In later sections data will be quoted on the basis employed in the original work and then translated into one of the other references to facilitate comparison.

$[K]_{ex}/[Na]_{ex}$ or $[K]_{ex}$ (for constant $[Na]_{ex}$) yields the value for n and provides a convenient means for analysing experimental data.

Three commonly used plots appear in Fig. 1. The computed values are based on equation (1) for $n = 3$ and $n = 1$ and $K_{K,Na} = 150$; $F_T^- = 100$ mEq/kg d.s.; $[Na]_{ex} = 150$ mEq/l. Plots 1A and 1B can be employed to estimate the saturation level from experimental data (corrected for dissolved components). The value for $K_{K,Na}$ can be determined from the value of $[K]_{ex}$ which results in half saturation ($K_{ad} = F_T^-/2$). This can be determined from the intercept of the line and $K_{ad}/(F_T^- - K_{ad}) = 1$ in Fig. 1C. At that point:

$$\frac{[K]_{ex}}{[Na]_{ex}} K_{K,Na} - 1 = 0 \tag{5}$$

and $K_{K,Na}$ can be readily computed from the known concentrations. On the basis of equation (4) the slope of the plot when $K_{ad}/(F_T^- - K_{ad}) = 1$. (Fig. 1C) yields the value for n.

Experimental data for smooth muscle will be evaluated by means of these plots.

Biological control

The concepts involved in the formulation of the *cooperative* adsorption isotherm can be extended to the interpretation of biological control. The location of highly polarizable sites at strategic locations in the matrix can account for the ability of a small amount of one molecule (bioregulant) to influence the interaction of many sites with other molecules. Such alteration can result from shifts in electron density induced throughout the backbone of the protein by the bioregulant.

One such regulating substance produced within the cell is ATP. The adsorption of ATP into controlling sites is an important factor for maintenance of K^+ selectively over Na^+ (Chapter 9 in Ling, 1962). The hydrolysis of ATP would result in a reversible shift in selectivity toward Na^+. The concept of metabolic regulation via the adsorption energy of an intermediary such as ATP constitutes a fundamental difference from the membrane models. In the latter system the continued hydrolysis of ATP via ATPase is considered to provide the energy for the maintenance of a relatively high cell K^+ and low cell Na^+. According to the 'association-induction hypothesis' the activation of ATPase activity underlies much of the reversible shifts in selectivity of sites falling under metabolic control, but does not provide an important energy source for ion accumulation.

Regulating substances produced outside the cell can also operate via strategically placed sites or, according to common terminology, receptors. The metabolic and hormonal control of smooth muscle electrolyte accumulation will be discussed later. Only a few of the many agents known to impinge on smooth muscle cells will be considered as examples of a general mechanism.

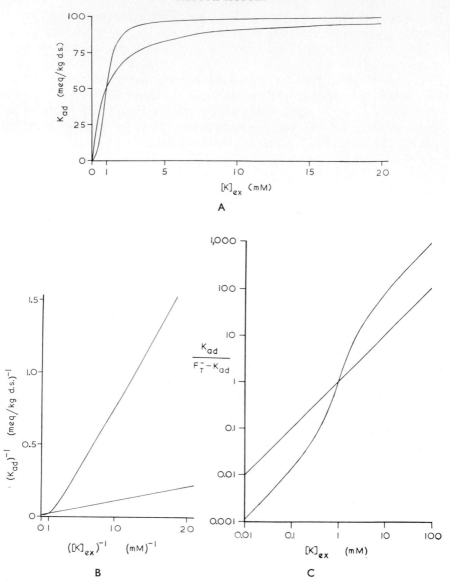

FIG. 1   Plots of the adsorption isotherms based on equation (1) for $n = 3$ and $n = 1$.
A. Adsorbed potassium versus extracellular concentration. The curve for $n = 3$ exhibits
a steeper slope at the midpoint ($[K]_{ex} = 1$ mM) than that for $n = 1$.   B. Reciprocal
of adsorbed potassium versus reciprocal of external concentration. The curve results
from the case of *cooperative* interaction ($n = 3$), while a straight line represents the
case of no interaction ($n = 1$).   C. Log–log plot of the ratio of adsorbed potassium to
adsorbed sodium versus external concentration. The curve results from the case of
cooperative interaction ($n = 3$) and a straight line for the case of no interaction ($n = 1$).
The slopes at the midpoint were 3 and 1 respectively.

Distribution of ions in cell water

Solute dissolved in the cell water and not associated with a fixed site is referred to as interstitial. Evidence has been put forward for the existence of cell water in a physical state different from that in the extracellular solution (Ling, 1962; Ling, 1965a, b, 1966a). The interstitial water is pictured as existing in polarized multilayers about the protein. Ions which enter such a phase are subject to restrictions in movement especially in rotational motion and to some extent translational. This consideration leads to the conclusion that such ions will undergo a decrease in entropy leading to a positive change in free energy (the change in enthalpy being less than the entropic changes). The mean equilibrium distribution coefficient between intracellular and extracellular water ($q^{\infty \to \text{ins}}$) would therefore be less than one ($q_{\text{Na}}^{\infty \to \text{ins}} \simeq 0.1$ to 0.2 in frog sartorius, Ling 1965b). At equilibrium, the concentration of an extracellular ion would exceed that in the interstitial water and is represented by the following relation for Na^+:

$$[\text{Na}]_i = q_{\text{Na}}^{\infty \to \text{ins}} [\text{Na}]_{\text{ex}} \qquad (6)$$

The presence of water in polarized layers throughout the cytoplasm provides a basis for an ion exclusion model. Ions are excluded from the cell water under equilibrium conditions because it is energetically less favourable for an ion to enter a structured phase. Relatively low intracellular concentrations can therefore be maintained under conditions in which intracellular and extracellular ions are free to exchange (p. 131). If one assumes exclusion mechanisms based on a membrane exhibiting impermeable or short-circuiting characteristics, it is difficult to account for the exchanging electrolyte contained in the cytoplasm. Exclusion as well as selective accumulation can both be satisfactorily treated in terms of a cytoplasmic model.

It might be pointed out that factors which will alter the cytoplasmic structure would be reflected in changes in $q^{\infty \to \text{ins}}$. A more open structure with relatively less polarized water would be expected to lead to values of $q^{\infty \to \text{ins}}$ approaching one. Such conditions would result in the net uptake of material because of the increased capacity to dissolve ions. Increased capacity can also result from an increase in cell volume.

Volume control

The volume of the cell under a given set of conditions is taken to be a function of the degree of crosslinking in the protein matrix in addition to the water present in polarized multilayers. The equilibrium volume depends on a balance between cohesive forces, e.g., S—S bonds, hydrogen bonds, and salt linkages; and opposing forces such as electrostatic interaction between similarly charged groups. The reversible forming and breaking of salt linkages between carboxyl and amino groups is of special importance in controlling cell volume. Whether a carboxyl group forms such a linkage or associates with a free counter ion such as K^+ depends on the associational properties. As noted above these fall under biological control.

Such a system would also undergo change in volume as a result of changes in osmolarity (or water activity) of the extracellular solution. Whether such a system behaved as an ideal osmometer depends on such interacting factors as: the value $q^{\infty \to \mathrm{ins}}$ of the solutes, the degree and stability of crosslinking, and changes in the physical state of water induced by volume shifts or by the entry of an interacting solute. The measurement of osmotic shifts of water alone does not provide a particularly critical means of determining whether the living cell behaves as membrane-enclosed solution or as a network of structured water and crosslinked proteins in equilibrium with the extracellular fluid.

Permeability to ions

Selective ionic permeability is regarded as a reflection of selective ion accumulation, but not as its underlying cause. This represents a sharp diversion from membrane models. It was noted above that the electrolyte contained within the cell is in two states, either dissolved in the interstitial water or associated with fixed sites. Likewise the entrance of ions into such a system would take place via two routes; diffusion through channels of polarized water (saltatory route), or adsorption–desorption to fixed charges (adsorption–desorption route) (Ling et al., 1967; Ling & Ochsenfeld, 1965; Ling, 1965a, 1966b). Ions which are weakly adsorbed would enter such a system primarily through the saltatory route. Selectively adsorbed ions would enter via both routes, the adsorption–desorption route being quantitatively more prominent. Changes in the selectivity of the cell surface for a given ion would be reflected in an altered permeability. Therefore references to changes in 'membrane permeability' according to the membrane models are analogous to changes in 'surface selectivity' according to the 'association–induction hypothesis'.

The kinetics applied to the entry of ions via a saltatory route are those of a bulk diffusion process occurring throughout the cytoplasm. The adsorption–desorption route would be described by a reversible reaction, assuming the inward movement from site to site not to be rate limiting. The differential equations for the simultaneous occurrence of these processes (Crank, 1956) have been applied to the description of the time course of entry of ions and water into frog sartorius and eggs (Ling, 1965a; 1966b; Ling et al., 1967). Bulk diffusion is governed by the equation

$$\frac{\partial c}{\partial t} = D \frac{\partial^2 c}{\partial x^2} - \frac{\partial S}{\partial t} \tag{7}$$

with a simultaneous reaction of the type:

$$\frac{\partial S}{\partial t} = \lambda c - \mu S \tag{8}$$

where D = diffusion coefficient, cm^2 sec^{-1}
 λ = rate constant for forward reaction, sec^{-1}
 μ = rate constant for backward reaction, sec^{-1}
 c = concentration of solute free to diffuse, mEq/kg d.s.
 S = concentration of adsorbed solute, mEq/kg d.s.
 x = space coordinate, cm.

Most of the entry of sodium was governed by the bulk diffusion process (equation 7), whereas almost all the potassium exchange followed the adsorption–desorption route (equation 8), (Ling, 1966b).

Under conditions in which there is no net change of adsorbed ion ($dS/dt = 0$) the ion exchange of the adsorbed potassium, $V_{K_{ad}}$, equals μS or μK_{ad} where K_{ad} is defined by equation 1 or 3 depending on the degree of *cooperative* interaction. For the case in which there is no *cooperative* interaction (equation 3) the formulation results in Michaelis–Menton kinetics in which:

$$\frac{1}{V_{K_{ad}}} = \frac{1}{V_{max}} (K_m) \frac{1}{[K]_{ex}} + \frac{1}{V_{max}} \tag{9}$$

where V_{max} = maximum rate, mEq/kg d.s./sec
 K_m = Michaelis constant, mEq/l

This formulation has also been put forward for membrane models in which the adsorption is onto a carrier molecule (Wilbrandt & Rosenberg, 1961). Kinetics which follow equation 9 are therefore consistent with both approaches. Ling & Ochsenfeld (1965) showed that a number of interactions can occur between two different ions entering the cell which cannot be explained by a 'pump' and 'leak' model. They could be predicted in terms of a fixed-charge system by taking into account several steps for adsorption–desorption. The equation (equation 5 in Ling & Ochsenfeld, 1965) for these interactions will not be included here because this approach has not been tested on smooth muscle. Another approach to testing the appropriateness of a 'pump' model will be reviewed, in which predictions of $V_{K_{ad}}$ based on equation 9 will be compared to those resulting from the product of the experimentally-derived rate constants and equation 1 which predicts *cooperative* interaction.

A crucial difference between the 'association–induction hypothesis' and membrane theories concerns the nature of the transitions made by a permeating ion when it crosses the *inside* of the membrane and passes into the cytoplasmic phase. The 'association–induction hypothesis' is an equilibrium model which considers the membrane and cytoplasmic phases as part of the same continuum. The interactions an ion undergoes in the membrane are undergone to much the same extent throughout the cytoplasm. In the membrane models a number of modes of ion interaction in the membrane have been suggested. However, when the ion leaves the inside of the membrane, it is considered to pass into a phase having properties similar to the extracellular solution. The ion therefore undergoes relatively few restrictions. This proposition has been supported by the observation that the translational movement of $^{42}K^+$ in squid axons was similar to that in the extracellular water (Hodgkin & Keynes, 1953). It is the restrictions on the rotational motion however which are of first importance in determining the behaviour of ions in a fixed-charge system (Chapter 4 in Ling, 1962,). This property is not readily measured by radio-isotope experiments. Recent developments in nuclear magnetic resonance (NMR) techniques have led to a direct assessment of those properties in biological systems.

Complexing of Na$^+$

The nuclear magnetic resonance spectrum of Na$^+$ has been shown to be suitable for direct analysis of Na$^+$ in tissues (Cope, 1967). Such analyses are conducted on whole tissues placed in a magnetic field and do not require impalement or otherwise damaging the tissues during measurements. The comparison of spectra obtained in living tissues, ashed specimens, and solutions of known composition led to the conclusion that approximately 70% of the skeletal muscle sodium did not contribute to the NMR spectrum (Cope, 1967). Similar findings were noted for kidney, brain, homogenates and actomyosin gel. In a recent study (Ling & Cope, 1969) NMR measurements were conducted on Na$^+$ loaded, K$^+$ depleted frog skeletal muscle. Most of the Na$^+$ taken up in exchange for K$^+$ was in a complexed state. These two studies provided direct evidence for the complexing of Na$^+$ in muscle cells.

Similar evidence for the complexing of Na$^+$ and K$^+$ in barnacle muscle has been presented by Hinke & McLaughlin (McLaughlin & Hinke, 1966, 1968; Hinke & McLaughlin, 1967) employing cation-sensitive glass electrodes. This technique has the disadvantage of disrupting the cytoplasmic structure in the environment of the glass. Also, the argument can be put forward that the low Na$^+$ activity may result from much of the fibre Na$^+$ being compartmentalized and therefore inaccessible to the electrode. The measurement of the NMR spectrum of skeletal muscle confirmed the Na$^+$ glass electrode studies under conditions in which the cells remained intact and in which all Na$^+$ was accessible to the magnetic field. NMR studies of water have also indicated structuring of this component (Bratton, Hopkins & Weinberg, 1965; Chapman & McLauchlan, 1967; Fritz & Swift, 1967; Hazlewood, Nichols & Chamberlain, 1969; Swift & Fritz, 1969). The study of Swift & Fritz (1969) indicated a decrease in the percentage of structured water (from 65% to 44%) which was associated with depolarization of the nerve. Such shifts in water structure may play an important role in the exchange of ions associated with nerve depolarization.

The future of the NMR technique holds many possibilities in advancing our understanding of the state of ions, water and protein in the living cell. The development of more powerful magnets may allow the measurements of the NMR spectrum of K$^+$. Investigation of transitions accompanying excitation will also help in understanding the mechanisms underlying ion movements and potential changes. From the material to be reviewed in the following sections, it would appear that smooth muscle would warrant investigation with the NMR technique.

HETEROGENEOUS EXCHANGE AND DISTRIBUTION

The assumption that the intracellular electrolyte is a uniform pool, comprises a basic premise for membrane models. The deviation of ion exchange and distribution from such behaviour can be termed heterogeneous.

There are a number of ways by which the uniformity of intracellular electrolyte can be tested. One is to conduct a profile analysis of the time course of isotope washout under steady-state conditions. According to theory based on a rate-limiting membrane such analysis should demonstrate that a single exponential is approached (the reader is referred to Solomon, 1960, for the description of the method of exponential analysis of such data). Studies conducted on sodium washout from smooth muscles have indicated that the process does not approach a single exponential under the experimental conditions (Durbin & Jenkinson, 1961a; Burnstock, Dewhurst & Simon, 1963; for review, Goodford, 1968). The mathematical description of such curves requires more than the two exponential functions needed for a simple model based on an extracellular and an intracellular component. As pointed out by Goodford (1968) this is probably much too simple an approach for smooth muscle. Complexities in the extracellular structure can splay the efflux kinetics. Also a profile analysis requires data with very good resolution. Because of the errors present in isotope studies, e.g. random counting errors, the resolution obtained toward the lower part of the washout curves usually does not allow easy differentiation between a single or multiple exponential processes. Quite commonly single exponentials are fitted to such curves when improved techniques might have indicated more to be present (Fig. 4A).

Another approach is to compare the distribution of electrolyte based on kinetic analysis to that based on chemical dissection techniques. The basis of the dissection approach has been established for many years (Hastings, 1940). The tissue is divided into cellular and extracellular components on the basis of chemical analysis. Since it is not yet possible to analyze smooth muscle cells directly for water and electolyte composition, the dissection technique approaches this problem by correcting total tissue contents for the quantity associated with the extracellular constituents. The principle constituent is water. Methods for measuring this component in smooth muscle have varied and have been extensively reviewed (Goodford, 1968; Burnstock, Holman & Prosser, 1963; Kao, 1967; Bohr, 1964a; Villamil, Rettori, Barajas & Kleeman, 1968). There is a wide range of values and, as noted by Goodford (1968), the volume of distribution is an inverse function of molecular size. Therefore in order to estimate the composition of the smooth muscle component with the objective of testing the kinetic uniformity of this material, markers which yield relatively high values have to be employed.

The extracellular solid also sequesters some electrolyte. This has been considered an especially important component in the vascular wall (Headings, Rondell & Bohr, 1960; Garrahan, Villamil & Zadunaisky, 1965; Friedman, Gustafson, Hamilton & Friedman, 1968*a*; Friedman, Gustafson & Friedman, 1968*b*). Analysis of adventitia (mostly collagen) from dog carotid has not given evidence of significant adsorption to this component (Jones & Karreman, 1969*a*). The degree of sodium association to acid mucopolysaccharides can be estimated from the chemical analysis of arterial mucopolysaccharide sulphate (28 m-Mole SO_4/kg d.s.) (Kirk & Dyrbye, 1956) and sodium adsorption onto

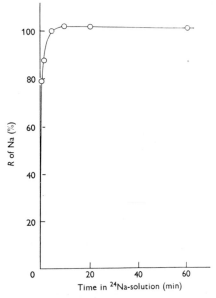

FIG. 2   Uptake of ^{24}Na by 6 pieces of taenia coli from one guinea-pig. (Goodford & Hermansen, 1961)

mucopolysaccharides (0·85 Na/SO_4, Farber & Schubert, 1957). Multiplying yields 24 mEq Na/kg d.s. in dog artery. Comparison of the collagen content of myometrium (Kao & Nishiyama, 1964) with that of dog carotid (Fischer and Llaurado, 1966) indicates that the acid mucopolysaccharide component of myometrium would be less than 20% that of artery, or 5 mEq Na/kg d.s., or 1 mEq Na/kg wet wt.

One of the first studies in which the sodium distribution based on chemical dissection techniques was compared with that based on kinetic analysis was conducted by Goodford & Hermansen (1961) on the guinea-pig taenia coli. The data for ^{24}Na entry under steady state condition appear in Fig. 2. The uptake was essentially complete in 5 min. These authors also noted that a 5 min washout removed 97% of the total activity. Given the uncertainties of this experiment and those noted above for kinetic analysis, approximately 5% of the total activity could be relegated to a slow component. On the basis of the extracellular measurement (inulin), 50% of the sodium should have exchanged with kinetics other than an extracellular diffusive process. Employing the sorbitol space (estimated to be $1 \cdot 15 \times$ inulin space from Goodford & Leach, 1966) decreases the estimate to 43%. Later studies confirmed the observation that smooth muscle sodium is not simply distributed between cellular and extracellular components (Freeman–Narrod & Goodford, 1962; Goodford, 1962). In a recent study in which simultaneous measurements were made of extracellular space (^{60}CoEDTA) and ^{24}Na washout in taenia coli and myometrium, it was concluded that the kinetics for 98 to 99% of the exchanging sodium were similar to those for ^{60}CoEDTA, yet 12 to 15 mEq Na/kg wet wt was in excess of that contained in the extracellular space (Brading & Jones, 1969).

The measurement of chloride exchange and extracellular distribution in the taenia coli indicated that some of the fast component exceeded the inulin space (Goodford, 1964) and the estimated sorbitol distribution.

The measurement of a fast-exchanging component of cell potassium in the taenia coli is made difficult by the large amount of slowly-exchanging material. Comparison of ^{42}K entry under conditions in which the sorbitol distribution was measured indicated an excess of rapidly-exchanging potassium (Goodford, 1966; Goodford, 1967). Because of interaction with other ions, e.g. calcium, this fraction was thought to be associated with negative sites on the membrane contributing 3–4 mEq/kg fr wt (Goodford, 1967; Goodford, Chapter 3). This concentration is insufficient to account for the quantity of excess fast-exchanging sodium present in normal Krebs solution (Brading & Jones, 1969).

Estimations of electrolyte exchange and distribution in the extracellular constituents have been carried out on the dog carotid artery (Jones & Karremen, 1969a). The difference between the fast-exchanging electrolyte and that calculated to be in the sucrose space or adsorbed onto mucopolysaccharides was 30 mEq/kg d.s. or $7 \cdot 5$ mEq/kg wet wt. There was an additional 20 mEq/kg d.s. characterized by relatively slow exchange. Both fast and slow components were observed for chloride as well. This has also been noted by Villamil, Rettori, Yeyati & Kleeman (1968) who studied chloride exchange and distribution in the same preparation. The quantity of excess fast-exchanging potassium was relatively small compared with the total cell pool and was less precisely characterized (Jones & Karreman, 1969a, b).

From the evidence available it appears that the electrolyte contained in smooth muscle cells exchanges not as a uniform pool, but with both fast and slow components. The former appear to fuse with the exchange of extracellular electrolyte, while the slowly-exchanging components often follow more than one exponential function. Sodium has been studied to the greatest extent, but similar behaviour has been noted for chloride. Under the conditions studied (releatively low [K]$_{ex}$) the presence of an intracellular fast-exchanging potassium fraction was not readily identified.

INTERSTITIAL COMPONENT

As pointed out above the 'association–induction hypothesis' provides a basis for describing the heterogeneous exchange and distribution of cell electrolyte. One component is dissolved in the interstitial water. It was noted that the mean equilibrium distribution coefficient between interstitial and extracellular

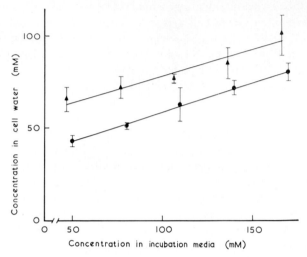

FIG. 3   Calculated concentration of Na (●) and Cl (▲) in cell water of rabbit myometrium versus the concentration of Na and Cl in the incubation media at 37°C. Each point is the average of 6 pieces ± S.E. (Jones, in press)

water, $q^{\infty \rightarrow \text{ins}}$, would be less than 1. This proposition has been tested in rabbit myometrium (oestrogen dominated) (Jones, in press). Tissues were equilibrated in Krebs solution of varying sodium chloride concentration and adjusted with sucrose to keep constant tissue water contents. Electrolyte and water were partitioned on the basis of the ^{60}CoEDTA distribution measured in each tissue. The calculated concentrations of $[\text{Na}]_i$ and $[\text{Cl}]_i$, appear in Fig. 3. A line was fitted to the data on the basis of a least square analysis. From equation 6 the slope of the line,

$$\frac{[\text{Na}]_i}{[\text{Na}]_{\text{ex}}} \quad \text{or} \quad \frac{[\text{Cl}]_i}{[\text{Cl}]_{\text{ex}}} = q_{\text{Na}}^{\infty \rightarrow \text{ins}} \quad \text{or} \quad q_{\text{Cl}}^{\infty \rightarrow \text{ins}},$$

which for the two ions was 0·30. An estimate of $q_{\text{Na}}^{\infty \rightarrow \text{ins}}$ and $q_{\text{Cl}}^{\infty \rightarrow \text{ins}}$ in vascular smooth muscle yielded 0·24 (Jones & Karreman, 1969a). These values are slightly greater than those estimated for frog sartorius (Ling, 1965b, 1966a). The observations provide a basis for considering cytoplasmic ion-exclusion as a mechanism underlying the maintenance of relatively low intracellular concentrations in smooth muscle.

It may also be noted that some cell sodium and chloride in Fig. 3 did not change in a proportional manner with extracellular concentration. Extrapolation to zero yields an estimate of $[Na]_i = 28$ mM (8 mEq/kg wet wt or 50 mEq/kg d.s.) and $[Cl]_i = 50$ mM (13 mEq/kg wet wt or 90 mEq/kg d.s.). This fraction

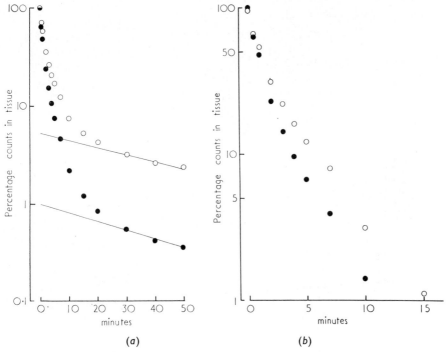

FIG. 4   Washout of ^{24}Na (●) and ^{60}CoEDTA (○) from oestrogen-dominated rabbit myometrium versus time at 37°C.   a. Total counts. The curves approach a single exponential indicated by the line.   b. Counts in the tissue after subtraction of the slow component. Ordinate: Counts expressed as percentage of the initial tissue content, log scale. Abscissa: Time in minutes after starting washout. (Brading & Jones, 1969)

has characteristics of an associated component (p. 140), and may explain the apparent reversal of the sodium 'equilibrium potential' at low $[Na]_{ex}$ observed by Brading & Tomita (1968b). The presence of the two components shown in Fig. 3 emphasizes the point made on p. 133 concerning heterogeneous distribution.

Since the interstitial electrolyte makes up a dissolved phase, the exchange with extracellular components is proposed to be via a diffusional or saltatory route (p. 130). The findings of Brading & Jones (1969) support this concept for the taenia coli and myometrium in Krebs solution at 37°C. The results of a study in which the washout of an extracellular marker, ^{60}CoEDTA, and ^{24}Na were measured appear in Fig. 4. The washout of total activity (Fig. 4a) was corrected for the slowly exchanging fraction and replotted (Fig. 4*b*). Both isotopes exhibited similar kinetics and the fast exchange did not follow a single exponential function. The data of Fig. 4*b* and for a similar run on a strip of taenia coli are plotted in Fig. 5 with the curve representing a bulk diffusion process (Crank, 1956). In these experiments 12 to 15 mEq Na/kg wet wt exchanged with bulk diffusion kinetics, yet this amount was located outside of the [60]CoEDTA

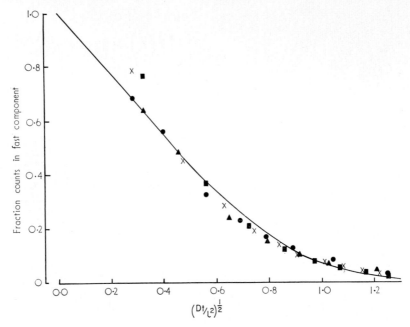

FIG. 5   Kinetic analysis of fast exchanging [24]Na and [60]CoEDTA. Ordinate: Fraction of counts in fast exchanging component. Abscissa: Non-dimensional time parameter (Crank, 1956). The curve represents the theoretical relation based on bulk diffusion kinetics through a plane sheet. ×, [60]CoEDTA, taenia coli; ●, [60]CoEDTA, myometrium; ■, [24]Na, taenia coli; ▲, [24]Na, myometrium. (Brading & Jones, 1969)

distribution. At $[Na]_{ex} = 137$ mM in Fig. 3, the dissolved component was equivalent to 12 mEq/kg wet wt which represented reasonable agreement between the two studies.

The possibility must be considered that a membrane transport system may be present and 'pumping' sodium at a very rapid rate under metabolically-supported conditions at 37°C in Krebs solution. Thus the cellular transport and extracellular exchange might be fused. Further experiments were designed to stop or slow down such a membrane transport process. The acute application of metabolic poisons at 37°C; or K-free Krebs solution, or solutions of high and low $[Na]_{ex}$ had no apparent effect on [24]Na washout at 37°C (Jones, in press). Another approach was to place tissues, previously equilibrated in normal

Krebs solution at 37°C (with ^{24}Na and ^{60}CoEDTA), into non-radioactive solutions at temperatures ranging from 0·5 to 42°C (Jones, in press). Thermal equilibrium occurs within seconds and the slowing of a temperature dependent process should be readily demonstrated. (See Buck & Goodford, 1966, for time course of response to acute temperature changes under different conditions than employed here). The slope of the slow component (see Fig. 4a) exhibited a relatively large temperature dependence. A range of 1 to 6 mEq Na/kg wet wt was associated with this component. The exchange of 12 to 18 mEq Na/kg wet wt, contained outside of the ^{60}CoEDTA space, exhibited little temperature

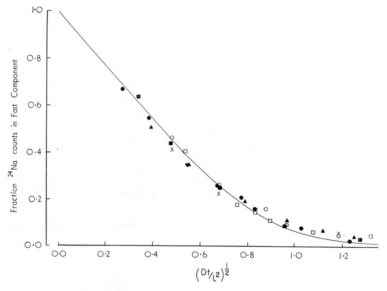

Fig. 6   Kinetic analysis of fast exchanging ^{24}Na in rabbit myometrium at various washout temperatures. Data and curve plotted as in Fig. 5. ●, 0·5°C; ■, 10°C; ▲, 20°C; ×, 30°C; ○, 37°C; □, 42°C. (Jones, in press)

dependence. The time course of the entire fast-exchanging sodium pool is plotted in Fig. 6 with a bulk diffusion curve. Since the runs determined at various temperatures fall along the curve in a uniform manner, it may be concluded that changes in temperature do not effect fast intracellular exchange differently from that of the extracellular space. An Arrhenius plot of temperature and D/a^2 ($a \simeq 0·03$ cm) for ^{24}Na and for ^{60}CoEDTA exchange indicated a low activation energy and yielded a value for the Q_{10} of 1·3 and 1·4 respectively. Similar temperature effects have been noted for the excess fast-exchanging sodium in dog carotid (Jones & Karreman, 1969a; Garrahan et al., 1965). The fusion of the fast-exchanging intracellular component with the extracellular fraction does not appear to be subject to a metabolically-dependent process, e.g. a membrane 'pump', but rather to a diffusive or saltatory route.

The evidence available supports the concept that an interstitial fraction of electrolyte is present in smooth muscle. The size of this component under normal *in vitro* conditions is about 12 to 15 mEq Na/kg wet wt for taenia coli and myometrium. In dog carotid which contains more connective tissue the value is about half. There is also a slowly exchanging fraction (1 to 6 mEq/kg wet wt) which exhibits properties of an associated component. The characteristics of this constituent will be the subject of the next section.

ASSOCIATED COMPONENT

According to the 'association-induction hypothesis' the second component which contributes to the heterogeneous behaviour of smooth muscle electrolyte results from the interaction of ions and protein. As pointed out on p. 125 such interaction can provide a basis for interpreting selective accumulation of ions with the *cooperative* adsorption isotherm describing the analytical relations.

Smooth muscle preparations exhibit some advantages in testing such a formulation. The potassium contained in fresh specimens can be reversibly removed by a period of overnight cold storage in K-free solution (Jones & Karreman, 1969*a*; Jones, 1968). Subsequent incubation in solutions of varying $[K]_{ex}$ lead to a stable accumulation, thus allowing investigation of the equilibrium relations between $[K]_{ex}$ and cell K and ^{42}K exchange. Similar experiments performed on frog sartorius would require tissue culture techniques and prolonged incubation periods (Ling & Ochsenfeld, 1966).

Cooperative adsorption

The relation between $[K]_{ex}$ and the slowly-exchanging potassium has been investigated in canine carotid arteries (Jones & Karreman, 1969*b*). Tissue potassium corrected for the quantity exchanging in three minutes is plotted in Fig. 7. The accumulation of potassium-exhibited properties consistent with those predicted on the basis of the *cooperative* adsorption isotherm. Saturation behaviour is noted, in that an upper level is approached, $F_T^- = 119$ mEq/kg d.s. or about 30 mEq/kg wet wt. The relatively large increase in uptake between $[K]_{ex} = 1$ to 2 mM indicates interaction. Curves were calculated on the basis of the *cooperative* adsorption isotherm (equation 1) for no interaction ($n = 1$) and interaction ($n = 2\cdot7$). There was good agreement between experimental and theoretical findings which took *cooperative* interaction into account. Since the value of n is positive, the interaction was an *auto-cooperative* process. That is, the adsorption of K^+ at one site made the adsorption of a succeeding K^+ at a nearest neighbour energetically more favourable. From equation 2 the free energy of nearest neighbour interaction is $-\gamma/2 = 0\cdot61$ kcal/mole; and from equation 5, $K_{K,Na} = 93$ (for half saturation at $[K]_{ex} = 1\cdot56$) representing $\Delta F_{K,Na}^\circ = -2\cdot8$ kcal/mole ($K_{K,Na} = \exp -\Delta F_{K,Na}^\circ/RT$). Because of *auto-cooperative* interaction the free energy for the first K^+—Na^+ exchange

$([K]_{ex} \to 0) = -2 \cdot 2$ kcal/mole $(-2 \cdot 8 + 0 \cdot 61)$, and for the last $([K]_{ex} \to \infty)$ $= -3 \cdot 4$ kcal/mole $(-2 \cdot 8 - 0 \cdot 61)$. Such self reinforcement can go a long way accounting for threshold and all or none behaviour.

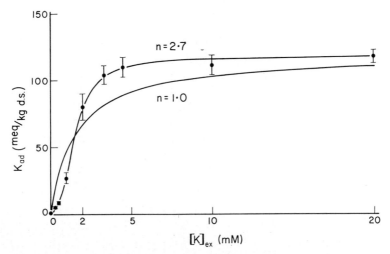

FIG. 7   Adsorbed potassium contents of carotid arteries in equilibrium with various external concentrations at 37°C. Each point represents the average of 13 dogs $\pm$ S.E. Theoretical curves were calculated on the basis of equation (1) for $n = 2 \cdot 7$ and $n = 1$.
(Jones & Karreman, 1969b)

The applicability of the *cooperative* adsorption isotherm was further tested by estimating the slowly-exchanging sodium at various $[K]_{ex}$ (Jones & Karreman, 1969b). In general the sodium levels followed the relation, $Na_{ad} = F_{T}^{-}$ $- K_{ad}$ where K_{ad} was calculated on the basis of equation 1. The relatively large correction employed for fast-exchanging sodium limits the resolution of this approach.

Good agreement has also been noted between potassium accumulation in rabbit myometrium and the theoretical values computed on the basis of the *cooperative* adsorption isotherm (see Fig. 10). The value found for F_{T}^{-} was 366 mEq/kg d.s. (55mEq/kg wet wt) and as a result of hormone treatment (p. 145) $K_{K,Na}$ was 142 to 170 and n was 2·2 to 2·0. The application of the *cooperative* adsorption isotherm to ion accumulation in other smooth muscles awaits future development. One aspect of the approach, saturation behaviour, is apparent in the studies of potassium contents of rat myometrium and taenia coli (Casteels & Kuriyama, 1965, 1966). Cell potassium values (calculated on the basis of the ethanesulphonate distribution) increased by less than 15% over a 20 fold range of $[K]_{ex}$. Unfortunately, steady-state relations were not determined for low $[K]_{ex}$ (<2 mM) which would have allowed the computation of $K_{K,Na}$ and n for the accumulation processes.

6+

From the information available, selective accumulation of potassium by smooth muscle exhibits properties which are consistent with those of an ion protein interaction, and follows a *cooperative* adsorption isotherm derived for such a system.

Fixed charge content

The amino acid composition of the tissue is an important consideration in applying a fixed-charge model. It is essential to establish that sufficient fixed charges are available to accumulate ions. Measurements of amino acid composition have been conducted on rabbit myometrium (Evans Electrosolenium Amino Acid Analyser, Department of Pharmacology, Oxford). The data for the myometrium from oestrogen treated and from oestrogen plus progesterone treated rabbits are summarized in Fig. 8. Hormonal treatment had little

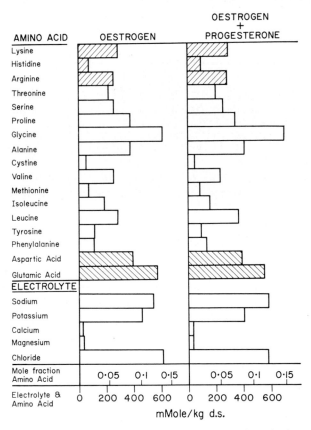

FIG. 8   Amino acid and electrolyte analysis of rabbit myometrium. Amino acid contents are represented in terms of mole fraction and m-mole/kg d.s. Electrolyte is expressed in terms of m-mole/kg d.s. only. Amino acids having a net charge are indicated by cross hatching.

effect on the amino acid composition or recovery (50% of dry solids analysed). There was an average of 951 mEq/kg d.s. of β and γ carboxyl groups (aspartic and glutamic acids) in the two muscles. This is well beyond the average F_T^- determined for this tissue (above) or, more specifically, for the intracellular K determined on the two muscles analysed. There was an average of 653 mEq/kg d.s. of positive charge available from lysine, argenine, and histidine; sufficient to accommodate the 90 mEqCl/kg d.s. estimated to be associated (p. 137). Although other inorganic anions were not evaluated (e.g., PO_4^{3-}, SO_4^{2-}, HCO_3^-) their contribution would not be expected to go beyond the fixed positive charge available. The amino acid analysis also shows an excess of fixed negative charge amounting to 300 mEq/kg d.s., indicating that the net charge of the protein matrix is negative.

It appears therefore that there are sufficient charges in smooth muscle to account for the quantity of electrolytes proposed to be selectively accumulated. This extends the observations noted for other preparations (Chapter 3 in Ling, 1962; Ling & Ochsenfeld, 1966), and provides a firmer basis for establishing this as a general property of tissues.

Competitive inhibition

Competitive inhibition between similarly-associated ions represents another property of selective accumulation in a fixed-charge system. This feature was examined in vascular smooth muscle by measuring the accumulation of potassium in the presence of a similarly associated ion, Rb^+ (Jones & Karreman, 1969b). The rubidium and potassium contents of dog carotid (corrected for fast exchanging components) are plotted in Fig. 9. With increasing $[K]_{ex}$ there is an uptake of K_{ad} and a loss of Rb_{ad}. The sodium levels were low and relatively independent of the value of $[K]_{ex}$ employed. Theoretical values were computed on the basis of equation 3 (no *cooperative* interaction) with $F_T^- = 126$ mEq/kg d.s. and $K_{K,Rb} = 0.68$, and exhibited good agreement with the experimental results.

It can be deduced from the values employed for $K_{K,Rb}$, and $K_{K,Na}$ that the order of selectivity exhibited by vascular smooth muscle for these three ions is $Rb^+ > K^+ \gg Na^+$. A similar order was observed in skeletal muscle (Ling & Ochsenfeld, 1966). Under conditions in which K^+ selection is much greater than the alternate ion, e.g., Na^+, *cooperative* interaction was apparent. When K^+ selectivity was similar to the alternate ion, e.g. Rb^+, little *co-operative* interaction was apparent indicating the inductive effects of these two ions to be similar (Chapter 5 in Ling, 1962).

The competitive inhibition between Rb^+ and K^+ is consistent with the ion accumulation behaviour predicted on the basis of the 'association–induction hypothesis'. The studies need to be extended to more ions, e.g. Cs^+, more concentration ranges, and other types of smooth muscle, to establish the generality of the observations made on vascular smooth muscle.

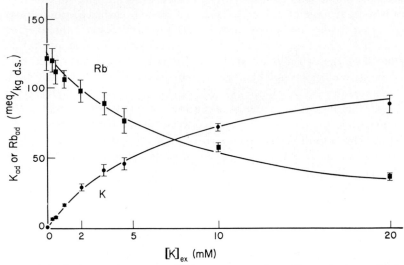

FIG. 9   Adsorbed potassium and rubidium contents of carotid arteries in equilibrium with various $[K]_{ex}$, and $[Rb]_{ex} = 5$ mM at 37°C. Average values for 4 dogs $\pm$ S.E. Theoretical curves calculated from equation (3) with Na term replaced by Rb. (Jones & Karreman, 1969*b*).

Control of selective accumulation

According to the 'association–induction hypothesis' the relative selectivity exhibited towards ions provides information about the physico-chemical properties of the proteinaceous cytoplasmic network. This information can provide a basis for interpreting other properties of the network such as bio-electric potentials and contractile responses. Also, shifts in the physico-chemical properties of the protein matrix can be reflected in the relative selectivity toward ions exhibited by the system. Such shifts could also underlie changes in other behaviour, e.g. excitation, contraction. The control of shifts in selectivity by bioregulants may take place by a number of routes, two of which will be discussed with respect to smooth muscle.

1 *Metabolic control*

There is ample evidence that metabolic inhibition, at 35 to 37°C, of various types of smooth muscle results in a loss of K^+ and a gain of Na^+ (Daniel, Sehdev & Robinson, 1962; Daniel & Robinson, 1960; Jones and Karreman, 1969*a*; Friedman *et al.*, 1968). Under these conditions $K_{K,Na}$ approaches one. However, to get a maximum effect (lowest tissue potassium levels) it is necessary to block both aerobic and anaerobic pathways. It would therefore appear reasonable to assume that a common product of aerobic and glycolytic metabolism is involved, with ATP of primary importance.

The effect on ATP and CP levels of aerobic and anaerobic incubation in the presence and absence of substrate has been studied in taenia coli (Bueding and Bülbring, 1964; Bueding, Bülbring, Gercken, Hawkins & Kuriyama, 1967). The lowest levels of ATP and CP were measured after substrate depletion under anaerobic conditions. The presence of glucose under anaerobic conditions led to a 50% reduction of the control level, as did the absence of substrate under aerobic conditions. In these experiments, the levels of CP were most affected by alterations of metabolic support, but parallel changes in ATP also occurred.

On the basis of the 'association–induction hypothesis' there should be a quantitative relationship between ATP and K contents of the cells, with more than 30 moles K being accumulated for each ATP (Chapter 9 in Ling, 1962). Qualitatively, this relationship appears to hold in smooth muscle. There is need, however, to measure ATP levels and smooth muscle potassium contents under conditions of varying metabolic activity to test the presence of a quantitative relation.

2 Hormonal control

Other examples of biological control are those mediated via regulating substances which impinge on the cell from external sources such as nerve terminals, endocrine glands, or syringe. Because of the rather long incubation periods required to derive the data for an adsorption isotherm, substances having a rather long action are more readily studied. One example is the effect of oestrogen and oestrogen plus progesterone treatment on rabbit myometrium (Jones, 1968). The accumulation of potassium was measured at various $[K]_{ex}$ in myometrium from rabbits treated with oestrogen alone (25 μg per day for 10 to 12 days) or with oestrogen and additional progesterone (4 mg per day for the last 4 days). As noted on p. 141 hormonal treatment did not effect F_T^- or amino acid composition (p. 142. From the data presented in Fig. 10 it is apparent that the effect of additional progesterone was to shift the *cooperative* adsorption curve to the left by increasing the intrinsic equilibrium constant, $K_{K,Na}$, from 142 to 170. The muscle therefore become more selective toward K^+ over Na^+. The differences in uptake at the two lowest $[K]_{ex}$ were significant at the 1% level. There was a slight decrease in n under progesterone treatment, but more studies are required to establish the significance of this observation.

The shift in selectivity of the muscle towards potassium would represent a change in the physico-chemical properties of the tissue protein. Such shifts would be related to changes in other behaviour. For instance the observation of Casteels & Kuriyama (1965) that the resting potentials recorded in the myometrium from the pregnant rat followed changes in $[K]_{ex}$ more closely than that from non pregnant rat is consistent with an increase in K^+ selectivity. According to the 'association–induction hypothesis', the resting potential results from a phase boundary potential existing at the junction of the fixed-charge system (surface of membrane) and the aqueous solution (Chapter 10 in Ling, 1962; Ling, 1966c; Karreman, 1964). The magnitude of the potential depends on the concentration of surface-fixed anionic sites and their selectivity towards counter ions. A shift towards K^+ from Na^+ would result in a more negative potential under conditions where $[K]_{ex}$ is $< [Na]_{ex}$ (normal Krebs solution) and the potential would follow more closely changes in $[K]_{ex}$. Both these observations have been made in pregnant rat uterus (Casteels & Kuriyama, 1965).

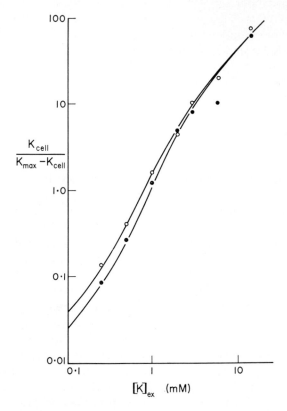

FIG. 10   Log–log plot of cell potassium ratio versus external concentration. ●, oestrogen treated myometrium; ○, oestrogen plus progesterone treated myometrium. Each point represents the average of 13 rabbits. Theoretical curves were calculated on the basis of equation (1).

The poor correlation between hormonal state and electrolyte accumulation reported by Casteels & Kuriyama (1965) and others (Kao, 1961; Kao & Siegman, 1963; Bitman, Cecil, Hawk & Sykes, 1959; review, Kao, 1967) probably resulted from conducting measurements at normal or high $[K]_{ex}$. Because of the saturation behaviour exhibited by myometrium, measurements at low $[K]_{ex}$ are required to demonstrate a shift in the selective accumulation properties.

It was pointed out on p. 130 that selective ion permeability was a reflection of selective accumulation. It may be noted that the increase in K$^+$ selectivity under the influence of progesterone was also related to an increase in K$^+$ permeability as measured by steady-state exchange of ^{42}K and by the net loss of accumulated potassium into a K-free solution (Jones, 1968).

Selective permeability

The accumulation of slowly-exchanging potassium was noted to exhibit properties which are consistent with an adsorption process (p. 140). The entrance of a selectively accumulated ion into the cell was proposed to occur via two routes, saltatory and adsorption–desorption, the latter being predominant (p. 130). Two general methods are available for investigating this proposition.

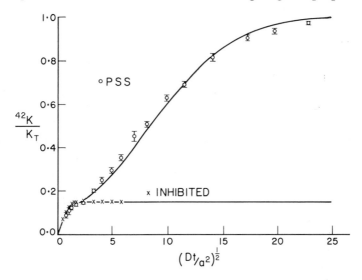

FIG. 11. Steady-state exchange of potassium in carotid arteries with ^{42}K under metabolic-ally-supported conditions (PSS) and inhibited conditions at 37°C. Ordinate: fraction of total potassium which exchanged with isotope under supported conditions. Abscissa: Non-dimensional time parameter (Crank, 1956). ○, supported conditions (9 dogs ± S.E.); ×, inhibited conditions (2 dogs). Theoretical curves were calculated on the basis of equations (7) and (8). (Jones & Karreman, 1969a)

One method of analysis is to compare the time course of isotope exchange with that predicted on the basis of a kinetic analysis which explicitly takes into account the diffusive exchange and the simultaneous adsorption–desorption processes (equation 7 and 8). This analysis has been carried out for the steady state entry of ^{42}K into dog carotid (Jones & Karreman, 1969a). The influx profiles for metabolically supported (PSS) and metabolically inhibited conditions at 37°C appear in Fig. 11. There is an initial rapid uptake following bulk diffusion kenetics. Under inhibited conditions only this component was apparent (the uptake data were normalized on the basis of total K$^+$ present under supported conditions). Under metabolically-supported conditions there was a plateau corresponding to the time for the diffusion limited exchange to reach equilibrium (3 to 5 min). This was followed by a slower uptake of ^{42}K which approximated adsorption–desorption kinetics. The curve in Fig. 11 was calculated on the basis of equation 7 and 8 employing the following values: $D = 1 \cdot 1 \times 10^{-5}$ cm^2 sec^{-1}; $a = 0 \cdot 035$ cm; $\mu = 7 \cdot 5 \times 10^{-5}$ sec^{-1}; $\lambda = 52 \cdot 5 \times 10^{-5}$ sec^{-1}. Good agreement was found between the time course of ^{42}K entry and the theoretical values. Good agreement was also noted for ^{22}Na and ^{36}Cl entry under similar conditions (Jones & Karreman, 1969a). In potassium-depleted tissues (K-free Krebs; 37°C) the additional sodium taken up exhibited adsorption–desorption kinetics (Jones & Karreman, 1969a). This fraction, in contrast to the interstitial component, was also sensitive to temperature changes and metabolic inhibition at 37°C (Jones, in press).

A second method of analysis is to measure the steady state exchange as a function of external ion concentrations. Under these conditions the rate of exchange, $V_{K_{ad}}$ equals μK_{ad}. One formulation leads to equation 9 (no *cooperative* interaction). The predictions of this approach can be compared to an approach which includes *cooperative* interaction and concentration effects on μ. The steady state entry of ^{42}K has been studied in dog carotid under similar conditions employed for the evaluation of the *cooperative* adsorption isotherm (p. 140) (Jones & Karreman, 1969b). The rate constant describing the process was dependent on $[K]_{ex}$ following the relation:

$$\mu = 0 \cdot 88 \times 10^{-4} \text{ sec}^{-1} + 3 \cdot 8 \times 10^{-4} \text{ sec}^{-1} \exp [K]_{ex} \tag{10}$$

Values for experimental $V_{K_{ad}}$ at various $[K]_{ex}$ are plotted in Fig. 12. Three equations were applied to the results, line A was calculated from equation 9 ($V_{max} = 10 \cdot 5 \times 10^{-3}$ mEq/kg d.s./sec and $K_m = 0 \cdot 47$ mEq/l) Curve B was based on the relation:

$$\frac{1}{V_{K_{ad}}} = \frac{1}{\mu K_{ad}} \tag{11}$$

where: K_{ad} was calculated from equation 1 employing the parameters noted on p. 140 for dog carotid and $\mu = 0 \cdot 88 \times 10^{-4}$ sec^{-1}. Curve C was calculated from the product of equation 1 and equation 10. Therefore line A took into account saturation behaviour; curve B, saturation behaviour, *cooperative* interaction; and curve C, saturation behaviour, *cooperative* interaction and the effect of $[K]_{ex}$ on μ. At high $[K]_{ex} > 3 \cdot 33$ mM all 3 relations exhibited good agreement with experimental results. Beyond this range it was necessary to take into account *cooperative* interaction between K^+ and Na^+ as well as concentration effects on the rate constants.

These permeability data alone do not rule out a surface limited exchange mechanism in smooth muscle (versus exchange limited at individual sites distributed throughout the cytoplasm). They do indicate that a membrane

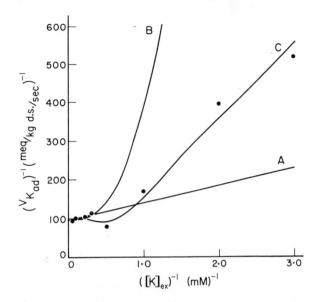

FIG. 12   Reciprocal plot of potassium exchange in carotid arteries and external concentration. Each point represents the average of 4 dogs. Theoretical curve A computed from equation (9), B from equation (11), and C from the product of equations (1) and (10). (Jones & Karreman, 1969*b*)

transport model employing Michaelis–Menten kinetics would not be generally applicable over the concentration range studied.

CONCLUDING REMARKS

This review has dealt primarily with only one approach to ion accumulation and permeability behaviour of smooth muscle. The objective was not to exclude other possibilities, but to indicate that much of the known behaviour is consistent with that of an equilibrium model based on interactions of ions-water-proteins distributed throughout the cell. For instance, it was noted that ion-accumulation processes exhibited saturation behaviour, *cooperative* interaction, and competitive inhibition; all properties of a fixed-charge system. Several contrasts were also drawn between exhibited behaviour and that predicted on the basis of a membrane-limited model. The exchange of cell sodium, for example, did not occur in a uniform manner.

It is not within the scope of this review or the state of knowledge in this field to draw firm conclusions as to which approach will ultimately be the most productive in furthering understanding of smooth muscle behaviour. Much work has been designed on the basis of membrane models and many current concepts are founded on this approach. A model which considers both surface

6*

and bulk properties, however, provides a basis for integrating a wide range of smooth muscle activity. It can be concluded that the problem of sorting out the relevant approaches still remains a challenge to present and future investigators of smooth muscle physiology.

ACKNOWLEDGEMENT — The author would like to thank Dr. G. Karreman for advice during the preparation of this manuscript.

5

CALCIUM FLUXES AND CALCIUM DISTRIBUTION IN SMOOTH MUSCLE

H LÜLLMANN

In a book which is concerned with all aspects of physiology and pharmacology of many kinds of smooth muscle, a review of the role of cellular calcium metabolism in the function of mammalian smooth muscle should play an important part. I started this work with enthusiasm, but soon realized that there are not many papers in which modern methods have been used to investigate this problem, and the results reported are often rather conflicting. The same holds true for pharmacological studies. Our own investigations have not yielded as much information on the role of Ca in smooth muscle function as we had hoped. This report will, therefore, be restricted to an account of the actual experimental observations and some tentative conclusions. Their interpretation must, however, be left open for discussion and speculation by the reader. An example of such a stimulating speculation has been published by Daniel (1965a).

METHODS FOR DETERMINING Ca IN SMALL MUSCLE PIECES

The difficulties involved in the quantitative determination of Ca in smooth muscle are caused by the small size of the tissue samples and the presence of other ions, which may interfere with the Ca determination. Two main principles have been used successfully for estimating the tissue Ca content:

1. Flame photometry, e.g. atomic absorption flame photometry,
2. Fluorescence photometry.

It is necessary, first, to mention a few points concerning the preparation of the tissue. The extraction of Ca from the muscle cannot be adequately performed by soaking the tissue in distilled water, since part of the Ca is firmly bound to muscular protein and is not extractable under such conditions. Homogenization of the muscle pieces in alcoholic trichloracetic acid yields better results. According to our experience the following procedure proved to be optimal: 0·15 ml of a mixture of 65% HNO_3 and 60% $HClO_4$ were added to 20 mg wet wt of muscular tissue in a tube of special glass (Schott G 20 or Duran, width 13 mm, length 130 mm). The tubes were placed in an aluminium block with holes of 125 mm depth to cover the entire glass tubes. The block was slowly heated up to 140°C for 2 hr and then to 210°C until the residue became dry. The latter temperature proved particularly critical, since above 215°C polyphosphates are formed from the muscular phosphates. The polyphosphates possess a high affinity towards Ca and interfere with other complexing agents as well as with the emission during the flame photometry. The dry residue can easily be dissolved in HCl.

An important factor in flame photometrical procedures is the flame temperature which should be maintained as hot as possible. This reduces the formation of Ca complexes in the flame which would otherwise depress the sensitivity of both emission and absorption spectrophotometry. Another problem in emission work is the high background, resulting from the sodium and potassium contained in the muscle sample. Compensation for this effect cannot easily be achieved by adding appropriate amounts of sodium and potassium to the standard solution.

The following procedure for Ca determination with a fluorescence method has proved to be the most reliable in our experience during many years: The complexing agent calcein (fluorescein-2·7-bis-methyleniminodiacetic acid) forms in a strongly alkaline solution (pH $\simeq$ 12·5) a calcium complex, which can be maximally excited by a wave length of about 480 mμ and displays a maximal emission at a wave length of about 540 mμ. Other ions such as Mg^{2+}, Zn^{2+} and PO_4^{3-} present in muscular tissues do not interfere with the Ca determination even in concentration 5 to 10 times higher than the Ca concentration. The smallest amounts of Ca that can be determined correctly, lie in the range of 5 to 7·5 n-mole Ca. Depending on the Ca concentration in the tissue, 5 to 10 mg wet wt of muscular tissue is sufficient for a determination. The details of the technique, briefly described above, has been published by Zepf (1966). The method can be modified for the determination of Mg as well. At a pH of 7, the affinity of Mg towards the complexing agent calcein becomes much stronger than that of Ca in the same sample.

Ca CONTENT OF SMOOTH MUSCLE

The tissue Ca concentration shows a remarkable variation although the analytical procedures are sufficiently accurate to yield results with a low scatter.

The variations occur between different animals, but they are still more pronounced between different animal batches and different seasons. Although this variation of the control values is troublesome for every scientist working in this field, the causes for the observations have not yet been investigated systematically. It seems that the Ca content of muscular tissue need not be kept as constant as the K- or Na-concentration in tissues and blood. In muscular tissue, a fundamental difference exists between Ca on the one hand and K and Na on the other hand: while a large proportion of Ca is bound and only a small fraction is ionized, K and Na are almost completely dissociated. Small changes in the K- or Na-concentrations usually influence the function of organs to a large extent, whereas changes in the bound Ca fraction (Ca store) do not noticeably interfere with it. As long as a certain Ca store is available to maintain a low but decisive Ca-ion concentration in muscular tissue, the size of the Ca store is more or less unimportant and it may vary greatly with environmental conditions. The situation of the tissue Ca may in some way be comparable to that of the catecholamines: the amount of stored noradrenaline is subject to a very large variation in a particular tissue of different individuals, but the normal function depends on a small available fraction.

Apart from this individual variation of the Ca content of smooth muscle samples, the tissue content depends largely on the Ca concentration of the medium used for the dissection and incubation of the isolated organs. For this reason it is difficult to compare the values for tissue Ca reported in the literature. To give some information the following figures may be mentioned which are listed in Table 1.

Bauer, Goodford & Hüter (1965) investigated the influence of dissection and incubation upon the tissue Ca of taenia coli. The Ca content of muscle pieces

TABLE 1. Tissue Ca concentrations for different smooth muscles and conditions

Organ	Ca concentration of the medium (m-mole/l)	Tissue Ca concentration (m-mole/kg wet wt)	Authors
Taenia coli of guinea-pig	0·85	0·65–1·2	von Hattingberg et al., 1966
Taenia coli of guinea-pig	1·8	2·5	Urakawa et al., 1964
Taenia coli of guinea-pig	1·8	2·7	Herrlinger et al., 1967
Taenia coli of guinea-pig	2·5	4·2	Schatzmann, 1961
Taenia coli of guinea-pig	2·5	2·8	Bauer et al., 1965
Longitudinal muscle of guinea-pig's small intestine	1·8	2·6	Lüllmann et al., 1968
Circular muscle of cat ileum	1·8	1·4	Sperelakis, 1962
Longitudinal muscle of cat small intestine	2·0	1·5	Potter et al., 1968
Rat uterus	1·5	2·2	van Breeman et al., 1966
Aortic strips of rabbit	2·4	3·1	Briggs, 1962

in vivo amounted to 1·8 m-mole/kg wet wt. After a 12 to 18 min period used for preparation in a glucose free dissection solution at room temperature the tissue Ca rose to about 5·2 m-mole/kg wet wt. During the following equilibration period of 60 min at 35°C the calcium content fell to 2·6 m-mole/kg wet wt. The same authors investigated other unfavourable conditions which resulted in an elevated Ca content of taenia coli: anoxia, low temperature (4°C) and crushing of the muscle pieces. These findings suggest, that any impairment of muscle function may be accompanied by a net uptake of Ca, and led the authors to postulate the existence of a calcium extrusion mechanism. Also Potter & Sparrow (1968) reported an increase of the Ca content of cat intestinal smooth muscle at low temperature or by damaging the tissue.

As mentioned above, the tissue Ca content depends upon the extracellular Ca concentration. Schatzmann (1961) published the following figures for taenia coli: at 0 m-mole Ca/l, incubation medium 2·8 m-mole Ca/kg wet wt; at 2·5 m-mole/l, 4·2 m-mole/kg wet wt; at 5·0 m-mole/l, 7·0 m-mole/kg wet wt; and at 7·5 m-mole/l, 8·5 m-mole/kg wet wt. In each case the tissue Ca level lies well above the Ca concentration of the medium. Comparable results have been found for longitudinal muscle strips of guinea-pig small intestine by Lüllmann & Siegfriedt (1968), namely at 0·6 m-mole Ca/l tyrode solution 1·0 m-mole Ca/kg wet wt; at 1·8 m-mole/l, 2·6 m-mole/kg wet wt; and at 2·7 m-mole/l, 3·6 m-mole kg wet wt. In experiments with isolated rat uterus, van Breeman, Daniel & van Breeman (1966) investigated the tissue Ca content in relation to the Ca concentration of the medium in the range from about 1 to 40 m-mole Ca/l. In order to elucidate the relationship between the extracellular and cellular Ca concentration more closely, Grosse & Lüllmann (unpublished) estimated the time course that governs the adaptation of the tissue Ca content to changes in the Ca concentration of the medium. Figure 1 represents the results indicating an adaptation process with an exponential time course. The experiments demonstrate, furthermore, the existence of a firmly bound Ca fraction, that is kept by the smooth muscle even in a Ca-free solution. This confirms the findings of Schatzmann (1961), who extracted homogenized taenia coli with distilled water and found residual calcium in the homogenate, the amount of which depended upon the Ca concentration of the initial incubation medium.

So far, two factors on which tissue Ca depends have been mentioned: the individual variations among animals, and the dependence on the extracellular Ca concentration. Still more factors may be involved. As Goodford (1967) has shown recently, the pH of the medium has no influence on the tissue Ca of taenia coli in the absence of phosphate but, if the PO_4^{3-}-concentration is raised to 1·2 m-mole/l in alkaline pH, the tissue Ca increases. One may wonder if this is a biological effect or simply a precipitation of Ca-phosphate on the muscle surface, as suggested by the author. In order to avoid this possible influence on the Ca metabolism, the PO_4^{3-}-concentration in the incubation medium should be kept low.

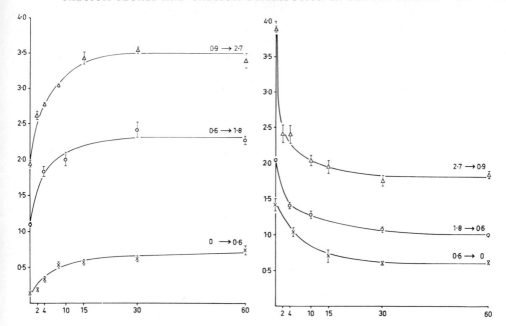

FIG. 1   Adaptation of tissue Ca level to a new extracellular Ca concentration. Ordinate: tissue Ca in m-mole/kg wet wt. Abscissa: time in minutes. The changes of the Ca concentration (m-mole/l) are indicated by the figures on top of each curve. The points represent means ± S.E.M.

It is necessary to consider yet another factor which complicates the work on Ca metabolism in smooth muscle. The tissue Ca can vary considerably in the same muscle according to the anatomical location of the piece chosen for study, as was found for the longitudinal muscle of guinea-pig small intestine (Lüllmann & Siegfriedt, 1968). In our experiments the control values showed an unusual scatter when we did not take into account the site in the small intestine from which the pieces were obtained. A systematic study of this problem revealed the results presented in Fig. 2. The muscle pieces that were obtained close to the stomach or close to the colon showed a considerably higher Ca content than those from intermediate sites. Since the flux rates also vary in a similar manner to the tissue Ca values, the control and experimental pieces of smooth muscle have to be selected very carefully in order to get matched pairs of muscle pieces.

Ca FLUXES IN SMOOTH MUSCLE

Determination of Ca influxes

The first paper, which deals with Ca fluxes in smooth muscle, was published in 1961 by Schatzmann. The uptake of ^{45}Ca, however, was measured only at two different incubation times, namely 20 and 50 min after adding the isotope.

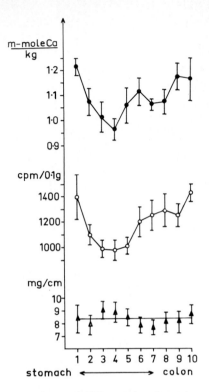

FIG. 2 Tissue Ca concentration and ⁴⁵Ca uptake, dependent upon the anatomical localization of pieces of longitudinal muscle of guinea-pig small intestine. Abscissa: origin of the smooth muscle pieces, 1 marks adjacent to the stomach, 10 adjacent to the colon. Ordinates: above, tissue Ca concentration in m-mole/kg wet wt; middle, ⁴⁵Ca uptake after 8 min of incubation, in c.p.m./0·1 g tissue (wet wt); below, size of the muscle pieces in mg (wet wt)/cm. (From Lüllman & Siegfriedt, 1968)

The exchange process was found to be almost finished after 20 min, which means that the half time of the uptake process had to be shorter than 5 min. At that time only 65% of the tissue Ca had been exchanged, this value seems remarkably low when compared with the results of other authors. On the other hand, the tissue Ca level given by Schatzmann is by far the highest so far reported for taenia coli. The quantity of exchangeable Ca, however, agreed well with that found by other authors.

Under comparable experimental conditions Bauer *et al.* (1965) determined the ⁴⁵Ca uptake by taenia coli of guinea-pigs with higher time resolution. Five minutes after adding ⁴⁵Ca the exchange process had already reached equilibrium. The time constant involved corresponded to a half time of about 1·2 min. In contrast to the results of Schatzmann (1961) the tissue Ca had exchanged completely, there was no unexchangeable fraction detectable. The repetition of uptake experiments at low temperature yielded different results which were in better agreement with the findings of Schatzmann (1961). The tissue Ca content rose to about 4 m-mole/kg wet wt and was partly inexchangeable. Beside the fast uptake process, which was characteristic for the experiments at 35 to 37°C, a second slow component of the uptake became measureable. As mentioned above, crushing of the muscle pieces leads to an increased tissue Ca content. Under this condition the tissue Ca was again not completely exchangeable (Bauer et al., 1965).

The time constant of the ^{45}Ca uptake process, at a lower extracellular Ca concentration than that used in the experiments of Schatzmann (1961) and Bauer et al. (1965), has been determined by von Hattingberg, Kuschinsky & Rahn (1966). At 0·85 m-mole Ca/l Tyrode solution the half time amounted to about 2·5 min. This was somewhat longer than the half times obtained with higher extracellular Ca concentrations. Nevertheless, the tissue Ca exchanged completely. In rat uterus, however, about 20% of the tissue Ca proved to be unexchangeable (van Breeman & Daniel, 1966).

In order to detect a possible correlation between the extracellular Ca concentration and the time course of the ^{45}Ca uptake we investigated this point in more detail (Lüllmann & Siegfriedt, 1968). The steady-state ^{45}Ca uptake was determined for three extracellular calcium concentrations. The results, obtained with longitudinal muscle of guinea-pig small intestine, are summarized in Table 2. The figures clearly demonstrate the dependence of the half time on the Ca concentration. The lower the external Ca concentration (and therefore the tissue Ca content) the slower the ^{45}Ca uptake and vice versa. Tentatively, it might be concluded that the Ca permeability is a function of the extracellular Ca concentration: it decreases in a condition of Ca deficiency and increases when Ca ions are present in excess. It may also be mentioned that in our experiments with smooth muscle of small intestine the total tissue Ca was completely exchangeable.

The half times shown in Table 2 are noticeably longer than those obtained for taenia coli. The difference may be explained by the fact that the longitudinal muscle of the small intestine is completely relaxed and inactive, while in the taenia coli there is always some tension and electrical activity under the experimental conditions used.

The amount of tissue Ca which does exchange with ^{45}Ca, may still be controlled by another factor. According to recent experiments, the tissue Ca concentration rose from 1·0 to about 1·4 m-mole/kg wet wt, when the smooth muscle pieces were incubated for 2 hr in a Mg^{2+} free solution. Under this condition, however, the tissue Ca became partly inexchangeable. The inexchangeable fraction amounted to about 25% of the total tissue Ca (Lüllmann & Siegfriedt, 1968).

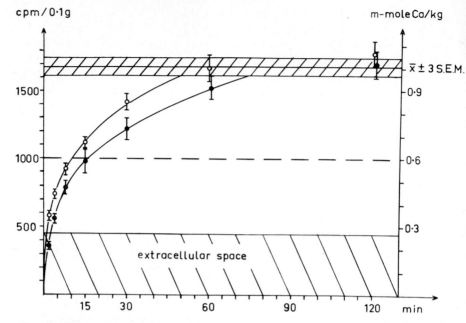

FIG. 3 ^{45}Ca uptake by longitudinal muscle strips of guinea-pig small intestine in normal and K rich tyrode solution. The K concentration was raised at zero time of the Ca uptake process. Abscissa: time in minutes. Left ordinate: ^{45}Ca content of the muscle pieces in c.p.m./0·1 g tissue, the radioactivity of the tyrode solution was 1000 c.p.m./0·1 ml. Right ordinate: tissue Ca concentration in m-mole/kg wet wt, the tyrode solution contained 0·6 m-mole Ca/l. The tissue Ca concentration is marked by the dashed curve ($\bar{x} \pm 3$ S.E.M.). ($\bigcirc$): control value in tyrode solution, containing 2·7 m-mole K/l. ($\bullet$): values in tyrode solution containing 40 m-mole K/l. The points represent the means $\pm$ S.E.M. (From Lüllmann & Siegfriedt, 1968)

TABLE 2.   Time course of ^{45}Ca uptake and exchangeability of tissue Ca of longitudinal muscle of guinea-pig small intestine at different extracellular Ca concentrations (Lüllmann & Siegfriedt, 1968)

Ca concentration in the medium (m-mole/l)	Tissue Ca concentration (m-mole/kg wet wt)	$t_{\frac{1}{2}}$ of the ^{45}Ca uptake (min)	Exchangeable tissue Ca
0·6	1·0	12	compl. exchang.
1·8	2·6	7	compl. exchang.
2·7	3·6	4·5	compl. exchang.

The experiments discussed above have been performed exclusively with smooth muscle of the gut or the uterus. To my knowledge no comparable results are available, obtained with mammalian smooth muscle of different origin, which allow a quantitative conclusion concerning the cellular exchangeability or the time course of cellular labelling with a Ca isotope.

Determination of Ca effluxes

The loss of ^{45}Ca from muscle pieces (taenia coli), which have been 'loaded' previously with the isotope and are in an ionic equilibrium, is not governed by a single exponential function. The analysis published by Schatzmann (1961), for example, revealed three different fractions with the following half times and relative sizes: (I) < 3 min, 44%; (II) 3 min, 42% and (III) 31 min, 14%. Chujyo & Holland (1963) found 4 discernible fractions, although one of these was rather small. In a further series of experiments with taenia coli, Goodford (1965) was able to measure two fast phases for the loss of ^{45}Ca. The two phases showed half times of 0·5 and 2·7 min and amounted to about 90% of the total ^{45}Ca of the muscle pieces. The remaining amount exchanged very slowly and with a declining rate constant. At low temperature (4°C) the processes were retarded.

TABLE 3. ^{45}Ca efflux in longitudinal muscle of guinea-pig small intestine, relative size and half times of the fractions involved (Lüllmann & Siegfriedt, unpublished)

Ca concentration	0·6 m-mole Ca/l		1·8 m-mole Ca/l		2·7 m-mole Ca/l	
	relat. size	half time	relat. size	half time	relat. size	half time
fraction 1	31%	1 min	34%	1 min	34%	1 min
fraction 2	49%	7 min	50%	4·5 min	52%	4 min
fraction 3	20%	39 min	16%	25 min	14%	22 min

Lüllmann & Siegfriedt (1968) have analysed the ^{45}Ca efflux curves of longitudinal muscle strips of guinea-pig small intestine at different Ca concentrations of the medium. Again, three fractions could be established. While the relative sizes of the fractions were almost unchanged, the half times were dependent upon the Ca concentration as shown in Table 3. The assumption that fraction 1 reflects the ^{45}Ca of the extracellular space is consistent with the half time of about 1 minute and the relative size (taking into account the different Ca concentrations of the extracellular and the cellular space).

In experiments with circular muscle of cat's ileum Sperelakis (1962) described at least two exponentials composing the efflux curve. The half times amounted to about 8 and 60 min, a more detailed analysis of the curves was not possible due to the experimental procedure (the first determination was performed 10 min after starting the wash out).

The ^{45}Ca efflux curves of taenia coli or isolated muscle layers of small intestine are rather complex although these preparations are mainly composed of muscle cells. It is not surprising to find still more complicated results with isolated organs which consist only partly of muscular tissue. Examples are given by Briggs (1962) in experiments with isolated rabbit aorta and by van Breeman *et al.* (1966). According to the latter authors the ^{45}Ca efflux curves obtained with rat uterus cannot be fitted to a unique sum of first order processes.

CORRELATION BETWEEN THE FUNCTIONAL STATE OF SMOOTH MUSCLE AND THE CELLULAR Ca METABOLISM

Effects of drugs on cellular Ca metabolism

The modern conceptions concerning muscle physiology grant Ca (bound or ionized) an important role. If this also holds true for smooth muscle, one would expect a significant influence on the cellular Ca metabolism by drugs, e.g. catecholamines, cholinergic drugs, which give rise to fundamental changes in the functional state of smooth muscle. Although some publications are devoted to these problems, the results are not quite as clear as one would expect.

The action of adrenaline upon the tissue Ca content, Ca-influx and -efflux was investigated by von Hattingberg *et al.* (1966). Neither the Ca content nor the Ca fluxes were changed in taenia coli, although adrenaline was applied in a concentration as high as 10^{-5} g/ml, which leads to a complete relaxation of the muscle pieces. These findings are in agreement with the results of Schatzmann (1961).

The Ca metabolism of smooth muscle in the presence of parasympathomimetic drugs has been investigated more intensively than that in the presence of catecholamines. Using pilocarpine (10^{-5} g/ml) von Hattingberg *et al.* (1966) were not able to demonstrate any change in the Ca content and Ca influx of taenia coli of the guinea-pig, but the Ca efflux was accelerated by a just detectable amount. Earlier, Schatzmann (1961) had published a similar result: acetylcholine (10^{-5} g/ml) applied during the slow phase of the efflux curve increased the ^{45}Ca loss by a very low percentage. However, van Breeman *et al.* (1966) in experiments with rat uterus could not detect any changes of the Ca fluxes in the presence of acetylcholine. Recent investigations have again emphasized the difficulties involved: we were unable to find evidence for an action of carbachol (2×10^{-7} g/ml) upon tissue Ca or Ca influx or Ca efflux in longitudinal muscle of guinea-pig small intestine (Lüllmann & Siegfriedt, 1968), although in an earlier paper Herrlinger, Lüllmann & Schuh (1967) described an increase in tissue Ca of taenia coli of about 25% upon exposure to acetylcholine (10^{-5} g/ml) for 8 minutes. It should be mentioned that the contractile response evoked by acetylcholine is Ca-dependent (Hurwitz, Joiner & von Hagen, 1967). Nevertheless, in experiments with intestinal smooth muscle, Potter & Sparrow (1968) could not establish a correlation between the tissue Ca content and the contractile response to acetylcholine under different experimental conditions.

Although the cholinomimetic drugs used possess a strong action upon the functional state of smooth muscle, the changes in Ca metabolism reported in the literature are surprisingly small or even conflicting. The reason for this may be sought in the assumption that only a small fraction of the cellular Ca is involved in the drug action and that the change in this particular fraction may be masked by the much greater amount of Ca which does not contribute to the response.

To exclude some of the experimental uncertainties, Durbin & Jenkinson (1961a) measured the Ca fluxes of taenia coli in a depolarized state induced by a high external potassium concentration. The application of carbachol (3×10^{-7} g/ml), which gave rise to an increase in muscle tension even under this condition, failed to produce a convincing change of the Ca fluxes.

In their investigations, von Hattingberg et al. (1966) also included papaverine, a smooth muscle inhibiting drug. The ^{45}Ca uptake as well as the ^{45}Ca loss were enhanced by papaverine, while the tissue Ca concentration remained unaltered.

An interpretation of the results in terms of the importance of Ca for muscle function is not possible at present.

Influence of a high external K-concentration on Ca metabolism

High external K-concentrations induce an excitation of smooth muscle, which is dependant upon the presence of Ca (Durbin & Jenkinson, 1961b; Urakawa & Holland, 1964). It was therefore appropriate to investigate the Ca metabolism of smooth muscle in the presence of an elevated K-concentration in the incubation medium. It has, however, to be kept in mind, that the functional state of smooth muscle in a medium with elevated K-concentration depends on how much the K-concentration is increased: about 40 m-mole K/l causes a partial depolarization, which leads to a sustained hyperactivity, whereas an isotonic K-concentration completely depolarizes the muscle membrane and keeps the muscle, after a short contraction, in a relaxed state.

According to Schatzmann (1961), a K-concentration of 100 m-mole/l exerted a marginal effect, namely an enhancement of ^{45}Ca uptake and loss. Sperelakis (1962), in experiments with circular muscles of cat's ileum, found an increased ^{45}Ca uptake in the completely depolarized muscle. Unfortunately, the significance of this effect cannot be evaluated because, after the loading period, the author incubated the muscle for 60 minutes in a non-radioactive solution to remove the 'extracellular ^{45}Ca'. Chujyo & Holland (1963) reported an enhanced uptake of ^{45}Ca in taenia coli incubated in Tyrode solution with 40 m-mole K/l, whereas the rate constant of the efflux curves did not change at all. These experiments were extended by Urakawa & Holland (1964). The authors differentiated between phasic and tonic contractions. They found that the ^{45}Ca uptake was more pronounced during a tonic and sustained shortening than during a phasic contraction of the smooth muscle pieces.

In experiments with longitudinal muscle of guinea-pig small intestine Lüllmann & Siegfriedt (1968) determined the ^{45}Ca fluxes and the Ca content of the tissues in an incubation medium with a high K-concentration. As seen in Fig. 3, the ^{45}Ca uptake was significantly delayed in the presence of 40 m-mole K/l, whereas the tissue Ca content remained unchanged (1·02 and 1·01 m-mole/kg wet wt respectively). The latter result agrees well with the findings of Potter & Sparrow (1968). A tentative analysis of ^{45}Ca effluxes in the presence of 40 m-mole K/l yielded the following relative sizes for the three fractions of the efflux curve: In the presence of normal K: (I) 34%, (II) 52%, (III) 14%; in the presence of 40 m-mole K/l: (I) 38%, (II) 34%, (III) 29%. This change in the relative size of the two cellular fractions in the presence of a high potassium concentration might reflect a transfer of Ca from a loosely bound fraction to a more firmly bound one. It may be mentioned that the muscle pieces were highly active under these conditions (40 m-mole K/l), whereas in the control experiments the longi-tudinal muscle strips stayed completely relaxed. The delayed ^{45}Ca uptake in the presence of 40 m-mole K/l could be interpreted in a similar manner to the explanation of the results of the efflux experiments, namely a firmer binding of the cellular Ca. The change might be due to a binding of Ca to active sites of the activated contractile proteins, but this assumption is merely speculative. A complete replacement of Na by K in the incubation medium did not influence the Ca metabolism of our preparations.

According to the findings of van Breeman *et al.* (1966) the influence of a high potassium concentration (93 m-mole K_2SO_4/l replacing 115 m-mole NaCl/l) upon the Ca fluxes of isolated rat uterus seems to be rather complex. Although the influx of ^{45}Ca was delayed, the efflux of ^{45}Ca was enhanced only during the last phase of the total efflux curve. A proper interpretation of these results is made more difficult by the complexity of the tissue used and by the high SO_4^{2-} concentration.

In rabbit aortic strips, which were soaked for 2 hr in a Ca-free solution, Briggs (1962) found an enhanced uptake of ^{45}Ca in an isotonic K_2SO_4-medium, as compared with controls in normal saline.

The conflicting results gained with different types of smooth muscle are rather surprising if not disappointing. A possible explanation may be found in the very different behaviour of the muscle species in the control state: e.g. the taenia coli is more or less contracted and electrically active, whereas the longi-tudinal muscle is completely relaxed. Thus, the excitatory effect of a high potas-sium concentration is much larger in longitudinal muscle pieces than in the taenia coli.

It is difficult to interpret in a simple way the results concerning ^{45}Ca fluxes in the presence of elevated potassium concentrations. Two different processes may be involved:

1. an increase of the Ca permeability, and
2. a displacement of cellular Ca from a loosely bound fraction towards a more firmly bound fraction, which would imply a decrease of the exchange rate.

The interferences between these two could account for the experimental results observed with different kinds of smooth muscles.

Influence of electrical stimulation on Ca metabolism

In order to investigate whether electrical stimulation influences the Ca metabolism of smooth muscle, Lüllmann & Mohns (1969) compared the tissue

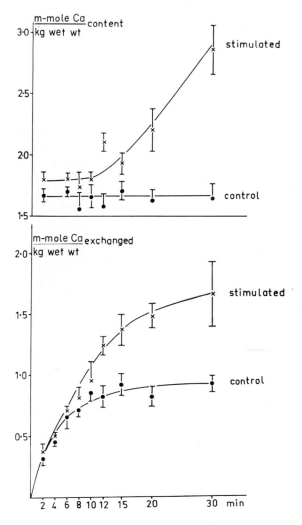

FIG. 4   Tissue Ca content (above) and ^{45}Ca uptake (below) of electrically stimulated longitudinal muscle strips of guinea-pigs small intestine as compared with non-stimulated control muscles. Ordinates: tissue Ca content in m-mole Ca/kg wet wt, and ^{45}Ca uptake in m-mole ^{45}Ca/kg wet wt. Abscissa: time in minutes, at zero time the stimulation and the exposure to ^{45}Ca are beginning.

Ca content and the Ca fluxes of longitudinal muscle strips of guinea-pig small intestine during stimulation and at rest. The muscle pieces were stimulated every 10 sec by rectangular pulses of 250 msec duration and supramaximal voltage. Each contraction developed a tension between 5 and 7 g. The stimulation periods could be extended over more than 60 min without a marked decrease of the height of contraction. The tissue Ca content and the ^{45}Ca uptake of stimulated and resting muscle pieces is shown in Fig. 4. The tissue Ca content increased from about 1·7 to about 2·8 m-mole Ca/kg wet wt in 30 min. Consequently, the net up-take of Ca was accompanied by an accumulation of ^{45}Ca during the entire period, whereas in the control muscle the ^{45}Ca uptake reached the maximal value in about 10 min. A closer analysis of the results revealed that the increased ^{45}Ca accumulation by the stimulated muscles was entirely due to the net uptake but not caused by any change of the influx as such. This holds true also for the effluxes, for which no difference could be measured between resting and stimulated muscles.

Based upon these results the net Ca uptake per stimulus was calculated. It amounts to 5×10^6 Ca atoms per single cell or about 10^{-12} mol Ca/cm^2 cell surface. Similar figures have been reported by Sunano & Miyazaki (1968) using taenia coli of guinea-pigs. These figures gain still more importance, if the surface/volume ratio of the muscle cells is taken into consideration. As compared with skeletal muscle, the smooth muscle cells take up much more Ca per unit volume and stimulus (for details see Lüllmann & Mohns, 1969).

CONCLUSIONS

All kinds of muscular tissue have in common the properties of excitability and of contractility. There is good experimental evidence that in skeletal muscle and in heart muscle these two fundamental processes are coupled by calcium. Unfortunately, smooth muscle has not been investigated as extensively as the two other muscular tissues mentioned.

The cellular Ca metabolism seems to display marked differences. In skeletal muscle only about 20% of the cellular Ca content exchanged with the extra-cellular ^{45}Ca when equilibrium was reached (isolated rat diaphragm, Lahrtz, Lüllmann & Reis, 1967) whereas, under identical experimental conditions, 50 to 70% of the Ca of guinea-pig heart muscle exchanged (Klaus & Lüllmann, 1964; Hoditz & Lüllmann, 1964). On the other hand, all the cellular Ca of smooth muscle exchanges rather quickly with the extracellular Ca under comparable circumstances. In the same order, the endoplasmic reticulum of the different muscle types is of less importance, i.e. of least importance in smooth muscle. It can be assumed that Ca which is bound into the endoplasmic reticulum does not exchange with the extracellular Ca at least during the resting state. In skeletal muscle, due to the larger size of the cells (surface/volume ratio about 1:15) and the higher speed of shortening, the contractile proteins have to be activated by an intracellular Ca release mechanism. On the other hand, in smooth muscle (surface/volume ratio approaching 1 : 1) the actomyosin may be activated sufficiently fast by a simple penetration of Ca through the cell membrane, or by a transfer of membrane-bound Ca towards the contractile proteins. However, the nature of this mechanism in smooth muscle is still not completely understood, as demonstrated by the results discussed above.

It can be stated that smooth muscle cells take up a relatively large amount of Ca compared with other muscular tissues when stimulated electrically. It is tempting to assume that excitation-contraction-coupling in smooth muscle is also brought about by Ca, but additional evidence is required for a proper evaluation of this hypothesis.

6

OSMOTIC PHENOMENA IN
SMOOTH MUSCLE

ALISON F. BRADING

INTRODUCTION

The most widely-accepted model proposed to explain the properties of excitable tissues is the membrane theory, which postulates that a semi-permeable membrane surrounds the cell, and divides the internal cellular environment from the extracellular fluid. The internal electrolytes are assumed to be dissolved in the intracellular water which is free to act as a solute to all the electrolytes. The distribution of electrolytes between the intra- and extracellular solutions in this model, is imposed on the cell by the properties of the membrane, and the concentration of non-permeant intracellular anions. Another model, the fixed charge, or adsorption theory, has been proposed by several authors. The most complete treatments of this type of model are given in the books by Troshin (1966) and Ling (1962). This model basically proposes that the cells are not bounded by semipermeable membranes, but that the membrane is permeable to solutes, and the distribution of water and electrolytes is determined by the colloidal properties and structural arrangements of the living protoplasm. These properties depend on the metabolism of the cell. Much of the cell water is bound in some form, and also many of the electrolytes, in particular potassium ions, and the concentration of ions in the free water of the cell is proportional to that in the external bathing medium.

In the field of smooth muscle research, similar models to these two have been used to explain the properties of the cells, although at the present time it is doubtful whether many of the workers would accept either view in its entirety. In this chapter the two models will be referred to as the membrane and non-membrane theories.

The osmotic behaviour of excitable tissues has often been studied, and the early work of Overton (1902a, b, 1904) on the osmotic behaviour of frog striated muscle forms the basis for the theory that cell membranes are semi-permeable. Meigs & Ryan (1912) and Meigs (1912) carried out duplicate experiments on striated muscle and smooth muscle (stomach) of the frog, and although they confirmed Overton's observations that the striated muscle did behave as if bounded by a semi-permeable membrane, the authors could not accept this model for the smooth muscle cells. Bozler (1959, 1961a, 1961b, 1962, 1965) in a series of papers describing the results from experiments that repeated and extended Meig's work with frog stomach muscle, also concluded that the cell cytoplasm was more important than the membrane in determining the distribution of water and electrolytes in this tissue.

Electrophysiological studies, on the other hand, have been carried out more extensively on mammalian smooth muscle, and have tended to demonstrate that the phenomena observed can be explained by the membrane theory. Tomita (1966a, 1967a) and Abe & Tomita (1968) have shown that the taenia coli muscle and the vas deferens of the guinea-pig, exhibit cable properties similar to those described by Hodgkin & Rushton (1946) for the nerve fibre. This is of course very strong evidence in support of the membrane theory.

Interpretation of work carried out on ionic distribution and fluxes of ions across the cell membranes, in particular in mammalian smooth muscle (see Chapters 2, 3 and 4), tends to reveal a system that is more complicated than would be expected if the intracellular electrolytes were simply dissolved in the cell water, with the distribution limited by a selectively permeable membrane. The models used to interpret these results vary between the extreme views.

This article will examine in more detail the osmotic properties of smooth muscle that have been described in the literature, with particular reference to the effects of tonicity on the tissue weight, electrolyte distribution, and ion fluxes. Some new work with guinea-pig taenia coli muscle will be described, and it is hoped that some conclusions may be reached concerning the applicability of the two theories to smooth muscle.

The results are also helpful in elucidating the effects of tonicity on the electrical and mechanical behaviour of excitable tissues, since hypertonic solutions are often used by electrophysiologists to minimize contractile responses in tissues, whether spontaneous, as in taenia coli (Tomita, 1966a) or evoked, as in striated muscle (Hodgkin & Horowicz, 1957). These effects may also be relevant to naturally-occurring changes in the osmotic pressure of the blood, particularly with respect to vascular smooth muscle.

TERMINOLOGY

Before describing any results of experimental work, it is necessary to define some of the terminology in use. At this stage however, it should be emphasized that osmosis is a physicochemical phenomenon, and the equations and terminology described are strictly applicable only to the simple and established conditions defined in their formulation. Cells and cell membranes are highly complex, and the application of osmotic equations to cells stems from a time when little was known about them except that they did in many cases (and particularly plant cells) shrink or swell in proportion to the osmotic pressure of the bathing solution. This observation led to the formulation that cells in general were bounded by semi-permeable membranes, and that cell volume was determined by the osmotic pressure of the bathing medium.

The fact that the membrane of cells is now known to be permeable to a greater or lesser extent, to a large number of solutes, both electrolyte and non-electrolyte, has meant that many other factors have to be considered when the control of cell volume is investigated. Nevertheless, studies of the osmotic effects of penetrating solutes can be used to deduce the relative permeability of the membrane to the solute in question. The study of the relationship between the concentration of non-penetrating solute in the external solution and the volume of the cell, can yield valuable information as to the nature of the cell. This can be done by comparing the behaviour with that predicted from simple osmotic considerations, although the conditions for the strict application of these equations do not hold.

Osmosis is the phenomenon of movement of a solvent from a solution of low concentration to one of a higher concentration, through a membrane permeable only to the solvent molecules. To express this in a different way, the solvent moves from an area of high solvent activity to an area of low solvent activity. The addition of solute effectively lowers the activity of the solvent. Osmosis can be prevented by applying pressure to the solution into which the solvent is moving. If the semi-permeable membrane is separating pure solvent from a solution, then the pressure applied to the system to just prevent osmosis is the osmotic pressure of the solution, and is usually denoted by the symbol Π.

In ideal or infinitely dilute solutions, where there is no interaction between the ions or molecules, the osmotic pressure can be derived from van't Hoff's equation

$$\Pi = RTC \tag{i}$$

where R is the gas constant, T the absolute temperature and C the concentration of particles in the solution. A similar expression can be derived from thermo-dynamic considerations (Dick, 1959). This derivation is more relevant, since van't Hoff's equation stems from the gas laws, and considers the pressure exerted by the solute particles, whereas it is in fact the solvent molecules that move. Nevertheless the equations derived are identical.

The concentration of the solution is in fact the particle concentration, so that, for electrolytes, a molar solution will have a greater osmotic pressure than a molar solution of non-electrolyte, since the electrolyte can dissociate into two or more particles.

Thus the osmotic pressure of a solution is, ideally, proportional to the con-centration of solute particles and, in an unrestricted system, the volume of solvent flowing from a solution of low to high osmotic pressure across a semi-permeable membrane will be directly proportional to the difference in osmotic pressure between the solutions, and equilibrium will be reached such that

$$\Pi V = K \tag{ii}$$

where V is the volume of a compartment, and K is constant.

Since solutions do not behave ideally, and are never infinitely dilute, there is interaction between the ions or molecules in solution, and the osmotic pressure measured is not directly proportional to the concentration. The osmotic co-efficient of a solution φ is the correction factor that has to be included in equation (i) to make the calculated and observed pressures agree:

$$\Pi = \varphi RTC \tag{iii}$$

Concentration is expressed either as molal concentration (moles/kg solvent) or molar concentration (moles/litre solution) and the coefficients can be found in prepared tables, for example in Robinson & Stokes (1955), but they apply to solutions of single solutes. When considering mixed solutes, other interactions occur and the osmotic coefficient can only approximately be derived from the coefficients of individual solutes. The osmotic pressure of any solution can however be determined from estimations of the depression of freezing point of the solution.

$$\Pi = \frac{L\rho T}{T_0} \tag{iv}$$

where L is the latent heat of fusion, T_0 is the freezing point of the pure solvent,

T is the depression of the freezing point of the solution and ρ is the density of the solvent.

The osmotic pressure of a solution is measured in mm of Hg or in atmospheres but, for biological purposes, the unit osmole is more convenient and is defined (Dick, 1966) as the osmotic pressure of a 1 M solution of an ideal solute. For non-ideal solutes, it will be obtained by multiplying the molar or molal osmotic coefficient by the molar or molal concentration. This unit is independent of temperature.

CELL VOLUME

On the assumption that the cell membrane is permeable to water, and separates an intracellular from an extracellular solution, the cell volume will be affected by the osmotic pressure of the bathing solution. Under normal conditions it is assumed that the cell is in equilibrium with its environment, and that the osmotic pressure of the internal solution will be equal to the osmotic pressure of the external solution. Any changes in the osmotic pressure of the external solution will disturb the equilibrium, and lead to changes resulting in the re-establishment of osmotic equilibrium. If the cell membrane were completely impermeable to all particles except water molecules, then the addition to or removal of solutes from the external solution would result in movement of water into or out of the cell as predicted from the osmotic equations, with the resultant changes in the cell volume.

It is, however, certain that many solutes can and do penetrate the cell. The mechanism of the distribution of electrolytes between the cell and its environment is also known to depend on the consumption of metabolic energy. Under steady-state conditions the influx and efflux of ions are equal, and there is a drop between the inside and outside of the cell. Thus it can be seen that a complicated state of equilibrium exists between the cell and its environment.

The membrane theory explains the distribution of ions and the membrane potential by postulating that a Donnan distribution of permeable ions exists across the membrane, so that the product of diffusible ions on each side remains equal. The high intracellular potassium is maintained to balance intracellular non permeating anions, the sodium ions being pumped out of the cell in exchange for potassium, which involves the utilization of energy. The diffusible ions are distributed passively in accordance with their chemical and electrical gradients. The cell membrane is considered to be effectively impermeable to sodium ions, through possession of the sodium pump. On the membrane theory, increase of the extracellular osmotic pressure will result in either movement of water out of the cell, or movement of the added solute into the cell if it is permeable or, more probably, a combination of the two, depending on the relative permeability of the membrane to water and the solute. Any change in volume of the cell will, however, disrupt the Donnan equilibrium by altering the concentration of the intracellular ions, and this may result in movement of these ions leading to secondary volume changes. The theory also predicts changes in the membrane potential of the cell, which will, in time, lead to redistribution of the ions and possible volume changes. Bearing these considerations in mind, it is possible to predict to a certain extent the changes in volume and ionic content that would be expected, according to this theory, when the external osmotic pressure is changed, provided that something is known of the permeability of the membrane to the solute. It is also possible to draw some conclusions as to the permeability of the membrane to particular solutes from the volume changes induced.

According to the non membrane theory, the distribution of electrolytes is proposed to be due to the selective absorptional properties of the cell protoplasm. The selectivity is related to energy consumption, through high energy phosphate compounds. The regulation of the water content of the cell protoplasm on this hypothesis is not osmotic but colloidal, depending on the hydration of the cell colloids and the change in this hydration as a result of the penetration of solutes into the cells. It is difficult to predict the changes in volume that will occur on changing the osmotic pressure of the external solution, but the theory offers much more scope for explaining results that cannot be predicted from the membrane theory.

It is proposed in this article to consider the results described initially in terms of the membrane theory, since the predictions of this theory are much better formulated.

Equations and theory

When osmotic equations are applied to cells, it is normally assumed that the membrane is effectively semipermeable and that, in equilibrium, the osmotic pressure of the cytoplasm is equal to the osmotic pressure of the external medium.

On these assumptions, the change in volume of a cell with changes in the osmotic pressure of the bathing medium can be predicted from the following equation given by Lucké & McCutcheon (1932)

$$\Pi(V - b) = \Pi_o(V_o - b) = K \tag{v}$$

Where V_o is the original volume of the cell, and V the final volume, Π_o is the original osmotic pressure and Π the osmotic pressure of the experimental medium, b is the non-solvent volume of the cell and is presumed to remain constant. If the cell behaves as a perfect osmometer, then a plot of V against $1/\Pi$ will yield a straight line, with an intercept, b, at $1/\Pi = 0$.

One can only assume an effective semipermeability since it is known that the cell membrane is permeable. Nevertheless, within the terms of the equation, it is only necessary that there should be no net loss or gain of solutes and, therefore, that any fluxes should be equal in both directions across the cell membrane.

Attempts to demonstrate the equality of osmotic pressure of the internal and external solutions have often failed. Investigations of the ionic content of smooth muscle by Daniel (1958) led him to conclude that there was apparently a hyperosmotic concentration of monovalent cations in the cell. This high concentration of monovalent cations has been found repeatedly in smooth muscle, and has led to the postulation that some of these cations are bound to proteins, since it appears highly unlikely that the cell can in fact maintain a difference in osmotic pressure across its membrane. However, if the intracellular anions are multivalent, the osmotic activity of the anions will be lower than that in the external solution, and the total internal osmotic pressure may be equal to the external solution.

Tonicity is a term frequently used when osmotic principles are applied to living systems. When a cell or tissue is placed in a solution which causes no change in cell volume, the solution is said to be isotonic, and it is presumed to have an osmotic pressure equal to the osmotic pressure of the intracellular solution. Tissues will swell in hypotonic solutions, and shrink in hypertonic ones. Relative tonicity has been defined by Ponder (1948), as the ratio of the freezing point depression of the suspending medium to the freezing point of normal mammalian plasma. Thus the relative tonicity of a solution is equal to Π/Π_o, where Π is the experimental solution osmotic pressure, and Π_o the osmotic pressure of the original medium. However the actual tonicity of the solution depends not only on the osmotic pressure, but also on the permeability of the membrane to the solute.

Another equation predicting changes in cell volume with the tonicity of the medium is given by Ponder (1948)

$$V = W\left(\frac{1}{T} - 1\right) + 100 \tag{vi}$$

where V is the volume of the cell, W the percentage water content of the cell in isotonic solution (when $T = 1$) and T is the relative tonicity of the solution (Π/Π_o). The application of this expression requires previous determination of the water content of the cell.

Hypertonic solutions
Non permeating solutes

The simplest way to look at the effect of non permeating solutes on the volume of cells is to apply equation (v), and plot cell volume against the reciprocal of the osmotic pressure of the external medium. Such plots have been made by Dydyńska & Wilkie (1963) and Blinks (1965) for frog striated muscle. In both investigations the osmolarity of the solution was increased by the addition of sucrose. A straight line relationship was observed by both authors, leading them to the conclusion that this muscle behaves as a perfect osmometer. With Blinks' results in single muscle fibres, the intercept where $1/\Pi = 0$ gave a value for nonsolvent volume of 33% and thus a solvent volume of 67% which was considerably less than the cell water, estimated by other means to be 80%.

In many tissues a straight line relationship has been found between $1/\Pi$ and the volume of the cells, indicating behaviour as predicted for an osmometer. However, a discrepancy between the solvent volume and the estimated cell water, and between the non-solvent volume and the estimated dry weight of the cell, has often been found. This has lead to the conclusion in many studies that a small amount of the cell water is not available as a solvent. Ponder (1948) clearly demonstrated this discrepancy in the erythrocyte, and introduced the letter **R** (the ratio of the found/calculated dry weight) to represent the deviation of the cells from ideal osmotic behaviour. Dick (1966) gives a list of the values of **R** calculated by different workers for many tissues. A ratio of less than one is often taken to indicate the presence of 'bound water' in the cell but, as pointed out by Olmstead (1966), the value **R** is the sum of all the factors causing the cells to deviate from ideal osmotic behaviour and would include (a) resistance of the cell framework to volume change, (b) deviations from isosmotic values of the osmotic coefficients of intracellular proteins with concentration and dilution of the cell solution, (c) immobilization of any of the cell water, (d) ion binding to protein, (e) net transport of ions across the cell membrane, and other factors.

Investigations of the effect of hypertonic solutions on smooth muscle were carried out by Meigs (1912), who followed the time course of loss of weight of frog stomach muscle exposed to a double strength Ringer's solution. He compared the results with those obtained with frog striated muscle and found that both tissues shrank with a similar time course. Bozler (1965), who exposed frog stomach, heart and striated muscle to solutions made hypertonic with addition of NaCl, obtained similar results. He found that all three types of muscle shrank in proportion to the tonicity of the medium.

The effects of hypertonic Krebs solution (sucrose added) on the volume of the smooth muscle of the guinea-pig taenia coli, have been investigated by Brading & Setekleiv (1968). Figure 1 shows the effect of tonicity on the volume of the muscle cells. This volume was estimated by subtracting the extracellular space, measured with ^{14}C Sorbitol, from the wet weight of the tissue, after exposure for 30 min to solutions made hypertonic with sucrose. Also included in this figure are the effects of diluting the bathing medium with distilled water. The curve is calculated from equation (vi) using a figure of 74% for intracellular water, derived from measurements of dry weight and extracellular space. It can be seen that in hypertonic solution the points fit reasonably well to the predicted curve, and it can be concluded that the cells are behaving as osmometers. If the points are plotted in accordance with equation (v), the intercept gives a value for the non-solvent volume that is approximately the same as the measured dry weight of the tissue, and thus the value for **R** is unity. However this figure assumes that the whole dry weight of the tissue is derived from the cells, and does not allow for a possible extracellular component. If there is a significant amount of extracellular dry weight, for instance from collagen, then the **R** value of this tissue would also be less than unity.

7+

The behaviour of cells so far described, in physiological solutions made hypertonic with non-permeating solutes, can be interpreted by the membrane theory. There is however similar work described in the literature that cannot be so interpreted. For example, Nasonov & Aizenberg (1937), as quoted in Troshin (1966) investigated the change in volume of frog striated muscle in a series of solutions made hypertonic with isosmolar amounts of different non electrolytes. It would be expected from the membrane theory that, if the solutes were permeable, the weight of the tissue would eventually return to its original value, and

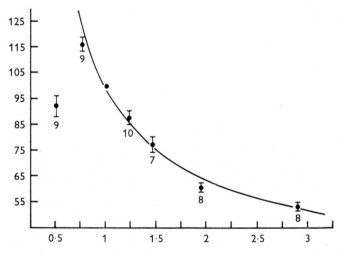

FIG. 1 The influence of tonicity on the intracellular space. The points plotted are values obtained by subtraction of the extracellular space from the wet weight of the tissue. The continuous line is drawn from Ponder's equation for a perfect osmometer: $V = W[(1/T) - 1] + 100$, where V = intracellular space, T = tonicity and W = cell water (taken as 74% of the fresh weight). Ordinate: intracellular space expressed as a percentage of the value in normal solution. Abscissa: tonicity expressed with the normal Krebs solution taken as 1. (From Brading & Setekleiv, 1968)

that, if the solutes were non permeable, the tissue would shrink by equal amounts and remain shrunk. The authors used urea, glycerol, alanine, glucose, galactose, sucrose, lactose, dextran, gum arabic, egg albumin and peptone, In every solution the muscle shrank and remained shrunk for up to $2\frac{1}{2}$ hr, but the volume change was different in each solution, and the effectiveness of the solute for removing water from the tissue increased in the order listed; thus the greater the molecular weight of the solute the greater the volume change induced.

Similar observations have been made on the taenia coli, where preliminary experiments (Brading, unpublished observations) have shown that the final volume of the taenia, when the Krebs solution is made hypertonic, depends on the substance used to increase the tonicity. Approximately isosmolar amounts of sucrose, erythritol and sodium chloride were used to double the osmolarity of the solution, and the tissue lost most weight in the sucrose solution, and least in the solution with sodium chloride.

The behaviour of tissues deviates even further from predicted behaviour when the tissues are exposed to pure solutions of 'non-permeating' solutes.

Meigs (1912), studying the effect on frog stomach muscle of isosmotic solutions of pure electrolytes, found that the muscle gained up to 25% in weight during the first 4 hr of exposure to isosmotic NaCl solutions, and then gradually lost weight, returning to the original weight in about 24 hr. In isosmotic solutions of sucrose, lactose and dextrose the tissue also gained weight and in hyperosmotic solutions of sucrose and dextrose as well. The gain of weight in these latter solutions was not consistent or predictable.

Bozler & Lavine (1958) confirmed that frog stomach muscle gains weight in isosmotic sucrose, and does so even in solutions of up to 0·1 M sucrose. The authors showed, however, that there is escape of electrolytes under these conditions, and probably penetration of sucrose. It would appear that the semipermeability of the membrane depends on the presence of the correct external environment, and in pure solutions these properties are modified or lost.

It is interesting that in mammalian smooth muscle (Brading & Jones, unpublished observations on guinea-pig taenia coli), even after 1 hr, there is no significant increase in weight if the tissue is exposed to an isosmotic pure sucrose solution, although a considerable amount of electrolyte has been lost. In contrast, if only the sodium chloride in normal Krebs solution is replaced by isosmotic sucrose, there is again a loss of intracellular electrolytes, but the tissue actually loses weight. When all the sodium is replaced, the tissue shrinks to 77% of its original volume, but if 10 mM sodium is left and the rest replaced with sucrose, the tissue only shrinks to about 83% of its original volume.

It can be concluded at this stage, that although smooth muscle and other tissues, often appear to behave as perfect osmometers with respect to the addition of non-permeating solutes to the normal medium, further investigations reveal deviations from the expected behaviour which are difficult at present to explain simply from the normal membrane theory.

Permeating solutes

Theoretically it would be expected that, when a permeating non electrolyte is added to the medium bathing a cell, thus increasing the osmotic pressure of the medium, osmotic equilibrium will be brought about by movements of both water and the solute. Initially the cell may shrink, but as penetration of the solute proceeds until its intracellular concentration is equal to that in the bathing medium, the water will enter the cell again, and equilibrium will be reached with the cell at the same volume as before addition of the solute. The amount of shrinkage observed and the time taken for equilibrium to be reached, will depend on the relative permeability of the cell membrane to water and the solute.

Bozler (1959) studies the osmotic effects of urea and thiourea on frog stomach muscle. In a Ringer solution containing 0·24 M urea, the muscle shrank by about 10% and then gradually the weight increased, although the original weight was not completely regained. The lack of complete recovery of weight was unexpected, since Bozler showed that, after equilibrium, the urea space was 88% of the cell volume, a space significantly higher than the water space, indicating that part of the urea was adsorbed. Initially the persistence of the shrinkages in urea was taken to indicate that despite its large volume of distribution, urea was excluded from some of the cell water, and could thus exert an osmotic effect. Later (1961b) Bozler showed that the shrinkage could be accounted for by loss of electrolytes from the tissue. Another interesting observation was that the equilibrium of urea in the tissue was complete even before any recovery of weight was apparent.

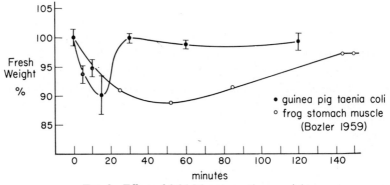

FIG. 2   Effect of 0·24 M urea on tissue weight.

In the taenia coli of the guinea-pig urea acted more predictably, causing a transient shrinkage and full recovery of weight. Figure 2 compares the effect of 0·24 M urea added to the bathing medium, on taenia coli and frog stomach muscle (the latter points taken from Bozlers' results).

Further studies with non-electrolytes and their volume of distribution in frog stomach muscle lead Bozler (1961a, b) to conclude that some non-electrolytes penetrate passively into muscle fibres, but are excluded from part of the fibre water, and are therefore osmotically active. Erythritol distributed itself in a space nearly twice as great as that of inulin, but smaller than that of urea, thiourea of glycerol, whereas the space occupied by sucrose was normally slightly larger than the inulin space, but smaller than the erythritol space. Effluxes of sugars indicated that there was significant penetration of the molecules into the fibres. Bozler proposed that

'as an alternative to the generally accepted view that distribution is determined by a semipermeable membrane... most of the fibre water is contained in segregated or narrow spaces, pores or layers, so that diffusion of molecules of moderate size is hindered. Semi-permeability, therefore, is considered a bulk property of the muscle fibres'.

This type of explanation would also be applicable to the results of Nasonov & Aizenberg (1937) described earlier, and such results are important evidence for those who advocate the non-membrane theory of the distribution of ions. They are certainly difficult to reconcile with the membrane theory, unless the membrane is assumed to present relatively little obstruction to the passage of non-electrolytes, showing its selective permeability properties to charged particles only.

Hypotonic solutions
Frog stomach
Using half strength Ringer solution, Meigs (1912) compared the swelling of frog striated and frog stomach muscle. He showed that during the first 8 min after transfer to the hyposmotic solution, the striated muscle swelled five times as rapidly as did the smooth muscle. Whereas the swelling of the former was more or less exponential in time course, the smooth muscle increased in weight more or less linearly with time, and the final gain in weight was much less than that observed in the striated muscle.

Bozler also found that the swelling of frog stomach and heart muscle in hypotonic solutions were less than that of sartorious. On return to the normal solution after exposure to hypotonic Ringer solution the stomach muscle lost more weight than it had gained, indicating a possible loss of electrolytes.

Taenia coli
The effect of tonicity on the tissue weight and extracellular space of the taenia coli of the guinea-pig can be seen in Fig. 3.

In hypotonic solutions there was a relatively small increase in the weight of the tissue. In 0·75 tonicity this increase in weight was accompanied by a decrease in the sorbitol space, which means that the swelling of the cells was greater than appeared from this figure, but not as large as would be expected from a perfect osmometer (see Fig. 1). In 0·5 times tonicity, there was a considerable increase in the sorbitol space, and also a much greater scatter of the experimental results. The line on the graph is drawn plus or minus the standard error of the mean, but the actual range was from 99·6 to 150·87% of the normal extracellular space. It is likely that in many cases there was some damage to the cell membranes, allowing penetration of the sorbitol into the cell. If this were so, and if the real extracellular space was 95% of the original (from the trend exhibited in Fig. 2), then the actual cell volume or intracellular space would be little changed from normal. As will be discussed later, there was possibly some loss of electrolytes from the cells, although the calculated amount lost depends on the value taken for the extracellular space.

Further investigations into the effect of 0·75 times tonicity indicated that the

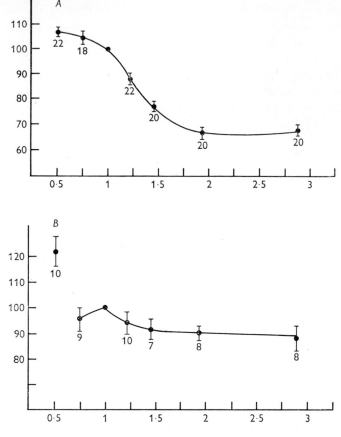

FIG. 3   Change in weight (A) and extracellular space (B) of tissue in hypo- and hypertonic solutions. Ordinate: weight expressed as a percentage of the weight in normal
Krebs solution. Abscissae: tonicity expressed with the normal Krebs solution taken
as 1. (From Brading & Setekleiv, 1968)

changes in weight with time of exposure were rather complicated and, as can
be seen in Fig. 4, the secondary loss and gain in weight seemed to be associated
with similar changes in tissue potassium.

The behaviour of smooth muscle in hypotonic solution deviates from the
predicted behaviour to a greater extent than that of striated muscle. The increase in tissue volume is normally less than predicted. There may be some
active control involved, since swelling in hypotonic solutions is much greater
at 4°C than at 37°C (Brading, unpublished observations).

Osmotic effects of electrolytes

The membrane theory of the distribution of ions across the cell membrane

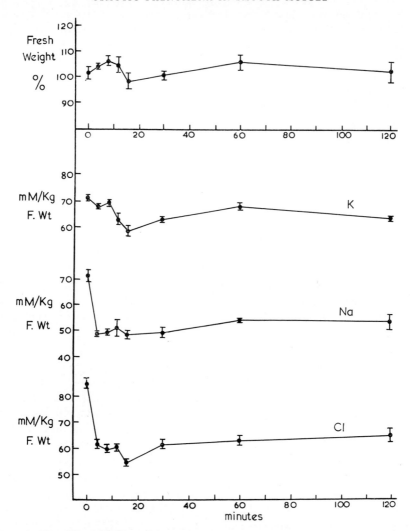

FIG. 4 The effect of 0·75 tonicity Krebs solution on tissue weight and ion content. Ordinates: Fresh weight (as percentage of weight in normal solution) and m-moles/kg fr wt. Abscissa: Time in minutes after transfer to the 0·75 tonicity Krebs solution.

supposes that a double Donnan equilibrium exists between the intracellular and extracellular compartments, by virtue of the non-permeating anions in the cell, and the effective impermeability of the membrane to the extracellular sodium. From the membrane theory it would be predicted that an electrolyte with one ion to which the membrane is effectively impermeable should act as a non-permeating solute, whereas electrolytes in which all the ions can cross the cell membrane should act as permeating solutes, when added to the external bathing solution. Sodium chloride can thus be considered to be non-permeating, as has been described earlier, but potassium chloride should be a permeating electrolyte.

Meigs (1912) has shown that, after a slight initial shrinkage, frog stomach muscle does in fact swell considerably in isosmotic KCl. Swelling of this muscle in isosmotic KCl has also been reported by Bozler (1962) although the time course was much slower than that for the swelling of striated muscle in the same solution. (Bozler also reported that the addition of a small amount of calcium chloride to the isosmotic KCl solution prevented the stomach muscle swelling).

Replacement of NaCl in the bathing solution by isosmotic KCl would, theoretically, be expected to upset the Donnan equilibrium, and penetration of KCl into the cells should occur, followed by water, due to the increase in internal osmotic pressure. Thus the tissue would be expected to swell, and such swelling has been observed in striated muscle (Boyle & Conway, 1941). Addition of KCl to the bathing medium, making it hypertonic, would be expected to cause an initial shrinkage; but a re-distribution of KCl and penetration of water back into the tissue should again occur, so that its weight would be expected to return to the original value.

In the taenia coli of the guinea-pig, Casteels & Kuriyama (1966) have shown that, contrary to expectation, no tissue swelling occurs when the NaCl in the Krebs solution is replaced by KCl. Moreover, addition of KCl to normal Krebs solution causes a shrinkage of the tissue, but no recovery of weight, and no penetration of KCl occurs.

Further investigations of the osmotic effects of potassium salts of different anions on the volume of the taenia coli of the guinea-pig have been recently carried out (Brading & Tomita, 1968a, c). Three anions were used, chloride, nitrate and benzenesulphonate. Two types of experimental solutions were used, isosmotic solutions in which the NaCl in the medium was replaced with the potassium salt, and hyperosmotic solutions made by adding the potassium salt to normal medium to double the osmolarity. If Krebs solution was used as the normal medium, in the isosmotic KCl solution, no swelling of the tissue occurred, a slight swelling occurred in KNO_3 solution, but much less than predicted by the membrane theory, and in the benzenesulphonate solution a shrinkage of the tissue was seen, that appeared to be due to loss of KCl from the tissue. In the hyperosmotic solutions, the three potassium salts had the same effect, causing a shrinkage of the tissue, but no recovery of tissue weight was observed up to 2 hr in the solutions. Removal of calcium from the medium had no effect on the behaviour of the tissue in the hyperosmotic solutions, but in the isosmotic solutions a greater differentiation between the three anions was seen. If Locke solution was used as the normal solution (see Table 1), swelling of the tissue was greater in isosmotic KCl and KNO_3 solution than in the corresponding Krebs solution, and this swelling was greatly enhanced in calcium free Locke solution. Table 2 summarizes the tissue weights after 1 hr exposure to the various solutions. In Fig. 5 the recovery of the weight of tissues in hypertonic

TABLE 1. Composition of media

(mM)	NaCl	KCl	CaCl₂	NaHCO₃	MgCl₂	NaH₂PO₄	Glucose	O₂/CO₂ (%)
Krebs	120·4	5·9	2·5	15·5	1·2	1·2	11·5	97/3
Locke	154·0	5·6	2·2	1·8	—	—	5·6	100/0

TABLE 2. Tissue weight (%) after 1 hr exposure (*30 min)

	Krebs solution		Locke solution	
	Normal Ca²⁺	0 Ca²⁺	Normal Ca²⁺	0 Ca²⁺
	100	104 ± 1·5*	100	107 ± 0·9*
Isosmotic solution (NaCl replaced by K salts)				
NO₃	109 ± 0·8	119 ± 2·8	117 ± 1·2	155 ± 3·0
Cl	103 ± 2·0	106 ± 2·9	118 ± 1·7	136 ± 2·7
C₆H₅SO₃	85 ± 0·4	80 ± 1·8	91 ± 1·7	99 ± 1·2
Hyperosmotic solution (154 mM K salts added)				
NO₃	78 ± 3·0	75 ± 1·7	90 ± 1·7	98 ± 2·6
Cl	79 ± 0·5	79 ± 1·9	84 ± 1·6	93 ± 2·3
C₆H₅SO₃	76 ± 1·9	72 ± 0·9	78 ± 1·3	88 ± 1·9
Na-free (NaCl replaced by sucrose)				
—	—	—	81 ± 1·3	119 ± 5·4

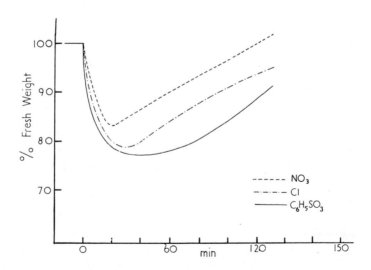

FIG. 5 Changes in the weight of tissues exposed to Ca-free Locke solution of twice normal osmolarity, made by addition of KNO₃, KCl or KC₆H₅SO₃.

7*

calcium free Locke solution is illustrated. The permeability of the membrane to the different anions appears to be in the order of $NO_3 > Cl > C_6H_5SO_3$. Of the various differences between Krebs and Locke solution one factor which appears to be responsible for the different tissue behaviour, is the bicarbonate content (Fig. 6). The behaviour of the tissue in HCO_3-free Krebs solution was very similar to that in Locke solution, in spite of a low pH, and the behaviour in Locke solution with added HCO_3 was very similar to that in normal Krebs solution. However, in other experimental conditions it was found that the external pH, the CO_2 as well as the Ca concentration in the medium could all

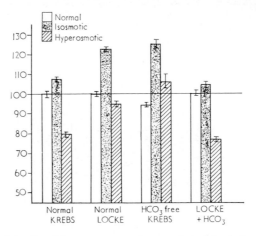

FIG. 6   Tissue weight after 1 hr exposure to the experimental solution. Isosmotic solution was made by replacing the NaCl in the medium with KNO_3, and hyperosmotic solution by the addition of KNO_3 to the normal Locke or Krebs solution, in an amount sufficient to double the osmolarity. Ordinate: Tissue weight as a percentage of the weight in normal Krebs or Locke solution.

influence the volume changes of the tissue. The mechanism by which these factors can change the behaviour of the tissue from one type that cannot be explained by the membrane theory, to the other type which can, is not yet understood. There are two main possibilities which might explain the osmotic effects of potassium salts on tissue volume. (1) In normal Krebs solution the membrane permeability to potassium salts may be so low that osmotically significant amounts cannot penetrate. However, in experimental conditions in which the predicted volume changes are observed, the membrane permeability could be increased by one or several of the factors mentioned above. (2) Some change in internal structure may occur which, in some situations, can limit the amount of swelling the tissue can undergo.

The osmotic effect of electrolytes on smooth muscle is more complicated than that on striated muscle, although it is not known whether, in fact, similar conditions exist that could prevent the predictable behaviour of striated muscle.

The evidence described above could possibly be explained by either the membrane or non-membrane theory, although it seems most likely that both the membrane and the internal structure of the cells are involved.

Summary of volume changes induced osmotically

It may be concluded that, when non-permeating solutes are added to physiological bathing solutions, amphibian smooth muscle, mammalian smooth muscle and striated muscle all exhibit volume changes which are consistent with behaviour as an osmometer. Differences between the behaviour of striated muscle and smooth muscle are reported concerning the osmotic effects of electrolytes, but these appear to be due to secondary factors in the medium and, under some conditions, the smooth muscle behaves in accordance with the membrane hypothesis.

If the bathing solution is diluted, smooth and striated muscle can again behave as osmometers for relatively small dilutions but, with further dilution, smooth muscle in particular appears to lose the semi-permeable nature of the membrane, which becomes less selective in its properties.

In solution of pure solutes, the properties of both mammalian and amphibian smooth muscle can no longer be adequately explained within the framework of the membrane theory and, in addition, differences between the two types of smooth muscle can be observed, as is shown by the response of the tissues to isotonic sucrose solutions.

The factors affecting tissue volume in conditions where the membrane appears to be permeable to solutes, will be discussed more fully later, but there seems to be a range of tonicity over which smooth muscle behaves as if it were a relatively good osmometer, and the next sections will discuss tissue electrolytes under these conditions.

IONIC DISTRIBUTION

If the smooth muscle cell does in fact behave as an osmometer over a certain range of tonicity of the external medium, its intracellular electrolyte content should, theoretically, remain constant, and thus become more or less concentrated as the cell shrinks or swells.

Such a study of the ionic content of taenia coli in relation to tonicity of the medium has been carried out by Brading & Setekleiv, (1968). Figure 7 shows the concentrations of sodium, potassium and chloride ions in the cell water at different tonicities. The lines represent the calculated ion concentrations if the cell had not lost any of its internal ions. This was calculated using the sorbitol space and, since in 0·5 times tonicity the sorbitol probably penetrated the cell, the calculated concentration of ions at this tonicity may be inaccurate. The symbols show the experimentally-determined concentrations, and it can be seen that there is in fact a loss of ions at the higher tonicities. It can be calculated that, under the experimental set up employed for these measurements, at twice the normal tonicity, a loss of sodium plus potassium ions of 24 m-moles/kg intracellular space has occurred after 1 hr (some 10% of the expected total), whereas the cell lost 44 m-moles of chloride, which is about 50% of the expected concentration.

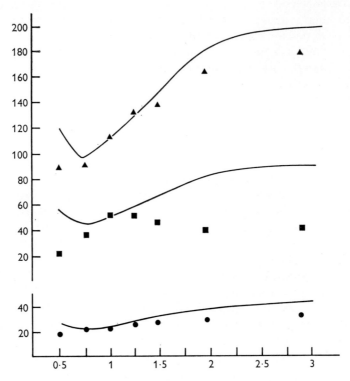

FIG. 7   The expected intracellular concentration of ions, assuming no loss from the cells (continuous line), and the observed concentrations in hypo- and hypertonic solutions. ▲, Potassium concentration m-moles/kg intracellular space; ■, chloride concentration m-moles/kg intracellular space; ●, sodium concentration m-moles/kg intracellular space. Ordinate: concentration (mM). Abscissa: tonicity. (From Brading & Setekleiv, 1968).

A similar loss of chloride ions unaccompanied by loss of sodium or potassium has been observed in the taenia coli by Casteels (1964) when the tissue was placed in a solution in which chloride had been replaced by ethanesulphonate. In this case it would be expected from the Donnan distribution that the cell would loose potassium and chloride, and shrink, but in fact, although slight shrinkage was observed, there was no loss of cation, and the penetration of ethanesulphonate into the cell was insufficient to replace the chloride lost. Casteels postulates the replacement of intracellular chloride ions by bicarbonate. Another explanation could be that changes in protein configuration could occur resulting in the unmasking of hitherto unavailable anionic sites in the cell. Changes in protein configuration are particularly likely to occur in hypertonic solution, where there is a very considerable increase in the ionic strength of the intracellular environment. Unmasking of anionic groups in plasma albumin has been described by Vijai & Foster (1967).

Thus, although the volume change of the cells of the taenia coli is consistent with the membrane hypothesis, nevertheless, on further examination, the distribution of the ions is not exactly predicted by this hypothesis, and other factors obviously play some part in the process.

ION FLUXES

A comparison of the effluxes of sodium, potassium and chloride ions in normal and twice tonicity Krebs solution has been made for the taenia coli (Brading, unpublished). The object was to see if there were any obvious changes in the rates of efflux that would necessitate postulating changes of the membrane permeability to these ions in hypertonic solution. If the membrane of the cells is significantly elastic, then, when shrinkage occurs, the cell membrane area might decrease, the membrane might become thicker and the permeability of the ions might decrease. On the other hand, the cell membrane area might remain constant and be thrown into folds on shrinking.

The complexity of the efflux curves that were obtained in normal as well as hypertonic solutions, made it difficult to compare the effluxes under the two conditions. It was found however, that the curves in every case could be reasonably fitted by the sum of three exponential terms, and followed the equation:

$$P = Ae^{-\lambda_1 t} + Be^{-\lambda_2 t} + Ce^{-\lambda_3 t} \tag{vii}$$

where P is the counts in the tissue, and t the time in minutes from the beginning of the wash out. A, B and C are the intercepts at zero time of the three component terms, and λ_1, λ_2 and λ_3 are the rate constants (min^{-1}) determined from the half time of the components in minutes (k) such that $\lambda = 0.693/k$.

The six variables could be estimated for each experiment, and these values were compared for tissues in normal and hypertonic solutions. In Table 3 can be seen the average values of the variables, and in Fig. 8 the method of graphical analysis used, and the computed curves of the average results for ^{42}K, ^{36}Cl and ^{24}Na in the two solutions.

In the potassium efflux, the two faster phases of the efflux are only a small proportion of the whole, and the errors involved in the calculations are large. There was no significant detectable difference in these two phases between normal and hypertonic solution. The slow phase on the other hand, which extrapolates to over 90% of the total tissue counts in both conditions, was

TABLE 3. Constants from three-compartment analysis of flux data

$$P = Ae^{-\lambda_1 t} + Be^{-\lambda_2 t} + Ce^{-\lambda_3 t}$$

The half times for each component $t_{1/2} = \dfrac{0 \cdot 693}{\lambda}$

		Normal			Hypertonic		
		Na	K	Cl	Na	K	Cl
A	m-moles	$45 \cdot 1 \pm 2 \cdot 9$	$3 \cdot 0 \pm 0 \cdot 5$	$35 \cdot 2 \pm 2 \cdot 0$	$35 \cdot 5 \pm 3 \cdot 2$	$4 \cdot 3 \pm 1 \cdot 2$	$38 \cdot 1 \pm 1 \cdot 7$
B	kg fr wt	$18 \cdot 9 \pm 1 \cdot 8$	$2 \cdot 8 \pm 0 \cdot 3$	$13 \cdot 8 \pm 1 \cdot 7$	$27 \cdot 1 \pm 2 \cdot 7$	$3 \cdot 1 \pm 0 \cdot 5$	$15 \cdot 6 \pm 1 \cdot 1$
C		$1 \cdot 77 \pm 0 \cdot 5$	$58 \cdot 9 \pm 3 \cdot 2$	$15 \cdot 0 \pm 2 \cdot 0$	$1 \cdot 78 \pm 0 \cdot 5$	$61 \cdot 9 \pm 2 \cdot 9$	$5 \cdot 9 \pm 1 \cdot 2$
λ_1 min^{-1}		$0 \cdot 96 \pm 0 \cdot 04$	$1 \cdot 21 \pm 0 \cdot 08$	$1 \cdot 10 \pm 0 \cdot 05$	$1 \cdot 01 \pm 0 \cdot 04$	$0 \cdot 98 \pm 0 \cdot 17$	$0 \cdot 95 \pm 0 \cdot 1$
$t_{1/2}$		$0 \cdot 72$ min	$0 \cdot 6$ min	$0 \cdot 63$ min	$0 \cdot 68$ min	$0 \cdot 7$ min	$0 \cdot 73$ min
λ_2 min^{-1}		$0 \cdot 29 \pm 0 \cdot 01$	$0 \cdot 21 \pm 0 \cdot 01$	$0 \cdot 19 \pm 0 \cdot 01$	$0 \cdot 35 \pm 0 \cdot 02$	$0 \cdot 20 \pm 0 \cdot 02$	$0 \cdot 21 \pm 0 \cdot 01$
$t_{1/2}$		$2 \cdot 4$ min	$3 \cdot 3$ min	$3 \cdot 62$ min	$2 \cdot 0$ min	$3 \cdot 4$ min	$3 \cdot 3$ min
λ_3 min^{-1}		$0 \cdot 035 \pm 0 \cdot 004$	$0 \cdot 011 \pm 0 \cdot 001$	$0 \cdot 055 \pm 0 \cdot 004$	$0 \cdot 031 \pm 0 \cdot 005$	$0 \cdot 015 \pm 0 \cdot 001$	$0 \cdot 055 \pm 0 \cdot 008$
$t_{1/2}$		$19 \cdot 5$ min	$64 \cdot 7$ min	$12 \cdot 69$ min	$22 \cdot 5$ min	$46 \cdot 6$ min	$12 \cdot 58$ min

significantly faster in the hypertonic solution, having an average half time of 65 min in normal, and 47 min in hypertonic solution.

The slow component of the sodium efflux was small and not significantly different in the two media, being 2% to 3% of the counts. The middle phase however was slightly, but consistently and significantly faster in the hypertonic solution, the half time being 2·4 min in normal solution, and 2·0 min in hypertonic solution. The rate and extrapolation of the fast compartment was not significantly changed in hypertonic solution.

Analysis of the efflux curves for chloride showed no significant difference between the rates for the three components in the different solutions, but there was a very significant difference in the size of the slow component, which was 25% of the total efflux (15 m-moles/kg fr wt) in normal solution, and 11% (6 m-moles/kg fr wt) in the hypertonic solution.

The ionic content of the tissues estimated from the sum of the three components at zero time, shows that, under these conditions, there is no significant difference in the cell content of sodium and potassium ions, but there is a significant loss of chloride ions in the hypertonic solution.

Since the description of the fluxes as the sum of three exponential terms is a mathematical convenience only, the interpretation of the results is very difficult. Even if the three components had their physical counterparts, the size of the components would not be given by the extrapolations to zero time of the components, due to back diffusions of radioactive material; and although mathematical corrections can be applied, these are very long, and can introduce inaccuracies resulting from the arbitrary assumption of the arrangement and interconnections of the compartments. The actual permeability of the membrane to the ions cannot be calculated unless the concentration of the ion on each side of the membrane involved is known, and as the curves are so complex, it is not

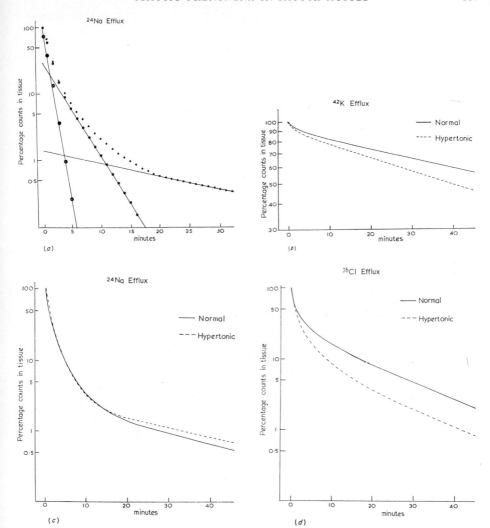

FIG. 8   (a) Method of analyses of efflux curve. ($\bullet$) Experimental results, i.e. percentage radioactivity in the tissue on a logarithmic scale plotted against time after beginning the washout of the activity. After 20 min the points can be considered to fall on a straight line. The middle component of the curve is obtained by subtracting the line from the experimentally determined points. The values of this second component ($\bullet$) can also be considered to fall on a straight line after about 5 min, and subtraction of this line from the non linear points gives the third component, the values for which ($\bullet$) can all be considered to fall on a straight line. The slopes of the three lines represent the rate constant for the three terms, and the extrapolation of the line to zero time gives the values A, B and C. Efflux curves of ^{42}K (b), ^{24}Na (c), ^{36}Cl (d) in normal and hypertonic solution. These curves are average curves and have been computed from the equation $P = Ae^{-\lambda_1 t} + Be^{-\lambda_2 t} + Ce^{-\lambda_3 t}$ where P is the percentage of the counts in the tissue at time t, and the values of the six variables are the average values shown in Table 3.

appropriate to assume that all the intracellular ions are dissolved in all the intracellular water. Nevertheless, the fact that the curves are so little changed in hypertonic solution, might suggest that there is no drastic change in the membrane permeability properties. The increase in rate of flux of the slow component for potassium efflux, and of the middle component of sodium efflux, could be explained by the increase in concentration of the ions that must occur within the cell. The loss of chloride from the slow compartment reflects the loss that was noted in the total ions, as described earlier.

CYTOPLASMIC STRUCTURE LIMITING VOLUME CHANGE

Under certain conditions the smooth muscle of the frog stomach does not behave as predicted for an osmometer. Bozler has investigated the behaviour of this muscle in some detail (1959, 1961a, b, 1962, 1965), and the following paragraphs summarize some of his findings.

As has been mentioned previously, in most hypotonic solutions and in solutions of pure non-electrolytes, the membrane seems to lose its selective permeability; it thus no longer governs the volume of the cell. In hypotonic solutions whose sole constituent is 2 mM $CaCl_2$ or $MgCl_2$, another phenomenon is apparent. The tissue only swells one fifth the amount that it swells in distilled water, or even in equivalent concentrations of monovalent cations. The increase in volume appears to be limited by the presence of the divalent cations. A study of the tissue ionic content in these solutions shows that the volume becomes constant at a time when the concentration of cations in the muscle is many times higher than in the outside medium. The inulin and sucrose spaces in these muscles are small, and the muscles will shrink on addition of 0·22 M sucrose (about isosmotic with the Ringer) indicating that the sucrose is largely excluded from the majority of the cell water.

If tissues, that have been exposed to these hypotonic divalent cation solutions, are returned to normal Ringer solution they swell further, despite the increase in tonicity of the medium. The 'selective permeability' appears to be lost as the sucrose and inulin spaces increase dramatically.

In pure isosmotic sucrose solutions, the frog stomach muscle swells, as has been mentioned. Addition of even less than 1 mM $CaCl_2$ or $MgCl_2$ to the sucrose solution causes a shrinkage of the tissue.

These solutions of low concentrations of $CaCl_2$ and $MgCl_2$ also induce striking changes in the physical properties of the muscles. The tissues become opaque and stiff, and the stress relaxation becomes many times slower, indicating that these cations can transform the fibres into a stiff gel.

From these results, Bozler concludes that under conditions of low external osmotic pressure (where it appears that many of the semi-permeable properties of the cell membrane are lost) divalent cations accumulate in the fibres and produce changes in the physical properties which indicate increased internal cross linking. Swelling is then limited as in a gel, and considerable hydrostatic pressure may develop within the fibres.

Bozler's results have been confirmed and extended by Sparrow, Mayrhofer & Simmonds (1967). Figure 9, reproduced from their paper, shows the effect of deionized sucrose solutions at different concentrations on the volume of frog stomach muscle, with and without addition of 1 mM CaCl to the solution.

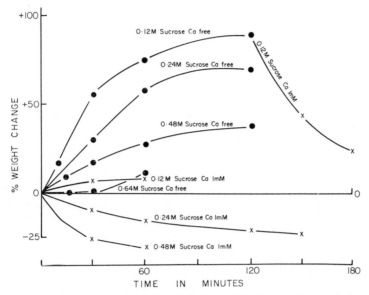

FIG. 9   Mean percentage weight changes of strips of toad stomach muscle in sucrose solutions of varying tonicity, ● = deionized sucrose, × = sucrose + Ca 1 mM. Weight change expressed as percentage weight increase or decrease. (From Sparrow, Mayrhofer & Simmonds, 1967)

It can be seen that in order to prevent swelling in deionized sucrose, a concentration of 0·64 M sucrose had to be used, that is about 2·5 times the osmotic strength of the normal Ringer solution, and even in this solution the tissue tended to gain weight after 30 min exposure. In the sucrose solution containing Ca, swelling was prevented. The cells accumulated calcium to a greater concentration than in normal Ringer, and this accumulated calcium appeared to be unavailable for activating the contractile machinery of the cell, making it likely that Ca was bound to sites not involved in the contractile response. If Mg was added to the deionized sucrose, the effect was similar to that of adding Ca, and the cations were accumulated to the same extent.

These authors also showed that in 0·5 isosmotic sucrose solution, although the tissue lost most of its sodium, only 30% of the potassium was lost even after 2hr exposure to the solution. This result was not affected by the addition of 1 mM CaCl to the sucrose, although without the calcium the tissue swelled by about 70% of its original weight, and only about 30% in the presence of calcium. Thus the volume of the tissue was not related to its ionic content.

EFFECT OF TONICITY ON ELECTRICAL AND MECHANICAL ACTIVITY

Striated muscle

It is perhaps relevant, before considering smooth muscle, to summarize the work done on the effects of hypertonic solutions on the electrical and mechanical activity in striated muscle. Most of the work stems from the observations of Hodgkin & Horowicz (1957) that hypertonic solution differentially affected the electrical and mechanical activity in striated muscle. In 2·5 times osmotic pressure (solutions made hyperosmotic with NaCl) the twitch of the muscle in response to applied electrical stimulation declined to zero, whilst the membrane was still excitable, and conducted normal action potentials. Howarth (1958) investigated the mechanical properties of frog sartorius muscle in hypertonic and hypotonic solutions. His results showed that the velocity of shortening of the muscle and the maximum tetanic tension developed were both reduced in hypertonic solution, but unaffected in hypotonic solutions. Although the twitch of the muscle to a single stimulus was abolished in hypertonic solutions of 2·5 to 3 times the normal concentration, he found that a series of action potentials still provoked activity in the contractile proteins, as shown by an increased resistance of the muscle to stretch. Howarth concluded that the failure to develop tension was due to the slowness of the shortening compared with the brief duration of the active state. He suggested that this slowness could be due to increased intracellular viscosity, or to some effect of the increased intracellular ionic concentration (the muscle shrinks) on the contractile proteins. Hill (1958a) working on the relationship between heat production and twitch tension in muscles in hypertonic solutions, showed that with hypertonicities of 1·9 times that of Ringer solution, the onset of the tension response was markedly delayed although otherwise not readily distinguishable from muscle responses in normal solution. The onset of heat production preceded the tension response in both cases, and occurred with about the same latency from the stimulus. If the solution was made more hypertonic (3 times that of Ringer solution) the mechanical response to a single stimulus was abolished, but heat production still occurred, the total being about 32% of that produced in normal solution. This heat was equal to Hill's 'activation heat', the heat of shortening and relaxation being lost.

More recent work has however made it likely that some step in the E/C coupling is affected by hypertonic solutions, rather than the contractile apparatus. Yamaguchi, Matsushima, Fujino & Nagai (1962) showed that although Ringer solution made hypertonic with glycerol (420 mM) caused a shrinkage of striated muscle, and initially a reduction in twitch size, yet recovery of twitch response occurred while the muscle was still in the glycerol Ringer. A small amount of twitch recovery also occurred in muscles exposed to solutions made hypertonic with Urea. These results suggest that it is the E/C coupling rather than the contractile apparatus that is usually affected by hypertonicity. This is also suggested by the work of Fujino & Fujino (1964), who showed that pre-treatment of the muscle in Ringer solution where NaCl had been replaced by KCl or K_2SO_4, prevented the loss of the twitch response on subsequent immersion of the tissue in hypertonic Ringer with the normal amount of potassium. Caputo (1966) used caffeine and potassium to induce contractions in frog striated muscle. Caffeine acts directly on the contractile machinery without involving depolarization of the membrane, whereas potassium contractures are mediated by depolarization of the membrane. In solution made hypertonic by addition of NaCl or sucrose, Caputo found that caffeine-induced contractures were in fact potentiated, whereas the K-induced contractures were reduced. These results further support the view that the differential effect of hypertonic solution on the electrical and mechanical activity, is caused by some interference in E/C coupling.

In summary, it is clear that under certain conditions normal contractions may be elicited in striated muscle in hypertonic solutions. Thus the contractile apparatus is still functional, while some step in the E/C coupling appears to be affected by hypertonic solutions. The work of Howarth and Hill, however, shows that a considerable amount of the E/C coupling is still intact. Since the caffeine contracture is not reduced in hypertonic solution, and since it has been shown (Ashley, 1967) that caffeine induces the intracellular release of bound calcium, it is tempting to suggest that the release of calcium by depolarization of the membrane may in some way be reduced in hypertonic solutions.

Smooth muscle
Intracellular recording

The effect of tonicity on the electrical and mechanical activity of smooth muscle is complicated in many instances by the fact that spontaneous activity occurs, and that the tension response of the muscle is normally a function of the spike frequency (see chapter by Axelsson). Changes in mechanical activity may thus be caused by effects of tonicity on the spontaneous activity of the tissue.

Holman (1957) studied the effects on the guinea-pig taenia coli, of increasing the tonicity of the bathing solution by the addition of sucrose. The spontaneous activity was inhibited and with the decline of electrical activity the tension fell. Different results were obtained on increasing the osmolarity by addition of NaCl; this treatment resulted in an initial excitation of the tissue, probably caused by depolarization due to penetration of Na^+ and a reduction of internal K^+.

Tomita (1966a) used sucrose to double the osmolarity of the Krebs solution, and found that in this medium the smooth muscle cells of the taenia coli underwent a hyperpolarization of 10 to 15 mV (mean 12 mV), and lost spontaneous electrical activity. The effect was completely reversible. While the tissue was quiescent in the hypertonic solution, spikes could still be obtained by externally applied electrical stimulation, but were not accompanied by any appreciable tension response. The parameters of the spikes in normal and hypertonic solution are shown in Table 4. Tomita was of the opinion that the suppression of spontaneous electrical activity was mainly due to the hyperpolarization of the membrane, especially since depolarization of the tissue with current pulses, or by raising the external concentration of potassium in the hypertonic solution to 18 or 24

TABLE 4. Membrane potential, spike parameters, spontaneous spike frequency and conduction velocity in normal solution and after 15 min exposure to hypertonic solution (From Tomita, 1966a)

Solution	Membrane potential	Spike amplitude	Maximum rate of rise	Maximum rate of fall	Spike frequency	Conduction velocity
	($\pm$S.E.) (mV)	($\pm$S.E.) (mV)	($\pm$S.E.) (V/sec)	($\pm$S.E.) (V/sec)	($\pm$S.E.) (spike/sec)	(cm/sec)
Normal Krebs	51 ± 0.6	55 ± 0.7	7.0 ± 0.4	7.2 ± 0.5	1.2 ± 0.8	$6.7 - 8.8$
Hypertonic Krebs	63 ± 1.1	72 ± 1.6	10.5 ± 0.4	8.8 ± 0.7	0	7.3 ± 0.7

mM, caused the spontaneous electrical activity to return (although the mechanical response did not recover). Brading & Setekleiv (1968) have shown that the intracellular potassium concentration rises in two times hypertonic solution to an extent that would change the potassium equilibrium potential from -87.5 to -102 mV, which is quite enough to account for the observed hyperpolarization.

Other observations

Earlier, Barr, Dewey & Evans (1965) and Barr, Berger & Dewey (1968), observed that hypertonic solutions decreased the nexal area, and thus the electrotonic coupling between the smooth muscle cells in the taenia coli, and they concluded that the spike activity was in this way abolished. This conclusion has not been substantiated by Cobb & Bennett (1969), nor by Nishihara (1969) who found that the nexus is not affected by treatment with hypertonic solution. Moreover, Tomita (1966a, 1967b) showed that spikes could be evoked in taenia coli, and were propagated normally in hypertonic solution at a velocity of 7.3 cm/sec. Bülbring, Burnstock & Holman (1958) give values for the propagation in normal solution of 6.7 to 8.8 cm/sec. More recently Tomita (1969), from measurements of the longitudinal tissue resistance in normal and hypertonic solutions, considers that the function of the junctions is preserved in solutions of up to twice the normal osmolarity. The tissue shows cable properties in these solutions. The return of spontaneous activity in hypertonic solution with increased external potassium also strongly suggests that there is no marked disruption of the junctions between the cells.

The absence of tension development, though a normal action potential can be evoked in hypertonic medium, and the failure of the tension response to recover, though spontaneous spike activity returns in hypertonic medium containing excess K, indicates that in smooth muscle, as in striated muscle, hypertonicity does have a differential effect on the electrical and mechanical activity. Axelsson (1961b) has shown that dissociation between electrical and mechanical activity can be brought about by several means, but for a fuller description of this phenomenon, the reader is referred to the chapter by Axelsson in this book.

In hypertonic solutions the taenia coli becomes stiff and considerably less extensible than in the normal solution. It is possible that the increased intracellular ion concentration has some effect on the contractile proteins as was postulated by Howarth for skeletal muscle, and that this also accounts for the reduced mechanical response to spike activity.

In the smooth muscle cells of taenia coli, it is thought that calcium ions pass into the cells through the membrane during the spike, probably carrying the action current (see Chapters by Casteels, Goodford and Tomita). If the amount of Ca entering the cell is sufficient to activate the contractile mechanism it is less likely that in this tissue the dissociation of electrical and mechanical events is caused by some interference in a calcium release mechanism, since normal spikes can be produced without tension response. On the other hand, a calcium release from, and uptake into, cellular stores may also play a part in excitation contraction coupling of smooth muscle, and may be disturbed by treatment with hypertonic solution.

Extracellular records and vascular smooth muscle

Setekleiv (personal communication) has recorded the mechanical and electrical activity of the taenia coli in the sucrose gap, and studied the effect of changing from normal Krebs solution to hypotonic solution. There was an initial depolarization of the tissue associated with an increased spike activity and the development of a sustained contraction which was terminated later through depolarization block of the spike activity.

Similar sucrose gap recording of the mechanical and electrical activity of the smooth muscle of the rat portal vein have been made by Johansson & Jonsson (1968). These authors investigated the effect of increasing the tonicity of the bathing solution with sucrose, xylose and urea, and of decreasing the tonicity by ommission of part of the NaCl from the medium. The preparation is spontaneously active in the physiological salt solution, contractions occurring at the rate of about 2/min, associated with bursts of electrical activity (see Chapter by Golenhofen). Addition of sucrose or xylose to the medium, reduced the frequency of the spontaneous activity, and thus the contraction (Fig. 10). On washing out the frequency returned to normal. Addition of urea, on the other hand, caused transient reduction in frequency but this returned to normal during exposure, and when the normal solution was returned, there was a transient excitation, and increased frequency of the spontaneous activity. A similar excitation was observed in hypotonic solution. The results could be linked to the volume changes that are presumed to occur in the various solutions. Excitation is associated with situations in which swelling of the tissue might occur, and inhibition when shrinkage might occur. There was also evidence that in conditions when shrinkage would be expected, a decrease in the size of the

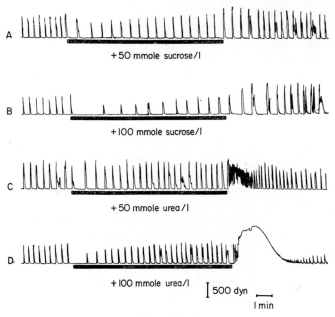

FIG. 10 Isometric recordings of mechanical activity in isolated rat portal vein. Responses to increased osmolarity produced by adding sucrose (A and B) and urea (C and D) to the physiological salt solution (PSS). (From Johansson & Jonsson, 1968)

mechanical response occurred. The authors do not consider that the changes in concentration of the intracellular ions would be sufficient to explain the change in electrical activity, and presume that ionic membrane permeability changes must also have occurred. The authors, however, made no measurements, and their theoretical treatment took no account of the dry weight of the cells, and so probably underestimated the magnitude of the changes.

Mellander, Johansson, Gray, Jonsson, Lundvall & Ljung (1967) attempted to correlate the vasodilation of vascular smooth muscle in skeletal muscle in response to exercise hyperaemia with changes caused by hypertonic solutions. Intra-arterial infusion of hypertonic glucose or xylose solutions into resting skeletal muscle at rates producing similar changes in regional osmolarity to that found in exercise, elicited a pattern of vascular response resembling that seen during work. Records of isolated vascular smooth muscle showed that increased osmolarity exerted its effect by inhibiting myogenic pacemaker activity, possibly by hyperpolarization of the membrane due to the increased intracellular potassium concentration. The authors also postulate an interference of the hyperosmolarity with propagation and excitation contraction coupling.

Summary

In striated muscle the study of the effect of tonicity has largely been confined to the differential effect of tonicity on the spike and the twitch tension. In spontaneously active smooth muscle however, study of the effects of tonicity is more complicated, due to the fact that tonicity may affect the membrane potential, spike initiation and propagation, any of which may result in alteration of the frequency of spontaneous activity, and thus tension. Nevertheless, it appears that there is also a differential effect on the mechanical and electrical activity similar to that described in striated muscle, although little in the way of quantitative work has been carried out on this aspect of smooth muscle activity.

FINAL DISCUSSION

It is apparent that changes in the osmolarity of the medium bathing smooth muscle, have many more effects than simply movement of water into and out of cells. Dilutions of the medium beyond a certain point alter the membrane properties. Alterations in the internal or external concentration of electrolytes result in changes of membrane potential which themselves may alter the mechanical properties of the muscle, and possibly have effects on metabolism which may also cause changes in membrane properties.

These secondary effects may be relatively more important in smooth muscle than in striated muscle, because they may affect processes maintaining the spontaneous activity of smooth muscle. Moreover, there is a great difference in the surface area volume ratio between the two tissues. The smooth muscle, due to the smallness and the shape of the cells, has a relatively much larger amount of membrane than striated muscle, and this means that the majority of the cell cytoplasm in smooth muscle is much nearer to the cell membrane than it is in striated muscle.

In spite of all the complexities of the effects of osmotic pressure, many of the results described in this article, if considered in isolation, could be used as evidence for the membrane theory, and many as evidence for the non-membrane theory. Neither of these two theories can satisfactorily explain all the results, and it would seem more profitable to propose models combining features of both theories, than to decide in favour of one or the other.

Under certain conditions, and particularly when the osmolarity of the physiological bathing solutions is raised by addition of non-permeating non electrolytes, smooth muscle behaves very much as predicted for a perfect osmometer. This is strong evidence in favour of the membrane theory but does not necessarily preclude the presence of structured water or bound potassium, since no assumptions are made as to the strength or stability of binding or structure inside the cell. The 'bound' water and potassium may be in equilibrium but freely exchanging with that in solution, and changes in the free solution due to osmotic effects could alter the bound substances in such a way that all the water in the cell becomes osmotically available.

In hypotonic media, on the other hand, and in conditions where the cells are exposed to an abnormal extracellular ionic environment, the smooth muscle cell membrane appears to lose its selective permeability, and under these conditions, the intracellular substance becomes much more important in determining the volume and ion content of the tissue.

The shrinkage of smooth muscle induced osmotically can most easily be attributed to a 'semi-permeable' cell membrane, but the swelling of the tissue seems to be limited by mechanisms which could easily be intracellular, and are affected by the amount of available energy, the calcium concentration, the pH, buffers and other factors.

7

ELECTRICAL PROPERTIES OF MAMMALIAN SMOOTH MUSCLE

T. TOMITA

INTRODUCTION

It has been known for some time that intestinal smooth muscle behaves like a syncytium (Bozler, 1948). Since smooth muscle consists of many short fibres (100 to 400 μ in length, see Chapter 1) in series and in parallel, the excitation necessary for physiological function must be transmitted from cell to cell.

Prosser (1962) proposed the following possibilities for spike propagation through smooth muscle. (1) Conduction mediated through nerve fibres which run parallel to the muscle fibres, (2) Conduction by mechanical pull from muscle fibre to muscle fibre, (3) Conduction by a chemical transmitter between muscle cells, and (4) Electrical conduction through the smooth muscle as in a syncytium. The last possibility is generally thought to operate in visceral smooth muscle, i.e. spike propagation by local circuit current through low resistance connections between cells (Prosser, 1962; Barr, 1963; Nagai & Prosser, 1963a, b; Barr, Dewey & Evans, 1965).

In electrophysiological terms, propagation (or conduction) is used when the excitation propagates within the same cell while transmission is used for the spread of excitation from one cell to another cell through a junction (or a synapse). Therefore, strictly speaking, the term transmission should be used in smooth muscle, especially when the significance of the junction between cells is emphasized. However, since the tissue behaves as a single cell in many respects, in this chapter the term propagation (or conduction) is used instead.

It is now generally believed that the 'nexus' or 'tight junction' is responsible for electrical coupling between cells in many smooth muscles (see Chapter by Burnstock). In many other tissues, in which cells are connected with relatively low resistance, the existence of the tight junction has been proved (cardiac muscle: Weidmann, 1952; Barr & Berger, 1964; Barr, Dewey & Berger, 1965; nerve cells: Furshpan & Potter, 1957, 1959; Watanabe & Grundfest, 1961; Hagiwara & Morita, 1962; Eckert, 1963; Bennett, Aljure, Nakajima & Pappas, 1963; Bennett, M.V.L., 1966; other tissues, e.g. epithelial tissue: Loewenstein, 1966; 1967a, b).

In order to understand the electrophysiological properties, passive or active, of smooth muscle, it is most important to study cell to cell interaction. This will be the main topic of this chapter. The description will be mainly confined to observations on the taenia coli and the vas deferens of the guinea-pig because they are the most thoroughly investigated. The technical difficulties in dealing with smooth muscle are such that it is not easy to obtain convincing results, and it is unfortunate that much of the published data cannot be taken as convincing evidence for the electrical properties although the publications contain interesting hypotheses or conclusions.

PASSIVE PROPERTIES

Membrane polarization and spike activity

It is well known that intestinal smooth muscle behaves like a syncytium in that excitation starts at the cathode when current is passed through the tissue, as Bozler showed (1938a.) The phenomenon has been further investigated using intracellular recording of the responses to external stimulation. The difficulty, when using the taenia coli, is that the shape of the electrotonic potential is distorted by the spontaneous spike activity and that the evoked spike differs depending on the refractoriness of the membrane following the preceding

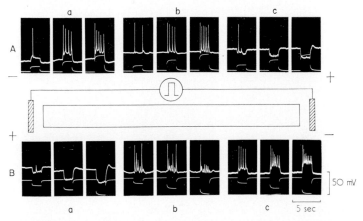

FIG. 1   Electrical responses of guinea-pig taenia coli recorded intracellularly (upper trace) from three different cells to stimulating currents with three different intensities (lower trace) applied externally. (Arrangement of the stimulating electrodes shown in the inset). The records of the upper row A were taken when the left stimulating electrode was the cathode and the right electrode the anode; the records of the lower row B were taken at reversed polarity. In each row, the three records on the left (a) were taken from the same cell near the left stimulating electrode, the three middle records (b) from a cell near the centre of the tissue, and the three right records (c) from a cell near the electrode on the right. The relative current intensities are shown in the lower trace. (From Tomita, 1966a)

spontaneous spike. This can be avoided by using hypertonic solution, made by the addition of sucrose (292 mM) to the medium, which suppresses the spontaneous activity (Holman, 1957). Spikes can still be produced by depolarization of the membrane and can propagate along the tissue (Tomita, 1966a). The possible effect of hypertonic solution is discussed in the section dealing with the longitudinal internal resistance (p. 208).

Figure 1 shows intracellular recording of the responses of the membrane of the guinea-pig taenia coli to stimuli of 2 sec duration at three different intensities and at three different sites when the tissue is placed between two stimulating electrodes, one at each end. The cells near the cathode are depolarized, those near the anode are hyperpolarized, while those in the middle part of the tissue show no polarization. With a weak stimulating current spikes are recorded over the whole length of the tissue. As the stimulating current intensity is increased the spike frequency increases near the cathode and in the middle part of the tissue, but spikes are abolished near the anode. Here small potential changes can still be observed at a frequency which is the same as that of the spikes in the cells near the cathode.

The only possible explanation of the relation between membrane polarization and the polarity of the stimulating electrodes is that the smooth muscle is composed of a series of cells connected with relatively low electrical resistances. The spike is evoked by depolarization of the membrane where current passes

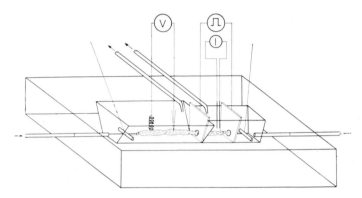

Fig. 2   Recording and stimulating arrangements. The muscle chamber is divided into two compartments, stimulating (*right*) and recording (*left*), by one of the stimulating electrodes. The surface exposed to the recording chamber is coated with araldite which acts as an insulating partition. The compartments are irrigated separately with hypertonic Krebs solution (arrows show direction of the flow). (From Abe & Tomita, 1968)

outward across the surface membrane; it propagates along the tissue and its propagation can be blocked when the membrane is hyperpolarized (Tomita, 1967b).

The responses to extracellular membrane polarization can be investigated more precisely by intracellular recording from the extrapolar region of the tissue which is separated from the stimulating site by an insulating partition (Tomita, 1966a; Abe & Tomita, 1968). The two stimulating electrodes are placed close together (5 to 10 mm apart) at one end of the tissue and an insulating partition is placed between stimulating and recording chamber as shown in Fig. 2.

With this method the results are similar to those described above. Figure 3a shows the responses to cathodal stimulation (the electrode near the recording chamber being negative) and, Fig. 3b, to anodal stimulation, recorded from cells close to the partition. The number of spikes produced by a stimulus increases with its intensity up to a certain limit, beyond which increasing the intensity reduces the number of spikes to one. The peak of the spike remains almost constant when the current intensity is increased. This is one of the indications that the electrotonic potential is actually produced across the membrane. During a long anodal current pulse small repetitive potential changes are often seen. This problem will be described further in the section concerning the spike propagation.

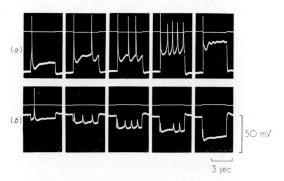

FIG. 3   Intracellular records of the responses of the taenia coli to external polarization. (*a*) Cathodal stimulation. (*b*) Anodal stimulation. The upper trace in each record is the stimulus, and the lower trace the response. All records are from the same cell close to the partition (about 200 μ distant). (From Tomita, 1966*a*)

Cable properties

Shuba (1961, 1965) has shown, with external recording, that an electrotonic potential spreads with an exponential decay along the smooth muscle of frog stomach, suggesting cable-like properties. It has also been shown that not only the taenia coli, which is spontaneously active, but also the vas deferens of the guinea-pig, which is quiescent and physiologically activated through the nerve, have cable-like properties (Tomita, 1966*a*, 1967*a*; Abe & Tomita, 1968). This result is surprising because the structure of the tissue is very different from that of a simple core conductor, e.g. a nerve or a skeletal muscle fibre, even if low-resistance connexions between cells are taken into account.

As will be described, smooth muscles can be represented electrically by a series of independent cables formed by end-to-end connexions between cells. Although interconnexions exist not only in the longitudinal direction but also in the transverse direction, the latter can be disregarded in external polarization, since the tissue is at equi-potential in the transverse direction.

Properties of electrotonic potentials

Spatial decay of the electrotonic potential. By using the method shown in Fig. 2, electrotonic potentials can be recorded intracellularly at different distances (ranging from 0·2 to 10 mm) from the insulating partition. Figure 4 shows examples of the electrotonic potential produced by anodal current pulses

(400 msec) of the same intensity and recorded at three different distances from the partition in the guinea-pig taenia coli.

Current-voltage relations during the steady state of electrotonic potentials at different distances from the insulating partition are plotted on a graph. In this relationship the longitudinal potential field along the tissue in the stimulating chamber is taken for the current axis, assuming that the intensity of the current which flows through the tissue is proportional to the applied potential field.

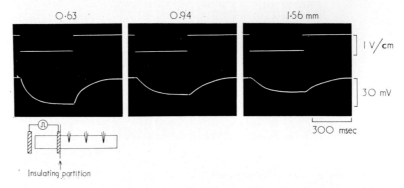

FIG. 4   Electrotonic potentials (lower trace) produced in the taenia coli by externally applied anodal current (upper trace) of constant intensity, and recorded intracellularly at three different distances from the stimulating electrode, as shown in the inset.

From the current-voltage relations, the electrotonic potentials produced by a given current intensity are replotted on a logarithmic scale against the distance from the partition, as shown in Fig. 5. This relation is roughly linear, indicating that the electrotonic potential decays exponentially along the tissue. The average value of the space constant is 1·5 mm in the taenia and 2·1 mm in the vas deferens, which is the distance at which the electrotonic potential decays to $1/e$ (37%).

Time course of the electrotonic potential. The time course of the electrotonic potentials in the taenia (Tomita, 1966a; Abe & Tomita, 1968) and in the vas deferens (Tomita, 1967a) has been compared with the theoretical time course using the following cable equation (Hodgkin & Rushton, 1946):

$$V_{x,t} = V_{x=0,t=\infty} \frac{1}{2} \left\{ e^{-x/\lambda} \left[1 + \mathrm{erf} \left(\frac{x/\lambda}{2\sqrt{t/\tau_m}} - \sqrt{t/\tau_m} \right) \right] \right.$$
$$\left. + e^{x/\lambda} \left[1 - \mathrm{erf} \left(\frac{x/\lambda}{2\sqrt{t/\tau_m}} + \sqrt{t/\tau_m} \right) \right] \right\} \quad (1)$$

where $V_{x,t}$ is the electrotonic potential which is a function of distance (x) from the partition and of time (t) after current is applied ; λ is the space constant; τ_m is the membrane time constant; erf is the error function; $V_{x=0,t=\infty}$ is the potential during the steady state at the partition ($x = 0$). $V_{x=0,t=\infty}$ is obtained by multiplying $V_{x,t=\infty}$ (the potential during the steady-state at a given distance (x)) with $e^{x/\lambda}$, since equation (1) becomes as follows when $t = \infty$,

$$V_{x,t=\infty} = V_{x=0,t=\infty}e^{-x/\lambda} \qquad (2)$$

The membrane time constant (τ_m) is obtained by inserting the experimentally obtained λ into equation (1) and by finding the best theoretical curve fitting the experimental results. The theoretical curve usually fits the experimentally obtained electrotonic potentials fairly well at any distance as shown in Fig. 6. The average value of the time constant obtained by this method is 107 msec in the taenia and also about 100 msec in the vas deferens.

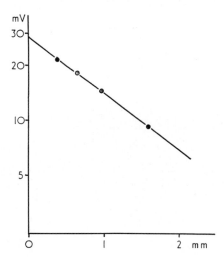

FIG. 5 Spatial decay of the electrotonic potential of the taenia coli. Points were obtained from the current-voltage relations at a constant current intensity (0·95 V/cm) at different distances. Ordinate: intracellularly recorded potential on a logarithmic scale. Abscissa: distance from the stimulating site. (From Abe & Tomita, 1968)

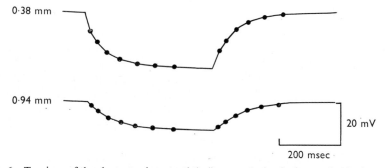

FIG. 6 Tracings of the electrotonic potentials (hyperpolarization) recorded in the taenia coli at 0·38 mm (*upper*) and 0·94 mm (*lower*) from the stimulating site and produced by nearly the same current intensity. Dots were calculated from the cable equations by inserting $\lambda = 1·31$ mm and $\tau_m = 110$ msec. (From Abe & Tomita, 1968)

As pointed out by Hodgkin & Rushton (1946) and Katz (1948), the time to reach the half amplitude of the electrotonic potential increases linearly with distance along the cable, the slope of which is expressed by $\tau_m/2\lambda$. In the taenia coli the relation is also linear and the time constant obtained by this method is 103 msec.

Electrical model of the smooth muscle. As will be described later, the contribution of the junctional resistance between cells measured longitudinally is of a similar order of magnitude to that of the myoplasmic resistance (Jones &

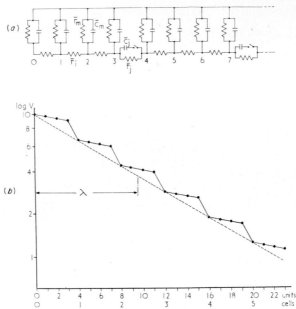

FIG. 7 (*a*): circuit diagram of a model of the smooth muscle (only two cells are shown). $\bar{r}_m$ = membrane resistance, 100 kΩ; $\bar{c}_m$ = membrane capacity, 0·1 μF; $\bar{r}_i$ = myoplasmic resistance, 330 Ω; $\bar{r}_j$ = junctional resistance, 3300 Ω and $\bar{c}_j$ = junctional capacity, 0–0·1 μF. External resistance is neglected. Cells ($4\bar{r}_m$, $\bar{c}_m$, $3\bar{r}_i$) are connected with $\bar{r}_j$, $\bar{c}_j$, (*b*): logarithmic decay of the electrotonic potential (*V*) along the model. Current was applied at O point in (*a*), and units on abscissa correspond to numbers shown in (*a*). λ = length, expressed either in units or in cells, at which the potential decays to $1/e$ (37%). (From Abe & Tomita, 1968)

Tomita, 1967). The question therefore arises as to the applicability of the cable equations to the taenia which is not a uniform cable but consists of short cables of individual cells connected with junctional resistances and capacities located between cells. In order to test this question, a simple model, shown in Fig. 7*a*, was constructed (Abe & Tomita, 1968).

A single cell is represented by four membrane resistances ($\bar{r}_m$) of 100 kΩ and four capacities ($\bar{c}_m$) of 0·1 μF (time constant: 10 msec), and three myoplasmic resistances ($\bar{r}_i$) of 330 Ω. Cells are connected by a longitudinal junctional resistance ($\bar{r}_j$) of 3300 Ω and capacity ($\bar{c}_j$) of 0·1 μF (time constant: 0·33 msec).

The model consists of 12 cells in series. Square current pulses are applied intracellularly at the end of the model and a transmembrane potential is recorded along the model. Figure 7b shows the decay of the electrotonic potential along the model. The decay is in a staircase-like manner being slow along a cell and sharp at the junction. If the average decay along the model is measured as shown by the dotted line, the space constant (λ) can be calculated from the ordinary cable equation ($\lambda = \sqrt{r_m/r_i}$). In this calculation, either (a) the membrane resistance ($\bar{r}_m$) and average internal resistance $[(3\bar{r}_i + \bar{r}_j)/4]$, or (b) the membrane resistance per cell ($\bar{r}_m/4$) and overall internal resistance ($3\bar{r}_i + \bar{r}_j$) are taken.

$$\lambda(a) = \frac{100{,}000}{1{,}075} = 9 \cdot 65 \text{ unit-length}$$

$$\lambda(b) = \frac{25{,}000}{4{,}300} = 2 \cdot 4 \text{ cell-length}$$

This model experiment suggests that in the taenia the decay of the electrotonic potential should be smoothly exponential even though the junctional resistance between cells is high compared with the myoplasmic resistance, since the cell length (about 200 μ) is much shorter than the space constant of the tissue (about 1·5 mm). The same argument can probably be applied to the vas deferens whose cell length is about 450 μ and the space constant is about 2 mm, although the junctional resistance is not known.

The time course of the electrotonic potential at any point in the model also fits the theoretical time course, when the overall space constant obtained above and the time constant of the membrane ($\bar{r}_m \times \bar{c}_m$) are inserted into the cable equation (equation 1). The time course of the electrotonic potential is scarcely affected by the capacity component at the junction ($\bar{c}_j$) provided that the time constant of $\bar{c}_j \times \bar{r}_j$ is less than 5% of the membrane time constant. In the taenia the time constant of the junctional membrane is less than 3% of the membrane time constant (Tomita, unpublished observation). Therefore it may be concluded that in the taenia the cable equations can be applied to obtain the membrane parameters (the membrane resistance and capacity) if the internal resistance including the junctional resistance is known (see p. 209 to 213).

Cable properties examined with alternating current and spike progagation

Responses to alternating current. Tasaki & Hagiwara (1957) slightly modified the cable equations and applied them to the responses to alternating current instead of square current pulses in frog skeletal muscle. When similar experiments are carried out in the taenia, it is found that the relationship between the frequency of alternating current and the potential change in the membrane fits the theoretical predictions, as in frog skeletal muscle (Abe & Tomita, 1968). From the analysis of the relationship the product of the specific internal resistance (R_i) and the membrane capacity per unit area (C_m) is calculated to be 0·5 msec/cm.

Spike propagation. In a uniform cable-like fibre the foot of a propagating spike rises exponentially and its time constant is determined by the cable properties of the fibre at rest and also by the conduction velocity (Tasaki & Hagiwara, 1957). The rising phase of the spike in the taenia is exponential as shown in Fig. 8 (Tomita, 1966*b*). This suggests that spike propagation is due to a local

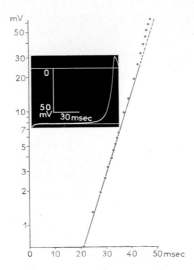

FIG. 8   The rising phase of the spike in the guinea-pig taenia coli. A spike recorded at about 6 mm from the stimulating electrode is shown in the inset. The spike was propagated at a speed of about 7 cm/sec. The line is the rate of rise of the foot of the spike shown in the inset, plotted on a semilogarithmic scale. It shows that the spike rises exponentially with a time constant of about 6 msec. (From Tomita, 1966*b*)

circuit current in a cable-like structure. The time constant of the foot of the spike is about 7 msec and the conduction velocity is about 6 cm/sec, measured in hypertonic solution.

Van der Kloot & Dane (1964) have examined the time course of the foot of the action potential in the frog ventricle and found that its properties fit the predictions made from the cable theory. It looks as if in both the smooth muscle of the taenia and the cardiac muscle of the frog ventricle the action potentials travel along the column of cells with cable-like properties.

Longitudinal resistance, membrane capacity and membrane resistance
Longitudinal internal resistance

In order to obtain quantitative values of the membrane parameters from the cable equation, it is important to know the value of the longitudinal internal resistance of the cell and also that of the junctional resistance between the cells.

It has been reported that the specific resistance of the myoplasm in the circular and longitudinal muscle of the cat small intestine is 225 Ω cm and the resistance of the junction between two cells is 17 to 18 MΩ (Nagai & Prosser, 1963b; Kobayashi, Prosser & Nagai, 1967). The myoplasmic resistance was measured by means of double micro-electrodes in one arm of a bridge circuit. The total resistance of the double micro-electrodes, when the electrodes are outside the cell, must be subtracted from the total resistance, when the electrodes are inside the cell, to obtain the resistance between the two intracellular electrodes. However, this could easily produce an error because the resistance of the micro-electrode might change when it is inserted into the cell (Schanne, Kawata, Schäfer & Lavallée, 1966; Tanaka & Sasaki, 1966; Wardell & Tomita, 1967). Moreover, the specific myoplasmic resistance was calculated by assuming that the cell is a cylinder of 3 μ diameter, although the cell was described as a double cone 100 μ long and 5 μ maximum diameter. Furthermore, in the calculation of the junctional resistance between cells, short circuits by the plasma membrane and by the connections with many surrounding cells were neglected (Vayo, 1965).

In the taenia coli of the guinea-pig, the tissue impedance along the direction of fibre orientation has been measured by applying alternating current to the two ends of the tissue, 2 to 3 cm long (Jones & Tomita, 1967). The impedance measurement was carried out in air after soaking the tissue in isotonic sucrose solution for 3 min, in order to eliminate electrical shunting by the extracellular fluid. Water and ion analysis indicates that this period is sufficient to remove the electrolytes from the extracellular space. The longitudinal impedance of a unit volume was calculated from the total impedance, tissue dimension and extracellular space. The tissue dimensions were calculated from its weight and specific gravity (1·06). The impedance is frequency dependent, as in cardiac and longitudinal intestinal muscle observed by Sperelakis & Hoshiko (1961). The value is 320 Ω cm at 10 c/s; this is reduced to 100 Ω cm at frequencies greater than 10 kc/s.

However, the tissue impedance is progressively increased by repeated washing with sucrose solution and can reach more than 10 times that measured after 3 min washing. The impedance is not constant but decreases with time of exposure to air. The change of the impedance is probably due to a change in the shunting resistance which is determined by the rate of leakage of ions from the cells and the rate of removal of ions from the extracellular space. It is also likely that pure sucrose solution which is free of ions has a direct effect on the junctional resistance. Soaking the tissue in sucrose solution for a long time probably increases the tissue impedance to a value much higher than the physiological value. This seems to be the reason why Sperelakis & Hoshiko (1961) obtained rather high values of about 2000 Ω cm for low frequency tissue impedance in the longitudinal intestinal muscle, since they immersed the tissue in sucrose solution containing only one tenth of the normal sodium concentration for 2 hr.

Since the impedance depends on the time of exposure to sucrose solution, it is difficult to obtain the true physiological value. Therefore, in another series of experiments, the impedance was measured when the tissue was suspended in a narrow tube and the external shunting resistance was altered by replacing half the sodium chloride with sucrose. The tissue impedance was calculated on the basis of the difference between the total impedance in normal Krebs solution, the impedance when half the sodium chloride was replaced with sucrose, and the specific resistance of the two solutions. It was assumed that the effect on the tissue impedance of reducing the sodium concentration to half is negligible. The values obtained by this method were 370 Ω cm to 10 c/s and 190 Ω cm at 10 kc/s, and they are similar to those above. Since the junctional resistance between cells is probably in parallel with the capacity component located at the junctional membrane, and also at the adjoining cell membrane surrounding the tight junction, it is reasonable to assume that the impedance of the junction is decreased with increasing current frequency due to a bypass of current through the capacity. The impedance measured at high frequency is, therefore, probably due to the resistance of the myoplasm, while the impedance which disappears with increasing frequency might be attributed to the junctional membrane. It may be concluded that in normal Krebs solution, the longitudinal tissue impedance is 300 to 400 Ω cm, which is the sum of the junctional resistance, about 200 Ω cm, and the myoplasmic resistance, 100 to 200 Ω cm (Tomita, 1969).

It is still difficult to correlate the electrical observations with histological findings so that the specific resistance of the junctional membrane could be calculated. We do not know whether the tight junction is the only junction responsible for electrical coupling and how large is the area of contact between cells, which has a low electrical resistance in physiological conditions.

Effects of hypertonic solution on the junctional membrane. Barr, Dewey & Berger (1965) observed a block of spike propagation in cardiac muscle immersed in a solution of twice normal tonicity. They explained this in terms of a disruption of the 'nexuses'. However, Dreifuss, Girardier & Forssman (1966), though agreeing that the nexus is a low resistance electrical pathway, believe that the blockage of the spike propagation in cardiac muscle is not due to separation of the nexus but to swelling of the sarcotubular system.

In the smooth muscle of the taenia, the tissue impedance is increased in hypertonic solution exceeding 1·5 × normal osmolarity, the increase being about 50% at 2 × normal osmolarity (600 milliosmolar). This suggests that there may be some effect on the junction. However, the effect is not dramatic. In Krebs solution of twice normal osmolarity the taenia shows proper cable-like properties and spike propagation (Tomita, 1966a, b; Tomita, 1967b; Abe & Tomita, 1968). Therefore, the function of the junctions seems to be preserved in a solution of up to twice normal osmolarity, although there is a report that hypertonic Krebs–Henseleit solution (630 milliosmolar) made by the addition of sucrose disrupts the nexus (Barr, Berger & Dewey, 1968).

The spontaneous spike activity is suppressed in hypertonic solution

(Holman, 1957; Tomita, 1966a). This seems to be mainly due to hyperpolarization of the membrane, because depolarization, whether by the addition of excess K or sustained depolarizing current, produces spontaneous repetitive discharge of spikes which propagate along the tissue (Tomita, 1966a, 1967b). Barr et al. (1968) explain the abolition of spontaneous spike activity by disruption of the nexus. However, this observation was not confirmed by Cobb & Bennett (1969), nor by Nishihara (1969) who found that in Krebs solution of twice normal osmolarity the nexuses had a normal appearance. For the interpretation of the electrophysiological results of Barr et al. (1968) the possibility cannot be excluded that the block of the spike propagation is due to hyperpolarization of the membrane (Tomita, 1967b). The degree of hyperpolarization of the membrane (about 10 mV) is expected from the loss of intracellular water and the increase in $[K]_i$ in the hypertonic solution (Brading & Setekleiv, 1968) (see Chapter 6).

Effect of Ca on the junctional membrane. In the functional state of the junctional complex, the ions Ca and Mg may play an important role. They determine the permeability of the junctional membrane and that of the perijunctional barrier which insulates the interspace of the junction from the external medium in the salivary gland and the sponge cell (Loewenstein, 1967a). The cytoplasmic surface of the junctional membrane is normally in contact with a medium containing low Ca and low Mg. If the Ca or Mg concentration is raised on either face, the permeability of the junctional membrane is reduced and it approaches that of the nonjunctional membrane (Loewenstein, Nakas & Socolar, 1967). When Ca is removed from the external medium, the perijunctional barrier becomes leaky.

The high permeability of the junctional membrane seems to be maintained by the Ca and Mg pump which keeps their intracellular concentration low, the energy being supplied by cell metabolism. Therefore, the permeability of the junctional membrane gradually falls when the metabolism is inhibited by cooling (6 to 8°C) or by metabolic inhibitors (2,4-dinitrophenol or N-ethyl maleimide) (Politoff, Socolar & Loewenstein, 1967).

Although the above results are obtained on the salivary gland of the midge and sponge cell, it is quite possible that the permeability of the junctional membrane in all tissues, including smooth muscle, is similarly controlled by Ca. However, the ease of changing the actual amount of Ca at the junction may differ from one tissue to another. In leech giant cells, unlike epithelial cells, reduction of the external Ca concentration to 10^{-5} M causes little change in the junctional resistance (Payton & Loewenstein, 1968). Simply omitting Ca from Krebs solution or adding Ca does not significantly change the longitudinal impedance of the guinea-pig taenia at least within 10 to 15 min (Tomita, 1969).

Membrane capacity

According to Tasaki & Hagiwara (1957) the product of the internal resistance

(R_i) and the membrane capacity of unit area (C_m) can be obtained by using the following equation

$$C_m R_i = \frac{a}{2v^2 T}$$

where a is the radius of the fibre, v is the conduction velocity and T is the time constant of the foot of the propagating spike. The value of $C_m R_i$ for the taenia is calculated to be 0·4 msec (based on the average values for conduction velocity = 6 cm/sec, for the time constant of the foot of the spike = 7 msec, and for the fibre radius = 2 μ). The product of C_m and R_i can also be obtained from the responses to alternating current as described above (0·5 msec) and is in reasonably good agreement with that obtained from the foot of the spike.

If $C_m \times R_i$ is taken to be 0·45 msec and R_i (including R_j) is 300 Ω cm, the membrane capacity in the smooth muscle of the taenia, C_m, is 1·5 μF/cm^2. From slightly different parameters, the value of about 3 μF/cm^2 (Tomita, 1966b) and of 2 to 3 μF/cm^2 (Abe & Tomita, 1968) have been obtained.

It is known that the frog skeletal muscle fibre (Katz, 1948; Fatt & Katz, 1951) and the crustacean muscle fibre (Fatt & Katz, 1953; Fatt & Ginsborg, 1958) have larger membrane capacities than nerve fibres. According to Falk & Fatt (1964), the capacity of the muscle membrane consists of a small capacity in the plasma membrane (2·6 μF/cm^2 in frog, 4·1 μF/cm^2 in crayfish) and a large capacity in the tubular system (3·9 μF/cm^2 in frog, 17 μF/cm^2 in crayfish).

The idea that the muscle membrane of the frog sartorius may be represented by a circuit with two time constants, one representing that of the plasma membrane and the other that of the tubular system, is supported by Freygang, Rapoport & Peachey (1967) who have correlated the dilation of the transverse tubular system induced by hypertonic sucrose solution with the change in the electrical constants, which are obtained from the measurements of the phase angle, the conduction velocity of the spike and the time constant of its foot.

Adrian & Peachey (1965) found that, in frog iliofibularis, the membrane capacity of the slow fibre (1·6 to 2·5 μF/cm^2) is one-third of that of the twitch fibre (6·8 μF/cm^2). They explained this difference by a lack of the transverse tubular system. Therefore, it is tempting to correlate the small membrane capacity of 1 to 3 μF/cm^2 in the smooth muscle with poorly-developed transverse tubules and sarcoplasmic reticulum (Peachey & Porter, 1959; Rhodin, 1962; also see Chapter by Burnstock).

The capacity in the cardiac Purkinje fibre is similar to that in the skeletal muscle fibre. According to the earlier analysis of the cable properties of the Purkinje fibre using square current pulses, the fibre has a membrane capacity of 10 to 12 μF/cm^2 (Weidmann, 1952; Coraboeuf & Weidmann, 1954). However, from the difference between the rate of rise of the computed spike and that of the experimentally observed spike, Noble (1962a) suggested the possibility of two capacity components in the Purkinje fibre. Only a small part of the membrane capacity (2 μF/cm^2) is discharged by the sodium current during the spike and the remaining capacity (10 μF/cm^2) has a resistance in series so that it is discharged mainly during the plateau. This idea was confirmed by Fozzard (1964, 1966) who calculated the membrane capacity from the propagation velocity and the time course of the foot of the action potential. This method gave a value of 2·4 μF/cm^2 for the capacity which is discharged by the foot of the action potential. The second component of the capacity which is in series with the resistance, has been estimated from voltage clamp experiments to be about 7 μF/cm^2.

Membrane resistance

Since the time constant of the electrotonic potential in a cable-like structure is the product of the membrane resistance and capacity, one of them could be calculated if the other was known. Taking the capacity of 1·5 to 3 μF/cm^2 and the membrane time constant of 100 msec, the membrane resistance is calculated to be 30 to 60 kΩ cm^2 (Tomita, 1966b; Abe & Tomita, 1968).

However, there is a little doubt whether one can calculate the membrane resistance from the value of the membrane capacity, obtained from the propagating spike, and from the membrane time constant, obtained from the analysis of the response to square current pulses. The capacity component in the electrotonic potential produced by a square current pulse could be larger than the capacity which is discharged by the foot of the spike, as it is in skeletal muscle and in the cardiac Purkinje fibre.

The fact that intracellular polarization, which has a very sharp spatial decay, affects the spike amplitude (Kuriyama & Tomita, 1965) suggests that every muscle fibre is involved in spike propagation. Therefore, the membrane capacity of all cells may be discharged by the upstroke of a propagated spike. However, it is also possible that some part of the membrane, e.g. overlapping end parts of the muscle fibre and vesicles along the surface membrane, behave electrically similarly to the sarcoplasmic tubules in the skeletal muscle fibre. Then, the capacity involved in the electrotonic potential would be larger than 1·5 to 3 μF/cm^2, which is obtained from spike propagation, and the calculated membrane resistance would be smaller than 30 to 60 kΩ cm^2. Nevertheless, this factor seems to be small, because the membrane resistance calculated from the data obtained by the double-sucrose gap method is also of the same magnitude (Bülbring & Tomita, 1969a) as those just described.

In the double sucrose-gap method, a current pulse of 1·5 $\times$ 10^{-7} A produces an electrotonic potential of 10 to 15 mV. Therefore, the total membrane resistance of the tissue exposed to Krebs solution is 60 to 100 kΩ. The volume of the tissue is calculated to be about 1·9 $\times$ 10^{-4} cm^3 (0·4 mm diameter $\times$ 1·5 mm length), in which 35% is extracellular space. Then the cell volume would be about 1·2 $\times$ 10^{-4} cm^3. Taking a volume-surface ratio of 1·5 $\times$ 10^{-4} cm (Goodford & Hermansen, 1961), the surface area where the electrotonic potential is produced is about 0.8 cm^2. Thus, the specific membrane resistance is 50 to 80 kΩ cm^2 which is of course a very rough estimation but is nevertheless of the same order of magnitude as that described above.

As shown in Table 1, the twitch fibre of frog skeletal muscle and the cardiac Purkinje fibre have a membrane resistance of 2000 to 4000 Ω cm^2. Adrian & Peachey (1965) found that in the frog iliofibularis the membrane resistance of the slow fibre (29,000 Ω cm^2) is about 10 times higher than that of the twitch fibre (3140 Ω cm^2), but it is of the same order of magnitude as that of the smooth muscle.

TABLE 1. Electrical membrane parameters and cell diameter

	Capacity	Resistance	Time constant	Diameter	Reference
	(μF/cm^2)	(Ω cm^2)	(msec)	(μ)	
Skeletal muscle					
Crab					
Carcinus, Portunus	42	116	4·6	180	(1)
Crayfish					
Astacus	20	1,000	20	150	(2)
Frog					
Add. magnus	6	1,500	9	75	(3)
Ext. dig. iv	4·5	4,000	18	45	(3)
Sartorius	6·1	1,900	12	86–104	(4)
Sartorius	8	4,100	34·5	137	(5)
Iliofibularis					(6)
(i) Twitch muscle	6·8	3,100	21·5	50	
(ii) Slow muscle	1·6–2·5	29,000	46	50	
Purkinje fibre					
Kid	12·4	2,000	19·5	75	(7)
Calf, sheep	11	1,200	13	75	(8)

(1) Fatt & Katz (1953), (2) Fatt & Ginsborg (1958), (3) Katz (1948), (4) Ishiko & Sato (1960), (5) Fatt & Katz (1951), (6) Adrian & Peachey (1965), (7) Weidmann (1952), (8) Coraboeuf & Weidmann (1954).

When a long cylinder of 2 μ in radius (a) is taken as a single cable representing a series of cells in the taenia, the specific membrane resistance (R_m) is taken as 30 to 60 kΩ cm^2, and the internal specific resistance (R_i) as 300 Ω cm, then the space constant ($\lambda = \sqrt{a/2} \sqrt{R_m/R_i}$) is 1.0 to 1.3 mm, which is similar to the observed value (1.45 mm).

Similarly, Kobayashi et al. (1967) calculated the space constant for the cat circular muscle to be 1.1 mm (0.94 mm for the cat longitudinal muscle) by taking the bundle diameter as 100 μ, the specific membrane resistance as 1050 Ω cm^2 (780 Ω cm^2 for the longitudinal muscle) and the specific myoplasmic resistance as 225 Ω cm. However, the value of the specific membrane resistance had been obtained from a calculation based on the input resistance measured with a micro-electrode and on the assumption that cells were not electrically interconnected (uniform current flow across the cell membrane the area being 15×10^{-6} cm^2). The value obtained by Kobayashi *et al.* (1967) is therefore not the specific membrane resistance of a bundle of 100 μ diameter. Their calculated value fits accidentally the observed space constant which was measured using pressure electrodes for stimulating and micro-electrodes for recording.

The space constant can also be calculated from the membrane time constant (τ_m), the conduction velocity (v) and the time constant (T) of the foot of the conducted spike ($\lambda = \sqrt{\tau_m \cdot v^2 \cdot T}$.) This relation can be derived from equation (3). Taking τ_m as 100 msec, v as 6 cm/sec, and T as 6 msec, the space constant for the taenia is calculated to be 1·6 mm.

Effects of ions on the membrane resistance

K, Cl and Na. The membrane potential of the guinea-pig taenia coli is mainly determined by the K concentration gradient across the membrane. Depolarization of the membrane potential by a ten-fold increase in the external K is 43 mV when the sum of NaCl and KCl is constant, and 51 mV when the product of K and Cl is constant (Casteels & Kuriyama, 1966). There is some indication of a Cl contribution to the membrane potential, since, when Cl is replaced with SO$_4$ or C$_2$H$_5$SO$_4$, the membrane is depolarized by 7 to 18 mV, and depolarization by excess K becomes greater (Kuriyama, 1963a).

The relatively low membrane potential (about 50 mV, Bülbring 1954; Holman, 1958; Kuriyama, 1963a) of the taenia compared with frog skeletal muscle (about 90 mV, Adrian, 1956, 1960) has been explained by a high Na permeability of the membrane (Kuriyama, 1963a). Hyperpolarization of the membrane (about 10 mV) by replacement of Na with sucrose (Brading & Tomita, 1968b; Brading, Bülbring & Tomita, 1969a) may be taken as evidence for a relatively high Na permeability compared with K permeability, although replacement of Na with tris (-hydroxymethyl-aminomethane) slightly depolarizes the membrane after transient hyperpolarization (Kuriyama, 1963a).

A low potassium conductance is probably another reason for the low membrane potential. The rate constant of K-efflux in smooth muscle (Goodford & Hermansen, 1961; Barr, 1961a; Daniel, Sehdev & Robinson, 1962) is low compared with that of skeletal muscle. In general, the effects on the membrane potential of changing the external ion concentration suggest that the membrane is mainly permeable to K, but is also considerably permeable to Cl and Na.

In order to investigate changes in the membrane resistance of the guinea-pig taenia coli, the double sucrose-gap method (Berger, 1963) has been used (Bülbring & Tomita, 1968a; 1969a). Only about 1 to 1·5 mm of the tissue between the two sucrose channels is exposed to the test solutions. Constant current pulses are applied to one side of the tissue, and the electrotonic potential is recorded from the other side of the tissue. The potential recording is the same as described by Bülbring & Burnstock (1960).

8*

Excess K depolarizes the membrane with a reduction of the membrane resistance. This is also demonstrable in the absence of the external Cl. In low K solution, the membrane resistance becomes larger. When Cl is reduced from 134 to 7 mM by replacing the ethane-sulphonate or benzene-sulphonate, the membrane resistance increases gradually. Often the increase still continues 15 to 20 min after the replacement, and the electrotonic potential usually reaches more than twice the control size. Nitrate has the opposite effect to that of ethane- or benzene-sulphonate (Bülbring & Tomita, 1969a, b).

When Na is replaced by either tris chloride or sucrose, the membrane is hyperpolarized and the membrane resistance is at first decreased but it usually recovers gradually. The hyperpolarization and the reduction of the membrane resistance are less in tris solution than in sucrose solution.

These results are still too preliminary to draw any conclusion as to the extent to which Na and Cl contribute to the membrane conductance. However, it is clear that ethane- or benzene-sulphonate is less and nitrate is more permeant than Cl. This has also been demonstrated by their osmotic effects on tissue weight (Brading & Tomita, 1968a). The gradual reduction of the membrane conductance in benzene-sulphonate may be due to a slow decrease in the intracellular Cl. The other possibility is that substitution of Cl with a large foreign anion reduces K conductance as a secondary effect, as reported by Casteels & Meuwissen (1968) on the basis of flux measurements. The Na contribution to the membrane conductance is probably small, since the membrane resistance is usually reduced in Na-free solution.

Ca. When the external Ca concentration is increased the membrane is hyperpolarized (Bülbring & Kuriyama, 1963a). This hyperpolarization is accompanied by a reduction of the membrane resistance (Bülbring & Tomita, 1968a, b, 1969c). This was confirmed with intracellular recording and external polarization (Brading et al., 1969b).

Reduction of the external Ca concentration depolarizes the membrane (Bülbring & Kuriyama, 1963a). At a low concentration of Ca the membrane resistance is usually increased. However, when the concentration is reduced further the membrane resistance becomes smaller. A concentration which produces a transition from an increase to a decrease of the membrane resistance depends on many factors, but it is usually between 0·2 and 0·6 mM. The critical Ca concentration is higher in Locke solution than in Krebs solution and also higher at high temperature than at low temperature. When Na is reduced by replacement with tris-Cl or sucrose after the membrane has been depolarized in Ca-free solution, the membrane is repolarized and the membrane resistance is increased (Brading & Tomita, 1968b; Brading, et al., 1969b).

According to the ionic theory (Hodgkin, 1951), the hyperpolarization of the membrane by excess Ca may be due to either an increase of K conductance or to a decrease of Na conductance, or to both, if a passive distribution of Cl is assumed. Since excess Ca reduces the electrotonic potential in the smooth muscle of the taenia, an increase of K conductance seems to be the main factor. The depolarization in low Ca may be mainly due to a decrease in K conductance, since the membrane resistance increases, but the depolarization in zero Ca is probably due to an increase in Na conductance, since the depolarization caused by removal of Ca is small in Na-free solution.

A similar conclusion has been reached from ion flux experiments in frog skeletal muscle in which the K permeability decreases to about half normal and the Na permeability increases to eight times normal in Ca-free Ringer solution containing EDTA (Kimizuka & Koketsu, 1963).

Temperature. When the temperature is reduced from 37° to 20°C, the size of the electrotonic potential is increased. This is accompanied by a small depolarization. The temperature effect is smaller in a solution in which Cl is substituted with benzene-sulphonate than in normal Krebs solution. This indicates that not only the K conductance but also Cl conductance decreases with the lowering of temperature (Brading, Bülbring & Tomita, 1969a).

The depolarization and the reduction of the membrane resistance produced by removal of Ca at normal temperature are reversed by cooling. A similar phenomenon has been reported in frog skeletal muscle. It has been postulated that this may be due to a higher amount of Ca in the membrane at low temperature than at high temperature (Apter & Koketsu, 1960). In cardiac muscle, the positive inotropic effect of cooling and the recovery of the contraction in a low Ca solution by reducing the temperature from 32° to 22°C have also been explained by a greater availability of Ca at low temperature (Kaufmann & Fleckenstein, 1965; Šumbera, Braveny & Kruta, 1967).

However, in the smooth muscle of the taenia, it would seem that Ca is lost from the membrane at low temperature, since in the normal solution the effect of reducing the temperature can be mimicked by a reduction of the external Ca concentration. This apparent contradiction could be overcome by postulating two different sites for the action of Ca and an independent control of the membrane permeability to ions. At the outer layer of the membrane Ca might control the permeability to Na, as postulated by Frankenhaeuser & Hodgkin (1957) for squid giant axon. An increase of Ca at this site would reduce the membrane permeability to Na. At the inner layer of the membrane Ca might control the permeability to K. An increase at this site would increase K permeability.

Cooling may increase the Ca at the outer layer as postulated in frog skeletal muscle and cardiac muscle, and decrease the Ca at the inner layer at which the Ca accumulation might be metabolically controlled as in the sarcoplasmic reticulum of skeletal muscle (Hasselbach, 1966). It is quite possible that this site of Ca action is not at the inner layer but at a different part of the membrane, e.g. the vesicles seen on the membrane surface. When the Ca concentration of the external solution is reduced, both membrane sites lose Ca, but the change in K conductance would be the dominant factor determining the membrane resistance, especially at lower temperature, because the P_{Na}/P_K ratio is probably in the order of 0·1; the membrane resistance therefore increases. However, at high temperature, in a very low external Ca concentration, the contribution of the Na conductance becomes relatively large and the membrane resistance is therefore reduced.

Interpretation of responses to intracellularly applied current

The observations on the electrotonic potential produced by large external stimulating electrodes, and also on the propagating spike, strongly suggest that the smooth muscle of the taenia coli and the vas deferens of the guinea-pig have nearly perfect cable-like properties (Tomita, 1966a, b, 1967a; Abe & Tomita, 1968). On the other hand, when intracellular stimulation is used for studying the electrotonic potential, its behaviour is quite different from that described above.

In the circular intestinal muscle of the cat, Nagai & Prosser (1963b) applied current intracellularly and recorded an electrotonic potential in another cell nearby. They concluded that there were relatively low-resistance interfibre junctions. However, Sperelakis & Tarr (1965) using the same preparation reported that current flow through one cell did not have any substantial effect on the transmembrane potentials of adjacent cells. Therefore, they concluded that electrotonic coupling between cells was weak. In similar experiments performed in the taenia coli and the vas deferens of the guinea-pig, Nagai & Prosser's result could not be confirmed (Tomita, 1967a, Holman, personal communication; Kuriyama & Tomita, unpublished observation).

In Nagai & Prosser's experiments, the responses to depolarizing or hyperpolarizing pulses, applied with an 'intracellular electrode', were reduced only by less than 3% when recorded in an apparently adjacent cell. According to Table 3 in their paper, the resistance between adjacent cells is 20 to 30 $M\Omega$ and the resistance between the outside and the inside of the cell (the input resistance) is 69 $M\Omega$. The reduction of the electrotonic potential should therefore be about 30%, and it is difficult to explain only 3% reduction. Nagai & Prosser (1963b) also reported that the space constant obtained from responses to stimulation with an intracellular micro-electrode was approximately the same as that obtained with a pressure electrode. When one considers the difference between the spread of the current applied by an intracellular electrode and that applied with a pressure electrode, as will be described later, it is surprising that both measurements gave approximately the same space constant, although there was a difference of the time constant (31 msec with intracellular and 133 msec with pressure-electrode).

In measurements on the longitudinal muscle of the cat intestine described recently (Kobayashi et al., 1967), the decay of the electrotonic potential produced intracellularly is much sharper than that previously reported for the circular muscle. The decline in amplitude is 20 to 30% in adjacent fibres and the electrotonic potential nearly disappears at a separation of 200 μ.

In Table 2 the time constants obtained by recording with a single intracellular

TABLE 2. Membrane time constant (msec)

		Function used*	Reference
Intracellular polarization			
Cat			
Circular muscle	31	—	Nagai & Prosser, 1963b
Circular muscle	21	erf.	Kobayashi, Prosser & Nagai, 1967
Longitudinal muscle	10	erf.	Kobayashi, Prosser & Nagai, 1967
Guinea-pig			
Taenia coli	2–4	exp.	Kuriyama & Tomita, 1965
Taenia coli	10	erf.	Kuriyama & Tomita, 1965
Vas deferens	2–3	exp.	Hashimoto, Holman & Tille, 1966
Vas deferens	1·8	exp.	Bennett, 1967b
Vas deferens	1·5–3	exp.	Tomita, 1967a
Vas deferens	3·5–5	erf.	Tomita, 1967a
Longitudinal muscle	3·5	—	Kuriyama, Osa & Toida, 1967b
External (including pressure electrode) polarization			
Cat			
Circular muscle	133	—	Nagai & Prosser, 1963b
Longitudinal muscle	92	erf.	Kobayashi, Prosser & Nagai, 1967
Guinea-pig			
Taenia coli	60–100	erf.	Tomita, 1966a
Taenia coli	107	erf.	Abe & Tomita, 1968
Vas deferens	100	erf.	Tomita, 1967a
Longitudinal muscle	14–100	erf.	Kuriyama, Osa & Toida, 1967b

* The time constant is taken as the time at which the electrotonic potential reaches either 63% (exponential function, exp.) or 84% (error function, erf.) of the steady level.

micro-electrode, through which also the current was applied using a bridge circuit, and those obtained by polarization with a large external electrode are listed for comparison. In many reports, it is not clearly stated whether the electrotonic potential has an exponential or error-functional time course in order to justify the definition of the time constant. Nevertheless, there is a clear difference between the time constants obtained by the two different methods of polarization: the time constant obtained with intracellular polarization is much shorter than that obtained with external polarization.

The differences between the spatial decay and between the time course of the electrotonic potentials obtained by intracellular and external polarization are explained by the hypothesis that the current applied by an intracellular electrode spreads in three dimensions, since the muscle fibres are electrically interconnected through low resistances, while the current applied externally spreads only in one dimension longitudinally, because the fibres run parallel and are polarized to the same extent, so that there is no current flow through the transverse interconnections (Tomita, 1966b, 1967a).

The electrical properties which depend on the geometry of the tissue have been discussed by Noble (1966). In cable-like preparations the membrane area increases linearly with the distance from a given point. In a thick preparation with three-dimensional interconnections between cells, like those in smooth muscle, the membrane area increases more rapidly with distance from a given point, probably as the cube of the distance. Therefore, the current density in the vicinity of the current-supplying micro-electrode decreases rapidly. In other words, in the region of the micro-electrode impalement, the membrane is shunted by a low electrical resistance due to the large membrane area surrounding the electrode. This explains why the spatial decay is much sharper and the

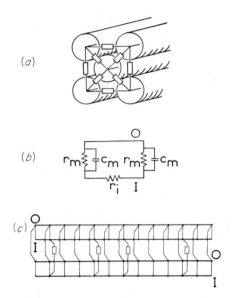

FIG. 9   *a* Schematic drawing of the model which represents five columns of muscle fibres connected in series. Each column is connected with the next through 10 kΩ resistances which represent interconnexions between fibres (see also *c*).   *b* The circuit of the unit, r_m = the membrane resistance, 100 kΩ: c_m = the membrane capacity, 0·1 μF; r_i = the resistance of myoplasm, 10 kΩ. The resistance of extracellular space was neglected. O, the outside and I, the inside of the fibre.   *c* Examples of interconnexions between two columns through 10 kΩ resistances at two to four unit intervals. Each square represents the unit shown in *b*. O, outside and I, inside of the fibre. (From Tomita, 1966*b*)

time course is much faster with internal polarization than with polarization with a large external stimulating electrode.

The difference between intracellular and external polarization can easily be demonstrated by the following model experiments (Tomita, 1966*b*). The model is represented by five columns of longitudinally connected fibres as shown in Fig. 9*a*. A column of single fibres connected in series is represented by 12 repeated units in series. The circuit of the unit is shown in Fig. 9*b*. The values of the resistances and the capacity are chosen arbitrarily. The inside of some cells is connected across 10 kΩ to the inside of some other cells in an adjacent column at random, but only four in each column as shown in

Fig. 9a and c. The outside of all cells is directly connected assuming that the resistance of the extracellular space is very low.

Current for 'intracellular polarization' is applied between the inside and the outside of one cell in the centre of the model. Current for 'external polarization' is applied between the inside and the outside of five cells, one in the middle of every column. The electrotonic potential is measured in the centre column.

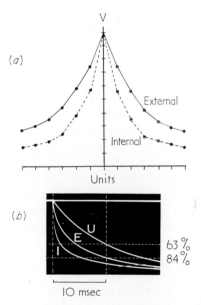

FIG. 10   a Spatial decay of the electrotonic potential measured in the centre column of the model (see Fig. 9a). Ordinate: the potential expressed as percentage of the potential at the unit. where the current is applied. Abscissa, the distance expressed by numbers of units. External = external polarization, and internal = intracellular polarization. b The time course of the electrotonic potentials recorded at the unit where the current is applied. E, with external polarization; I, with intracellular polarization; U, the electrotonic potential produced with one membrane resistance (100 kΩ) and with one parallel capacity (0·1 μF), but without any other connexion. Note that the electrotonic potential (U) decays to 37% and the potential (E) to 16% in 10 msec; the potential (I) decays most quickly. (From Tomita, 1966b)

As shown in Fig. 10a, the spatial decay of the electrotonic potential along the column was much sharper with intracellular stimulation than with external stimulation. The time course of the electrotonic potential produced by external polarization could be expressed by an error function, while the time course of the electrotonic potential produced by intracellular polarization could neither be expressed by an error function nor by an exponential function. As seen in Fig. 10b, the rate of change of the electrotonic potential produced by intracellular polarization was quicker at first, then became slower than the exponential curve and this tendency was also observed in the living tissue of the taenia coli or of the vas deferens.

Although the analogue is an over-simplification, the results of intracellular and external polarizations are clearly different, and qualitatively as observed in the living tissue.

As mentioned above, when the muscle membrane is not uniformly polarized, for example by transmitter action, the time course of the response must be different depending on the spatial current spread.

In the fast skeletal muscle fibre, the miniature end-plate potential has the same characteristics as the end-plate potential, i.e. the same localization and the same time course, except for its spontaneous occurrence and small amplitude (Katz, 1962; Dudel & Kuffler, 1961). Its falling phase is determined by the muscle time constant (Fatt & Katz, 1951, 1952; Dudel & Kuffler, 1961). However, Burke & Ginsborg (1956a, b) reported that in the slow skeletal muscle fibre which has cable-like properties and which receives distributed innervation, stimulation of the small nerve fibres produces a uniform depolarization which decays more slowly than a focally-produced electrotonic potential.

The time course of the junction potentials in the guinea-pig vas deferens also depends on their spatial current spread (Tomita, 1967a). The junction potential evoked by nerve stimulation is recorded with similar amplitude in every cell, i.e. the whole tissue is depolarized by more or less the same magnitude. Therefore, little current spreads from cell to cell. The amplitude of the junction potential evoked by field stimulation of the end of the tissue decreases with distance. This indicates that there is a slight current spread in one direction. The spatial decay of the spontaneous junction potential is very much sharper (Burnstock & Holman, 1966a; Kuriyama, 1964a), suggesting that it is due to depolarization of a small spot of the tissue and that current spread is three-dimensional as with intracellular polarization.

The time course of the falling phase of the junction potential evoked by hypogastric nerve stimulation is slightly slower than that evoked by field stimulation, whereas the spontaneous junction potential decays much faster than the evoked junction potential (Burnstock & Holman, 1961; 1962a, 1963). The half-decay time of the electrotonic potential recorded close to the stimulating electrode (40 to 50 msec) lies between that of the evoked junction potential (200 to 300 msec) and that of the spontaneous junction potential (20 to 30 msec). There is a good correlation between the spatial current spread and the time course of the potential change (Fig. 11). When there is no current spread, the rate of decay is slow while when there is three-dimensional current spread the rate of decay is fast. In addition, there are many other factors affecting the junction potential (e.g. duration of the transmitter action), (see Chapter 8 by Holman).

Membrane parameters obtained by intracellular polarization

As shown in Table 3, many investigators have calculated the specific membrane

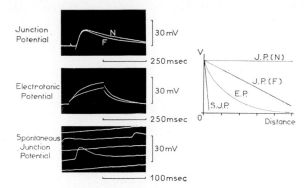

FIG. 11 The time course of the junction potential in the vas deferens produced by hypogastric nerve stimulation (N) and field stimulation (F) (upper record), of the electrotonic potential (middle record) and of the spontaneous junction potential (lower record). The spatial decay is shown schematically in the graph. (From Tomita, 1967a)

TABLE 3. Membrane resistance and capacity measured with intracellular electrode

	Input resistance (MΩ)	Specific resistance (Ω cm²)	Specific capacity (μF/cm²)	Time constant (msec)	Dimension (μ)
Cat intestinal circular muscle	(a) 104	980	—	—	100 × 6
	(b) 70	560	56	31	100 × 5
	(c) 66–71	1050	21	21	200 × 5
	(d) 7·5	377	40	5–30	200 × 8
Cat intestinal longitudinal muscle	(c) 51–54	780	13	10	200 × 5
Guinea-pig taenia coli	(e) 40	320	10	3	100 × 5
	(f) 40	600	5	3–4	100 × 5*
Guinea-pig jejunum and rectum	(f) 40–45	600–650	4·5–5	3–4	100 × 5*
Guinea-pig ureter	(g) 20	160	15	2·5	100 × 5
	(g) 15–23	300	8	2–3	100 × 5*

* Calculation by the short-cable equation.
(a) Barr (1961a); (b) Nagai & Prosser (1963b); (c) Kobayashi, Prosser & Nagai (1967); (d) Sperelakis & Tarr (1965); (e) Kuriyama & Tomita (1965); (f) Kuriyama, Osa & Toida (1967b); (g) Kuriyama, Osa & Toida (1967a).

resistance from the input resistance obtained by intracellular polarization. In most of the calculations, the input resistance is simply multiplied with the cell surface area, assuming that current spreads homogeneously over the whole surface membrane (i.e. neglecting the internal myoplasmic resistance). However, this assumption cannot be true even for a single cell neglecting the syncytium because the cell dimensions (5 μ diameter and 100 μ length) imply a high internal longitudinal resistance (5 MΩ/100 μ assuming a cylinder of 5 μ in diameter whose specific resistance is 100 Ω cm). In some calculations the fibre is treated as a short cable. However, if current spreads three-dimensionally, the error in both calculations may be serious, since the membrane near the intracellular micro-electrode is shunted by the low resistance of the surrounding membrane (Vayo, 1965).

In the electrical model shown in Fig. 9, the effective membrane resistance of the model, R_e (calculated by dividing the voltage, measured at the unit where the current is applied, by the intensity of the applied current) is about 16 kΩ when external polarization is used. When the resistance values of the units are inserted into the cable equation (Hodgkin & Rushton, 1946) the same effective resistance is calculated.

$$R_e = \frac{\sqrt{r_m \cdot r_i}}{2} = \frac{\sqrt{100 \text{ k}\Omega \times 10 \text{ k}\Omega}}{2} = 16 \text{ k}\Omega$$

However, when intracellular polarization is used, the effective membrane resistance is only about 6·5 kΩ.

When the model is polarized externally, i.e. when current is passed into all cables simultaneously, the time course of the electrotonic potential can be expressed by an error function and the time constant, i.e. the time to reach 84% (erf 1) of the steady level according to the cable theory, is 10 msec. This is the actual time constant of the unit (100 kΩ $\times$ 0·1 μF = 10 msec). However, the time course of the electrotonic potential produced by intracellular polarization cannot be expressed by an error function or by an exponential function and the time to reach 84% (about 5 msec) or 63% (1 − 1/e; about 1·5 msec) is much shorter than the actual value (10 msec).

From the results of the model experiments, it can be said that, if intracellular polarization is used in a muscle with three-dimensional interconnections between cells, the membrane parameters cannot be derived by the calculation applicable to an ordinary cable-like fibre.

Resistance change measured by intracellular polarization

As described above, in the measurement of the membrane resistance, its absolute value or its change, it is very important to take the geometrical factor into account (i) When current flows uniformly across the whole membrane of unit length, the current-voltage relation shows directly the membrane resistance ($V/I = r_m$). In some tissues, e.g. the squid giant axon, this condition can be achieved by inserting a long stimulating electrode inside the fibre. (ii) If current is applied at a point of a cable-like fibre using a micro-electrode, the current density decreases exponentially along the fibre. As seen from the cable equation (Hodgkin & Rushton, 1946) the current-voltage relation at the point of current injection is now proportional to the square-root of the membrane resistance ($V/I \approx (r_m)^{1/2}$). For example, 75% reduction of the membrane resistance gives only 50% reduction of the current-voltage relation. (iii) When current is applied by an intracellular micro-electrode in 'closed' syncytium (a close-packed hexagonal array) the relation is expressed by $V/I \approx (r_m)^{1/4}$ if the branches of the syncytium are much shorter than the space constant. For example, 75% reduction of the membrane resistance (r_m) gives only 30% reduction of the input (or effective) resistance (V/I). In a three-dimensional syncytium, the input resistance would thus be much less dependent on the membrane resistance than in the system described above. In an open syncytium (cable repetitively branching into two at some distance) the input resistance is independent of the membrane resistance, if branches are shorter than the space constant (George, 1961).

In the smooth muscle, no change or only a very small change in the input resistance has been reported during the spike (Kuriyama & Tomita, 1965; Bennett, 1967a, b; Nagai & Prosser, 1963b; Kobayashi et al., 1967) or during the junction potential (Bennett & Merrillees, 1966; Holman, 1967). These negative results do not necessarily mean that the actual change in the membrane conductance is very small. It is very likely that in the smooth muscle, intracellular polarization is not an adequate method to pick up the change of membrane conductance. The input resistance and its dependence on the membrane resistance are probably mainly determined by the extent of electrical interconnections between cells. In order to explain failures to record a resistance change during the spike, Nagai & Prosser (1963b) have considered the arrangement in three dimensions of many interconnected fibres. However, in the same paper and also in the paper by Kobayashi et al. (1967), they have ignored these interconnections in the calculation of the membrane parameters, as pointed out by Vayo (1965).

The linearity of the current-voltage relation is also different with intracellular and external polarization. With intracellular polarization, the relation is more or less linear (Kuriyama & Tomita, 1965; Hashimoto, Holman & Tille, 1966; Bennett & Merrillees, 1966; Bennett, 1967b; Tomita 1967a) while with external polarization it is often non-linear (Tomita, 1966a, 1967a). The cable properties have the effect of reducing non-linearity in the current-voltage relation (Cole & Curtis, 1941; Burke & Ginsborg, 1956a; Hutter & Noble, 1960) when this is obtained with intracellular polarization. This results from the fact that areas of the membrane near the electrode, which have been polarized into the non-linear region of the current-voltage relation, are shunted by areas of the membrane at a distance from the electrode which, because of the decrement of current with distance, are much less polarized and so more nearly linear. In a preparation with three-dimensional interconnections, non-linearity should be even more reduced because the membrane areas at a given distance from the electrode are very much larger than in a cable-like preparation (Noble, 1962b, 1966).

ACTIVE PROPERTIES

Spike initiation

When the active properties of a smooth muscle are examined, it is very important to take the geometrical factors of the preparation into consideration, as discussed by Noble (1966). If the membrane of a cable-like fibre is stimulated uniformly, the whole area produces an action potential simultaneously. There is no potential difference along the fibre and therefore no current flows longitudinally. However, when one point of a cable-like fibre is stimulated, the evoked action potential propagates along the fibre and the action current is influenced by a local circuit current flowing between active and surrounding resting areas of the membrane. As a result, the rate of rise of the propagating spike is slower and its refractory period is longer than that of the spike produced uniformly in the whole membrane. In a tissue in which cells are electrically interconnected in three dimensions the current flow may be very complicated.

As described in the section on passive properties, the smooth muscles of the taenia coli and the vas deferens of the guinea-pig behave like a cable when large polarizing electrodes are used, but they behave like a three-dimensional syncytium when an intracellular micro-electrode is used for polarization of the membrane. Since the density of the local circuit current depends on the ratio of the active area to the surrounding passive area, the active response shows quite different properties depending on the way of stimulation.

Intracellular stimulation

In smooth muscles it is difficult to initiate a spike by intracellular electrical stimulation. Since the resistance of the micro-electrode used for experiments on smooth muscle has to be very high (30 to 100 MΩ) and this value is easily changed by passing current or by penetrating a cell, the experiments are usually carried out by trial and error, selecting a suitable electrode. It is, therefore, rather difficult to estimate what percentage of failure in producing the spike is due to a bad electrode, to a poor penetration perhaps damaging the cell, or to the real properties of the cell.

In most cells of the taenia coli (Kuriyama & Tomita, 1965) and the ureter (Kuriyama, Osa & Toida, 1967a) of the guinea-pig, no spike could be evoked by intracellular stimulation even if the membrane was depolarized by more than 50 mV. In some cells a graded response was recorded, and only in a few cells the spike was triggered in an all-or-none manner. The spike produced in an all-or-none manner had a shape similar to that of the spontaneously generated spike (Fig. 12).

In the guinea-pig vas deferens, initiation of the spike by intracellular stimulation seems to be slightly easier than in the taenia coli. 20 to 25% of the cells gave an active response including a graded response (Bennett, 1967b; Tomita, 1967a). According to Hashimoto et al. (1966), graded responses were observed in most cells, and spikes of the all-or-none type were very rare. Even when the spike was evoked at the stimulating site, it did not propagate and thus caused no measurable contraction of the tissue (Bennett & Merrillees, 1966; Tomita, 1967a). Figure 13 shows examples of a cell which produced a spike (a) and one which did not produce a spike (b) by intracellular stimulation. In a cell which

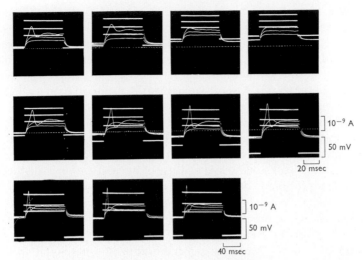

FIG. 12 Spikes triggered in the taenia coli by intracellular stimulation. In each record three depolarizing pulses of increasing intensity are superimposed. Interrupted line = original membrane potential. *Top — left to right*: effect of increasing conditioning depolarization. *Middle*: effect of increasing conditioning hyperpolarization. *Bottom*: records taken after 1, 3 and 5 sec constant conditioning hyperpolarization. (From Kuriyama & Tomita, 1965)

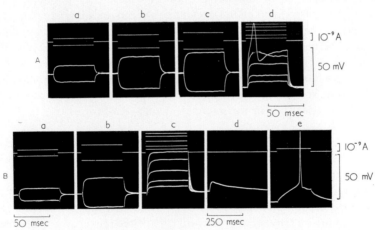

FIG. 13 Guinea-pig vas deferens, intracellular records (lower trace). A. Membrane polarization and spike in response to intracellular stimulation (current monitored in upper trace), a–c: depolarization (upward) and hyperpolarization were superimposed; d: responses to outward current at five different intensities were superimposed. Note short time constant, no rectification, high threshold and short latency for spike. Resting potential: −67 mV. B. An example of failure to evoke an active response to intracellular stimulation (a–c); a junction potential evoked by nerve stimulation (d), and the spike evoked by external polarization with a large electrode (e). Resting potential: −63 mV; base line was shifted downward in (e). Calibration for stimulating current is only applicable to (a–c), but not (e). Note different time scales. (From Tomita 1967a)

only produced a passive response, a junction potential and also a spike were evoked by external current application.

From the reports just mentioned, comparing the spike evoked by intracellular stimulation with the spike evoked by external stimulation, the following characteristics may be summarized. (1) A tendency to graded activity. (2) A high critical firing level. (3) A short latency. (4) A localized activity, i.e. almost no propagation.

In the propagating spike in a cable-like fibre, the inward current is drawn away from the active membrane to the passive membrane to form a local circuit current, thereby reducing the current discharging the membrane capacity of the active region. This slows the rate of rise of depolarization and allows more time for the inactivation of the carrier system for the inward current. Thus the rate of rise of the spike is slower when the spike propagates than when the spike is initiated uniformly in the whole membrane. For example, in squid nerve, the computed values for the maximum rate of rise are 431 V/sec for a propagated action potential and 564 V/sec for a membrane action potential (Hodgkin & Huxley, 1952d).

In smooth muscle, the depolarization of the membrane at a distance from the stimulating micro-electrode is very small as a result of the sharp spatial decay, as already described in the section on passive properties. Therefore, the inward ionic current generated by the active membrane near the micro-electrode will have to supply a large local circuit current to the surrounding areas in order to depolarize their membrane to a firing level by discharging the membrane capacity. Even if the membrane near the intracellular electrode is depolarized more than to the threshold for uniform polarization, this is still not enough to produce a spike, because more current will flow from the areas of the membrane supplying outward current, so that there is an excess of outward ionic current that recharges the membrane capacity near the micro-electrode (Rushton, 1937, 1938; Noble, 1966; Noble & Stein, 1966).

This may be the reason for the difficulty in triggering the spike in most smooth muscle cells when intracellular stimulation is used, and also the reason why the threshold is higher than with external stimulation when the spatial decay of polarization is less. The fact that the spike evoked intracellularly cannot propagate while external stimulation produces a propagated spike can also be explained by the disproportion between the low intensity of the local current and the large surrounding area which has to be excited (Bennett & Merrillees, 1966; Tomita, 1967a). This situation is probably similar to the failure in producing a propagated response by a very local stimulation in the skeletal muscle fibre (Rushton, 1937; Huxley & Taylor, 1958).

According to this idea, a spike is more easily evoked in a cell which has fewer connections with surrounding cells. However, another possibility, e.g. different excitability of individual cells, cannot be excluded in order to explain why some cells give an active response and others do not (Bennett, 1967b).

External stimulation

When current is applied to the taenia coli or the vas deferens of the guinea-pig in a longitudinal direction using large external stimulating electrodes, the tissue shows cable-like properties. Therefore, with this method of stimulation, the fundamental mechanism of spike initiation is the same as that in other excitable tissues, such as nerve or skeletal muscle fibres which have cable-like properties. Thus, also in smooth muscle, as already described (see *Membrane polarization and spike activity*), when the surface membrane is depolarized to a critical firing level by an outward current crossing the membrane, a spike is evoked and it propagates along the tissue.

The longitudinal potential gradient at the rheobasic intensity is 10 mV/mm for excitable uterine strips (Bozler, 1938a). A gradient of about 20 mV/mm is necessary in the taenia coli of the guinea-pig in hypertonic solution. Since a single fibre is about 0·1 mm long, the potential gradient is only a few mV which is too small to depolarize the membrane by 10 to 20 mV to the firing level, if a single cell were the unit of excitation and if cells were not interconnected. However, when a tissue length of more than the space constant (1·5 to 2·0 mm) is considered as the unit of excitation, the smooth muscle can be treated similarly to a nerve or a skeletal muscle fibre. Thus, when the tissue length, or the distance between the stimulating electrodes, is reduced to less than the space constant, the current intensity must be increased to produce the response (Bülbring & Tomita, 1967) as in nerve fibre (Rushton, 1927; 1938).

Spontaneous spike generation

Most of the intestinal smooth muscles, especially longitudinal muscles, have spontaneous spike activity. The origin of the spikes could be either myogenic and/or neurogenic. Myogenic activity is suggested by the fact that spontaneous activity can be observed after treatment with atropine, hexamethonium, or cocaine (Bülbring, 1955; Bülbring, Burnstock & Holman, 1958; Bülbring & Kuriyama, 1963a; Kuriyama, Osa & Toida, 1967c), and also with tetrodotoxin which blocks the response to nerve stimulation (Nonomura, Hotta & Ohashi, 1966; Kuriyama, Osa & Toida, 1966; 1967c; Hashimoto, Holman & McLean, 1967; Bülbring & Tomita, 1967; Tomita, 1967a). Therefore, myogenic activity seems to be the main underlying mechanism for the spontaneous activity in many smooth muscles. However, there is also evidence for a strong neurogenic influence on spike activity in the same tissues, since atropine and tetrodotoxin modify the spontaneous activity by reducing the slow changes in the membrane potential (Kuriyama, Osa & Toida, 1967c; Hukuhara & Fukuda, 1968). Since the nervous influence on the electrical activity is treated in other chapters (see those by Kuriyama and Holman), only myogenic activity will be described here.

The mechanism of automaticity in smooth muscle, is still not known but the pacemaker potential must be the most important factor. Since depolarization increases the frequency of the pacemaker potential and hence of the spike discharge, and hyperpolarization decreases them when external polarization is used (Bülbring, 1955; 1957a), the pacemaker activity depends on the membrane potential (Fig. 14). Every cell may have a tendency to produce a pacemaker potential at random. However, since the potential change in only one cell cannot affect the others, as deduced from intracellular stimulation, and since the shape of the prepotential varies from time to time, the existence of a specific pacemaker cell is doubtful. It is very likely that many cells are involved in forming a pacemaker region in the tissue. Consequently, there must be some mechanism by which the cells in the pacemaker region are first synchronized. The following possibilities may be considered.

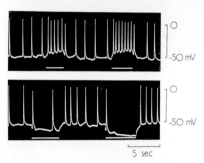

FIG. 14   Intracellular records of responses of taenia coli to externally applied polarizing current. *Top*: depolarization. *Bottom*: hyperpolarization. Note the frequency modulation. (From Bülbring, 1967)

1. Every cell has the ability to produce a pacemaker potential. When the number of cells, which are producing a pacemaker potential, exceeds a critical value at a particular time in some part of a bundle, the potential change in individual cells tends to synchronize and to affect the cells in other parts of the bundle. This change is similar to a transition from point polarization to space polarization, or to that from a spontaneous junction potential to an evoked (synchronized) junction potential.

2. Stretch depolarizes the membrane and increases the spontaneous activity (Bozler, 1938b; Bülbring, 1955; Bülbring & Kuriyama, 1963c; Mashima, Yoshida & Handa, 1966; Mashima & Yoshida, 1965) (Fig. 15). A thinner part or a branching part of a bundle may be more affected by mechanical stretch than a thicker and straight part of the bundle. As a result, this particular part of the bundle becomes the pacemaker region.

3. The rate of metabolism of the cells may differ in different parts of the tissue, according to the ease of supply of oxygen or glucose. A change in metabolism may cause depolarization of the membrane (the pacemaker potential). It is also possible that a local circuit current, flowing within a bundle due to differences in resting potential in various parts, stimulates the most excitable part of the bundle (probably a branching part, like the hillock of neurone).

If a propagated spike has once been initiated by the pacemaker potential, then the recovery process (the refractory period) determines the next pacemaker potential in the region where the activity of the cells is already synchronous.

Possible ionic basis for the pacemaker potential. The pacemaker potential (depolarization) may be due to (a) an increase in Na permeability, (b) an increase in Ca permeability, (c) a decrease in K permeability, or (d) an increase in Cl permeability, or a combination of the above factors.

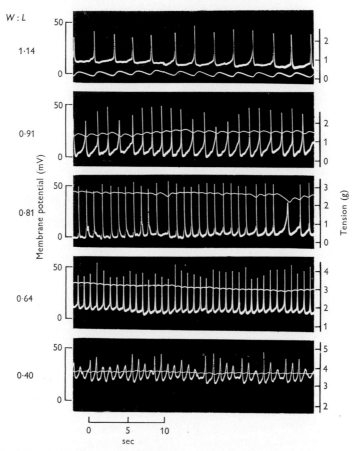

FIG. 15   Records from one experiment showing the effect of stretch on membrane potential, membrane activity, and tension in the taenia coli. The degree of stretch is indicated by the values of $W:L$. (From Bülbring & Kuriyama, 1963c)

Na excess increases the amplitude and the duration of the local potentials (the slow waves), while they are abolished in the absence of Na (tris chloride substitution) in the guinea-pig taenia coli (Bülbring & Kuriyama, 1963a). However, Na is probably not essential for the pacemaker potential, since spontaneous spikes with a nearly normal frequency can still be observed in Na-free solution. When external NaCl is reduced by replacing with sucrose, the spontaneous activity is always suppressed (Holman, 1957; Brading *et al.*, 1969*b*). However, the spontaneous spike generation can be started again by increasing the external KCl. Therefore, the suppression of the spontaneous activity in low Na solution may be due either to hyperpolarization of the membrane or to a lack of Cl, or to both.

If the spike is due to an entry of Ca ions rather than Na ions, the pacemaker potential (the local potential) could also be due to an increase of the membrane permeability to Ca, in the same manner as Na is responsible for the local potential as well as the spike in nerve fibres. However, there is no convincing evidence in smooth muscles to support this idea.

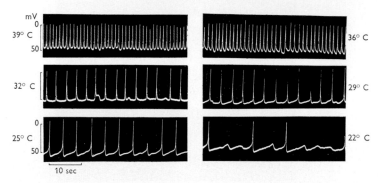

FIG. 16   Effect of temperature on membrane potential and membrane activity, in the taenia coli. (From Bülbring & Kuriyama, 1963*c*)

The pacemaker potential could be due to a reduction of the membrane permeability to K. As described above (see *Effects of ions on the membrane resistance*), it has been postulated that an increase of Ca at the inner layer (or the vesicles) of the membrane may increase K permeability. Then a removal of Ca from this site, probably by actively transporting Ca to the outside of the membrane, would decrease K permeability. This outward transport of Ca may be coupled with an uptake of K, corresponding to the Na-K pump in nerve fibres in which the spike is due to an inflow of Na ions. This hypothesis may also explain the high sensitivity of the spontaneous spike frequency to temperature change (Bülbring & Kuriyama, 1963*c*) (Fig. 16).

Another possibility is that the pacemaker potential is due to a leakage of anions (Cl ions in normal solution). The internal Cl concentration is probably too high to be passively distributed, the equilibrium potential being about -20 mV (Casteels & Kuriyama, 1966). Replacement of Cl with a less permeant anion (ethane or benzene-sulphonate) usually suppresses the spontaneous electrical activity and also the mechanical response, while replacement of Cl with a more permeant anion (nitrate) potentiates the spontaneous activity and also the mechanical response (Axelsson, 1961*a*; Kuriyama, 1963*a*; Bülbring &

Tomita, 1969a). It is also known that many factors which reduce the spontaneous activity reduce the internal Cl concentration (low K, low temperature and hyperosmolarity, A. F. Brading, unpublished observations). It is tempting to speculate that the vesicles seen on the membrane in the taenia have an anion permeable membrane and that this is an important site for excitation–contraction coupling, as postulated for the transverse tubular system in crab muscle (Girardier, Reuben, Brandt & Grundfest, 1963; Brandt, Reuben, Girardier & Grundfest, 1965; Reuben, Brandt, Garcia & Grundfest, 1967; Selverston, 1967). According to this idea, an outward movement of anions (Cl or nitrate) through the vesicle membranes would produce the pacemaker potential leading to the spike in the plasma membrane, and an inward movement of the anions through the vesicles, as a result of the depolarization of the plasma membrane (the spike) would release Ca from the vesicle membranes.

Spike propagation

Since the guinea-pig taenia coli has cable-like properties, it is natural to suppose that propagation of the spike is brought about by a local circuit current which flows between active and resting regions, as in the nerve fibre. Propagation cannot take place if the spike is initiated only in a single cable which is surrounded by many interconnected cables as already discussed, because the local circuit current spreads into surrounding cells in three-dimensional directions and becomes too weak to excite surrounding cells. However, when spikes are generated simultaneously in many cables (a cable being composed of a series of cells) at one region in a bundle, then there is little current flow between cables in a transverse direction, and the local circuit current flows only in the direction of the fibre axis, and the spike can propagate along the cable.

It has actually been shown, as described above in *Cable properties examined with alternating current and spike propagation*, that the early time course of the conducted spike in the taenia coli fits the prediction based on the cable theory (Tomita, 1966b; Abe & Tomita, 1968). Supporting evidence for assuming that spike propagation is brought about by a local circuit current is also found in the following observations:

(a) A diphasic potential can be recorded with external bipolar electrodes (Bülbring, Burnstock & Holman, 1958), (see Chapter 10 by Golenhofen)

(b) A triphasic potential can be recorded with a monopolar external electrode having an indifferent electrode at a distance in a large volume of medium (Nagai & Prosser, 1963b). The membrane current, which is recorded with a monopolar electrode as a potential drop outside the membrane, should be proportional to the second derivative (triphasic shape) of the propagating spike in a uniform cable (Katz & Schmitt, 1940; Hodgkin & Huxley, 1952d).

(c) The conduction velocity is reduced (Nagai & Prosser, 1963b) or the conduction is reversibly blocked (Barr et al., 1968) by increasing the electrical resistance of the fluid outside of the tissue.

(d) Conduction through a sucrose gap can be restored when the shunting resistance is reduced by connecting the chambers separated by the gap with a low electrical resistance (Barr et al., 1968).

Spike propagation appears to be facilitated by nervous influence. Tetrodotoxin (10^{-7} g/ml), which blocks nervous activity without affecting smooth muscle activity, slows the conduction velocity in the longitudinal muscle of the guinea-pig jejunum and rectum (Kuriyama, Osa & Toida, 1967b). The conduction velocities were 3·2 cm/sec in the jejunum and 4·8 cm/sec in the rectum. These were reduced to 2·1 cm/sec and to 4·0 cm/sec respectively in the presence of tetrodotoxin.

Conduction velocity

Conduction velocity in smooth muscle is less than 10 cm/sec. The velocity in a cable-like structure depends on the rate at which the membrane capacity ahead of the spike is discharged. In smooth muscle which conducts fastest, the extracellular space is small, the fibres are closely packed, and the longitudinal resistance of a strip in the sucrose-gap is low compared with that of slow conducting muscle (Burnstock & Prosser, 1960b; Prosser, Burnstock &Kahn , 1960). Thus, the conduction velocity may be closely related to the longitudinal internal resistance (myoplasmic and junctional resistances) of the tissue.

According to Huxley (1959), conduction velocities (θ) in a cable-like fibre under different conditions can be compared by the following equation when the spike propagates at a constant velocity.

$$\theta = \sqrt{\frac{a}{2R_i\gamma C_m}} \cdot \varphi \cdot f\left(\frac{\eta}{\gamma\varphi}\right) \tag{4}$$

where a is the radius of the fibre; R_i is the specific internal resistance; γ is a factor by which the membrane capacity is increased; φ is a factor by which all permeability changes are accelerated; η is a factor by which the absolute value of all ionic currents is increased. η can roughly be estimated by the change in the maximum rate of rise of the spike which is indicative of inward ionic currents. $f(\eta/\gamma\varphi)$ is the function of η, γ and φ. As seen from Table 4, a slow conduction velocity in smooth muscle is probably related to a small fibre diameter, a high internal resistance, and a slow rate of rise of the spike. Table 4 shows a clear relation between the maximum rate of rise and the conduction velocity. The ratio of these parameters is roughly constant in all muscle fibres. This ratio (the maximum rate of rise/the conduction velocity) is the maximum potential gradient along the cable produced by the propagating spike.

Slow component and spike component of the electrical activity

The slow components and spike components generated spontaneously in the taenia coli usually occur in close relation to each other (Bülbring, 1957c; Bülbring et al., 1958; Holman, 1958; Bülbring & Kuriyama, 1963a) (Fig. 17).

TABLE 4

	Radius	R_i	C_m	Max. rate of rise	Conduction velocity	Max. rate of rise / conduction velocity
	(μ)	$(\Omega\ cm)$	$(\mu F/cm^2)$	(V/sec)	(cm/sec)	(mV/mm)
Loligo axon (a)	250	30	1·1	630 (b)	2500	25
Sepia axon (c)	100	60	1·2	840	1150	73
Frog sartorius (d)	30	230	2·6 (e)	450 (f)	160	280
Purkinje fibre (g)	40	120	2·4	570 (h)	260	220
Terminal Purkinje (i)	15	120	2·4	530	120	440
Ventricular fibre (i)	8	120	2·4	230	88	260
Atrium (j, k)	7·5 (l)	120	2·4	75(65)	58(42)	130(150)
Taenia coli	2	300	1·5	7	6	120

(a) Hodgkin (1939); (b) Hodgkin & Katz (1949); (c) Weidmann (1951); (d) Katz (1948); (e) Falk & Fatt (1964); (f) Nastuk & Hodgkin (1950); (g) Fozzard (1966); (h) Weidmann (1955); (i) Matsuda (1960); (j) Vaughan Williams (1958a); (k) Vaughan Williams (1958b); (l) Woodbury & Crill (1961).

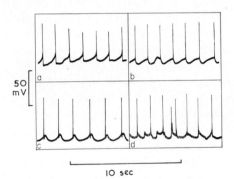

FIG. 17 Patterns of spontaneous activity in the taenia coli (a) spikes of 'pacemaker' type; (b) spikes occurring on the falling phase of the slow waves; (c) spikes occurring on the rising phase of the slow waves; (d) spikes occurring at any point. (From Bülbring, 1957c)

The slow component usually precedes a spike, triggers it and may persist as a depolarization delaying the full recovery of the membrane potential. However, sometimes the slow component follows a spike and is seen as a clearly different activity from the spike, or it appears on the falling phase as a notch.

When the tissue has been kept at low temperature (5°C) spontaneous activity is absent. During the recovery process at normal temperature (36°C) at first only small irregular slow components appear, either spontaneously or in response to electrical stimulation. They become gradually larger and regular, and the spike component is produced if the slow component reaches a critical level. The slow components and spikes are also differently affected by changes in the ionic composition of the external solution (Bülbring & Kuriyama, 1963*a*), or by intracellular polarization of the membrane (Kuriyama & Tomita, 1965). In hypertonic solution, or Cl-deficient solution, long-lasting slow components can be produced triggering repetitive spikes (Bülbring & Kuriyama, 1963*a*).

At least one type of the slow component, A, may be a pacemaker potential, since the tissue has spontaneous electrical activity. This component A may be produced in a special region of the tissue and trigger the spike there as already described. The spike is then conducted to other parts of the tissue, while the slow component A spreads electrotonically in a limited portion of the tissue, so that in the non-pacemaker regions the relation between the slow component and spike changes. At the same time, the slow component A can be produced independently in another part of the tissue (at another pacemaker region). However, it may not trigger a spike if it occurs during the long refractory period of the preceding spike or if its amplitude is too small. At other times it may succeed in setting the pace and thus the pacemaker region may shift from one part of the tissue to another. The slow component A is important for controlling the spontaneous spike frequency.

It has been shown that, in the taenia coli, there is a good correlation between the spike frequency and the tension development (Bülbring, 1954; 1955). The spike appears to be the initial event in the excitation–contraction coupling (Holman, 1958). Since depolarization itself does not cause any rise in tension and the tension rises only when spikes occur (Bülbring, 1957*c*), slow components observed without a mechanical response (Burnstock & Prosser, 1960*b*) may be the type A which is in no way connected with spike activity in neighbouring cells.

On the other hand, it is likely that there is also another type of slow component, B, the underlying mechanism of which is quite different. Since the electrical activity of the cells lying closely together is well synchronized, and the electrotonic potential produced by external current application is detected intracellularly over a distance of many cell lengths, there must be some electrical interaction between cells. The electrotonic spread (due to the local circuit current) of the electrical activity in neighbouring cells may initiate an active response (a local response and a spike). The most frequently observed slow component is probably this type of the slow component, B, i.e. the electrotonic spread of the spike and the local response, and this is important for the spike propagation. The size and shape is very variable and may depend on the condition of the cells. Sometimes it is very difficult to distinguish between the spike and the slow component. Some of them could be abortive spikes on large slow components.

The spike in the taenia coli conducts probably along functional bundles, as expected from the cable-like properties. However, the bundle, unlike a single skeletal muscle fibre, is not a simple cylinder but branches out and joins other bundles forming a mesh. Therefore, the geometrical factor of branching may produce some complications in the spike propagation. The local circuit current of a spike which propagates along a bundle must excite a larger area at the branching point. Hence the safety factor would be low and the spike would be more easily blocked than in a simple cylinder-like bundle.

There seems to exist a critical amplitude of the spike for normal propagation. A spike larger than the critical amplitude increases in amplitude to full size during propagation along the tissue. A spike smaller than the critical amplitude decreases in amplitude during propagation and finally becomes only a passive electrotonic potential (Tomita, 1967b). The critical amplitude of the spike which can propagate without decrement cannot be determined exactly because it

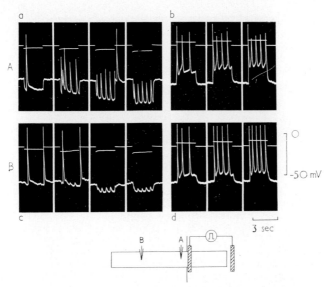

FIG. 18   Intracellular records (lower trace) from two different cells of the taenia coli at distances of 0·2 mm (A) and 2 mm (B) from the partition. (a) and (c): responses to hyperpolarizing currents, intensity increased stepwise (monitored in upper trace); (b) and (d) responses to depolarizing current. In hypertonic solution containing tetrodotoxin (10^{-6} g/ml). (From Tomita, 1967b)

depends mainly on the geometry of cell arrangement and the excitability of the region. Furthermore, the active component of the spike cannot easily be discriminated from the passive electrotonic component. However, the conduction probably becomes decremental when the spike amplitude is reduced to about half the normal amplitude or the membrane is hyperpolarized by about 10 mV (Fig. 18).

The conduction velocity of the spike may be slightly different in individual bundles and may be affected differently by membrane polarization. When two cells, probably in different bundles, are impaled simultaneously by microelectrodes, the records show that spontaneous spikes are not necessarily synchronous although (see Fig. 15) they are of the same frequency (Bülbring, Burnstock & Holman, 1958). Therefore, the shape of the spike recorded from the part where different bundles join might be altered depending on the time lag between the arrival of the spikes conducted along the different bundles. This may be an explanation for a notch on the spike. Notch formation on the spike or change of the spike shape are known to appear frequently during spontaneous electrical activity observed in normal Krebs solution (Bülbring, 1957a; Holman, 1958; Tomita, 1967b). A notch sometimes appears, or the spike is split into two small spike-like potentials when the membrane is hyperpolarized as shown in Fig. 19.

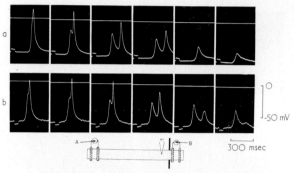

FIG. 19   Effects of external polarization on the conducted spike in the taenia coli (upper trace = current monitor, lower trace = intracellular record). The records were taken at about 0·3 mm from the partition. Conditioning polarization (by electrodes B) was started 300 msec before the stimulus (by electrodes A) used to evoke a response and then was maintained through the period of recording. Distance between the nearest stimulating electrode and the recording site was about 8 mm. The nearest stimulating electrode was the anode in (a) and the cathode in (b): the intensity was kept constant, slightly above threshold. In both (a) and (b), the anode of the polarizing electrodes was near the micro-electrode (ME). In hypertonic solution containing tetrodotoxin (10^{-6} g/ml). (From Tomita, 1967b)

Spike mechanism

The guinea-pig taenia coli has been most frequently used for the study of the spike mechanism. Therefore, in this section descriptions will be mainly confined to the taenia cell.

Effects of Na-deficiency

In the guinea-pig taenia coli, the amplitude of the spontaneous spike remained normal when Na in the external medium was reduced to as little as 20 mM by replacement with choline chloride; below this concentration the spike amplitude was reduced and, in the presence of 2 mM Na, spontaneous electrical activity was abolished (Holman, 1957; 1958). Bülbring & Kuriyama (1963a) confirmed that the size and the overshoot of the spontaneous spike were not dependent on the external Na concentration between 0 and 137 mM when Na was substituted with tris chloride; on the other hand, they observed a reduction of the rate of rise (from 8 to 5 V/sec) with lowering the external Na concentration.

Since the above observations were made on spontaneous activity, it is possible that the decrease of the rate of rise was due to an impairment of spike propagation, and that the loss of activity was due to the suppression of the pacemaker potential in low Na solution. Thus, in Na-free solution (hydrazine chloride substitution) some tension response can be recorded using the sucrose-gap method in response to electrical stimulation while no spikes were conducted, indicating that the cells were not inexcitable but that conduction had failed (Axelsson, 1961b).

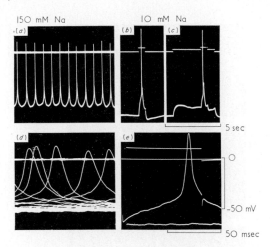

FIG. 20   Effects of a low external Na concentration on the action potential in the taenia coli. Spontaneous spikes recorded intracellularly in normal Krebs solution with (a) slow sweep and (d) fast sweep. After 1 hr in 10 mM Na (sucrose substitution), the spike was evoked by externally applied current in (b) by a 600 msec pulse taken at low sweep speed, in (e) evoked by a 75 msec pulse and taken by fast sweep, (c), spike evoked during conditioning depolarization. Upper trace shows zero potential level and relative current intensity in (b), (c) and (e). Note hyperpolarization of the membrane, suppression of spontaneous activity, increase in amplitude and rate of rise of the spike in low Na. Also note little change in overshoot by conditioning depolarization. (From Brading, Bülbring & Tomita, 1969b)

When NaCl was replaced with sucrose, leaving 10 to 15 mM Na contained in the buffer, the membrane was hyperpolarized by about 10 mV and spontaneous spikes stopped after about 10 min (Holman, 1957). However, spikes were easily evoked by a depolarizing current pulse, even after two or three hours exposure to the low Na solution (Brading & Tomita, 1968b; Brading et al., 1969b). (Fig. 20). In the low Na solution, the spike amplitude was larger (by about 20 mV) than in normal solution, and the maximum rate of rise was increased from 7 to 13 V/sec.

Effects of changing the external Ca concentration

Reduction of the external Ca from normal (2·5 mM) to 0·25 mM depolarized

9+

the membrane, and decreased the spike amplitude and its rate of rise (Holman, 1958; Bülbring & Kuriyama, 1963a; Brading et al., 1969b). The effects of changing the external Ca concentration in Krebs solution differed slightly from those in Locke solution in which the changes of the membrane potential and of the spike were more pronounced.

Excess Ca increases the amplitude and the rate of rise of spontaneously generated spikes (Holman, 1958; Bülbring & Kuriyama, 1963a) and of evoked spikes (Brading et al., 1969b). Excess Ca also increased the threshold, partly as

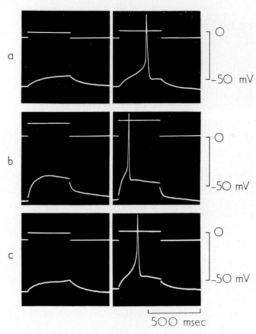

FIG. 21   Effect of excess Ca (12·5 mM) on subthreshold and threshold responses of the taenia coli obtained in twice hyperosmotic solution. *Top and bottom*: in normal Ca solution: *middle*: in excess Ca. Note the shape of the subthreshold response, the increase in threshold and amplitude of the spike and the short latency in excess Ca. (From Brading, Bülbring & Tomita, 1969a)

a result of the membrane hyperpolarization and partly due to a shift of the firing level towards a more positive inside potential. In the presence of a normal Ca concentration (2·5 mM) the latency could easily be more than 200 msec at threshold stimulation while, in the presence of excess Ca, it was relatively short (less than 100 msec). (Fig. 21).

Effects of low Na and low Ca

In a Na deficient solution (10 mM, sucrose substitution), the depolarization of the membrane caused by reducing the external Ca from 2·5 to 0·2 mM was very small, but the rate of rise of the evoked spike decreased from 13 to 6 V/sec. When the Ca concentration was reduced to 0·5 mM in Krebs solution containing the normal Na concentration, the membrane was depolarized by as much as 10 mV and the spike became small (Holman, 1958; Bülbring & Kuriyama, 1963a). Thus, the effect of reducing Ca was much greater in the presence of the normal Na concentration than with low Na (Fig. 22). Moreover, when the external Na concentration was reduced from 137 mM to 15 mM after the membrane had been depolarized in low Ca (0·5 mM), the membrane was gradually repolarized, the spike frequency became low and the amplitude of the spike increased. The spontaneous activity finally stopped after further repolarization. In this condition, a spike with nearly normal amplitude and rate of rise could be evoked by depolarizing current.

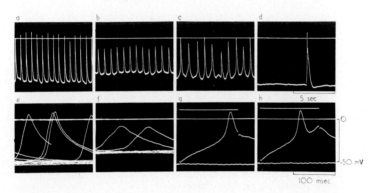

FIG. 22   Effects of reducing Na after the preceding exposure to low Ca. Intracellular recordings.   (a) and (e): spontaneous spikes in normal Krebs solution at two different sweep speeds; (b) and (f): in 0·5 mM Ca solution containing normal Na; (c) after 5 min; (d), (g) and (h) after 30 min in 15 mM Na and 0·5 mM Ca. Note recovery of amplitude and rate of rise of the spike by reducing Na concentration in the presence of low Ca. (From Brading, Bülbring & Tomita, 1969b)

In the presence of the normal Na concentration, exposure to Ca-free Locke solution caused membrane depolarization, it abolished all activity, and the membrane resistance became very low. However, after complete removal of Na from the medium (replacement of NaCl with sucrose or tris chloride), repolarization of the membrane and an increase in the membrane resistance were observed (Fig. 23). Spike activity returned after addition of only one-tenth of the normal Ca concentration, in the absence of Na. Further increase of the Ca concentration to 1 mM suppressed the spontaneous activity, although the spike could be evoked by a depolarizing current (Brading et al., 1969b).

Possible ionic basis for the spike

The amplitude of the spike in the smooth muscle of guinea-pig taenia coli is relatively insensitive to the external Na concentration. However, a contribution

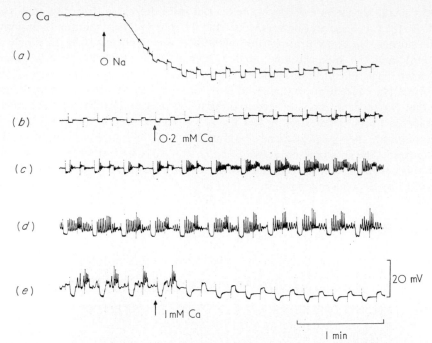

FIG. 23   External recording from taenia coli with double sucrose-gap method. Constant current pulses with alternating polarities (3 sec duration, every 10 sec) applied throughout.   (*a*) State of depolarization and low membrane resistance after 15 min exposure to Locke solution containing zero Ca. At arrow, removal of Na (sucrose substitution) causing hyperpolarization and recovery of membrane resistance.   (*b*) After 7 min in zero Na and zero Ca, at arrow 0·2 mM Ca added.   (*c*) and (*d*) Continuous records following (*b*).   (*e*) After 20 min in 0·2 mM Ca and zero Na, at arrow 1 mM Ca added. Note recovery of spike activity by 0·2 mM Ca but suppression of spikes by 1 mM Ca in zero Na. (From Brading, Bülbring & Tomita, 1969*b*)

of Na to the spike has been suggested by the finding that the maximum rate of rise of the spontaneous spike, which is indicative of the inward ionic current, is decreased by reducing the external Na concentration (Holman, 1958; Bülbring & Kuriyama, 1963*a*).

The reduction of the rate of rise may not be simply due to a lack of Na ions, and replacement of Na with choline or tris might have some secondary effect on the membrane. Depolarization in tris solution, which is the opposite effect to what one would expect from the relatively high Na permeability of the membrane, may suggest a toxic effect. When NaCl was substituted with sucrose, the membrane was actually hyperpolarized and the rate of rise of the evoked spike was increased.

A definite conclusion regarding the spike mechanism can as yet not be drawn from the available data. Na may carry part of the total current during the spike. When Na is reduced by substitution with sucrose, the effect is probably the same as that of increasing Ca, if Ca and Na compete at the membrane (Lüttgau & Niedergerke, 1958). Then, an increase of Ca at the membrane potentiates the regenerative process of the spike, and this potentiation overcomes the effect of Na deficiency. Similarly, although the effect of changing the external Ca concentration on the spike suggests a direct contribution of Ca to the spike, it is not clear whether the effect is actually due to a change in the Ca current or, more indirectly, to a variation of the permeability of the membrane to other ions. One of the known effects of excess Ca is the increase of the availability of membrane sites for Na inward movement during the spike (Weidmann, 1955; Frankenhaeuser & Hodgkin, 1957).

Nevertheless, the Na contribution to the spike, if any, is probably relatively small, because not only are the overshoot and the rate of rise of the spike actually increased in low Na solution, but the spike can also be evoked in zero Na solution containing Ca. These observations strongly suggest that the spike in the taenia is due to Ca entry. However, the possibility that Ca ions may replace Na ions and carry the charge to generate the spike only when the external Na concentration is reduced, as postulated by Bülbring & Kuriyama (1963a), cannot be neglected. In the cat ureter, a contribution of both Na and Ca to the spike has been suggested (Kobayashi & Irisawa, 1964; Kobayashi, 1965).

The idea that Ca is the main ion involved in the spike mechanism of the taenia, is based on considerable evidence:

(a) Even though the normal solution contains only 2·5 mM Ca, the Ca equilibrium potential may be over $+100$ mV which is sufficient for spike production, since the free intracellular Ca is probably less than 10^{-6}–10^{-7} M, as in other muscles (Caldwell, 1968). The maximum rate of rise of the smooth muscle spike is slow (about 10 V/sec) (see Table 4) probably because the external Ca concentration is only 2·5 mM. The maximum rate of rise of the spike in nerve or skeletal muscle, which is due to influx of Na ion, is about 500 V/sec and is roughly proportional to the external Na concentration (Hodgkin & Katz, 1949).

(b) In low Na solution (10 mM), the intracellular Na concentration is only reduced from 35 to 24 mM, reversing the equilibrium potential from $+34$ to -22 mV, while the overshoot of the spike is increased from $+10$ to $+20$ mV (Brading & Tomita, 1968b; Brading et al., 1969b). It is unlikely that reduction of the external Na decreases the intracellular Na so much that the concentration gradient becomes actually larger than before and causes an increase of the overshoot of the spike.

(c) In low Na solution (which increases the spike amplitude and its rate of rise) Ca deficiency causes only a small depolarization but decreases the amplitude and the rate of rise of the spike considerably, suggesting an increase in Ca permeability of the membrane during the spike. The increase in the rate of rise and the spike amplitude in low Na solution, and the decrease in low Ca solution, are more than can be explained only by the shift of the membrane potential.

(d) The following observations are similar to those made on crustacean muscle (Fatt & Ginsborg, 1958; Hagiwara & Naka, 1964; Hagiwara & Nakajima, 1966) whose spike is due to Ca entry: (i) Ineffectiveness of tetrodotoxin on the spike (Kuriyama et al., 1966; Nonomura et al., 1966) although it blocks the electrical response produced by nervous elements in the muscle (Bülbring & Tomita, 1967; Kuriyama et al., 1967c). (ii) Spike generation in Ca-free solution containing Ba (Hotta & Tsukui, 1968; Bülbring & Tomita, 1968b; 1969c).

The negative effect of tetrodotoxin on the spike may not be taken as unequivocal evidence for a Ca spike, because the muscles in taricha newt and tetrodon fish are resistant to tetrodotoxin, even though the spike in the muscles is sensitive to a change in the external Na concentration (Kao, 1966). Similarly, tetrodotoxin had no effect on the membrane activity of the guinea-pig ureter while 3 mM $MnCl_2$ blocked its electrical activity (Washizu, 1966, 1968; Kuriyama et al., 1967a). On the other hand, the spike of the cat ureter depends on the external Na concentration (Kobayashi & Irisawa, 1964; Kobayashi, 1965).

As a result of observing the effects of Na and Ca on the spike, Bennett (1967a) also reached the conclusion that in the guinea-pig vas deferens, part of the current responsible for the rising phase is carried by Ca. Tetrodotoxin also had no effect on the spike although it blocked the electrical response caused by nervous elements (Tomita, 1966c, 1967a; Hashimoto et al., 1967).

Calcium may contribute to the spike in two different ways: One, as a current carrier, the amount of which is determined by the external free Ca ion concentration, and the other, as a stabilizing agent, the amount of which is determined by its adsorption at the membrane. As a stabilizing agent, Ca reduces Na permeability and activates the regenerative process for the spike generation (Frankenhaeuser & Hodgkin, 1957). The stabilizing action of Ca is known not only in nerve, in which the spike is due to Na influx, but also in barnacle muscle, in which the spike is due to Ca influx (Hagiwara & Takahashi, 1967). The weak effect of calcium deficiency on the spike of taenia in Krebs solution (Holman, 1958) and the dramatic effect of calcium deficiency in Locke solution (Brading et al., 1969b) may be due to the fact that Ca binds more favourably at the membrane in the buffer system in Krebs solution.

The effect of reducing the external Na concentration on the membrane potential and on the spike is similar to that of increasing the external Ca concentration. The reduction of ionic strength by replacing NaCl with sucrose may increase Ca adsorption at the membrane (see Chapter by Goodford) and potentiate the stabilizing action of Ca. The degree of depolarization in low or zero Ca solution depends on the external Na concentration, suggesting an increase of Na conductance of the membrane in low Ca solution. Therefore it is likely that Na influences the spike activity indirectly through competition with Ca as in the cardiac muscle (Lüttgau & Niedergerke, 1958; Niedergerke & Orkand, 1966a) or by competing with Ca in the control of the membrane potential, which in turn affects the spike activity.

CONCLUSION

It has been stressed that, in smooth muscles which have electrical interconnections between cells, knowledge of the current distribution in the tissue is very important, not only for the evaluation of the membrane resistance and capacity, but also for the interpretation of the responses to electrical stimulation. The taenia coli and the vas deferens of the guinea-pig are typical examples of tissues with strong electrical cell to cell connexions. However, the degree of interconnections is probably quite different from one smooth muscle type to another, or even from one part of the same tissue to another part.

The taenia coli and the vas deferens of the guinea-pig have cable-like properties. The smooth muscle cells are so arranged that they form functional bundles. Electrically these can be represented by a series of independent cables formed by end-to-end connexions between cells. Although there are also many transverse interconnexions these can be disregarded when external polarization is applied. The current spreads only in the longitudinal direction, because the fibres run parallel and are at a uniform potential, so that there is no transverse current flow. In contrast, when intracellular polarization is applied, the current is dissipated in three dimensions by the shunting of the membrane near the point of current application by the large membrane area surrounding it. Consequently, the membrane properties cannot be derived by applying cable equations to responses to intracellular polarization.

Effects produced by changes in ionic environment have been described and discussed only from observations on the guinea-pig taenia coli. The conclusions cannot be generalized to include other smooth muscles which may have different properties. For example, it is known that various smooth muscles respond differently to acetylcholine and to catecholamines. It is very likely that the differences in the responses to drugs are related to differences in the permeability of the muscle-cell membrane to ions.

The hypothesis concerning the cellular function of calcium, both in controlling membrane permeability and in the spike mechanism of the taenia, is very tentative. In order to substantiate it further careful studies are necessary. Finally, it must be emphasized that it is very important to employ an adequate method for the measurement of changes in membrane conductance in future investigations of the membrane properties of different smooth muscles.

8

JUNCTION POTENTIALS IN
SMOOTH MUSCLE

MOLLIE E. HOLMAN

INTRODUCTION

Views on the innervation of smooth muscle have varied between the two extremes illustrated diagrammatically in Fig. 1. In Fig. 1A individual smooth muscle cells are independent of each other. Transmitter is liberated from nerve terminals into extracellular spaces, eventually reaching all cells by diffusion. (For details of this model see Rosenblueth, 1950.) In Fig. 1B, neighbouring cells are able to interact with each other. Transmitter is liberated at sites where nerve terminals form morphologically distinct neuromuscular junctions and the effect of transmitter is passed on to neighbouring 'non-innervated' cells by electrotonic spread or by the propagation of action potentials (see Eccles, 1936). This vastly simplified diagram points to the need for an understanding of two aspects of the innervation of smooth muscle before any attempt can be made to interpret new electrophysiological data:

1. The morphology of autonomic nerve terminals and their relationship with smooth muscle.

2. The electrical properties of the smooth muscle cell membrane and the nature of the interaction between individual cells. These questions have been discussed in earlier chapters by Burnstock and Tomita. Only the briefest outline of present concepts will be given here.

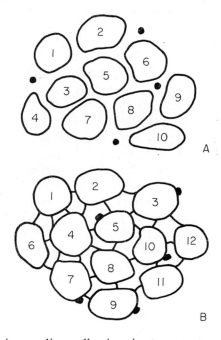

FIG. 1 Diagram of transverse sections through two bundles of smooth muscle cells. In A, the cells (1 to 10) are independent units. Nerve terminals (●) release transmitter into the extracellular space surrounding the cells. In B, the cells (1 to 12) are able to interact with each other. Nerve terminals (●) make discrete neuromuscular contacts (five such profiles are shown).

The nonmyelinated axons of autonomic ganglion cells give rise to an extensively branching system of terminal axons which ramify within the effector organ for distances of several mm (Malmfors, 1965a). The ultrastructure of the terminal axons has been clarified by a large number of electron microscope studies, especially that of Merrillees (1968), who followed the ramifications of several bundles of axons throughout a strand of smooth muscle (guinea-pig vas deferens) over a distance of 500 μ. This marathon task required a total of 3200 sections. Terminal axons resemble a string of beads, with thick portions, 0·5 to 1·0 μ in diameter, 1 to 2 μ in length, separated from each other by narrow regions 0·1 to 0·2 μ in diameter. Beads or varicosities, occur at a frequency of about 20 per 100 μ, single varicosities or small clusters of them undoubtedly correspond with the rows of bright green spots characteristic of the appearance

9*

of noradrenergic axons when they are observed in the fluorescence microscope following the histochemical procedure of Falck & Hillarp (Falck, 1962). Axons are accompanied by Schwann cells until close to their actual termination. Many axon varicosities however, are only partially covered by Schwann sheath and regions of 'naked' axon membrane are common, especially in varicose regions. Single axons, as opposed to those found in bundles, may be associated with only a fine Schwann cell process.

Axon varicosities contain neurotubules, neurofilaments, mitochondria and vesicles of various kinds. Small synaptic vesicles, indistinguishable from those of the skeletal neuromuscular junction, are predominant within cholinergic varicosities. Noradrenergic terminal axons also possess small vesicles of similar size (about 500 Å) but many of these have an extremely electron-dense core (small granular vesicles). Large granular vesicles (about 1000 Å) with a somewhat more diffuse core can be seen in both types of varicosity. It is generally considered that transmitter can be synthesized, stored and released from all the varicosities of terminal axons (Malmfors, 1965a).

The distribution of the terminal axons and the distances separating them from smooth muscle cell membranes vary greatly from one tissue to another and even for the same organ from different species (for a detailed account, see Chapter by Burnstock). In Fig. 2A and B, two extreme forms of innervation pattern are indicated diagrammatically. In both cases, the smooth muscle cells are arranged in bundles, from 100 to 200 μ thick. In Fig. 2A, terminal axons are confined to the extracellular space outside the bundle of cells whereas in Fig. 2B, terminal axons penetrate the bundle, pushing their way through the narrow spaces between neighbouring smooth muscle cells. The pattern of innervation shown in Fig. 2A is usually associated with a fairly wide gap between axon and muscle membranes (greater than 0·1 μ though there are exceptions), and this pattern is typical of intestinal smooth muscle. The pattern of innervation of Fig. 2B is usually associated with closer neuromuscular contacts, from 0·1 μ to less than 200 Å, and is most exaggerated in tissues like the vas deferens and other pelvic viscera (Nagasawa & Mito, 1967).

The consequences of these different patterns of innervation in relation to the transmission of excitation and inhibition will depend on the electrical properties of the smooth muscle cells and the interaction between them. Recent work on a number of visceral smooth muscles has confirmed Bozler's observations that smooth muscles can behave as electrical syncytia. The judicious use of a combination of external and intracellular electrodes by Bülbring, Kuriyama, Tomita and others has made it possible to give a quantitative description of the linear electrical properties of smooth muscle (see Chapter by Tomita).

Provided one is dealing with a sufficiently large aggregation of cells (e.g. bundles greater than 100 μ in diameter), it is possible to show that smooth muscles have electrical properties which are equivalent to a single 'core-conductor' or cable with a length constant of 1 to 2 mm and a time constant of the order of 100 msec (see Chapter 7). The parameters for electrical stimulation with large *external* electrodes and the velocity of conduction of an action potential so initiated can be explained in terms of this model (Tomita, 1966*a*, *b*; Kuriyama, Osa & Toida, 1967*a*, *b*; Abe & Tomita, 1968). One might look therefore, for a pattern of innervation in visceral smooth muscle which utilizes these characteristics of *bundles* rather than *single* smooth muscle cells (Bennett & Rogers, 1967).

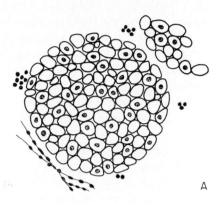

A

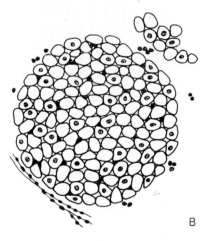

B

FIG. 2   Diagram of transverse sections through two smooth muscles with contrasting patterns of innervation.   A. Terminal axons, shown in cross-section as black dots and in longitudinal section as a group of varicose fibres (*lower, left-hand corner*) are confined to the extracellular space surrounding the bundle.   B. Terminal axons penetrate the bundle of smooth muscle cells and can be seen lying in some of the spaces between the tightly packed smooth muscle cells.

    The morphological basis of electrical interaction between individual cells which enables bundles of smooth muscle to behave as single units (Bozler, 1948), is probably a form of tight junction or nexus (Barr, Berger & Dewey, 1968). This is a region where the membranes of neighbouring cells appear to fuse with each other to form a five-layered complex. There have been extensive studies on the properties of similar regions of membrane fusion in epithelial and glandular tissues (Loewenstein, 1966). Here, it has been demonstrated that fusion alters the permeability of the cell membrane so that ions and other particles can move with ease from one cell to another. The low electrical resistance of these structures can probably account for the syncytial nature of many visceral smooth muscles.

The electrical properties of smooth muscle have also been studied with various bridge circuits which enable the cell membrane potential (henceforth MP) to be recorded while current is injected through the same electrode (e.g. Kuriyama & Tomita, 1965; Bennett & Merrillees, 1966; Hashimoto, Holman & Tille, 1966; Tomita, 1967a). The input resistance observed in these experiments is relatively high compared with that of other muscles, ranging from 15 to 100 MΩ (Holman, 1968). On the other hand, the time course of the changes in MP at the onset and cessation of a current pulse are very much faster than those observed during polarization with external electrodes (Tomita, 1966a, b). It is likely that current flow from an intracellular electrode causes a fairly localized change in MP since it is ineffective in altering patterns of spontaneous electrical activity (Kuriyama & Tomita, 1965).

Differences in the temporal and spacial characteristics of electrotonic potentials induced by current flow from an intracellular electrode compared with larger external electrodes can be explained in terms of the syncytial nature of many visceral smooth muscles (Tomita, 1966a, b, 1967a). This problem first arose some years ago in relation to cardiac muscle. As a result of studies on the steady state changes in MP caused by intracellular current flow in atrial muscle, Woodbury & Crill (1961) proposed that a syncytial structure such as the atrium could be represented electrically as a continuous sheet of membrane with the current-passing electrode at its centre. Since the area of membrane available to current flow increases rapidly with distance away from its point of application, changes of MP will be limited to a region close to the current-passing electrode. The linearizing effects of such a syncytial structure on the relation between current and membrane potential have been emphasized by Noble (1966). Although the time course of the changes in MP evoked by a square pulse of current have yet to be worked out mathematically for the sheet model, it is clear that the initial phase of the rise or decay of potential will be very much faster than the membrane time constant (the product of the resistance, R_m, and capacitance, C_m, of unit area of all membrane) (Jack & Redmond, personal communication).

Arguments in favour of a similar explanation for the effects of intracellular polarization of smooth muscles of the single unit type are difficult to refute. Convincing evidence for this view has been found for the taenia coli, and other intestinal muscles, guinea-pig vas deferens and ureter (see Chapter by Tomita).

It has, however, proved to be extremely difficult to demonstrate single-unit characteristics such as the longitudinal spread of electrotonus in the *mouse* vas deferens — a preparation which will be referred to elsewhere in this chapter. This was so, in spite of the use of a variety of external electrodes and much larger currents than those needed to demonstrate conduction of excitation or electrotonus in parallel experiments on the guinea-pig vas deferens (Yonemura, unpublished work). This has led us to question whether the sheet model of Woodbury & Crill (1961) should be used to account for the effects of intracellular polarization in the mouse. The time course of the changes in MP following intracellular current injection, is somewhat slower in the mouse than in the guinea-pig preparation, the time constant of the initial exponential phase being 5 to 10 msec, compared with 2 to 3 msec in the guinea-pig (Hashimoto *et al.*, 1966).

If one assumes that the dimensions of the smooth muscle cells of the mouse vas deferens are such that current density is uniform across the entire cell membrane and, further, that each cell is completely isolated from its neighbours, then it is possible to calculate the time constant from the values of membrane resistance (R_m) and membrane capacitance (C_m). If R_m is approximately 50 KΩ cm^2 and C_m is 2 μF cm^{-2} (see Chapter by Tomita), then the time constant should be equal to 100 msec.

The rapid time course of the electrotonic potentials observed during intra-cellular current injection in the mouse vas deferens suggests that there must be some electrical coupling between neighbouring cells though this may be less extensive than in the guinea-pig vas deferens. The smooth muscle cells of the mouse vas deferens may be coupled into small groups or bundles whose geometry is inappropriate for the longitudinal spread of electrotonus. The extent and the geometry of the electrical coupling in different smooth muscles may show a continuous range of variation from that of a three-dimensional syncytium like cardiac muscle to a relatively uncoupled situation where only a fraction of injected current passes across the membrane of cells other than that impaled with the intracellular electrode.

In all smooth muscles studied so far, action potentials (APs) initiated by intracellular stimulation are not conducted throughout the whole smooth muscle or even along the bundle in which the stimulated cell lies (Bennett & Merrillees, 1966; Tomita, 1967a, b). Such APs are probably localized to one or a very small number of cells. Again this could be explained in terms of a tightly-coupled syncytium since the inward current flowing across the membrane at the site undergoing an active change in permeability would be insufficient to de-polarize the large area of surrounding inactive membrane to threshold (Tomita, 1966a; Bennett, 1967b). This could also come about if there was very little interaction between cells. In either case, a pattern of innervation leading to the excitation of individual cells, separated from one another by a number of non-innervated neighbours, would be ineffective in bringing about excitation of the whole organ.

Many visceral smooth muscles show spontaneous rhythmic contractions which persist in the presence of tetrodotoxin, a drug which blocks nerve excitation (Kao, 1966). Such contractions are generally triggered by APs of spike form. Before discussing the effects of nerve stimulation it is important to have some idea of the factors which determine the pattern of firing of APs in these muscles. (This question will be discussed in detail in the Chapter by Kuriyama.) The following is a brief account of our own view (Holman, 1968) which will be referred to again later on:

1. All regions of spontaneously-active smooth muscles have low RMPs and are probably in a state of excitability which is close to that at which APs arise spontaneously.

2. If the MP of the muscle is decreased, for example by stretch, some cells or groups of cells may begin to fire at their own intrinsic rate. Such APs may be conducted to neighbouring regions — the most excitable region becoming the pacemaker for the neighbouring regions of the tissue.

3. Many smooth muscles are characterized by slow rhythmic fluctuations in MP and excitability — the so-called 'slow-waves'. These may occur in the absence of APs and if so, they do not trigger contractions (see Holman, 1968).

It is not yet clear whether slow waves are generated by all cells or by specialized regions of the tissue. The frequency of the slow waves is remarkably constant for any particular tissue — usually of the order of 1 to 20 per minute. The action of the slow waves is to modulate the frequency of firing of APs which, again, are probably initiated first in the most excitable pacemaker regions. Thus, although all cells may show similar slow waves, some cells are characterized by APs of the pacemaker type whereas others show conducted APs ('driven-cells'— Bennett, Burnstock & Holman, 1966a). The relationship between pacemaker regions and driven regions in smooth muscle resembles that of cardiac muscle. Smooth muscle differs from cardiac muscle however, in that the frequency of its pacemakers is modulated by inherent fluctuations in excitability and MP — the slow waves.

Smooth muscle is infamous for its diversity. Electrophysiological studies on neuromuscular transmission have been limited so far to relatively few examples, and it is probably premature at this stage to attempt to generalize. However, a general rather than a particular account of the junction potentials of smooth muscle follows, since it is hoped that a certain amount of theorizing might help to stimulate further experimentation.

NATURE OF EXCITATORY AND INHIBITORY TRANSMITTERS

It is well known that both acetylcholine (ACh) and noradrenaline (NA) may fill the role of either an excitatory or an inhibitory transmitter and that a number of smooth muscles receive both an excitatory and an inhibitory innervation which are cholinergic or noradrenergic respectively (or vice versa). However, the generalization so popular with students, that the two types of nerve fibre are always antagonistic is far from true. The innervation of the pelvic viscera, for example, includes a number of examples of the synergistic action of cholinergic and noradrenergic nerves (see Chapter 15). The question as to whether or not these two neurohormones exert their inhibitory or excitatory effect through the same membrane mechanism will be discussed in later sections.

The only junctions studied so far at which neither NA nor ACh appears to be the transmitter are those between the axons of intrinsic inhibitory neurones of the gastro-intestinal tract. Transmission at these junctions is resistant to both atropine and guanethidine (Bennett et al., 1966b). So far no specific blocking agent for this inhibition has been found and the identity of the transmitter is still unknown.

SPONTANEOUS JUNCTION POTENTIALS

Characteristics

Before giving an account of the responses to nerve stimulation it is of interest to see what can be learned about neuromuscular transmission from studies on the spontaneous discharge of junction potentials (SJPs) which have been recorded in some smooth muscles (vas deferens of guinea-pig, rat and mouse, Burnstock & Holman, 1966a; dog retractor penis, Orlov, 1963a). These are all noradrenergic excitatory junctions. Although spontaneous inhibitory JPs have been described for the guinea-pig taenia coli (Bennett et al., 1966b), it is possible that these might have been due to mechanical stimulation of pre-junctional nerve fibres by the micro-electrode.

There is evidence that the excitatory SJPs are due to the spontaneous release of transmitter from axon varicosities in a packaged manner — a feature in common with skeletal neuromuscular junctions, ganglionic and central nervous system synapses (Martin, 1966). The discharge of SJPs in smooth muscle is unaffected by concentrations of tetrodotoxin which block nerve impulses and evoked neuromuscular transmission (Tomita, 1967a; Hashimoto, Holman & McLean, 1967). In the dog retractor penis they are abolished by denervation (Orlov, 1963a) — a procedure which cannot be readily undertaken for the vas deferens. However SJP frequency is markedly reduced by reserpine in the guinea-pig vas deferens where the majority of terminal axons are noradrenergic (Burnstock & Holman, 1962b).

Figure 3 illustrates the characteristics of the SJPs recorded from the mouse vas deferens. Amplitudes are variable, ranging from 20 mV to less than 1·0 mV (or less than the noise level of the recording system). Amplitude histograms are skewed towards low amplitude SJPs, as are those for the guinea-pig vas deferens (Burnstock & Holman, 1966a). Figures 4A and B show scatter diagrams in which the amplitude of successive SJPs has been plotted as a function of rise time (resting MP to peak) and half-decay time (peak to half of total amplitude) for an intracellular recording from mouse vas deferens (Holman & Yonemura, unpublished work). The most obvious feature of this and similar scatter diagrams from other cells is the small variation in rise time of SJPs of variable amplitude

(2·5 to 15 mV). Some of the smaller SJPs may have longer rise times but the errors involved in their measurement make this point uncertain. The scatter diagram for half-decay times (Fig. 4B) shows a tendency for larger SJPs to decay more slowly than smaller SJPs.

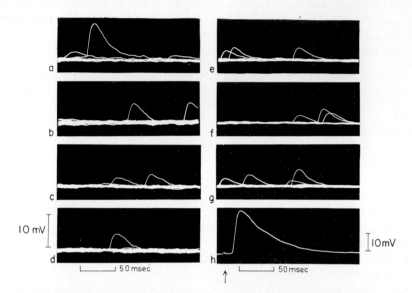

FIG. 3   Intracellular records from mouse vas deferens. Superimposed traces (a to g) showing the time course of SJPs of varying amplitude. Record h, *bottom right*, shows JP evoked by nerve stimulation (indicated by arrow). (Burnstock & Holman, 1966a)

Burnstock & Holman (1962a) suggested that the largest SJPs in the guinea-pig vas deferens were due to transmitter action at a site close to the recording electrode (focal SJPs) whereas the smaller SJPs were initiated by transmitter action at distant sites from which depolarization spread to the region of the intracellular electrode by electrotonus. Evidence for the electrotonic spread of SJPs in the guinea-pig preparation has since come from experiments in which two cells, less than 50 μ apart, were impaled simultaneously. Although the discharge of SJPs occurred independently in many such pairs of cells, on some occasions a pair of cells was found where *some*, but not all, of the SJPs were synchronous. In such cells synchronous SJPs occurred too frequently to be explained as random coincidences (see Burnstock & Holman, 1966a, Fig. 2). If the two cells were 100 μ apart, SJPs were completely independent (Tomita, 1967a).

In a tightly-coupled syncytial structure depolarization evoked by transmitter action on a *small area* of smooth muscle membrane would be expected to decay rapidly with distance from the site of transmitter action (see Chapter by Tomita). Hashimoto *et al.* (1966) and Tomita (1967a) have pointed out that SJPs are probably quite localized events compared with evoked JPs (see below). Nevertheless,

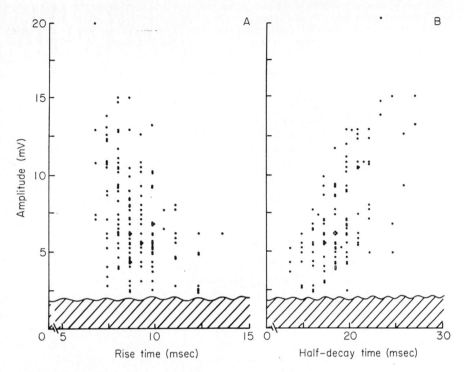

FIG. 4 Scatter diagrams showing the relationship between the amplitude of a series of SJPs recorded from mouse vas deferens and their rise times (A) and half-decay times (B). The wavy line indicates the noise level recorded throughout this impalement.

the finding that synchronous SJPs *do* occur occasionally in different cells suggests that there may be some degree of electrotonic spread which could contribute to the variation in their amplitude. Variation in amplitude may be partly due to variation in the concentration of NA at the smooth muscle membrane, resulting from the release of packets of NA from different regions of terminal axon varicosities situated at variable distances from the smooth muscle membrane (see below). The possibility that the packets of NA are not of uniform size must also be considered.

It is of interest to compare the time course of the SJPs with that of 'local' electrotonic potentials of similar amplitude, evoked by intracellular current injection. If the inward current at the site of transmitter action is restricted to a small area of membrane, comparable with that affected by local current injection, one might expect the rising phase of the SJP to be faster than that of an electrotonic potential — provided the SJP was recorded at the site of transmitter action where membrane conductance was increased. The maximum rate of rise of a few of the larger SJPs in the mouse vas deferens did exceed that of a local electrotonic potential of similar amplitude, recorded in the same cell (Yonemura & Holman, unpublished work). In most instances however, the maximum rate of rise of the SJP was similar to or slower than that of the electrotonic potential.

The rising phase of the SJPs of the guinea-pig preparation is also slower than that of the 'local' electrotonic potential (Hashimoto *et al.*, 1966; Tomita, 1967a). This probably indicates that a larger area of membrane is depolarized during an SJP than during a local electrotonic potential (see Chapter by Tomita). In both preparations the decay phase of the SJP is very much slower than that of the electrotonic potential.

The duration of transmitter action which gives rise to the SJPs is unknown. If it could be demonstrated that the decay phase of the SJPs was determined by the passive electrical properties of the tissue, it would be reasonable to conclude that transmitter action was brief — possibly only amounting to a fraction of the rise time of a 'local' SJP. Although we now have an equivalent circuit which can explain the longitudinal spread of electrotonus in single unit smooth muscles when many cells are polarized simultaneously, and we also have the 'sheet' model to account for the effects of intracellular current injection at a point, the effects of current flow across a finite area of a syncytial structure have yet to be worked out.

Furthermore, we have no idea of the dimensions of the area undergoing transmitter action following the release of a single packet of NA. It is possible that the slower rate of decay of the SJPs in the guinea-pig vas deferens compared with those of the mouse, might be a consequence of the fact that a greater area is involved in the transmitter action of a packet of NA in the guinea-pig preparation than in the mouse, but evidence for this view is purely circumstantial.

In some preparations of mouse vas deferens which have low RMPs, SJPs may initiate APs, as shown in Fig. 5 (Holman, unpublished work). It can be seen that following the repolarization phase of the AP, the MP falls again, eventually returning to the resting RMP with a time course resembling that of the SJP. At first sight, this observation could be taken as evidence for continuing transmitter action during the time that the AP was generated and for some tens of msec afterwards. However, as pointed out by Dr. T. Tomita (personal communication) this sequence of changes in MP could also arise if the area of membrane undergoing transmitter action, for say 10 msec, was much greater than that involved in generating the AP.

Ogston (1955) calculated that the time course of action of ACh in sympathetic ganglia could be due simply to the diffusion of ACh away from its site of liberation and action, and that there was no need to propose a mechanism for the inactivation of transmitter (ACh esterase) to account for the time course of the post-synaptic potential. For Ogston's models, the time taken for the transmitter concentration to decay to $1/e$ ranged from 1 to 6 msec. The diffusion coefficient for NA may be a little lower than that estimated for acetylcholine but probably not much (Ogston — personal communication). If the NA is liberated into the narrow gap corresponding to a 'close contact' between axon and muscle membrane one would expect the most intense transmitter action to be complete within 10 msec. This is time enough for diffusion of NA over a considerable area of smooth muscle. However it must be emphasized that, at this time, we have no way of deciding whether the transmitter action of a single packet of

NA in the vas deferens lasts for 1, 10, or more than 100 msec. The need to measure the duration of the total inward current which gives rise to the SJPs is
obvious.

Although these experiments have provided evidence for the spontaneous
release of NA in discrete packets, they have not helped to solve the problem of
whether such packets of NA are of constant size. Even if it were possible to

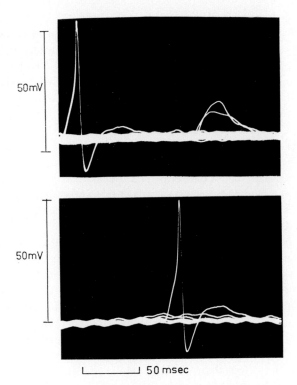

FIG. 5   Superimposed records taken with an intracellular electrode from a preparation
of mouse vas deferens with a relatively low RMP (about 50 mV). Note marked under-
shoot of spontaneous APs.

measure the inward current generated by a single packet of transmitter, variations in current intensity would not necessarily indicate a variation in the
amount of NA released from a terminal axon. The distance separating the
membrane of an axon varicosity from the muscle cell membrane varies from
less than 200 Å up to 1·0 µ or more. (see Chapter by Burnstock). If a fixed number of molecules of NA were deposited instantaneously into the extracellular
space near the axon membrane the maximum concentration occurring in the
vicinity of the smooth muscle membrane will be variable, depending on the
dimensions of the gap. If the size of the packages of NA is comparable with

those estimated for ACh (e.g. a few thousand molecules, Katz & Miledi, 1965a), and if such a package was dispersed into a narrow 'junctional cleft' of 300 Å, concentrations of up to 0·1 mM would occur in the gap. Although the shape of the dose-response curve for NA versus membrane depolarization is unknown, one might expect a considerable difference in the effectiveness of packets of equal size released into gaps ranging from 200 Å up to 1000 Å (Bennett & Merrillees, 1966).

The synthesis, storage and release of NA at sympathetic nerve terminals have been the subject of extensive biochemical and ultrastructural studies (Potter, 1967) which suggests that NA is stored within the small granular vesicles of the axon varicosities. The NA content of an average varicosity has been calculated from fluorescence studies to be about 5×10^{-3} pg (Dahlström, Häggendal & Hökfelt, 1966). If there are about 1500 vesicles per varicosity, the NA content of a vesicle is about 3×10^{-6} pg or 10,000 molecules; i.e. of the same order of magnitude as that discussed by Katz & Miledi (1965a) for the number of molecules of ACh per package at the skeletal neuromuscular junction. The large amplitude of some of the SJPs recorded from vas deferens could be accounted for if the effectiveness of such a packet of NA, released within a few hundred Å from the smooth muscle membrane was similar to that of a 'quantum' of ACh — since the input resistance of this smooth muscle is so much higher than that of skeletal muscle (Katz & Thesleff, 1957).

As previously pointed out by Burnstock & Holman (1966a) it is not possible at this stage to give an accurate estimate of the distance separating an axon varicosity from the smooth muscle membrane which would enable a single package of NA of these dimensions to cause a detectable change in muscle MP. Unfortunately SJPs have only been recorded from a few smooth muscles and of these, only the vas deferens has been subjected to detailed electronmicroscopic studies. It is tempting to suggest that SJPs will only occur in those smooth muscles where packages of NA are released at sites relatively close to the smooth muscle membrane (e.g. less than 1000 Å). However such a conclusion must await further studies on other smooth muscles.

Factors affecting SJPs

Nerve stimulation

The mean frequency of discharge of SJPs in the vas deferens is variable and sometimes quite low. One of the most convenient means of increasing their frequency is by stimulation of the hypogastric nerve (Burnstock & Holman, 1966a). The time during which discharge frequency is increased, following a single stimulus to the nerve, has yet to be worked out but preliminary experiments suggest that this effect lasts for many seconds. A relatively long-lasting increase in the frequency of SJPs, compared with the duration of the evoked JP in response to pre-junctional nerve stimulation has also been reported for the chick ciliary ganglion (Martin & Pilar, 1964b). It is likely that the effect of the pre-junctional nerve impulse on the release of transmitter greatly outlasts the time during which an axon terminal is in a depolarized state (Katz & Miledi, 1965b, c).

α-Adrenergic blocking drugs

Pharmacological studies on the mechanical responses of the vas deferens to nerve stimulation indicate that this preparation is remarkably resistant to the blocking action of drugs such as phentolamine, phenoxybenzamine and other α-blockers. It is not surprising that the SJPs themselves can still be observed in preparations exposed to high concentrations of these drugs (Burnstock & Holman, 1964).

This resistance to blockade has led, from time to time, to the suggestion that NA may not be the transmitter involved in transmission of excitation in the vas deferens. However, all other methods for identifying the transmitter point to NA. For example, enzymes necessary for its synthesis and degradation have been demonstrated in this preparation. Monoamine oxidase (MAO) is present, but little or no catechol-O-methyl transferase (Austin, Livett & Chubb, 1967). The concentration of NA, around 10 µg/g wet wt of tissue (Sjöstrand, 1965) is perhaps higher than that of any other tissue innervated by post-ganglionic sympathetic nerves.

Although α-blockers are ineffective, transmission is readily blocked by guanethidine and other pre-junctional sympathetic blockers, in concentrations which are likely to be specific for noradrenergic nerves. The discharge of SJPs continues in the presence of these drugs.

Attention has already been drawn to the high density of close contacts between axon and muscle membranes in the vas deferens and the high local concentration of NA which would be likely to exist, for a short period of time, within these junctional gaps. It may be that the α-blocking drugs cannot be given in sufficiently high doses to counteract such high local concentrations of NA.

Drugs which block the uptake of NA by terminal axons

The conservation of NA, by its uptake into noradrenergic terminal axons is now generally accepted (Iversen, 1967a). It seemed possible that this mechanism might be important in determining the duration of transmitter action at noradrenergic junctions; i.e. that the uptake mechanism might be responsible for the inactivation of NA and thus play a role in transmission at noradrenergic junctions comparable with that of ACh esterase at some cholinergic junctions.

Cocaine, in concentrations ranging from 10^{-6} g/ml to 5×10^{-5} g/ml had no effect on the time course of the SJPs in the mouse vas deferens although it caused marked potentiation and prolongation of the effects of added (exogenous) NA (Holman, 1967). In these experiments there was some indication that the amplitude of the SJPs increased in the presence of cocaine, but, since amplitude-distribution curves were skewed, it was difficult to obtain quantitative evidence on this point. Furthermore, cocaine caused an increase in resting membrane potential, and this effect would be expected to lead to an increase in SJP amplitude.

In the cat spleen, cocaine does not increase the overflow of NA into the venous effluent, during repetitive nerve stimulation (Blakeley, Brown & Ferry, 1963) in spite of the fact that it causes a marked increase in the recovery of infused NA (Gillespie & Kirpekar, 1965). Phenoxybenzamine (PBA), on the other hand, causes a marked increase in the overflow of NA during nerve stimulation (Brown, 1965). It therefore seemed possible that PBA might be more potent than cocaine in blocking the uptake mechanism in the vas deferens. Contractile responses to nerve stimulation are, in fact, potentiated by low doses of PBA in this preparation. Preliminary experiments on the mouse vas deferens suggest that PBA may cause a small increase in the amplitude or frequency of the SJPs, though a detailed statistical analysis will be needed to check this point. There was no evidence for a marked increase in the duration of the SJPs.

These negative results may be taken as evidence that the time course of the SJP depends on factors other than the uptake mechanism for NA. As already suggested, transmitter action during the SJP may be determined by the rapid decay of NA concentration as a result of passive diffusion. Under normal conditions the uptake mechanism could still be important in removing this much diluted NA from the extracellular space. If the uptake mechanism was blocked, this might have little effect on the time course of diffusion of NA. However, under these conditions spontaneously-released NA would be expected to accumulate and eventually reach a sufficiently high concentration to depolarize the smooth muscle (provided that the drug used to block uptake did not interfere with the action of NA on the smooth muscle membrane). Studies aimed at testing this hypothesis are in progress in this laboratory.

It seems likely that further work on the properties of the SJPs may provide important clues to the nature of the transmission process. The effect of a single packet of transmitter released at a site close to the effector cell membrane would seem to provide a simpler system for analysis than the sequential release of an unknown number of packets from the long series of varicosities which constitute the terminal axons of autonomic nerves.

EXCITATORY JUNCTION POTENTIALS EVOKED BY SINGLE STIMULI

The earliest published tracings of evoked excitatory JPs (EJPs) uncomplicated by the presence of action potentials are probably the N waves of Eccles & Magladery (1937a, Fig. 15) recorded from the nictitating membrane of the cat. The latency, time course, gradability and facilitation of the N wave is typical of the EJPs that have been recorded more recently with intracellular electrodes from a variety of tissues. It is interesting that the first clear demonstration of graded, subthreshold junction-potentials in smooth muscle preceded their observation with micro-electrodes by more than 20 years.

The characteristics of the EJPs recorded from a number of smooth muscles in response to a single stimulus are summarized in Table 1. The implications of these findings will now be considered.

Latency

Latencies of the EJPs of the oesophagus, taenia coli and colon are very long, ranging from 80 to 220 msec. In the colon and possibly the taenia-coli this may be partly due to the stimulation of pre-ganglionic nerves and to a delay in ganglionic transmission. In the oesophagus however, ganglion blockers do not alter latency (Ohashi & Ohga, 1967). It is interesting that transmission at all these junctions involves the interaction of ACh with muscarinic receptors. Similar very long latencies have also been recorded for muscarinic junctions in salivary glands (Lundberg, 1958; Creed & Wilson, 1969) and in cardiac muscle (del Castillo & Katz, 1955c).

It is difficult to account for the latency of 400 msec observed by Gillespie (1962b) in his first series of experiments on the rabbit colon. Perhaps this may reflect several ganglionic relays during the transmission of excitation through Auerbach's plexus.

Stimulation of post-ganglionic nerve trunks to the vas deferens, nictitating membrane, bladder, retractor penis and small arteries evoked JPs after somewhat shorter latencies ranging from 10 to 50 msec, depending on the distance between the stimulating and recording electrodes. Minimum latencies of 6 to 10 msec have been recorded from the vas deferens of the guinea-pig (Kuriyama, 1963b) and mouse (this laboratory), in response to stimulation of intramural nerve fibres.

During the latent period following nerve stimulation several events must take place:

1. Excitation and conduction of the nerve impulse along the terminal axons.
2. Release of transmitter from successive varicosities.
3. Diffusion of transmitter across the junctional gap.
4. Interaction of transmitter with receptors on the smooth muscle cell.
5. Changes in the membrane properties of the smooth muscle membrane.

Kuriyama's (1963b) experiments with field stimulation suggest that the conduction velocity of terminal axons may be very slow — of the order of 0·1 m/sec. Conduction time along a 5 mm terminal axon could therefore take as long as 50 msec. This could account for the longer latencies of responses to stimulation of extramural post-ganglionic nerve fibres compared with responses to transmural or field stimulation. At the frog neuromuscular junction (room temperature) there is a delay of less than 1 msec following invasion of the axon terminal by a nerve impulse before release of ACh takes place (Katz & Miledi,

TABLE 1. Characteristics of JPs recorded from vertebrate smooth muscles

Preparation	Nerve stimulated	Recording method**	Latency (msec)	Estimated junctional delay (msec)	Rise time (msec)	½ Decay time (msec)	Total duration (msec)	Reference
A. Excitatory Junctions								
Cat nictitating membrane	Post-g	External electrodes	20	10–15	50–80	—	≈ 500	Eccles & Magladery, 1937a
Rabbit detrusor muscle	Post-g	CME	35–100	—	< 100	—	≈ 1000	Ursillo, 1961
Guinea-pig small mesenteric arteries	Pre-g	CME	145–175	—	⩽ 100	—	< 1000	Speden, 1964
Rabbit arteries (ear, mesenteric)	Post-g (field)	CME	12–40	< 12	min 70 variable ≈ 100	—	500–1000	Speden, 1967
Chick oesophagus	Post-g (field)	CME	90–160	—	150–250	—	700–950	Ohashi & Ohga, 1967
Guinea-pig taenia coli	?Post-g (field)	CME	100–200	—	200–400	—	500–800	Bennett, 1966b
Rabbit colon	?Pre-g Post-g	CME + pressure electrodes	400	—	—	—	600	Gillespie, 1962b
Rabbit colon	Pre-g (field)	CME	220	—	≈ 250	—	—	Gillespie, 1964
Guinea-pig vas deferens	Pre-g	CME	20 (minimum)	10	[½ rise time 15–20]	≈ 150	< 1000	Burnstock & Holman, 1961

Guinea-pig vas deferens	Post-g (field)	CME	6 (minimum)	—	≈ 40	—	≈ 500	Kuriyama, 1963b
Guinea-pig vas deferens	Pre-g	CME	—	—	45–100	135–300	—	Tomita, 1967a
Guinea-pig vas deferens	Post-g (field)	CME	—	—	45–100	≈ 200	—	
Mouse vas deferens	Post-g (field)	CME	10–20 (minimum < 10)	—	10–20	—	100–250	This laboratory unpublished work
Rat vas deferens	Post-g (field)	CME	10–20	—	10–20	—	≈ 150	
Dog retractor penis	Post-g	CME	—	—	—	—	300–350*	Orlov, 1962
B. Inhibitory Junctions								
Guinea-pig taenia coli	Post-g ? (field)	CME	80	—	200–500	—	< 1·0	Bennett, Burnstock & Holman, 1966b
Guinea-pig taenia coli	Post-g ? (field)	CME	140	—	200–300	260–590	—	Bülbring & Tomita, 1967
Guinea-pig jejunum	Post-g ? (field)	CME	50–80	—	120–280	—	< 1·0	Kuriyama, Osa & Toida, 1967c

*Orlov's (1962) published records suggest that duration may be greater than this.
**CME: capillary microelectrode

1965c). It is possible that the release of ACh and NA at autonomic nerve terminals may be a more time consuming process though it is difficult to imagine why this should be so.

In densely-innervated smooth muscles, it is unlikely that a significant fraction of the junctional delay could be due to diffusion of transmitter. If the *minimum* distance separating axon and muscle membranes is of the order of 1 μ the time taken for the concentration of transmitter to reach its maximum level at the muscle cell membrane would only amount to a fraction of a msec. However, in less densely-innervated smooth muscles, especially those whose innervation pattern in comparable with that of Fig. 2A, it may be necessary for the transmitter to diffuse over considerably longer distances in order to reach a sufficiently large number of cells to ensure effective transmission (see p. 275). Diffusion over distances of the order of 50 μ might be expected to take as long as 100 msec.

It is possible that the interaction of transmitter with the smooth muscle membrane may take several msec to occur. It is usually assumed that the receptors for neurohormones are on the surface of cells. This proposition was elegantly confirmed for ACh at the skeletal neuromuscular junction by del Castillo & Katz (1955a), but direct evidence for this has yet to be found for smooth muscle.

The latency of secretory potentials in the cat submaxillary gland has been analysed by Creed & Wilson (1969) who concluded that at least 100 msec must elapse *after* ACh reached the cell membrane before the initiation of a secretory potential. This may also be true for muscarinic junctions in smooth muscle. At nonmuscarinic junctions, however, it would seem that the latency of the JP is mainly due to the slow conduction velocity of terminal axons and, in some instances, to the diffusion of transmitter.

Time course

Rise times (from resting MP to peak amplitude) of the EJPs recorded so far, fall into 3 groups:

   i Cholinergic JPs of the gastro-intestinal tract (chick oesophagus, 150 to 250 msec; taenia coli, 200 to 400 msec; rabbit colon, about 250 msec).

   ii Adrenergic JPs of the guinea-pig vas deferens, blood vessels and the nictitating membrane (40 msec to about 100 msec).

   iii EJPs of rat and mouse vas deferens (10 to 20 msec).

In all cases the decay phase of the JP is much longer in duration and more variable than the rising phase (see Table 1).

If the time course of the EJPs in response to nerve stimulation is compared with the time course of the SJPs in the same tissue it is clear that, although the minimum rise times of both potentials may be comparable, the decay time of the evoked JP is much longer than that of spontaneous JPs (see Fig. 3). Tomita (1967a) has suggested that this may be partly due to the long time course of the electrotonic potentials set up by passing current through a large population of cells compared with the electrotonic potentials observed during intracellular polarization. Since the branching terminal axons of a single post-ganglionic fibre probably release transmitter near many hundreds of smooth muscle cells, *passive decay* of the depolarization induced by evoked transmitter action might be expected to mimic the decay of electrotonus induced by external electrodes rather than that set up by current flow from an intracellular electrode. It is interesting that the half decay time of the evoked JPs in mouse and rat vas deferens (40 to 50 msec; Holman, unpublished work) is of the same order of magnitude as that of externally-evoked electrotonic potentials in the guinea-pig vas deferens (Tomita, 1967a). However, it must be remembered that the degree of coupling between the smooth muscle cells of the mouse vas deferens is unknown and that it is difficult to demonstrate the longitudinal spread of electrotonus in this preparation. It is probably unwise to extrapolate data from the guinea-pig vas deferens to other preparations and certainly it cannot be assumed at this stage, that the decay of the evoked JP in the mouse vas deferens reflects the passive electrical properties of this tissue.

The evoked JPs of the guinea-pig vas deferens decay much more slowly than externally-evoked electrotonic potentials (Tomita, 1967a). This suggests that transmitter action may continue throughout the evoked JP. Further evidence for this view can be found from the relation between APs and JPs in this tissue (see below).

The duration of transmitter action following the release of a single packet of transmitter cannot be any longer than the total duration of a spontaneous JP. What mechanisms can explain the prolongation of transmitter-action during an evoked JP? The spread of excitation along the terminal axons of the ground plexus may take several tens of msec and this process could be significant in determining the time course of the initial stages of the JP. It may also contribute to the differences in time course of the JPs recorded in response to 'field' stimulation compared with those in response to stimulation of the hypogastric nerve (Holman, 1964; Kuriyama, 1963b). There is another prejunctional phenomenon which might also contribute to the prolongation of evoked transmitter action. We have already noted that a single nerve impulse can cause long term changes in the frequency of the SJPs. A prolonged increase in the probablility for release of transmitter following excitation may be a characteristic feature of all autonomic terminal axons.

Finally it may be argued that the time taken for the concentration of transmitter to fall to a level at which it can no longer cause a detectable change in membrane properties may be longer, if transmitter is released, more or less synchronously from many hundreds of varicosities, as opposed to the release of single packets which cause SJPs (Tomita, 1967a).

Can these arguments be extended to other smooth muscles in which EJPs have been recorded in response to single stimuli? The EJPs of muscular arteries have a time course similar to that of the guinea-pig vas deferens (Speden, 1967). Here the pattern of innervation differs greatly from the vas deferens with the varicose terminals confined to a region at the medioadventitial border, where there are few, if any, contacts between axon and muscle membranes which are closer than 1000 Å (a situation analogous to that of Fig. 2A). Nevertheless, the concentration of NA which builds up following a single stimulus is sufficient to cause several mV of membrane depolarization. This probably depends on the release of a number of packets of NA from different varicosities so that there is an overall increase in the concentration of NA in the extracellular space bathing the smooth muscle. No SJPs comparable with those of the vas deferens have been observed in these preparations — probably because the concentration of NA following the release of a single packet is too low to cause a detectable change in MP.

Since terminal axons are confined to the adventitia and outer layer of media the question arises as to whether transmitter action involves the diffusion of NA throughout the media or whether the EJPs initiated in the cells of the outer media lead to APs which then propagate throughout the media.

There are great technical difficulties in obtaining intracellular records from these preparations owing to the tough external elastic lamina. It is still not clear whether the individual cells are independent units (multi-unit smooth muscle) or coupled together electrically (single unit smooth muscle).

If arterial smooth muscle *is* of the multi-unit type it is obviously necessary to postulate that the diffusion of NA throughout the media is necessary for complete activation of the smooth muscle. Such diffusion has been demonstrated by histochemical methods during repetitive nerve stimulation (Gerova, Gero & Dolezel, 1967). It is questionable whether diffusion through the narrow channels between the smooth muscle cells of the media could account for the time course of the response of this tissue to single stimuli. These arguments will be considered in detail in Speden's chapter.

The relatively slow time course of the 'cholinergic' JPs of the gastro-intestinal tract deserves comment. This may be partly due to the need for extensive diffusion of ACh in order that the activity of a sufficiently large number of cells can be increased to ensure effective transmission. As already noted, in a tightly coupled three-dimensional syncytium, changes in the activity of a single or small group of cells — such as those evoked by intracellular stimulation — will have no effect on the overall activity of the preparation (see below). Alternatively, the interaction of ACh with muscarinic receptors may be the rate determining process underlying these slow excitatory JPs.

Effect of current flow on junction potentials

Transmitter action at the vast majority of synapses and neuromuscular junctions studied so far involves an increase in membrane conductance for one or more ions. At the skeletal neuromuscular junctions of the 'twitch' fibres of amphibia the excitatory transmitter (ACh) causes an increase in conductance for both Na and K ions (g_K and g_{Na}) (Takeuchi & Takeuchi, 1960). Under the continuing action of ACh the MP of the endplate region tends to be 'clamped' at about -15 mV. It follows that the amplitude of the endplate potentials caused by a given amount of ACh will vary, depending on the MP from which they arise — the more negative the MP, the larger the EPP.

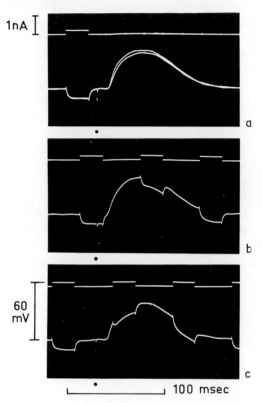

FIG. 6 Records from an experiment on mouse vas deferens in which a bridge circuit was used to pass current across the cell membrane while the MP was recorded with the same electrode (see text). The upper trace in each record monitors current (nA) and the lower trace shows changes in MP induced before (a) and during (b and c) a JP evoked by stimulation of intramural nerves (indicated by •).

Figure 6 shows the results of an experiment in which short pulses of current were injected by a microelectrode at various times before, during and after an evoked JP in the mouse vas deferens (Dorward & Holman, 1967; Holman, 1967). The time course of the electrotonic potentials recorded in this impalement were rather faster than usual. It can be seen that the amplitude of the electrotonic potential was significantly reduced during the early part of the JP — indicating that membrane conductance was increased at this time. Only some of the cells of the mouse vas deferens behaved in this way. In most instances it was not possible to detect such a large change in amplitude of the electrotonic potential. Furthermore, it was often uncertain whether or not the change in conductance was due to transmitter action or to the development of a 'local response' which was superimposed on the JP.

In a further series of experiments on the mouse vas deferens the MP was displaced by injection of conditioning current for longer times. Special attention was paid to the effects of hyperpolarizing currents, since local responses and APs developed as a result of the interaction of JPs with depolarizing pulses. It was apparent that the amplitude and time course of the majority of the JPs were independent of MP when this was changed from -100 mV to -50 mV. It must be emphasized that sub-maximal nerve stimulation was used in these experiments, in order to prevent the initiation of APs and contraction. It is likely that only a small fraction of total population of terminal axons were excited.

During similar experiments on the guinea-pig vas deferens, Bennett (1967b) was unable to increase the amplitude of the EJP as a result of hyperpolarization by injected current though he reported a decrease in amplitude of the EJP during depolarization of *some* cells.

The simplest explanation for these predominantly negative results would be to assume that the cells under study were coupled into three-dimensional syncytium. Under these conditions it would be difficult to detect changes in membrane conductance by observing changes in MP induced by point polarization with an intracellular electrode. It can be shown theoretically (George 1961) that input resistance when measured in this way may be independent of membrane resistance (R_m). This argument has been developed by Tomita (1967a) and Bennett (1967b) to account for their findings for the guinea-pig vas deferens.

Assuming that sites of transmitter action are scattered throughout a syncytial structure, an individual cell could be depolarized by:

1. outward current arising from transmitter action elsewhere, and,
2. inward ionic current due to transmitter action on the cell itself.

Bennett suggested that the inward current generated in a cell undergoing transmitter action might be relatively small compared with outward passive current. During changes in MP induced by intracellular polarization, passive outward current would give rise to an EJP whose amplitude was independent of MP. Any increase or decrease in *inward current* due to an altered driving force on Na and K ions might make a negligible contribution to the amplitude of the EJP.

It would be interesting to check this explanation with a suitable electrical model. Theoretical analysis of this problem may be very difficult, especially since the area of membrane undergoing transmitter action is unknown. It may, however, be possible to detect the inward current at sites of transmitter action with extracellular electrodes. Small spontaneous potentials (about 1 mV), of brief duration (10 to 20 msec) and negative polarity can often be recorded by micro-electrodes in the mouse vas deferens, immediately before the electrode enters a cell. Unfortunately it is difficult to distinguish these random events from the general level of background 'noise' which often increases when the electrode enters the tissue.

Although it seems possible that the sheet model for a tightly-coupled syncytium can account for the properties of the EJPs in the guinea-pig vas deferens, one feels a little hesitant in applying these arguments to the mouse vas deferens. At this time however, we have tentatively concluded that the cells of the mouse preparation are coupled electrically to some extent and that the sheet model can be used to account for the lack of effect of current injection on the EJP. However, coupling may be inadequate to endow this smooth muscle with the passive electrical properties needed for single unit function.

Summing up, one could say that it is difficult to detect changes in membrane conductance during the EJP as a result of the syncytial nature of these smooth muscles. It is rather more difficult to explain why it has not been possible to find any evidence for inward ionic current. Perhaps an alternative explanation to that of Bennett (1967b) may be considered — at least for the mouse vas deferens. Regions of transmitter action evoked by submaximal stimulation may be small and only involve a fraction of the total surface area of the smooth muscle. Such regions may not have been impaled with the micro-electrode during these experiments.

Grading

The EJPs of Table 1 vary in amplitude with changes in the strength or duration of the stimulus. In the vas deferens, if extramural nerve fibres are stimulated (e.g. the hypogastric nerve or its branches) it is difficult to detect any significant difference in the amplitude of the EJPs recorded from cells scattered over a distance of up to 1 cm (Burnstock & Holman, 1961; Tomita, 1967a). This is true, whether the stimulus is just supra-threshold, or supra-maximal. If intramural nerve fibres are excited by field stimulation, EJPs can still be graded. However, their amplitude falls as the distance between the stimulating and recording electrodes is increased (Kuriyama, 1963b; Tomita, 1967a). In the guinea-pig vas deferens EJPs decay in a roughly linear fashion, falling to half-maximum at a distance of 3 to 4 mm. This probably reflects the spacial distribution of the pre-terminal axons and their terminal extensions which are excited by the stimulating electrodes. The spread of terminal axons over distances of this order of magnitude is in accordance with Malmfors' observations on the iris (Malmfors, 1965a).

In the dog retractor penis the amplitude of the EJPs can be graded with strength of stimulation of the extramural post-ganglionic nerve trunk (Orlov, 1962). For any given stimulus, however, the amplitude of the EJP depends on the site of recording — the largest JPs being recorded from the region of the muscle which receives the densest innervation. The frequency of SJPs is also highest in the most densely-innervated region.

The relation between strength of stimulation and response was one of the key experimental procedures which led to debates about transmission in smooth muscle during the 1930s (Eccles 1936; Rosenblueth, 1950; Hillarp, 1960). In the absence of an adequate description of the innervation of smooth muscle, Rosenblueth and his colleagues doubted that it was possible to account for the grading of responses in terms of the recruitment of discrete sites of transmitter release in close proximity with the smooth muscle membrane, analogous with the skeletal neuromuscular junction. Instead, they put forward the view that transmitter was released from autonomic nerve fibres into the extracellular space of the effector organ and that the response of the effector organ was related to the overall concentration of transmitter (Fig. 2A). Eccles (1936) argued against this interpretation. Studies by Eccles & Magladery (1937a), on the cat nictitating membrane, led to the suggestion that transmission from autonomic nerves to smooth muscle was basically similar to that of autonomic ganglia, central synapses and the skeletal neuromuscular junction. Grading of response was considered to be due to recruitment of additional sites of transmission. However, Eccles & Magladery (1937b) also found evidence that NA released as a result of motor nerve stimulation could cause long term rhythmic changes in the activity of the nictitating membrane, presumably as a result of diffusion to more remote sites of action.

Much confusion about the innervation of smooth muscle was swept aside during the 1950s by the work of Hillarp and his collaborators. Hillarp's view of the 'autonomic ground plexus', formulated largely as the result of careful studies with the light microscope, has since been confirmed with the electron-microscope. Hillarp suggested that transmission from nerve to autonomic effector organs occurred where apposing membranes were separated by a 'short diffusion distance' rather than by a generalized increase in the concentration of transmitter in the extracellular space (diffusion of transmitter over 'long' distances as proposed by Rosenblueth). He felt there were a sufficiently large number of intimate contacts between axons of the ground plexus and the effector tissues whose physiology he investigated, to explain the grading of response as a function of the number of post-ganglionic axons stimulated. Recent evidence for the extensive electrotonic spread of the EJPs in some smooth muscles makes this view seem very reasonable.

It has often been suggested that the density of innervation of a smooth muscle is inversely related to its ability to function as an electrical syncytium (see Burnstock, Holman & Prosser, 1963). Since the *patterns* of innervation of different smooth muscles are turning out to be so variable it is beginning to look as though this generalization might not be a very useful one. Thus there is no doubt that the guinea-pig vas deferens has single unit properties *and* a dense innervation. Grading of response in this muscle appears to be due to recruitment of additional sites of release somewhat akin to the discrete neuromuscular junctions envisaged by Eccles and Hillarp. The density of innervation of the vas deferens and other smooth muscles of the pelvic viscera, the nictitating membrane and possibly the dilator muscles of the iris may vary in accordance with the degree with which their responses to nerve stimulation can be graded.

Later on we shall discuss examples of single unit muscles where transmission would seem to depend on a more general diffusion of transmitter, and grading of response on an overall increase in the concentration of transmitter — as proposed by Rosenblueth.

Initiation of APs by JPs

It is generally considered that the AP is the electrical event which triggers contraction in smooth muscle. In the guinea-pig vas deferens when the EJP reaches sufficient amplitude (around 30 mV), an AP is initiated and contraction occurs. Such APs are initiated at a threshold similar to that of the APs in response to electrical stimulation with external electrodes (Tomita, 1967a). They are usually all-or-none. During their repolarization phase the MP moves towards the resting level but subsequently undergoes a further depolarization whose decay may mimic the time course of the falling phase of the EJP. This suggests that the duration of transmitter action in this smooth muscle may continue, to a progressively diminishing extent, throughout the falling phase of the EJP. Even if the EJP giving rise to the AP is due to passive depolarization rather than local transmitter action it may still be argued that the 'post-spike' depolarization is re-established by electrotonic spread of the depolarization which is maintained by transmitter action elsewhere.

In the mouse vas deferens graded increases in stimulus strength do not always lead to the initiation of APs in such an all-or-none manner. The records of Fig. 7 show a smooth transition from the small JP of Fig. 7a to the large depolarization of about 60 mV of Fig. 7e. By contrast, Fig. 8 shows the initiation of an AP by a JP of 30 mV amplitude, this record resembling those from the guinea-pig vas deferens (except for the difference in time course). It seems possible that records such as those of Fig. 7 come from regions of smooth muscle cells directly influenced by NA. It may be that the action of the transmitter is sufficiently intense to prevent the development of a regenerative response in such a cell and, further, to counteract the effect of an increase in g_K which normally brings about the repolarization of the AP. In other words, transmitter action has 'clamped' the MP at about -10 mV. We have already mentioned that it was possible to detect a decrease in input resistance during the initial stages of the EJPs in some cells of this preparation. These EJPs were always of the type illustrated in Fig. 7. No change in input resistance has been detected so far during an EJP like that of Fig. 8, which *did* fire an AP.

10+

Changes in MP such as those of Fig. 8a and b were usually recorded at distances of 2 mm or more distant from the field stimulating electrodes whereas records like that of Fig. 7 came from cells close to the electrodes. There were no obvious differences in the excitability of cells giving these differing responses to nerve stimulation, when their excitability was tested with intracellular polarization.

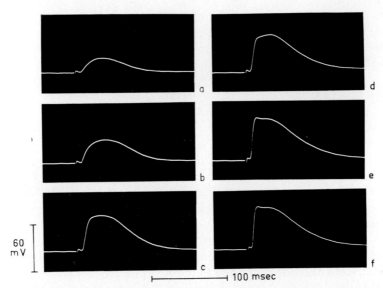

FIG. 7   JPs recorded from mouse vas deferens in response to progressively increasing strengths of stimulation (a to f) applied to intramural nerves. (Holman, 1967)

The relation between APs and EJPs in the other smooth muscles studied so far appears to resemble that of the guinea-pig vas deferens. In the chick oesophagus, under normal conditions, *in vitro*, JPs always give rise to APs (Ohashi & Ohga, 1967). The level of excitability of this smooth muscle may be such that APs can be triggered by the smallest possible EJPs in response to threshold stimulation of intramural cholinergic nerves. Similar results were obtained by Gillespie (1962b) from rabbit colon during stimulation of pelvic nerve. Most, if not all the EJPs recorded in his experiments were associated with all-or-none APs, or with smaller depolarizations of similar time course. These latter potentials probably represent passive depolarization of the impaled region due to electrotonic spread of APs generated in neighbouring regions. (In a further series of experiments, nerve fibres within the wall of the colon were stimulated. Under these conditions Gillespie (1964) was able to observe graded EJPs, uncomplicated by APs. Unfortunately these records have yet to be published.)

Gillespie (1962b) noticed that the ability of successive EJPs to initiate APs fluctuated in accordance with the changes in excitability associated with the slow waves of the colon. At the peak of the depolarization phase large, all-or-none APs were initiated whereas during the troughs of the slow waves only small 'electrotonic' spikes could be seen.

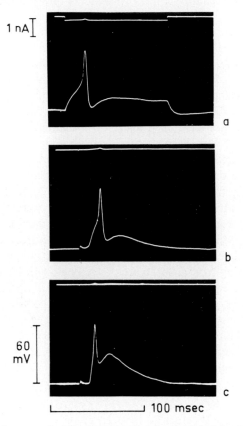

FIG. 8 Intracellular records from mouse vas deferens. In a the AP was initiated by intracellular stimulation. In b and c the AP was initiated by nerve stimulation. Strength of stimulation was increased between b and c. (The upper trace in these records monitored current.) (Holman, 1967)

In the colon it is possible to trigger APs from successive EJPs at frequencies of up to 2/sec. At higher frequencies (up to 12/sec) the 1 : 1 relationship between stimuli and APs is lost. The amplitude of successive APs falls, the electrical activity changes to a damped oscillation and eventually to a steady depolarization which is maintained throughout the period of stimulation (Gillespie 1962b).

In contrast with these findings for the rabbit colon, Bennett (1966b) was unable to initiate APs at a frequency greater than 1/sec in the taenia coli during repetitive stimulation of intramural cholinergic nerves. This was probably due to the simultaneous excitation of intrinsic inhibitory nerves. Most regions of the taenia coli are under the influence of inhibitory innervation whereas only some regions may have an excitatory supply (Bennett & Rogers, 1967). Until some means is found for blocking the intrinsic inhibitory system of the gastrointestinal tract, analysis of the excitatory effects of 'field' or 'transmural' stimulation will be a difficult task.

Although EJPs give rise to APs in small muscular arteries (guinea-pig), *in vivo* (Speden, 1964), APs have yet to be demonstrated for arterial smooth muscle *in vitro* (Speden, 1967). The possibility exists that the smooth muscle cells closest to the dense network of terminal axons at the medio-adventitial border are under such intense transmitter action that their AP mechanism is short-circuited. However, such an interpretation awaits further studies on this preparation.

Relatively few studies have been made on the initiation of APs during prolonged stimulation of densely-innervated smooth muscles like the vas deferens. In the guinea-pig, repetitive stimulation at frequencies greater than 5/sec gives rise to a maintained depolarization on which is superimposed a repetitive discharge of APs. Initial frequencies of discharge of up to 35/sec have been observed during stimulation at 20 to 50/sec but impalements were not maintained for more than 1 sec (Burnstock & Holman, unpublished work).

Takeda & Nakanishi (1965) used extracellular microelectrodes to record from the guinea-pig vas deferens during and after long trains of stimuli (up to 30 sec) at frequencies ranging from 1 to 20/sec. They demonstrated a maintained discharge of APs for frequencies of stimulation greater than 2 to 3/sec. The frequency of the discharge increased with increasing rates of stimulation. At a repetition rate of 20/sec, APs continued to occur for some seconds after the cessation of stimulation. Takeda & Nakanishi (1965) used an AC-amplifier during their experiments and it is important that their results be followed up by DC recording with intracellular or sucrose-gap techniques.

EFFECTS OF REPEATED STIMULATION

Facilitation

Although APs and mechanical responses to single stimuli can be recorded from some smooth muscles (e.g., vas deferens of mouse and rat, small arteries, taenia coli, colon, oesophagus, etc.) this is not true for all smooth muscles. In the guinea-pig vas deferens, summation and facilitation are usually required for a mechanical response, *in vitro*. Thus the effects of low frequency repetitive stimulation on neuromuscular transmission can be studied in the guinea-pig vas deferens without the complication of movement artefacts. Facilitation of successive EJPs is very marked in this preparation (see Fig. 9). A six-fold increase in EJP amplitude occurs at frequencies as low as 2/sec. Burnstock, Holman & Kuriyama (1964) put forward indirect evidence that this was due to an increase in the amount of transmitter liberated by successive nerve impulses. Throughout their experiments, sub-maximal nerve stimuli were used to avoid contraction. The possibility that successive stimuli recruited an increasing number of nerve fibres could not be ruled out though this would seem unlikely.

A similar form of facilitation in response to relatively low frequency stimulation has been observed at the frog skeletal neuromuscular junction in the presence of high Mg–low Ca solutions (del Castillo & Katz, 1954). Under these conditions the probability for release of quanta of ACh by the nerve impulse is greatly reduced so that the number of quanta released per nerve impulse is small compared with the total number of quanta available for release.

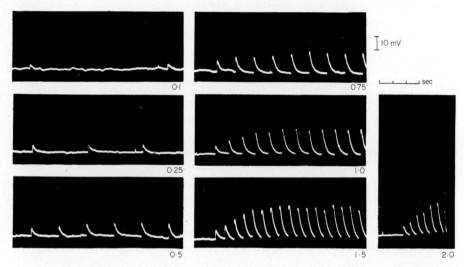

FIG. 9 Facilitation of JPs in response to varying frequencies of stimulation/sec (as indicated). Intracellular records from guinea-pig vas deferens. (Burnstock, Holman & Kuriyama, 1964)

It is tempting to suppose that this situation might exist under normal conditions in the guinea-pig vas deferens. There is no direct evidence for this idea though it is certainly in accordance with the finding that the form of successive JPs fluctuates considerably, especially in response to sub-maximal stimulation.

The maximal amount of NA liberated by a single nerve impulse from the constrictor fibres of the cat's hind limb has been measured by Folkow, Häggendal & Lisander (1967). When this is expressed in terms of the fraction of the total NA contained within a single varicosity it turns out that either each nerve impulse must release only a fraction of NA content of a *single vesicle* or alternatively, that the probability for release of the entire NA content of a single vesicle is very low (3%). When similar data become available for other noradrenergic junctions it will be interesting to see if there is a correlation between the fraction of transmitter released by a single nerve impulse and the degree of facilitation of transmitter output by successive impulses.

The mechanism which underlies facilitation is unknown. Since it can be observed at frequencies as low as 0·3/sec it is unlikely to be due to long lasting changes in the MP of the terminal axons (Burnstock *et al.*, 1964). Perhaps the nerve impuse leads to the redistribution of vesicles within an axon varicosity so that a larger number of them are strategically placed near the axon membrane for subsequent release.

Facilitation of successive JPs also occurs in small mesenteric arteries, dog retractor penis and nicitating membrane. It has also been observed in the chick oesophagus after transmission had been depressed by high Mg solution. It is premature at this stage to say whether or not it is a more conspicuous feature of noradrenergic than cholinergic junctions.

EXCITATORY EFFECTS UNACCOMPANIED BY INDIVIDUAL EJPs

In general, the excitatory effects of nerve stimulation in smooth muscle are associated with EJPs in response to single stimuli, which can be recorded from some, or all of the smooth muscle cells of the effector organ [see Table 1 and Gillespie (1962*b*); Bennett, (1966*b*)]. Transmission of excitation from sympathetic nerves to the smooth muscle of portal and mesenteric veins may be an exception to this rule. Holman, Kasby, Suthers & Wilson (1968) were unable to detect any depolarization in response to a single stimulus ('field' electrodes) during their studies on the rabbit portal vein with the sucrose gap. However, this point needs to be checked with intracellular electrodes. Likewise no change in the pattern of spontaneous electrical or mechanical activity could be recorded in response to single stimuli. The minimum frequency of stimulation for a detectable response varied from 1/sec to 5/sec (Holman *et al.*, 1968; Johansson & Ljung, 1967*a*), as is also found for the inhibitory effects of stimulation of the sympathetic nerve supply to the gastrointestinal tract which can only be observed in response to repetitive stimulation (see below).

The spontaneous activity of strips of portal and mesenteric veins, set up under minimal tension (Fig. 10A), consists of bursts of spikes superimposed on 'slow waves' of depolarization, (Nakajima & Horn, 1967; Holman *et al.*, 1968). Stimulation of sympathetic nerves causes an increase in the frequency of these 'multi-spike complexes'. Increasing strengths or frequencies of stimulation cause the continuous firing of spikes, as shown in Fig. 10B. The electrical response of the rabbit portal vein, to a 10 sec train of stimuli, reaches its peak intensity from 15 to 20 sec after the onset of stimulation. The increased frequency of firing of spikes or spike complexes is maintained for periods of up to 2 min. A prolonged excitatory response to sympathetic stimulation (5 sec train) lasting for several minutes was also reported for rat mesenteric arterioles by Steedman (1966).

It is possible to mimic all the features of the response of the portal vein to nerve stimulation by perfusing a strip of muscle with a small quantity of NA

(Fig. 10C, see also Holman, 1967). Pharmacological blocking agents are effective against both responses at similar concentrations, both *in vitro* and *in vivo* (Johansson & Ljung, 1967a). As a result of these experiments, and concurrent electronmicroscope studies on the distribution of terminal axons in the portal

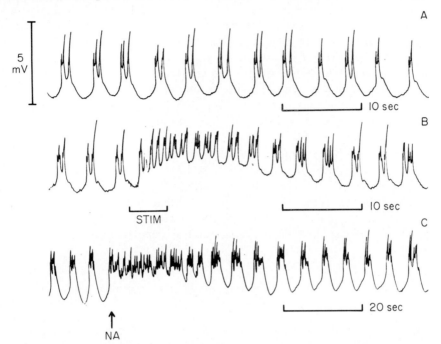

FIG. 10   Records taken with sucrose-gap from a strip of rabbit portal vein.   A. An example of normal activity.   B. Response to stimulation of intramural noradrenergic nerves (indicated by horizontal bar).   C. Response to 0·1 μg of noradrenaline.

vein, Holman *et al.* (1968) suggested that the NA released upon nerve stimulation probably reaches the smooth muscle membrane at a fairly low concentration, comparable with that of the perfusion experiments. Since terminal axons are largely confined to the extracellular space surrounding bundles of smooth muscle cells rarely approaching to within 1000 Å of the smooth muscle membrane, this interpretation seems reasonable. In this case the mechanism of transmission probably involves the diffusion of transmitter over relatively 'long diffusion distances' — in accordance with the model of Rosenblueth.

Smooth muscle cells situated at the periphery of a bundle will obviously be influenced by nerve-released NA in preference to those situated deeply within the bundle. The action of NA will be to increase the firing of APs in these cells, and they will become the dominant 'pace-maker' for the remainder of the muscle bundle. (The 'synchronizing' effects of threshold doses of NA is a common feature of sucrose-gap records). The *key* cells of Rosenblueth's model would thus correspond to those closest to a group of terminal axons. However, in contrast with Rosenblueth's model, it seems likely that excitatory effects may spread throughout the tissue by the conduction of APs rather than by diffusion (Holman *et al.*, 1968; Johansson & Ljung, 1967*b*). The time course of the onset of the excitatory response may be partly determined by the time needed for synchronization of the activity of all the smooth muscle cells within a bundle at a frequency dictated by the action of NA on the outermost cells of the bundle.

Again, it is likely that if the nerve-released NA only affected one or a few cells there would be little chance of effective transmission. The passive electrical properties of the portal vein are probably similar to those of visceral smooth muscle (Funaki, 1967; Johansson & Ljung, 1967*b*) and again it is necessary to postulate that the activity of a relatively large group of cells must be influenced by the transmitter.

The portal vein is remarkably sensitive to exogenous NA. Some preparations give detectable responses to doses as low as 10^{-10} g/ml (Holman *et al.*, 1968). Threshold doses of NA increase the frequency of the spike complexes. Higher doses cause changes in the electrical activity which are similar to those described for the action of ACh in intestinal smooth muscle, i.e., continuous spiking at increased frequencies and finally, a maintained depolarization (Nakajima & Horn, 1967).

ACTION OF EXCITATORY TRANSMITTERS ON THE SMOOTH MUSCLE MEMBRANE

There are many difficulties involved in measuring the changes in membrane properties which might be expected to occur during transmitter action, even in densely innervated smooth muscles. It would therefore seem desirable to investigate the ionic basis of externally applied (exogenous) transmitter. As a first step, it is important to be sure that the two sources of transmitter give rise to similar changes in MP. Although this may be true for some muscles (e.g. portal vein, see p. 274) it is more difficult to mimic the effects of nerve stimulation with exogenous NA in densely-innervated smooth muscles like the vas deferens (Holman, 1967). Relatively high concentrations of NA (10^{-6} to 10^{-5} g/ml) are needed to elicit reproducible contractions of the vas deferens and tachyphylaxis occurs if the preparation is exposed to these concentrations for periods of 1 min or more. Although it is dangerous to generalize one is tempted to suggest that the sensitivity of a smooth muscle to exogenous transmitter is inversely related to the density of its innervation. No pharmacologist would choose the vas deferens as a tissue for assaying NA or the circular muscle of the intestine for assaying ACh.

A partial explanation for this may follow from the means by which ACh and NA are removed from their site of release. It is likely that smooth muscles receiving a dense cholinergic innervation also have a high concentration of cholinesterase.

Inactivation of exogenous NA depends mainly on the neuronal uptake mechanism (Iversen, 1967*b*). The denser the innervation, the more effective the uptake mechanism is likely to be. The sensitivity of the nictitating membrane to exogenous NA is increased 100 fold, 2 days after section of its post-ganglionic nerve supply (Langer, Draskóczy & Trendelenburg, 1967). [A further increase in sensitivity (800 fold) occurs during the next 24 days but this is post junctional in origin.] The sensitivity of the rabbit ear artery to exogenous NA is much greater if it is applied intraluminally compared with its effect when added to the solution bathing the adventitia — in this case the NA must pass by the medioadventitial plexus of noradrenergic fibres before reaching the smooth muscle of media (de la Lande & Waterson, 1967).

Numerous other examples could be quoted in support of the view that noradrenergic terminals are extremely effective in mopping up exogenous NA, so preventing its access to the smooth-muscle receptors. Similarly, it is well known that the sensitivity of a smooth muscle to ACh can be greatly increased by the use of an anti-cholinesterase drug.

The first difficulty that must be overcome in comparing the effect of exogenous transmitter with transmitter released upon nerve stimulation is that of building up a sufficiently high concentration of exogenous transmitter throughout the smooth muscle. If some of the release sites are buried deeply within a tightly-packed smooth-muscle bundle (see Fig. 2B) this may be an impossible task unless some means is taken to block the inactivation mechanism. On the other hand, if the sites of release are fairly distant from the smooth muscle membrane and limited to the larger extracellular spaces between bundles (see Fig. 2A) it should be possible to mimic the effects of nerve released transmitter more readily — as is the case for the portal vein.

It seems reasonable to assume that the ionic effects of transmitter action in the less densely-innervated smooth muscles is similar to that caused by exogenous transmitter since the two cause similar changes in MP. In the case of the more densely-innervated muscles where it is difficult or impossible to mimic all the characteristics of transmitter action with exogenous transmitter one should be more cautious in coming to such a conclusion.

Where NA is the transmitter a further complication arises in attempts to make this comparison. This follows from the presence in many smooth muscles of both α and β adrenergic receptors. For example, exposure of the guinea-pig vas deferens to low concentrations of NA may depress transmission. This effect is markedly potentiated in the presence of an α-blocking drug. Holman & Jowett (1964) suggested that this was probably due to the existence of β-receptors which caused stabilization of the cell membrane. Excitatory effects mediated by α-receptors and inhibitory effects mediated by β-receptors are well-known characteristics of many types of vascular smooth muscle.

Inhibitory and excitatory effects of ACh on the same smooth muscle have yet to be confirmed. There is evidence for both excitatory and inhibitory effects of

10*

ACh on the smooth muscle of portal and mesenteric veins (Funaki & Bohr, 1964; Nakajima & Horn, 1967; Holman & McLean, 1967). However one of the actions of ACh on these preparations is to release NA from nerve terminals, possibly as a result of depolarization and the initiation of nerve impulses, and it is difficult to be sure that excitatory effects attributed to the direct action of ACh on the smooth muscle membrane are not due to the release of NA.

The excitatory action of ACh on smooth muscle is associated with the initiation of APs or an increase in their frequency. In the taenia coli, ACh causes depolarization and the continuous firing of APs of diminished amplitude (Bülbring & Kuriyama, 1963b). The repolarization phase of the AP is particularly sensitive to ACh and some smooth muscles generate prolonged APs of several seconds duration in the presence of this neurohormone (see Burnstock & Holman, 1966b). There have been extensive studies on the mechanism of the action of ACh, especially in the taenia coli (Bülbring & Kuriyama, 1963b). Although direct evidence for an increase in membrane conductance has yet to be found, all the available evidence points in this direction (see Chapter by Kuriyama). An increase in conductance for Na, K and Ca ions probably occurs. In common with its action at the skeletal neuromuscular junction ACh may act on smooth muscle

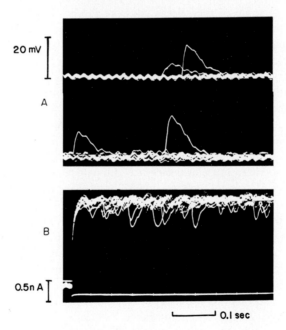

FIG. 11   SJPs recorded from mouse vas deferens. In A, superimposed records show SJPs in normal solution, at normal RMP. In B, the preparation was exposed to K_2SO_4 (see text). A bridge circuit was used to displace the MP in a positive direction (the lower trace of B monitored current). SJPs reappear at a high frequency but their duration is shortened and their polarity reversed.

by opening up negatively-charged pores in the membrane which increase permeability for cations but not for anions.

Although there is an impressive volume of literature concerned with the excitatory effect of exogenous NA on vascular smooth muscle, few of these experiments have helped to clarify the ionic basis of this effect. It is generally accepted that an increase in conductance for cations is involved, as is the case for the excitatory action of ACh. However other effects, including the release of Ca ions, may be involved (see Chapter by Speden). The smooth muscle of veins could provide a useful tool for further investigation of this problem.

Since the smooth muscle of the vas deferens is relatively insensitive to the action of exogenous transmitter it seemed important to look for changes in membrane properties in response to nerve released transmitter, in spite of the difficulties mentioned earlier.

Preliminary experiments have been undertaken in an attempt to modify the amplitude of the SJPs of the mouse vas deferens by changes in MP. Following the experiments of del Castillo & Katz (1955b) the preparation was exposed to concentrations of K_2SO_4 which reduced the resting MP to around 0 mV. Under these conditions SJPs were no longer observed (Dorward & Holman, 1967). It was found that the resistance to current flow in an outwards direction was high compared with resistance to current flow in the opposite direction (anomalous rectification). It was possible therefore, to set the MP at about $+20$ mV with quite small currents. During this procedure, SJPs were again observed but their polarity was reversed (Fig. 11). Although many more experiments along these lines need to be carried out to confirm this result it would seem that transmitter action in mouse vas deferens may resemble that of ACh at the skeletal neuro-muscular junction.

INHIBITORY JUNCTIONS

Characteristics of inhibitory junction potentials (IJPs)

The nerve fibres and intrinsic neurones of the gastrointestinal tract can be excited by means of transmural electrodes (Paton, 1955) or field stimulation (Kuriyama, 1963b). Changes in membrane potential in response to stimulation of these nerves have been recorded from the guinea-pig taenia coli and the longitudinal muscle of the small intestine (Bennett et al., 1966b; Bülbring & Tomita, 1967; Bennett & Rogers, 1967; Kuriyama, Osa & Toida, 1967c; see also Chapter by Kuriyama).

In the absence of pharmacological blocking-agents the response to field stimulation of the smooth muscle is complex, consisting of a mixture of excitatory and inhibitory effects. In the presence of atropine, however, field stimulation causes large and abrupt waves of hyperpolarization such as those shown in Fig. 12. Figure 12d shows the time course of the individual waves of hyperpolarization in response to stimulation at low frequency. Repetitive stimulation

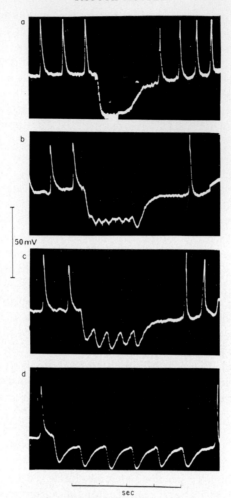

FIG. 12   Effect of stimulating across the taenia coli with repetitive pulses of maximal strength at different frequencies. Frequencies of stimulation 60, 4, 2 and 1 pulses/sec in a, b, c and d, respectively. Pulse duration 200 μsec. Note the increase in amplitude of the action potential after the IJP in b, c and d, and the increased rate of firing after the IJP in a and c. Electrode was dislodged from the cell in the record at the end of b.

(From Bennett, Burnstock & Holman, 1966b)

at high frequencies causes summation of successive responses, as in Fig. 12a, b, c.

These changes in membrane potential differ in many ways from those recorded during sympathetic nerve stimulation of the guinea-pig taenia (Bennett, Burnstock & Holman, 1966a; also see Chapter by Campbell). For example, a hyperpolarization caused by sympathetic nerve stimulation appears only during repetitive stimulation, whereas a large wave of hyperpolarization can be elicited by a single transmural stimulus. Secondly, no matter how intense and prolonged the sympathetic nerve stimulation, the MP of the taenia does not increase by more than 10 to 20 mV, whereas hyperpolarization of up to 50 mV can be reached in response to transmural stimulation. Thirdly, the intrinsic inhibitory system is unaffected by sympathetic blocking drugs (Bennett et al., 1966b).

During the last five years it has become apparent that this non-sympathetic inhibitory mechanism is a ubiquitous feature of the gastrointestinal tract of many species. Its significance in the functioning of the intestine will be discussed in the Chapters by Campbell, Kottegoda and Crema.

Several characteristics of the inhibitory response which were noted in early electrophysiological studies suggested that it was probably due to the release of an inhibitory transmitter from axons arising in Auerbach's plexus. For example, stimuli of brief duration (less than 0·1 msec), which would be unlikely to stimulate smooth muscle directly, caused a large hyperpolarization. The latency of the response was long (about 100 msec) and rather variable but was not significantly different if the recording electrode was moved from 2 to 8 mm from the stimulating electrode (Bennett et al., 1966b). More recently Bülbring & Tomita (1967) were able to measure an increase in latency at increasing distances between stimulating and recording electrodes and their results suggested a conduction velocity of 10 to 20 cm/sec.

Perhaps the most convincing evidence that the hyperpolarization caused by transmural stimulation is indeed an inhibitory junction potential has come from experiments with tetrodotoxin (Bülbring & Tomita, 1967; Kuriyama et al., 1967c). Tetrodotoxin has no effect on the spontaneous or on the evoked action potentials of the taenia coli but completely blocks the inhibitory potential, as shown in Fig. 13. Unfortunately, the nature of the inhibitory transmitter is still unknown. Attempts to find a specific blocking-agent for the inhibitory junction potential, which might antagonize the action of the inhibitory transmitter, have been unsuccessful. It is of interest that the IJPs recorded in guinea-pig intestinal muscle are much more resistant to the blocking action of solutions containing high Mg and low Ca concentration than the EJPs recorded from the same tissue.

The morphology of the inhibitory axons is unknown, though a combination of electronmicroscopy with electrophysiological studies by Bennett & Rogers (1967) suggests that the terminal axons of these neurones may extend as far as the serosa of the taenia coli. It is also likely that they penetrate bundles of smooth

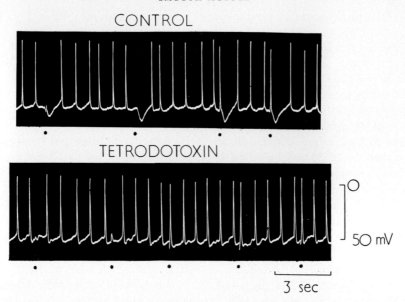

FIG. 13 Effect of tetrodotoxin on the inhibitory potential and spike activity of guinea-pig taenia coli. *Top*: spontaneous spike discharge in normal Krebs solution and responses evoked by two anodal and two cathodal pulses (●) of 1 msec duration. *Bottom*: after 15 min exposure to tetrodotoxin (5×10^{-7} g/ml) the spontaneous spike discharge was unchanged, but anodal pulse (1 msec, first dot) and four cathodal pulses (1, 2, 3 and 1 msec) failed to produce inhibitory potentials, though spikes were evoked by cathodal stimulation. (From Bülbring & Tomita, 1967)

muscle cells and are not confined to the larger extracellular spaces of this tissue. However, it is interesting to note that Bennett & Rogers were unable to find any examples of 'close contacts' between axon and muscle membrane in the taenia coli, the closest approach between them being 1000 Å. If excitation of the intrinsic inhibitory nerves occurred close to their origin in Auerbach's plexus, 10 msec or more might elapse before excitation could reach the axon terminals close to smooth muscle cells on the serosal surface of the taenia which are normally studied with intracellular electrodes. However, it seems unlikely that conduction time could explain the 100 msec latency of the IJPs recorded in cells within 1 mm of the stimulating electrodes. This long latency is comparable with that of the excitatory JPs recorded in this tissue (see Table 1).

Bülbring & Tomita (1967) compared the spatial spread of the IJPs of the taenia coli with the longitudinal spread of the electrotonic potential in the same preparation (Fig. 14). Whereas the electrotonic potentials decayed exponentially with distance, the IJPs decayed in a linear way and could be recorded at much longer distances from the stimulating electrodes than the electrotonic potentials. They deduced that the inhibitory axons were probably a few mm long, with a space constant of 0·1–0·2 mm and an absolute refractory period of 3–4 msec.

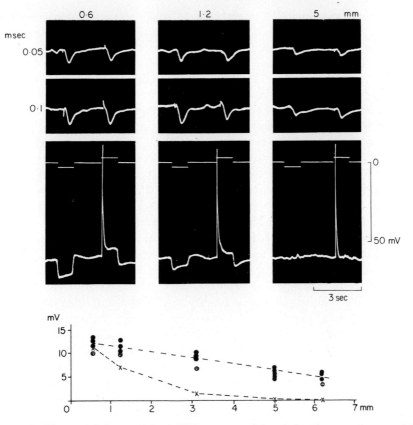

Fɪɢ. 14   The spatial decay of the inhibitory potential and the electrotonic potential. The inhibitory potentials (*shown in the upper and middle row*) were taken from 3 different cells at 0·6, 1·2 and 5 mm from the partition, in response to an anodal (first) and cathodal (second) pulse, of 0·05 (*top row*) and 0·1 msec duration (*middle row*). The bottom records show the membrane polarization produced by an anodal (first) and cathodal (second) pulse of 1 sec duration recorded from the same cell as the inhibitory potentials at 0·6, 1·2 and 5 mm from the partition. The cathodal pulse produced a conducted spike. In the graph the amplitude of the inhibitory potential (●) and of the electrotonic potential (×) is plotted against the distance from the partition. Note that the inhibitory potential decays linearly along the tissue and more slowly than the electrotonic potential which decays exponentially. (From Bülbring and Tomita, 1967)

The rise times of the IJPs are variable (usually about 200 msec). The time to decay to half-maximum ranges from 200 to 400 msec. The decay phase is roughly exponential with a time constant varying from 250 to 500 msec (Bennett *et al.*, 1966*b*; Bülbring & Tomita, 1967). The time constant of the decay phase is very much longer than the time constant of the electrotonic potentials induced by external polarizing electrodes in this smooth muscle. It is interesting that the time course of the IJPs in the taenia closely resembles that of the Type I secretory potentials of cat salivary glands (Lundberg, 1958).

The amplitude of the IJPs in response to single stimuli can be graded with increasing strength of stimulation up to a maximum amplitude of 35 mV. Repetitive stimulation at low frequencies (up to 10 sec) causes summation of individual IJPs (see Fig. 12). The maximum hyperpolarization observed during repetitive stimulation (10 sec) may be as large as 50 mV.

The time taken to reach maximum hyperpolarization during repetitive stimulation at 10/sec or more is slightly shorter than that of the rise time of a single IJP. The MP does not remain at this peak level but begins to decay, in spite of continuing stimulation. This feature of the inhibitory mechanism has yet to be explained. Electrophysiological studies have been confined so far to short periods of stimulation (less than 10 sec). Throughout this time the level of hyperpolarization, although decreasing, is still sufficient to block any spontaneous APs. Studies have been made on the mechanical activity of segments of small intestine of rat and rabbit during more prolonged periods of repetitive stimulation in the presence of atropine (Holman, unpublished work). These confirm the findings for the taenia and clearly demonstrate that the inhibition is a transient event which cannot be maintained by electrical stimulation for more than 20 sec. The significance of this effect *in vivo* is unknown.

Whereas it is relatively simple to modify the frequency of firing of APs of the taenia coli by sympathetic stimulation (see p. 285) the response to just suprathreshold stimulation of intramural inhibitory neurones at low frequencies is the immediate and complete abolition of APs and full relaxation of the muscle.

One of the most characteristic features of the IJPs in the taenia coli is the rebound excitation which always occurs towards the end of the recovery phase of the IJP (Bennett, 1966a). As the MP returns to its resting level a series of APs are initiated at a higher frequency than during the control period. In quiescent preparations APs occur as a result of rebound excitation and the *mechanical* effect of the IJP is therefore one of contraction rather than relaxation (see Chapter by Campbell).

The nature of the spontaneous activity of visceral smooth muscles was mentioned briefly on p. 250. It is our impression that owing to the relatively low value of their resting MP, most smooth muscle cells of the longitudinal muscle layer of the gastrointestinal tract are poised at, or close to, the point of spontaneous activity. Hyperpolarization may increase the magnitude of the voltage dependent increase in conductance for inward current in response to a subsequent depolarization, by reversing the 'inactivating' effects of the low resting membrane potential. This could lead to a lowered threshold for the initiation of APs (anode-break excitation). As the MP returns to its normal level during the decay phase of the IJP this new threshold for excitation would be exceeded and one or more APs could be generated. Rebound excitation has been noted in the taenia coli following hyperpolarizations induced by both external and intracellularly applied currents (Kuriyama & Tomita, 1965; Tomita, 1966a).

INHIBITORY EFFECTS UNACCOMPANIED BY INDIVIDUAL IJPs

With the exception of the IJPs of the gastrointestinal tract described above, all the inhibitory effects of autonomic nerve stimulation on smooth muscle reported so far, have only been demonstrated in response to repetitive nerve stimulation. These include the inhibitory effects of sympathetic stimulation of the gastrointestinal tract (Gillespie, 1962a; Bennett et al., 1966a) and the parasympathetic inhibition of the retractor penis (Orlov, 1963b).

Although the intrinsic inhibitory neurones of some regions of the gastrointestinal tract may be excited by extrinsic autonomic nerves, this does not seem to be the case for taenia coli or any regions of the gastrointestinal tract beyond the uppermost segments of small intestine (see Chapter by Campbell). Stimulation of periarterial nerves to the taenia coli, in the presence of atropine (Bennett et al., 1966a) and sympathetic nerves to the colon (Gillespie, 1962a), causes inhibition which is readily abolished by sympathetic blocking drugs. Thus it is possible to distinguish the inhibitory effects of sympathetic stimulation from those of the intrinsic inhibitory neurones of Auerbach's plexus.

In the rabbit colon, optimal frequencies of sympathetic nerve stimulation (50/sec) cause the MP to stabilize at a level similar to that observed between slow waves of depolarization (and APs) in unstimulated preparations. In stretched preparations which tend to fire APs continuously, Gillespie (1962a) was able to demonstrate hyperpolarization — but again only to a level comparable with that observed *between* slow waves in unstretched preparations. The effects of stimulation outlasted the duration of stimulation — as noted previously for the sympathetic excitation of venous smooth muscle.

In the taenia coli (Bennett et al., 1966a) all cells studied with intracellular electrodes showed a decrease in frequency of APs during sympathetic stimulation but detectable increases in MP were only recorded from cells of the pacemaker type. These authors drew attention to two types of activity which could be recorded from this preparation. Pacemaker regions could be distinguished by a phase of slow depolarization which developed from the maximum level of repolarization of one AP and appeared to lead directly to the initiation of the next — a pattern of activity resembling that of the cardiac pacemaker. 'Driven' regions were characterized by an abrupt threshold level of MP at which the AP appeared to be initiated (see Bennett et al., 1966a). Inability to detect hyperpolarization of 'driven' regions during sympathetic stimulation could be due to their relatively high membrane potential. Abolition of spikes in 'driven regions' could then clearly be explained by blockade of the spontaneous activity in 'pacemaker' regions.

The latency of the hyperpolarization observed in some of the cells of the taenia coli depends on the frequency of sympathetic nerve stimulation, decreasing from about 250 msec at 50/sec to 150 msec at 80/sec. The rate and amplitude of the hyperpolarization increases with increasing frequencies. These observations suggest that it may be necessary to build up the concentration of NA by repetitive stimulation for the effective transmission of inhibition. An explanation for the minimum latency of 150 msec remains to be found. During this time the terminal axons would be re-excited several times during high frequency stimulation, and presumably the concentration of NA in their vicinity would be steadily increasing. However, the sympathetically-induced hyperpolarization of some cells of the taenia is characterized by a very sharp 'take-off' whose timing varies with stimulus frequency rather than a gradual increase in MP that one might expect if latency was determined by the diffusion of NA over long distances.

As previously noted for the sympathetic excitation of blood vessels, the effects of sympathetic inhibition of the taenia coli and colon are prolonged for some time after the cessation of stimulation. This may be due to the sparsity of the noradrenergic innervation of these tissues and thus to an ineffective uptake mechanism. A further similarity between these two situations is the ease by which sympathetic nerve stimulation can be mimicked by exogenous transmitter.

ACTION OF INHIBITORY TRANSMITTERS ON SMOOTH MUSCLE

The transmitter responsible for the IJPs of the taenia coli and small intestine of the guinea-pig is unknown. It is therefore of special interest to study the ionic basis of these potentials — since this might provide a clue as to the nature of the transmitter. The maximum level of MP at the peak of transmitter action is probably close to -100 mV. The K equilibrium potential (E_K) for the taenia coli is about -100 mV and the simplest explanation for the IJP would seem to be that the transmitter causes marked increase in K conductance (Bennett, 1966c). This conclusion is supported by experiments in which the ionic composition of the solution bathing the taenia was varied. No evidence has been found for an increase in anion conductance (Bennett, Burnstock & Holman, 1963).

Direct evidence that transmitter action causes an increase in membrane conductance has yet to be found. Attempts to alter the amplitude of the IJP by changes in MP induced by current flow from an intracellular electrode have given negative results (Bennett & Rogers, 1967). This could be due to the syncytial properties of the taenia coli, as described previously for the EJPs recorded from the vas deferens.

The inhibitory action of exogenous NA and adrenaline on the gastrointestinal tract appears to be similar to the effects of sympathetic nerve stimulation (compare Fig. 10 of Bülbring & Kuriyama, 1963b with Fig. 4 of Bennett et al., 1966a). It is reasonable to expect that both sources of NA might have the same effect on membrane properties. There have been extensive studies on the mechanism of action of adrenaline on the taenia coli which have been reviewed in detail elsewhere (Burnstock & Holman, 1966b; Chapter by Kuriyama and Chapter by

Setekleiv). At the present time it appears that both types of adrenergic receptors are involved but that the mechanism of inhibition depends on whether the NA (or adrenaline) combines with α-receptors or β-receptors.

Combination with α-receptors causes an increase in K efflux and influx in depolarized smooth muscle (Jenkinson & Morton, 1967b). No increase in the rate of exchange of Cl was detected in these experiments. This suggested that the mechanism of inhibition might be an increase in K conductance similar to that caused by ACh in cardiac muscle. More recently Bülbring & Tomita (1968a, 1969b) were able to show, by the use of a double sucrose-gap method, that NA causes an increase in membrane conductance of normal taenia coli. The effect of changes in ionic environment on the changes in MP due to pulses of externally applied current suggested that both K and Cl conductances were increased (Chapter 12 by Kuriyama).

The hyperpolarization caused by catecholamines or by sympathetic stimulation of the taenia coli is generally much smaller than the hyperpolarization in response to intrinsic inhibitory nerves. This might be expected if the former is due to an increase in both g_K and g_{Cl} and the latter to an increase in g_K only. The Cl equilibrium potential (E_{Cl}) is probably relatively low in this smooth muscle (possibly even less negative than the RMP) (Casteels & Kuriyama, 1966). Thus the MP might be clamped somewhere between E_K and E_{Cl} under the action of NA but at, or close to E_K during the action of the unknown inhibitory transmitter.

Combination of adrenaline with β-receptors (i.e. in the presence of α-blockers) does not cause any detectable changes in MP, or conductance (Bülbring & Tomita, 1968a, 1969b). Nevertheless, spontaneous activity is abolished due to suppression of the generator potentials. Bülbring and Tomita made the interesting suggestion that the generator potentials could be due to a decrease in passive K conductance and depolarisation caused by the activity of a process removing Ca from binding sites in the cell membrane. They suggest that catecholamines may *inhibit* this Ca removal.

Some years ago an alternative suggestion came from Edith Bülbring's laboratory which still seems worthy of consideration. It was proposed that the inhibitory effects of catecholamines might involve the *stimulation* of an ion pump. If Ca as well as Na ions were handled by such a pump, an increase in its activity could lead to membrane stabilization as the result of an increase in the Ca concentration immediately outside the cell membrane. Previously we drew attention to the similarities and differences between the effects of adrenaline and high Ca on the taenia coli (Burnstock & Holman, 1966b). At that time we noted that high Ca did not seem to be nearly as effective as adrenaline in blocking spontaneous activity. However, it might be worthwhile to reconsider this possibility. Evidence — admittedly circumstantial — has been accumulating for vascular smooth muscle which suggests that β inhibition may be associated with an increase in the activity of an ion-pump (see Somlyo & Somlyo, 1968a, b).

CONCLUSION

Recent electrophysiological studies on transmission from autonomic nerves to smooth muscle have been confined to muscles with single-unit properties. The mouse vas deferens may be an exception, but even in this tissue it seems likely that there is considerable electrical interaction between neighbouring smooth muscle cells.

In order to influence the activity of such a syncytial structure it is necessary for a transmitter to act on a large number of cells, more or less simultaneously. It seems that this can come about by two alternative mechanisms:

1. The diffusion of transmitter liberated into the extracellular space from a number of axon varicosities situated at a distance of 1000 Å or more from a bundle of smooth muscle cells.

2. A series of close contacts (less than 1000 Å separation) made 'en passage' by a terminal axon which penetrates into the depths of a tightly-packed bundle of smooth muscle cells.

In the former case it should be possible for circulating transmitter to mimic the effects of autonomic nerve stimulation. The cholinesterase content of blood makes it unlikely that circulating ACh would ever be significant in influencing the activity of smooth muscle. On the other hand circulating NA and adrenaline are clearly important in the control of some autonomic effector organs. The dense adrenergic innervation of some smooth muscles, which is implied by the second of these alternatives, necessarily involves an effective uptake mechanism for catecholamines and, as a consequence, reduces sensitivity to circulating transmitter. This may be one of the reasons why smooth muscles such as those of pelvic viscera, have evolved with a pattern of innervation which enables local rather than generalized responses to sympathetic stimulation.

ACKNOWLEDGEMENT — I would like to thank Dr Y. Hashimoto for permission to describe some of his unpublished work.

9

MECHANICAL PROPERTIES OF SMOOTH MUSCLE, AND THE RELATIONSHIP BETWEEN MECHANICAL AND ELECTRICAL ACTIVITY

J. Axelsson

INTRODUCTION

The task of a muscle is to develop tension. This sometimes results in physical work.

The information given in the preceding chapters about the ionic and electrical characteristics of smooth muscle is highly relevant to an understanding of how activation of the contractile machinery leading to release of energy is initiated in this type of muscle.

A short introductory section summarizing a few well-known facts concerning striated-muscle contraction, drawing comparison with that of smooth muscles, might help to orientate the reader in the confusing situation regarding smooth-muscle mechanics. This is bound to be highly superficial, merely pointing out the few apparent similarities and dissimilarities between the different muscle types. For example, one should remember that there are just as many differences between muscle types known as 'smooth muscles', as between any of them and striated muscles. Even among striated muscles there is great variety in behaviour. For example, there exist the 'slow fibres' which are physiologically activated not by action potentials but by graded depolarization. Their mechanical response is a contracture (defined below) and not the usual twitch of striated muscle.

Structural basis of contraction

The structural basis for the sliding-filament theory in vertebrate striated muscle is well established (for review see Huxley, A. F., 1964, and Huxley, H. E., 1965). This theory can also be extended to invertebrate muscles, both striated and smooth, on the basis of Hanson & Lowy's findings (1964a, b) that there exist in these muscles two types of filaments sliding in relation to each other. In the case of vertebrate smooth muscle, including taenia coli, the picture is still not as clear as one could wish. It is known, however, that these muscles contain both actin and myosin (see review by Needham & Shoenberg, 1967). Further there is electron microscopic evidence that the actin is present in filament-form. Evidence indicating that the sliding filament theory may also be applicable to vertebrate smooth muscles is summarized in two of Shoenberg's communications (Needham & Shoenberg, 1967 and Shoenberg, 1969). Another model for smooth muscle contraction based on sliding filaments has been presented by Panner & Honig (1967). Details of the structural basis of contraction in smooth muscle are given in Chapter 1.

Activation processes

One step in the physiological activation of the contractile mechanism in striated muscle is usually a propagated change in transmembrane potential (action potential). I have, however, just referred to the exception of the 'slow fibre system'. Although striated in structure the slow fibres are physiologically activated by graded depolarization (Kuffler & Vaughan Williams, 1953a, b).

Although a change in transmembrane potential is generally considered an essential factor for the activation of the contractile mechanism in striated muscle (Katz, 1950; Huxley, 1957) exceptions have been found. Axelsson & Thesleff (1958) showed that caffeine induced tension in both polarized and depolarized striated muscle, without any reduction in membrane potential, whether propagated or non-propagated. Such forms of activation may not play any physiological role in striated muscle. Their discovery has, however, prompted a great deal of work on the mechanism of this type of activation (see Caldwell & Walster, 1963; Miyamoto, 1963; Hasselbach, 1965a, b; Herz & Weber, 1965; Ashley, 1967).

In the majority of smooth muscles studied, an early step in the physiological activation of the contractile mechanism appears to be similar to that of striated muscles, i.e. a change in transmembrane potential. There are, however, numerous deviations some of which may be of physiological importance. These will be systematically discussed below.

We shall now consider briefly the later steps involved in the activation of the contractile machinery. Here again the picture is much clearer in the vertebrate striated muscle than it is in smooth muscles. In striated muscles at rest, vesicles of the endoplasmic reticulum store calcium. When the muscle depolarizes the calcium is released and leads to contraction. When the depolarization is over the calcium is pumped back. There are, of course, other modes of 'activating' or releasing calcium than depolarization. For further description, the reader is referred to the excellent review by Hasselbach (1965b).

The development of the endoplasmic reticulum in different muscle types varies greatly. In many smooth muscles it appears to be very poorly developed or even absent (see Burnstock, Chapter 1). In spite of this there is strong evidence, to be presented below, that in various types of smooth muscles calcium, in one way or another, plays a determining part in the activation of the contractile system.

Mechanical properties

The work of A. V. Hill, Wilkie and others on the frog sartorius yielded a classical model of striated muscle mechanics. For a detailed picture of the present situation in this work, the reader is referred to Wilkie (1966) and A. V. Hill (1965). So far most experiments on intestinal smooth muscle mechanics have been repetitions of the classical mechanical experiments. One could, however, hardly expect this model to provide an adequate account of mammalian smooth muscle which is frequently spontaneously active and therefore difficult to maintain in steady state. In fact one is surprised whenever any rule in its mechanical behaviour is discovered. A good deal of work towards bringing the mechanics of smooth muscles on a quantitative basis is in progress. But continued efforts will be required.

We may now proceed to an enquiry into the connections between the electrical and mechanical phenomena in smooth muscles — the correlation between transmembrane potential on the one hand and tension or changes in length on the other.

INITIATION AND CONTROL OF 'TONE'

The relationship between membrane potential, frequency of spike discharge and tension in 'physiological condition'

Since 1931 extracellular electrodes have been used in attempts to record simultaneously mechanical and electrical activity in various types of smooth muscles (Hasama, 1931, 1933, 1935; Berkson, Baldes & Alvarez, 1932; Puestow, 1932; Berkson, 1933 and 1934). Bozler (1948) concluded that the tone in smooth muscles was maintained by slow but continuous spike discharge. This was confirmed for taenia coli by Bülbring (1955). She used intracellular recording and found that tension was directly proportional to the frequency of spike discharges which in turn was inversely related to the transmembrane potential. This inverse relation, however, only holds when the transmembrane potential falls to a certain limit. In that case the tension (in 'isometric' condition) rises as a consequence of increased frequency of discharge. If the intervals between

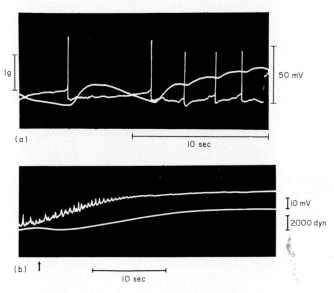

FIG. 1   Simultaneous records of electrical and mechanical activity from guinea-pig taenia coli.   (a) Intracellular record of spontaneous spike discharge (kindly supplied by H. Ohashi). Each action potential is followed by a phasic tension response.   (b) Extracellular record (sucrose-gap technique). The arrow indicates an increase in [K]$_0$ from 5·9 mM to 128 mM. This resulted in a sustained depolarization and the development of a contracture.

action potentials become sufficiently short the tension does not revert to resting level but summates. With these increases in frequency of discharge which, at the present stage of knowledge, are assumed to occur spontaneously, the pattern of mechanical activity in taenia coli comes to resemble tetanus in striated muscle, complete or incomplete. This pattern is called 'tone' which may thus be defined as myogenic mechanical activity initiated by action potentials arising spontaneously in pacemaker regions. Questions regarding the sources of these action potentials are dealt with in the Chapters 7 and 12 on the origin and control of electrical automaticity in this book. Numerous investigators have since confirmed and extended the original observations of Bülbring, both for taenia coli and other types of smooth muscle.

Each action potential or transient depolarization is normally followed by a brief mechanical response referred to as 'phasic response' (Fig. 1a). The mechanical response due to sustained depolarization, consisting in reversible shortening of the fibres or the development of tension, is termed 'contracture' (Fig. 1b). In both cases depolarization, whether transient or sustained, leads to or is associated with the activation of the contractile mechanism and a conversion of chemical energy into mechanical activity. In this connection the reader is referred to the outstanding work of Kuriyama (1961b), Marshall (1962) and others on the uterine muscle (references to this work can be found in Chapter 13). A similar relationship has recently been established in different kinds of vascular smooth muscle (for references see Chapter 20).

General hypothesis of a causal connection between changes in transmembrane potential and tension

On the basis of the numerous observations, just referred to, of a close connection between electrical and mechanical activity the general hypothesis may be advanced that in smooth muscles changes in transmembrane potential are the sole cause — sufficient and necessary — of contraction. The hypothesis thus postulates that the change in potential, be it by transient or sustained depolarization, initiates contraction without exception, and that neither phasic response nor contracture is ever initiated without change in membrane potential. Thus formulated the hypothesis lends itself easily to critical testing.

Although, under physiological conditions, the electrical and mechanical processes are obviously linked in the majority of smooth muscles studied, they are probably only steps in a chain of events or processes. If this may be assumed we should enquire into the possibility that 'intermediate steps' in the chain can be affected in such a way that the relation between transmembrane potential and tension is disrupted. Should this turn out to be so, our hypothesis, at least in its most general form, would be refuted, and then we are faced with the possibility of such atypical forms of activation playing a physiological part in certain tissues.

In the remainder of this section I shall describe briefly experimental results incompatible with the hypothesis.

Electrical activity without tension response

Metabolic factors. In taenia coli electrical activity is maintained without tension response in the presence of various metabolic inhibitors (Bülbring & Lüllman, 1957). In glucose-free solution the frequency of discharge is increased while the phasic response to each action potential is gradually abolished (Axelsson, Bueding & Bülbring, 1959; Axelsson & Bülbring, 1961 (see Fig. 2). A more quantitative study of the effects of the removal of glucose revealed a fall in both physical work and tension in response to general depolarization obtained

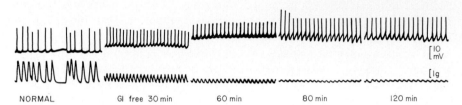

NORMAL   GI free 30 min   60 min   80 min   I20 min

FIG. 2   The partial dissociation obtained in guinea-pig taenia coli by removing glucose from the superfusing solution. Electrical and mechanical activity were recorded simultaneously by the sucrose-gap method. The upper row shows the electrical, the lower the mechanical activity. (From Axelsson, Högberg & Timms, 1965)

by raising the $[K]_0$ to 128 mM (Axelsson, Högberg & Timms, 1965). The hypothesis will therefore have to be restated more carefully and claim that changes in transmembrane potential are sufficient and necessary only to trigger the processes involved in the activation of the contractile mechanism. It then becomes another question whether contraction always follows the triggering. And this, it now appears, will depend on numerous factors.

In the examples quoted above the failure to evoke contraction may be due mainly to energy shortage of the contractile mechanism, although it is not excluded that removal of glucose may also affect essential steps in the activation processes in spite of the continued changes in transmembrane potential.

Ionic environment. Various degrees of dissociation of maintained electrical activity from mechanical response are also observed in experiments involving changes in the ionic composition of the external medium.

Na and K deficiency. Such dissociations in taenia coli were described and discussed by Axelsson (1961*b*, 1962) who substituted sodium chloride with lithium chloride, hydrazine chloride, choline chloride and lithium ethanesulphonate and obtained a high degree of dissociation. It was also found that removal of external potassium reduced the phasic response to action potentials of increased amplitude. Prolonged exposure to potassium-free solution also abolishes the mechanical response to induced depolarization (Axelsson, Holmberg, Högberg & Wahlström, 1969). Most of these findings await further examination.

Ca. In calcium-free or calcium-deficient solution the phasic response to continued action potentials in taenia coli is greatly reduced or abolished (Axelsson & Bülbring, 1959; Axelsson, 1961*b*) (see Chapters 7 and 12).

Cl. The dissociation was easily demonstrated in a solution in which chloride was replaced by nitrate which by itself causes a great increase in the frequency of discharge and excitability (Axelsson, 1961c). If calcium was reduced below 10% of its normal concentration in nitrate solution, the tension rapidly fell in spite of persistent spike activity of high frequency.

The mechanical response of taenia coli to depolarization by high $[K]_0$ is also dependent on calcium (Durbin & Jenkinson, 1961b). The same is true of uterine smooth muscle (Edman & Schild, 1961, 1962) and vascular smooth muscle, both for the mechanical response to action potentials and the contracture in response to sustained depolarization (Axelsson, Johansson, Jonsson & Wahlström, 1966; Axelsson, Wahlström, Johansson & Jonsson, 1967). Again one may insist that the change in potential is sufficient to activate the contractile machinery. Calcium would then be considered essential for the contraction process itself and its removal would only dissociate the mechanical response from the potential change in the same way as, for example, shortage of energy would.

Osmolarity. Dissociation of electrical from mechanical activity in hyperosmotic solutions has been shown for striated muscle by Hodgkin & Horowicz (1957). Howarth (1958) concluded that in frog striated muscle this was mainly due to increased viscosity. Similar phenomena have been found in taenia coli. The effects of osmotic changes in taenia coli are discussed in detail in Chapter 6. In the portal vein small changes in osmolarity caused changes in mechanical activity 'which can be ascribed essentially to changes in the pattern of electrical activity' (Johansson & Jonsson, 1968). With greater changes in osmolarity, however, dissociation is observed (Mellander, Johansson, Gray, Jonsson, Lundvall & Ljung, 1967).

Dissociation of the tension from electrical activity in spontaneously-firing taenia coli in normal solution has been reported by Mashima, Yoshida & Handa (1966). After stimulation by 'optimum a.c. field' they recorded relaxation while discharge of action potentials continued at high frequency (the a.c. stimulation also relaxed contractures of depolarized muscles). The authors concluded that this was due to decreased viscosity.

In spite of numerous examples of electrical activity, transient or sustained, without a mechanical response, I have failed to refute the first part of the hypothesis, i.e. that changes in membrane potential are sufficient for triggering contraction (where 'triggering' means one of the intermediate steps in the chain between potential change and mechanical response). However, in order to be considered a cause of activation, the changes in potential have to be also necessary for triggering contraction. Here the refutation is easier.

Changes in tension without changes in transmembrane potential (depolarized muscles)

Observations of dissociation of electrical and mechanical activity in different kinds of depolarized muscles are now so numerous that a detailed account would fill pages. Therefore only a few references will be given which by no means cover this field.

Drug action. In completely depolarized chick amnion tension was increased by drugs which increase membrane permeability, without any change of potential (Evans, Schild & Thesleff, 1958).

Edman & Schild (1961) showed that ACh applied to a depolarized rat uterus produced an increase in the contracture, while adrenaline relaxed it.

Durbin & Jenkinson (1961*a*) found that carbachol increased both inward and outward fluxes of various ions in a depolarized taenia coli and simultaneously increased the degree of contracture.

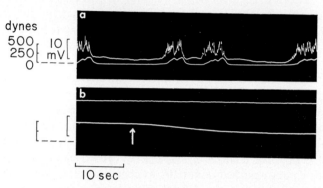

FIG. 3   Relaxing effect of isoprenaline on contracture of the rat portal vein without a measureable change in membrane potential. a Spontaneous activity in normal Krebs solution.   b Depolarization and contracture produced by increasing the $[K^+]_o$ to 128 m-mole/l. Isoprenaline 1 mg/l was added at the arrow. (From Johansson, Jonsson, Axelsson & Wahlström, 1967)

Waugh (1962) demonstrated that adrenaline and noradrenaline could induce contractions in depolarized arterial muscles. In the depolarized portal vein noradrenaline increased, and isoprenaline in low concentrations decreased, contracture without measureable change in potential (see Fig. 3) (Axelsson, Johansson *et al.*, 1966; Johansson *et al.*, 1967).

Axelsson, Holmberg & Högberg (1965) showed that chlorbutolum, which is a common ingredient in many commercially-obtainable drugs, relaxed contracture of taenia coli reversibly at 37°C without measurable change in transmembrane potential. This was later confirmed by Imai & Takeda (1967*a*) who also showed that papaverine and theophylline relaxed contracture in muscles depolarized by K_2SO_4. Axelsson & Högberg (1967) showed that caffein relaxed contractures of depolarized taenia reversibly without change in transmembrane potential.

Axelsson Holmberger & Högberg. (1965, 1966) found that the contractures maintained by taenia coli which had been depolarized by raising the $[K]_o$ to 128 mM were after one hour 58 ± 5% of the maximum mechanical response to this type of depolarization. At this stage the contractures were relaxed, without a measurable change in transmembrane potential, by ATP, AMP, adenosine and adenine. The lowest effective concentration was similar for the various compounds (about 0·1 mM). In the higher concentrations, however, the order of effectiveness was adenine > adenosine > AMP ≃ ATP.

Changes in ion concentration. As mentioned previously, Edman & Schild (1961, 1962) demonstrated that changes in the extracellular calcium concentration greatly affected the contractures maintained by depolarized rat uterus. The same was found in depolarized taenia coli by Durbin & Jenkinson (1961b). These findings have since been confirmed and extended by many authors working with various tissues. The dependence of the contractures of the portal vein on the external calcium ion concentration has recently been demonstrated (Axelsson, Wahlström, Johansson & Jonsson, 1967).

Changes in pH. Axelsson, Holmberg & Högberg (1965) found that a decrease in pH increased the contracture tension in taenia coli. This was confirmed and extended by Åberg, Mohme-Lundholm & Vamos (1967).

Some authors have not actually measured the membrane potential but, being satisfied that the muscles under examination were fully depolarized, assumed that the potential did not change.

The establishment of the fact that the contractile machinery can be affected in various kinds of depolarized smooth muscle without a change in potential refutes the hypothesis that a potential change is necessary for contraction. We must therefore, turn our attention to the ion-movements or activities which are associated with or constitute an action potential or a graded depolarization and work on the assumption that these may be induced (or inhibited) in various ways without a change in transmembrane potential.

Polarized muscles in which the usual relationship between the frequency of spike discharge and tension is not maintained

There are polarized smooth muscles which exhibit 'only limited quantitative correlation between drug-induced tension development on the one hand, and extent of depolarization and/or increase in spike frequency on the other' as cautiously put by Somlyo & Somlyo (1968b). In this category are the 'atypical' forms of activation already referred to which, I believe, may prove to be of physiological importance. Adrenaline and noradrenaline contract circular smooth muscle in rabbit stomach without initiating action potentials or depolarization (Furchgott, 1960, 1962). In pulmonary artery Su, Bevan & Ursillo (1964) obtained tension responses to drugs without change in potential. Niu (personal communication) has made similar observations in various other vascular muscles. Tension responses to drugs which could apparently not be accounted for by changes in electrical activity have been reported and discussed by Cuthbert & Sutter (1965), Axelsson & Högberg (1967), Somlyo & Somlyo (1968b)

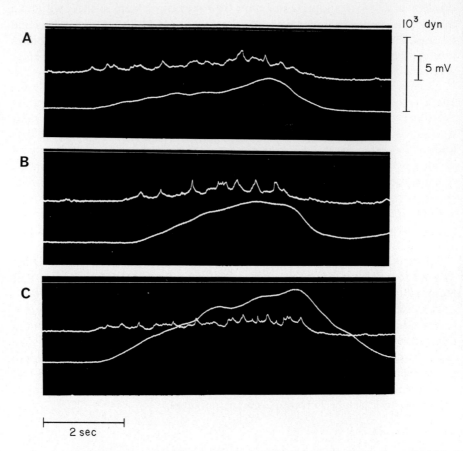

FIG. 4 Simultaneous recording of electrical and mechanical activity from the rat portal vein (sucrose-gap method). The upper row in each panel shows the electrical, the lower row the mechanical activity. The figure illustrates the partial dissociation of electrical activity and mechanical response during exposure to noradrenaline. (A) Spontaneous activity in normal solution. (B) Activity after 2 min exposure to noradrenaline 10^{-7}g/ml. (C) After 5 min exposure to noradrenaline. (Axelsson & Wahlström, unpublished)

and for partly 'desensitized' taenia coli, by Axelsson & Wahlström (1966). Figure 4 is typical for this kind of record. The danger of accepting partial dissociation of electrical and mechanical activity as a proof of mechanical response in these conditions being independent of changes in membrane potential has been pointed out (Axelsson, 1962; Cuthbert & Sutter, 1965; Somlyo &

Somlyo, 1968*b*). The objections to such an acceptance as proof of independence are equally valid for intracellular recordings as for sucrose gap recordings, except if electrical and mechanical activity were recorded from the same cell.

In an attempt to obtain a higher degree of objectivity in our methods we have translated our records of electrical as well as mechanical activity into numerical values and are now investigating the statistical properties of the variations in membrane potential and the transfer function between the electrical and mechanical activity by means of Fourier — and spectral analyses (Axelsson, Gudmundsson & Wahlström, 1968). Figure 4 is typical for the kind of records on which our quantitative analysis is based. Methods and results of this analysis are further discussed in the Appendix to this chapter. The analysis inspired confidence in the recording method and demonstrated dissociation of electrical and mechanical activity in the presence of, and after administration of noradrenaline. The result is consistent with the idea that a form of activation of the contractile mechanism not mediated via changes in transmembrane potential may play a physiological part in the portal vein.

Meanwhile there is a wealth of data which needs no laborious mathematical treatment to demonstrate dissociation in polarized smooth muscles. Decreased frequency of discharge was observed simultaneously with increase in tension when the external calcium concentration was raised from 0·1 to 2·5 mM in partly 'desensitized' taenia coli as well as in the portal vein (Axelsson & Wahlström, and Axelsson, Johansson, Jonsson & Wahlström, unpublished data). Several observations of increased electrical activity and failing tension have been discussed earlier in this chapter.

AN ESSENTIAL STEP IN THE ACTIVATION OF THE CONTRACTILE ELEMENTS

As mentioned in the introductory section above, the role of calcium in the mechanical activation of striated muscles is fairly well known. In spite of the poorly-developed endoplasmic reticulum in most smooth muscles there has been great reluctance to give up the idea of calcium playing a similar part in smooth muscles.

Bauer, Goodford & Hüter (1965) pointed out that many smooth muscle cells have a very small diameter (3 to 4 μ). Consequently, 'extracellular calcium is not far from the contractile mechanism'. Ebashi (1961) proposed that contraction and relaxation of smooth muscles might be related to a shift of calcium towards and away from the contractile elements.

There is good evidence for increased inflow of calcium during electrical activity in various tissues. In the squid axon Hodgkin & Kenyes (1957) found an increased inward movement of calcium during the action potential and increased inflow of calcium during spike activity was observed in frog skeletal muscle by Bianchi & Shanes (1959). Urakawa & Holland (1964) who studied ^{45}Ca uptake during potassium induced contraction of taenia coli found it increased. Furthermore, Lüllmann has found increased uptake of calcium during electrical stimulation of taenia coli (see Chapter 5).

Fatt & Ginsborg (1958) suggested that inward movements of calcium and other divalent cations might account for the spikes of crustacean muscle fibres in sodium-free solution. In taenia coli an inward movement of calcium during an action potential was suggested on the basis of experiments in which conducted spikes were obtained in isotonic sucrose solution only containing 2·5 mM CaCl$_2$ (Axelsson, 1962). And now the latest work of Brading & Tomita (1968b) and Brading, Bülbring & Tomita (1969b) strongly suggests that the action potential in taenia coli even in normal solution may be mainly a 'calcium spike' (see Chapter 7).

In fact, calcium plays a prominent part in most attempts to 'explain' the different phenomena, when tension changes are dissociated from changes in membrane potential. Durbin & Jenkinson (1961b) advanced the hypothesis that 'the carbachol contracture of depolarized smooth muscle was a consequence of net movement of calcium into the cells, following the increase in permeability produced by the drug.' Imai & Takeda (1967a) suggest that the relaxing effects of papaverine and theophylline on contractures of taenia coli were due to inhibition of calcium fluxes into the muscles.

While it is accepted that calcium ions, contained in both extracellular and intracellular compartments, normally play a part in the mechanical activation processes, the relative importance of an inflow of extracellular calcium and the release of intracellular calcium for the initiation of contraction is not known. Imai & Takeda (1967b) conclude 'that the phasic response and a part of the tonic response in taenia coli depend upon the extracellular calcium for their initiation (and also for the maintenance of tension in the case of the latter) and the rest of the tonic response, here named contracture, draws on a store of "bound" calcium for its evolution.' It has also been suggested that 'bound' calcium plays a part in the contraction of uterine smooth muscle (Edman & Schild, 1962; Daniel, 1963a). Regarding vascular smooth muscle there are findings which favour the view that bound calcium may be liberated by drugs and play a part in the physiological activation process. In the portal vein noradrenaline can repeatedly restore both spike discharge and mechanical response when these have been abolished in calcium-free solution in which the muscles do, of course, retain Ca^{2+}. In the presence of EGTA the restoring effect of noradrenaline is abolished while it still initiates both spike discharge and tension in control preparations exposed to calcium-free solution without EGTA for the same length of time (Axelsson, Johansson et al., 1966; Axelsson, Wahlström et al 1967). Noradrenaline may still initiate spikes and tension in calcium-free solution when graded depolarization by raising the $[K]_0$ fails to do so. Similarly, after short exposure to noradrenaline, an increase in $[K]_0$ which before failed to initiate activity may now restore it. Even contracture in response to depolarization which was abolished in calcium-free solution may redevelop after addition of noradrenaline. All this suggests that intracellular or membranal calcium has been liberated. Other drugs such as isoprenaline (in high concentrations), acetylcholine and carbachol may also restore activity to some extent in calcium-free solution. All these restoring effects are cancelled by administration of EGTA. Hinke (1965) working on perfused isolated artery segments suggested that noradrenaline maintained tone in calcium-free solution by liberating bound calcium.

Assuming that intracellular as well as extracellular calcium is involved in the physiological activation of the contractile elements it would be, as already mentioned, of great interest to know the relative importance of the inflow of extracellular calcium and the proposed intracellular or membrane liberation. This may, of course, vary in various smooth muscles. For a detailed discussion of the possible situation in taenia coli see Imai & Takeda (1967b). Using standardized depolarization as stimuli we are collecting quantitative information about the dependence of the contractile response in the portal vein on the $[Ca^{2+}]_0$ in the absence and presence of noradrenaline, relating all responses to muscle volume and diameter, in the hope that this may throw some light on the relative importance of liberation of intracellular calcium versus changes in permeability towards extracellular calcium in this tissue.

Doubts about the role of calcium in the activation process in smooth muscle have been expressed. Isojima & Bozler (1963) working on frog stomach concluded that calcium did not participate in the activation process in that muscle. This conclusion was based on the observation that calcium did, under certain circumstances, enter the fibres without causing contraction. It seems, however, that there are alternative interpretations of these results (see Wilkie, 1966).

SMOOTH MUSCLES AS MECHANICAL SYSTEMS

It strikes one as strange with a muscle as thoroughly investigated from as many points of view as taenic coli how little attention has been paid to its mechanics. In the meagre literature on the subject, including the present author's contributions, there is little of basic interest. What is known about its structure, the foundation of any mechanical system, is described in Chapter 1 by Burnstock. The very completeness of Burnstock's review only reveals the incompleteness of our knowledge on this vital point. Heat measurements are just beginning to be made. Yet no values of its mechanical efficiency (defined as the ratio $W/(H + W)$, i.e. the ratio of work performed to the total energy used) have been obtained. Collection of biochemical data continues, but only in a few cases

11+

have they been systematically related to mechanical events. Analyses relating mechanical output to heat produced and ATP and CP used have only recently been undertaken.

It appears that the time is not yet ripe for developing theories or models of contraction in this muscle. But one may hope that the aforementioned gaps in our knowledge will be gradually filled.

As already pointed out in the introduction, most of the mechanical experiments in taenia have been attempts to repeat those of classical striated muscle mechanics (see Åberg & Axelsson, 1965). They have the character of a mere collection of facts rather than of problem-solving. They are valid only for the defined experimental situations to be described below. The 'simplest' mechanical studies are hampered by variables still imperfectly controlled. Many of these derive from the prominent visco-plasticity of smooth muscle and consequent difficulties in standardizing the initial experimental conditions. Emphasis on the numerous uncertainties may save us from undue generalizations and oversimplifications.

As the terms plasticity, viscosity and elasticity will occur throughout, definitions of these terms, as used in this context, may help the reader. For further references see *Tissue Elasticity* (1957).

'Elasticity is that property of a material which determines the tendency of the stressed material to return to its unstressed geometrical configuration.
Viscosity is that property of a material which tends to retard deformation of the stressed material.
Plasticity is that property of a material by which it may be subjected to stresses of less than a critical or "yield" value without suffering permanent set, but will flow at stresses above the yield value. When the stress applied to such a material is removed, the material does not return to its original geometrical condition, provided the stress has exceeded the yield value.'

Standardizing the initial conditions and controlling variables

Elastic, plastic and viscous properties determine, to varying degrees, the mechanical behaviour of taenia coli depending on the experimental conditions. This does not sound encouraging and may explain the reluctance to investigate its mechanics. The prime difficulties in obtaining a reliable reference length for taenia were discussed in detail by Åberg & Axelsson (1965). There they describe the method finally adopted for determination of L_0, which gave the most comparable and reproducible results. L_0 was defined as the greatest length of an inactive muscle when the force applied to straighten it did not exceed 50 dynes. In later studies it was found desirable to add the following pretreatment. The pieces of taenia were dissected at 37°C in a solution containing adrenaline. After determination of the inactive 'unstressed' length of adrenaline-relaxed pieces *in vitro*, the muscle pieces were depolarized and allowed to shorten. After recovering in normal solution, the pieces were once more relaxed by adrenaline and their L_0 (original) determined (Axelsson, Högberg, & Löwenhard, unpublished). This provided more comparable initial conditions for the various pieces and improved the consistency of the results obtained during the first application of stress. Although others may have the good fortune to obtain reproducible and comparable mechanical results without this laborious pretreatment to standardize the initial conditions — we did not.

Let us now briefly consider a few of the variables to be taken into account during the simplest mechanical measurements of taenia coli. (1) The effect of stretching a visco-plastic system depends on the rate of deformation. An apparatus was therefore constructed which allowed one to stretch and relax the preparation at a constant rate which could be varied over a wide range. (2) It is well known that this muscle behaves like a stretch receptor. It responds to stretch by a series of action potentials that normally induce the mechanical responses. Therefore, while stretching an adrenaline relaxed muscle one may be recording a mixture of active and passive responses. The elimination of this last variable was attempted in different ways: (a) By abolishing the contractile response by using calcium-free EGTA solution. Although stretch neither induces spikes nor phasic tension in the calcium-free EGTA solution, the muscle membrane may, nevertheless, be depolarized. It is not known whether this depolarization and the consequent changes in membrane permeability might affect the mechanical properties. This possibility of an uncontrolled variable is now being investigated in experiments in which stress relaxation and creep in polarized and depolarized muscles in calcium-free EGTA solution are compared. The binding of calcium will most likely, by itself, introduce changes in 'passive properties'. For example, Alexander (1967), working on venous smooth muscle, has described the effect of changing the $[Ca^{2+}]_0$ on the 'plastic resistance to stretch' of these muscles. (b) By abolishing mechanical responses with metabolic inhibitors. (c) By depolarizing the muscle, thus inducing a permanent state of activation where action potentials and phasic tension development would not interfere. (d) By using glycerol-extracted pieces of taenia.

We have still a troublesome variable: the prominent plasticity of the muscle. All mechanical qualities of the system change with the unavoidable plastic deformation. Some of the qualities lying at the root of the instability of taenia as a mechanical system have been subjected to systematic study. A few results obtained during controlled stretch for the purpose of studying creep and stress–relaxation together will be presented.

Visco-plastic properties

Creep (increase in length during constant stress) and stress relaxation (decreased stress during constant strain) are prominent features of this system. Both precede plastic changes — but appear not to be a sufficient cause of these as both may occur without measurable plastic deformation.

Both creep and stress relaxation, when plotted versus the logarithm of time, revealed a linear phase and could be subjected to mathematical treatment earlier applied to dead material. Figure 5 shows the stress-relaxation of two different muscles plotted as logarithmic functions of time.* Curve (a) shows the stress relaxation of a glycerinated piece after the first stretching while curve (b) shows that of the same muscle after the sixth stretching. Curve (c) shows the stress-relaxation of an intact muscle after the first stretching. This muscle was relaxed and immersed in Ca^{2+}-free EGTA solution, thus it was incapable of an active mechanical response. It appears that the decrease in stress is linearly proportional to the logarithm of time during approximately 150 sec. ($\ln t$ ca. 5) for

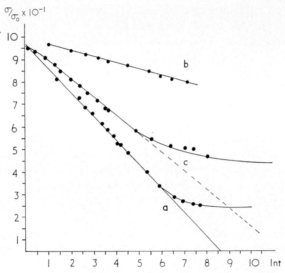

FIG. 5 Stress relaxation. Relative stress as a logarithmic function of time. The ordinate represents the normalized stress $\sigma/\sigma_0 \times 10^{-1}$. The stress values are given as proportional parts of the original (starting) stress σ_0. is the actual measured stress. at $t = 0$, $\sigma = \sigma_0$. The abscissa shows natural logarithms of time in seconds. For further description see text.

both muscles. The slope φ is different however. Compare the slope $\varphi = -1\cdot125 \times 10^{-1}$ for glycerinated muscle with $\varphi = -0\cdot80 \times 10^{-1}$ for intact muscle. After the sixth strain and consequent plastic deformation the slope φ for the glycerinated muscle was $-0\cdot278 \times 10^{-1}$. Note, however, that the ratio of $\varphi/\Delta\sigma$ was similar for the glycerinated and intact muscle: $0\cdot148$ and $0\cdot140$ respectively ($\Delta\sigma = \sigma_0 - \sigma_\infty$). σ_0 (absolute) was in this case 5000 dyne/mm² for the glycerinated muscle and 3680 dyne/mm² for the intact one. This agreement between the values of $\varphi/\Delta\sigma$ was a constant finding and is borne out in Table 1 which summarizes the results obtained from 13 muscles yielding $\varphi/\Delta\sigma = 0\cdot13 \pm 0\cdot014$. This value appears to be independent of the absolute initial stress (σ_0). Zatzman,

* These results are from unpublished work by Axelsson, Högberg & Löwenhard. Some of the results have been communicated to the *First International Congress of Biophysics in Vienna* (1966).

TABLE 1

Muscle No.	σ_0 (dyn/mm²)	σ_∞/σ_0	Slope φ (1) = dσ/d ln t	$\varphi/\Delta\sigma$	Limit of linearity (ln t)	Slope φ (6) = dσ/d ln t
1	3,672	0·125	−1·1292	0·1477	4·5	−0·0390
2	4,788	0·275	−0·0969	0·1234	5·5	−0·0440
3	11,400	0·150	−0·1022	0·1202	5·8	−0·0223
4	5,000	0·245	−0·1023	0·1355	5·8	−0·0278
5	1,235	0·065	−0·1220	0·1305	4·9	−0·0538
6	1,080	0·375	−0·1000	0·1600	4·6	−0·0273
7	6,400	0·150	−0·1083	0·1274	5·3	
8	4,730	0·245	−0·0906	0·1200	5·5	
9	5,520	0·180	0·0859	0·1048	5·3	
10	5,460	0·250	−0·1007	0·1343	5·6	
11	4,000	0·250	−0·0986	0·1315	6·0	
12	4,230	0·130	−0·1160	0·1333	5·5	
13	3,490	0·150	−0·1250	0·1471	4·7	−0·0286
Mean value:			−0·106	0·132	5·3	−0·0347
total range of variation:			±0·023	±0·028	±0·75	±0·0157
standard deviation:			±0·013	±0·0136	±0·47	±0·01

Table 1 shows the following measured values: (a) The stress at the moment stretching was stopped (σ_0). (b) The equilibrium stress when changes in stress with time were negligible (σ_∞), expressed as proportional parts of σ_0 (σ_∞/σ_0). (c) The slope φ of the linear part of the curve. φ (1) is the slope after the first straining, φ (6) the corresponding slope after the 6th straining. (d) The quotient $\varphi/\Delta\sigma$ (it has been shown empirically that this quotient, at least concerning dead material, is more stable than the slope φ). (e) Limits of linearity in terms of ln t.

Stacey, Randall & Eberstein (1954) studied stress-relaxation in isolated arterial segments. They concluded that stress–relaxation 'is a phenomenon associated with the smooth muscle component of the artery'. They found it non-exponential but 'a successful plot was obtained when pressure (or wall tension) was plotted vs. the logarithm of time. On such a plot, the time course followed a straight line relationship over two or more log cycles of time'. In this respect there appears to be a similarity between taenia coli and arterial segments.

Both creep and stress-relaxation in taenia coli can be quantitatively described and, as claimed above, subjected to mathematical formalism applied to dead material (Trouton & Rankine, 1904; Feltham, 1961; Kubát, 1953, 1954, 1965). During stretching internal stress gradients are built up, mainly due to visco-elastic properties of the system. In general terms stress-relaxation may be described as a slow decrease of internal stress gradients due to the action of local shear forces which in turn depend on the total stress. A formalized description based on these considerations would predict: (1) A general type of stress-strain behaviour for a wide class of material. (2) A linear phase for stress-relaxation. (3) A linear phase for creep. (4) A sigmoid shape of stress-relaxation curves. Although our results so far appear to be in good agreement with these predictions it must be remembered that they only describe the behaviour of a system as a function of experimental variables. No assumptions are made about intrinsic properties.

Attempts have been made to analyse more closely the events during the first seconds after stretch was terminated. Experiments were performed on both active muscles, i.e. muscles which were depolarized and allowed to shorten before stretching was started (Fig. 6a) and on inactive muscles (calcium-free EGTA solution) at L_0 (Fig. 6b). Here the tension was plotted on a logarithmic

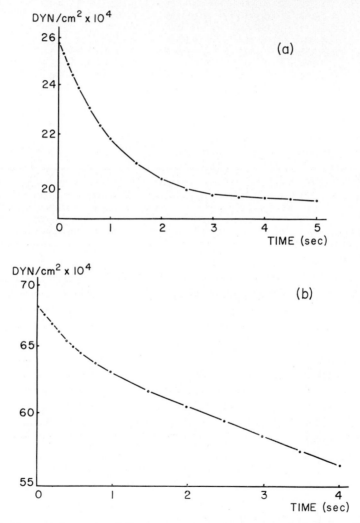

FIG. 6   Decrease in tension following stretch at constant speed. Ordinate: Tension plotted on a logarithmic scale. Abscissa: Time after stretch in sec.   (a) Active muscle.
(b) Inactive muscle.

scale and followed for five seconds. In all muscles studied the decrease in tension could be expressed as the sum of two exponential curves.

Figure 7 shows the analysis of the time course of the two exponential components. The slow component was obtained from the slope of the linear portion of the decay curve. This was extrapolated back and the tension values subtracted from those in the earlier part of the decay curve to give the time course of the first component (Högberg, unpublished result).

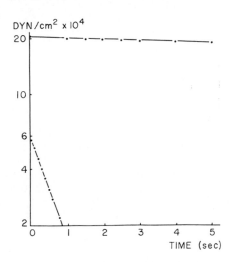

FIG. 7   Analysis of curve a in Figure 6a. The ordinate represents tension plotted on a logarithmic scale. The abscissa is time after end of stretch in seconds (see text).

It has been emphasized that the same general behaviour holds for creep and stress–relaxation. Furthermore that both preceded plastic deformation. Figure 8 gives a quantitative presentation of these characteristics. Figure 8a shows results obtained from a glycerinated muscle piece, Figure 8b the results from an intact piece in which the mechanical response was abolished by exposure to calcium-free EGTA solution. During repeated stretches up to a certain maximum stress, there was a progressive loss of plasticity as a function of the number of stretches. The 'viscous processes' became, so to say, more reversible. Finally both creep and stress–relaxation occurred without measurable plastic deformation at lengths between 130 to 145% L_0 (original). After three stretches the muscle in Fig. 8a showed almost purely visco-elastic behaviour.

Elasticity

It is of course meaningless to make calculations of 'elastic moduli' while plastic changes are still occurring. Even at stages when the muscles show almost purely visco-elastic behaviour the nonlinearity of the stress-strain diagrams, found by all workers using widely-different conditions, makes the commonly

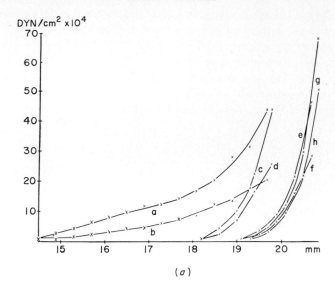

(a)

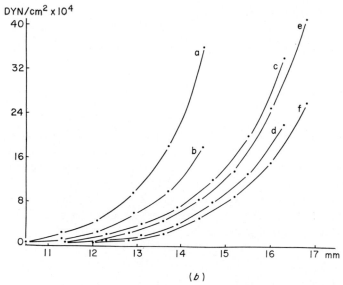

(b)

FIG. 8   Stress-relaxation and plastic changes during repeated stretches. (a) Glycerinated muscle. (b) Intact muscle with mechanical response abolished by exposure to calcium-free EGTA solution. On the abscissa the actual length of the muscle is plotted in millimetres. Stress is plotted on the ordinate. At each nominal strain two stress values are plotted. The upper shows the stress at the moment elongation was stopped. The lower shows the stress 30 min later. Curves a and b show the strain–stress relationship during the first stepwise stretch. At the end of curve b the muscle was relaxed until approximately the original stress was reached. Then from this new L_0, the muscle was again stretched stepwise (curves c and d). This was repeated (curves e–f and g–h).

used definition of Young's modulus $E_Y = (F/A)/(\Delta L/L_0)$ not applicable. One might of course derive a specific function $(dF)/A \ |(dL)/L = G(E)$ for calculations of 'moduli' for each given length. The interpretation of this 'modulus' would still be questionable.

Surely the only thing that can be said with certainty is that this muscle is capable of storing elastic energy. Whether it is homogeneously stored or located in specific structural components is not known. With regard to the elastic behaviour of taenia coli, references will be made to some values from recent work of Åberg (1967). These values give information about the elastic behaviour of taenia in the given experimental situation but, like those from our earlier work, do not lend themselves to generalizations. To avoid the theoretical difficulties attached to the term 'elastic moduli', Åberg omits cross-sectional area and absolute length used in the formula for Young's modulus; and due to the nonlinearity of the load extension curve he uses dT/dL, which he calls 'stiffness' according to Buchthal & Rosenfalck (1957), i.e. the compliance of a material as exhibited in the variable slope of the load extension curve. Using the method of quick release he obtained non-linear force-extension curves which he supposed to be determined by what he calls 'series elastic components of the muscle'. In agreement with Aubert (1955), working on striated muscle, and Parmley & Sonnenblick (1967), working on heart muscle, he found that the curve had an exponential shape within the investigated range. He also shows force-extension curves obtained by stretching intact muscle pieces at a constant rate. This rate was chosen in accordance with the maximum velocity of shortening found by Åberg & Axelsson (1965), i.e. 30% of L_0 per sec. For numerous reasons descussed earlier one would not expect them to follow Hook's law. Comparison of 'stiffness' thus measured with that calculated from the quick release experiments do not contribute to a clearer picture of the mechanical properties of taenia coli.

Having thus stressed the need for careful consideration of the inherent variables of our mechanical system, one can proceed to the attempts to repeat the experiments of 'classical striated muscle mechanics'. Most of the results to be described are those of Åberg & Axelsson (1965); Axelsson, Högberg & Timms (1965); Åberg (1967) and Beviz, Mohme-Lundholm & Vamos (1969).

Attempts to make 'classical' mechanical measurements on taenia coli

All the values referred to are obtained *in vitro* in what we call either 'isotonic' or 'isometric' conditions. It will prove useful to consider the limitations of these terms — both those deriving from our method of measurements and those that are inherent and unavoidable (A. V. Hill, 1965).

Let us first consider how the requirements of 'isotonicity' are methodically fulfilled in our experiments. Neglecting the frictional forces and the mass of the muscle, the general equation for the shortenings described here can be written $P = F + M\dot{v}$. (M = effective mass. $\dot{v} = d^2x/dt^2$ where x = amount of

11*

shortening and t = time). To approach the conditions of isotonic shortening where $P = F$, the term $M\dot{v}$ must be small compared to F (here the terms P and F are used in the conventional sense: P = muscular force, F = load). In the experiments described here the forces due to inertia would be small compared with P_0 at the velocities present (P_0 = maximum isometric force at L_0). During acceleration from zero velocity to the velocity of a few millimeters per second however, $M\dot{v}$ may sometimes become sufficiently great compared with a small F to cause a deviation from the conditions of 'isotonic' contraction. Besides these deviations one may recall Hill's comments on the inherent and unavoidable deviation from truly isotonic contraction, which really never occurs. At the same time he stresses that a contraction is never truly isometric, not even in an idealized muscle (Hill, 1965), and taenia coli is far from being 'ideal' in Hill's sense.

The reader realizes of course, how misleading the use of such terms as 'in physiological condition' may be when what is actually meant, is the isometric condition *in vitro*. The similarities between this condition and that *in situ* must be limited indeed. In the body the taenia contracts neither isotonically nor isometrically, not even in the sense determined by our recording methods *in vitro*.

After these reservations the following measurements may be listed:

The amount and velocity of isotonic shortening. Taenia coli loaded with 100 dynes shortens by 60 to 80%. The maximum velocity of shortening of an unloaded taenia was extrapolated to 30% of L_0 per sec (Axelsson, 1964).

The relationship between external force and velocity of shortening. Åberg & Axelsson (1965) finally obtained curves which obeyed Hill's equation: $(P + a)$ $(v + b) = (P_0 + a)b$ = constant. Graphical determination according to Katz (1939) of the maximum velocity of shortening against different afterloads of 19 mm long muscle gave 800 dynes for a, and 0·24 mm/sec for b. In our opinion the many failures and deviations from predicted behaviour could often be traced back to changes in the metabolic rate in the course of an experiment. The view that generation of energy may be a factor limiting the rate of contraction is not new: see Fenn (1924, 1957), Hill (1938), Wilkie (1954). These ideas were supported by the findings of Axelsson, Högberg & Timms (1965) who studied quantitatively the effects of the removal of glucose on the velocity of shortening under 1000 dynes load and on the rate of isometric tension development.

Maximum isometric tension at L_0 and velocity of shortening. Hill's equation rearranged into the form $V = (P_0 - P)b(P + a)$ suggests that the velocity of shortening depends on the ratio P_0/P. (P_0 means here and elsewhere in this chapter the greatest load that the muscle can lift, i.e. the maximum isometric tension at L_0). That this holds true was clearly demonstrated for taenia coli in experiments in which isometric and isotonic recordings were obtained from the same muscle (Fig. 9).

Muscle length and rate of isometric tension development. The tension developed per unit time was found to increase with the force on the muscle and thus to increase with muscle length up to a given limit for each muscle.

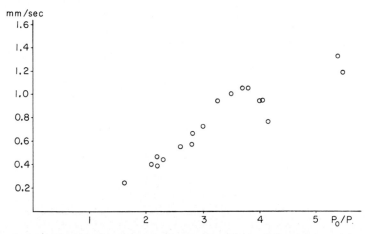

FIG. 9   The relationship between velocity of shortening and P_0. Abscissa: the tension in each contraction as a fraction of the maximum tension developed by that muscle at L_0. Ordinate: maximum velocity of shortening. (Åberg & Axelsson, 1965)

Work. Maximum isotonic or maximal physical work was performed with a constant load of 30% of $P_0 \pm 5\%$ (Fig. 10). The maximum physical work varied in these experiments between $(P_0L_0)/7\cdot0$ and $(P_0L_0)/8\cdot2$. Åberg (1967) tried to make corrections for the internal work and found that the isometric work performed at $P_0/3$ was approximately 10% of the maximum physical work. (Maximum isometric work is defined by him as the isometric work performed at L_0 when the series elastic elements exert maximum isometric force.) The total work (the sum of isometric and isotonic work) performed by taenia at $P_0/3$ would accordingly be between $(P_0L_0)/6\cdot4$ and $(P_0L_0)/7\cdot5$. Csapo (1954) found a value $(P_0L_0)/6$ for rabbit uterus which was the same as that for striated frog muscles.

Length-tension relationship. The length-tension curves in taenia were in principle similar to those obtained in other muscles (see Åberg & Axelsson, 1965; Mashima & Yoshida, 1965; Csapo, 1962; and for vascular smooth muscle Lundholm & Mohme-Lundholm, 1966).

Attempts to relate the supply and use of energy to mechanical activity

Attempts to relate mechanical and electrical activity of taenia coli quantitatively to its content of glucose and glycogen have been referred to above.

Normally the tissue glucose ranges from $0\cdot7$ to $1\cdot1$ mg/g wet wt. In glucose-free solution the fall in tissue glucose is fast, more than 75% disappears in

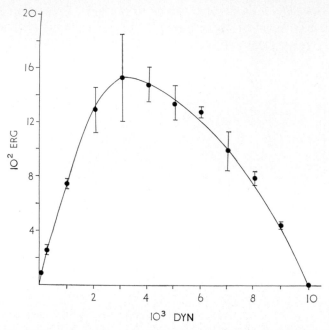

FIG. 10   Physical work. Abscissa: load. Ordinate: work. Vertical lines show standard deviations. Each period is the mean of four observations. (From Åberg & Axelsson, 1965. By courtesy of *Acta physiol. scand.*)

3 min. Changes in the glucose concentration of the bathing solution are immediately reflected in the glucose content of the muscle and simultaneously in the frequency of discharge of action potentials. The dissociation of spontaneous electrical and mechanical activity in glucose-free solution (described above) does not occur until the glucose content of the muscle is extremely low. The decrease in mechanical response is thus not parallel to the fall in glucose. This did not, however, appear to be sufficient evidence to rule out the possibility of parallelism between tissue glucose and mechanical activity. Consequently the effects of removing and readmitting glucose on the mechanical response to supramaximal stimulation (128 mM $[K^+]_0$) were studied. In isotonic conditions the physical work performed by muscles mounted under a 5000 dynes load fell gradually during 3 hr exposure to glucose-free solution from approximately 5×10^3 erg to less than 1×10^3 erg, i.e. the work was reduced by more than 80%. The effects of removing glucose on maximum shortening, velocity of shortening and the subsequent recovery of the mechanical response in the presence of glucose have all been described in detail (Axelsson, Högberg & Timms, 1965). Parallel measurements were also made in isometric conditions. Although no parallelism between the fall in mechanical response and tissue glucose was found, the possibility remained that such parallelism might be demonstrated

with respect to the glycogen content of the muscle. The rate of glycogen depletion under glucose-free aerobic conditions was therefore determined. The glycogen content of the muscle appeared to fall slowly, i.e. in the course of 1 hr by 15 to 50%. There was, however, a more rapid initial phase in the fall of glycogen content during which the rate appeared to be similar to that of the fall in mechanical response.

The idea that 'the chemical reactions which produce muscular energy are themselves controlled by the force on the muscle' (see Wilkie, 1957) seems to gain support from recent experiments in which pieces of taenia coli were depolarized and allowed to shorten until they only maintained slight force. The muscles were then stretched to varying degrees at a constant rate. When the stretch had been terminated, the usual stress–relaxation began — but continued only for a few seconds, to be gradually reversed into active tension development. The active development of tension at the new stress or force was abolished by metabolic inhibitors (Axelsson & Högberg, unpublished). Furthermore, Åberg & Axelsson (1965) found (within certain limits) a great increase in work per unit time with increasing force on the muscles.

From determinations of hydrolysis of energy-rich phosphate compounds and of lactate formation, the energy used by taenia during an isometric contraction could be calculated (Beviz et al., 1969). Using the relationship LP/H (Hill, 1958) (L = muscle length in cm; P = increase in tension in g wt/cm^2 of the cross-sectional area; H = energy used expressed in g cm), they obtained a maximum value of 2·4. This value is much lower than the value, 4·8, obtained by the same group of workers for mesenteric arteries (Beviz & Mohme-Lundholm, 1969), which in turn is only about half that which Hill (1958) found for striated muscles using thermoelectric measurements.

Mechanical measurements *in vivo*

Continuous recording of gastro-intestinal mechanics in conscious animals using an extra luminal strain-gauge force transducer was described and discussed by Reinke, Rosenbaum & Bennett (1966). The same authors (1967) have used the method for analysis of wave shape, force, frequency and velocity of single contractions. For a description of the results obtained, the reader is referred to the quoted papers. The authors stress the advantage of having succeeded in quantifying gastro-intestinal muscle force *in vivo* thus making it possible to compare 'these data with other measurable physiological properties of such muscle *in vitro*'.

They have made comparisons between their results and those obtained *in vitro* by Åberg & Axelsson (1965). The similarities are encouraging.

APPENDIX
by G. Gudmundsson

A linear model of the relationship between electrical and mechanical activity is given by the convolution integral

$$Y(t) = \int_{-\infty}^{\infty} W(t - t') \, X(t') \, dt' \tag{1}$$

where Y is the mechanical, X the electrical response and $W(t - t')$ is the impulse response function describing the influence of $X(t')$ on $Y(t)$.

The records are digitized and the Fourier transforms of X and Y estimated numerically by methods described by Gudmundsson (1966, 1968). Denoting the Fourier transforms of X, Y and W by x, y and w respectively where

$$x(u) = \frac{1}{2\pi} \int_{-\infty}^{\infty} X(t) \exp(-iut) \, dt, \text{ etc.} \tag{2}$$

we obtain from (1)

$$w(u) = \frac{y(u)}{2\pi \, x(u)} \tag{3}$$

where $w(u)$ is the transfer function.

In practice the integration in equation (2) is carried out for each burst separately, assuming that X and Y are zero except for the duration of the burst.

Biophysical interpretation of the results is yet at an early stage, but estimates of the Fourier transforms in the lowest frequencies demonstrate the dissociation of the electrical and mechanical activity under the influence of noradrenaline.

The Fourier transforms are complex values and can be written:

$$x(u) = z_x(u) \exp(i\varphi_x(u)) \tag{4}$$

with an analogous notation for y and w. Here φ and z are real functions and z is positive. From equation (3) follows:

$$z_w(u) = \frac{z_y(u)}{2\pi z_x(u)} \quad \text{and} \quad \varphi_w(u) = \varphi_y(u) - \varphi_x(u) \tag{5.6}$$

Table 2 shows estimates of $x(0)$, $y(0)$ and $w(0)$ from one experiment. The values of $x(0)$ and $y(0)$ are multiplied by 2π whereby they provide the values of the integrals of X and Y over one burst according to equation (2).

The increase in tension due to noradrenaline is thus not accompanied by a corresponding increase in the electrical activity. The average value of the standard deviation within each group for $z_x(0)$, $z_y(0)$ and $z_w(0)$ are 12, 11 and 10% of the average value respectively. Estimates of z at wavelengths of the order of the duration of one burst yield similar results, but at shorter wavelengths the variation within each group increases sharply. It is worthy of attention that the

TABLE 2

Group	$2\pi x(0)$ mV sec	$2\pi y(0)$ dyn sec	$w(0)$ dyn/mV
I	11·7	640	8·7
II	13·9	955	10·9
III	10·5	865	13·1
IV	9·8	825	13·4

Explanations: I. Normal Krebs; II. Immediately after administration of noradrenaline; III. 5 min after administration of noradrenaline; IV. 10 min after administration of noradrenaline. Each value is the average of four consecutive bursts.

variance of z_w is similar or smaller than that of z_x and z_y. If X and Y were independent apart from the fact that the bursts occur simultaneously, the variance of z_w would be greater according to equation (5). The estimates of the moduli of the Fourier transforms thus indicate that within the time interval of one group, supposedly short enough that only a negligible change in the drug effect takes place, there is proportionality between the electrical and mechanical activity.

Noradrenaline has not been found to affect $\varphi_w(u)$ nor the shape of $z_w(u)$ apart from the multiplicative factor, changing slowly with time according to the Table 2. Variation in the estimates of $w(u)$ within the groups is large except at the lowest frequencies. An eventual change in the shape of the transfer function, caused by noradrenaline, might therefore not be detectable without analysing a larger number of bursts than we have had an opportunity to carry out so far.

Department of Mathematics,
University of Manchester,
Institute of Science & Technology

10

SLOW RHYTHMS IN SMOOTH MUSCLE (MINUTE-RHYTHM)

K. GOLENHOFEN

The rhythmical activity of smooth muscle varies in frequency over a wide range. Even in smooth muscle of homoiotherms, to which the present section is limited, the duration of a period may range from a fraction of a second to many hours. For the analysis and the classification of these rhythms it is necessary to consider not only the duration of the period but also additional functional characteristics.

Modern techniques in electrophysiology allow, firstly, a separate treatment of the fastest oscillation, i.e. the short membrane depolarizations known as 'spikes' or 'action potentials'. These spikes obviously trigger the development of mechanical tension, and thus provide the basic element of every mechanical activity. However, at the usual frequency of about 1/sec, these individual twitches merge into a tetanic contraction, and in the mechanical record the spike frequency is usually not evident. Thus, the tension changes in smooth muscle are generally modifications of tetanic activity (Fig. 1).

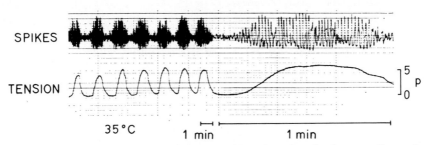

FIG. 1 Electrical activity (extracellular records) and tension development (in ponds; 1 pond = 981 dynes) of isolated smooth muscle (taenia coli of the guinea-pig), recorded at low speed on the left and at higher speed on the right side. The picture shows the two main rhythms of the spontaneously active tissue: the spike rhythm with a frequency of about 1/sec and the minute-rhythm of about 1/min. Tissue in Krebs solution with 5% CO_2. (Golenhofen & von Loh, unpublished)

In contradistinction to the high frequency of rhythmical spike discharge the rhythmical movements could then be called 'slow rhythms'. Older observations, however, which were necessarily limited to mechanical recordings, called the most frequent discernible rhythms 'fast rhythms'. Even today this classification is still in use; and on the basis of visible rhythmical movements waves of type I, II and III are distinguished (Truelove, 1966). According to this classification, the 'slow rhythms' begin with the 'minute-rhythm' (type 2 contractions). This will be the main topic of this section.

It seems necessary, however, to draw a certain line of demarcation also with regard to the very slow rhythms, such as diurnal and annual rhythms, which are processes involving the entire organism (see p. 323, Fig. 8). They include, of course, the smooth muscle, but only in a secondary manner. Thus 'long-wave rhythms' are not specific, intrinsic rhythms of the smooth muscle itself. The same is true for the submultiples of the diurnal rhythm (periods of 6, 8 or 12 hr).

SLOW RHYTHMS OF SMOOTH MUSCLE *in situ*

Digestive tract
Stomach
The fastest movements of the human stomach are peristaltic waves, which show a remarkable regularity and vary only slightly in frequency. The older

literature gives the following normal values for the interval between peristaltic waves (cited from Klee, 1927): 18 to 21 sec (Dietlen, 1913), 22 sec (Kaestle, 1913), 20 sec (Kaufmann & Kienböck, 1911; Fraenkel, 1926). An apparent deceleration can be produced if individual waves fail to occur. The continuation of the basic frequency under these conditions is shown by the fact that all intervals are multiples of 20 to 22 sec (Davenport, 1966; Code, Hightower & Morlock, 1952).

Slower fluctuations of tone, with period durations in the minute range, are superimposed on the peristaltic waves (Fig. 2). These slow rhythms have also been observed at an early date, but Weitz & Vollers (1925; 1926) were the first to examine the minute-rhythm systematically and to recognize its significance. Since the peristaltic waves pass across the stomach, they show a certain phase shift if recorded simultaneously with several small balloons. On the other hand, if a single large balloon is placed in the stomach, the minute-rhythm can be

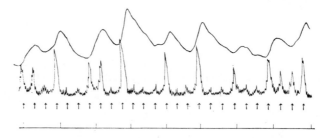

FIG. 2   Fluctuations in the activity of the human stomach: minute-rhythm above, peristaltic waves below. The peristaltic waves are marked by arrows; they are sometimes abortive. Time in minutes. Minute-rhythm recorded with a balloon filled with about 150 ml water, peristalsis with a balloon filled with 10 to 30 ml air. (Weitz & Vollers, 1925)

demonstrated in the integrated total tone of the stomach (see Fig. 2). It is irrelevant whether the minute-rhythm is interpreted as a fluctuation of the basic tone on which more or less uniform peristaltic waves are superimposed, or whether it is regarded as a summation of peristaltic waves with rhythmic variations in intensity.

Very slow fluctuations with a periodicity of 1 hr or more are often superimposed on the minute-rhythm of the stomach. Such hour-rhythms were observed by Boldireff (1904) in starved dogs. He found phases of activity in the stomach at intervals of 1·5 to 2·5 hr; each phase lasted for 20 to 30 min and consisted of 10 to 20 rhythmic contractions. The findings of Weitz & Vollers (1925) in humans are in close agreement with Boldireff's. However, they found that hunger and an empty stomach were not prerequisites for these fluctuations.

Small intestine

In the small intestine, rapid segmental movements predominate. In the human duodenum, their frequency is 11/min, with less than 1% variation (Davenport,

1966). The frequency varies from one species to another, and appears to be correlated with the size of the animal (Trendelenburg, 1927). In the dog it is 18/min, and decreases to 5 to 10 per min in the direction of the ileum (Daniel, Honour & Bogoch 1960; Bass, Code & Lambert, 1961). The segmental movements are associated with fluctuations of the membrane potential (basic electrical rhythm, slow waves), upon which spikes are superimposed. The number of spikes in a basic electrical wave determines the intensity of the contraction (Daniel, Wachter, Honour & Bogoch 1960; Bortoff, 1961a; Bass *et al.*, 1961; Davenport, 1966). Also in the small intestine, slow fluctuations in tone with a minute-rhythm are superimposed upon the rapid segmental movements, as described above for the stomach.

Large intestine

The reports on the rhythmic activity of the colon are not in agreement (cf. Truelove, 1966; Connell, 1968; Gillespie, 1968). Apart from differences in experimental procedures, the reason for this lack of agreement may be the

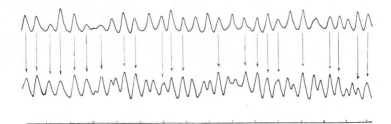

Fig. 3   Minute-rhythm of the human descending colon, recorded with two water-filled balloons 20 cm apart. Time in minutes. (Weitz & Vollers, 1926)

occurrence of several intermediate forms of different rhythms. Since the investigations of Weitz & Vollers (1926), however, there can be no doubt that the colon also possesses a marked minute-rhythm. Moreover, this has sometimes been found to be co-ordinated with the minute-rhythm of the stomach. In simultaneous recordings of fluctuations in tone with two balloons 20 cm apart in the descending colon, there is generally co-ordination as shown in Fig. 3. Frequently, the peaks of the curves coincide exactly, but at times the patterns diverge. Such 'relative co-ordination' (von Holst, 1939) is typical for autonomous rhythms which are connected by coupling mechanisms. In conscious dogs a marked co-ordination of the minute-rhythm in the colon (type 2 contractions) was seen by Templeton & Lawson (1931). The procedure used by these authors also revealed a faster rhythm (type 1 contractions) with a frequency of about 6 to 8 per min.

There are apparently no differences in minute-rhythm from one species to another. Figure 4 shows a minute-rhythm in the fluctuations of activity in the caecum of the guinea-pig *in situ*.

Very slow fluctuations of activity with a period duration in the range of hours have also been observed in the intestine (type 3 contractions, Templeton & Lawson, 1931).

Ritchie, Ardran & Truelove (1962) classified the intrinsic rhythms of the human colon as follows: (1) 5 sec-rhythms, (2) 30 sec-rhythms, (3) slow basic pressure changes. The 30 sec-rhythms correspond to the minute-rhythms of the present discussion.

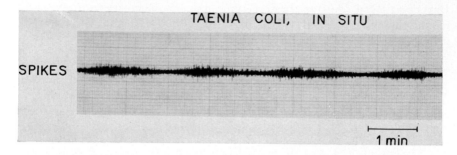

FIG. 4   Electrical activity in the taenia coli of an anaesthetized guinea-pig, measured extracellularly with platinum wires in the tissue. Activity is indicated by superimposed spikes, producing a broadening of the record. (Golenhofen & Wienbeck, unpublished)

Urogenital tract

Minute-rhythm is also found in the smooth muscle of the urogenital tract. In the fluctuations of uterine activity it is particularly strong. Figure 5 shows this in a pregnant rabbit uterus. Weitz & Vollers (1926) observed minute-rhythms also in the human uterus during pregnancy; in the urinary bladder, the scrotum, and the penis.

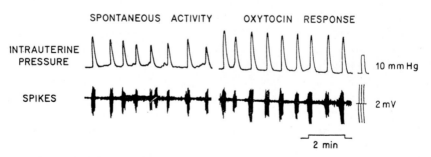

FIG. 5   Electrical activity (spikes) and intrauterine pressure in a pregnant rabbit, 12 hr after removal of the ovaries and placentae. (According to Takeda & Csapo, 1961; from Csapo, 1962)

Circulation

Circulatory rhythms, evident as fluctuations in blood pressure, are usually classified as follows: (1) The fastest, i.e. the pulse waves, as blood-pressure waves of the 1st order; (2) the respiratory waves as blood-pressure waves of the 2nd order; (3) blood-pressure waves of the 3rd order. These are usually slower than the respiration. In humans they have a frequency of 5 to 6 per min (10 sec-rhythm; Golenhofen & Hildebrandt, 1958). They are also called Traube-Hering-Mayer waves (cf. Koepchen, 1962).

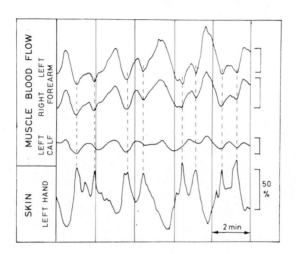

FIG. 6   Spontaneous fluctuations of muscle and skin blood flow in human limbs; simultaneous measurement in four areas. For rough calibration, 50% deflections from the average resting blood flow are indicated. (Golenhofen, 1962a)

In addition, a distinct rhythm is found in measurements of organ blood flow. An example is given in Fig. 6. These fluctuations might be called circulatory rhythms of the 4th order. The splenic rhythms which have been examined in animal experiments (Roy, 1881; Barcroft, Khanna & Nisimaru, 1932; Barcroft & Nisimaru, 1932a, b) belong to this area.

In records of human blood pressure the minute-rhythm is normally not perceptible, since the fluctuations of blood flow through various organs run in opposite direction and thus compensate each other (Golenhofen & Hildebrandt, 1957a, b). This compensation takes place mainly between muscle and skin, as shown in Fig. 6. The blood flow through the skin changes in the opposite direction to that through three muscle areas. The slow fluctuations in skin blood flow are already described by Burton (1939), Burton & Taylor (1940), and Aschoff (1944). The rhythms of intestinal blood flow are not always strictly synchronous with those of the skin and muscle. If there is co-ordination, they generally run

in the same direction as those of the skin, and opposite to that of the muscle
(Graf, Graf & Rosell, 1958, 1959).

Since the minute-rhythm fluctuations in blood flow are not caused passively
by fluctuations of the blood pressure, they must be attributed to fluctuations in
vascular tone; this means that they are smooth-muscle rhythms. Vascular tone
is likewise involved in the blood-pressure waves of the 3rd order, whereas the
fast waves of the 1st and 2nd order can no longer be considered to be vascular
rhythms. Burch, Cohn & Neumann (1942) and Burch & De Pasquale (1960),
also begin their classification of vascular rhythms with waves of the 3rd order
(α waves), and call the slower waves with a frequency of 1 to 2 per min β waves.
Accordingly, the α waves of the blood vessels correspond to type 1 contractions
of the intestine and the β waves to type 2 contractions of the intestine.

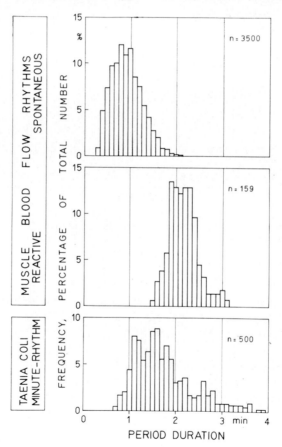

FIG. 7   Histograms of the period durations of muscle blood-flow rhythms in man
(spontaneous minute-rhythm (*top*) reactive fluctuations (*middle*)) and of the minute-
rhythm of isolated taenia coli (*bottom*). (Upper part from Hildebrandt & Golenhofen,
1958; middle part from Golenhofen, 1962*a*; lower part from Golenhofen, 1966*a*)

In addition to the spontaneous rhythmical changes in blood-flow, there are even slower fluctuations which appear as damped oscillations after a sudden change in blood flow and which can be included with the minute-rhythm on the basis of their period duration (Golenhofen, 1962). These reactive oscillations are particularly clearly seen in human muscle blood flow; while they are scarcely perceptible in skin blood flow. In contrast to spontaneous rhythms, these reactive oscillations are confined to the site of action, and are obviously of a local nature. The period duration is approximately twice as long as that of the spontaneous rhythm of the muscle, as shown by the histograms in Fig. 7.

Comparisons with heart rhythm

Because of its close functional relation with smooth muscle, the heart will be discussed briefly here.

Periodic fluctuations in the spontaneous activity of the heart were first noted by Luciani (1873), although in a frog's heart, and under rather 'non-physiological' conditions. Schellong observed such Luciani periods in a dying human heart (Rothberger, 1926). In normal conditions the heart is only slightly involved in the minute-rhythm of the circulation. If it is involved, the minute-rhythm fluctuations in heart frequency run concurrent with the fluctuations in muscle blood-flow. The amplitude may amount to as much as $\pm 10\%$ of the average (Golenhofen & Hildebrandt, 1957a). The marked periodic phenomena of heart activity, which can be evoked experimentally, will be discussed on p. 328.

Conclusions

The rhythms of smooth muscle can be classified as follows:

1. A high-frequency spike rhythm, the frequency of which varies around 1/sec over a relatively wide range. It is determined mainly by the characteristics of the excitable membrane, and is usually not apparent in records of mechanical activity.

2. A rapid movement rhythm (type 1 contractions, α waves in blood vessels), which is specific for particular organs and species, and which covers roughly the frequency range of 3 to 20 per min.

3. A slower rhythm (type 2 contractions, β waves in blood vessels), which is similar in all smooth muscle and in all species, and which has a preferred frequency of ca. 1/min. This 'minute-rhythm' is an important basic rhythm in the whole smooth muscle system.

4. A very slow rhythm with period durations in the range of hours (type 3 contractions), about which relatively little is known.

SMOOTH-MUSCLE RHYTHMS AND THE TOTAL SPECTRUM OF BIOLOGICAL RHYTHMS

Figure 8 shows the rhythms found in humans, represented in the order of their period duration. The smooth-muscle rhythms are in the middle part of the total spectrum. It is noteworthy that the rhythms become more and more complex with increasing period duration, from the rhythmic action of individual cells ('nervous action') to rhythms which include the whole organism ('organism' and 'complex regulations' in Fig. 8). In addition, according to Hildebrandt (1961, 1967) there are systematic changes in frequency variability throughout the spectrum. For the high-frequency rhythms, a continuous variability over a wide range, often without any preferred frequency, is characteristic. On the other hand, the frequency of the very slow diurnal and annual rhythms is rigidly

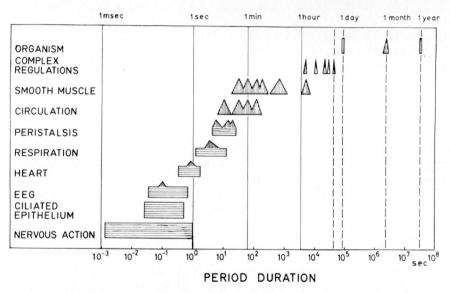

PERIOD DURATION

Fig. 8 Spectrum of rhythmic functions in man. Period duration on a logarithmic scale. Ranges with continuous variability in frequency are indicated by horizontal hatching. The vertically-hatched triangular peaks indicate a preferred or 'normal' frequency. Several preferred frequencies with whole-number relations are typical for the middle range and are indicated by multipeaked areas. (Hildebrandt, 1967)

fixed. The middle range covers all conceivable intermediate steps between the two extremes. Heart and respiration have a distinct frequency norm (represented in Fig. 8 as small triangular peaks), from which the organism can, however, deviate during stress (represented as wide rectangular area in Fig. 8). Smooth muscle belongs to the next group with a pronounced constraint in frequency. Multi-peaked histograms are typical; although the frequency may change over a relatively wide range, the basic intrinsic frequency is so strong that simple whole-number multiples are preferred (in Fig. 8 indicated by multipeaked areas).

It can be demonstrated that a whole-number frequency co-ordination exists between the different rhythms of the middle range, as shown in Fig. 9. The minute-rhythm of smooth muscle may well serve as a guide line in this co-ordinated middle range of biological rhythms and may play an important role in 'biological clock' mechanisms.

The characteristics which apply to the whole spectrum are also found in the narrower field of smooth-muscle rhythms. The spike rhythm, as the fastest oscillation, is in principle a cellular rhythm, the frequency of which can be changed continuously over a relatively wide range in order to produce a graded

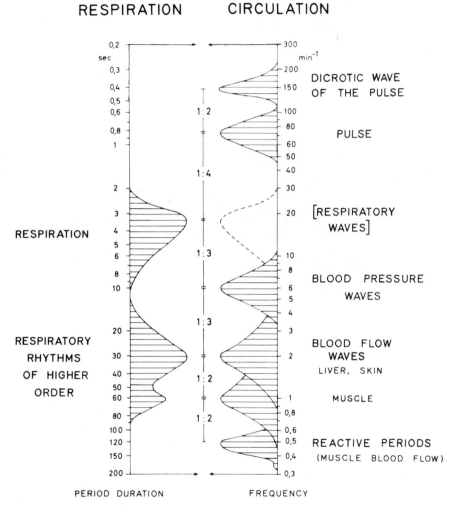

FIG. 9 Semidiagrammatic histograms of respiratory and circulatory rhythms in man, calibrated in period duration (*left*) and in corresponding frequency (*right*). All rhythms are co-ordinated in such a manner that the normal frequencies of neighbouring rhythms have a simple whole-number relation, as shown in the middle. 'Respiratory rhythms of higher order' indicate periodic fluctuations of the respiration (analogous to Cheyne-Stokes-Respiration). (Hildebrandt, 1967)

force of contraction. The rapid movement rhythms (type 1 contractions) involve parts of the organ either in a particular sequence, or alternately. The frequency may be quite constant for one special area, but in relation to the entire organism, and even more so comparing different species, the frequency variability is still relatively large. In contrast, the frequency of minute-rhythm (type 2 contractions) is very rigidly fixed; whole organs, and even groups of organs, are synchronized in the minute-rhythm oscillations, and there is almost no difference between different species. Greater changes in frequency are possible only by whole-number multiplication of the basic frequency as will be discussed in the following sections.

MECHANISMS OF MINUTE-RHYTHM IN SMOOTH MUSCLE

Intestinal smooth muscle

It was demonstrated long ago that even isolated parts of the intestine can carry out rhythmic contractions. The possible influence of surviving nervous elements can be eliminated by drugs. Of late, tetrodotoxin is preferred for this

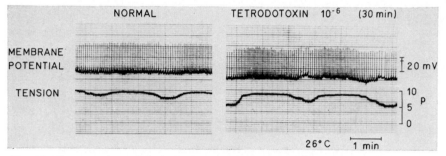

FIG. 10   Spontaneous activity in the isolated taenia coli. Spike- and minute-rhythm are not influenced by tetrodotoxin (10^{-6} g/ml). Intracellular record of the membrane potential above, and tension development below, calibrated in ponds.

purpose; it selectively blocks the excitability of nerve fibres, but does not influence the smooth muscle itself (Mosher, Fuhrman, Buchwald & Fischer, 1964; Nonomura, Hotta & Ohashi, 1966; Bülbring & Tomita, 1967). The example in Fig. 10 shows that not only the spike rhythm is of myogenic origin, as has been repeatedly demonstrated (cf. Bülbring, 1955; Schatzmann, 1964a), but the minute-rhythm as well.

The longitudinal muscle of the guinea-pig (taenia coli), which was used for the experiment shown in Fig. 10, is, however, a preparation with a particularly marked automatic myogenic activity, and the findings may therefore not be completely applicable to all parts of the intestine.

The question remains, whether the minute-rhythm *in situ* is entirely based on fundamental myogenic processes, or whether neural and hormonal influences are involved. The synchronization, which is sometimes quite strict, might suggest a neural control. However, the relative freedom of individual parts (see Fig. 3) indicates a decisive involvement of myogenic automaticity.

If the myogenic component predominates, the coupling mechanisms which link the myogenic rhythms of the various parts of an organ, must be explained. First, there is the possibility of mechanical coupling effects. It has been shown that rhythmic stretching of the taenia coli can induce a predetermined rhythm in the tissue; this was the easier the closer the applied extrinsic frequency agreed with the intrinsic frequency (Golenhofen, 1964, 1965). Since there is probably a close agreement of intrinsic frequency in various parts of the intestine, mechanical coupling effects alone could provide quite good synchronization. The electrical transmission of excitation in smooth muscle as well as local reflexes could augment this coupling. In addition, a central control through neural and hormonal mechanisms must also be assumed, since vegetative nervous activity exhibiting minute-rhythm is known to exist in the circulatory rhythms (see below). The normal minute-rhythm *in situ* is probably based on a synergistic action of all these factors.

Uterine smooth muscle

If there is any spontaneous activity present in isolated strips of uterine smooth muscle, there are always fluctuations with a very strong minute-rhythm (Marshall, 1959; Csapo & Kuriyama, 1963; Kuriyama, 1964b). Thus, the remarks on intestinal muscle in the previous section are also applicable here; the normal minute-rhythm of the uterus is inherent in the muscle as a fundamental characteristic and is modified *in situ* by humoral and neural influences.

Vascular rhythm

The minute-rhythm of peripheral blood flow stops if the innervation of the area under investigation is blocked (Golenhofen & Hildebrandt, 1957b). The synchronization of the blood-flow waves in different parts of the body make the existence of a central control mechanism very probable, since mechanical coupling effects and myogenic conduction alone would not suffice as coupling mechanisms over such great distances. However, it should not be concluded that the minute-rhythm is merely imposed upon the blood vessels, and that they have no rhythm of their own. Even in blood-flow rhythms, the individual parts have a certain freedom; the picture that emerges is that of a co-ordination which is only relative. Sometimes, even if two curves run synchronously, the maximal amplitudes measured from the two sites alternate (Golenhofen & Hildebrandt, 1957a). This indicates that here nervous activity interferes with local factors, and that the individual vascular sections respond differently to the nerve impulses, depending upon the phase of their intrinsic rhythm at the time.

It is a well-known physiological fact that isolated blood vessels exhibit spontaneous rhythmicity (Atzler & Lehmann, 1927). However, the older investigations, which were carried out on strips of larger vessels, and often under rather 'non-physiological' conditions, permit only limited conclusions as to the minute-rhythm *in situ*. More recent investigations, however, have shown conclusively that a spontaneous rhythm is part of the normal behaviour of blood vessels (Somlyo, Woo & Somlyo, 1965; Johansson & Bohr, 1966; Biamino, Vorthaler & Thron, 1967.

Of especial importance for our question are the studies by Seller, Langhorst, Polster & Koepchen (1967), who perfused an isolated flap of canine skin and found spontaneous fluctuations in resistance, with a mean period duration of 25 sec (Fig. 11) and, in the histogram, a secondary peak at twice the period duration. Continuous stimulation of the vasomotor nerves intensified the spontaneous fluctuations, or evoked them if they had not existed before.

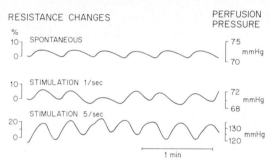

FIG. 11   Fluctuations of resistance in an isolated perfused skin area of a dog. Upper record: spontaneous. Two lower records: during stimulation of the vasoconstrictor nerves at different frequencies. (Seller *et al.*, 1967)

Rhythmic stimulation synchronized the oscillations within a limited range of frequencies above and below the spontaneous frequency. Independent oscillations of different vascular regions could be synchronized by synchronous rhythmic stimulation of adequate frequency. In these investigations it was also found that the spontaneous minute-rhythm is one of the most sensitive of functional phenomena. During prolonged perfusion it gradually disappeared before an impairment of the reaction to nerve stimulation could be detected.

In connection with the findings of Seller *et al.* (1967), we should interpret our findings in humans as follows. The minute-rhythm is inherent in both the circulatory centres and in the blood vessels themselves. The functional state of the vessels immediately after nerve block is probably a 'latent rhythmicity'. In particular, the fact that a sudden dilatation is followed by a reaction consisting of periodic contractions of local origin, indicated that the normal behaviour of the vessels is determined by an intrinsic minute-rhythm (Golenhofen, 1962a, b).

Comparisons with heart rhythm

It has already been mentioned that Luciani (1873) described periodic fluctuations in the spontaneous activity of the frog heart. Langendorff (1908) evoked these Luciani-periods in curarized frogs by functional isolation of the cardiac ventricle from the rest of the heart by crushing. Within a few hours, the heart, which had been left *in situ*, developed a very regular periodic activity consisting of a few regular heart beats followed by a longer interval. The duration of the periods, in six experiments, was between 115 and 222 sec, and remained quite constant within each individual experiment. Langendorff also tested excitability during the intervals, and found that it gradually increased during the pause. However, even at the very beginning of an interval completely normal contractions could be evoked; thus, excitability was always present.

In analogy with the findings on smooth muscle, I should like to interpret the observations on the heart as showing that the minute-rhythm is also one of the basic properties of heart muscle. However, in the primary pacemaker area, the spontaneous generation of excitation is, under normal conditions, so strong that minute-rhythm fluctuations are completely masked (cf. p. 339, Fig. 21). Sometimes, periodic fluctuations can be detected in isolated pacemaker tissue; these are generally evaluated as a certain deterioration of the functional condition.

Conclusions

It has been demonstrated that the activity of isolated smooth muscle also has a minute-rhythm which is thus inherent in the basic functional processes of smooth muscle. Under normal conditions *in situ*, this myogenic minute-rhythm is modified by nervous and hormonal influences which either augment, or depress, or co-ordinate and synchronize it at the proper frequency. No general answer can be given to the question whether myogenic or neural factors are more important here; since they are synergistic, the question becomes irrelevant. In any case, the minute-rhythm is manifest mainly in the activity of smooth-muscle organs, and we are therefore justified in accepting it as a functional rhythm of smooth muscle. The latent presence of minute-rhythm in the activity of the heart emphasizes the close functional relationship with that of smooth muscle. In addition, it should be mentioned that minute-rhythm in the central nervous system is not limited to the circulatory centres; the respiration can fluctuate with a minute-rhythm, which is co-ordinated with that of skin blood flow (Hildebrandt, 1963; see Fig. 9).

MYOGENIC MINUTE-RHYTHM OF SMOOTH MUSCLE

For a closer analysis of the myogenic minute-rhythm, the taenia coli of the guinea-pig was chosen. This preparation is always taken from the caecum, since in the guinea-pig the longitudinal muscle is concentrated in taenia only there. Various findings indicate that the caecum is a centre of myogenic automaticity. The studies on dogs by Templeton & Lawson (1931) showed, for example, that the caecum is generally the pacemaker for the activity of the adjacent parts of the colon. Thus, guinea-pig taenia coli can be taken as the prototype of spon-taneously-active smooth muscle and is particularly suitable for an investigation of the fundamental processes underlying the myogenic minute-rhythm.

The author's experiments which will be described below, were done with two different techniques. (1) Extracellular recording of spike activity with simul-taneous records of tension development. Preparations were mounted at *in situ*-length of 15 to 20 mm in a glass capillary perfused with Krebs solution and containing platinum wire electrodes (Golenhofen, 1966a). (2) Intracellular measurement of the membrane potential, with simultaneous records of tension development. For this purpose slightly shorter and more strongly stretched preparations were generally used in order to reduce the movements. Usually, the *in situ*-length was here also more than 10 mm.

Conditions for the manifestation of minute-rhythm

In isolated preparations, depending on the experimental conditions, the minute-rhythm will appear stronger, or weaker or not at all. Observations indicating that it represents the 'normal' functional state are as follows:

1. The minute-rhythm is found *in situ* under normal conditions, and can be recorded *in situ* in the taenia coli of anaesthetized guinea-pigs (Fig. 4 p. 319).
2. In isolated preparations in Krebs solution, at their *in situ* length and at a temperature of 35°C, the minute-rhythm will be the more marked the longer the preparation. As a rule, preparations of less than 5 mm length are continuously active. Long preparations, however, reflect the natural conditions more closely (Golenhofen, 1965).
3. Strong stretching, considerably beyond the normal *in-situ* length, favours the transition to continuous activity. If one wishes to suppress the minute-rhythm, e.g. for an electrophysiological analysis of the spike mechanisms, it is advisable to use short and markedly stretched preparations (Bülbring, 1954, 1957a; Holman, 1958; Gillespie, 1962a; Kuriyama, 1963a).
4. Various deviations from the normal situation damp or suppress the minute-rhythm, e.g. lowering the temperature (p. 334), changes in CO_2 tension and pH value (p. 334), etc. The disappearance of minute-rhythm and appearance of continuous spike discharge is often the first sign of a dis-turbance. Bacterial growth in the nutrient solution can apparently cause such disturbances, glucose-stock solution being particularly susceptible. At the end of a long experiment transition to continuous spike discharge is regularly found before the activity begins to die out.

A number of additional factors, such as the age and sex of the animals, may play a part in the manifestation of minute-rhythm. If the taenia is placed in a glass capillary for extracellular measurement of electrical activity, as described above, the minute-rhythm is especially well marked; this is possibly connected with the long length of tissue and with a better electrical conduction of excitation, and therefore a better synchronization of activity.

Period duration of minute-rhythm

When spontaneous activity is measured in isolated preparations over several hours, it is found that the period duration of the minute-rhythm does not remain completely constant. Figure 12 shows sections of an experiment which

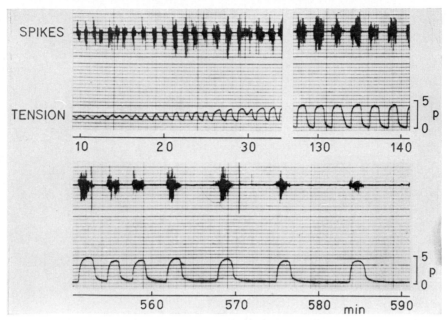

FIG. 12   Electrical activity and tension development of isolated taenia coli. Sections from an experiment in which spontaneous activity was recorded for 18 hr under constant conditions. Tissue in a glass capillary perfused with Krebs solution with a bicarbonate-5% CO_2 buffer, containing penicillin, at 35°C. Spikes measured with extracellular electrodes (platinum wires). Tension in ponds. Time in minutes after the tissue was introduced into the chamber. (Golenhofen & von Loh, unpublished)

lasted 18 hr under constant conditions (with addition of penicillin to the solution). At first the fluctuations were, typically, of relatively high frequency, with a weak over-all tension development; with increasing peak activity, the periods became longer. After the phase of adaptation during the first hour the pattern of activity changed but little during the following hours. From the 10th to the 12th hour, the period duration increased, with extreme values of up to 10 min. In the further course of the experiment the minute-rhythm decreased again and finally disappeared to be replaced by continuous spike discharge during the last hour before activity stopped. The histogram (Fig. 13) of all period durations found in this experiment, shows that the spontaneous changes of the period duration did not take place continuously but rather as whole-number multiples of a basic period which appeared to be quite constant. This means that, in spite of a relatively large variation in frequency of the manifest rhythm, the basic frequency is very stable, and that the minute-rhythm actually possesses the characteristics of long-wave rhythms (see p. 324). In addition, it follows that the basic process of the minute-rhythm is a sinusoidal rather than a sawtooth oscillation.

The principle, recognizable in Fig. 13, of a stepwise whole-number change in period, appears to be applicable to the minute-rhythm in general. Weitz & Vollers (1926) already mentioned half-minute, one-minute, and two-minute

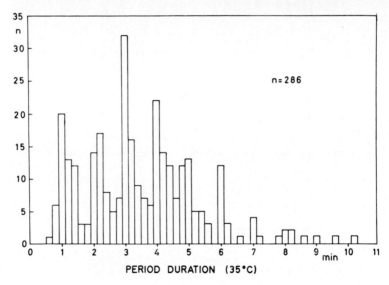

FIG. 13   Histogram of the period durations of all waves from the experiment shown in Fig. 12. (Golenhofen & von Loh, unpublished)

fluctuations, which could alternate. Although the authors presented no histograms, it can be assumed from their description, together with the original curves, that these frequency changes were not smooth, but stepwise. In measuring blood-flow waves, we likewise found multipeaked histograms, with simple whole-number relations between the maxima (Golenhofen & Hildebrandt, 1957a). On the other hand, in the summation of results from many experiments the multi-peaked nature of the individual histograms becomes naturally more and more obscured (see Fig. 7). The fact that the preferred period duration for muscle blood flow was found to be 1 min, but for skin and liver blood flow only 0·5 min (see Fig. 9), may be interpreted as showing that, in accordance with the rules of whole-number frequency multiplication, a uniform rhythm can manifest itself differently at different sites, depending upon the characteristics of the organ involved.

At the beginning of the experiment shown in Fig. 12 a direct relation is seen between the extent of tension development and the period duration of the manifest rhythm; this is a general rule. Accordingly, during substrate-depletion (see p. 334) and in Ca-free solutions (see p. 335), the period duration decreases with a decrease in tension development.

The effect of stretching and loading on minute-rhythm

An extensive series of experiments was carried out to test the effect of stretching (with isometric measurement) and loading (with isotonic measurement) on the frequency of the minute-rhythm (unpublished observations). The results were not uniform, and can best be explained by assuming that the basic period duration remains constant, while only the form of manifestation changes. This

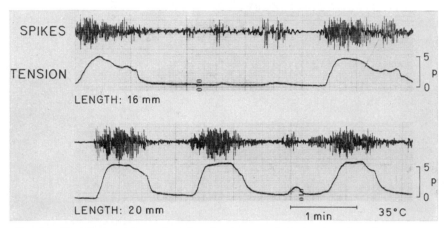

FIG. 14   Electrical activity and tension development in isolated taenia coli. Technique as in Fig. 12. Between the upper and lower part, the tissue was stretched from 16 to 20 mm. Steady-state conditions. (Golenhofen & von Loh, unpublished)

is supported by observation of preparations with a pronounced and regular minute-rhythm, which remained unchanged in frequency over a wide range of length and tension changes. Figure 14 shows an example of an increase in the frequency of a manifest rhythm of contractions after stretching. However, the record of electrical activity shows that the higher frequency was already present before the stretch, the intermediate waves being unable to spread through the entire preparation. In the mechanogram (at a length of 16 mm) these intermediate waves were only just perceptible and would not have been taken into account. After excessive stretching, individual activity waves may merge and thereby simulate a decrease in frequency. Under such conditions, it was impossible either to demonstrate or to rule out an influence of stretching on the frequency of the basic rhythm.

12+

Substrate depletion and minute-rhythm

If the taenia is placed in a substrate-free solution, the spontaneous activity gradually decreases. In the final stage it stops temporarily so that only sporadic bursts of activity appear (Golenhofen, 1966a; Bülbring & Golenhofen, 1967). The activity-free intervals between bursts become longer and longer with continuing substrate depletion. As a rule, the individual bursts continue to appear in a minute-rhythm form and there is evidence that the minute-rhythm continues in a sub-threshold manner even during the activity-free intervals (Golenhofen, 1966a). The frequency of the manifest minute-rhythm increases with a decrease in tension development during substrate depletion, and the principle of whole-number changes in frequency can also be observed in this condition.

Dependence of minute-rhythm on temperature

If the temperature is lowered, the frequency of the minute-rhythm remains remarkably constant, as Fig. 15 shows. With decreasing temperature, the

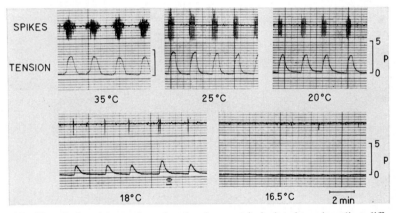

FIG. 15   Electrical activity and tension development in isolated taenia coli at different temperatures. All records from the same experiment. Technique as in Fig. 12. (Golenhofen & von Loh, unpublished)

number of spikes per wave decreases with a Q_{10} of about 3, until finally (at 18°C in Fig. 15) there is only one spike per wave. Thus, in the critical-temperature range spike rhythm and minute-rhythm coincide. Below this temperature range, spontaneous activity ceases entirely. In many cases, the minute-rhythm disappears in an intermediate temperature range below 30°C, and reappears with further cooling (Golenhofen & von Loh, 1966).

Effect of changes in CO_2 and pH

Figure 16 shows the effect of changing the partial CO_2 pressure. The changes were sudden, since a two-way tap was used to switch to a solution which had been aerated for at least 10 min with a different CO_2 concentration. An increase in CO_2 pressure, at a constant extracellular pH (i.e. a corresponding increase in the bicarbonate ion concentration) inhibits activity; a decrease in CO_2 increases activity. The transition to lower CO_2 tension caused an initial period of continuous activity which was frequently longer than that shown in Fig. 16; at 1·25% CO_2 it persisted. Conversely, in a tissue with continuous spontaneous activity, an increase in CO_2 tension can lead to the appearance of minute-rhythm (Wienbeck, Golenhofen & Lammel, 1968).

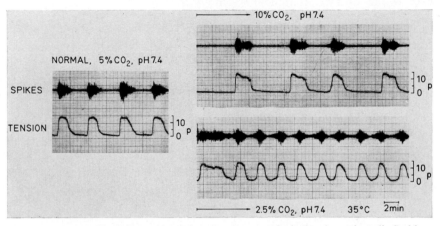

FIG. 16 Electrical activity and tension development in isolated taenia coli. Sudden change in CO_2-tension at constant pH. Control on left side. The records on the right start with the transition from 5% CO_2 to 10% (*top*), and from 5% CO_2 to 2·5% CO_2 (*bottom*). Technique as in Fig. 12. (Golenhofen & Wienbeck, unpublished)

Lowering the pH value at 5% CO_2 produces the same effect as decreasing the CO_2 tension at constant pH, i.e. it increases the activity; raising the pH value has the opposite effect. Changes in O_2 pressure accompanying those in CO_2 pressure were excluded as cause for the observed effects. It is possible that the effect of CO_2 at constant extracellular pH may be exerted, in part, via changes in the intracellular pH, since CO_2 can easily diffuse through the cell membrane.

Other effects on minute-rhythm

Because of the myogenic nature of the minute-rhythm in isolated tissue, drugs which affect neuromuscular transmission have no influence (atropine 10^{-5}, phentolamine 10^{-4}). The β-receptor blocking agent propranolol inhibits spontaneous activity in concentrations exceeding 10^{-5} g/ml. Before activity ceases, there is usually a regular spike discharge with strongly altered spike shape, but no minute-rhythm. The inefficacy of tetrodotoxin has already been mentioned (see Fig. 10).

In this connection, the effect of calcium ions is noteworthy. In Ca-free solution the spontaneous spike discharge usually stops within 10 to 20 min. As shown by the example in Fig. 17, the minute-rhythm persisted during Ca-depletion, as long as any activity could still be detected. Intracellular measurements of membrane potential at the time corresponding to Fig. 17(b), have shown that membrane function was already severely impaired and that the slight development of tension was associated with small abortive oscillations in the membrane potential [Fig. 17(c)]. The frequency of the minute-rhythm generally increases with the decrease in tension development. Thus, Ca-depletion is one of the conditions which lead to a dissociation of spike- and minute-rhythm.

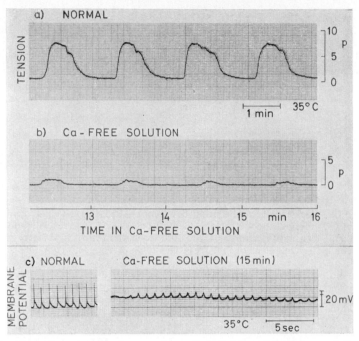

FIG. 17 Normal tension development in taenia coli (a), and declining activity of the same preparation in a calcium-free solution, with persisting minute-rhythm (b). (c) Electrical activity of the same preparation, measured intracellularly, under the same conditions as (a) and (b). Note different time scales in the three records.

According to the latest investigations it is possible to suppress the minute-rhythm selectively with metabolic inhibitors. The tissue than exhibits continuous spike activity (Golenhofen & Petrányi, unpublished).

Membrane potential and minute-rhythm

The minute-rhythm fluctuations in the activity are accompanied by fluctuations in the membrane potential which may have an amplitude of as much as 20 mV (Bülbring, 1957b, c, 1961b, 1962). If the fluctuation is marked, as in Fig. 18, spike discharge is absent in the potential troughs (maxima of the membrane potential). If it is less marked, as in Fig. 19, a slow spike discharge remains even during the activity minima. The increase in tension is then connected with depolarization and increase in spike frequency. The frequency of the generator oscillations (often called 'slow waves'), which initiate the spikes, is usually difficult to assess. In Fig. 19, the generator oscillations at the activity minimum are clearly defined and regular. With increasing activity, they become less clear and start to merge, so that there are usually two or more spikes per

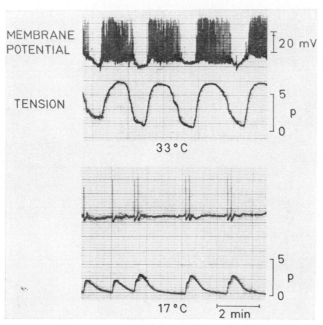

FIG. 18   Spontaneous activity in taenia coli at 33°C and at 17°C. Intracellular records of membrane potential (upper trace) and tension (lower trace). Tissue in Krebs solution with bicarbonate-5% CO_2 buffer. (Golenhofen & von Loh, unpublished)

oscillation. An increase in the frequency of the generator oscillations, therefore, cannot be derived from this picture. On the contrary, the intrinsic frequency of the slow waves appears to remain fairly constant. This would agree with the finding that the first resonance frequency during rhythmic stretching which coincides with the intrinsic frequency of the generator oscillations is quite stable (Golenhofen, 1965).

The normal correlation described between basic membrane potential, spike frequency, and mechanical tension is not always strictly true. This may result from the fact that measurements of membrane activity with glass micro-electrodes can be made only in individual cells, whereas the mechanical tension is measured in the whole tissue whose activity is not always fully synchronized. Figure 20 shows a case in which membrane potential and mechanical tension reveal synchronous fluctuations, but in which only every other wave in the membrane potential is sufficient to produce spikes in the area covered by the micro-electrode. The weaker waves remain subthreshold.

Other experiments raise the question whether the fluctuations in the basic membrane potential are really an essential element in, or even the cause of,

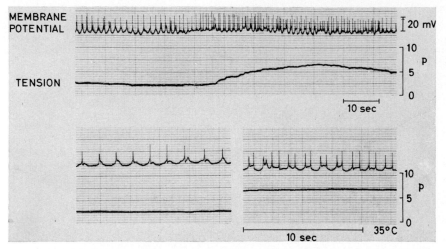

FIG. 19   Spontaneous activity in taenia coli at 35°C. Technique as in Fig. 18. The records below were taken at a higher speed from a phase with low activity (*left*) and high activity (*right*) during a minute-rhythm cycle. All parts from the same experiment.

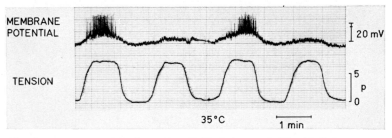

FIG. 20   Spontaneous activity in taenia coli. Technique as in Fig. 18. Only every second minute-rhythm wave reaches the threshold in the area picked up by the microelectrode.
(Golenhofen & von Loh, unpublished)

minute-rhythm, as might be indicated by the findings described so far, or whether they are merely a secondary concomitant. At low temperatures, for example, we no longer found any minute-rhythm fluctuations in the membrane potential (Fig. 18), although the minute-rhythm was still quite clear in spike discharge and in tension development (cf. p. 334). In view of the difficulties mentioned above in evaluating measurements in individual cells, no hasty conclusion should be made. However, the finding presented in Fig. 18 was regularly confirmed, and thus must be taken into consideration in dealing with the total concept of minute-rhythm (p. 340).

Basic and manifest minute-rhythm

Some observations have already indicated that the frequency of the minute-rhythm seen in the mechanogram does not always agree with that of the basic membrane oscillations. The schematic diagram in Fig. 21 is intended to show that this can easily be understood by assuming, as a basic process, an oscillation on which its next harmonics are superimposed, and by further assuming that the excitation threshold can be varied independently.

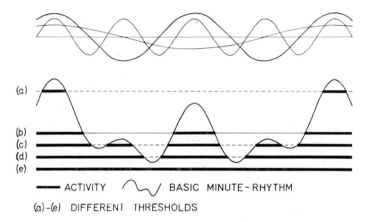

FIG. 21   Diagram showing different forms of manifestation of a minute-rhythm with a constant basic oscillation. *Upper part*: Main component with the next two harmonics. *Lower part*: The basic oscillation of the minute-rhythm around (b) as the zero line was obtained by summing the oscillations around the upper zero line. (a)–(e) indicate various excitation thresholds. Activity (thick lines) is present when the oscillation exceeds the threshold. The whole figure represents a time of ca. 2 min.

One way of causing a general shift in the threshold is stretching. The stretch effects described on p. 333 can be well explained by the scheme shown in Fig. 21. If the threshold is lowered by stretching, e.g. from (a) to (b) and then from (b) to (c), weaker intermediate waves, which previously had been subthreshold, become visible. Thus, with each further stretch, the frequency of the manifest rhythm increases, without any change in the basic minute-rhythm oscillation. Further stretch, lowering the threshold to (e), causes the manifestation of continuous activity.

Not all influences on the minute-rhythm are included in the diagram of Fig. 21. It is possible, for example, that individual components of the complex basic oscillation might change their amplitudes independently, so that one of the next harmonics would become dominant. In this way stepwise whole-number changes of frequency, which are observed again and again in minute-rhythm, can easily be explained even if there is no shift in threshold.

The assumption in Fig. 21 that the threshold remains constant, is, of course, a simplification. There is evidence that the manifest activity influences the threshold so that, following activity, the threshold is raised. This would seem to be an expression of a general law which, during spike rhythm, reveals itself in the hyperpolarization following the spike. Such a mechanism can also promote whole-number period changes and can also lead to extremely long intervals between individual waves of activity.

Furthermore, if it is remembered that changes in the extent of synchronization of the tissue mean an additional complication, the variable nature of the phenomenon of minute-rhythm becomes comprehensible.

The multiplicity of intrinsic and extrinsic factors influencing the manifest frequency renders the determination of the 'normal' basic frequency difficult if not impossible. The question as to the norm can be answered satisfactorily only by considering the functional relations in their entirety. However, an upper limit for 'normal' period duration can be given with considerable certainty. After a series of experiments without measurement of intracellular potentials, we listed the normal range as being from 1 to 3 min; 70% of all the periods were between 1 and 2 min (Golenhofen, 1966; Fig. 7). Measurement of intracellular potentials undoubtedly comes somewhat closer to the basic process. Considering the fluctuations of the basic membrane potential, it can often be shown that periods of more than 2 min must be interpreted as double waves; this is also true of the example given in the upper part of Fig. 18. The main component of the basic oscillation probably lies always in the narrow range between 1 and 2 min, and certainly not in a period duration of more than 3 min. Longer intervals, such as those shown in parts of Fig. 12 and Fig. 16, result from the fact that certain waves remain below the threshold.

The scheme presented in Fig. 21 also contains the changes in membrane excitability by external stimuli, which Kuriyama (1964b) demonstrated in uterine muscle. As early as 1908, Langendorff described such fluctuations in excitability for the Luciani periods in the heart.

Conclusions

Important characteristics of the minute-rhythm *in situ* can be traced back to basic myogenic processes, especially the strict fixation of a preferred frequency with the possibility of whole-number multiplications of frequency. This results in the typical multi-peaked spectrum for the frequency distribution of period duration of manifest rhythm (Fig. 13). The basic process can be considered as a complex oscillation consisting of a main component with its superimposed harmonics. Changes in the basic oscillation via changes in the intensity of the partial components, shifts in the excitation threshold, effects of activity on the excitation threshold, and changes in the degree of synchronization in the tissue can, if taken together, explain the manifold nature of manifest rhythm. In comparison, continuous changes in the basic period duration undoubtedly play only a secondary part.

Under various conditions there may be a dissociation of spike- and minute-rhythm, e.g. if the temperature is lowered, during Ca-depletion, if the CO_2 tension is changed, or metabolic inhibitors are added. Thus, in contrast to the spike rhythm, the minute-rhythm should not be considered as being mainly connected with the membrane; it should rather be taken as a comprehensive process involving the entire cell or group of cells, and it is probably closely connected with metabolic processes. In this context, there are new biochemical findings of special interest, which are briefly discussed below.

SOME NEW BIOCHEMICAL ASPECTS OF MINUTE-RHYTHM

Continuous oscillations of metabolic processes have recently been measured in a cell-free extract of *Saccharomycis carlsbergensis*, with a period duration in the minute range (Pye & Chance, 1966; Hess, Brand & Pye, 1966; Hess, 1968).

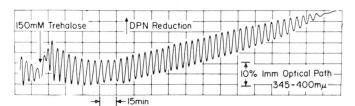

FIG. 22   Continuous glycolytic oscillations in a cell-free extract of yeast after addition of trehalose. Room temperature. (Hess *et al.*, 1966)

An example is given in Fig. 22, where the oscillations of the concentrations of DPN/DPNH are recorded. The proper concentration of substrate is a prerequisite for such continuous oscillations in yeast extracts. Trehalose, which was used in Fig. 22, is a natural substrate of yeast. The frequency was ca. 0·2/min at 25°C, and 0·55/min at 35°C (Hess *et al.*, 1966). That is surprisingly similar to the minute-rhythm of smooth muscle. In addition, frequency changes to the higher harmonics, in the ratio 1:2:4:8 to the fundamental frequency of 0·2/min (at room temperature), were occasionally observed in this system.

This agreement in frequency must not tempt us to consider the two processes as being simply identical. There are definite differences, e.g. in temperature dependence. The Q_{10} value for the glycolytic oscillations is about 3; but for the minute-rhythm of taenia coli it is near 1 (p. 334). Nevertheless, the demonstration of minute-rhythm oscillations in isolated metabolic systems support the assumption that oscillations in cellular metabolism are an important factor in the basic process of normal myogenic minute-rhythm.

12*

SUMMARY

A survey of the rhythms of smooth-muscle organs *in situ* shows that a rhythm with a period duration in the minute range (ca. 0·5 to 2 min) is common to the entire smooth-muscle system in mammals. This minute-rhythm also appears in isolated, spontaneously-active smooth muscle, in preparations of intestine and uterus and in isolated blood vessels, and remains after all peripheral nervous activity has been blocked. The minute-rhythm is thus one of the basic processes of automatic myogenic activity. Under normal conditions *in situ*, myogenic minute-rhythm is modified by neural and hormonal influences; in particular, it is synchronized. This has been demonstrated for the minute-rhythm of peripheral blood flow in man.

In addition, the study reports new investigations of the mechanisms involved in minute-rhythm. Changes of frequency which occur in the manifest rhythm are mainly due to the fact that a basic oscillation exists which is quite stable in frequency but which can apparently vary its manifest frequency via whole-number multiplications. Stretching does not influence the frequency of the basic oscillations, but merely changes the pattern in which it appears.

The frequency of the basic minute-rhythm remains remarkably constant when the temperature is lowered from 35°C to the temperature at which spontaneous activity ceases below 20°C. In contrast, the spike frequency is markedly dependent upon temperature. Besides temperature, there are other procedures which influence spike-rhythm differently, and apparently independently, from the minute-rhythm. For example, during calcium depletion the minute-rhythm remains unaltered until activity ceases, in spite of extreme changes of the spikes. On the other hand, metabolic inhibitors selectively suppress the minute-rhythm and lead to continuous spike discharge. CO_2 also has a strong effect on minute-rhythm. Thus, in contrast to spike rhythm, which is mainly connected with the membrane, the minute-rhythm must be thought of as a comprehensive process involving the entire cell or group of cells, and which is closely connected with metabolic processes of the tissue.

11

EFFECT OF DRUGS ON ION DISTRIBUTION AND FLUX IN SMOOTH MUSCLE

JOHANNES SETEKLEIV

INTRODUCTION

Electrophysiological techniques have introduced a new approach to the cellular mechanisms of drug action in smooth muscle. In terms of the ionic hypothesis, the electrical events are due to ion movements across the cell membrane. Estimation of drug effects on ion distribution and flux should therefore allow a more direct examination of the ionic mechanisms underlying the tissue response to drugs. In recent years growing interest has been paid to these aspects in the evaluation of drug action. So far the investigations have been mainly concerned with the action of the cholinergic and the adrenergic neurotransmitter. Other agonists and antagonists have, however, also been applied and quantitative analyses of the results have been attempted in studies of drug-receptor interaction.

The techniques employed in these investigations are essentially the same as those used in studies of ion content and tracer kinetics of smooth muscle. Either the cellular ion concentration is determined before and after application of a drug, or the change in the rate of uptake and loss of radioactive isotopes upon drug addition is followed.

At present the methods are not perfect. The most difficult technical problems are due to the secondary effect produced by the action of drugs on the smooth muscle, i.e. changes in electrical and mechanical activity, which make the interpretation of the data obtained very difficult. The present article includes a discussion of the technical problems involved in using ion contents and ion flux as parameters in the evaluation of drug action. This will provide a background for the review of the experimental results available.

FACTORS AFFECTING ION DISTRIBUTION AND FLUX

Extracellular space

The main difficulty in studies of the effects of drugs on intracellular ion content is the uncertainty in obtaining an accurate measurement of the extracellular space. An estimation of the effect on the total ion content is not adequate because the drug might have changed the size of the extracellular space as well.

With the present methods, using inulin, sorbitol, ethanesulphonate or other molecules which should be restricted to the extracellular space, the scatter of the values obtained for extracellular space is usually considerable. Small and rapid variations in the ion content have therefore been difficult to detect.

The kinetics of tracer exchange

Investigations of drug action on cellular uptake and loss of radioactive tracer ions aim at detecting permeability changes induced by the drugs studied. The interest of the investigator is therefore focused on those components of the flux curves which can be related to the transfer of ions across the cell membrane. At present, however, the different components of the flux curves are only partly explained in relation to the anatomical cell structures, tissue organization, diffusion barriers, etc. (see Chapters 2 and 3).

The uptake of a radioactive isotope is usually estimated by soaking the tissue in Krebs solution containing the isotope under standardized conditions allowing the ion studied to exchange with its corresponding radioactive isotope. By measuring the radioactivity in the muscle at certain time intervals the amount of radioactivity transferred to the muscle per unit time can be determined. The uptake of tracer is rapid in the beginning due to diffusion into the extracellular space; later the uptake depends on the rate of penetration through the cell membrane. Drugs might increase or decrease the rate of penetration by affecting passive or active transport of that particular ion through the membrane. It is, however, obvious that if the drug affects the size of the extracellular space, the uptake curve will be affected. The uptake into the extracellular space therefore ought to be verified by simultaneous determination of the rate of uptake of a non-permeant ion. This has as yet not been done. Changes in cell-surface/cell-volume ratio could also affect the ionic transfer. At present it is not possible to estimate such changes.

The efflux curve of a tracer is usually determined by transferring the tissue, after loading in radioactive solution, through a series of tubes containing inactive solution. After a given time in the inactive solution, the tissue — mounted on a steel rod — is moved to the next tube in the series. The radioactivity transferred from the tissue to the solution in each tube is then estimated. The method is simple and easy to perform but, because of the relatively long time periods needed, especially if the radioactivity is low, small and rapid changes in ionic flux are difficult to detect. Another method is to mount the tissue in a chamber which is perfused with Ringer solution at a constant flow rate. The radioactivity is then continuously transferred to the passing Ringer solution and, by collecting constant volumes of the effluent at constant time intervals, the rate of loss of radioactivity from tissue can be determined. A great advantage of such a method is that the mechanical activity can be recorded simultaneously (Born & Bülbring, 1956; Jenkinson & Morton, 1967a; Brading, 1967). The effluent might also be led by tubes directly to a detector. Combined with a ratemeter and a recorder a continuous record of the tracer efflux can be obtained (Spero, 1967; Burgen & Spero, 1968).

The radioactivity remaining in the tissue is measured at the end of the experiment. By summing the amounts lost to the remaining activity in reversed order, the total radioactivity remaining in the tissue during the washout is calculated and plotted against time.

The typical efflux curve usually has three components, an initial fast, a second transitional and a third slow loss of radioactivity. The fast component corresponds to the clearing of radioactivity from the extracellular space. If drugs which influence the size of the extracellular space are administered during this phase, they may change the curve in a way which have nothing to do with the effect on the membrane. Drugs are therefore given in the *third* phase when the rate of tracer loss is slower and is believed to represent mainly the transfer of tracer through the cell membrane.

Usually an estimate of the tracer lost per unit time gives a better picture of the changes in efflux than the estimate of the radioactivity remaining in the tissue. Since the efflux is dependent on the amount of radioactivity left in the tissue, the loss is usually calculated as a fraction of the remaining radioactivity lost per unit time of washout.

Electrical and mechanical activity

The most obvious difficulties on the evaluation of drug effects on the ionic level are caused by the close relationship between electrical and mechanical phenomena in smooth muscle with the possibility of interfering with, or masking, ion movements across the membrane. Several means have been applied to dissociate or eliminate one of these two factors in order to study the actual effect of drugs on the cell membrane.

Regarding the electrical phenomena two components have to be considered in smooth muscle, i.e. the spontaneous discharge of action potentials and the resting membrane potential.

The action potentials

Excitation of smooth muscle is accompanied by an increased spike discharge, and inhibition by a decrease or cessation of spike activity. One might expect that, during the action potentials, an increased membrane permeability to sodium as well as calcium and potassium occurs and between the spikes, an increased ion flux in the opposite direction, but experimental verification is still at an early stage (see Chapter 5). Recent investigations indicate that calcium is the most important ion for spike generation (Kuriyama, Osa & Toida, 1966; Nonomura, Hotta & Ohashi, 1966; Brading, Bülbring & Tomita, 1969*b*).

The membrane potential

A change in membrane potential influences the ion fluxes and can obscure the drug effect proper on the cell membrane. The exact correlation between membrane potential and ion fluxes in smooth muscle is, however, not known.

The membrane potential can be altered by changing the external potassium concentration, and, in a series of experiments, we have studied how the external potassium concentration influenced the rate of ^{42}K-loss. Pieces of taenia coli from the same guinea-pig were equilibrated and the rate of ^{42}K-efflux determined in normal Krebs solution. In a series of subsequent experiments the muscles were transferred during the washout, to bathing solutions with different potassium concentration (Fig. 1). After exposure to solutions with higher potassium concentration the ^{42}K-efflux increased sharply; later, the efflux declined gradually and reached a relatively stable level (Setekleiv, 1967). When potassium was removed from the solution, the efflux was depressed.

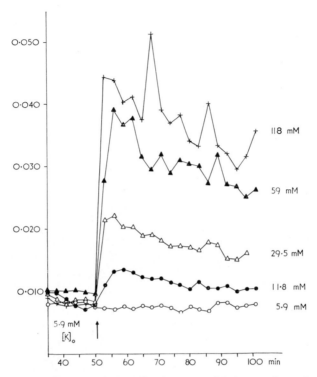

FIG. 1   Potassium efflux (fraction of ^{42}K lost per min) from guinea-pig taenia coli before and after changing the external potassium concentration, from normal (5·9 mM) to 11·8, 29·5, 59 and 118 mM (2, 5, 10 and 20 times increase).

When the potassium efflux is plotted on a logarithmic scale against the potassium concentration, the values fall on a straight line (Fig. 2). The change of K-efflux is thus proportional to the logarithm of the external potassium concentration (Setekleiv, 1967).

The muscle cells are depolarized in solutions containing a high potassium concentration. The relationship between external potassium concentration and the membrane potential has been studied by Casteels & Kuriyama (1966) in pieces of taenia coli in a solution of the same ionic composition as that used in the experiment referred to above. Their results are also plotted in Fig. 2. The decrease in membrane potential seems to be well correlated to the increase in ^{42}K-efflux.

These results indicate a strong relationship between the potassium efflux and the membrane potential. The external potassium concentration itself will, however, also influence the potassium fluxes. To investigate whether the potassium concentration or the membrane potential is the most important factor in determining the rate of ^{42}K-loss, a high calcium concentration was used in the solutions, since calcium is known to hyperpolarize the cell membrane (Kuriyama, 1963a). After the first washout in a solution with normal calcium and high potassium, the muscles were reloaded in radioactive solution and another washout was performed in a solution with three times the normal calcium

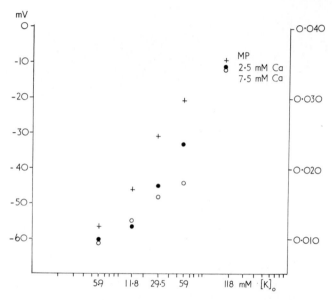

FIG. 2   Efflux of ^{42}K (right ordinate) from guinea-pig taenia coli in solutions containing different potassium concentrations (abscissa) and containing normal (2·5 mM) Ca (●) and 3 times normal (7·5 mM) Ca (○). Corresponding values of membrane potential (+) (left ordinate) measured by Casteels & Kuriyama (1966).

concentration. Small or no changes in potassium efflux were obtained in normal and 2·5 times the normal potassium concentration. However, in 5 and 10 times potassium the efflux was lower. In 20 times potassium, where one would hardly expect any change in the membrane potential by increasing the calcium concentration, the potassium loss, in fact, was the same as in the control series. Therefore, if other effects of calcium on the ^{42}K-efflux can be excluded, the results indicate that the membrane potential is the most important factor in determining the potassium efflux.

The membrane can be completely depolarized by soaking the tissue in a bathing solution containing high potassium, e.g. the K_2SO_4-Ringer solution used by Evans & Schild (1957b). In this preparation the membrane potential is not measurable (Evans, Schild & Thesleff, 1958). The rate of potassium efflux is about doubled under these circumstances (Bülbring, Goodford & Setekleiv,

1966), but the rate can be reduced to about normal by lowering the temperature to 20°C. In the depolarized preparation the effect of drugs on membrane permeabilities can be studied without interference by alterations in membrane potential. In spite of the unphysiological treatment of the tissue, recent studies (see below) seem to indicate that valuable information about drug effects on ion transfer can be obtained with this preparation. However, it remains to be investigated how this treatment influences the intracellular ionic composition, and the membrane properties. The effect of drugs on the contractile mechanisms in depolarized muscles is not abolished, and their mechanical response is qualitatively the same as that of smooth muscle in normal solution (Evans et al., 1958). A possible influence of the drug induced tension on the ion fluxes is thus not eliminated. Spero (1967) and Burgen & Spero (1968) have pointed out that, when the membrane is completely depolarized by high external potassium, the potassium efflux is only one fifth of the maximum efflux obtained by carbachol stimulation of the muscle. This observation provides the possibility to separate the two factors by comparing the two dose response curves. In this way the effects observed by changing the external potassium concentration might be helpful in differentiating between the genuine drug effect and secondary effects on the fluxes due to change in the membrane potential.

The muscle tension

The close correlation between the tension and the electrical activity of the smooth muscle was first demonstrated by Bülbring (1955). She found that stretch depolarized smooth muscles and increased the spike activity whereas relaxation had the opposite effect. The smooth muscle should therefore be regarded not only as a contractile tissue, but also as a receptor organ, sensitive to change in length. This change in electrical activity resulting from change in tension also affects the ion fluxes, further complicating the evaluation of drug effects on the ionic level. Since little is known about this subject, a study of the effect of tension of the ^{42}K-efflux was undertaken.

Manual shortening of the muscle. Before starting the experiments, the tissue loaded with radioactive potassium was washed with inactive solution for 35 min. Decrease of tension by allowing the muscle length to shorten from weight/ length ratio $(W/L) = 1$ to $W/L = 1.5$ or 2 was usually followed by a decrease in ^{42}K-efflux (Setekleiv, 1967). The fall in potassium loss was more pronounced if the initial tone was high due to stretch of the muscle. Sometimes the potassium efflux settled on a new (lower) level (Fig. 3). In other experiments the decrease was transient, and after a few minutes the efflux became more variable, sometimes returning to the level before the shortening, indicating reappearance of rhythmic tension changes. A decreased ^{42}K-efflux was also sometimes seen after shortening of muscles in solutions where the potassium concentration was increased 2.5 and 5 times (14.5 and 29.9 mM respectively) but not in 10 (59 mM) or 20 (118 mM) times potassium concentration.

The influence on the rate of sodium loss by shortening of the muscle has been explained as being due to an increased ratio of cell surface area to cell volume (Freeman-Narrod & Goodford, 1962). A similar mechanism may play a part in the decrease in the rate of ^{42}K-loss after shortening of the muscle. It can, however, only play a minor role, since the efflux increases again when the rhythmic activity reappears and, furthermore, since shortening has little or no effect on the K-efflux from muscles in increased external potassium concentration.

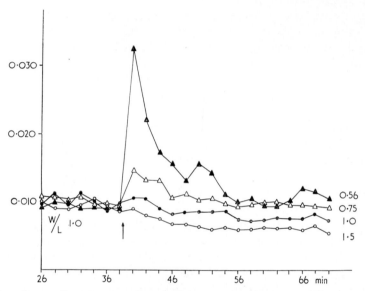

Fig. 3   Effect on ^{42}K-efflux (fraction lost per min) of manual shortening (weight/length ratio increased from 1 to 1·5) or stretch (W/L ratio decreased from 1 to 0·75 and 0·56).

The results indicate that the ^{42}K-efflux is mainly determined by the tension in the muscle which again is correlated to the membrane potential (Bülbring, 1955).

Stretch of the muscle. In a study of the relationship between muscle tension and K-efflux it was found that an increase in muscle length by rapid stretch, resulted in a transient increase in potassium efflux (see Fig. 3). This was proportional to the degree of stretch. In normal Krebs solution, stretching the muscle from $W/L = 1$ to $W/L = 0·75$ produced an increase in potassium efflux to about 150%, and a stretch from $W/L = 1$ to about $W/L = 0·6$ increased the efflux in a series of experiments up to 350%. The increased efflux was, however, transient and after a period, which lasted from 5 to 20 min depending on the degree of stretch, the ^{42}K-loss decreased to a level only slightly higher than the initial ^{42}K-efflux (Setekleiv, 1967). Moderate stretch is known to increase tension, to produce depolarization of the cell and to increase spontaneous electrical activity. A rapid and more pronounced stretch induces a high initial tension, which, however, declines again (Bülbring & Kuriyama, 1963c). Born & Bülbring (1956) first observed that development of tension increased potassium efflux. The present findings demonstrate that the ^{42}K-loss is well correlated to the time course of the change in tension after stretch.

Several factors may be involved in the production of this increased potassium efflux. Firstly, the change in the shape of the tissue might, by producing higher internal pressure, squeeze out intracellular potassium. Secondly, the potassium in the extracellular space might diffuse more rapidly into the bathing solution because of a distortion of the extracellular space. Thirdly, one must also consider the change in the ratio of cell surface area to cell volume. However, since the changes in ^{42}K-efflux seem to follow the tension development, it is most likely that they are due to the increase in tension and the depolarization associated with stretch.

When the external potassium concentration is raised, the resting ^{42}K-efflux is increased (from about 0·01 to about 0·03 min^{-1}). Under such circumstances it was found that after moderate stretch ($W/L = 0.75$) the increase of ^{42}K-loss was small or absent. Nor did stronger stretch ($W/L = 0.5$) produce as high an increase of ^{42}K-efflux as the same stretch in normal Krebs solution (190 to 280% increase). This may be due to the already existing depolarization of the cells. However, an influence of the W/L ratio was encountered when the tissue was transferred from normal solution to a high potassium concentration. In a series of experiments, pieces of taenia coli from the same guinea-pig were mounted isometrically at different degrees of stretch ($W/L = 0.56$, 1.5 and 2.0). After a 40-min washout period in normal solution, the potassium concentration was raised from normal (5·9 mM) to ten times (59 mM). Figure 4 shows that the increase in potassium efflux is most pronounced in the most stretched preparation which indicates again the correlation between ^{42}K-loss and tension (Setekleiv, 1967).

Muscle contraction

Determination of the dose-response relationship for the contraction as well as for the increase in potassium efflux has been used to dissociate these two parameters. Spero (1967) and Burgen & Spero (1968) have shown that a maximal contraction caused by carbachol produces only a two- to threefold increase in the potassium efflux, whereas carbachol can increase the potassium efflux by at least a hundred times. Furthermore, the doses of carbachol which elicited maximal contractions were lower than those causing maximum potassium efflux (Fig. 5). A reservation to these results is, however, that isotonic muscle shortening was used as indicator of the contractile response. Isotonic recording might be misleading because only partial activation of the muscle is necessary to produce a maximal shortening (Csapo, 1954; Setekleiv, 1964). The result therefore ought to be confirmed by isometric technique.

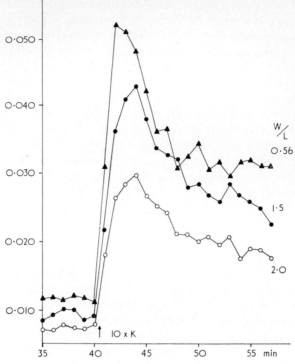

FIG. 4   Influence of the degree of stretch (weight/length ratio) on the increase of ^{42}K-efflux in response to change from the external concentration (5·9 mM), to 10 times normal (59 mM) potassium.

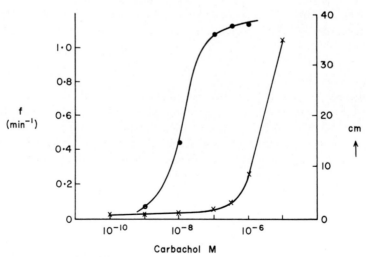

FIG. 5   The dose response curves of contraction (●) and K-efflux (×) to carbachol. (Spero, 1967)

Temperature

The contractile response can be abolished by lowering the temperature and this procedure might therefore be used to dissociate the tension from ion fluxes. However, since the membrane potential depends on the internal and external ion concentration, on the ion permeability of the cell membrane as well as on the temperature, a change in temperature might have consequences for the ionic fluxes.

In the cold, the active processes which maintain the ionic distribution across the cell membrane are less effective. Thus taenia coli looses potassium and accumulates sodium when the temperature is lowered (Freeman-Narrod & Goodford, 1962). The consequence is a decrease in the membrane potential (Axelsson & Bülbring, 1961). In guinea-pig taenia coli Bülbring & Kuriyama (1963c) found that the Q_{10} for the change in membrane potential and spike amplitude was 1·3 and 1·5 respectively. Similarly, the Q_{10} for the rate of rise and fall of the spikes was 2·6 and 2·7 respectively. These changes probably reflect the influence of alterations in temperature on the ionic permeability. By measuring the membrane conductance with the double sucrose-gap technique Brading, Bülbring & Tomita (1969a) found that between 19° and 37°C the membrane conductance increased with rising temperatures and decreased with cooling. The mean Q_{10} was 2·5. Several attempts have been made to evaluate how the ion fluxes are influenced by change in temperature.

Potassium. In normal Krebs solution it is not likely that the potassium *efflux* is influenced by active processes. To evaluate if that could be the case in a high external potassium concentration, (in which adrenaline affects the ^{42}K-loss, see later), we have studied the effect of different temperatures on potassium efflux from muscle suspended in ten times normal potassium concentration (59 mM).

In a series of experiments the rate of ^{42}K-efflux was measured in taenia coli at different temperatures (4, 10, 15, 20, 25, 30, 32·5, 35, 37·5 and 40°C). The measurements were started after 35 min washout to secure a steady rate of efflux. Change in temperature was obtained by transferring the muscle to tubes maintained at the wanted temperature. Usually 3 to 4 min elapsed before the potassium efflux was stabilized at the new temperature. Figure 6 shows that the increase in ^{42}K-efflux is proportional to the increase in temperature at temperatures from 10 to 40°C. The Q_{10} was calculated to be 1·94 (Setekleiv, 1967). A decrease in temperature thus reduces the rate of ^{42}K-efflux.

The *uptake* of potassium is also decreased by lowering the temperature. Freeman-Narrod & Goodford (1962) found that the rate of uptake of radioactive potassium was much higher at 20°C than at 4°C. In these experiments the rate of uptake of radioactive potassium from the bathing solution at 35°C was about the same as at 20°C.

Sodium. The Q_{10} for sodium efflux and uptake has not been estimated, but the changes in these parameters as a consequence of alteration in temperature are large (Freeman-Narrod & Goodford, 1962). By increasing the temperature from 4° to 35°C the Na-efflux increased more than one hundred times (Buck & Goodford, 1966). The high exchange rate of sodium in smooth muscle makes it difficult to differentiate between the various components in the flux curves. Lowering the temperature reduces the transfer rate. The slow phase of Na-efflux is, however, not significantly altered by temperature variations (Brading, Bülbring & Tomita, 1969a).

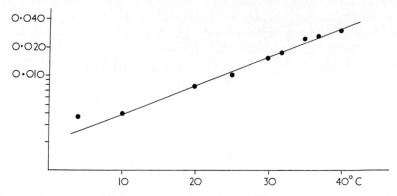

FIG. 6.   Fraction of ^{42}K lost per min (ordinate) at different temperatures (abscissa).

Chloride. ^{36}Cl-efflux is decreased considerably by lowering the temperature from 37·5° to 19·5°C. Changing the chloride in the bathing solution with benzenesulphonate, the Q_{10} of the membrane conductance has been found to decrease from about 2·5 to 1·5 (Brading, Bülbring & Tomita, 1969a).

Calcium. The effect of temperature on calcium permeability in smooth muscle has been studied by Goodford (1965) who found that low temperature (4°C) decreased the calcium efflux.

The results so far obtained regarding the effect of temperature on ion movements in smooth muscle indicate that the influence of cooling is rather complex. Several functional and structural parameters are probably interfered with. Therefore, although cooling effectively abolishes the mechanical and electrical activity, possible secondary effects and artifacts have to be carefully controlled.

Possible mechanical artifacts in cardiac muscle

The effect of changes in tension has been neglected in many previous investigations of ion fluxes in smooth muscle. On the other hand, some investigators of ion fluxes in heart muscle have abandoned their previous findings on cation fluxes because they think that they are secondary to the mechanical events. Pak, Walker, Greene, Loh & Lorber (1966) observed that in heart tissue the efflux of cations, as well as of labelled substances known to be largely restricted to the extracellular space, was synchronous with the systole, and therefore regarded the increase in potassium efflux during contraction as an artifact.

Lamb & McGuigan (1968) reached similar conclusions from ion flux measurements in beating frog ventricles stimulated electrically. They found that during the period of contraction there was an overall reduction of both potassium influx and efflux compared with the resting period. The decreased K-efflux occurred at the same time as the mechanical twitch. However, there was a transient increase in the efflux of Na, Ca, K, Cl, SO_4, sorbitol and erythritol with the peak occurring just after the point of maximum rate of contraction. They concluded that all these effects were mechanical artifacts.

Results derived from heart muscle cannot, of course, be applied directly to smooth muscle, but for further research it will be necessary to take these factors into consideration.

Conclusions

Before we are able to distinguish between a genuine drug effect and an effect secondary to a change in mechanical or electrical activity, only few conclusions can be drawn from the results obtained regarding the effect of drugs on ion permeabilities. These non-specific effects must be controlled, and corrected for, in future experiments. However, although previous investigations of drug effects on the ion distribution and permeability have only partly fulfilled these requirements, results have been obtained which have supported or substantiated ideas reached by electrophysiological techniques.

In smooth muscle, the same transmitter can produce excitation in one tissue and inhibition in another. These opposite effects have been explained by an action on a receptor, still a hypothetical structure, which receives the transmitter at the membrane level and determines the effect exerted on the cell membrane.

Although our present knowledge is still incomplete, it will be attempted here to cover the transmitter action on the ion level under two headings, the excitatory and the inhibitory responses. The similarity between the responses initiated either by cholinergic or adrenergic agents might justify such an approach.

EXCITATORY RESPONSES TO DRUGS

Acetylcholine

Acetylcholine initiates contractions in smooth muscle from a variety of visceral organs such as intestines, bladder, uterus and vas deferens as well as in smooth muscle with other origin such as the nictitating membrane, iris and chick amnion. Electrical recording shows that in taenia coli the excitatory response to acetylcholine is associated with a fall in membrane potential, initiation or increased rate of firing of action potentials and prolongation of the spike duration (Bülbring, 1957a, b, c). It has been suggested that an increase in Na- and K-conductance could account for many of the excitatory actions of acetylcholine in smooth muscle (Burnstock, 1958a; Bülbring & Burnstock, 1960; Bülbring & Kuriyama, 1963b).

Effect on ion distribution

Rather few studies have been concerned with the effect of acetylcholine on the ion distribution. Banerjee & Lewis (1964) observed a significant fall of potassium content in isolated strips of guinea-pig ileum treated with acetylcholine, carbachol and histamine. Paton & Rothschild (1965) estimated the content of potassium, sodium and calcium in isolated longitudinal muscle of the guinea-pig ileum. Using graded doses of acetylcholine, they found a decreased content of K in the tissue and a quantitatively similar rise in Na content, but no effect on the Ca content. Extracellular space was not estimated in these studies and the effect on the intracellular concentration could therefore not be calculated.

The effect of excitatory drugs on the intracellular ion content of smooth muscles which are stimulated by the adrenergic transmitter has not been investigated.

Effect on ion flux

Potassium. Acetylcholine and other cholinomimetic stimulants have been found to increase the potassium efflux (Born & Bülbring, 1956; Lembeck & Strobach, 1956; Hurwitz, 1960; Durbin & Jenkinson, 1961a; Weiss, Coalson &

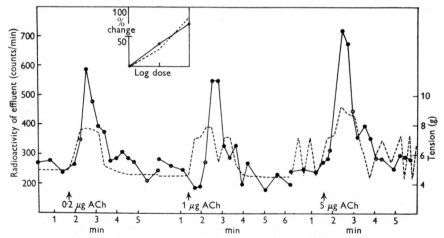

FIG. 7   Effect of increasing doses of acetylcholine (0·2, 1 and 5μg) on tension (– – –) and ^{42}K-efflux (———) from guinea-pig taenia coli. *Inset*: Log dose plotted in relation to percentage change in tension and efflux. (From Born & Bülbring, 1956)

Hurwitz, 1961; Banerjee & Lewis, 1964; Spero, 1967; Fig. 7). Most investigators have found that the uptake of radioactive potassium is reduced after acetylcholine (Hurwitz, 1960; Durbin & Jenkinson, 1961a; Weiss *et al.*, 1961; Banerjee & Lewis, 1964).

In order to eliminate the effects of alterations in the membrane potential, Durbin & Jenkinson (1961a) studied the potassium uptake and loss in muscle depolarized in K_2SO_4-Ringer. Under these conditions an increase in uptake as well as in efflux was obtained by application of carbachol (3.10^{-7} g/ml), indicating an increased permeability to potassium. Depolarized smooth muscle still develops tension after application of acetylcholine (Evans et al., 1958). The tension development can be abolished by omitting calcium from the bathing fluid. The effect of acetylcholine on the K-efflux also requires a normal calcium concentration. A decrease in external Ca concentration reduces the acetylcholine induced increase in K-efflux (Weiss & Hurwitz, 1963; Schatzmann, 1961; 1964a, b; Banerjee & Lewis, 1963). Durbin & Jenkinson (1961b) observed that the acetylcholine effects on K-efflux and on tension can be dissociated, because the development of tension is more sensitive to calcium lack than is the increase in K-efflux. By gradual reduction of the external calcium they obtained an increased K-efflux without concomitant increase in tension.

Cocaine has been found to block the acetylcholine contraction selectively (Hurwitz, 1965). The opposite effect is exerted by desoxycorticosterone which reduces the large increase of potassium efflux caused by pilocarpine, but only slightly affects tension development in isolated ileal muscle (Bass, Hurwitz & Smith, 1964). Spero (1967) differentiated between the increase in K-efflux caused by carbachol and that due to the contraction in response to carbachol by studying the dose-response relationships (see Fig. 5). He found that the contractile response was a more sensitive parameter than the potassium efflux. Further, the rise in K-efflux produced by a maximal contraction or complete depolarization was much lower than the K-efflux after maximal carbachol stimulation. He also found that various agonists influenced the contraction and the K-efflux in a different manner. After application of some agonists the contraction was accompanied only by a small increase in ion permeability, while others induced a much larger increase, indicating that there probably exist two pathways leading to contraction. The effects on K-efflux were blocked by atropine-like antagonists, but not by hexamethonium.

The results support the view that cholinomimetic agents increase the potassium permeability. They also stress the important role played by calcium.

Calcium. Carbachol has been found to increase the calcium content in taenia coli (Herrlinger, Lüllmann & Schuh, 1967). Other investigators have, however, not been able to detect any significant increase in the intracellular Ca-concentration upon addition of cholinomimetic drugs (Breeman, Daniel & Breeman, 1966; Lüllmann & Siegfriedt, 1968). The efflux of calcium is increased by acetylcholine and cholinomimetic agents (Durbin & Jenkinson, 1961b; Schatzmann, 1961, 1964a, b; Hattingberg, Kuschinsky & Rahn, 1966). The influx of Ca was also found to increase after acetylcholine in depolarized muscle (Robertson, 1960; Durbin & Jenkinson, 1961b) but not in normal Ringer solution (Schatzman, 1961, 1964a, b). Banerjee & Lewis (1963) found that acetylcholine increased the uptake as well as the loss of ^{47}Ca in guinea-pig ileum whereas carbachol increased the uptake, but not the efflux. For more details see Chapter 5.

Since Ca^{2+} is indispensable for the transmitter effect, it has been postulated that calcium is released by the transmitter, either from an intracellular store, probably the cytoplasmatic reticulum (Edman & Schild, 1962; Schatzman, 1961, 1964a, b), or that calcium enters the cell from the extracellular space due to an increased calcium permeability of the membrane (Durbin & Jenkinson, 1961b).

Sodium. The effect of acetylcholine on sodium distribution has been difficult to investigate because of the high transfer rate of this ion. Durbin & Jenkinson (1961a) found an increased uptake after acetylcholine, but no significant effect on the efflux. From electrophysiological studies Bülbring & Kuriyama (1963b) suggested that the permeability to sodium is increased during depolarization by acetylcholine.

Other ions. The inward and outward movement of ^{36}Cl and ^{82}Br has also been found to be increased by acetylcholine (Durbin & Jenkinson, 1961a).

The effect of cholinomimetic stimulating drugs seems thus to be a non-selective increase in permeability to several ions.

Catecholamines, oxytocin and angiotensin

Adrenaline and noradrenaline produce an excitatory response in several types of smooth muscle, such as the vas deferens, the oestrogen dominated rabbit uterus, vascular smooth muscle and cat nictitating membrane. From electrophysiological studies similar ionic mechanisms as those involved in excitation by acetylcholine have been proposed for the excitatory effect at adrenergic nerve–smooth muscle junctions (Burnstock, 1960), but tracer studies have not been performed to substantiate this view.

Oxytocin has an excitatory effect on the uterus (Marshall, 1964). Higher doses of oxytocin increase the sodium efflux (Türker, Page & Khairallah, 1967). This increase is blocked by ouabain in a similar manner as the *angiotensin*-induced increase. Because of the sensitivity to ouabain the authors postulate that angiotensin stimulates the sodium pump. Bradykinin which also contracts myometrial cells, is not associated with an increased sodium efflux.

INHIBITORY RESPONSES TO DRUGS

Catecholamines

Inhibitory responses of smooth muscles to drugs have been most thoroughly investigated in taenia coli where adrenaline causes an arrest of spontaneous spike discharge, hyperpolarization and relaxation (Bülbring, 1957a, b, c). Regarding the underlying ionic mechanisms which initiate these electrical and mechanical changes, effects on several ions have been suggested (see also Chapter 12.)

Ion contents and fluxes

Sodium. Theoretically, the hyperpolarization could be explained by a decrease in sodium permeability. This would reduce the intracellular content of sodium provided sodium was still being pumped out of the cell. The transfer of positive charge by an electrogenic pump would also hyperpolarize the cell. Burnstock (1958b) suggested that the effect of adrenaline in taenia coli might be caused by stimulation of an electrogenic sodium pump. From electrophysiological studies Bülbring & Kuriyama (1963b) concluded that adrenaline affected the Na-conductance during the active and resting state of the membrane, modifying the movement of sodium across the membrane. However, Bülbring, Goodford & Setekleiv (1966) obtained only small and inconsistant effects of adrenaline on the intracellular content of sodium in taenia coli, changes which could be accounted for by the recorded increase in extracellular space.

As already mentioned, changes in the transfer of sodium are difficult to measure in taenia coli because of the high exchange rate (Goodford & Hermansen, 1961; Durbin & Jenkinson, 1961a). An increase of the rapid as well as of the slow phase of sodium efflux after adrenaline was obtained by Bülbring et al. (1966), but the effect was small and could well be due to a change in extracellular space, to slight movements of the tissue, or to a change in the ratio of the cell volume to cell surface area. On the other hand, the consistent increase of the slow phase of Na-efflux by adrenaline might be the result of the competition between Na and Ca for anionic membrane sites (Bülbring, Goodford & Setekleiv, 1966) so that, in the presence of adrenaline, Na is dislodged and Ca fixation is favoured (see below, and Chapter 12). In a modified K_2SO_4-Ringer solution with 51 mM Na and 190 mM K at 10°C, Jenkinson & Morton (1967b) were not able to increase the uptake of ^{24}Na in guinea-pig taenia coli with noradrenaline $(3.10^{-7}$ g/ml), although a significant increase in sodium uptake was obtained with this method after carbachol $(5.10^{-7}$ g/ml).

The results so far regarding the effect of adrenaline on sodium permeability are not conclusive and the sodium hypothesis remains to be proven.

Calcium. Excitatory responses in smooth muscle are believed to be associated with mobilization of calcium (see p. 346). Inhibitory responses, on the other hand, have been thought to be associated with fixation of calcium at membrane sites (see Chapter 12). Bülbring & Kuriyama (1963a, b) concluded that the low membrane potential in taenia coli and the slow rate of rise of the action potential might both be due to a poor fixation of calcium at the membrane, resulting in a high Na-permeability of the resting membrane and a limited availability of Na carrier for the spike. Adrenaline would then act by increasing calcium fixation and, like excess calcium in the external solution, would reduce sodium permeability, stabilize the membrane, and increase the rate of rise of the action potential by activation of the Na carrier. (The function of calcium at the cell membrane is discussed in detail in Chapters 7 and 12.)

Schatzmann (1964b) studied the effects of adrenaline on the uptake of Ca in taenia coli, but found no increase, and Banerjee & Lewis (1963) observed no effect on the ^{47}Ca-efflux. The same results were obtained by Hattingberg et al.

(1966). They found, however, that papaverine, which also relaxes smooth muscle, increased the uptake as well as the efflux of ^{45}Ca, but had no effect on the cellular calcium content. At present, therefore, little can be deduced from the effect of adrenaline on Ca^{2+}-movements during the inhibitory response.

Potassium. Since the measured membrane potential is less than the potassium equilibrium potential, an increased potassium permeability would hyperpolarize the cells, and that could be a possible mechanism underlying the hyperpolarizing effect of adrenaline in taenia coli. By studying the effects of changing the external concentration of potassium and sodium, Kuriyama (1963*a*) and Bülbring & Kuriyama (1963*b*) obtained some evidence that adrenaline increased the K-conductance. By estimating the intracellular content of potassium before and after adrenaline (10^{-7} g/ml) Bülbring *et al.* (1966) found a small increase in the intracellular potassium but, as previously mentioned, simultaneous change in extracellular space and tension make this observation less reliable.

In studies of potassium fluxes in taenia coli, several investigators have found that adrenaline increases the rate of uptake of radioactive potassium, in normal Krebs solution (Born & Bülbring, 1956; Hüter, Bauer & Goodford, 1963; Bülbring *et al.*, 1966) as well as in tissue depolarized with a high external potassium concentration (Jenkinson & Morton, 1967*b*). However, the increase in the efflux of ^{42}K after adrenaline application in normal Krebs solution at 35°C is so small and variable (Born & Bülbring, 1956; Bülbring *et al.*, 1966) that one has to consider whether it is due to an artifact or not. As first shown by Jenkinson & Morton (1965), the increase in K-efflux by noradrenaline becomes consistent in muscle depolarized with K_2SO_4-Ringer.

We have studied the effect of different external potassium concentrations on the adrenaline-induced increase in ^{42}K-efflux in a series of experiments. In normal Krebs solution, after transferring the muscle to inactive solution, the radioactive potassium is washed out of the extracellular space in about 5 min, after which the ^{42}K-efflux proceeds at a constant rate for 1 to 2 hr. The muscles were, therefore, washed for at least 35 min to ensure that they were equilibrated and that a constant efflux of radioactive potassium per unit time was obtained. Adrenaline was then added to each test-tube just before the muscle was transferred to the tube, making a concentration in the bathing solution of 10^{-7} g/ml. In normal Krebs solution adrenaline sometimes increased the potassium efflux, sometimes decreased it and sometimes had no effect, which confirms the findings reported previously.

Figure 8 represents the results of four muscles from the same guinea-pig incubated in four solutions with different external potassium concentrations (0, 5·9, 29·5 and 59 mM). In this particular experiment (Setekleiv, unpublished data), adrenaline depressed the ^{42}K-efflux from the muscle in normal Krebs solution ($[K]_0$ 5·9 mM). No effect was seen on the muscle in the bathing solution containing zero potassium. However, in solutions with high potassium ($[K]_0$ 29·5 and 59 mM) adrenaline increased potassium efflux.

The variable response to adrenaline in solutions with normal potassium concentration has been attributed to different factors. It has not been possible to evaluate the effect of the cessation of spike activity on K-efflux. However, it has been shown that the increase in the rate of K-loss after adrenaline is less if the tissue (guinea-pig taenia coli) exhibits spontaneous contractions and responds

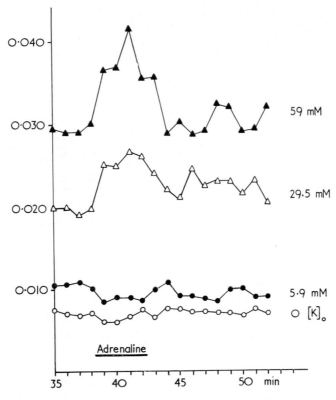

FIG. 8   Effect of adrenaline (10^{-7} g/ml) on the potassium efflux (fraction, lost per min) from guinea-pig taenia coli in bathing solutions with different potassium contractions (0, 5·9, 29·5 and 59 mM).

with a relaxation than when the mechanical response is blocked by lowering the temperature (Jenkinson & Morton, 1967a). That the relaxation itself is not the most important factor for the reduced response has been shown by application of isoprenaline. This relaxes taenia coli in a similar way to noradrenaline, but there is no simultaneous increased efflux of potassium (Fig. 9). Since the potassium efflux seems to be closely linked to the membrane potential, the rise in membrane potential probably opposes the increase in K-efflux produced by adrenaline in normal Krebs solution at normal temperature (Setekleiv, 1967).

It is therefore concluded that adrenaline and noradrenaline induce an increase in potassium influx as well as in efflux, i.e. an increased potassium permeability of the membrane. This is the most likely explanation for the ionic mechanisms underlying the hyperpolarization, a view which is strongly supported by electrophysiological evidence (Bülbring & Tomita, 1969a, b, c). The increase in potassium permeability is the main component of the inhibitory action of adrenaline and noradrenaline — but not of isoprenaline — on taenia coli.

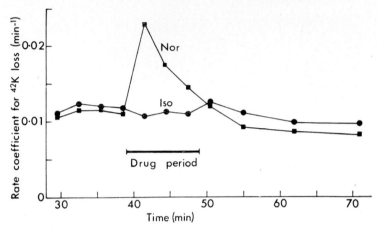

FIG. 9   Effect of 3.10^{-7} g/ml isoprenaline (●) and 3.10^{-7} g/ml noradrenaline (■) on the ^{42}K-efflux from guinea-pig taenia coli at 20°C depolarized by high external potassium concentration (235 mM). (From Jenkinson & Morton, 1965)

Chloride. This is ascribed an important role in inhibitory processes at nervous and nerve skeletal muscle junctions. Although not much studied in smooth muscle, it has been shown that adrenaline has no effect on the intracellular concentration of chloride (Bülbring *et al.*, 1966) nor on the chloride efflux in depolarized muscle (Jenkinson & Morton, 1967b). Measurements of membrane conductance, however, indicate that adrenaline increases Cl conductance in taenia coli (Bülbring & Tomita, 1969a).

Adrenergic receptors

Additional information regarding the effect of catecholamines on K-efflux has been established by Jenkinson & Morton (1967c) with adrenergic blocking agents (Fig. 10). They found that the α-blocking agent phentolamine abolished the increase in potassium efflux caused by noradrenaline. The β-blocking agent pronethalol failed to do this. The increase in potassium efflux is therefore mediated by α-receptors. The ionic mechanisms involved in the inhibition mediated by β-receptors are not known. The inhibitory effect of adrenaline is accompanied by an increase in 3'5'-cyclic AMP and other energy-rich phosphate compounds in the guinea-pig taenia coli. This indicates an increased metabolic energy supply which has been proposed to induce changes in ionic permeability, possibly mediated via an influence on calcium in the membrane (Bueding, Bülbring, Gercken, Hawkins & Kuriyama, 1967).

Since the metabolic effect of adrenaline in other tissues is mediated through β-receptors, and since the β-action does not affect the potassium permeability in taenia coli (Jenkinson & Morton, 1965), the metabolic action might influence other inhibitory mechanisms than the potassium permeability.

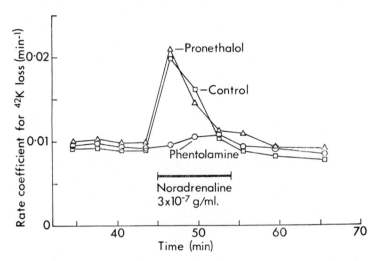

FIG. 10   The different effect of α- and β-adrenergic blocking agents on the increase in ^{42}K-efflux induced by noradrenaline. Phentolamine abolishes the increase while pronethalol has no effect. (From Jenkinson & Morton, 1965)

By measurement of membrane resistance in taenia coli, Bülbring & Tomita (1968a, 1969a, b) found that the reduction of the electrotonic potential by adrenaline and noradrenaline was abolished by an α-blocker (phentolamine). The suppression of the spikes, however, was not abolished until a β-blocker (propranolol) was applied in addition.

By changing the external ion concentration Bülbring & Tomita (1969a, b, c) found that an increased permeability to potassium and chloride was the most likely explanation for the reduced membrane resistance. These changes were attributed to an increased Ca-binding in the membrane mediated by α-adrenergic receptors. They propose that the generator potential is due to removal of calcium from the cell and that catecholamines, especially isoprenaline, oppose this removal. The effect is believed to be mediated by their action on the β-receptors (see Chapter 12).

Hormonal inhibition of uterine activity

Progesterone, which hyperpolarizes the myometrial cell membrane and abolishes mechanical and electrical activity, induces no changes in the intracellular ion concentrations (Casteels & Kuriyama, 1965), but increases the potassium permeability of the membrane. By studying the K-uptake and K-efflux in myometrial tissue depleted of potassium, Jones (1968) found that progesterone increased the rate of K-uptake as well as the K-efflux (see Chapter 4). The results indicate that the inhibitory effect of progesterone on the myometrium is mediated by an increased potassium permeability.

SUMMARY

Conclusive evidence regarding the effect of drugs on ion distribution and flux in smooth muscles has been difficult to establish, (i) because the present methods of intracellular ion determination are inaccurate, (ii) because the kinetics of the ion fluxes are not adequately related to the ion movement through the cell membrane and (iii) because non-specific effects due to changes in electrical and mechanical activities are difficult to eliminate. In smooth muscle the influence of the latter parameters are especially complicated since smooth muscles are not only contractile units, but are also sensitive to stretch. A review of our present knowledge of the influence of electrical and mechanical activity on the ion movement in smooth muscle has therefore been presented, and experiments have been described which have been carried out to elucidate these aspects further. The results show a close correlation between potassium efflux and the membrane potential and also between potassium efflux and muscle tension.

In the analysis of the ionic basis of the responses to drugs, the interference by changes in electrical activity has been eliminated by studies on depolarized muscle; the mechanical activity has been abolished by lowering the temperature or by removal of calcium. The consequences of these procedures on the ion exchange and the influence on the effect of drugs on ion fluxes has been reviewed. It is essential to distinguish between the genuine drug effect on the ion permeabilities of the cell membrane, and the secondary effects resulting from changes in electrical activity and tension.

Since the ionic mechanisms involved in nervous excitation and inhibition of smooth muscle are still insufficiently known, the studies of drug effects on ion distribution and flux have so far been restricted mainly to the action of the cholinergic and adrenergic neurotransmitter. The same transmitter may elicit in smooth muscle both excitatory and inhibitory responses. Since similar mechanisms may be activated during the same type of response, regardless of the initiating agent and of the origin of the tissue, the present review has been arranged under these two headings, i.e. as ionic mechanisms involved in excitatory and inhibitory responses.

The results obtained by the administration of drugs indicate that the ionic mechanisms involved in excitatory responses are a non-selective increase in membrane permeability to several ions. The ionic mechanism underlying inhibition of smooth muscle activity is not yet fully understood. Although an increased permeability to potassium after adrenaline administration is well established in intestinal smooth muscle, influences on chloride, sodium and calcium permeabilities, as well as on metabolic processes, seem to be involved. Selective blockade of adrenergic receptors indicate that several mechanisms are activated in the inhibitory response to catecholamines.

Future experiments must aim at controlling non-specific interference with ion flux, and at developing more quantitative techniques.

13+

12

EFFECTS OF IONS AND DRUGS ON THE ELECTRICAL ACTIVITY OF SMOOTH MUSCLE

H. Kuriyama

INTRODUCTION

Most of the early research into smooth muscle electrophysiology was done by Bozler who recorded electrical activity with extracellular electrodes. This technique has produced a wealth of information about the physiology of smooth muscles and, in 1948, Bozler summarized his ideas on the functional behaviour of smooth muscle in an article 'Conduction, automaticity and tonus of visceral muscles'. Most of his early work has since been confirmed by workers using different techniques.

The introduction of the micro-electrode made possible more precise measurements of electrical activity. This technique was first employed for mammalian smooth muscle by Bülbring & Hooton (1954), Greven (1954) and Woodbury & McIntyre (1954). Since then the electrophysiology of smooth muscle has progressed a great deal especially in the past decade. However, most publications deal mainly with phenomenological observations, and there are still many problems to be solved before the basic mechanisms of physiological function or of drug action are understood.

In 1954, Bülbring reported a membrane potential of 60 mV in the smooth-muscle fibres of the taenia coli of the guinea-pig, measured with micro-electrodes. In 1955 and 1957a, she demonstrated a close correlation between the membrane potential and both the frequency and configuration of the action potential, from investigations of the effects of applied current and chemicals. Subsequently the excitation and conduction in smooth muscle were investigated (Bülbring, Burnstock & Holman, 1958). Since then, Bülbring and her co-workers have continued to study the membrane properties and the mechanisms of drug action on various smooth muscle types (see reviews by Bülbring, 1962; 1964).

Another group that have studied the electrophysiology of smooth muscle over this period are Prosser and his co-workers. They have concentrated on the comparative physiology of smooth muscle tissues and on the mechanism of the propagation of excitation (see review by Prosser, 1962).

Since it is much easier to study the electrical activity of the tissue *in vitro*, most of the available information is based on *in vitro* experiments. However, the activity of smooth muscles is influenced by many factors, for example stretch, temperature, composition of the medium, and also by the nervous activity within the tissue. These factors can alter the muscle behaviour and may produce a large variation in the results. Therefore, when one considers the physiological properties of a smooth muscle, the results obtained in *in vitro* experiments must be interpreted with great care. Unfortunately, it is usually even more difficult to draw definite conclusions from the results obtained in *in vivo* experiments, which involve many uncertainties because of the technical limitation in controlling the conditions.

This article summarizes the electrophysiological observations obtained since 1960 from smooth muscles, mainly by the micro-electrode technique. The description is limited, with few exceptions, to intestinal smooth muscle.

GENERAL FEATURES OF RESTING POTENTIAL AND ACTION
POTENTIALS RECORDED FROM DIFFERENT MAMMALIAN
SMOOTH MUSCLES

Resting membrane potential

The membrane potentials of smooth muscle fibres differ from one tissue to another, as shown in Table 1. For example, a non spontaneously active smooth muscle, e.g., sphincter pupillae and vas deferens, has a higher membrane potential than a spontaneously active smooth muscle, e.g., intestine, and pregnant myometrium. In spontaneously active preparations, the resting membrane potential is defined by most investigators as the maximum polarization between spikes recorded by a micro-electrode. If the tissue shows bursts of spontaneous activity, the value obtained during the quiescent period is taken as the membrane potential.

There are several difficulties in comparing the data reported by different investigators even on the same tissue. One problem is the junction potential at the tip of a micro-electrode. Even in the frog skeletal muscle fibres, large variations (about 30 mV) of the recorded resting potential can be introduced by the tip potential, and electrodes of high resistance are more likely to give rise to this trouble than electrodes of low resistance (Adrian, 1956). Since it is essential to have a very fine electrode tip to avoid damage to the small smooth muscle fibres, resistances of electrodes usually range between 40 and 100 MΩ. Therefore, the resting potentials recorded from smooth muscles must be interpreted with some reservation.

Another problem is that many factors affect the membrane potential. For example, stretch of a tissue reduces the membrane potential (Bülbring, 1955; Bülbring & Kuriyama, 1963c). In the taenia, the highest membrane potential is recorded between 29°C and 32°C (Bülbring & Kuriyama, 1963c). In the vas deferens, the membrane potential is reduced by lowering the temperature below 32°C (Kuriyama, 1964a).

Action potential

Configuration of the action potential

Variability of the action potential is characteristic of smooth muscles. The shape differs greatly from one tissue to another, and even in the same tissue, it varies depending on conditions (or it may be different at different sites of recording). In some tissues, the action potential is a simple spike-like potential, as in the taenia; but in other tissues, it consists of spike-like potentials superimposed on a slow plateau potential, as in the ureter. Some tissues have spontaneous activity as the taenia; and other tissues are quiescent but may be activated through nervous activity, as vas deferens.

The configuration of the action potential also differs with the method of recording, i.e. external recording with wick electrodes, pressure electrodes, sucrose-gap method or intracellular recording with micro-electrodes.

It is difficult to be certain, even with micro-electrodes, that the true trans-membrane potential is really recorded. Some methods of extracellular measurement (e.g. pressure electrode) produce records that are very similar to those observed with micro-electrodes, and it may be that in a tissue with such small cells as smooth muscle, a significant amount of damage is done to the cell membrane even with very fine electrodes. Records obtained with a micro-electrode in a smooth muscle cell may vary considerably depending on the degree of damage done to the cell, from a perfect penetration recording a true transmembrane potential to a badly damaged cell virtually recording extra-cellularly (Bortoff, 1961a; Gillespie, 1962b).

Table 1 contains values for the amplitude of the action potential measured in various smooth muscles of several species, with micro-electrodes. The mean overshoot potential usually does not exceed 10 mV in spontaneously active tissue. However, in quiescent smooth muscles, e.g. vas deferens, the overshoot potential of spikes triggered by nerve stimulation may exceed 10 mV.

The maximum rate of rise of the spike in all smooth muscles is very slow compared with that in other excitable tissues, such as frog skeletal muscle (Nastuk & Hodgkin, 1950) or squid giant axon (Hodgkin & Katz, 1949) which have a rate of rise of about 500 V/sec. The values in the smooth muscle never exceed 30 V/sec, and the rate of fall often exceeds the rate of rise of the spike.

Holman (1958) postulated that the taenia coli may have a limited carrier for Na ions, partly due to inactivation of the current carrying system at the low membrane potential. An increase of the external Ca concentration or condition-ing hyperpolarization by inward current increases the rate of rise (Bülbring & Kuriyama, 1963a; Kuriyama & Tomita, 1965). This may be explained by a reduction of inactivation in the Na carrying system, as in the nerve fibre (Hodgkin & Huxley, 1952c; Frankenhaeuser & Hodgkin, 1957).

More recent experiments (Brading, Bülbring & Tomita, 1969b) have suggested that the action potential in the taenia coli is due to an entry of Ca ions rather than of Na ions (see Chapter 7). The effects of excess Ca or of hyperpolarization indicate that the Ca carrier system is affected similarly to the Na carrier system in nerve. The maximum rate of rise of the action potential in squid axon is about 500 V/sec in the presence of the physiological concentration of 450 mM Na, and is roughly proportional to the external Na concentration (Hodgkin & Katz, 1949). One of the reasons for the slow rate of rise of the smooth muscle spike, which utilizes Ca, is probably that the external solution contains only 2·5 mM Ca.

The action potential produced spontaneously or elicited by electrical, chemical or mechanical stimulation is associated with the development of tension (see Chapter 9). A single spike triggers the phasic twitch tension development, and

TABLE 1. Membrane potential and action potential recorded from various smooth muscle tissues and species using microelectrode technique (except uterine and vascular smooth muscle)

(OS = overshoot)

Species	Tissues	MP(mV)	AP(mV)	Observers
dog	small intestine	35–50	5–30	Daniel, Honour & Bogoch (1960)
cat	jejunum (longit)	52	max. 68	Kuriyama & Osa (unpublished)
guinea-pig	stomach antrum	58·1	OS	Hidaka, Kuriyama & Tasaki (1969)
guinea-pig	stomach pylorus	61·2	OS	Hidaka, Kuriyama & Tasaki (1969)
guinea-pig	stomach greater curvature	53·2	OS	Hidaka, Kuriyama & Tasaki (1969)
guinea-pig	stomach lesser curvature	59·33	OS	Hidaka, Kuriyama & Tasaki (1969)
guinea-pig	bile bladder	52	42	Hidaka, Kuriyama & Tasaki (1969)
guinea-pig	common bile duct	54	58	Hidaka, Kuriyama & Tasaki (1969)
guinea-pig	duodenum (longit)	54·3	62·5	Kuriyama, Osa & Toida 1967b)
guinea-pig	jejunum (longit)	56·0	64·2	Kuriyama, Osa & Toida (1967b)
guinea-pig	jejunum (circular)	60·4		Kuriyama, Osa & Toida (1967b)
guinea-pig	ileum (longit)	54.0	61.8	Kuriyama, Osa & Toida (1967b)
guinea-pig	ileum (circular)	57.4		Kuriyama, Osa & Toida (1967b)
guinea-pig	caecum (circular)	57·8	64·3	Kuriyama, Osa & Toida (1967b)
guinea-pig	rectum (longit)	55·6	60·8	Kuriyama, Osa & Toida (1967b)
guinea-pig	taenia coli	51·5	59·3	Holman (1958)
guinea-pig	taenia coli	53	61	Bülbring & Kuriyama (1963a)
guinea-pig	taenia coli	55·4	57·6	Casteels & Kuriyama (1966)
guinea-pig	taenia coli	53·5	63·2	Kuriyama, Osa & Toida (1967b)
cat	ureter	45	32	Kobayashi & Irisawa (1964)
guinea-pig	ureter	60	70–75	Bennett, Burnstock, Holman & Walker (1962)
guinea-pig	ureter	50–65	OS	Kuriyama, Osa & Toida (1967a)
rabbit	urinary bladder	40	20–70	Ursillo (1961)
guinea-pig	vas deferens	50–80	57–90	Burnstock & Holman (1961)
guinea-pig	vas deferens	62	77	Kuriyama (1963b)

repetitive spikes elicit either incomplete or complete tetanus. Figure 1 shows an example of such a relationship between the spike and tension obtained from the guinea-pig taenia coli.

Besides the action potential just described, smooth muscle normally has another type of electrical activity, the slow potential (or slow wave) which will be described below.

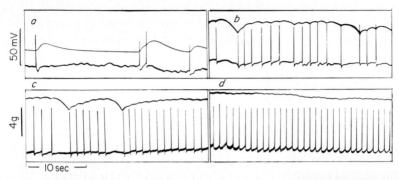

FIG. 1   Intracellular electrical recording from single cells and tension record from a 5 mm length of guinea-pig taenia coli. Stretch in steps from *a* to *d* causes depolarization, acceleration of spike discharges, and consequently an increase in tension. The height of the tone is determined by the spacing of the spikes. (Bülbring, 1962)

Slow potential changes

Recordings of spontaneous activity in smooth muscle in physiological solutions show that the patterns of electrical activity depend very much on slow transmembrane potential changes. Several types of slow potential changes can be seen, and three types will be defined below for convenience in further descriptions.

Type I. Slow potentials with short duration (100 to 1000 msec). These may be divided into the following two categories.

(*a*) Produced near or at the recording site and usually present with a single spike, e.g. pacemaker potentials (Fig. 2*a*), negative and positive after-potentials. (The negative after-potential may trigger a second spike) (see Chapter 7).

(*b*) Produced by electrotonic spread of spike activity in another part of the tissue: can trigger spikes at recording site.

Type II. Synaptic potentials produced by a single stimulus of nerve fibres (usually a few hundred msec) (see Chapter 8). (*a*) Excitatory (Fig. 2*e*). (*b*) Inhibitory; this could lead to after-excitation. (*c*) Combination of excitatory and inhibitory potentials.

Type III. Slow potentials normally associated with a train of action potentials (Fig. 2*b*, *c* and *d*). Gradual development of the slow wave usually leads to firing of the spikes, but sometimes a train of spikes appears abruptly. Termination of the slow wave is usually very slow. The duration is about 5 to 10 sec. Individual spikes may have a type I slow potential. The number of spikes on the slow wave may vary and the slow wave may appear without spikes. Some of this type of slow potential could be due to transmitter action resulting from nervous activity. When the slow potential is small in amplitude the spike amplitude remains more or less constant during the slow potential; but when the slow wave is large the spike amplitude decreases as the slow potential grows.

These three classifications are only tentative. There may be slow potentials with intermediate properties between the types described. The type III slow potential is usually clearly distinguishable from a plateau potential, which occurs in the ureter, and which has a larger amplitude (more than 30 mV) and

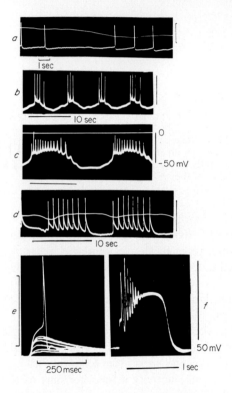

FIG. 2 Various patterns of the membrane activity recorded from various smooth muscle tissues of the guinea-pig. *a* Taenia coli.   *b* Jejunum.   *c* Stomach. *d* Uterus.   *e* Vas deferens.   *f* Ureter. (Bülbring & Kuriyama, 1963*a*; Kuriyama, Osa & Toida, 1967*a*, *b*; Hidaka, Kuriyama & Tasaki, 1969)

a much shorter duration (up to 1 sec). Here, the plateau is initiated by a spike (Fig. 2*f*). and a train of spikes (often decreasing in amplitude) appears only at the beginning of the slow potential; termination of the potential is abrupt.

Slow potentials (type III) in intestinal muscle

Although every type of muscle has a tendency to produce one particular type of slow potential, the tissue is also able to vary its pattern of activity, as seen in Fig. 3. This is especially true for the longitudinal muscle coat of guinea-pig jejunum where no characteristic pattern of the electrical activity can be recorded from any particular part (Kuriyama, Osa & Toida, 1967*b*). A train of spikes can sometimes appear without a slow potential (see top record in Fig. 3). This may be due to the fact that spikes can propagate with a larger safety factor than the slow potentials and, occasionally, the slow potential propagation may be blocked at some point in the tissue, while the spike propagation is unaffected. Gonella (1965) also described two types of slow waves in the longitudinal muscle of rabbit small intestine: one, in which the spike preceded the slow wave and a second in which the slow depolarization preceded the spike, which was only triggered when depolarization reached firing level.

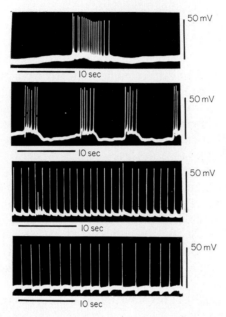

Fig. 3 Various patterns of the spontaneously generated spikes recorded from the smooth muscle cells of guinea-pig jejunum. (Kuriyama, Osa & Toida, 1967*b*)

Stretching the tissue modifies the pattern of spike discharge by reducing the interval between the bursts of discharge or producing continuous firing without silent periods (Mashima & Yoshida, 1965; Mashima, Yoshida & Handa, 1966; Hukuhara & Fukuda, 1968).

At the beginning of *in vitro* experiments, the electrical activity of the longitudinal muscle coat usually consists of bursts of spikes, but after prolonged exposure to Krebs solution the amplitude of the slow potential (type III) often increases or, sometimes, the activity becomes continuous. In this condition, treatment with atropine (10^{-5} g/ml) increases the membrane potential, lowers the spike frequency, and increases the amplitude of the spike. Therefore, it may be that this activity pattern is at least partly due to the spontaneous release of

13*

acetylcholine from nervous elements in the tissue. The release of acetyl-choline is probably increased when the tissue is stretched (Kuriyama *et al.*, 1967*b*, *c*; Hukuhara & Fukuda, 1968).

When the longitudinal muscle is ablated from the circular muscle layer it may loose its bursts of activity and generate spikes irregular in amplitude and frequency. This suggests that a component responsible for the slow potential (type III) is eliminated or damaged. The slow potential (type II) probably pro-duced by electrical stimulation of nervous elements becomes also much smaller than in the intact longitudinal muscle (Kuriyama *et al.*, 1967*b*).

The circular muscles are quiescent and a spike appears only in response to electrical or chemical stimulation, except in the caecum where spontaneous spikes appear regularly, but without a pattern of periodical bursts (Kuriyama *et al.*, 1967*b*). This suggests that the slow potentials (type III) originate in the muscle cells of the longitudinal layer (Sperelakis & Prosser, 1959; Bortoff, 1961*a*). A strip of the longitudinal muscle carefully removed from the circular layer can show the slow wave (type III) (Tamai & Prosser, 1966), while the circular muscle shows the slow wave only when connected to the longitudinal layer. At least in the cat intestine, interaction seems to be electrotonic through strands of muscle which pass diagonally between the two layers. (Bortoff, 1965; Kobayashi, Nagai & Prosser, 1966).

Bi-polar recordings of the slow potential (type III) in the longitudinal muscle show a wave form which is approximately the first time derivative of the monopolar or intracellular recording. When the current generated by the slow potential is measured in the volume conductor, its form is roughly the second time derivative of the slow potential. These observations are similar to those in nerve fibre and thus suggest that the slow potential is propagating at more or less constant speed in a core conductor-like structure (Bortoff, 1961*a*).

The conduction velocity of the slow potential decreases from the duodenum to the ileum. In the upper small intestine, it is 8 to 22 cm/sec and in the lower small intestine, 0·2 to 0·7 cm/sec (Armstrong, Milton & Smith, 1956; Daniel & Chapman, 1963; Bass & Wiley, 1965).

The frequency of slow potentials (type III) decreases also along the small intestine from the duodenum aborally to the ileum (Alvarez & Mahoney, 1922). The frequency of the slow waves in a given region of the intestine is influenced by the frequency of the waves in the region above it. After applying a clamp, the frequency in the region below the clamp is reduced while that of the region above the clamp is unaltered (Milton, Smith & Armstrong, 1955; Milton & Smith, 1956).

Slow potentials (type III) in the stomach

In general the slow potentials are larger in stomach than in the intestine, though there is no basic difference in configuration or in the spikes appearing on the slow potential. Figure 4 shows examples of the spontaneous electrical activity obtained from the guinea-pig stomach. The amplitude of the slow potential may exceed 40 mV and the duration is usually longer than 5 sec. A train of spikes appears mainly during the early phase of the slow potential. The spikes occasionally have an overshoot. The slow potential can appear without triggering the spikes.

Papasova, Nagai & Prosser (1968) classified the slow potential in the cat stomach into two components: the initial rapid component and the second slow component. The spikes may appear only on the second component. The total

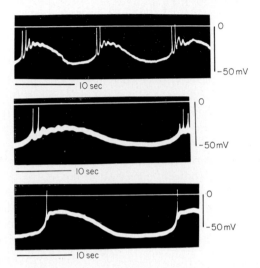

FIG. 4   Spontaneous discharges recorded intracellularly from the longitudinal muscle
of guinea-pig stomach. (Hidaka, Kuriyama & Tasaki, 1969)

duration of the slow potential is 5 to 10 sec and the intervals between the potentials are 12 to 20 sec. In the guinea-pig stomach, the two components of the slow potential could not be differentiated (Hidaka, Kuriyama & Tasaki, 1969).

In the cat stomach, the origin of the slow potential is in the longitudinal muscle layer, as in the intestine, and the slow potential seems to spread electrotonically to the circular muscle (Papasova et al., 1968). Contraction is correlated with the second component of the slow wave (Daniel, 1965b; Papasova et al., 1968). When spikes are present on the slow component, contractions are enhanced. The conduction velocity of the slow wave is 4 to 14 mm/sec and is maximal near the pyloric sphincter. Conduction is faster on the side of the greater curvature than on the side of the lesser curvature. The first component conducts, but the second component may not propagate, at least not in the absence of the first component (Papasova et al., 1968).

EXTERNAL IONIC COMPOSITION AND RESTING MEMBRANE
POTENTIAL

The membrane potential of many tissues can be described in terms of the distribution of K, Na and Cl ions across the cell membrane and of the relative permeability of the membrane to these ions.

In order to avoid repetition, the influence of Na ions on the membrane potential will be discussed in connection with their effect on membrane activity, and the role of Ca in connection with the responses to drugs. This section will therefore be restricted to the importance of K and Cl for the membrane potential.

From the effects observed after changing the external ionic concentrations, it is found that in smooth muscles K is the most dominant ion determining the membrane potential, as in most other tissues. The effects of altering the ionic environment can often be interpreted in terms of membrane permeability and ionic distribution. However, there is evidence which suggests that the mechanisms of regulating ionic distribution and cell volume in smooth muscle differ from those which apply in other excitable tissues, such as frog skeletal muscle. Casteels & Kuriyama (1966) and Brading & Tomita (1968a) found that addition of KCl to Krebs solution produces a loss of intracellular water in the taenia, which persists, and no significant penetration of KCl occurs. Replacing the NaCl of Krebs solution with KCl, however, has no effect on the tissue weight and again no penetration of KCl occurs. This suggests that under these conditions the Donnan distribution of ions, does not apply to the smooth muscle of the taenia (see Chapter 6). This should be borne in mind when trying to interpret the results in terms of the ionic theory of the membrane potential (Hodgkin, 1958).

Effects of K ions on resting membrane potential

In a solution with a constant sum of NaCl and KCl, the line relating the membrane potential to the logarithm of the external K concentration has a maximum slope of 43 mV for a tenfold change of K ions. In a solution with a constant product of K and Cl, the maximum slope is 51 mV (Casteels & Kuriyama, 1966).

If the membrane potential were solely determined by the K diffusion potential, the maximum slope of change in the membrane potential per tenfold change in the external K concentration would be 61 mV. The difference between this theoretical value and the observed value may be due either to a change in the intracellular K concentration or to a contribution by other ions, or to both factors. According to ionic analysis, the intracellular K concentration remains more or less constant under these conditions (Casteels & Kuriyama, 1966). Therefore, it is probable that diffusion potentials of other ions are affecting the relationship. At an external K concentration below 30 mM, the slope deviates from a linear relationship and the membrane potential is less sensitive to changes in the external K concentration. This may also be explained by a relatively high permeability of the membrane to Na and Cl ions.

Effects of substitution of Cl with other anions

When Cl in the external solution is replaced with foreign anions the membrane is transiently depolarized, independent of the substitute for Cl (Kuriyama, 1963a). Within 5 to 10 min the membrane is gradually repolarized reaching a new steady level, which is different for the different anions. The membrane potential at this steady state is still lower with SO_4 (7 mV) and with Br (3 mV) than the normal resting potential, but it is higher with NO_3 (6 mV) and with I (8 mV).

Depolarization caused by substitution of Cl with SO_4, $C_2H_5SO_3$ or Br may be explained by the fact that the substitutes are less permeant than Cl, thus moving the Cl equilibrium potential towards more positive inside. Hyperpolarization by NO_3 and I is due to the fact that these are more permeant than Cl and thus their equilibrium potentials, which are more negative than the membrane potential, affect it more strongly than the Cl equilibrium potential. Repolarization of the membrane within 5 to 10 min after the initial depolarization in Cl deficient solution may be due, partly, to a loss of the intracellular Cl and partly to a secondary effect of an increased spike activity, especially when the substitutes are NO_3 and I.

Figure 5 illustrated the relationship between the membrane potential (at the steady state after 30 to 60 min exposure to a solution) and the logarithm of the external K concentration (K salts added to the solution) in the presence of various foreign anions. The maximum slope per tenfold change in $[K]_0$ is 49 mV with SO_4, 47 mV with $C_2H_5SO_3$, 42 mV with Br, 38 mV with Cl, and 15 mV with NO_3 or I.

Thus, when Cl is replaced with SO_4 or $C_2H_5SO_3$, the changes of the membrane potential caused by increasing the external K concentration come closer to those predicted from a K diffusion potential. This observation, and also the fact that Cl deficiency (in the presence of a normal K concentration) changes the membrane potential, both suggest that Cl-ions contribute to the membrane potential of the taenia. This conclusion is supported by the fact that replacement of Cl with ethane- or benzene-sulphonate decreases the membrane conductance while it is increased by replacement Cl with NO_3 (Bülbring & Tomita, 1969a).

When the concentration of one particular ion in the external solution is changed, it is possible that the membrane permeability to other ions is affected, due to some interaction between different ions in the membrane. If this is true, especially between Cl and K as suggested by Casteels & Meuwissen (1968) based on ion flux experiments, then the interpretation of a change in resting potential or membrane resistance becomes very difficult.

The degree to which Cl ions contribute to the membrane potential is larger in muscles than in nerves, but seems to be very different from one tissue to another. In some muscle, such as the somatic muscle of *Ascaris* (Del Castillo, de Mello

& Morales, 1964) and the twitch fibre of the seawater elasmobranch (Hagiwara & Takahashi, 1967) Cl ions are more dominant than K ions in determining the resting potential. It may well be that the Cl permeability in mammalian smooth muscles also differs from one tissue to another, and this could partly be responsible for producing different responses to drugs (see Chapter 13).

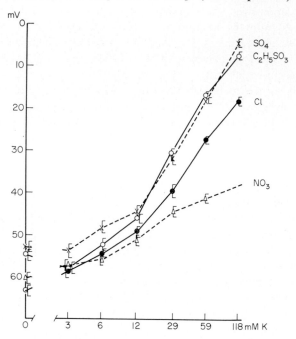

FIG. 5   The relation between membrane potential and the logarithm of the external K concentration in the presence of different anions (guinea-pig taenia coli). (Kuriyama, 1963a)

EFFECTS OF CHANGING THE EXTERNAL IONIC COMPOSITION ON THE ACTION POTENTIAL

Change in external Na and Ca concentrations

When the effect of different Na concentrations was first studied on the spontaneous spike in taenia coli it was found that the amplitude was not much affected by Na deficiency, although the rate of rise was reduced (Holman, 1957; 1958; Bülbring & Kuriyama, 1963a). In a solution containing less than 5 mM Na the spontaneous electrical activity was abolished, but it could be restored by excess Ca. From these observations, it was suggested that, if the action potential was due to Na entry, the inward movement was rather limited, and that Ca could replace Na when the external Na concentration was reduced.

It was therefore of great interest to study the action potential of smooth muscle in Na deficient solution in order to obtain information on the contribution of Na ions which, in most other excitable tissues, is the main ion involved in carrying the inward current. Most of the work has been done on the taenia coli. However, since the observations were made on the spontaneously generated spikes it was difficult to reach a conclusion concerning the underlying mechanism by which Ca and Na ions modified the activity. The effects may be on the slow potentials (type I, II, or III) which evoke the spikes, or on the threshold for triggering the spike, or on the regenerative process for the spike itself.

When the spikes evoked by electrical stimulation are investigated in various Na and Ca concentrations, the results suggest strongly that the Ca contribution to the action potential in the taenia is more important than the Na contribution (Brading, Bülbring & Tomita, 1969b). (see Chapter 7).

One difficulty in investigating the effects of Na reduction is the uncertainty about the effects produced by the substitute itself on the membrane properties. For example, the depolarization of the membrane when Na is reduced is difficult to explain. If tris is the substitute in Na-free solution, the membrane is slowly depolarized by 15 mV, after a transient hyperpolarization (Bülbring & Kuriyama, 1963a). Li substitution produces a larger depolarization than tris and increases the spontaneous spike activity (Axelsson, 1961b; Bülbring & Kuriyama, 1963a; Bülbring & Tomita, 1969b). Choline (Holman, 1957; 1958; Axelsson, 1961b) and tetraethyl ammonium (Bülbring & Kuriyama, 1963a; Suzuki, Nishiyama & Inomata, 1963) have a strong stimulating action which is reduced by a high concentration of a cholinergic blocking agent, atropine (10^{-4}–10^{-5} g/ml).

If sucrose is used as a substitute, the effects of Cl deficiency must be taken into account. Furthermore, reduction of the ionic strength in the external solution might affect the membrane, probably by increasing fixed sites in the membrane for other ions (especially Ca). Na may be important for the regulation of the intracellular Ca concentration. In squid giant axon, the Na-pump is utilized for pumping Ca across the membrane (Baker, Blaustein, Hodgkin & Steinhardt, 1969; Blaustein & Hodgkin, 1969). Therefore, if a similar mechanism is involved in smooth muscle, substitution of Na with either tris or Li may disturb the pump, thereby modifying the Ca distribution. This may lead to some change in the response even if Na ions are not directly contributing to the inward current responsible for the spike.

The ionic basis of the action potential in different smooth muscle types
Intestinal muscle

The guinea-pig taenia coli has been most extensively studied, and a detailed account will be found in Chapter 7. The following observations which support the possibility of a Ca spike should however be mentioned briefly, since they resemble those made on crustacean muscle (Fatt & Ginsborg, 1958; Hagiwara & Naka, 1964; Hagiwara & Nakajima, 1966) where the spike is due to Ca entry across the membrane.

(i) Tetrodotoxin has no effect on the spike (Kuriyama, Osa & Toida, 1966; Nonomura, Hotta & Ohashi, 1966; Kao, 1966; Bülbring & Tomita, 1967) (see Fig. 6).

(ii) Mn blocks spike generation (Nonomura, Hotta & Ohashi, 1966; Brading *et al.*, 1969*b*; Bülbring & Tomita, 1969*c*).

(iii) Ba can substitute for Ca in spike generation (Hotta & Tsukui, 1968; Bülbring & Tomita, 1968*b*, 1969*c*).

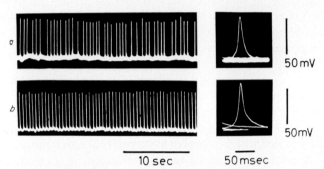

FIG. 6　Effect of tetrodotoxin (5×10^{-7} g/ml) on the spontaneous discharges of the smooth muscle cells of guinea-pig taenia coli. *a* Control. *b* After 20 min exposure to tetrodotoxin. Note increase in spike frequency. (Kuriyama, Osa & Toida, 1966)

Vas deferens

Tetrodotoxin (10^{-6} g/ml) does not appear to have any effect on the excitability of either rat, mouse or guinea-pig vas deferens when the action potential is studied by intracellular stimulation although tetrodotoxin blocks the response to nerve stimulation (Hashimoto, Holman & McLean, 1967; Tomita, 1967*a*). In the mouse vas deferens, Mn (5 mM) abolishes the action potential evoked by intracellular stimulation (Hashimoto & Holman, 1967).

Bennett (1967*a*) studied the effects of Na and Ca on the action potential evoked by intracellular stimulation and by hypogastric nerve stimulation in the guinea-pig vas deferens. Reduction of the extracellular Na ions to less than 30 mM did not change the action potential. Reduction of Ca to one-tenth of the normal concentration decreased the resting potential by 25 mV and the peak of the action potential by 22 mV (Fig. 7). The threshold for initiation of the action potential was also reduced in the presence of low Ca. If the Na ions were also reduced, keeping the ratio of Ca/Na^2 constant, low Ca did not depolarize the membrane, but it decreased the action potential by 20 mV per tenfold change. These observations suggest that Ca ions may carry the current to generate the action potential.

Ureter

In the cat ureter, both the amplitude and the rate of rise of the action potential decreased as the external Na concentration was reduced (Kobayashi & Irisawa, 1964). In Na-free solution (tris-Cl substitute), spontaneous action potentials

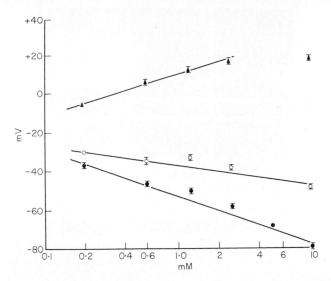

FIG. 7   Guinea pig vas deferens. Effect of external calcium on resting potential (●), peak of the action potential (▲) and threshold for initiation of the action potential (○). Abscissa: calcium activity (log scale). Ordinate: membrane potential (negative inside with respect to outside). The overshoot changes by 22 mV per tenfold change in calcium activity in the range 0·2–2·5 mM. Vertical bars give twice the standard error of the mean, $n = 7$. (Bennett, 1967a)

stopped but they were restored by addition of excess Ca (Kobayashi, 1965). These observations are very similar to those on the guinea-pig taenia coli (Bülbring & Kuriyama, 1963a). Although further experiments, using a proper stimulation, seem necessary to analyse the spike mechanism in the cat ureter, it is possible that both Ca and Na contribute to the action potential even in physiological solution. As shown in Fig. 8, in Na-free solution, the action potential looses its plateau and excess Ca enhances the earlier phase of the action potential. The results indicate that Ca entry may cause the earlier part and Na entry the later plateau of the action potential.

Similar observations have been made on the guinea-pig ureter. The amplitude of spikes became smaller when the external Ca was reduced, while the spike became larger in excess Ca (Bennett, Burnstock, Holman & Walker, 1962). Recent experiments (unpublished), in which the influence of Na and Ca was studied further suggest that the spike component is due to Ca entry and the plateau component is due to Na entry. Both appear to be controlled by Ca bound in the membrane. It is interesting to note that, although the plateau of the action potential is probably due to an increase in Na conductance of the membrane, tetrodotoxin (10^{-7} g/ml) has no effect on the plateau nor on the spike component and Mn (3 mM) blocks the action potential (Kuriyama, Osa & Toida, 1967a).

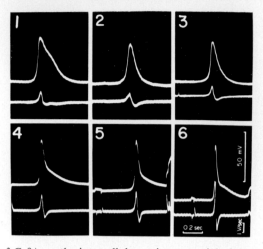

FIG. 8   Effects of Ca^{2+} on the intracellular action potential of the cat ureter in Na^+-free solution. In each recording, the upper tracing is the action potential and the lower is the electrical time derivative of the upper.   (1) Control in Krebs solution (2·5 mM Ca^{2+}), and responses in   (2) Na^+-free, 2·5 mMCa^{2+};   (3) Na^+-free, 5 mMCa^{2+};   (4) Na^+-free,   7·5   mMCa^{2+};   (5) Na^+-free,   12·5   mMCa^{2+};   (6) Na^+-free,   25 mMCa^{2+} (3–6 are evoked responses). (Kobayashi, unpublished observation)

Effects of Ba and Sr

The addition of Ba ($2·4 \times 10^{-5}$ to $2·4 \times 10^{-4}$ M) to Krebs solution depolarizes the membrane in taenia coli and prolongs the spike duration, mainly by delaying the falling phase (Bülbring & Kuriyama, 1963a). Similar results have been obtained in cat circular intestinal muscle (Burnstock & Prosser, 1960c). In the longitudinal muscle of the guinea-pig jejunum, a higher concentration of Ba (0·5 to 1·0 mM) prolongs the action potential to form a very long plateau (up to 10 sec). Treatment with atropine (5×10^{-5} g/ml) nearly abolishes the plateau produced by Ba. Therefore, the effects are probably due to the release of cholinergic transmitter by increased nervous activity evoked by the presence of Ba (Kuriyama et al., 1967c).

Ba also seems to have a direct action on the smooth muscle. In taenia coli, in Ca-free solution, when the membrane is depolarized and spikes are very small or abolished, the addition of Ba (0·25 to 0·5 mM) caused gradual repolarization of the membrane and recovery of spikes (Bülbring & Tomita, 1968b; 1969c). With higher Ba concentrations (1·0 to 2·5 mM) the spike amplitude and duration increases further forming a long-lasting plateau or a maintained state of depolarization. Recent experiments suggest that the plateau of the spike (and the maintained state of depolarization) is due to an increase in the Na conductance of the membrane, since it is abolished in absence of Na (Bülbring & Tomita, unpublished). However, in the presence of Ba a spike can still be produced when Na and Ca are absent, indicating that Ba is able to substitute for Ca in generating the action potential (Hotta, & Tsukui, 1968; Bülbring & Tomita, 1968b; 1969c).

In the guinea-pig ureter, the addition of Ba (10^{-5} M) to normal Krebs solution depolarizes the membrane and produces spontaneous activity. Ba reduces the number of the spikes on the plateau and prolongs the plateau (Bennett et al., 1962; Kuriyama et al., 1967a).

Sr can replace Ca for spike generation in the taenia without changing the membrane potential, the membrane resistance or the spike configuration. However, in the ureter Sr prolongs the plateau. In the absence of Na and Ca, the spikes are produced, but without plateau, when the solution contains Sr (unpublished). Sr seems to have an intermediate effect between Ca and Ba. Divalent cations probably compete with each other for a site in the membrane, because the effects of Ba or Sr are larger when the external Ca concentration is reduced.

Effects of foreign anions on the action potential

In the guinea-pig taenia coli, the spontaneous spike frequency is increased during the initial phase of depolarization when Cl is replaced with another anion (Kuriyama, 1963a). This effect is especially large when the substitute is NO_3. The frequency, compared with that in normal solution, is increased 1·7 times with $C_2H_5SO_3$, 1·9 times with SO_4, 1·5 times with Br, 2·9 times with NO_3, and 2·0 times with I.

The pattern of spike activity differs depending on the Cl substitute. When Cl is replaced with a less permeant anion, such as ethane- or benzene-sulphonate, the spike frequency (after the transient increase) becomes lower than in normal solution; sometimes bursts of spikes appear with quiescent periods and sometimes spontaneous activity ceases. On the other hand, if Cl is reduced by replacing with NO_3, there is usually a continuous high frequency activity (Axelsson, 1961a; Bülbring & Tomita, 1969a).

Modification of the spike activity in Cl deficient solution may partly be due to a change in the membrane potential caused by a shift of the Cl equilibrium potential, or by the diffusion potential of a foreign anion. The other possible mechanism is that the foreign anions have some effect on the Ca permeability of the membrane, as is known for skeletal muscle fibres (Shanes, 1958; Bianchi & Shanes, 1959; Frank, 1960).

EFFECTS OF CHANGING THE EXTERNAL IONIC COMPOSITION ON THE SLOW POTENTIAL

In Na-free solution (tris substitute), the slow potential in the taenia (type I and probably type III) is suppressed, while in excess Na the amplitude of the slow potential is enhanced and repetitive spikes are superimposed on the slow potential (Bülbring & Kuriyama, 1963a; Fig. 9).

In the longitudinal muscle of jejunum or duodenum, reduction of the external

Na concentration by replacing Na with Li decreases the amplitude of the spikes, and in 20% of the normal concentration (50% sucrose replacement) the spike disappears. The slow potential (type III) is reduced below 20 to 30% of the normal Na concentration, and blocked in Na-free solution (Tamai & Prosser, 1966).

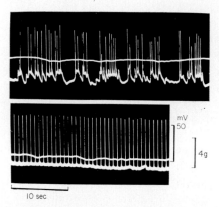

FIG. 9   Effects of excess sodium (1·5 times hyperosmotic) and sodium free (substituted by tris) solution on the membrane activity of the guinea-pig taenia coli. *Upper*: excess sodium (205 mM Na$^+$). *Lower*: sodium free. (Bülbring & Kuriyama, 1963a)

In the cat stomach, reduction of the external Na concentration to 38 mM by sucrose, Li or tris replacement, has little effect on the spontaneous activity. Further reduction of Na (below 15 mM) suppresses the first rapidly rising component of the slow potential, which is abolished in Na-free solution, leaving the second slow component and the spikes intact (Papasova, Nagai & Prosser, 1968). The second component is more sensitive to the external Ca concentration than the first component. In low Ca solution (25 to 75% of the normal concentration of Ca) the first component of the slow potential is not affected, but the second component is reduced. The frequency of the slow potentials is reduced in low Ca solution and is increased in excess Ca. In a low Na solution, the frequency of the activity becomes insensitive to a change in the Ca concentration. Adding Mn produces effects similar to reducing Ca in the solution. Mg ions can be omitted without effect.

In the guinea-pig stomach, the slow potential is reduced in a low Na solution, and Mn does not affect the slow potential at a concentration of less than 1 mM (Hidaka, Kuriyama & Tasaki, 1969).

DRUG ACTIONS ON MAMMALIAN SMOOTH MUSCLE

There are numerous publications on the pharmacology of smooth muscle, describing the effects of various drugs and chemical agents. Classification of the observed effects is extremely difficult, since they vary not only from tissue to tissue, but can also be modified by different experimental conditions. The results may also differ depending on the electrical recording technique. Many smooth muscles are densely innervated, others are sparsely innervated so that the effects of drugs may be due to a direct action on the smooth muscle cells, or mediated indirectly via effects on the nerves.

In this chapter, only the electrical responses of smooth muscle to the application of a few drugs, i.e. acetylcholine and catecholamines will be discussed. The effects of drugs on the uterine myometrium are described in Chapter 13 and the effects of drugs on the transmission of excitation and inhibition from autonomic nerves to smooth muscle cells are described in Chapters 15 and 16.

Acetylcholine
The effects on the resting and action potential
There is a great deal of conclusive evidence that acetylcholine is the excitatory transmitter in most visceral smooth muscle. Details of this evidence are given in Chapter 16. With the exception of resistance blood vessels (see Chapter 20) and

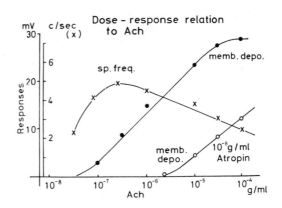

FIG. 10 Dose response relation to acetylcholine measured from single smooth muscle cells of guinea-pig jejunum. ($\times$) spike frequency; ($\bullet$) grades of the membrane depolarization; ($\bigcirc$) depolarizations of the membrane by acetylcholine after treatment with 10^{-8} g/ml atropine. (Hidaka & Kuriyama, unpublished observation)

possibly some circular intestinal muscles (Kottegoda, 1969), the majority of smooth muscle tissues are depolarized when acetylcholine is applied in physiological solution, and spontaneous activity is initiated or, if present, the spike frequency is increased (see Figs. 10 and 11). These effects are observed in the presence or absence of a functional cholinergic innervation (guinea-pig taenia coli (Bülbring, 1954; 1955; Bülbring & Burnstock, 1960; Burnstock & Straub, 1958; Burnstock, 1958a; Axelsson, 1961b; Bülbring & Kuriyama, 1963b), toad stomach (Sato, 1960), guinea-pig stomach (Hidaka, Kuriyama & Tasaki, 1969), pig oesophagus muscle (Burnstock, 1960), cat small intestine (Burnstock & Prosser, 1960c; Bortoff, 1961b), guinea-pig jejunum (Kuriyama et al., 1967c), rat bladder (Carpenter, 1963), rat ureter (Prosser, Smith & Melton, 1955), cat

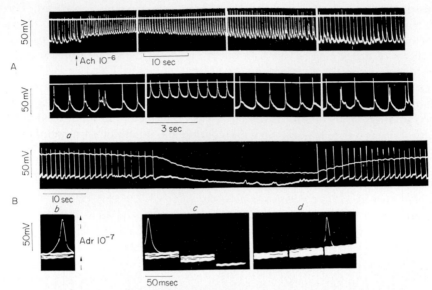

FIG. 11 Effects of acetylcholine and adrenaline on the membrane activity and tension of the guinea pig taenia coli. (A) Acetylcholine 10^{-6} g/ml. (B) Adrenaline 10^{-7} g/ml. (Bülbring & Kuriyama, 1963*b*)

ureter (Irisawa & Kobayashi, 1963), pig sphincter muscle of iris (Schaeppi & Koella, 1964), cat nictitating membrane (Burnstock, Holman & Prosser, 1963), chick ammion (Kuschinsky, Lüllman & Muscholl, 1954; Cuthbert, 1962).

In spontaneously active smooth muscle, the increased spike frequency induced by acetylcholine may be due to its effect on the pacemaker mechanism at some particular part of the tissue, or due to its depolarizing action which may convert non-pacemaker into pacemaker cells. For example, in the rat ureter, local application of acetylcholine to any region of the tissue can cause this area to become a pacemaker, whereas normally impulses originate only from the renal end (Burnstock, Holman & Prosser, 1963).

In intestinal smooth muscle it is not possible to differentiate morphologically between pacemaker and non-pacemaker cells. A functional differentiation can be made from electrophysiological observations. Cells showing spikes with well developed pre-potentials (type I slow potential) as observed in cardiac pacemaker cells, or showing slow depolarization (type III slow potential) triggering a burst of spikes, might be called pacemaker cells, and the other cells might be called propagating cells. The effect of acetylcholine on the pacemaker cells is to increase the slope of the prepotential or the rate of the slow depolarization, thus causing an increase in spike frequency. This increase in spike frequency is also recorded in the propagating cells. In addition, these cells are normally depolarized by acetylcholine, although in some cells of the guinea-pig taenia coli and jejunum an increased spike frequency is sometimes recorded without any significant depolarization. The changes are probably direct effects on the muscle cells, since they are unaffected by treatment with tetrodotoxin, which blocks nervous activity.

The depolarization of the smooth muscle cells by acetylcholine normally observed, is probably due to an effect on the cell membrane permeability. There is little evidence that the acetylcholine receptors are concentrated at a particular site on the cell membrane, and many investigators have postulated that the receptors are distributed over the whole cell surface. There may be differences in sensitivity from cell to cell. The mechanism of action of acetylcholine is thought to resemble its action on the end-plate of frog skeletal muscle. Takeuchi & Takeuchi (1959; 1960) demonstrated that at this site acetylcholine caused an increase in G_{Na}, G_K and G_{Ca}, but had little effect on G_{Cl}.

In the guinea-pig jejunum it was shown that iontophoretic application of acetylcholine depolarized the membrane and increased the membrane conductance (Hidaka & Kuriyama, 1969). In the taenia coli, in which acetylcholine depolarizes the cells in normal solution (see Fig. 11) the effect was reversed when the external potassium concentration was increased, i.e. the membrane was hyperpolarized on addition of acetylcholine. Burnstock (1958a) and Bülbring & Kuriyama (1963b) observed that acteylcholine did not depolarize the membrane in a solution containing 40 mM K, and that, when the membrane was depolarized to − 10 or − 15 mV (from − 55 mV in normal solution), the effect on the membrane potential was reversed. Recently, Bennett (1966c) calculated an 'acetylcholine equilibrium potential' from the results of Bülbring & Kuriyama (1963b) using an equation derived by Eccles (1961). He estimated that the ACh equilibrium potential might be in the range of − 20 to − 26 mV.

In sodium free (tris) solution, acetylcholine slightly depolarized the membrane, but no change in spike frequency was observed (Bülbring & Kuriyama, 1963b). In contrast, excess sodium (1·5 × normal) potentiated the membrane depolarization and the increase in spike frequency caused by acetylcholine. Excess calcium also enhanced the action of acetylcholine, whereas in the absence of calcium, acetylcholine was ineffective.

The contraction produced by low doses of acetylcholine in normal physiological solution appears to follow the spike activity (Bülbring, 1957a), i.e. each spike produces an increase in tension and the steps fuse to give an overall contraction of the muscle. With higher acetylcholine concentrations, the membrane is depolarized to a greater extent, the spikes deteriorate, and there may be a decline in frequency (see Fig. 10). Occasionally only oscillatory potential changes are superimposed on the depolarization. Under these circumstances, the tissue is probably in a state of contracture, triggered by the overall membrane depolarization.

When smooth muscle is completely depolarized by placing it in high potassium solutions, there is an initial contracture, which is followed by partial or complete relaxation (Durbin & Jenkinson, 1961b). In this condition the muscle will still produce a fully reversible contractile response to the administration of acetylcholine (Schild, 1964). This response cannot be mediated by depolarization, but may still be mediated via a change in the membrane permeability. Durbin & Jenkinson (1961a) have shown that there is an increase in influx and efflux of ^{42}K, ^{36}Cl and ^{82}Br, and also changes in ^{24}Na and ^{45}Ca fluxes on addition of carbachol to depolarized smooth muscle, suggesting an increased membrane permeability to all these ions. Calcium is very important for the initiation of contraction. In the depolarized muscle, after 40 to 60 min exposure to calcium-free solution, the application of acetylcholine fails to elicit a response, but addition of calcium to the medium will quickly restore the response (Schild, 1964). The observation indicates that acetylcholine produces its contractile effect partly by releasing bound calcium from a cellular store and partly by increasing the Ca permeability of the membrane (see Chapter 9 and 11). Also in normal conditions an increased Ca-efflux (Schatzmann, 1961) and an increased K-efflux (Born & Bülbring, 1956) have been observed in the presence of acetylcholine (see Chapter 11).

The effects of acetylcholine on the slow potentials

There are some indications that nervous factors are involved in the spontaneous and evoked electrical activity of smooth muscles. Since the responses to nerve stimulation (type II) are discussed in Chapter 8, the description here will be confined to the spontaneous slow waves (type III).

It has been shown that tetrodotoxin blocks the responses mediated by nervous elements when field stimulation is applied to the taenia (Bülbring & Tomita, 1967). Also in the vas deferens nervous effects are abolished by tetrodotoxin (Hashimoto, Holman & McClean, 1967; Tomita, 1967a).

In the longitudinal muscle of the guinea-pig jejunum tetrodotoxin (10^{-7} g/ml) and atropine (10^{-6} g/ml) reduced the slow potential (type III) (Kuriyama et al., 1967b, c). When the tissue had been depolarized by prolonged exposure (several hours) to Krebs solution, it was repolarized by atropine and spike activity recovered. These results suggest that the normal membrane activity of the smooth muscle is influenced by a release of acetylcholine from nerve terminals and that the slow potential (type III) in this intestinal muscle is at least partly due to such a mechanism.

However, the slow potential (type III) in the cat stomach does not seem to be caused by nervous activity, since neither atropine (10^{-6} g/ml), nor procaine (10^{-5} g/ml) nor tetrodotoxin (10^{-6} g/ml) have any effect on its shape or on its conduction (Papasova et al., 1968).

In longitudinal intestinal muscle of the cat, acetylcholine (10^{-6} g/ml) actually produces slow potentials (type III) with trains of spikes on their crests (Burnstock & Prosser, 1960c).

In the guinea-pig jejunum (Kuriyama et al., 1967c), treatment with prostigmine (5×10^{-5} g/ml) increases the spike frequency, and augments the amplitude of the slow potential (type III) (up to more than 35 mV in amplitude and 2 to 3 sec in duration). When the slow potential becomes very large, the repetitive firing of spikes is suppressed and only a single spike appears at the beginning of the slow potential. If the concentration of prostigmine is increased to 10^{-4} g/ml, the membrane remains depolarized by more than 15 mV, and both the spike and the slow potential are blocked. Atropine (5×10^{-6} g/ml) abolishes the slow potential when it had been enhanced by prostigmine, and in the presence of atropine the spikes appear without the slow potential.

The nervous participation in the pattern of spontaneous activity in other parts of the intestine requires further investigation.

Catecholamines

The relaxation of intestinal smooth muscle produced by catecholamines is brought about by the cessation of spontaneous spike activity which normally maintains the tone.

In the guinea-pig taenia coli the block of spike discharge caused by adrenaline is usually accompanied by hyperpolarization of the membrane. The threshold concentration blocking spike generation is about 10^{-9} g/ml and a strong effect is obtained with 10^{-7} g/ml (see Fig. 11).

The hyperpolarization is not always observed since its occurrence and its magnitude depends on the height of the membrane potential at the time when adrenaline is administered. For example, the absolute increase of membrane potential produced by a given dose of adrenaline was directly proportional to the degree of depolarization caused by stretching the preparation (Bülbring & Kuriyama, 1963c). On the other hand, the duration of the block of spontaneous activity was shortest in the stretched preparation and longest in the unstretched muscle.

The increase in membrane conductance

The rise in membrane potential produced by adrenaline appears to be mainly due to an increase of the membrane permeability to potassium.

An increase of ^{42}K efflux, and also of ^{42}K uptake, by noradrenaline, was first demonstrated on the depolarized taenia by Jenkinson & Morton (1965; 1967b, c). This effect is also produced by adrenaline (Bülbring, Goodford & Setekleiv, 1966). It is not readily detectable under normal conditions, probably because the hyperpolarization of the membrane reduces the outward movement of K (see Chapter 11).

The effect of catecholamines on the membrane conductance of the taenia coli has recently been examined with the double sucrose-gap method (Bülbring & Tomita, 1968a; 1969a, b, c). The effect on the membrane potential, on spontaneous and evoked spike activity, and on the electrotonic potential in response to constant current pulses was observed in different ionic environments. The results showed that the hyperpolarization produced by adrenaline and

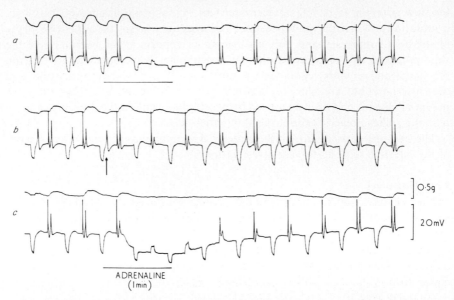

FIG. 12   The effect of removing potassium. (The external Cl was replaced with a large anion, benzene-sulphonate, to study the effect of adrenaline on K-conductance) 24°C. Constant current pulses of alternating polarity every 15 sec.   (a) the effect of adrenaline (2×10^{-7} g/ml) in benzene-sulphonate solution containing the normal potassium concentration (5·9 mM).   (b) potassium was reduced to zero at the arrow. Note the increase in the electrotonic potential, block of anodal break excitation and weak contraction in potassium-free solution.   (c) The effect of adrenaline (2×10^{-7} g/ml) after 8 min exposure to potassium-free solution. Note the increase in hyperpolarization but less reduction of the electrotonic potential by adrenaline. (Bülbring & Tomita, 1969a)

noradrenaline in taenia coli was associated with an increased membrane conductance (Fig. 12), mainly to K, but also to Cl. The hyperpolarization was converted to depolarization when the membrane was polarized to a level about 20 mV more negative than the resting potential by applying conditioning current.

Excess K in the external solution decreased the size of the electrotonic potential while low K enhanced it. The magnitude of the hyperpolarization caused by adrenaline depended mainly on the external K concentration. In low K the hyperpolarization became larger while the reduction of the electrotonic potential became smaller (Fig. 12). In excess K, the effect of adrenaline on the electrotonic potential was enhanced (Fig. 13).

When Cl was replaced with a less permeant anion, benzene-sulphonate, the electrotonic potential was increased, while Cl substitution with nitrate decreased it. Accordingly, the increase of membrane conductance by adrenaline was greater in the presence of NO_3 than in the presence of benzene-sulphonate. Adrenaline often caused depolarization in low Cl (benzene-sulphonate substitution) and excess K (Fig. 13). These results indicate that also the Cl ion contributes to the adrenaline effect.

The suppression of spontaneous spike activity

In the presence of phentolamine, which blocks the α-effects of catecholamines, the effect on membrane conductance was abolished, but adrenaline and noradrenaline still suppressed spontaneous spike activity (Fig. 14). Though spikes

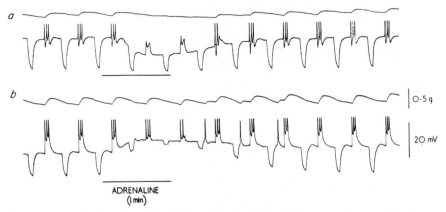

Fig. 13   The effects of (*a*) low ($\frac{1}{5}$ normal) and (*b*) high (4 × normal) external potassium on the response to adrenaline (2 × 10^{-7} g/ml) in benzene-sulphonate solution. 20°C. Constant current pulses of alternating polarity applied every 15 sec.   (*a*) 7 min after the external potassium concentration was reduced from 5·9 mM to 1·2 mM.   (*b*) 20 min after the top record and 5 min after the potassium concentration was increased from 1·2 mM to 24 mM. (Top part of the spikes off scale). Note the reduction of the electrotonic potential and abolition of the positive after-potential of the spikes in high external potassium. Note also that, in high potassium, adrenaline caused depolarization, a greater reduction of the electrotonic potentials but a weaker suppression of the evoked spikes and the anode break excitation. (Bülbring & Tomita, 1969*a*)

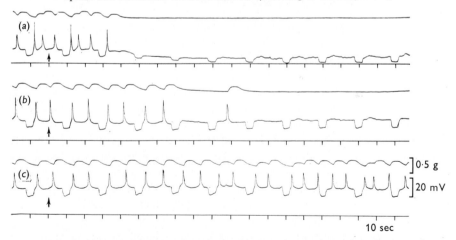

Fig. 14   Taenia coli, double sucrose-gap, 28°C. *Top*: isometric tension. *Bottom*: electrical record. Electrotonic potentials evoked by hyperpolarizing current pulses of 4 sec duration every 20 sec. Effect of noradrenaline 2 × 10^{-7} g/ml added at arrow, (*a*) before (*b*) in the presence of phentolamine 10^{-5} g/ml, (*c*) propranolol 10^{-5} g/ml in addition. (Bülbring & Tomita, 1968*a*)

could be evoked by electrical stimulation, the slow depolarization (type I) or pacemaker potential no longer occurred. This suppression of the spontaneous spike generation is the main mechanism by which isoprenaline exerts its inhibitory action, and this is antagonized by propranolol, which blocks the β-effect of catecholamines (Bülbring & Tomita, 1969b).

A comparison, in the guinea-pig taenia coli, of the three catecholamines showed that their potency, in terms of α-activity, was adrenaline > noradrenaline > isoprenaline, whereas, in terms of β-activity, their potency was isoprenaline > adrenaline > noradrenaline.

This is in agreement with the observation that K efflux in depolarized taenia is only increased by adrenaline and noradrenaline, but not by isoprenaline, and that this effect is abolished by α-blockers (Jenkinson & Morton, 1967a, c).

In addition to the β-action on electrical activity, which is confined to the suppression of the spontaneous generator potential (type I) there is some evidence for an interference by isoprenaline with excitation-contraction coupling (Bülbring & Tomita, 1969b). In depolarized muscle, isoprenaline inhibits the contracture elicited by calcium (Jenkinson & Morton, 1967c).

The role of Ca

The presence of Ca in the external solution is essential for the action of catecholamines (Bülbring & Kuriyama, 1963b; Bülbring & Tomita, 1969c). In low external Ca the adrenaline effect is reduced, in excess Ca it is enhanced (Fig. 15). In spite of the fact that electrical activity can be maintained when the Ca in the external solution has been substituted with Ba, adrenaline fails to abolish this activity, nor does it change the membrane conductance in the absence of Ca.

The effect of raising the external Ca concentration on membrane conductance resembles that of adding adrenaline, in that the stabilization of the membrane is associated with a decrease of membrane resistance which is probably mainly due to an increase in K-conductance (Bülbring & Tomita, 1969c).

The resting potential of the taenia is low, mainly because the K-permeability of the cell membrane is low, and this appears to depend on the amount of Ca which can be bound at some membrane site. Ca may have two functions at the membrane (Brading, Bülbring & Tomita, 1969a), (1) to control the Na- and Ca-conductance (which is closely related to its function of carrying the action current) and (2) to control K-conductance. These two functions may be exerted at two different sites (see Chapter 7) but, nevertheless, the replenishment of the amount of bound calcium at the two sites might involve processes which are interconnected.

Adrenaline has no influence on the spike mechanism. In fact, a larger and faster action potential can be evoked in the presence of adrenaline if a sufficiently strong stimulus is applied, since the threshold is raised, as in excess Ca (Brading *et al.*, 1969b).

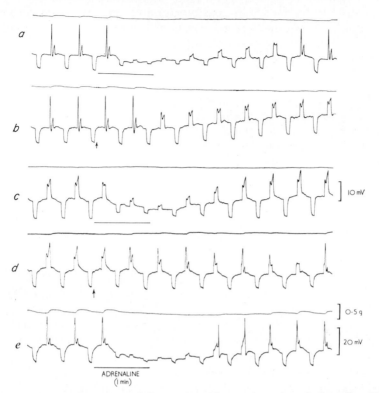

FIG. 15   Influence of the external Ca concentration on the action of adrenaline. Upper tracing tension, lower tracing electrical responses to constant current pulses of alternating polarity, every 15 sec. 20°C.   (*a*) Effect of adrenaline (2 × 10⁻⁷ g/ml) applied for 1 min, indicated by bar, in normal Krebs solution (2·5 mM Ca).   (*b*) At arrow, reduction of Ca concentration to 0·25 mM.   (*c*) Adrenaline effect after 8 min exposure to low Ca.   (*d*) At arrow increase of Ca concentration to 7·5 mM.   (*e*) Adrenaline effect after 18 min in high Ca. Note that Ca deficiency increased and Ca excess decreased membrane resistance. The adrenaline effect was weak in low Ca and it was potentiated in high Ca. Note that the gain was decreased in *d* and *e*. (Bülbring & Tomita, 1969*c*)

The action of adrenaline is more dramatic than that of excess Ca, which merely swamps the membrane while adrenaline apparently acts on processes taking place in the membrane. One of these might be an increased Ca uptake from the medium leading to a sudden accumulation of Ca at the binding site, hence a sudden increase in K-conductance and hyperpolarization (α-action). Removal of Ca from the binding site would reduce the K conductance and cause depolarization. This may be one factor in the generation of the pace-maker potential. It may be that adrenaline suppresses the pacemaker potential (β-action) by reducing the Ca removal (Bülbring & Tomita, 1969*c*).

Adrenaline possesses both α- and β-actions and therefore tends to increase and to preserve Ca stores at the membrane. Adrenaline and noradrenaline act predominately by their α-action, and cessation of spike activity is mainly the consequence of membrane hyperpolarization. Isoprenaline acts predominately by its β-action and cessation of spike activity is mainly due to the failure of the cell membrane to develop pacemaker potentials.

The metabolic action

How far the action of catecholamines depends on metabolic energy supply is still an open question. Adrenaline increases the formation of adenylcyclase, 3',5'-AMP, in the taenia (Bueding, Butcher, Sutherland & Hawkins, unpublished). Moreover, coinciding with the hyperpolarization and inhibition of activity, adrenaline increases the tissue concentration of ATP and creatine phosphate (Bueding, Bülbring, Gercken, Hawkins & Kuriyama, 1967). The effect of adrenaline does not coincide with an activation of phosphorylase and has no obligatory requirement for carbohydrate metabolism. However, the action of adrenaline does depend on metabolic processes generating energy-rich phosphate compounds. It is abolished by anaerobic substrate depletion and recovers after readmission of oxygen and/or substrate. When adrenaline is applied simultaneously with the readmission of substrate, already after 30 sec the ATP and CP content of the tissue is higher than that of the controls. This indicates an increased rate of ATP synthesis. Moreover, although adrenaline has no effect on oxygen uptake during the initial recovery period when glucose is readmitted to a depleted tissue, the presence of adrenaline shortens the time interval until spontaneous activity is resumed, indicating an acceleration of the recovery process (Bülbring & Golenhofen, 1967). The possibility that adenylcyclase may be the adrenergic receptor has been discussed by Robinson, Butcher & Sutherland (1967). They suggest that both α- and β-receptors might be different sites of adrenaline-enzyme interaction on the adenylcyclase molecule. The common factor, Ca, for the physiological α- and β-action may also indicate a very close relationship between the two receptors.

CONCLUSION

This chapter is meant to be a link between physiology and pharmacology. It is necessary to have more information about the physiology of smooth muscle before pharmacology can be properly understood, but pharmacological experiments are also needed in order to understand the physiology. Unfortunately, our knowledge of both is still at a rather too early stage, and the techniques have been too limited to obtain unequivocal results.

Most of the results presented are observations of phenomena and we still do not know how to interpret them properly.

Since many smooth muscles have spontaneous electrical activity, this has been the main object of most studies and the response to electrical stimulation has not been thoroughly investigated. However, we know from nerve physiology, that it is impossible to understand mechanisms of excitation without investigating the response when the membrane potential is changed artificially by current application. The current spread in the tissue is a very important factor which affects the electrical responses (see Chapter 7). Unless a proper method of current application is used, responses to applied current may be very difficult to interpret because of the electrical interconnections between smooth muscle cells. However, since this kind of investigation has just started, a great deal of progress may be expected in smooth muscle research.

In order to express the effects of drugs in terms of ionic conductances of the membrane, changes of the membrane resistance should be measured. But also in these experiments, the problem of current spread in the tissue must be taken into account. If one treats smooth muscles as functional bundles, not as a single cell unit, most of them have cable-like properties. These properties are very useful for the measurement of changes of membrane resistance in response to pharmacological agents and to changes in the ionic environment. Progress is also being made in this line of approach.

There is a large variability in the responses observed even in the same preparation, and the properties may differ from those of another preparation taken from the same tissue even in physiological condition. This may be expected from the very complicated arrangement of the cells in smooth muscles, and perhaps also from differences in the density of innervation. However, in *in vitro* experiments many more factors increase the variability. The history of a preparation after dissection is one very important factor and has to be carefully controlled to obtain reproducible results. Responses may also vary considerably with different recording and stimulating techniques.

It is known that the responses in different smooth muscles or in different species can be quite different. When they have different functions, their physiological and pharmacological properties would naturally be expected to be different. Therefore, it is dangerous to generalize from one conclusion, based on results obtained from one or two tissues, to include all other smooth muscles.

At the present time, the interpretation of many observations is bound to be still very speculative. However, even a tentative hypothesis usually stimulates research and leads to further experiments and ideas.

13

THE HORMONAL CONTROL AND THE EFFECTS OF DRUGS AND IONS ON THE ELECTRICAL AND MECHANICAL ACTIVITY OF THE UTERUS

Y. ABE

INTRODUCTION

Since the uterus is an organ of reproduction, the characteristics of the myometrium vary depending on hormonal influence. During the progress of gestation or during treatment with oestrogen and progesterone, not only does the cell size increase (hypertrophy) but also the number of cells increases (hyperplasia) and, after delivery, they decrease again, within a few days, to the same status as before (rat: Brody & Wiqvist, 1961). Chronic stretch also exerts effects on uterine growth and protein synthesis (rabbit: Csapo, Erdös, De Mattos, Gramss & Moscowitz, 1965). The ratio of RNA to DNA suggests that the growth under these conditions involves hypertrophy rather than hyperplasia.

The 'defence mechanism of pregnancy' (Csapo, 1956a, b, 1961b), i.e. the mechanism maintaining pregnancy, is mainly concerned with the prevention of placental ablation from the endometrium, which might be caused by contraction of the myometrium. On the other hand, moderate contractions seem to be required to provide nutrition to the myometrium and to the foetus via the circulation. These intricate mechanisms are apparently mainly controlled by ovarian and placental hormones.

In addition there occurs, in the course of pregnancy, an increment in the myometrial response to oxytocin. This is very specific, since no changes occur in the response to vasopressin as gestation progresses. Although the guinea-pig and human uterus respond to oxytocin in all stages of pregnancy, the sensitivity gradually increases and maximum oxytocin activity is reached just before term. The rabbit uterus, however, is at first insensitive and only on the last 2 or 3 days of pregnancy does the myometrium reacquire its responsiveness to oxytocin (Caldeyro-Barcia, 1965).

Much research has been carried out on the mechanical response under various conditions. However, the contraction of the muscle is physiologically achieved by a sequence of processes: spike initiation, spike propagation, excitation-contraction coupling and contraction. Each process might be affected differently and should therefore be studied separately.

The author intends to write this chapter not as an encyclopaedic review but as a description of recent electrophysiological investigations of the myometrium, i.e. of its membrane phenomena.

Most of the electrophysiological studies so far have been confined to observations on the resting membrane potential and spontaneous electrical activity. Measurement of the resting potential is of course, very important and fundamental for the analysis of the electrical properties of uterine muscle, but it is technically very difficult, as stressed by Kao (1967). So far, no measurements of the membrane parameters of the myometrium have been reported (i.e. membrane resistance, capacity, space constant, etc.), but some unpublished results will be given below. Moreover, it would be useful to have information on the changes in membrane resistance caused by alteration of the ionic composition of the external medium, or by drugs, and some preliminary results will also be reported.

Since the spike shape is probably influenced by processes involved in propagation within the tissue, it is important to differentiate whether the observed effects on the spike shape are due to changes in the regenerative mechanism directly involved in spike generation, or to changes in mechanisms involved in spike conduction, e.g. threshold. For example, as will be described later, in a low Na solution the shape of the spontaneously generated spike is quite different from that of the spike locally evoked by electrical stimulation. Although the conducted spikes are more important in the physiological function, it is easier to analyse the fundamental mechanism if one investigates spikes at a site where they are evoked. This information can then be applied to the analysis of the conducted spikes. A proper method of stimulation and a careful study of the spike at

14+

various distances from the site of stimulation will help in understanding the properties of myometrium.

CABLE PROPERTIES

The uterus, like many visceral smooth muscles, behaves like a syncytium. The propagation of contraction along the tissue and the co-ordinated contractions of the whole organ are well-known phenomena which suggest conduction of excitation between muscle cells.

For the guinea-pig taenia coli (Tomita, 1966a, b; Abe & Tomita, 1968) and the vas deferens (Tomita, 1967a) the cable properties have been demonstrated. Records obtained with the sucrose-gap method of spontaneous activity of guinea-pig and cat uterus (Bülbring, Casteels & Kuriyama, 1968) and of electrically evoked activity in rat uterus (Abe, 1968) suggest that the properties of uterine smooth muscle resemble those of the taenia and vas deferens.

Quantitative measurements of passive membrane parameters of pregnant rat uterus have recently been made (Abe, unpublished). The results indicate that the myometrium has cable-like properties. The spatial decay and the time course of the electrotonic potential, evoked by external stimulation and recorded intracellularly, could be expressed by the cable equation derived for nerve fibre by Hodgkin & Rushton (1946) (see Fig. 1). The space constant and the time constant of pregnant rat uterus were somewhat longer than in intestinal muscle. Between the seventh and twentieth day of pregnancy the membrane resistance decreased significantly.

It is interesting that the magnitude of the electrotonic potential and the time constant of the membrane of the uterus were affected by chloride replacement with foreign anions in the same manner as intestinal smooth muscle, (Kuriyama, 1963a; Bülbring & Tomita, 1969a). Chloride substitution with a large anion increased the size of the electrotonic potential and the time constant of the membrane. Benzene sulphonate appeared to be the least and iodide the most permeant anion.

RESTING MEMBRANE POTENTIAL

The values for the resting membrane potential of myometrium are different in different species and in different states of hormonal influence. The range of variation is between 35 and 65 mV (see Table 1). These values are much lower than those of skeletal muscle fibres. However, from the analysis of the intracellular K concentration, the K equilibrium potentials have been calculated to be 70 to 80 mV. (rabbit: Horvath, 1954; Kao & Nishiyama, 1964; rat: Cole, 1950; Casteels & Kuriyama, 1965; cat: Daniel & Singh, 1958.)

The low membrane potential has been explained by a relatively high Na permeability of the cell membrane. This explanation was based on the finding that the

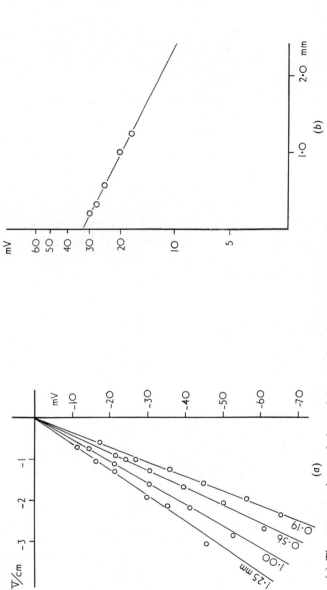

FIG. 1 (*a*) The current-voltage relations of the membrane obtained at four different distances from the stimulating electrode in twentieth day pregnant rat uterus. Ordinate: displacement of membrane potential by applied current. Abscissa: intensity of applied electrical field. (*b*) The spatial decay of the electrotonic potential. Each point was obtained from (*a*) at constant current intensity. Ordinate: recorded potential on logarithmic scale. Abscissa: distance from the stimulating electrode.

depolarization caused by a tenfold increase in the external K concentration is only between 32 and 52 mV, depending on the myometrial condition, a change which is smaller than would be expected from the potassium diffusion potential (Kuriyama, 1964b; Casteels & Kuriyama, 1965). A similar idea has often been put forward to explain the low resting potentials in other smooth muscles. However, this does not necessarily mean that the Na permeability in smooth muscles is higher than in skeletal muscles. It is also possible that the potassium permeability is lower than in skeletal muscles.

If the contribution of the Na conductance to the membrane potential is significant, one would expect hyperpolarization of the membrane in Na deficient solution. However, this has not always been observed. In postpartum rat myometrium, the membrane was slightly depolarized when NaCl was replaced with choline chloride or with sucrose (Kuriyama, 1961b), but Csapo & Kuriyama (1963) reported a transient hyperpolarization after 30 min followed by depolarization after 180 min. Marshall (1963, 1964) investigated the effect of reducing the external Na concentration to 10–50% of normal, using choline or LiCl, or sucrose as substitutes. When more than 50% of the NaCl was replaced with sucrose the membrane was hyperpolarized (Marshall, 1965). It may be that choline and Li exert an effect of their own and, therefore, may not be good substitutes for Na in studying the effect of Na deficiency. On the other hand, when NaCl is replaced with sucrose the concomitant Cl deficiency may oppose the effects of reducing the Na concentration on the membrane potential.

In low external Ca concentrations, the membrane is depolarized, and the membrane is hyperpolarized by an increase in the external Ca concentration (rat: Kuriyama, 1961b; Csapo & Kuriyama, 1963; Casteels & Kuriyama, 1965). Marshall (1965) reported that, in pregnant rat uterus, hyperpolarization by excess Ca was observed provided that the external Na concentration did not fall below 50% of normal. At a lower Na concentration (sucrose substitution), the membrane was already hyperpolarized and an increase in Ca had no significant effect. The change in the membrane potential produced by changing the external Ca concentration can be explained by a control of the passive Na permeability by Ca, as in nerve fibres (Stämpfli & Nishie, 1956; Huxley, 1959).

ACTION POTENTIAL

The electrical activity of the rat myometrium has been most frequently studied. There are some reports on mouse, rabbit and guinea-pig uteri but they have similar fundamental properties. Most excised uterine muscles exhibit spontaneous activity during the non-pregnant, pregnant and post-partum conditions. Exceptions are the early pregnant cat uterus and the rabbit uterus in the middle stage of pregnancy.

The spike amplitude depends on the hormonal condition. In non-pregnant uterus an overshoot is rarely observed, but in pregnant uterus the amplitude of the action potential may exceed 70 mV (see Table 1). The parameters of the spike are more or less similar to those in the guinea-pig taenia coli, e.g., the duration is 10 to 50 msec (Marshall, 1959; Kuriyama & Csapo, 1961*a*), and the maximum rate of rise is about 10 V/sec (Marshall, 1959; Goto, Kuriyama & Abe, 1961; Casteels & Kuriyama, 1965).

In most uterine muscles the spontaneously generated or the electrically evoked spike is followed by an after depolarization ('negative after-potential') throughout all stages of gestation, post-partum and in the non-pregnant condition. Exceptionally, the shape of the spike changed from the type with after depolarization in the non-pregnant cat myometrium to a spike with after hyperpolarization ('positive after-potential') in the early stage of gestation.

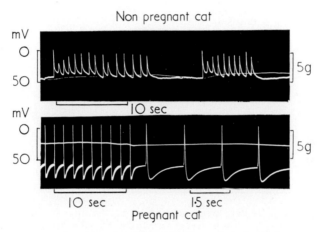

FIG. 2   Intracellular records of spontaneous electrical activity in virgin cat myometrium (upper record) and of evoked action potentials in seventeenth-day-pregnant cat myometrium, taken at two sweep speeds (lower record). (Kuriyama, unpublished record)

The after hyperpolarization, however, was abolished in excess K (Bülbring, Casteels & Kuriyama, 1968), indicating that it was the result of an increased K-conductance (Fig. 2).

A great deal of variation in values for the conduction velocity has been reported: in the guinea-pig, 1 to 3 cm/sec; in the rabbit, 1 cm/sec and in the cat, 6 cm/sec (Bozler, 1938*a*); in the rat, 0·26 to 6 cm/sec for spontaneous spikes and 0·3 to 9·5 cm/sec for evoked spikes (Melton, 1956), in the post-partum rat, 9·6 cm/sec for evoked spikes (Csapo & Kuriyama, 1963); in the mouse 10·2 cm/sec (Goto, et al., 1961). It is very likely that the conduction velocity is different in different species and that it is changed by the hormonal influence.

However, more careful studies of this problem seem necessary. For example, mechanical activity of rabbit uterus induced by stimulating one end of a uterine strip, was well conducted along the entire length of the post-partum uterine strip, but in early pregnancy the activity was restricted to near the stimulated portion, i.e. conduction of excitation was decremental (Kuriyama & Csapo, 1961*b*).

Spontaneous activity appears as a burst of spikes alternating with silent periods which vary usually between 10 and 60 sec. During the fluctuations from activity to the inactive state, a change of excitability occurs, so that during a silent period, a stimulus fails to evoke a spike until shortly before a spontaneous burst starts (Casteels & Kuriyama, 1965). The stimulus intensity has to be increased more than ten times to trigger spikes throughout the silent period. When the spikes are regularly triggered by electrical stimulation, spontaneous bursts of spikes are suppressed (Kuriyama, 1964*b*) (see Fig. 3).

Rat 21 days pregnant

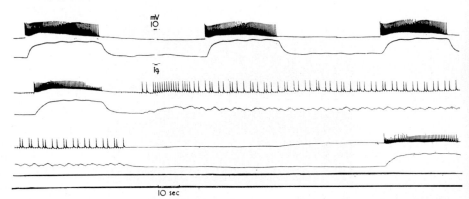

10 sec

FIG. 3   Rat uterus, 21 days pregnant. Sucrose-gap records of spontaneous electrical activity (upper trace) and tension (lower trace). The regular cycle of activity is abolished during electrical stimulation, applied at a rate of 1/sec, though not every stimulus evokes a spike. After cessation of stimulation there is a long delay before the rhythm of spontaneous activity returns. (Bülbring, unpublished record)

In the early pregnant rat myometrium, the membrane potential returned to more or less the original potential after each spike (Kuriyama & Csapo, 1961*a*), but in the late pregnant mouse myometrium the membrane remained slightly depolarized during a train of discharges, i.e., successive spikes were initiated before the repolarization of a previous spike was complete (Kuriyama, 1961*a*).

The non-pregnant guinea-pig myometrium showed spontaneous spikes with low frequency, thus the membrane was repolarized to the original level after each spike. In mid and late pregnancy, however the spike appeared as a train discharge with rather high frequency and the membrane remained slightly depolarized during discharges.

In early gestation, cat myometrium never generated spontaneous spikes, but in virgin cat, the spike appeared without electrical stimulation and sustained depolarization with low amplitude could be observed during the train discharges (Bülbring, Casteels & Kuriyama, 1968) (see Fig. 2).

The pacemaker potential, which is a gradual depolarization leading to a spike, may be found in any part of the tissue. The pacemaker region is not located in any particular site but often changes its site within the tissue (Kuriyama, 1961b; Marshall, 1962).

Effects of Na-deficiency are so far open to controversy. Goto & Woodbury (1958) observed in pregnant rat myometrium that partial substitution of NaCl with sucrose caused a large decrease in the amplitude of the spontaneous action potential. The magnitude of reduction of the action potential had a slope of 45 mV for a tenfold change in the external Na concentration. This is in agreement with the Na theory for the action potential. Marshall (1963) studying spontaneous activity in pregnant rat, and Kao (1967) in pregnant rabbit reached a similar conclusion.

Recently, the influence of Na ions on the rat myometrium have been investigated with the voltage-clamp method by Anderson & Moore (1968). In Na-free solution the transient current was greatly reduced and accompanied by a shift in the Na equilibrium potential along the voltage axis to a lower internal potential. These observations also support the Na theory for the spike generation.

According to Daniel & Singh (1958), reduction of the external Na-concentration to one-ninth of normal, replacing NaCl with choline chloride or with sucrose, did not reduce the action potential in the rat uterus. Kuriyama (1961b; 1964b) and Csapo & Kuriyama (1963) reported that in the pregnant and post-partum mouse and rat uterus an overshoot potential was recorded until the external Na concentration was reduced to 20 mM, substituting with choline or tris-(hydroxymethyl aminomethane). The rate of rise of the spike was gradually decreased when Na was reduced further. Kao (1967) argues that the small effect of Na-deficiency is due to very slow diffusion of Na out of a thick preparation.

Most of the above observations have been made on spontaneously generated spikes and probably on conducted spikes. The possibility exists, therefore, that Na-deficiency may suppress the pacemaker potential and block the spike propagation. In rat myometrium, when electrically evoked spikes were investigated (Abe, 1968), the amplitude and the maximum rate of rise of the spike were increased in low Na solutions (15 mM Na) (sucrose substitution), whereas the amplitude and the rate of rise of the spontaneously generated spikes were both reduced (Fig. 4). The maximum rate of rise was increased from 15·1 to 23·3 V/sec and the spike amplitude from 63·1 to 68·3 mV. The half-duration was decreased from 29·5 to 24·5 msec. Even after 3-hr exposure to low sodium, larger and faster spikes could be observed in response to electrical stimulation though spontaneous activity was absent. This experiment indicates that Na may not directly contribute to the spike generation. However, there is still a possibility that a small concentration of Na is necessary for regulating the Ca distribution across the membrane.

In contrast to Na, Ca seems to be the most important for determining the spike amplitude and its rate of rise, as shown in Fig. 5.

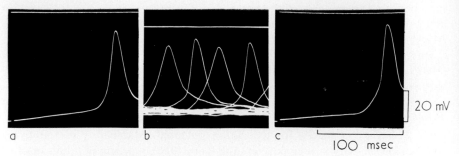

FIG. 4.   Rat uterus, 20 days pregnant. Intracellular records: (a) evoked spike in normal
Krebs solution; (b) spontaneous spikes, and (c) evoked spike, after 30 min exposure to
low Na (15 mM) (sucrose substitution).

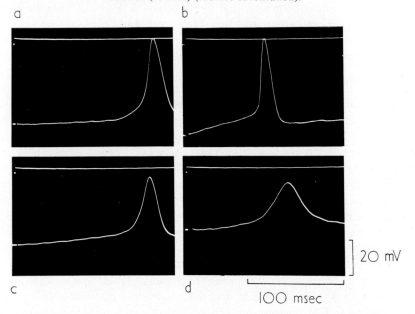

FIG. 5   Rat uterus, 20 days pregnant. Effect of changing the external Ca concentration
(in the presence of low Na) on the rate of rise and amplitude of the evoked action
potential. Intracellular records: (a) in normal Krebs solution containing 2·5 mM Ca and
137 mM Na; (b) 10 mM Ca and 15 mM Na; (c) 0·6 mM Ca and 15 mM Na; (d) 0 Ca
and 15 mM Na. Records b to d were taken about 35 min after changing the ionic composition of the solution.

The effects of changing the external Ca concentration on the spontaneous
action potential of the uterus has been described by Csapo & Kuriyama (1963)
and by Marshall (1965). The evoked spike was affected in the same way: when
the Ca concentration was increased the membrane was hyperpolarized and the
rate of rise and amplitude of the action potential were increased. Reduction
of the Ca concentration in the solution had the opposite effect (see Fig. 5).

It was noted by Marshall (1965) and Abe (1968) that the effect of Ca-deficiency was less if the Ca was reduced in a solution containing low Na than when the normal Na concentration was present.

In pregnant rat myometrium, Ca may carry most of the inward current during the spike. There is strong evidence supporting a similar conclusion in intestinal smooth muscle (see Chapters 7, 8 and 12). However, in uterine muscle, effects of Na-free solution should be thoroughly studied, and experiments employing a proper method for electrical stimulation should be extended to the myometrium of different species, before generalization of the theory is made.

Strontium can replace Ca in spike activity in rat myometrium near parturition, although the spike frequency is slightly reduced (Marshall, 1965). Manganese suppresses the spike, but tetrodotoxin has no effect on the spike activity in rat myometrium, as in intestinal smooth muscles (see Chapters 8 and 12).

CHANGES OF THE RESTING POTENTIAL DURING GESTATION AND UNDER HORMONAL TREATMENT

The non-pregnant uterus muscle has a low membrane potential (Table 1). During oestrus, it is about 40 mV and during the progress of gestation the membrane potential increases up to 65 mV (Kuriyama, 1964b). In rat myometrium, the membrane potential increases from 42 mV at the beginning to a maximum (60·5 mV) on the 15th day of pregnancy, and it remains high until the last day of the gestation when it begins to fall slightly (54 mV). After parturition the membrane potential declines rapidly, and 3 to 4 days after delivery the membrane potential is not different from that before gestation (Kuriyama, 1964b; Casteels & Kuriyama, 1965).

Cells at the placental site have a higher membrane potential than those of the non-placental site. The maximum potential difference between the two sites (up to 7 mV) appears on the seventeenth or nineteenth day of pregnancy. Similar observations have been made on different species: rabbit (Goto & Csapo, 1959; Kuriyama & Csapo, 1959), rat (Thiersch, Landa & West, 1959; Kuriyama, 1961b) mouse (Goto, Kuriyama & Abe, 1961).

Measurements of the intracellular ion concentrations of the myometrium in the rat (Casteels & Kuriyama, 1965) in rabbit (Kao, 1961; Kao & Siegman, 1963) and in guinea-pig (Bülbring, Casteels & Kuriyama, 1968) indicate that the ionic concentration remains constant throughout pregnancy. Only in the cat was the intracellular Cl content increased in the early stage of gestation (Bülbring, Casteels & Kuriyama, 1968). The increase in the membrane potential, is therefore, probably caused by a change in ionic permeability (an increase in the relative K permeability of the membrane). This explanation is supported by the finding that on the 20th day of pregnancy the maximum change of the uterine membrane potential produced by a tenfold change of the external K concentration

14*

TABLE 1. Resting potential and action potential of the uterus in different species

Condition	Resting potential	Action potential	Author
Rat			
Castrated			
Untreated	35·2	—	Marshall, 1959
Oestrogen-dominated	57·6	65·3	Marshall, 1959
Progesterone-dominated	63·8	—	Marshall, 1959
Non pregnant			
Anoestrus and oestrus	38–42	ca. 35	Casteels & Kuriyama, 1965
Oestrus	39	—	Kuriyama, 1964b
Pregnant			
6–9 days	63	—	Marshall & Miller, 1964
15–16 days	60·5	ca. 63	Casteels & Kuriyama, 1965
18–20 days	52	36–68	Kuriyama & Csapo. 1961a
18–20 days	ca. 57	ca. 64	Casteels & Kuriyama, 1965
20–21 days	54·5–58	ca. 57	Casteels & Kuriyama, 1965
20–22 days	62·8	71·8	Marshall, 1962
20–22 days	58	70	Marshall & Miller, 1964
Parturient	48	44	Kuriyama & Csapo, 1961a
Post partum			
6 hr	48	—	Kuriyama & Csapo, 1961a
12 hr	ca. 50	ca. 58	Casteels & Kuriyama, 1965
24 hr	40	—	Csapo & Kuriyama, 1963
2–3 days	ca. 42	ca. 40	Casteels & Kuriyama, 1965
Mouse			
Non pregnant Oestrus	38	—	Kuriyama, 1964b
Pregnant 18–20 days	53	42	Kuriyama, 1961a
Post partum 6 hr	46	41	Kuriyama, 1961a
Guinea-pig			
Non pregnant	38	15–45	Bülbring, Casteels & Kuriyama, 1968
Oestrus	40	—	Kuriyama, 1964b
Oestrogen + Progesterone treated (8 days)	58	59	Bülbring, Casteels & Kuriyama, 1968
Pregnant	58	ca. 63	Bülbring, Casteels & Kuriyama, 1968
Cat			
Virgin	48	—	Bülbring, Casteels & Kuriyama, 1968
Pregnant (17 days)	64	ca. 75	Bülbring, Casteels & Kuriyama 1968
Rabbit			
Non pregnant			
Oestrogen-dominated	ca. 43	—	Goto & Csapo, 1959
Oestrogen-dominated	38–46	9–25	Kuriyama & Csapo, 1961a
Oestrogen-dominated	42	—	Kuriyama, 1964b
Oestrogen-dominated	49·8	—	Kao & Nishiyama, 1964
Progesterone-dominated	ca. 55	—	Goto & Csapo, 1959
Progesterone-dominated	48·9	—	Kao & Nishiyama, 1964
Pregnant			
20–26 days placental	53·4	—	Goto & Csapo, 1959
20–26 days non placental	42·3	—	Goto & Csapo, 1959
30–31 days	54	33–66	Kuriyama & Csapo, 1961a
Post partum			
6–12 hr	50	35–65	Kuriyama & Csapo, 1961a

($[K]_0 \times [Cl]_0$ constant) was 51 mV while in the non-pregnant uterus it was only 38 mV (Casteels & Kuriyama, 1965).

It is likely that an increase in the membrane potential during pregnancy is due to effects of progesterone, because treatment with oestradiol and progesterone produces a change in the membrane potential similar to that during pregnancy: rabbit (Goto & Csapo, 1959; Kuriyama & Csapo, 1959), rat (Marshall, 1959; Jung, 1960).

In the castrated rat, injection of oestradiol for 5 days (6 µg each) increased the resting potential from 32·5 to 57·6 mV, but treatments with oestradiol for 3 days (6 µg each) followed by progesterone for 5 days (12 mg each) caused an increase to 63·8 mV (Marshall, 1959).

In the rabbit, the experimental results are contradictory. For example, Goto & Csapo (1959) found that in the castrated rabbit, oestradiol for 10 days (25 µg each) increased the membrane potential from 35 mV to 42 mV; additional treatment with oestradiol for 5 days only caused a slight further increase, but progesterone for 5 days (5 mg each) increased the membrane potential to 55 mV. In the rabbit, 3 days after delivery, the membrane potential of the myometrium was 35 mV; treatment with progesterone for 3 days (5 mg each) increased it to 55 mV, but oestradiol for 3 days (50 µg each) increased it only to 48 mV (Goto & Csapo, 1959). However, Kao & Nishiyama (1964) failed to find a difference between the resting potentials of the oestrogen-dominated and the progesterone-dominated non-pregnant rabbit myometrium. The rabbits were either treated with 25 µg oestradiol only for 10 to 14 days (oestrogen-dominated uterus), or additionally with 5 mg progesterone each day on the last 3 to 4 days (progesterone-dominated uterus). The resting membrane potential of the oestrogen-dominated uterus was 49·8 mV and that of the progesterone-dominated uterus was 48·8 mV. They also measured the ion content of uteri in the two different conditions of myometrium and found no difference. The discrepancy of the results obtained in rabbit uterus needs further study to be resolved.

Recently, Bülbring, Casteels & Kuriyama (1968) measured the resting membrane potential and ion content of guinea-pig and cat uterus, taking the tissue samples for electrical recording and ionic analysis from the same animal. In the guinea-pig the average membrane potential increased from 38 mV in the virgin uterus to 58 mV on the thirtieth day of pregnancy. A similar increase was produced by 8 daily injections of oestradiol (5 µg) and additional 1·5 mg progesterone on the last 4 days. The ionic contents (Na, K and Cl) remained unaltered during pregnancy and also during treatment with ovarial hormones. In the cat, the membrane potential was 48 mV in virgin uterus and 64 mV on the seventeenth day of pregnancy. When the ionic contents were measured, no significant change in K and Na content was detected, but there was an increased intracellular chloride content during pregnancy.

With the exception mentioned above, the results generally indicate that progesterone increases the membrane potential further when oestrogen is administered previously. The maximum membrane potential observed during gestation corresponds with the stage when the progesterone content in the blood is high, and the progesterone-treated myometrium has nearly the same membrane potential as that observed in the middle stage of gestation in rat, rabbit, guinea-pig and cat myometrium (see Table 1).

EFFECTS OF GESTATION AND OF HORMONE TREATMENT ON THE ACTION POTENTIAL AND CONTRACTION

The action potential

In non-pregnant rat uterus spontaneous spikes are not frequent, and in response to electrical stimulation the action potentials rarely have an overshoot potential. The non-pregnant uterus has a low excitability and its responses are irregular. During the progress of gestation, the spike amplitude becomes larger as the membrane potential increases and an overshoot potential is usually observed both in spontaneous and electrically evoked spikes (Casteels & Kuriyama, 1965).

The pregnant myometrium has a higher membrane potential, higher threshold of excitation, higher spike amplitude, and shorter refractory period than the post-partum myometrium (Kuriyama, 1964b). The maximum rate of rise of the spike in the rat was 3 to 8 V/sec in the non-pregnant uterus, 7 to 16 V/sec on the 15th day, 5 to 12 V/sec on the 19th to 20th day, and 2 to 6 V/sec on the 1st day after delivery (Casteels & Kuriyama, 1965).

Similar changes as those described for the rat occur in other species during pregnancy. Treatment of non-pregnant, immature or castrated animals with oestrogen and progesterone mimics the effects of pregnancy not only on the resting potential but also on the size of the action potential (see Fig. 6) while the shape remains essentially unaltered.

The cat differs in several respects from other species so far investigated. In early pregnancy the cat uterus (*in vitro*) is not spontaneously active. Moreover, the shape of the action potential is altered. In the virgin cat uterus (as in other uterine muscles under any hormonal condition) the action potential has a prolonged falling phase lasting 0·5 to 1·2 sec, but in early pregnancy the spike is followed by a marked after-hyperpolarization as shown in Fig. 2.

The contraction

There is active synthesis of proteolytic enzymes in the myometrium of the rabbit under the influence of oestrogen (Goodall, 1965; 1966). The presence of a high concentration of progesterone significantly increases the amounts of acto-myosin which can be extracted from the uterine muscle. Therefore, progesterone seems to stabilize the proteolytic enzymes, preventing the degradation of acto-myosin, and to permit continuation of uterine growth and pregnancy. At birth,

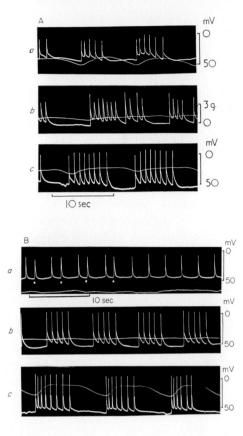

FIG. 6 (A) Mature guinea-pig uterus. Intracellular records of spontaneous activity in (a) non pregnant; (b) fifteenth day; (c) thirtieth day pregnant uterus. (B) Immature guinea-pig uterus. (a) Untreated control; (b) after 6 days oestradiol and 2 days progesterone injections; (c) after 8 days oestradiol and 4 days progesterone injections. (Bülbring, Casteels & Kuriyama, 1968)

the decrease in progesterone releases proteolytic enzymes which cause post-partum involution.

The uterus acquires the maximum working capacity already by mid-term (Csapo, 1960) due to an increase in actomyosin content during pregnancy (Needham, 1961). However, pregnancy is maintained by suppression of large contractions of the myometrium. Mechanical activity, increased by electrical stimulation of one end of the strip, was well-conducted along the entire length (8 cm) of the strips of post-partum uterus, while in the uterus in early pregancy the activity was restricted to the stimulated portion, suggesting impairment of spike propagation (Csapo, 1961b; 1962; Kuriyama, 1961a; Kuriyama & Csapo, 1961b).

Progesterone

The suppression of the uterine motility during pregnancy has been called 'inactivation' by Reynolds (1965) or 'progesterone block' by Csapo (1956a, b). However, in *in vitro* experiments, electrophysiological observations do not clearly support this idea. According to Casteels & Kuriyama (1965), non-pregnant rat uterus had long silent periods between bursts consisting of 3 to 18 spikes. During the progress of gestation, both the amplitude of the spikes and the number of spikes in a train increased. After delivery the number of spikes in a train decreased. The duration of stimulus required to trigger a spike was 5 msec for non-pregnant uterus, 2 to 3 msec for eighteenth to twentieth day of pregnancy, and 1 to 2 msec 6 to 8 hr after delivery. Thus there is no clear sign of suppression of electrical activity during pregnancy.

In vitro experiments probably do not completely reproduce the *in situ* conditions, and the hormonal influence which inhibits the membrane activity in the body may change after the myometrium is excised (Kuriyama, 1964b). A careful study of spike propagation seems to be necessary.

According to Kuriyama (1961a, b) and Kuriyama & Csapo (1961a) *in vitro* treatment of the pregnant and post-partum myometrium of the rat, mouse or rabbit with 10 µg/ml progesterone decreased the number of the spike discharge and often blocked the propagation of spikes, abolishing contractions.

In castrated rats, where the uteri are virtually inactive, injections of oestradiol caused spontaneous discharge of action potentials, and rhythmic muscle contractions occurred as a consequence of the regular membrane activity. In contrast, the progesterone-dominated uterus showed irregular amplitude of the action potentials and electrical stimulation to elicit the spike required stronger intensity than that in oestrogen-dominated uterus (Marshall, 1959).

The above results support the progesterone block hypothesis, but many questions are still not solved. Especially, the difference between the membrane activity of the non-placental and the placental region, postulated by Csapo (1956b), and the role of progesterone at the placental site during gestation, are still obscure.

EFFECTS OF OVARIAN HORMONES ON THE MYOMETRIUM IN
RELATION TO EFFECTS OF Ca AND Mg

Csapo (1956; 1961a) and Coutinho & Csapo (1959) found that the progesterone-dominated pregnant uterus of the rabbit was relatively resistant to immersion in Ca-free Krebs solution, so that the tension developed in response to electrical stimulation declined slowly. However, the oestrogen-dominated uterus lost the tension response rapidly in Ca-free solution. Similarly, in the pregnant rat (Kuriyama, 1964b) and guinea-pig (Schofield, 1964) Ca-lack produced a slow decline in tension of myometrium and, after restoration of the normal Ca concentration, the tension slowly recovered. In the post-partum myometrium

Ca-lack reduced tension much more rapidly and recovery was equally rapid. This suggests that adsorption of Ca in the post-partum myometrium is more labile than in the pregnant myometrium (Coutinho & Csapo, 1959; Kuriyama, 1964b).

According to Goto & Csapo (1959), the membrane potential of oestrogen-dominated rabbit uterus drops to a low level in Ca-free Krebs solution, whereas in the progesterone-dominated uterus there is only a small, transient decrease. In the rat, depolarization of the membrane takes longer in the pregnant myometrium than in post-partum myometrium (Kuriyama, 1961b). Excess Ca mimics the change during pregnancy in hyperpolarizing the membrane, and increasing the amplitude and the maximum rate of rise of the spike. The most probable ion involved during a long-term hormonal action is Ca (Kuriyama, 1964b).

A contribution by Mg is suggested by the following observations: Csapo & Corner (1952) showed that the staircase phenomenon in rabbit uterine muscle is influenced by the ovarian steroids. They found that the force developed by electrical stimulation in oestrogen-dominated uterus is increased when the frequency of stimulation is increased ('positive staircase'). The uterus dominated by progesterone showed an inverse relationship ('negative staircase'). These findings have been confirmed by Schofield (1954, 1955). The negative staircase, which is observed in the progesterone-dominated rat uterus in normal Krebs solution, is also shown in the oestrogen-dominated uterus when the external Mg-concentration is increased to 5·9 mM or more. The progesterone-induced negative staircase is absent in Mg-free solution. It is possible that substitution of Ca at the membrane by Mg is the mechanism by which the change from a positive to a negative staircase is produced in progesterone-dominated uterus (Coutinho, 1966).

However, according to Marshall (1965) Mg has rather weak effects on electrical properties of rat uterus near the end of the gestation period. Elevation of Mg to 10 mM caused no significant change either in the membrane potential or in the amplitude of the action potential, but it markedly reduced the spike frequency and the magnitude of the contraction. After an exposure of about 15 min to 10 mM Mg, the action potentials disappeared and the contractions stopped, but the resting potential was unaltered. In a Mg-free solution the behaviour of the muscle was essentially not different from that seen in normal solution.

EFFECTS OF OXYTOCIN

During the first part of gestation the rabbit myometrium usually does not respond to oxytocin, but towards the end of pregnancy (thirtieth to thirty-first day), the sensitivity to oxytocin increases. In a progesterone-dominated uterus (thirteenth to fifteenth day of pregnancy), oxytocin, 5 μU/ml, produced no effect, and 50 μU/ml slightly increased the active tension to 7 g. However, on the eighteenth day of pregnancy, oxytocin 5 μU/ml increased the active tension to 12 g. The maximum sensitivity to oxytocin of the mouse myometrium appears either during delivery or shortly afterwards (Kuriyama, 1961a).

A fresh preparation of the rabbit uterus on the 25th day of gestation had no spontaneous activity, even after treatment with high doses of oxytocin (50 mU/ml). However, after several hours incubation in oxygenated Krebs solution, the membrane activity was triggered by treatment with oxytocin (1 mU/ml), and after storage for more than 24 hr the tissue sometimes showed spontaneous activity (Kuriyama & Csapo, 1959; Csapo, 1961b). It seems that the suppressing effect of progesterone gradually disappears *in vitro*. This might account for some variability of the sensitivity to oxytocin in *in vitro* experiments.

Marshall (1964) has reported that oxytocin can produce spike discharges only when the membrane potential is near threshold for the discharge of pro- pagated action potentials.

When the Na concentration is increased to 1·25 times normal (from 143 to 178 mM), the effects of oxytocin were markedly enhanced. The stimulating action of oxytocin persisted when the external Na concentration was reduced to as low as 10% of normal, but below this concentration oxytocin was ineffective in the pregnant rat (Marshall, 1963; 1964). Oxytocin (50 μU/ml) increased the action potential to approximately normal amplitude in 10% Na solution (LiCl substitution).

A reduction of the external Ca concentration diminishes the effect of oxytocin (Berger & Marshall, 1961; Marshall & Csapo, 1961). According to Marshall (1968) the action of oxytocin on pregnant rat uterus at term is abolished within 15 min after removal of Ca. In the presence of excess Ca the membrane potential is high and a larger dose of oxytocin is required to stimulate the uterus.

Evans & Schild (1957a) have shown that even when the membrane in rat myometrium is completely depolarized by K, oxytocin can produce contractions, although oxytocin probably acts through a change in the membrane potential under physiological conditions. It is difficult to evaluate the underlying ionic mechanism for the oxytocin action in this condition. It may be that oxytocin increases Na permeability of the membrane in physiological conditions, but it is also possible that oxytocin mobilizes Ca ions from the membrane as carriers of positive charges across the membrane (Marshall, 1963). Kao (1967) presents a hypothesis that oxytocin shifts a curve relating the membrane potential Na inactivation towards depolarization, so that more Na carriers become available in the presence of oxytocin.

Csapo & Pinto-Dantas (1965) emphasize the importance of withdrawal of the progesterone block for the initiation of labour. Another hypothesis (Caldeyro-Barcia & Poseiro, 1959; Caldeyro-Barcia & Sico-Blanco, 1962–1963; Caldeyro-Barcia, 1965) proposes that the initiation of labour is due to the stimulating action of oxytocin, released by the posterior lobe of the hypophysis. They measured the progesterone concentration in the human myometrium during pregnancy and could not detect any difference of the concentrations between term and mid pregnancy. On the whole, it seems essential that the myometrium should have a high sensitivity to oxytocin at term; and labour may actually be initiated by oxytocin. But there are probably more factors determining the onset of labour.

EFFECTS OF CATECHOLAMINES

Responses of the myometrium to catecholamines vary with species and, in some species, with different conditions of hormonal influence (Table 2). Excitation is generally associated with activation of α receptors, and inhibition with activation of β receptors (Ahlquist, 1948, 1962, 1966). The uterus has usually both α and β receptors, but the final effects depend on the degree to which one or the other of these receptors is activated by catecholamines. Besides the hormonal influence, other factors can also affect, qualitatively, the actions of catecholamines. For example, small doses of adrenaline produced only inhibition of uteri from some rabbits, while higher doses caused stimulation followed by inhibition (Rothlin & Brügger, 1945). In isolated uteri from immature guinea-pigs, adrenaline caused relaxation at first, but later produced a biphasic effect and finally, 6 to 8 hr later, a contraction (Hermansen, 1961). The stretch of the preparation may also affect the responses (Miller, 1967).

The change of the uterine responses to catecholamines during pregnancy can be most clearly demonstrated in the cat. In 1906, Dale observed that electrical stimulation of the hypogastric nerve caused a relaxation of the uterus in the virgin cat, a contraction in the early pregnant uterus, and in late pregnancy the adrenaline effect became diphasic. These phenomena were called 'adrenaline reversal'. Graham & Gurd (1960) have suggested that a water-soluble substance is formed in the progesterone-dominated cat uterus which reverses the action of adrenaline on the non-pregnant uterus. However, Vogt (1965) studied the responses of the myometrium to hypogastric nerve stimulation and stated that there appears to be no change in the nature of the transmitter liberated when the uterine response to hypogastric nerve stimulation changes from relaxation to contraction as a result of pregnancy.

Bülbring, Casteels & Kuriyama (1968) confirmed Dale's observation. In the virgin cat the dose of adrenaline which produced a relaxation equal to that caused by maximal nerve stimulation was about 2 μg for a cat weighing 2·5 kg, but the dose of noradrenaline required to produce the same effect was 20 to 100 times larger. In the early pregnant cat, a contraction equal to that caused by maximal nerve stimulation was produced by adrenaline and noradrenaline in the same dose range of 5 to 10 μg. These observations agree with those of Vogt (1965).

Electrical stimulation of the hypogastric nerve elicits a contraction of the myometrium in oestrogen-dominated rabbits and the response is thought to be mediated by the adrenergic transmitter (Schofield, 1952; Setekleiv, 1964). Miller

TABLE 2. Effect of adrenaline on the uteri of different species in various hormonal states

Animal	Condition	Response	Reference
Rat	All states	Inhibition	Ahlquist, 1948
		Inhibition	Levy & Tozzi, 1963
	Oestrogen-dominated	Stimulation	Mann, 1949
		Stimulation	Mann & West, 1950
		Stimulation	Bonney & Ferguson, 1950
Guinea-pig			
Immature	Untreated	Inhibition	Greeff & Holtz, 1951
	Oestrogen-dominated	Stimulation	Holtz & Wölpert, 1937
		Stimulation	Greeff & Holtz, 1951
		Stimulation	Van der Pol, 1956
		Stimulation	Hermansen, 1961
Mature	All states	Biphasic	Bülbring, Casteels & Kuriyama, 1968
	Oestrus	Stimulation	Kochmann & Seel, 1929
Rabbit	Non-pregnant	Inhibition (small dose)	Rothlin & Brügger, 1945
		Biphasic (high dose)	Rothlin & Brügger, 1945
	Progesterone-dominated	Stimulation	Willems, Bernard, Delaunois & De Schaepdryver, 1965
Cat	Oestrogen-dominated (non-pregnant)	Inhibition	Robson & Schild, 1938
		Inhibition	Balassa & Gurd, 1941
		Inhibition	Tsai & Fleming, 1964
		Inhibition	Bülbring, Casteels & Kuriyama, 1968
	Progesterone-dominated	Stimulation	Robson & Schild, 1938
		Stimulation	Balassa & Gurd, 1941
	Pregnant	Stimulation	Tsai & Fleming, 1964
		Stimulation	Bülbring, Casteels & Kuriyama, 1968
Dog	Non-pregnant	Stimulation or Biphasic	Greeff & Holtz, 1951
	Labour or Post partum	Inhibition	Rudolph & Ivy, 1930
Sheep	Non-pregnant	Stimulation	Gunn, 1944
	Pregnant	Inhibition	Gunn, 1944
Human			
In vivo	Pregnant	Inhibition	Garrett, 1954
		Inhibition	Reynolds, Harris & Kaiser, 1954
		Inhibition	Zuspan, Cibils & Pose, 1962
		Inhibition (small dose)	Woodbury & Abreu, 1944
		Stimulation (large dose)	Kaiser & Harris, 1950
In vitro	All states	Stimulation	Garrett, 1955
	Non-pregnant	Inhibition	Kumar, Wagatsuma & Barnes, 1965

& Marshall (1965) studied effects of hypogastric nerve stimulation on the oestrogen-dominated (4 days 70 μg oestradiol every other day) and progesterone-dominated uterus (70 μg oestradiol every other day for 4 days then 5 mg progesterone daily for 5 days). Nerve stimulation contracted uteri from untreated and oestrogen-dominated rabbit. On the other hand, nerve stimulation inhibited the spontaneous contractions of progesterone-dominated uteri. The contractions of the oestrogen-dominated uteri were abolished by the adrenergic blocking agent phentolamine, and the relaxation of the progesterone-dominated uteri was abolished by propranolol. The catecholamine content of the uterus was unaltered by hormonal treatment, and reserpine greatly diminished uterine catecholamine content in all conditions. From the above observations, they proposed that adrenergic receptors in the oestrogen-dominated myometrium are predominantly excitatory (α-receptor) whereas those of progesterone-dominated uteri are inhibitory (β-receptor). Though normal labour can occur after the extrinsic nerve supply to the uterus has been cut, there is evidence that nervous impulses may play an important part in parturition (Theobald, 1965). Further experiments are necessary to elucidate the nervous influence and the influence of catecholamines on the uterine motility in relation to labour.

It has generally been assumed that the rat myometrium contains only beta receptors and that adrenaline relaxes it irrespective of its hormonal state (Rudzik & Miller, 1962; Levy & Tozzi, 1963; Marshall, 1959; Csapo & Kuriyama, 1963). However, there is now evidence that rat uterus has both alpha and beta receptors, and that the response to catecholmines, in some species, depends on the hormonal state of the muscle. It has been shown that adrenaline can produce excitation of the rat uterus after treatment with beta blockers, and that the stimulating effect is blocked by alpha blockers (Jensen & Vennerød, 1961; Brooks, Schaeppi & Pincus, 1965; Brody & Diamond, 1967; Diamond & Brody, 1966).

According to Marshall (1967), noradrenaline (4×10^{-7} to $1 \cdot 2 \times 10^{-5}$ M) relaxed the uterus from ovariectomized rats treated with a combination of oestrogen and progesterone as described previously (Miller & Marshall, 1965), but a beta blocker, propranolol, abolished the inhibitory action or unmasked the stimulating action of noradrenaline. The uterus from rats treated only with oestrogen was stimulated by noradrenaline and this action was converted to an inhibitory effect by an alpha blocker, phentolamine. Noradrenaline (10^{-7} to 10^{-5} M) generally relaxed the rat myometrium during early and mid pregnancy but in low concentration (around 10^{-7} M) stimulated the uterus at term (see Fig. 7). Moreover, after treatment with propranolol, the adrenaline action was also converted to a stimulation. From these results Marshall (1967) suggested that oestrogen may increase the sensitivity of the alpha receptors to catecholamines.

As already mentioned above, the Cl content of cat uterus was increased during pregnancy (Bülbring, Casteels & Kuriyama, 1968). It was therefore thought that the change in Cl content might be related to the reversal of the response to adrenaline. If it is remembered that it has been shown that adrenaline increases K- and Cl-permeability in the guinea-pig taenia coli (Bülbring & Tomita, 1969a)), one might assume that it has a similar action on pregnant cat uterus. However, in this tissue, the K-permeability is probably already high, so that the action of adrenaline on Cl- permeability becomes predominant, while in the virgin cat uterus the effect on K-permeability predominates. Further experiments, including measurement of membrane conductance, in different species are required. It should be mentioned

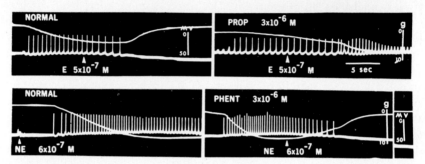

FIG. 7 Comparison of effects of adrenaline (E) and noradrenaline (NE) on uterine segments from two parturient rats, before (normal) and after adrenergic blocking agents. Top trace isometric tension (increase in tension downward); bottom trace membrane potentials. *Upper left frame*: Adrenaline given during spontaneous contraction of muscle, action potentials abolished, membrane potential hyperpolarizes to around 60 mV, muscle relaxes. Contractions ceased for about 10 min (not shown here). *Upper right frame*: same muscle after propranolol (PROP). Adrenaline given near end of spontaneous contraction, after about 10 sec action potential frequency increases, membrane depolarizes, contractile force increases and remains high for about 1 min. Thereafter (for about 15 min) rate of contractions is greater than during control period. *Lower left frame*: second muscle, noradrenaline given during silent period between contractions, action potentials initiated, membrane depolarizes, spike frequency increases, muscle contracts. *Lower right frame*: same muscle in presence of phentolamine (PHENT). Noradrenaline given during spontaneous contraction now abolishes spikes, muscle relaxes, membrane hyperpolarizes. Break between frames signifies 5 min interval. Muscle contractions inhibited for about 10 min (not shown here). (Marshall, 1967)

that adrenergic blocking agents themselves affect the membrane conductance of the pregnant rat myometrium. For example, di- chloroisoproteranol (DCl, 5 × 10^{-5} g/ml) and propranolol (5 × 10^{-5} g/ml) increased the membrane resistance measured by the extracellular polarizing method (Kuriyama, unpublished).

In general, it seems that the inhibitory action of adrenaline is mainly due to an increase in Ca binding at the membrane (Marshall, 1968). Catecholamines are known to produce relaxation in the depolarized rat uterus. Schild (1964, 1966) supposed that the underlying mechanism is the same in depolarized and normal tissues and postulated that catecholamines either prevent the entry of external Ca and/or activate a Ca-accumulating mechanism which lowers the free intracellular Ca concentration, thereby causing inhibition.

There are some indications that the action of catecholamines especially their beta action, is related to their ability to increase the formation of cyclic $3',5'$-AMP in many muscles (see Haugaard & Hess, 1965; and Chapter 12). In the myometrium, relaxation (rat: Brody & Diamond, 1967) and contraction (rabbit: Bartelstone, Nasmyth & Telford, 1967) may involve the cyclase system. Bülbring, Casteel & Kuriyama (1968) observed that a low concentration of ATP caused inhibition in the virgin cat uterus and excitation in the early pregnant cat uterus. In the guinea-pig uterus, when given during a quiescent period, ATP caused a burst of spikes and contraction like adrenaline and noradrenaline. When given during spontaneous activity the effect of ATP resembled that of noradrenaline more than that of adrenaline, for the stimulant component of its diphasic effect was more pronounced. The effect of ATP was, like that of catecholamines, greatly influenced by the external Ca concentration, being abolished by Ca deficiency and enhanced by excess Ca. Equiactive concentrations of catecholamines were about 100 times lower than those of ATP. A causal relationship between adrenaline and ATP on myometrium was not established.

CONCLUSION

Mechanical responses are easy to study, but it is nearly impossible to analyse the underlying mechanism from the mechanical activity alone. Electrophysiological investigations give more fundamental information and are useful for understanding the function of the myometrium, the hormonal influence on its function, and the mechanisms of drug action. Unfortunately, the results of electrophysiological investigation are still too limited for a convincing hypothesis. There are many problems to be solved. Among them the control of the free intracellular Ca concentration and the bound Ca at the cell membrane seem the most important. As discussed by Kuriyama (1964b), it may be that Ca adsorption to the membrane is related to the action of oestrogen and progesterone through a metabolic process. Ca adsorption to the membrane might also modify the responses to drugs.

ACKNOWLEDGEMENT — This article has been written in collaboration with Dr. H. Kuriyama and Dr. T. Tomita to whom the author wishes to express his gratitude.

14

5-HYDROXYTRYPTAMINE RECEPTORS

G. V. R. Born

INTRODUCTION

5-Hydroxytryptamine (5-HT) is found in numerous species of plants and animals (Erspamer, 1966; Welsh, 1968). An enormous amount of information is now available about the distribution and effects of 5-HT in both vertebrates and invertebrates. Much less is known about the biological reasons for its distribution and about the mechanism of the effects. This discussion is mainly about the mode of action of 5-HT. Its effects on mammalian smooth muscle are known in some detail and references to other biological systems will be made only when they throw light on the main topic.

The fact that 5-HT has a characteristic effect on a smooth muscle is taken to indicate that the muscle possesses specific chemical groupings with which the 5-HT interacts in some way. Information about these chemical groupings or 'receptors' has come mostly from *in vitro* experiments with freshly isolated preparations of smooth muscle. As well as smooth muscle cells, these preparations usually contain other types of tissue including nerve cells, connective tissue cells and fibres, and small amounts of vascular endothelium, erythrocytes and leucocytes, large mononuclear cells, etc. In attributing effects to 'smooth muscle' the assumption is implicit that actions on other types of cell are irrelevant or insignificant. This may seem reasonable when muscular activity is followed by recording contractions and relaxations of the preparation, but even these mechanical effects can be produced both by a direct action of 5-HT on the muscle cells and by an indirect action on the nerves. Furthermore, the presence of other tissues may affect the concentrations of 5-HT at the muscle receptors; thus connective tissue can retard the diffusion of 5-HT (Born, 1962). There is no evidence that the action of 5-HT on smooth muscle preparations involves cells other than smooth muscle and nerve so that, for the purposes of this discussion, they can be neglected.

Numerous analogues as well as 5-HT itself have been used to investigate 5-HT receptors of smooth muscle from three aspects, viz. (1) their pharmacological properties; (2) their location and numbers; and (3) their chemical identification. These three aspects will be considered in turn.

PHARMACOLOGICAL PROPERTIES

Effects of 5-HT on different smooth muscle preparations *in vitro*
Invertebrates

Smooth muscles in certain invertebrates respond to 5-HT in concentrations that are startlingly small; for example, the beating of cilia in embryos of nudibranch molluscs is accelerated by 5-HT in concentrations as low as 1×10^{-15} M (Koshtoyants, Buznikov & Manukhin, 1961). 5-HT increases the resting tone and the frequency and amplitude of the beat of the hearts of molluscs, e.g. *Venus mercenaria* (Welsh, 1957) and crustaceans, e.g. *Carcinus moenas* (Florey & Florey, 1954) at 10^{-9} M. The heart of *Venus mercenaria* is also stimulated by lysergic acid diethylamide (LSD) which antagonizes most of the other biological effects of 5-HT including smooth muscle contraction (see p. 421). LSD is astonishingly potent on the *Venus* heart since maximal stimulation is produced by only about six molecules per cell (Wright, Moorhead & Welsh, 1962).

5-HT causes relaxation of the anterior byssus retractor-muscle of the lamellibranch *Mytilus edulis*. When this muscle is stimulated by acetylcholine or by long cathodal pulses, tension is maintained after stimulation has ceased. The maintained tension which is known as 'catch' does not require a continuation of the active state of the contractile mechanism (Twarog, 1954) and is not accompanied by the breakdown of high energy, i.e. arginine, phosphate (Nauss & Davies, 1966). 5-HT added at low concentrations (about 10^{-9} M) causes catch tension to relax immediately. The relaxation is not accompanied by significant changes in the membrane potential; indeed, the catch mechanism does not depend on the level of the membrane potential. It seems that this relaxation is due to a direct action of 5-HT on the contractile mechanism, i.e. to an intracellular action.

Vertebrates

Many preparations of vertebrate smooth-muscles suspended in physiological saline solutions respond to the addition of 5-HT with *contraction*, measured either as decrease in length or as increase in tension. In mammals, 5-HT contracts the smooth muscle of blood vessels, the gastro-intestinal tract, the urogenital system and the bronchial tree. The sensitivity of different smooth muscles to 5-HT varies greatly (Table 1). This variation depends on several factors including species, organ, ionic environment and, sometimes, on other hormones. For example, the isolated uterus of the rat is highly sensitive to 5-HT during oestrus or pregnancy, and sensitivity diminishes during di-oestrus, post partum, or after removal of the ovaries (Erspamer, 1954a). The sensitivity can then be greatly increased by oestradiol but not by testosterone, progesterone, or other steroid hormones. The uterus isolated from a rat a day or so after injection of an oestrogen was for long the most widely used organ for the bioassay of 5-HT (Erspamer, 1954b). It has been largely superceded for this purpose by the fundus-strip preparation of the rat stomach (Vane, 1957). This preparation is more sensitive to 5-HT than the rat uterus and has several other advantages, viz. the animal needs no pretreatment with hormones; the fundus, unlike the uterus, works in various physiological solutions; and the dose-response curve is less steep. The fact that 5-HT acts on a particular smooth muscle provides by itself no clue to the biological significance of the effect. We do not know, for example, why the rat's uterus during oestrus or the fundal part of the stomach should be so sensitive.

Differentiation of the actions of 5-hydroxytryptamine on smooth muscle preparations

When a smooth muscle preparation contains nerve as well as muscle cells, 5-HT may stimulate both but the actions can be differentiated. This was first clearly demonstrated for isolated guinea-pig ileum (Gaddum, 1953). When the ileum was continuously exposed to 5-HT in comparatively high concentrations, initial contraction was followed after 10 to 15 min by relaxation; thereafter the preparation responded to other drugs but not to 5-HT or tryptamine; from the specificity of this tachyphylaxis it was concluded that the ileum contains a receptor specific for 5-HT and tryptamine. Similar exposure of other smooth muscles responsive to 5-HT such as isolated rat uterus did not cause such loss of response. It seemed, therefore, that the tachyphylaxis shown by guinea-pig ileum was due to an action of 5-HT on something other than the smooth muscle cells themselves.

This conclusion was supported by experiments with antagonists. On the one hand, substances such as lysergic acid diethylamide (LSD) and its analogues, which are most effective in antagonizing 5-HT on rat uterus, are comparatively feeble antagonists on guinea-pig ileum. Thus it was found that the effect of 5-HT on the ileum was halved by LSD at 2×10^{-8} M whereas concentrations a hundred times greater had no more effect (Gaddum & Hameed, 1954). On the other hand, several very different drugs including atropine, cocaine and morphine, which are inactive or non-specific antagonists on rat uterus, are highly potent antagonists to 5-HT on guinea-pig ileum (Rocha e Silva, Valle & Picarelli, 1953; Cambridge & Holgate, 1955a). Again, the inhibition of the ileum remains incomplete however high the concentration of these antagonists (Kosterlitz & Robinson, 1955). Furthermore, adenosine and the adenine nucleotides inhibit the action of 5-HT (but not of acetylcholine) on the guinea-pig ileum but not on rat uterus (Werle & Schievelbein, 1964).

Receptors on nerve cells

To explain these observations it was proposed (Gaddum & Picarelli, 1957) that in guinea-pig ileum 5-HT acts on two different sites, one of which is blocked by LSD and its analogues and the other by atropine and morphine. The site which is inactivated by morphine was called the M-receptor. Since atropine and morphine inhibit cholinergic nerve transmission in different ways it was concluded that the M-receptors are in the nervous tissue. The validity of this conclusion has been strengthened by many other observations. Isolated guinea-pig ileum can be prepared in such a way that the main intrinsic nervous system, i.e. Auerbach's plexus, is removed leaving the smooth muscle functional (Paton & Zar, 1968). With this preparation it has been shown that all the acetylcholine which the ileum contains or is capable of producing in response to stimulation originates in the nervous tissue, and that contractions produced by 5-HT (10^{-6} M) are about ten times greater in the presence than in the absence of the plexus. Clearly, therefore, 5-HT has a potent stimulating action on the nerves in guinea-pig ileum.

The nervous tissue in the ileum becomes rapidly refractory to 5-HT. This calls to mind the observation that platelet aggregation is induced by brief exposure to 5-HT in low concentrations but inhibited by prolonged exposure or by higher concentrations (Baumgartner & Born, 1968). This inhibition seems to be caused by the saturation with 5-HT of specific receptors in the platelet membrane. It remains to be seen whether a similar explanation can account for the rapid desensitization of the nerve cells to 5-HT. There is evidence that 5-HT released in the intestinal wall is involved in the peristaltic reflex (see Chapter 17) by acting on the intrinsic nerve plexuses (Bülbring, 1961a). 5-HT may act also as a ganglionic neuro transmitter in the vagal inhibitory pathway to the stomach

TABLE 1. Sensitivities of smooth muscles to 5-hydroxytryptamine

Species	Organ	Effect	Minimal effective concentration (order of magnitude) (M)	Effect of catecholamines	Reference
Helix pomatia	Stomach	Contraction	10^{-13}	Contraction	Gryglewski & Supniewski (1963)
Helix sp.	Heart	Increases in amplitude	10^{-11}	None	Kerkut & Cottrell (1963)
Hirudo sp.	Dorsal muscle	Relaxation	10^{-11}		Poloni (1961)
Cardium edule	Heart		10^{-10}		Erspamer (1966)
Spisula solida	Heart	Increase in amplitude	10^{-10}	Stimulate	Gaddum & Paasonen (1955)
Anodonta cygnea	Heart ventricle	Increases in amplitude and frequency	10^{-9}	None	Fänge (1955)
Mya arenaria	Heart		10^{-9}		Erspamer (1966)
Cyprina islandica	Heart		10^{-9}		Erspamer (1966)
Mytilus edulis	Anterior retractor muscle of byssus	Relaxation of tissue contracted by acetylcholine or electrical stimulus	10^{-9}		Cambridge & Holgate (1955b)
Frog	Lung	Relaxation	10^{-11}	Relaxation	Brecht & Jeschke (1960)
Rat	Stomach fundus strip	Contraction	10^{-10}	Inhibitory	Vane (1957)
Rat, ovariectomized plus oestradiol	Uterus	Contraction	10^{-8}	Inhibitory	Erspamer (1954a)
Rat, virgin plus stilboestrol	Uterus	Contraction	10^{-8}	Inhibitory	Gaddum & Hameed (1954)
Rat	Bronchial muscle	Contraction	10^{-8}		Brocklehurst (1957)
Rat	Colon	Contraction	10^{-7}	Inhibitory	Dagliesh, Toh & Work (1953)

Animal	Tissue	Effect	Concentration		Reference
Rat, castrated	Seminal vesicle	Potentiates contraction given by adrenaline noradrenaline ca. 10^{-5} M	10^{-6}		Picarelli, Hyppolito & Valle (1962)
Guinea-pig, fasted 18 hr	Ileum	Contraction	10^{-8}		Amin, Crawford & Gaddum (1954)
Guinea-pig	Uterus	Contraction	10^{-7}		McGovern, Ozkaragoz, Hensel & Burdon (1961)
Guinea-pig	Bronchiole muscle	Contraction	10^{-7}		Brocklehurst (1957)
Rabbit	Isolated ear refrigerated overnight	Vaso-constriction	10^{-12}	Contraction	Gaddum & Hameed (1954)
Rabbit	Bronchiole muscle	Contraction	10^{-5}		Brocklehurst (1957)
Cat	Bronchiole muscle	Contraction	10^{-8}		Brocklehurst (1957)
Cat	Nictitating membrane	Contraction	10^{-7}		Reid & Rand (1952)
Cat	Constrictor pupillae	Constriction of iris	10^{-6}		Koella & Schaeppi (1962)
Dog	Bronchiole muscle	Contraction	10^{-7}		Brocklehurst (1957)
Dog	Ureter	None			Borgstedt (1959)
Pig	Isolated ear	Vasoconstriction	10^{-11}		Grette (1957)
Horse	Constrictor	Constriction of iris	10^{-9}		D'Ermo & Bonomi (1961)
Rhesus Monkey	Bronchiole muscle	Contraction	10^{-4}		Brocklehurst (1957)
Man	Bronchiole muscle	Contraction	10^{-4}		Brocklehurst (1957)

(Bülbring & Gershon, 1968); a detailed description is found in Chapters 15, 16 and 17. Indeed, 5-HT stimulates or inhibits nerves in many situations from ganglia in molluscs (Gershenfeld & Stefani, 1966) to brain cells in mammals (Krnjević & Phyllis, 1963; Roberts & Straughan, 1967). As far as smooth muscle is concerned, guinea-pig ileum seems to be exceptional in so far as the effect of 5-HT is mediated principally by M-receptors in the nerves.

Receptors on smooth-muscle cells

After inactivation of the M-receptors, LSD and some other drugs including phenoxybenzamine (Dibenzyline) abolish the remaining effect of 5-HT on guinea-pig ileum. The sites which are inactivated by phenoxybenzamine were named the D-receptors and it was inferred that these were situated in the smooth-muscle cells themselves (Gaddum & Picarelli, 1957). Such a direct action of 5-HT on smooth muscle cells is clearly seen with other types of smooth muscle which contain little or no nervous tissue. The amniotic membrane of a 10 to 12-day-old chick embryo contains smooth muscle but no nervous tissue at all. 5-HT causes rhythmical contractions of the amniotic membrane which must be due to a direct effect on the muscle cells (Evans & Schild, 1953). The contractions of the rat uterus caused by 5-HT are also due to a direct action on the muscle cells. These muscle receptors have been characterized pharmacologically by their reactions to antagonists and to analogues of 5-HT.

Antagonists

When the rat uterus is exposed to LSD the contractions caused by 5-HT diminish progressively for about 2 hr. This antagonism by LSD is specific to 5-HT since the effects produced by acetylcholine, histamine, and the catechol-amines are unchanged (Gaddum, Hebb, Silver & Swann, 1953). Several compounds related to LSD, e.g. 2-bromo LSD and 1-acetyl LSD, antagonize 5-HT more strongly than LSD itself.

The antagonism between 5-HT and LSD in low concentrations is competitive, presumably for the D-receptor. This is probably accounted for by structural similarities in the molecules of the two substances (Fig. 1) and is in keeping with the observations that both substances act in *parallel* on certain tissues, e.g. on the heart of *Venus mercenaria* (Greenberg, 1960). When LSD is left in the bath, in time even the largest doses of 5-HT are unable to cause maximal contraction; the antagonism is said to be unsurmountable (Gaddum, Hameed, Hathway & Stephens, 1955). Furthermore, prolonged exposure to LSD may render the antagonism irreversible. The mechanism of these effects is not yet clear.

Dihydroergotamine and other ergot alkaloids antagonize almost as potently as LSD but less specifically. The antagonism by phenoxybenzamine and diben-amine is even less specific because they also antagonize the effects of acetyl-choline, histamine, and the catecholamines (Furchgott, 1954). Furthermore,

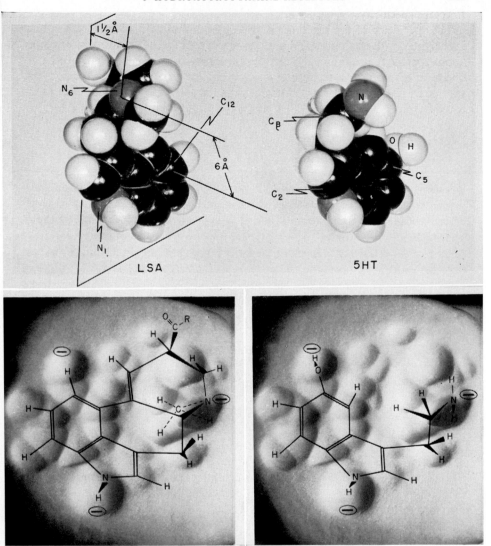

FIG. 1   (a) Taylor-Hirschfelder models of (+)-lysergic acid (LSA) and 5-hydroxytrypt-amine (5-HT)   (b) Composite negative imprint of the molecules in (a) as a model of 5-hydroxytryptamine receptor area. Structural formulae of (+)-lysergic acid diethyl-amide (*left*) and 5-hydroxytryptamine (*right*) show orientation of molecules. The H at C_5 of (+)-lysergic acid diethylamide is omitted; R is $-N(C_2H_5)_2$. Three proposed negative binding sites for the 5-hydroxyl group, 1-nitrogen atom and terminal amino-group of 5-hydroxytryptamine are indicated. (From Greenberg, 1960)

the antagonism becomes increasingly irreversible, presumably because of the alkylation of one or more essential groups in or near the receptor (Belleau & Triggle, 1962).

Receptor specificity

The specificity of the reacting sites in smooth muscle preparations has been investigated with analogues and antagonists of 5-HT. The resulting information about specificity requirements has been interpreted by some as indicating only one type of receptor and by others as indicating two different receptors, one specific for 5-HT and the other for tryptamine. The reasons for this uncertainty emerge from a brief review of the most relevant evidence.

Influence of monoamine oxidase

Before the relative activities of 5-HT and its analogues can be meaningfully compared, the effect of monoamine oxidase of smooth muscle on these activities has to be explained. It was found (Vane, 1959) that inhibition of monoamine oxidase present in the rat fundus-preparation increases the activities of tryptamine and 5-methoxytryptamine derivatives much more than those of the 5-hydroxylated tryptamines. This and subsequent work established that large differences in activity may be caused by differences not only of fit on the receptor but also of rates of access to the enzyme in the cell which inactivates some of these amines. This conclusion has other implications which will be discussed later. For the present, it is evident that comparisons of the potencies of the amines on the 5-HT receptor are valid only after inhibition of the monoamine oxidase.

Analogues

The effectiveness of different analogues of 5-HT has been compared on the uterus from the rat in oestrus (Erspamer, 1952, 1954a) and on the rat fundus strip (Vane, 1959). Table 2 shows the activities on the fundus strip of 30 analogues compared with that of 5-HT; the activities are expressed as the number of moles equal in potency to one mole of 5-HT. All these analogues caused the strip to contract but 5-HT itself was the most potent. The relative activities of the others varied from 3×10^4 for indolemethylamine which was the weakest to 1·4 for 5-hydroxy-α-methyltryptamine which was the strongest. The log dose-response curves of the more active substances were parallel to that of 5-HT, suggesting that all of them acted on the same receptor. The equipotent molar ratios varied as follows: for tryptamine and its derivatives (Nos. 1 to 12) from 32 to 33,000; for 5-hydroxylated tryptamines (Nos. 24 to 30) from 2·0 to 20; for other 5-substituted tryptamines (Nos. 19 to 23 and 31) from 1·7 to 16; and for other ring-substituted derivatives (Nos. 15 to 18) from 2·0 to 950.

These groupings permit some generalizations. First, all tryptamines substituted in position 5 are more active than those without substituents there: tryptamine had less activity than 5-hydroxy-, 5-methoxy-, 5-chloro- or 5-methyl-tryptamine. 4-hydroxytryptamine was almost as potent as 5-hydroxytryptamine whereas 6-hydroxytryptamine was more than 500 times weaker. This suggests that the 4- and 5- positions of the indole ring may be important points of attachment for tryptamine derivatives to the receptor site.

Secondly, the importance of the length of the side-chain is demonstrated by the comparative activities of indolemethylamine, tryptamine and homotryptamine (Nos. 2, 1 and 3). As would be expected, the order of potency is tryptamine, homotryptamine and, very much weaker, indolemethylamine. If the terminal amino-group is a second point of attachment of the molecule to the receptor site, the rotation of the longer side-chain of homotryptamine will permit attachment of this second point whereas the side-chain of indolemethylamine will be too short.

Thirdly, evidence about other points of attachment of tryptamine derivatives to the receptor site is provided by the relative activities of $N'N'$-dimethyltryptamine (No. 10) and $2:N'N'$-trimethyltryptamine (No. 13). The introduction of the methyl group at the 2-position of the indole ring is associated with a tenfold diminution in activity. The methyl group will limit free rotation of the side-chain by steric hindrance. If the reduction in activity is assumed to be due to steric hindrance, then it can be deduced that the point of attachment of the terminal amino-group fits best when the side-chain is rotated to a position fairly close to the indole ring, at a maximum distance away from the ring hydroxyl group. These two probable points of attachment of 5-HT to the receptor site may not, however, be the only ones.

The role of the indole nitrogen in the receptor reaction has been assessed by comparing the activity on the rat fundus strip of 5-HT and other tryptamines with the corresponding indene alkylamines in which the ring nitrogen is replaced by a methylene group (Winter, Gessner & Godse, 1967). For pairs of corresponding compounds both the intrinsic activity and the receptor affinity were similar. This indicates that the ring nitrogen is not essential for binding the active tryptamine to the receptor.

These conclusions on structure-activity relationships imply that, in smooth muscle, 5-HT and other tryptamines as well as these isosteric indene alkylamines act on the same receptor. The validity of this assumption has been questioned on the basis of various observations. Thus, aged strips of sheeps carotid artery lose their sensitivity to tryptamine before that to 5-HT (Woolley & Shaw, 1953). Furthermore, LSD and 2-bromo-lysergic acid diethylamide (brom-LSD) appear to antagonize the action of 5-HT on the rat fundus strip more than that of tryptamine (Barlow & Khan, 1959). However, this difference disappears if the preparation is pretreated with an amine oxidase inhibitor (Vane, unpublished results). Although brom-LSD has no effect on the rates of penetration for 5-HT or tryptamine into the fundus tissue (Handschumacher & Vane, 1967), the difference in its antagonizing potency could be explained by the existence of two receptors (Barlow, 1961).

Other evidence has come from experiments in which the antagonism by phenoxybenzamine to 5-HT, to other tryptamines, and to their indene isosteres has

TABLE 2. Equipotent molar ratios of tryptamine derivatives estimated on the isolated rat stomach strip, before and after amine oxidase inhibition (Vane, 1959)
The numerals in brackets refer to the number of experiments on each compound. 5-hydroxytryptamine was taken as unity.

No.	Ring substituent	Side-chain R	Trivial name used in text and systematic chemical name	Normal (A) Equipotent molar ratio Mean ± S.E. (No. of tests)	After amine oxidase inhibition (B) Equipotent molar ratio Mean ± S.E. (No. of tests)	Potentiation (A/B)
1	—	$-CH_2-CH_2-NH_2$	Tryptamine 3-(2-Aminoethyl)indole	408 ± 49 (15)	32 ± 4·8 (11)	12·7
2	—	$-CH_2-NH_2$	Indolemethylamine 3-Aminomethylindole	29,000 (2)	33,000 (2)	0·9
3	—	$-CH_2-CH_2-CH_2-NH_2$	Homotryptamine 3-(3-Aminopropyl)indole	1,920 (2)	460 (2)	4·2
4	—	$-CH_2-CH(CH_3)-NH_2$	α-Methyltryptamine 3-(2-Aminopropyl)indole	31 ± 5·7 (5)	40 (2)	0·8
5	—	$-CH_2-CH(C_2H_5)-NH_2$	α-Ethyltryptamine 3-(2-Aminobutyl)indole	4,600 (2)	—	—
6	—	$-CH_2-C(CH_3)_2-NH_2$	αα-Dimethyltryptamine 3-(2-Amino-2:2-dimethylethyl)indole	1,400 (1)	—	—
7	—	$-CH_2-CH_2-NH(CH_3)$	N'-Methyltryptamine 3-(2-Methylaminoethyl)indole	1,120 ± 192	126 (1)	8·9
8	—	$-CH_2-CH_2-NH(C_2H_5)$	N'Ethyltryptamine 3-(2-Ethylaminoethyl)indole	250 (2)	170 (1)	1·5
9	—	$-CH_2-CH_2-NH(C_3H_7)$	N'-Propyltryptamine 3-(2-Propylaminoethyl)indole	330 (2)	330 (1)	1·0
10	—	$-CH_2-CH_2-N(CH_3)_2$	N'N'-Dimethyltryptamine 3-(2-Dimethylaminoethyl)indole	196 ± 30 (3)	110 (1)	1·8
11	—	$-CH_2-CH_2-N(C_2H_5)_2$	N'N'-Diethyltryptamine 3-(2-Diethylaminoethyl)indole	83 (2)	87 (1)	1·0

No.	Substituent	Side chain	Name			
12	—	—CH₂—CH₂—N(C₃H₇)₂	N'N'-Dipropyltryptamine / 3-(2-Dipropylaminoethyl)indole	34 (2)	41 (1)	0·8
13	2—CH₃	—CH₂—CH₂—NH₂	2:N'N'-Trimethyltryptamine / 3-(2-Dimethylaminoethyl)-2-methylindole	1,200 (1)	—	—
14	5—OH	—CH₂—CH₂—NH₂	5-Hydroxytryptamine / 3-(2-Aminoethyl)-5-hydroxyindole	1·0	1·0	1·0
15	4—OH	—CH₂—CH₂—NH₂	4-Hydroxytryptamine / 3-(2-Aminoethyl)-4-hydroxyindole	1·8 (2)	2·0 (2)	0·9
16	6—OH	—CH₂—CH₂—NH₂	6-Hydroxytryptamine / 3-(2-Aminoethyl)-6-hydroxyindole	460 (2)	560 (2)	0·8
17	6—OCH₃	—CH₂—CH₂—NH₂	6-Methoxytryptamine / 3-(2-Aminoethyl)-6-methoxyindole	1,520 (2)	950 (2)	1·6
18	5:6—(OCH₃)₂	—CH₂—CH₂—NH₂	5:6-Dimethoxytryptamine / 3-(2-Aminoethyl)-6-methoxyindole	300 (2)	147 (2)	2·0
19	5—OCH₃	—CH₂—CH₂—NH₂	5-Methoxytryptamine / 3-(2-Aminoethyl)-5:6-dimethoxyindole	20 ± 4 (4)	1·7 ± 0·2 (4)	11·8
20	5—OCH₃	—CH₂—CH(CH₃)—NH₂	5-Methoxy-α-methyltryptamine / 3-(2-Aminoethyl)-5-methoxyindole	2·9 ± 0·6 (3)	2·5 (2)	1·2
21	5—Cl	—CH₂—CH₂—NH₂	5-Chlorotryptamine / 3-(2-Aminopropyl)-5-methoxyindole	100 (2)	7·6 (2)	13·2
22	5—CH₃	—CH₂—CH₂—NH₂	5-Methyltryptamine / 3-(2-Aminoethyl)-5-chloroindole	184 ± 42 (3)	9 (1)	20·0
23	5—CH₃	—CH₂—CH(CH₃)—NH₂	5:α-Dimethyltryptamine / 3-(2-Aminoethyl)-5-methylindole	14 ± 3·3 (3)	16 (1)	0·9
24	5—OH	—CH₂—CH(CH₃)—NH₂	5-Hydroxy-α-methyltryptamine / 3-(2-Aminopropyl)-5-methylindole	1·4 ± 0·1 (7)	2·0 (2)	0·7
25	5—OH	—CH₂—CH(C₂H₅)—NH₂	α-Ethyl-5-hydroxytryptamine / 3-(2-Aminopropyl)-5-hydroxyindole	7·0 (1)	—	—
26	5—OH	—CH₂—CH₂—NH(CH₃)	5-Hydroxy-N'-methyltryptamine / 5-Hydroxy-3-(2-methylaminoethyl)indole	6·3 (2)	9·5 (1)	0·7
27	5—OH	—CH₂—CH₂—N(CH₃)₂	5-Hydroxy-N'N'-dimethyltryptamine / 3-(2-Dimethylaminoethyl)-5-hydroxyindole	10 ± 3·6 (3)	8·0 (1)	1·2
28	5—OH	—CH₂—CH₂—N(C₂H₅)₂	N'N'-Diethyl-5-hydroxytryptamine / 3-(2-Diethylaminoethyl)-5-hydroxyindole	20 (1)	20 (1)	1·0
29	5—OH	—CH₂—CH₂—N(C₃H₇)₂	5-Hydroxy-N'N'-dipropyltryptamine / 3-(2-Dipropylaminoethyl)-5-hydroxyindole	4·0 (2)	6·0 (2)	0·7
30	5—OH	—CH₂—CO₂H	5-Hydroxyindoleacetic acid / 5-Hydroxyindol-3-ylacetic acid	Inactive	—	—
31	5—OCH₃	—CH₂—CO₂H	5-Methoxyindoleacetic acid / 5-Methoxyindol-3-ylacetic acid	Inactive	—	—

15+

been compared on the rat stomach strip (Winter & Gessner, 1968). In these experiments, both agonists and the antagonists were used at very high concentrations (10^{-3} to 10^{-5} M), particularly in attempts to protect receptors against phenoxybenzamine. The concentrations of 5-HT used (10^{-4} to 10^{-5} M) are 10^3 to 10^1 times greater than those which can cause considerable contractions of the fundus so that the receptors responsible for these contractions were certainly desensitized. Phenoxybenzamine totally blocked the response to 5-HT but only partially the responses to tryptamine and its isostere 3-(2-amino ethyl)-indene and to N,N-diethyltryptamine and 3-(N,N-diethyl-2-amino ethyl)-indene. This was taken as evidence that 5-HT acts on a receptor which is blocked totally by phenoxybenzamine whereas the other compounds act on the same receptor *and* on another which is not blocked. High concentrations of the agonists protected the 5-HT receptor against phenoxybenzamine block. This cannot be taken as unambiguous evidence that the contraction caused by tryptamine and the other substances was mediated by the 5-HT receptor (Waud, 1962). Nevertheless, it was proposed that the contractile response is mediated in part by a phenoxybenzamine-resistant tryptamine receptor which is distinct from another receptor powerfully activated by 5-HT and wholly blocked by phenoxybenzamine.

Furthermore, in a tissue responding maximally to an agonist, further contraction in response to simultaneous exposure to a second agonist acting on the same receptor would be expected only if the second agonist possesses greater intrinsic activity than the first (Ariens, 1964). In keeping with this, a rat fundus strip maximally contracted to 5-HT contracts further on the addition of 3-(N,N-dimethyl-2-amino-ethyl)-indene the intrinsic activity of which is greater than that of 5-HT. Similarly, a strip maximally contracted to 3-(3-amino-propyl)-indene, which has only half the intrinsic activity of 5-HT, contracts further when 5-HT is added. However, a strip maximally contracted to 5-HT also contracts further when 3-(3-amino propyl)-indene is added. This has been put forward as further evidence that the indene alkylamines act, at least in part, on a receptor distinct from that for 5-HT. This other receptor is neither that for acetylcholine nor that for histamine.

LOCATION AND NUMBERS OF 5-HT RECEPTORS

So far, smooth muscle receptors for 5-HT have been characterized by the specificity of antagonists and to some extent by the actions of analogues. The next problem is to determine, if possible, the location and numbers of the receptors. Experiments designed to solve this problem have to be considered against the following background.

1. The rapidity with which smooth muscle contracts when 5-HT is added and relaxes when it is washed out suggests that the primary site of action is on the surface membranes of the cells. This conclusion is supported by the slowness

with which 5-HT penetrates through the membrane (Born, 1962), presumably because of its low solubility in lipids (Vane, 1959). In acting on the cell surface, 5-HT would be like other local hormones including acetylcholine, histamine, and the catecholamines even in their effects on metabolism (Sutherland, Rall & Menon, 1962) as well as like some general hormones such as insulin. The evidence is, however, not absolutely conclusive. In some lower animals such as *Mytilus* 5-HT is known to act intracellularly (Twarog, 1968) and there may be unknown intracellular actions of 5-HT also in mammalian cells. Some cells have specialized mechanisms for the concentrative uptake of amines, e.g. platelets for 5-HT (Born & Gillson, 1959), adrenergic neurones for noradrenaline (Iversen, 1967*b*) and, possibly, smooth muscle cells also for catecholamines (Avakian & Gillespie, 1968). Although the existence of these mechanisms does not imply that the amines have intracellular functions in these cells, rigorous proof that the primary action of 5-HT on mammalian smooth muscle is on the cell membrane is still required.

2. Apart from active receptors which mediate the contractor response to 5-HT, smooth muscle cells may also possess 'silent receptors' (Cavallito, Arrowood & O'Dell, 1956). These are chemical groupings to which 5-HT binds without producing any effect. At physiological pH, cell surfaces carry a net negative charge whereas 5-HT is a monovalent cation; simple electrostatic attraction may, therefore, be the primary determinant in causing 5-HT to bind to silent sites. There is evidence that 5-HT is in ionic association with ATP in platelets (Born, Ingram & Stacey, 1958). However, any specificity of silent receptors for 5-HT would require other forces as well.

3. Smooth muscles contain the enzyme monoamine oxidase which inactivates 5-HT by oxidative deamination. The enzyme is found in many different tissues in association with intracellular granules (Blaschko, 1952) and this seems to be so also in smooth muscles (Vane, 1959). If, therefore, the contractor action of 5-HT is exerted on the outer membrane of the cell, the action cannot be influenced by the intracellular monoamine oxidase; if, on the other hand, the action occurs while there is movement of amine through the cell membrane, the monoamine oxidase inside the cell could influence the action by increasing the concentration gradient in the membrane.

4. The general effectiveness of LSD and its analogues in antagonizing the action of 5-HT suggests that 5-HT receptors in different types of smooth muscle are chemically very similar if not identical. That would imply that the differences in sensitivity of different smooth muscles to 5-HT cannot be accounted for by its initial reaction with the receptor, unless minor variations in the receptor were associated with major changes in affinity for 5-HT. It seems more likely that the observed differences are due to other variables of which the most obvious is the absolute number of receptors on each cell. Determination of this number is, therefore, essential for a complete quantitative description of the action of 5-HT on smooth muscle.

Attempts to determine the sites and numbers of receptors have made use mainly of autoradiographic techniques and of uptake measurements.

Autoradiography

Striated muscle takes up tritium-labelled (+)-tubocurarine which is seen on autoradiographs to be concentrated in the motor end-plate regions (Waser, 1966). Since (+)-tubocurarine is a specific competitive antagonist to the transmitter action of acetylcholine at the neuromuscular junction, the localization of the label is assumed to indicate the sites of the acetylcholine receptor. The technique has provided an estimate of about 4×10^6 (+)-tubocurarine molecules bound per end-plate; this should be the number of acetylcholine receptors. In principle, therefore, it is possible to determine both sites and numbers of a receptor by appropriate autoradiographic methods. However, the methods have inherent difficulties, e.g. the displacement or washing out of labelled substances during the preparative procedures, which can make interpretation hazardous. There is also the possibility that the labelled molecules are taken up by silent receptors, i.e. by sites with enough affinity to bind the molecule but not involved in its effect (Veldstra, 1956). This would clearly result in an over-estimate of the number of specific receptors responsible for the effect.

5-HT labelled with tritium has been localized by autoradiography on electron-micrographs of rabbit platelets (Davis & Kay, 1965; Davis & White, 1968). The label was seen to be concentrated in the storage organelles but occurred also scattered elsewhere over the platelets. It is indeed surprising that the highly-soluble labelled 5-HT was not lost from the platelets during their washing and dehydration. It seems reasonable to conclude that the labelled amine is firmly bound at least in the storage organelles; little can be concluded from the radio-activity distributed elsewhere because it may have been spread as an artifact. This work suggests that a similar technique, if not with 5-HT itself then with a more firmly bound specific antagonist, may show up the 5-HT receptors on smooth muscle.

Uptake

If a drug combines with receptor sites on a muscle cell membrane or penetrates through the membrane, changes in distribution of the drug between these different compartments with time should show themselves as a number of different rates and equilibria during the uptake of the drug by the muscle preparation. One of the equilibria may represent the equilibrium association between drug and active receptor. The possibility of demonstrating such a reaction increases when the dissociation constant of the drug-receptor complex is low (Paton, 1961). This tends to be the case less with agonist than with antagonist drugs. With the use of radioactive atropine an estimate has been obtained of the number of acetyl-choline receptors in the longitudinal muscle of guinea-pig small intestine (Paton

& Rang, 1965). A similar kind of investigation has apparently not been attempted for 5-HT although the properties of some of its specific antagonists, e.g. LSD or its more potent derivative 1-methyl lysergic acid butanolamide (methysergide), suggest that they could be used for this purpose.

Uptake by platelets

Measurement of the uptake of 5-HT by blood platelets, under conditions in which its active transport is abolished, has provided an estimate of the number of specific combining sites for 5-HT on the platelet membrane (Born & Bricknell, 1959). This was done by incubating human citrated platelet-rich plasma at 0 to 1°C with radioactive 5-HT added at increasing concentrations. At equilibrium, the concentration of 5-HT in the platelet water was almost exactly the same as that in the plasma, but the straight line relating the concentrations cut the Y axis a little above the origin. This was explained by assuming the existence on the platelet membrane of specific receptors with a high affinity for 5-HT which became saturated at a very low 5-HT concentration; the number of these receptors was calculated to be of the order of 10^4 per platelet. It was also suggested that these receptors might be part of the specific carrier mechanism responsible for transporting 5-HT actively through the platelet membrane at 37°C. Recently, other kinds of experiments have provided evidence in support of this proposition (Baumgartner & Born, 1968).

Uptake by smooth muscle

Measurement of the uptake of 5-HT and various analogues including tryptamine provided information about their distribution and fate in smooth muscle preparations from which interesting inferences have been made about the situation and nature of the specific receptors responsible for contraction. Like other primary amines, 5-HT is a monovalent cation at physiological pH so that it should either not penetrate through the membrane smooth muscle cells at all or only very slowly. If a smooth muscle preparation contained significantly more 5-HT at equilibrium than could be accounted for by the extracellular space, the excess 5-HT might be bound to extracellular sites at least some of which could be the specific receptors. This possibility was looked into by measurements of the uptake of radioactive 5-HT by guinea-pig taenia coli at 0 and 37°C. Radioactivity measurements combined with bioassay allowed the separate determination of intact 5-HT and its biologically inactive derivatives in the preparation; at the same time, the extracellular space was measured with inulin. The results have shown the distribution and fate of the amine in this smooth muscle.

The rate of movement of 5-HT through the muscle cell membranes is slow and similar at 0 and 37°C. The only difference is that, at physiological temperature, all the 5-HT that enters the cells is immediately broken down by monoamine oxidase. Observations on taenia loaded with 5-HT in the absence and presence of a monoamine oxidase inhibitor (phenyl isopropyl-hydrazine) showed that the rates of movement of intact 5-HT and its oxidation product through the cell membrane are very similar, with half-times of 32 and 29 min respectively. It seems, therefore, that 5-HT moves through the membrane solely by diffusion and that this is not significantly influenced by the presence or absence of the primary amino-group in the molecule. When 5-HT is added to a taenia coli preparation the tension reaches maximum after about 2 min. This is the time required for 5-HT to diffuse evenly throughout the extracellular space (Born & Vane, 1961). Furthermore, since 5-HT crosses the cell membranes so slowly, the amounts that enter the cells during tension development must be exceedingly small. These results support the conclusion that the site of the 5-HT receptor is on the outside of the cell membrane.

In the presence of phenyl isopropyl-hydrazine the concentration of intact 5-HT at equilibrium is about 2·5 times greater in the taenia than in the medium. This concentration difference is accounted for most simply by assuming that the amine distributes itself between cells and medium according to a Donnan equilibrium. Krebs bicarbonate solution, which was used in these experiments, has a pH of about 7·4. The internal pH of smooth muscle cells is not known but in striated muscle it is just over 7·0 (Caldwell, 1958). If the pH of smooth muscle is similar, the concentration of H^+ ions in them would be 2 to 3 times greater than in the bathing solution. Since 5-HT is almost entirely in the form of a cation at physiological pH (Vane, 1959) the concentration of 5-HT should also be 2 to 3 times greater in the muscle than in the external solution, and this is what was found experimentally. Thus, a mechanism for concentrating 5-HT is absent from smooth muscle cells and there is neither an active transport process for 5-HT in the membrane nor any significant binding inside the cells as there is, for example, in human erythrocytes (Born, Day & Stockbridge, 1967).

A particularly interesting result emerged from measurements of the rates at which intact 5-HT was lost from taenia in the absence of amine oxidase inhibitor. Most of the amine was lost rapidly at an exponential rate which was similar to that for the loss of inulin; this presumably represented diffusion out of the extracellular space. The phase of rapid loss was followed by one with a very much slower exponential rate of loss which continued over the whole period of observation, i.e. for nearly 2 hr. What are the possible sources of this slow-moving 5-HT? First, it could be coming out of the cells, like the breakdown product, which would imply that a small proportion of intracellular 5-HT is not broken down. This seems unlikely in view of the slowness with which 5-HT diffuses across the cell membranes and the high activity of the intracellular monoamine oxidase (Vane, 1959). Secondly, it may represent the effect of a barrier, possibly of connective tissue, which impedes the free diffusion of 5-HT from the extracellular space. This is improbable because the loss of inulin was complete long before that of 5-HT although inulin has much larger molecules. Thirdly, the 5-HT may be dissociating slowly from extracellular sites to which it is firmly bound; this is clearly the most interesting possibility. Which of these explanations is the right one has still to be decided.

This work has also provided a possible explanation for the slow decline in responsiveness of smooth muscles which shows itself in progressively smaller responses when the same dose of 5-HT is given repeatedly (Bülbring & Burnstock, 1960). The cause of this tachyphylaxis may be the persistence of small amounts of 5-HT on or near the receptors on the muscle cells.

The similarity in the rates of movement of 5-HT and its oxidation product out of taenia coli suggests that both diffuse through the same pores in the cell membrane. The following experiment (Born, unpublished) suggests that these pores can be obstructed by 5-HT.

Isolated taeniae coli were soaked for 2 hr at 0°C in Krebs bicarbonate solution containing radioactive 5-HT; then, at the same temperature, the loss of intact 5-HT and of its radioactive breakdown product were followed for 100 min. Almost all the intact 5-HT was lost with great rapidity, presumably from the extracellular space. Surprisingly, the rate of loss of radioactive breakdown product *increased* for the first 25 min; only then did it show the expected continuous decrease (Fig. 2). The initial increase indicated that the extracellular space must have been free of breakdown product to begin with. Therefore it seemed that, under the conditions of this experiment, the presence of comparatively high concentrations of 5-HT extracellularly prevented the diffusion of 5-HT breakdown products out of the muscle cells.

An intriguing explanation for this observation is that intact 5-HT molecules occupying receptor sites on the cell membrane cause a functional if not a structural closure of the pores or channels through which the deaminated 5-HT molecules diffuse from the interior of the muscle cells into the extracellular space.

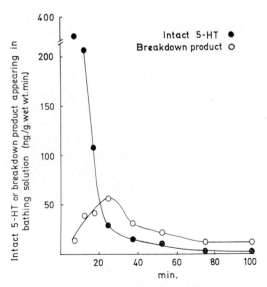

FIG. 2   Loss of intact 5-HT and 5-HT breakdown product at 0°C from isolated taenia coli which has been soaked in Krebs' solution containing 1·1 × 10⁻⁴ M 5-HT labelled with ¹⁴C.

This picture is made less fanciful by observations (Brown & Gillespie, 1957; Brown, Davis & Gillespie, 1958) which showed that, when cat spleen or intestine are stimulated via the sympathetic nerves, the amount of noradrenaline found in the venous blood coming from the organ is increased after the administration of the antagonists dibenamine or phenoxybenzamine. To explain this it was suggested that the mechanism for the destruction of noradrenaline at the nerve ending is linked to noradrenaline receptors so that blocking of the receptors by the antagonists results in the accumulation and overflow of noradrenaline in the blood. In the light of the smooth muscle experiment just described, it may be that the binding of the antagonists and, perhaps, of noradrenaline itself to the receptor sites hinders its movement back into nerve (and possibly into other cells) via pores in the cell membranes. The siting of a drug receptor at the mouth of a pore in a cell membrane has also been proposed to account for the effect of a drug such as acetylcholine on the movements of sodium and potassium through the membrane (Waud, 1968). In the light of that proposal it may be that, on smooth muscle, each 5-HT molecule becomes briefly attached to the receptor site after which there is a large chance of displacement outwards back into the medium and a small chance of displacement deeper into the pore.

Relation between uptake and contraction

More information about the location and properties of the 5-HT receptor came from experiments in which the uptake by rat fundus of 5-HT and trypt-amine was measured at the same time as the contractions caused by them. Contractions were maximal within 1 min, showing that diffusion of both sub-stances throughout the extracellular fluid was equally rapid. In homogenates of the tissue, 5-HT and tryptamine were inactivated by monoamine oxidase at about the same rate showing that, with free access to the enzyme, both were equally susceptible substrates. When the contractor potencies of 5-HT and tryptamine were compared on the intact fundus strip, the presence of mono-amine oxidase inhibitor (Iproniazid or phenyl isopropyl hydrazine) greatly potentiated the action of tryptamine but not that of 5-HT. The potencies of thirty analogues of 5-HT were compared with those of tryptamine and 5-HT itself in the absence and presence of monoamine oxidase inhibitor (Vane, 1959). The general rule emerged that amines with lipophilic substitutions were poten-tiated by the monoamine oxidase inhibitor, like tryptamine itself, whereas amines with a hydroxyl group on the indole ring or other hydrophilic substitu-tions were not. The results were explained by supposing that the lipophilic amines including tryptamine pass through the cell membrane into the muscle cells whereas the more hydrophilic amines, like 5-HT, do not, in accordance with the established generalization that cell membranes are more easily pene-trated by lipid-soluble molecules than by water-soluble molecules. The amines that enter the cells are inactivated by monoamine oxidase with the result that the effective concentration around the receptor on the cell membrane is diminished (Fig. 3); in contrast, the activities of 5-HT and other derivatives which do not penetrate through the membrane are unaffected by inhibition of the enzyme (Handschumacher & Vane, 1967). This hypothesis is supported by the demonstration that the oil/water partition coefficient is almost twenty times greater for tryptamine than for 5-HT (Vane, 1959) showing the major contribution made by the ring hydroxyl group to the extreme slowness with which 5-HT penetrates through a muscle cell membrane (Born, 1962).

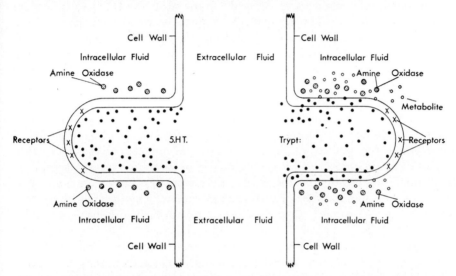

FIG. 3 Diagrams showing that, as the tryptamine molecules near the receptors, the concentration decreases because of diffusion into the cell. Prevention of this diffusion by inhibition of amine oxidase would increase the concentration of tryptamine at the receptor. Since 5-HT is excluded from the cell, amine oxidase inhibition does not influence the action of 5-HT. (From Handschumacher & Vane, 1967)

When the pH of the medium was decreased, the contractions of the rat fundus to 5-HT were unaffected whereas those to tryptamine were potentiated and its uptake into the cells was retarded (Handschumacher & Vane, 1967). Therefore, like inhibition of monoamine oxidase, acidification can be regarded as causing potentiation of tryptamine by increasing its concentration at the receptor.

It should be pointed out that the outer membrane of the cell is not the only possible diffusion barrier to polar molecules such as 5-HT. Indeed, it would be difficult to explain how some cells such as the platelets have both monoamine oxidase activity and a capacity for storing 5-HT if some intracellular mechanisms were not present to keep them apart.

This work also suggested a physiological function for monoamine oxidase (Blaschko, 1952). The enzyme may protect cells by diminishing the concentration of unwanted but active compounds around receptor sites on which they would otherwise act. This idea is made more plausible if the enzyme is unable to

15*

inactivate the naturally occurring, presumably wanted substrate, 5-HT, because of a diffusion barrier (Vane, 1959).

Restrictions on membrane diffusion

The more lipophilic tryptamines presumably diffuse rapidly through the lipid portions of the muscle cell membranes whereas the polar molecules of 5-HT presumably diffuse only through pores or channels which are lined by hydrophilic groups and contain water. Since this movement of 5-HT is so slow these channels must either be few in number or impose restrictions on free diffusion of 5-HT by being very narrow or by structural specificities in their walls. That such structural restrictions may occur is suggested by observations on the movement of amines into human erythrocytes (Born *et al.*, 1967). 5-HT, histamine and (−)-noradrenaline enter red cells by simple diffusion at different velocities which is highest for 5-HT, about four times less for histamine and about thirty times less for (−)-noradrenaline. It has been proposed (Schanker, Nafpliotis & Johnson, 1961) that these differences are determined by two physical properties of the amines, viz. lipid solubility and ionization. However, these properties are interdependent since the degree of ionization largely determines lipid solubility. Furthermore, the observed velocities of diffusion are not related simply to the concentrations of unionized molecules. Thirdly, human erythrocytes take up (−)-noradrenaline but not (+)-noradrenaline (Born *et al.*, 1967); this recalls the observation (LeFevre & Marshall, 1958) that the erythrocyte membrane can also discriminate between stereo-isomers of certain sugars in that those with positive specific rotations penetrate faster than those with negative. Differences of this kind cannot be explained on the basis of lipid solubility or ionization but must depend on structural specificities in the cell membrane. These specificities can pertain either to a receptor or carrier moving the molecules through the membrane for which, in the case of the amines, there is no evidence, or to fixed channels or pores through which the amines diffuse; this seems the more likely alternative. Such an explanation has to be invoked whenever differences in the rates of molecular diffusion through a cell membrane cannot be accounted for by simple differences in lipid solubility. Comparisons of the rates at which different hydroxytryptamines diffuse through the membranes of smooth muscle cells have apparently not yet been made. However, the failure of monoamine oxidase inhibitors to potentiate the action of any of the hydroxylated tryptamines (Vane, 1959) indicates that differences in the diffusion rates must be small and cannot account for the large differences in their observed activities which are, therefore, presumably due to differences in fit. The activities of the more lipophilic tryptamines are clearly limited by their rapid diffusion through the cell membrane. Neither monoamine oxidase inhibitors nor brom-LSD retard this diffusion (Handschumacher & Vane, 1967), but other drugs may be discovered which will potentiate the action of these tryptamines by impeding their passage through the cell membrane.

CHEMICAL IDENTIFICATION OF 5-HT RECEPTOR

The ultimate aims of receptor research are (1) the chemical identification of the binding site, which is necessary for understanding the specificities of drug action; and (2) the elucidation of the processes by which the combination of drug with receptor produces the observed effect; this is necessary for understanding the energetics of drug action, e.g. how contraction of a muscle results from the presence of a particular, comparatively simple substance in the environment. At present almost nothing is known about either. However, attempts have been made to identify the chemical nature of the 5-HT receptors of smooth muscles, based on (1) the intracellular effects of 5-HT on metabolism; (2) the nature of 5-HT binding in storage sites; and (3) the effect of chemical alterations of the cell membrane on the activity of 5-HT.

Intracellular effects of 5-HT

The relaxation by 5-HT of the catch tension in the contractor muscle of *Mytilus* has already been mentioned together with the evidence that this is an intracellular action. The gills of *Mytilus* contain 5-HT in considerable concentrations (Aiello, 1960). The gills possess cilia the movement of which is strongly stimulated by 5-HT, which also increases their anaerobic glycolysis and respiration (Moore, Milton & Gosselin, 1961). The movement of the cilia is also accelerated by acetylcholine which does not, however, stimulate metabolism (Bülbring, Burn & Shelley, 1953). 5-hydroxytryptophan acts like 5-HT but more slowly, presumably only after decarboxylation. These observations suggest that, in the gills, 5-HT acts intracellularly on metabolism. Effects of 5-HT on metabolism presumably imply that the amine is free inside the cell, at least temporarily. In human blood-platelets, the flux of potassium across the cell membrane is accelerated by 5-HT only while it is actively taken up into the intracellular storage organelles; once bound there, the effect disappears (Born, 1967). 5-HT free inside cells is presumably liable to inactivation by monoamine oxidase as has been shown for platelets (Bartholini, Pletscher & Bruderer, 1964) guinea-pig taenia coli (Born, 1962) and rat fundus strip (Handschumacher & Vane, 1967). The enzyme is absent from *Mytilus* gills (Blaschko & Hope, 1957; Blaschko & Milton, 1960) in which 5-HT appears to act intracellularly. The sites, presumably enzymic, with which 5-HT reacts intracellularly are still unknown so that the observations throw no light on specific chemical interactions.

This comment applies also to the observation (Mansour, 1959) that in the liver fluke *Fasciola hepatica* the formation of cyclic adenosine-3',5'-phosphate (cyclic AMP) from ATP is induced by 5-HT; cyclic AMP causes activation of phosphorylase which is the rate-limiting enzyme in glycolysis. In mammalian smooth muscle an increased formation of cyclic AMP is brought about not by

5-HT but by adrenaline. There is no activation of phosphorylase but the ATP plus creatine phosphate content of the muscle is increased and is believed to be used at the membrane, possibly as substrate for cyclic AMP synthesis (Bueding, Bülbring, Gercken, Hawkins & Kuriyama, 1967). Although the manifestation is metabolic, the sequence of biochemical events is initiated by an action of the amines on the cell membrane (Sutherland & Rall, 1960). The chemical nature of the site of this action is unknown.

Association of 5-HT with ATP in storage sites

Following the discovery that the catecholamines in the adrenal medulla are concentrated in intracellular organelles in association with adenosine triphosphate (ATP) (Falck, Hillarp & Högberg, 1956; Blaschko, Born, D'Iorio & Eade, 1956), it was surmised that the 5-HT present in blood platelets might also be associated with ATP (Born, 1956a). Considerable indirect evidence for this (Born, Ingram & Stacey, 1958; Born, Hornykiewicz & Stafford, 1958; Baker, Blaschko & Born, 1959) culminated in the isolation from platelets of a special type of organelle in which both substances are present together in extraordinarily high concentrations (Tranzer, Da Prada & Pletscher, 1966).

The site to which 5-HT is bound in the storage organelles may be called a *storage receptor*. Apart from the stoichiometric involvement of ATP, the chemical constitution of the storage receptor is still unknown, mainly because the isolation of the storage organelles from platelets has turned out to be more difficult than the isolation of other cytoplasmic organelles including the storage granules for adrenaline in the adrenal medulla. When human platelets are saturated with 5-HT the molar ratio 5-HT/ATP is about 2·8 or not much less than the stoichiometric ratio to be expected for ionic binding between the two types of molecule. Granules can be isolated from dog intestine which contain high concentrations of both 5-HT and ATP in a molar ratio of 2·6 (Prusoff, 1960) which is very similar to that found in the platelets. Simple electrostatic forces cannot account for the specificity for 5-HT in its association with ATP which is presumably determined by the un-ionized parts of the molecules, perhaps via a metal ion as has been suggested for the association of adrenaline with ATP (Belleau, 1960); or the ATP may be part of a larger complex the conformation of which favours the specific binding of 5-HT.

An attempt has been made to throw light on this problem by determining whether 5-HT forms a complex with ATP and how specific and strong such a complex might be. In aqueous solution, ATP forms a complex with 5-HT but not with tryptamine nor with 5-hydroxyindole, indicating that both the 5-hydroxyl group and the ethylamine side chain are involved in complex formation (Roberts, 1967). 5-methoxytryptamine, α-methyl-5-methoxytryptamine, α-ethyl-5-hydroxytryptamine, and N'N'-dimethyl-5-hydroxytryptamine all form complexes with ATP; but 6-methoxytryptamine does not. 5-HT complexes also with ADP but not with adenosine monophosphate (AMP); it seems that both the purine ring and at least two phosphate groups of ATP take part in the complex.

The complex between 5-HT and ATP is very weak: the stability constant of a one to one 5-HT:ATP complex in the absence of other ions was calculated to be 59 l/mole and in the presence of 0.1 M Na^+ only 26.9. This low value for the stability constant suggests either that the complex is, by itself, of little biological significance or that its demonstration in an aqueous medium is irrelevant to biological situations, i.e. to the protein- and lipid-rich interior of a storage organelle. The existence of this interaction may contribute to the stability of a ternary complex with a metal (Colburn & Maas, 1965) or with a macro-molecule (Weiner & Jardetzky, 1964). The results of titration experiments (Roberts, 1967) are compatible with the existence of ternary complexes between ATP, 5-HT and divalent metal ions; a possible complex of this kind is shown

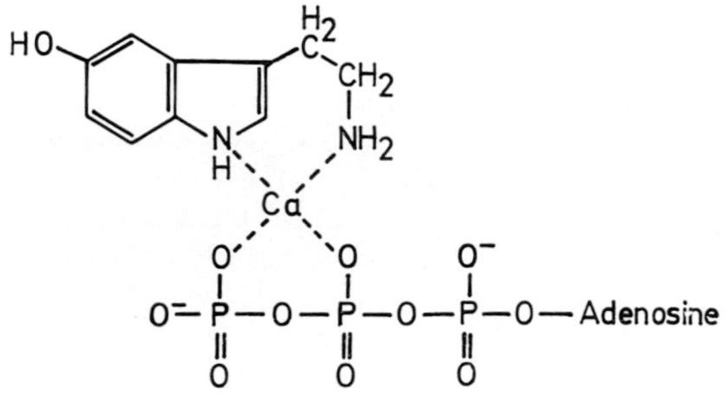

FIG. 4   Possible complex between 5-HT, calcium and ATP.

in Fig. 4. However, such ternary complexes would not explain the maximal molar ratio, 5-HT/ADP, in blood platelets. It is, of course, improbable that such simple complexes between adenine nucleotides, divalent cations and 5-HT may throw light on the interaction of 5-HT with other receptors, e.g. on those which mediate the activation of adenyl cyclase by 5-HT in lower animals. Whether the storage receptor and the smooth muscle receptor for 5-HT have any-thing in common remains to be seen.

Several other mammalian tissues including some endocrine glands (Falck & Owman, 1968) and the central nervous system (Fuxe, Hökfelt & Ungerstedt, 1968) contain 5-HT concentrated in intracellular granules or organelles. The mechanisms responsible for the localizations are entirely unknown. On the analogy with the inactivity of 5-HT stored in platelet organelles and with the inactivity of other agents such as adrenaline in other intracellular organelles, 5-HT localized in the granules of these different cells is presumably inactive until released from them; how the release is effected is also unknown.

Isolation of a ganglioside that binds 5-HT

This work began with the idea (Woolley, 1958) that 5-HT causes smooth muscles to contract by 'opening a special valve for the transport of calcium ions through the outer membrane of the muscle cells'. Like other hormones causing smooth muscle contractions, 5-HT was pictured as promoting this flow of calcium ions. Since the cell membrane, by virtue of its lipid constituents, is almost impermeable to calcium, it was suggested that 5-HT combines with a constituent which can then form a lipid-soluble complex with calcium. After diffusing through the membrane this complex would somehow be caused to disintegrate, releasing calcium ions and 5-HT. On the basis of these ideas, a substance was extracted from the tissues of several species including ape, rat, lamb and mussel, which had the property of making 5-HT soluble in a lipid solvent (a mixture of benzene and butanol), in the presence of calcium ions (Woolley, 1958). This substance occurred in tissues which responded to 5-HT and not in those which did not.

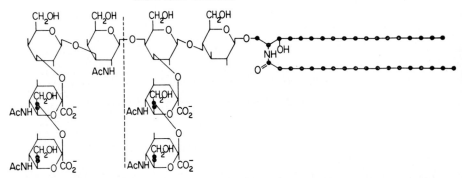

FIG. 5 Representative structure for the class of compounds called gangliosides. The 5-HT 'receptor' is shown to the right of the broken line: Sialyl $(2 \to 8)$ sialyl $(2 \to 3)$ galactosyl $(1 \to 4)$ glucosyl $(1 \to 1')$ ceramide. (From Rapport, 1968)

Thereupon, various purified complex lipids were tested for their ability to bind 5-HT (Woolley & Gommi, 1966). This led to the discovery that 5-HT was bound with remarkable specificity by a purified ganglioside (Gielen, 1966). Gangliosides are a family of lipid substances present mainly in cell membranes, particularly of nerve cells, which contain two long-chain fatty acids attached to an oligosaccharide moiety containing glucose, galactose, N-acetylglucosamine and N-acetylgalactosamine. The characteristic structural component of ganglio-sides is N-acetyl neuraminic acid, up to four of which may be attached to sugar residues. The particular ganglioside which binds 5-HT is shown in Fig. 5 to the right of the broken line: it is a di-sialobio-ganglioside (N-acylsphingosinyl-N-acetyl-sialyl-$(2 \to 8)$-N-acetyl-sialyl-$(2 \to 3$ Gal)-lactoside (Kuhn & Wiegandt, 1964); two neuraminic acid residues are attached to a ceramide lactoside. Dialysis experiments have shown that the complex between this ganglioside and 5-HT dissociated very slowly at 37°C (40 to 50% in 24 hr) and hardly at all below 4°C. When the ganglioside is exposed to neuraminidase one of the neuraminic acid residues is split off and simultaneously the high capacity for binding 5-HT is lost (Gielen, 1966). The binding of 5-HT to the ganglioside is prevented by both LSD and reserpine.

Although many vertebrate tissues that contain or react to 5-HT contain this ganglioside it does not meet the other requirements of Woolley's original hypothesis (Rapport, 1968). Thus, calcium displaces 5-HT from the ganglioside rather than forming a ternary complex with both. Furthermore, neuraminic acid has not been found in bivalve molluscs so that the ganglioside cannot be involved in any physiological function of 5-HT in *Mytilus*; these animals may have a different receptor substance for 5-HT.

Effect of neuraminidase on the action of 5-HT

The enzyme neuraminidase (N-acetyl-neuraminate glycohydrolase 3.2.1.18 from *Clostridium perfringens* or *Vibrio cholerae*) catalyses the hydrolytic removal of neuraminic acid from gangliosides and other molecules. When the oestrus uterus or the fundus strip of the rat is incubated with low concentrations of this enzyme in the presence of EDTA the 5-HT contractions are abolished while those due to acetylcholine or bradykinin are unaffected.

The abolition of the 5-HT response by neuraminidase does not occur in the absence of EDTA which suggests that a divalent metal, presumably calcium, is bound to the receptor in such a way as to protect it against hydrolytic attack by the enzyme. The inactivating effect of neuraminidase has been taken to indicate that neuraminic acid is part of an essential constituent of the binding site for 5-HT on the muscle cells, i.e. the D receptor (Woolley & Gommi, 1964). The muscles can be resensitized to 5-HT by the addition of gangliosides so that the essential constituent could be a ganglioside.

Other fundus strips have been incubated with neuraminidase together with N-acetyl neuraminic acid lyase which splits sialic acids into N-acetyl-D-mannosamine and pyruvate (Wesemann & Zilliken, 1968). When the enzymes are washed out, the response to 5-HT is first increased and then decreased, and almost abolished after 2 hr (Fig. 6). When this unresponsive preparation is incubated with N-acetyl neuraminic acid (10^{-5} M) for 15 min, the responsiveness increases again within 1 hr to the original value.

Dependence of the action of 5-HT on sialic acid metabolism

Further progress has been made by exposing isolated smooth muscles to neuraminidase together with either precursors or inhibitors of the biosynthesis of the N-substituted neuraminic acid derivatives known as sialic acids (Wesemann & Zilliken, 1967; 1968). Rat fundus strips from which the mucosa had been removed as far as possible were incubated in oxygenated Tyrode solution at 37°C and their contractions recorded isotonically. The sensitivity to 5-HT was measured by the cumulative dose-response curve, and the maximal contraction

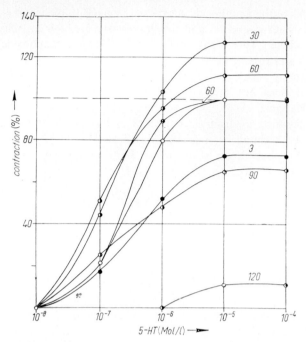

F<small>IG</small>. 6   Cumulative dose-response curves for 5-HT on rat fundus strips after simultaneous incubation with neuraminidase and N-acetyl neuraminic acid lysase. Abscissa: log 5-HT concentration (M); ordinate: isotonic contraction (per cent of maximal contraction of control). Control contraction before incubation ○—○. 30 60, 90 and 120 min after 15 min incubation with 0·03 units neuraminidase and 0·12 units lysase ◑—◑. 3 and 60 min after 15 min incubation with 10^{-5} M N-acetyl neuraminic acid 120 min after the incubation denoted by ●—● (compare the lowest curve). Curves marked ◑—◑ indicate time after incubation with the enzymes, and curves marked ●—● indicate time after incubation with N-acetyl neuraminic acid. (From Wesemann & Zilliken, 1968)

was also determined. When synthetic sialic acids or purified gangliosides (10^{-5} M) were added to the solution for 15 min and then washed out, the response to 5-HT was increased; after passing through a maximum in 60 min or so the response returned to that of the controls. When, on the other hand, strips were preincubated with inhibitors of the biosynthesis of N-acetyl-neuraminic acid, the 5-HT dose-response curves were shifted to the right and the maximal contractions were greatly diminished; these effects also were reversible (Fig. 7).

Recently, the responsiveness of rat fundus strips to 5-HT has been measured in relation to the rate of synthesis and degradation of sialic acids (Wesemann & Zilliken, 1968). The strips were preincubated for 15 min with different concentrations of precursors or inhibitors of the biosynthesis of sialic acid which were washed out before the effectiveness of 5-HT was determined. Both the sensitivity and the maximal contraction to 5-HT were considerably increased by N-acetyl

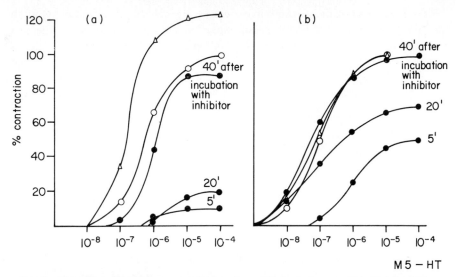

FIG. 7 The effect of 0·03 i.u. neuraminidase from *Vibrio cholerae* and bis-[2,4,5-trichlor-phenoxy]-acetic acid on the 5-HT log dose-response curve of the rat stomach fundus. (a) 5-HT log dose-response curve before ○—○ and after △—△ an incubation period of 15 min with 0·03 i.u. neuraminidase. ●—● 5-HT curves obtained 5, 20 and 40 min after incubation with 0·03 i.u. neuraminidase for 15 min followed by treatment with 1·5 × 10⁻⁵ M bis-[2,4,5-trichlor-phenoxy]-acetic acid for another 15 min (b) 5-HT log dose-response curve before ○—○ and after △—△ an incubation period of 15 min with 0·005% CaCl₂. ●—● 5-HT curves obtained 5, 20 and 40 min after incubation with 0·005% CaCl₂ for 15 min followed by treatment with a 1·5 × 10⁻⁵ M solution of the inhibitor bis-[2,4,5-trichlor-phenoxy]-acetic acid for another 15 min. (From Wese-mann & Zilliken, 1967)

D-mannosamine at 10^{-8} M (Fig. 8); N-acetyl glucosamine produced similar increases only at concentrations a thousand times higher. Substrates of energy metabolism including glucose and fructose had little or no effect. Certain nucleotides, i.e. cytosine triphosphate, adenosine 3'5'-monophosphate (cyclic AMP) and uridine diphosphate-N-acetyl-D-glucosamine increased the 5-HT contraction in concentrations as low as 10^{-16} M. In concentrations of 10^{-10} M or higher some other nucleotides viz. ATP, ADP and UDP were also effective but only some considerable time (about 60 min) after they were washed out.

Of six D-glucosamine derivatives which inhibit the biosynthesis of N-acetyl neuraminic acid (Boschman, 1968) only N-(phenylacetyl)-D-glucosamine, N-γ-(Z-glutaminyl-α-benzylester)-D-glucosamine, and N-α-(Z-glutaminyl-γ-benzylester)-D-glucosamine caused significant inhibition of the 5-HT contraction at 10^{-3} M; lower concentrations were ineffective. These results are more hazardous to interpret and by no means support the proposition unequivocally that the responsiveness of smooth muscle to 5-HT depends on the unhindered biosynthesis of sialic acid.

The observations have been interpreted as follows (Wesemann & Zilliken,

1968). Treatment with neuraminidase causes the release of components of the ganglioside receptor for 5-HT, presumably from the muscle surface. The early increase in the muscle's responsiveness to 5-HT is due to an increase in the rate of resynthesis or turnover of the 5-HT receptor. The subsequent loss of responsiveness is brought about by the depletion of precursor substances for receptor resynthesis, leading to the progressive diminution in the number of intact receptor

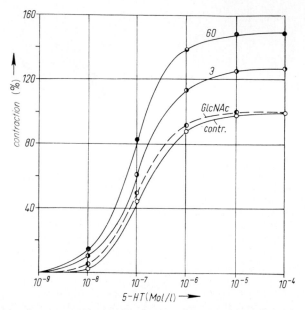

FIG. 8   Effect of preincubating rat fundus strips for 15 min with 10^{-8} M N-acetyl-D-glucosamine and N-acetyl-D-mannosamine on the cumulative dose-response curve to 5-HT. Abscissa: log 5-HT concentration (M); ordinate: isotonic contraction (% of maximal contraction of control). Control ○—○; 3 min ◑—◑ and 60 min ●—● after washing out the N-acetyl-D-mannosamine; 60 min ◑—◑ after washing out of the N-acetyl-D-glucosamine. (From Wesemann & Zilliken, 1968)

molecules. The restoration of the response to 5-HT by N-acetyl neuraminic acid indicates that this substance promotes the resynthesis and is an essential component of the 5-HT receptor.

This interpretation was supported by experiments in which the uptake of N-acetyl neuraminic acid labelled with ^{14}C was measured in fundus strips under conditions similar to those used for the determination of the dose-response curves to 5-HT. The radioactivity of the muscles increased for 15 to 20 min and then remained on a plateau. Pretreatment of the muscles with neuraminidase increased the uptake of radioactivity by about 60% within the first 15 min; thereafter the radioactivity diminished until, after about 60 min, it was similar to that of the untreated muscles. Other strips were incubated with radioactive N-acetyl neuraminic acid until their radioactivity was constant. After washing, the rate at which these preparations lost about 60% of their radioactivity into the medium was greatly increased in the presence of neuraminidase, although the enzyme did not increase the total loss. From this it was concluded that rather over half of the radioactive sialic acid had been incorporated into the 5-HT receptor on the cell surfaces from which it was slowly removed by a normal turnover and much more rapidly by the neuraminidase.

Preincubation with neuraminidase followed by incubation with radioactive N-acetyl neuraminic acid caused increased uptake of radioactivity which fell subsequently to that of muscles not exposed to the enzyme. This again could mean that the neuraminidase splits N-acetyl neuraminic acid from the 5-HT receptor where it is rapidly replaced by the radioactive molecules in the incubation medium. This incorporated radioactive N-acetyl neuraminic acid then exchanges more slowly with unlabelled N-acetyl neuraminic acid synthesized endogenously by the muscle itself.

The postulated connection between sialic acid metabolism and the 5-HT receptor of smooth muscle is summarized in the following reaction sequence (Wesemann & Zilliken, 1968) in which X represents the receptor precursor of the muscle cell membrane and the other reactants are abbreviated thus: ManNAc = N-acetyl-D-mannosamine; NeuNAc = N-acetyl-neuraminic acid; CMP = cytidine monophosphate; CTP = cytidine triphosphate, and Pi = inorganic phosphate:

1   ManNAc + ATP → ManNAc-6-P + ADP

2   ManNAc-6-P + P-Enolpyruvate → NeuNAc-9-P + Pi

3   NeuNAc-9-P → NeuNAc + Pi

4   NeuNAc + CTP → CMP-Neu-NAc + PPi

5   X + CMP-NeuNAc → Pi + $\boxed{\text{X-NeuNAc}}$

                          (= Receptor)

6    $\boxed{\text{X-NeuNAc}}$ + 5-HT → $\boxed{\begin{array}{c}\text{X-NeuNAc} \\ \hline \text{5-HT}\end{array}}$

                 (= 5-HT-Receptor-Complex)

7    $\boxed{\begin{array}{c}\text{X-NeuNAc} \\ \hline \text{5-HT}\end{array}}$ → X + NeuNAc + 5-HT

8    $\boxed{\text{X-NeuNAc}}$ → X + NeuNAc

9   NeuNAc → Pyruvate + ManNAc

The crucial reactions that mediate the effect of 5-HT are reactions 6 and 7. 5-HT-receptor-complex denotes an association between the receptor and 5-HT which is reversible and shows the specificities already described; nothing further is known about it yet. The initial increase in 5-HT contractions after simultaneous incubation with neuraminidase and N-acetyl neuraminic acid lyase suggests that the contraction-producing reaction is not the formation of the complex but its dissociation into receptor precursor X, N-acetyl neuraminic acid, and 5-HT (reaction 7). This recalls the evidence indicating that the conformational change in the membrane of the platelet which precedes their aggregation requires not only the association of 5-HT with its specific carrier in the membrane but also their subsequent dissociation (Baumgartner & Born, 1967; 1969).

Distribution of sialic acids in the smooth muscle membrane

The evidence that sialic acid is an essential part of the 5-HT receptor raises the question to what sort of macromolecule it is attached in the cell membrane. In an attempt to answer this, a method has been developed for isolating the membranes of the smooth muscle cells from rat uteri (Carroll & Sereda, 1968a). The membrane fraction, which was free of mitochondria, resembled a network of rings as would be expected for membrane sections which are held together by connective tissue fibres. This myometrial membrane preparation had a sialic acid content of 0·5% by dry weight of which only about one-seventh was lipid-soluble. This indicates that sialic acid of gangliosides contributes only a small proportion of the total sialic acid in the membrane. Indeed, there is only very little ganglioside in uterus altogether. Neuraminidase acts on the non-lipid sialic acids of the membrane fraction of rat uterus but not on the lipid-soluble sialic acid. These results suggest that the sialic acid of the 5-HT receptor is a constituent not of a lipid-soluble ganglioside but of a water-soluble glycoprotein (Carroll & Sereda, 1968b).

CONCLUSIONS AND COMMENTS

From the various observations that have been described, it is reasonable to draw the following tentative conclusions. Some, but by no means all, mammalian smooth muscles possess cell membrane structures capable of reacting specifically with 5-HT. These are the 5-HT receptors of the D type. The receptors of different muscles are chemically identical or at least very similar, but they differ from other sites with which 5-HT interacts, i.e. the M receptors for 5-HT on nerve cells; the storage organelles in various cells, particularly the platelets; and the catalytic site of the monoamine-oxidase molecule.

The 5-HT receptor on the muscle cell membrane is situated in a structural or functional trough which limits the concentration of lipid-soluble analogues of 5-HT capable of activating the receptor. The trough has the effect of a selective diffusion barrier. The receptor straddles or forms part of a pore through which 5-HT as well as its oxidation product 5-hydroxyindoleacetic acid can diffuse slowly through the membrane.

The receptor binds the 5-HT molecule with considerable specificity, and all its analogues have less affinity than 5-HT itself. When the indole ring is substituted other than in the 3 and 5 positions it fits the receptor less well, although indene derivatives are bound as strongly as the corresponding indoles. Binding is effected mainly through the 5-hydroxy and primary amino groups which must be at the maximum possible distance from each other. In this, 5-HT resembles other pharmacologically active molecules of similar size, such as acetylcholine and the catecholamines, which become attached to their respective receptors through two different groups, one of them polar, separated from each other by the greatest distance permitted by their structures.

The receptor groups involved in binding 5-HT are unknown. The evidence indicating that the sensitivity of smooth muscle cells to 5-HT depends, inter alia, on a sialic acid, leaves open several possibilities as to its site of operation. It is a big step to the proposition that sialic acid is an essential constituent of the specific membrane-receptor which mediates the pharmacological action of 5-HT. Apart from the criticism (Rapport, 1968) of the original ganglioside hypothesis (see p. 441) the more recent experimental evidence (Wesemann & Zilliken, 1967; 1968) also requires critical assessment. First, no substance other than 5-HT itself appears to have been tested so that the specificity for 5-HT of the changes in sensitivity has not been established. To conclude that a specific 5-HT receptor is affected it will be necessary to show that the different treatments do not alter the muscle's responses to, say, acetylcholine or prostaglandins, but that they do alter those to 5-HT analogues which, on the basis of pharmacological evidence, can be assumed to act on the 5-HT receptor. Secondly, explanations have still to be given for the increase in responsiveness to 5-HT of the smooth muscle immediately after its exposure to neuraminidase and for the fact that some substances known to inhibit the biosynthesis of sialic acids do not diminish the responsiveness whereas others do. Until these questions have been answered the observations are no more than consistent with the working hypothesis that the sialic acid is a component of the 5-HT receptor. Even if it is, the question remains whether the acid is part of the specific attachment site for 5-HT or whether it is required to maintain the rest of the receptor in a specific conformation required for binding the amine.

Many other interesting problems await solution. When means have been found for estimating the number of active 5-HT receptors on a smooth muscle cell, it will be possible to decide whether this number is related to the sensitivity to 5-HT of different smooth muscles and whether muscles that are wholly insensitive to 5-HT have no receptors at all. It will also be necessary to establish how the different 5-HT receptors that have been demonstrated pharmacologically differ in chemical constitution and whether the reactions of 5-HT with other cell types, such as platelets, are mediated by one of these receptors or by yet another kind.

Finally, there is the question as to the biological purposes of the sensitivity of smooth muscles and of other tissues to 5-HT. Evidence available now suggests that 5-HT has physiological functions in the intestinal peristaltic reflex and in central nervous systems. But its presence in and actions on many tissues remain mysterious. Nothing could illustrate this better than the high sensitivity of the chick amniotic membrane not only to 5-HT but to numerous other, very different substances as well (Evans & Schild, 1957): does this sensitivity have a purpose, or is it perhaps an accident of biochemical development?

15

AUTONOMIC NERVOUS SUPPLY TO EFFECTOR TISSUES

G. CAMPBELL

INTRODUCTION

Many of the early attempts to study the physiology of mammalian smooth muscle were made difficult by the presence of autonomic nerves within the muscular tissue. The nerves often could not be eliminated, as they had been in studies of skeletal muscle, by techniques of surgical denervation, because in many tissues the autonomic neurones in their entirety were embedded inaccessibly within the muscle bundles. For this reason, it has proved impossible to consider the physiology of smooth muscle without also considering the composition and distribution of the autonomic nervous system (ANS).

The accounts given of the autonomic nervous system in textbooks of general physiology or pharmacology are mostly brief and are typically based on the generalizations made by Langley in 1921. While Langley's rules are still to some extent convenient, there has in the past been a tendency to regard the rules as absolute and invariable. As a result the generalizations have attained an almost unassailable classical status and in far too many instances there have been attempts to fit what are truly deviant phenomena into the classical framework. Worse than this, some unproven and even disproven theories have at times achieved a classical status. For instance, as late as the 1930s, some authors still held that adrenaline acted by stimulating sympathetic nerve terminals as proposed by Brodie & Dixon (1904), even though at that time adrenaline was known to act on denervated muscles (Lewandowsky, 1899; Langley, 1901). Similarly, in quite recent papers, the view has been expressed that when a transmitter substance causes a tissue to respond, the tissue must be innervated in normal circumstances by nerves releasing that substance. This theory was originally put forward in 1905 by Elliott, but it was later found (Baur, 1928) that the chick amnion is responsive to acetylcholine although it is never at any time innervated.

In retrospect, the development of ideas about the autonomic nervous system is seen to have been retarded by an over-rigid adherence to theories formulated at the beginning of the century. To avoid perpetuating the tradition, this chapter will concentrate only on autonomic nervous pathways which do not fit the classical descriptions. Since we are mainly concerned with smooth muscle, only one type of deviation will be considered, namely situations in which the post-ganglionic neurones innervating an effector tissue do not appear to release the appropriate transmitter substance, as judged by the old rule that parasympathetic post-ganglionic autonomic neurones release acetylcholine whereas their counterparts in the sympathetic system mostly release an adrenaline-like substance. For immediate comparison, a brief account of the classical arrangement and composition of the autonomic nervous system is given; detailed descriptions are available in a number of textbooks (e.g. Kuntz, 1953; Mitchell, 1953, 1956).

CLASSICAL PICTURE OF THE AUTONOMIC NERVOUS SYSTEM

Definition and description of the mammalian ANS

The ANS may be defined as comprising all those efferent nervous pathways which have a ganglionic synapse outside the confines of the central nervous system. As such, the system includes the innervation of visceral, vascular and all other smooth muscle, of a number of secretory cell types and of cardiac striated muscle. In mammals, as described by Langley (1921), nervous pathways fitting this description leave the neuraxis in cranial nerves III, VII, IX, X and XI and in spinal nerves from upper thoracic to lower sacral levels.

The preganglionic autonomic nerve fibres which leave the thoracic and lumbar regions of the spinal cord, pass through white (i.e. myelin-rich) rami communicantes to the paravertebral chains, a bilateral system of ganglia joined together by longitudinal connectives. The ganglionated chains extend rostrally to the upper cervical region and caudally to the lower sacral level, but receive no efferent contribution from the spinal nerves at these extremes. In the chains, the preganglionic autonomic fibres make synaptic connections with post-ganglionic neurones situated within the paravertebral ganglia. The post-ganglionic axons then pass out of the chains in one of three ways. They may pass, via grey (i.e. myelin-poor) rami communicantes, to the spinal nerves at all levels and run with them to superficial structures such as pilomotor muscles and sweat glands and to peripheral vascular beds. They may pass via connectives to the cranial nerves, which they follow to their respective fields of innervation, e.g. heart, iris, salivary glands. Finally they may pass into nerves which leave the paravertebral chains directly and run to visceral organs. These nerves include the splanchnic nerves, supplying the digestive tract and associated glands, the cardiac nerves, supplying the heart, and the hypogastric nerves, supplying the internal genitalia and urinary system. However, many preganglionic autonomic nerves in the thoracic and lumbar outflows do not make their synaptic connections in the paravertebral chains. Instead, the fibres pass straight through the chains into the various visceral nerves, where they make synaptic connections with post-ganglionic neurones lying in the prevertebral ganglia, free-lying ganglionic masses such as the coeliacomesenteric and hypogastric ganglia. The prevertebral ganglia, like the ganglia in the paravertebral chains, are generally remote from the organs to be innervated, but, for instance, the hypogastric ganglia lie virtually on the surface of the innervated organs.

The adrenal medullary chromaffin cells are innervated by fibres which pass through the paravertebral chains to the adrenal gland without intervening synapses. If one regards the chromaffin cells as the target organs of the innervation, these fibres do not fit the definition of autonomic fibres as given above since the pathway lacks a peripheral ganglionic synapse. But there is a good argument for equating the chromaffin cells with post-ganglionic neurones, so that the innervation of the cells can be regarded as ganglionic rather than terminal, and the target organs now become those organs acted on by circulating adrenal medullary hormones.

The autonomic preganglionic fibres leaving the neuraxis in the cranial and sacral nerves do not pass through the paravertebral chains. The fibres in the cranial nerves run directly towards the organs to be innervated, making synaptic connections with post-ganglionic neurones either on the way, as in the ciliary ganglion on the oculomotor nerve, or within the substance of the innervated tissue, as in the vagal pathways to the heart, lungs and gastro-intestinal tract. The sacral preganglionic fibres run in the sacral spinal nerves for some way before leaving them to supply visceral and vascular structures in the pelvic region, where they synapse with post-ganglionic cells lying in the innervated organs.

The outlines of the arrangement just described are illustrated in Fig. 1.

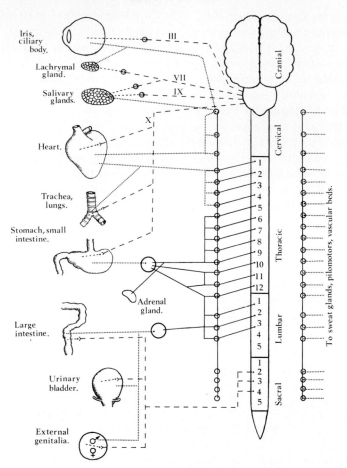

FIG. 1   Diagram of the distribution of autonomic nerves. Sympathetic preganglionic neurones (———) leave the neuraxis in thoracic and lumbar spinal nerves. They enter the paravertebral ganglionated chains (o—o) or the prevertebral ganglia (○) and make synaptic connections with post-ganglionic neurones (·····). The adrenal gland is provided with a preganglionic sympathetic innervation. Parasympathetic preganglionic neurones (— — — —) leave the neuraxis in sacral spinal nerves and in cranial nerves III, VII, IX and X. They make synaptic connections with post-ganglionic neurones (----) either in the innervated tissue (≺·) or in remote ganglia (o). (Simplified from Goodman, L. S. & Gilman, A., 1965, *The Pharmacological Basis of Therapeutics*, 3rd. Edn., Fig. 21-1)

Subdivision of the ANS: sympathetic and parasympathetic

Gaskell (1886) showed that there were gaps in the autonomic outflow from the neuraxis at the levels of the anterior and posterior limb outflows, dividing the autonomic outflow into cranial, thoracico-lumbar and sacral regions. Langley (1900), having noted that only the thoracico-lumbar outflow passed through the paravertebral chains and that the effects of this outflow were generally antagonistic to the effects of the cranial or the sacral autonomic nerves, suggested that the ANS could be divided into three functional parts, cranial, sympathetic (= thoracico-lumbar) and sacral. The term sympathetic was the old term for all visceral nerves. By 1905, Langley had produced pharmacological evidence which indicated that the cranial and the sacral outflows were similar, and that both were different from the sympathetic system. He therefore grouped the cranial and sacral outflows together as the parasympathetic system, the name being chosen to indicate the antagonism between these nerves and the sympathetic outflow. Langley (1900) had previously pointed out that there was no evidence to indicate whether all or only some of the neurones in the gastro-intestinal plexuses of Auerbach and Meissner were post-ganglionic neurones in cranial or sacral pathways, a comment which is still true today. He therefore proposed a third category of autonomic nerves, the enteric system, although it is now more usual to call these neurones parasympathetic. Langley's (1921) final classification appears as follows:

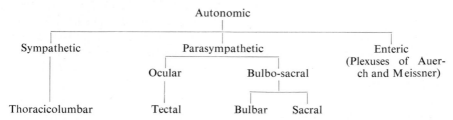

The features of the subsystems of the ANS may be summarized as follows. The parasympathetic system consists of autonomic pathways in the cranial and the sacral spinal nerves, running to the innervated organs without passing through the paravertebral chains and having ganglionic synapses in or near to innervated organs. The organs innervated by this subsystem include the intrinsic eye muscles, the salivary and lacrimal glands, the gastro-intestinal tract and the abdominal digestive glands, the pulmonary airways, the heart, the urinary system and a number of restricted vascular systems, e.g. in the lungs, the salivary glands and the external genitalia. The sympathetic system consists of a thoracicolumbar autonomic outflow, passing through the paravertebral chains and generally having ganglionic synapses rather remote from the organs innervated. The sympathetic innervates most of the organs already listed as receiving parasympathetic innervation, but in addition supplies the internal genitalia and a number of widespread systems which tend to respond to sympathetic stimuli in a rather massive way, such as systemic vascular beds and sweat glands and pilo-erector muscles in the dermis. Because of the widespread nature of many sympathetic nervous responses, often associated with the simultaneous effects of catecholamines released from the adrenal medulla, the sympathetic system has generally been regarded as mediating blanket-like reactions, especially of the vasculature. In contrast, the parasympathetic has been considered to give more precise control of limited regions of the organism. It should be noted that when both the sympathetic and the parasympathetic supply an organ, fibres from the two subsystems certainly interweave in the terminal plexus and they may become mingled in a nerve trunk some distance away from the organ.

Chemical transmission: adrenergic and cholinergic nerves

The phenomenon of chemical transmission of nervous activity across synaptic discontinuities is now so well established that no description of the process is needed here. Langley (1901) first pointed out that 'the effects produced by supra-renal extract are almost all such as are produced by stimulation of some one or other sympathetic nerve' and that 'the effect of supra-renal extract in no case corresponds to that which is produced by stimulation in normal conditions of a cranial autonomic or a sacral autonomic nerve', conclusions later confirmed for purified adrenaline (Elliott, 1905). With the discovery (Langley, 1905) that a number of drugs caused responses which usually mimicked the actions of parasympathetic, but not sympathetic, nerves, it became obvious that there were two pharmacologically different types of autonomic nerve fibre. Later, Langley (1921) was to term these two types adrenophil and cholinophil. In 1933, when the details of chemical transmission were clearer, Dale substituted the terms adrenergic, for nerves which released adrenaline or, as it later turned out, noradrenaline, and cholinergic, for nerves which released acetylcholine. Both Langley and Dale pointed out that this was a physiological classification which should, if necessary, cut across the anatomical categories of sympathetic and parasympathetic, although they both thought that the sympathetic post-ganglionic fibres were mainly adrenergic (adrenophil) and the parasympathetic mainly cholinergic (cholinophil). But, when Dale wrote, he was able to say that a few exceptional instances of cholinergic sympathetic post-ganglionic fibres had been discovered. It is the question of deviations from the general rule that parasympathetic post-ganglionic neurones are cholinergic, and sympathetic neurones adrenergic, that will be discussed in the remainder of the chapter. In addition to these considerations regarding the post-ganglionic fibres, Dale (1933) suggested that probably all preganglionic fibres, whether sympathetic or para-sympathetic, were cholinergic.

Summary

The classical picture of the ANS is that it consists of two subsystems. The sympathetic subsystem consists of a thoracico-lumbar outflow of pregangli-onic cholinergic neurones, making synaptic connections with post-ganglionic-neurones in the paravertebral or prevertebral ganglia. The post-ganglionic sympathetic neurones are mainly adrenergic but some are cholinergic. The para-sympathetic subsystem consists of a cranial and a sacral outflow of preganglionic neurones making synaptic connections with post-ganglionic neurones lying in or near the innervated organs. The post-ganglionic parasympathetic neurones are mainly, if not exclusively, cholinergic. When both subsystems innervate the same organ the two systems usually cause responses in opposing directions. Some of the actions of the two subsystems are indicated in Table 1.

TABLE 1. 'Classical' effects of sympathetic and parasympathetic nerves

	Sympathetic	Parasympathetic
Heart	+	− *
Limb vascular beds	+ *	○
Spleen	+	○
Gastro-intestinal muscle	− *	+ *
Internal genitalia	+ / −	○
Urinary bladder detrusor muscle	−	+
Constrictor pupillae	○	+
Dilator pupillae	+	○
Pilo-erector muscles	+	○
Sweat glands (some species)	+	○
Salivary glands	+ (some)	+

+, excitation; −, inhibition; ○, absence of innervation; *, deviations are discussed in detail below.

DEVIATIONS FROM THE CLASSICAL PICTURE

Non-adrenergic sympathetic post-ganglionic neurones

When Dale (1933) introduced the terms adrenergic and cholinergic, he was able to cite only three examples of effects apparently attributable to cholinergic post-ganglionic fibres in the sympathetic. These effects were the initiation of sweat secretion in certain species (Langley, 1922a; Dale & Feldberg, 1934), the dilatation of certain blood vessels and a paradoxical contraction of a number of denervated striated muscles, presumably caused by diffusion of acetylcholine from sympathetic vasodilator nerves (von Euler & Gaddum, 1931). In addition to these cholinergic effects and to a number of others discovered since then, a number of recent investigations have indicated that the sympathetic also contains post-ganglionic fibres which are neither adrenergic nor cholinergic.

Cholinergic fibres

In the last ten or so years, a considerable number of organs showing apparently cholinergic responses to sympathetic post-ganglionic nerve stimulation have been found. For instance, Burn & Rand (1960) showed that the cat spleen, which normally responds to splenic nerve stimulation with a constriction, responds with a dilatation after the neural noradrenaline has been depleted by reserpine treatment; the dilatation is inhibited by atropine, a drug which antagonizes direct actions of acetylcholine on smooth muscle. It has further been shown that acetylcholine is released from the spleen following stimulation of the splenic nerve (Brandon & Rand, 1961; Leaders & Dayrit, 1965) and that the acetylcholine content of the organ diminishes in parallel with the noradrenaline content after sympathetic denervation (Brandon & Rand, 1961). Similar observations have been made on the arteries of the rabbit ear (Armin, Grant, Thompson & Tickner, 1953; Burn & Rand, 1960; Holton & Rand, 1962), the guinea-pig vas deferens (Burn, Dromey & Large, 1963; Burn & Weetman, 1963; Birmingham & Wilson, 1963; Della Bella, Benelli & Gandini, 1964) and a number of other organs (see Burn & Rand, 1965). Not all of the postulated cholinergic effects have been verified by detailed study. For instance, it has been proposed that part of the innervation of the cat nictitating membrane is cholinergic, since atropine or hyoscine treatment reduces responses to sympathetic nerve stimulation (Bacq & Frédéricq, 1934; Burn & Rand, 1960; Burn, Rand & Wien, 1963), but the doses of atropine or hyoscine needed also reduce direct responses to noradrenaline (Cervoni, West & Fink, 1956; Mirkin & Cervoni, 1962; Reas & Tsai, 1966). However, enough verified responses remain to indicate that acetylcholine is often released following sympathetic post-ganglionic nerve stimulation.

Until quite recently it was generally accepted that cholinergic responses to sympathetic nerve stimulation could be explained by the presence of actual cholinergic fibres mixed with the more usual adrenergic fibres, although the sympathetic innervation of the sweat glands, in cats at least, was entirely cholinergic. However, in 1959 Burn & Rand suggested that acetylcholine plays an essential role in the normal release of transmitter by adrenergic nerves, and that the cholinergic responses to adrenergic stimulation are caused by acetylcholine reaching the effector more or less by accident. The theory suggests (see Burn & Rand, 1965; Burn, 1967) that acetylcholine is released when an action potential invades the terminals of an adrenergic nerve. The released acetylcholine re-acts on the fibre to initiate a release to the exterior of the true transmitter substance, noradrenaline. To explain the effects described above, which are apparently caused by a direct action of released acetylcholine on the effector cells, it is supposed that part of the released acetylcholine diffuses over to the effector tissue. A modification of the theory was put forward recently by Burn (1968b), who refers to electronmicroscopic pictures by Tranzer & Thoenen (1967a) which 'show two terminations in the same Schwann sheath, one with agranular vesicles and one with dense-cored vesicles, side by side. If that is the general rule it may be that acetylcholine is coming out of one fibre to release noradrenaline from the other.' It is not intended to carry out a detailed discussion of the original theory here (see Ferry, 1966), but simply to consider those specifically cholinergic responses to stimulation of largely adrenergic nerve trunks. The responses could be caused by the release of acetylcholine from three sources; from truly adrenergic nerve fibres, as suggested by Burn & Rand (1959), from truly cholinergic sympathetic post-ganglionic nerve fibres, or from parasympathetic cholinergic nerve fibres mixed with the sympathetic nerve supply. The two latter alternatives will be considered.

Admixed parasympathetic cholinergic fibres. One of the most convenient preparations for the routine demonstration of adrenergic innervation is an isolated segment of mammalian intestine, innervated by perivascular nerves, from which contractions of the longitudinal muscle are recorded (Finkleman, 1930). The perivascular nerve supply to all regions of the gut in mammals is usually regarded as being sympathetic in origin and there is much evidence to show that the nerve fibres are predominantly adrenergic and inhibitory, reaching the perivascular plexus from the splanchnic nerves. However, there are now a number of reports of cholinergic responses to stimulation of the perivascular nerves in Finkleman preparations from all regions of the gut. The cholinergic

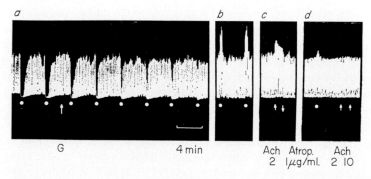

FIG. 2 Record of pendular activity. Finkleman preparation of rabbit ileum, to show cholinergic responses to perivascular nerve stimulation. Periarterial nerves stimulated at the white dots with 2 msec pulses for 14 sec every 4 min, in *a* at 50/sec, and in *b* and *d* at 10/sec. Guanethidine, 1 µg/ml at G, blocked the inhibitory response. In *b*, 120 min later, stimulation produced a motor response. In *c*, 2 µg acetylcholine (ACh) produced a motor response. Atropine 1 µg/ml in *d* blocked the motor responses to stimulation and to acetylcholine. (From Day & Rand, 1961. *Br. J. Pharmac. Chemother.*, **17**, 245–260)

excitatory responses may be seen in untreated preparations when the nerves are stimulated at low frequencies (less than 5 pulses/sec), as in preparations of the guinea-pig taenia coli (Ng, 1966). They may be seen after normal inhibitory responses have been abolished by treatment with adrenergic neurone-blocking drugs such as bretylium and guanethidine, especially if this treatment is combined with the application of an anticholinesterase drug such as physostigmine (Day & Rand, 1961; Bentley, 1962; Boyd, Gillespie & Mackenna, 1962; Akubue, 1966; Fig. 2). They may also be seen after treatment of the animal with reserpine, a drug which depletes the stores of noradrenaline in adrenergic nerve-terminals and therefore impairs adrenergic transmission (Gillespie & Mackenna, 1961; Bentley, 1962; Paton & Vane, 1963). The excitatory responses to nerve stimulation after reserpinization are greater than those seen after adrenergic neurone-blockade, but this difference has not yet been explained. All of the excitatory responses are inhibited by treatment with atropine or hyoscine and are clearly mediated by acetylcholine. It has also been shown that, after treatment with hyoscine, relaxations of Finkleman preparations caused by low frequencies of stimulation of the nerves are enhanced (Burn, Dromey *et al.*, 1963; Ng, 1966), suggesting that there is normally a masked antagonism of the inhibitory responses by simultaneously released acetylcholine.

The presence of cholinergic responses to stimulation of the perivascular innervation of Finkleman preparations has been taken to support the theory that there is a cholinergic link in the adrenergic transmission process, i.e. that the acetylcholine is being released from the same nerve fibres that are releasing noradrenaline (see Burn & Rand, 1965). However, it is also possible to suppose that part, if not all, of the excitatory response is caused by stimulation of parasympathetic cholinergic nerve fibres mixed in the perivascular nerve plexus. Firstly, many of the early experiments which demonstrated that the vagus nerves provide an excitatory innervation to the small intestine were actually carried out on segments of gut transected or ligated at both oral and anal ends (e.g. Bunch, 1899). In such preparations the vagal excitatory fibres cannot possibly run to the segment under study along the gut wall, and the only remaining pathway for the fibres is through the mesentery and along the blood vessels traversing the mesentery. The parasympathetic fibres must therefore be stimulated in a Finkleman preparation of the small intestine. Dissection reveals a branch of the vagus nerve running across to the region of the sympathetic ganglia of the coeliac plexus (see Mitchell, 1953), and it is probable that vagal efferent fibres are distributed to the perivascular plexuses by this route. Secondly, there is general agreement that there is a ganglionic synapse in the gut wall in the cholinergic pathways observed in Finkleman preparations, for the excitatory responses to perivascular nerve stimulation are inhibited by the application of drugs such as hexamethonium which prevent the nicotinic ganglionic actions of acetylcholine (Gillespie & Mackenna, 1961; Bentley, 1962; Paton & Vane, 1963; Akubue, 1966; Ng, 1966; cf. Day & Rand, 1961). The presence of ganglionic synapses in the gut wall would be expected if the pathways are parasympathetic and, classically, such synapses should be absent from sympathetic pathways, although this by no means proves that the fibres are not sympathetic. Finally, there is very clear evidence that the cholinergic excitatory fibres in the perivascular supply to the rabbit colon are provided directly from the parasympathetic pelvic nerves. Gillespie & Mackenna (1961) showed that the excitatory responses were absent in preparations taken from animals subjected to degenerative section of the pelvic nerves. They further showed that the excitatory responses declined after the direct supply of pelvic excitatory nerve fibres to the colon had been stimulated to a state of partial exhaustion, an observation which suggests that both the pelvic nerves proper and the perivascular excitatory nerves shared a common pathway. In contrast, Day & Rand (1961) found that cholinergic responses in Finkleman preparations of rabbit small intestine persisted after bilateral abdominal vagotomy, although the nerves may have been cut below the level of the branch to the coeliac plexus.

This example is a good illustration of the difficulties encountered if one assumes that the autonomic nervous system is perfectly ordered, with no admixture of sympathetic and parasympathetic pathways. But it should be noted that the problems posed by the presence of excitatory cholinergic responses to stimulation of perivascular nerves to the gut are by no means completely solved. Only in the rabbit colon has there been any direct proof that all of the cholinergic effects are mediated by separate parasympathetic fibres. In the small intestine the possibility still remains that part of the cholinergic response stems from fibres other than vagal fibres. An additional problem arises from the observations of Rand & Ridehalgh (1965), who found that the inhibitory response of guinea-pig colon to perivascular nerve stimulation was abolished by treatment with hemicholinium, a drug which is known to inhibit transmission in cholinergic nerves, and an excitatory response was revealed. The excitation was not cholinergic, since it was resistant both to the hemicholinium with which it was revealed and to atropine or hyoscine. This is the only report of such fibres to date and no comment can yet be made as to their nature or their origin.

There are no other clear examples of parasympathetic cholinergic fibres running in a sympathetic nerve trunk. It has been claimed that the vas deferens of the guinea-pig has a parasympathetic cholinergic innervation via the hypogastric nerve in addition to its sympathetic innervation via that nerve (Della Bella et al., 1964), but there is no evidence to support the conclusion that the cholinergic effects of hypogastric nerve stimulation are mediated by fibres of parasympathetic origin (cf. Langley & Anderson, 1895), and it appears that this is a case of confusion of the physiological term 'cholinergic' with the essentially anatomical term 'parasympathetic'. Similarly, in most other cases of cholinergic responses mediated by sympathetic nerves, it is highly unlikely that parasympathetic fibres are involved. For instance, although there is evidence for cholinergic fibres in the sympathetic nerve supply to the spleen, Masuda (1927) found that, provided vagal effects on blood pressure were avoided by using a blood-pressure compensator in the abdominal aorta, stimulation of the vagi had no effect on the spleen, and Utterback (1944) was unable to detect degenerating nerve fibres in the splenic nerve after vagotomy. The bulk of cholinergic effects of sympathetic nerve stimulation must therefore be mediated by sympathetic fibres.

Sympathetic cholinergic fibres. The presence of fibres which are cholinergic but which are anatomically sympathetic has received general acceptance in only one case, the innervation of the sweat glands in cats, until quite recently. But, as pointed out above, there are many instances of cholinergic responses to stimulation of sympathetic nerve trunks which are largely adrenergic. The most

16+

fully studied example of such effects is the sympathetic cholinergic vasodilator system.

Cholinergic vasodilator effects of sympathetic nerve stimulation were first noted in the facial skin of dogs (von Euler & Gaddum, 1931) and then in the muscles, but not in the skin, of the hindlimb of dogs and cats (Bülbring & Burn, 1935; Rosenblueth & Cannon, 1935). In the same year, Bacq (1935) suggested that the sympathetic vasodilators to the penis are cholinergic. The hindlimb vaso-dilator system was later studied in considerable detail by Folkow and his colleagues (Folkow, Haeger & Uvnäs, 1948; Folkow & Uvnäs, 1948a, b; 1949; Folkow, Frost, Haeger & Uvnäs, 1949). Burn & Rand (1959) suggested that cholinergic sympathetic effects on the vasculature were caused by the release of acetylcholine from sympathetic, adrenergic post-ganglionic nerves. But their speculation ignored two important observations made on the sympathetic vasodilator system. Firstly, Folkow & Uvnäs (1948a) had shown that the thres-hold stimulus for eliciting vasoconstriction was lower than that for vasodilator responses, revealing that two discrete populations of nerve fibres were involved, one dilator and the other constrictor. Secondly, it had been found that stimula-tion of brainstem centres could cause a selective cholinergic vasodilatation in hind limb muscles, without any involvement of adrenergic vasoconstrictor-responses (Eliasson, Folkow, Lindgren & Uvnäs, 1951; Eliasson, Lindgren & Uvnäs, 1952, 1954; Lindgren & Uvnäs, 1953, 1954, 1955; Lindgren, 1955; Lindgren, Rosen, Strandberg & Uvnäs, 1956). By similar stimulation of central nervous areas, it has now been shown that a discrete set of cholinergic sym-pathetic vasodilator fibres innervate the coronary blood vessels (Teplov & Vasil'eva, 1968). The differentiation between the fibres mediating vasodilatation and the fibres mediating vasoconstriction has been extended by showing that the dilatation caused by activity in sympathetic cholinergic fibres and the dilata-tion caused by decreased activity in the sympathetic adrenergic fibres occur at different loci in the muscle vascular bed (Rosell & Uvnäs, 1962; Renkin & Rosell, 1962; Bolme & Novotny, 1968).

While the presence of separate cholinergic sympathetic vasodilator fibres would explain most of the vasodilator effects of sympathetic stimulation, the possibility remained that part of the vasodilator response was caused by acetyl-choline released from adrenergic fibres in the manner proposed by Burn and Rand. However, there are circumstances in which this possibility can be ruled out. Dorr & Brody (1965) stimulated discrete vasoconstrictor areas in the central nervous system and found that the vasoconstrictor responses were abolished, but were not reversed to vasodilator responses, after treatment with adrenergic neurone-blocking drugs. It therefore seems that the cholinergic vasodilator response caused by stimulating peripheral sympathetic pathways may be medi-ated by discrete cholinergic fibres which fit the anatomical description of sym-pathetic fibres.

In view of the evidence that both the sweat glands and some vascular beds are innervated by discrete cholinergic sympathetic fibres, it seems highly likely that some of the other reported cholinergic responses to sympathetic nerve stimulation can also be attributed to separate cholinergic fibres rather than to the release of acetylcholine from adrenergic fibres. In fact, as Burn (1968a) has pointed out, the limited evidence available suggests that there is a much more extensive cholinergic component in the sympathetic system of lower vertebrates. It may be that the cholinergic sympathetic fibres in mammals are simply remnants of a more widespread primitive condition and that the primitive functions carried out by these fibres are in the process of being either discarded altogether or displaced to, say, hormonal systems in mammals. If the fibres are regarded as vestigial in an evolutionary sense, although not in the sense of their function in a particular species, it becomes easier to understand the extraordinarily random way in which sympathetic cholinergic supplies to both sweat glands and vascular beds occur in various mammalian species. For instance, the muscle vasodilator system is present in three canids (dog, jackal and fox) but is absent from two members of the quite closely related group of mustelids (badger and polecat); present in the cat and the mongoose, but absent from rabbits, hares, and a number of primates (see Uvnäs, 1967).

Non-cholinergic, non-adrenergic fibres

At the present time there is considerable evidence that the sympathetic subsystem contains post-ganglionic fibres which are neither adrenergic nor cholinergic. The same is true of the parasympathetic subsystem (see below, p. 484). But, as is suggested by the title of this section, one of the difficulties in presenting a coherent survey of such fibres stems from the very negativity of our recognition of the fibres, i.e. the fibres are only recognized as different because they do not fit into either of the two accepted patterns. In no case has a complete identification of the transmitter substances involved been made and it is therefore impossible to state at this stage whether we are seeing the effects of only one 'other' type of nerve fibre or, as has been claimed, of a number of different types. A further difficulty in many instances is that the anatomical origin of the fibres involved has not been worked out. This takes on a great importance, for it has been known for many years that vascular smooth muscle is inhibited by antidromic or axon-reflex activity in true sensory, i.e. non-autonomic nerve fibres. In the absence of anatomical information, we are unable to say whether nerves affecting a certain smooth muscle or gland are autonomic. To make things worse, there is still doubt about the vasodilator substance released from the sensory nerves, which has been variously proposed to be histamine, substance P, acetylcholine, a metabolite of adrenaline, or adenosine triphosphate (this subject is briefly reviewed by Beck & Brody (1961)).

One further difficulty has arisen with respect to the pharmacological analysis of responses mediated by nerves which are not adrenergic or cholinergic. A vast array of drugs is now available for use in analytical studies of the innervation of smooth muscle, but everything that we know about these drugs has been determined by experimentation on adrenergic and on cholinergic systems. Assuming that the nerves really are of a new type, we must be prepared for the normal range of autonomic drugs to do anything, or nothing, to them. In fact, it appears that the pharmacopoeia is about as useful in dealing with such a system as a book of incantations would prove to be in the transmutation of base metals. In a sense, one can only envy workers at the turn of the century, who had access to only a strictly limited range of natural pharmacological agents but who found that these agents were particularly apt for the study of the adrenergic and cholinergic systems.

Given that no consistent picture can yet be drawn up, only two representative examples of postulated non-adrenergic, non-cholinergic sympathetic post-ganglionic neurones will be discussed. The first is a system of sympathetic vasodilator-fibres which appear to act by releasing histamine. As a second example, the possibility of vasodilator and other nerves releasing a prostaglandin as a transmitter substance will be considered.

Histaminergic vasodilator fibres. The question of whether the sympathetic nerves supply histaminergic post-ganglionic fibres to blood vessels is still highly controversial. It has been known for many years that the induction of an increase in systemic blood pressure, by means of manoeuvres such as the rapid transfusion of blood or the injection of a pressor amine, is followed by a dilatation of peripheral vascular beds. The vasodilatation is a reflex, initiated by an increase in the firing rate of sensory nerve fibres arranged to detect vascular distension near the heart. Until quite recently it has been thought that the peripheral vasodilatation is a completely passive effect, caused by a diminution in the activity of vasoconstrictor nerves. The reflex does not involve the dorsal root vasodilator fibres, nor does it involve the cholinergic sympathetic vasodilators since it is not specifically reduced by atropine treatment (Folkow & Uvnäs, 1948b; Lindgren & Unväs, 1954). However, it has now been claimed that the baroreceptor reflex dilatation is an active process, involving increased activity in histaminergic vasodilator neurones (see Beck & Brody, 1961; Beck, 1964). Beck (1965) and Brody, Du Charme & Beck (1967) have shown that the vasodilator responses in dog hind limbs to systemic injections of pressor catecholamines or of veratrine are inhibited by a variety of drugs known to antagonize the actions of histamine. Tuttle (1966) has shown that the reflex vasodilatation in the cat tail caused by direct electrical stimulation of the baroreceptor nerve fibres in the carotid sinus nerve is also prevented by treatment with an antihistamine; furthermore, it is enhanced by isoniazid, a drug which inhibits the action of an enzyme, diamine oxidase, involved in the degradation of histamine. The responses were not greatly affected by treatment with histamine-releasers such as compound 48/80 (Beck, 1965) but it is known that the bulk of tissue histamine, except in the skin, is not released by such drugs (Smith, 1953a, b). Evidence has now been obtained that histamine emerges from the tail and limb during reflex vasodilatation. Brody (1966) has shown an increased efflux of ^{14}C-label during reflex vasodilatation of dog hindlimb after loading with ^{14}C-histamine, but a considerable proportion of the effluent label appears as methyl-histamine, a degradation product. In view of the high level of loading used and the proportion of degraded histamine emerging, Glick, Wechsler & Epstein (1968) have suggested that the increased efflux of label results from a perfusion during dilatation of vascular regions which are normally poorly perfused. On the other hand, Tuttle (1967) has shown an output of ^{14}C-histamine from the cat tail and gracilis muscle during reflex dilatation after the tissue has been loaded with a ^{14}C-labelled precursor of histamine, histidine. In his experiments, the bulk of the increased output of label during reflex stimulation occurs as histamine and not as unaltered histidine, so that the increased efflux does not seem to be due to enhanced perfusion of vascular backwaters.

Evidence has been presented that the histaminergic vasodilator nerves can be stimulated as they run in the sympathetic chains. Zimmerman (1966a) has reported that stimulation of the sympathetic chains in dogs treated with the adrenergic neurone-blocking drug bretylium causes vasodilatation in the skin of the paw. This response, like the vasodilator responses observed in the dog ear by Bülbring & Burn (1936), is not abolished by atropine. Zimmerman (1966b) showed that the combination of atropine and an antihistamine inhibits the vasodilator response. The vascular bed in the hind limb muscles shows both a cholinergic and an antihistamine-sensitive vasodilator process (Zimmerman, 1968). In contrast to this distribution of vasodilator fibres in the dog, it is found that most of the vasodilator response in the cat hind limb to stimulation of the sympathetic chain occurs in the muscle and it is usually completely eliminated by treatment with atropine (Zimmerman, 1968). But one must be cautious in accepting such reports of the absence of histaminergic responses obtained by experimentation on animals treated with adrenergic neurone-blocking drugs, for Sakuma & Beck (1961) found that guanethidine, bretylium and xylocholine (TM 10) inhibited the supposedly histaminergic reflex dilatation in the same doses in which they prevented adrenergic transmission. Only certain doses of the β-methyl derivative of xylocholine (β-TM 10) were found to eliminate the effects of adrenergic nerves selectively. The way in which the adrenergic neurone-blocking drugs achieve an inhibition of the histaminergic process is unknown, but the dangers of their use are obvious. A similar difficulty is found with the use of reserpine to expose vasodilator responses, for in animals chronically pretreated with reserpine to deplete tissue stores of catecholamines both the reflex constriction and the reflex dilatation are lost, while after acute treatment with reserpine there is a very rapid loss of the reflex dilatation before any diminution of the vasoconstrictor responses can be seen (see Beck & Brody, 1961). In addition, the ergot alkaloids, which may be used to reveal vasodilatations by eliminating the α-adrenergic vasoconstriction, prevent the reflex vasodilatation in doses so low that adrenergic function is not affected (Beck, 1961b; Wellens,

1964) and it has been suggested that these drugs also prevent the release of histamine (Beck, 1961*b*).

In view of the difficulty in finding drugs which selectively eliminate the vaso-constrictor effects of adrenergic nerves to reveal the histaminergic vasodilator-responses, it is clear that the most effective way to study the histaminergic

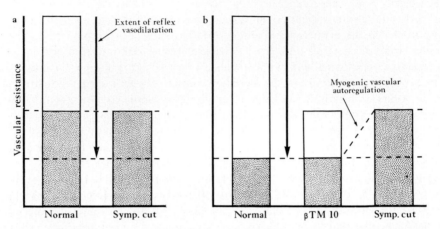

Fig. 3   The nature of baroreceptor reflex vasodilatation. In normal animals the resistance of the hind limb vascular bed comprises a myogenic component (stippled) and a neurogenic component caused by vasoconstrictor nerve activity (not stippled). *a* The Theory of Beck (1961*a*). After sympathectomy (symp. cut) the remaining resistance must be entirely myogenic. It is therefore assumed that the same level of myogenic resistance occurs in normal preparations (upper dotted line). However, during reflex vasodilatation (arrow) the resistance falls below this level (lower dotted line) and must therefore involve an active process mediated by vasodilator nerves. *b* The theory of Glick *et al.* (1968). It is argued that the resistance after sympathectomy (upper dotted line) does not reflect the myogenic component of resistance in normal preparations. Instead, after sympathectomy, myogenic autoregulation occurs so that the myogenic resistance of sympathectomized preparations is greater than the myogenic resistance component in normal preparations. Reflex vasodilatation is seen as a passive phenomenon, consisting solely of an obliteration of the neurogenic component of resistance by reducing the activity of vasoconstrictor nerves (lower dotted line). Beck *et al.*, have found that reflex vasodilatation persists in preparations treated with βTM 10, although sympathectomy causes no further decline in vascular resistance. It is suggested (middle column) that this could come about if βTM 10 reduces the neurogenic component of resistance to such an extent that autoregulatory processes can compensate completely for the abolition of the remaining neurogenic component by sympathectomy.

nerves is by creating a situation in which only these nerves are stimulated, a result which cannot be achieved by stimulation of the mixed nerve trunks in the periphery. It is in this respect that the nature of the baroreceptor-reflex dilatation becomes highly important to the argument. Beck and his colleagues believe that the reflex dilatation is caused actively by the release of a vasodilator compound because, during the reflex, the vascular resistance of the perfused dog hind limb falls to a level lower than that seen after inhibition of vasoconstrictor tone following either surgical sympathectomy or adrenergic neurone blockade. Since passive vasodilatation, caused by an inhibition of activity in vasoconstrictor nerves alone, cannot exceed the vasodilatation caused by total prevention of adrenergic activity, it seems that any excess dilatation beyond the levels existing after sympathectomy must be caused by the release of an actively vasodilator substance (see Fig. 3a). Since any reduction in vasoconstrictor tone will tend to depress the amplitude of vasodilator responses, it follows that the active component of dilatation in a normal animal will be of greater magnitude than the component remaining after inhibition of adrenergic tone by treatment with β-TM 10. Sakuma & Beck (1961) have calculated that between 80% and 90% of the normal baroreceptor reflex dilatation in the dog hind limb is caused by the active histaminergic vasodilators.

But a serious criticism of the argument outlined in the preceding paragraph has now been made by Glick et al. (1968). They point out that, when adrenergic vasoconstrictor tone is eliminated, the vasculature will not simply dilate to an extent predicted by the loss of nervous influences. Instead, the vascular bed will undergo the normal process of autoregulation, so that the resistance will return towards the level seen when neurogenic vasoconstrictor tone existed. Autoregulatory processes take a relatively long time to develop, so their effects would not be seen during brief falls of neurogenic vasoconstrictor tone but only during prolonged deprivation of adrenergic activity. It is therefore not correct to compare the minimal level of resistance observed during the reflex with the sustained resistance observed after sympathectomy or adrenergic neurone-blockade (Fig. 3b). This criticism is wholly correct, but it would be extremely hard to determine the extent of autoregulation and therefore the extent to which a rapidly occurring passive vasodilatation could decrease the resistance. A criticism along the same lines can be made about another feature of the reflex dilatation. Sakuma & Beck (1961) found that, after treatment with β-TM 10, the reflex still occurred but transection of the sympathetic chains no longer caused a fall in the sustained level of vascular resistance. Their explanation was that sympathetic adrenergic tone had already been selectively abolished by β-TM 10 and sympathectomy could cause no further decrease of neurogenic tone. But it is quite feasible that β-TM 10 had merely reduced the neurogenic vasoconstrictor tone to the extent where autoregulation could completely compensate for the further loss of tone on sympathectomy (Fig. 3b); if that were so, the reflex observed after treatment with β-TM 10 could still be passive. This would be completely consistent with the observation that higher doses of β-TM 10 or treatment with other adrenergic neurone-blocking drugs completely abolished the vasodilator response, an observation which can only be explained otherwise by postulating that the drugs act to prevent the release of histamine.

Assuming that the criticisms made by Glick et al., are correct, how then can one explain the actions of the antihistamines in blocking the dilatation? If it is assumed that the reflex is wholly passive, one must explain how the antihistamines can affect the adrenergic system. Glick *et al.* (1968) postulate that the antihistamines act like cocaine. Both cocaine and the antihistamine pyribenz-amine reduced the reflex dilatation in the hind limb in response to systemic injections of noradrenaline, enhanced the direct vasoconstrictor action of noradrenaline (see also Beck, 1965) and diminished the constriction caused by tyramine. Cocaine is thought to affect responses to sympathomimetic amines in these ways by inhibiting the neuronal uptake mechanisms for the amines, and the same action could be postulated for pyribenzamine. Glick and his colleagues suggest that both cocaine and the antihistamines inhibit the normally occurring re-uptake of noradrenaline released by nervous activity so that, when adrenergic nerve activity is reduced in a passive dilator reflex, it takes much longer than usual for the concentration of extracellular noradrenaline to fall to whatever value is dictated by the new rate of neurogenic release. This would have the effect of maintaining the vascular tone at high levels for longer periods than normal and would therefore, as observed, decrease the amplitude of the relatively brief reflex dilator-responses elicited by noradrenaline injections. But, in a preliminary report, Brody (1968) has now stated that cocaine does in fact decrease ^{14}C-histamine output as it blocks reflex vasodilatation, whereas the antihistamine pyribenzamine inhibits the reflex without preventing the con-comitant histamine release, both of which observations favour the theory of specific histaminergic nerves.

One other way of eliciting responses mediated purely by the histaminergic nerves would be to stimulate appropriate structures in the central nervous system. Tuttle & Connor (1965) found that stimulation of a small medial region of the posterior medulla oblongata in cats caused a vasodepression of such intensity that it was sometimes lethal. Tuttle (1965) found that this response was accom-panied by the release of a vasodilator substance into the circulation. The action of the vasodilator or the amount released was depressed by atropine and by an antihistamine and was enhanced by treatment of the animal with a diamine-oxidase inhibitor, aminoguanidine. Tuttle also found that ^{14}C-histamine, formed *de novo* from ^{14}C-histidine, was released into the blood stream during stimulation of this vasodepressor region. While Tuttle's experiments do not show the site of the release of histamine, there is a possibility that it does come from specific vasomotor nerves. This question may perhaps be solved by studies using a newly developed fluorescence histochemical test for histamine recently adapted for use with freeze-dried tissues (Ehinger, Håkanson, Owman & Sporrong, 1968). This technique may also help to resolve the very crucial problem which must be faced by any worker proposing histaminergic nerves, namely that sympathetic-denervation procedures have not been found to alter tissue histamine concentra-tions to any great extent (von Euler & Purkhold, 1951), although sympathetic nerve trunks contain higher levels of histamine than other autonomic nerves (von Euler, 1949).

The role of prostaglandins in autonomic transmission. Although the case for prostaglandins (PGs) as actual transmitter substances is, at least at the present time, weak, these compounds are currently receiving a great deal of attention, as must be true of any group of compounds which exert actions on a variety of glands and smooth muscle preparations in rather low concentrations and which also occur naturally in those organs. The arguments concerning the involvement of PGs in the autonomic transmission process are just as applicable to parasympathetic as they are to sympathetic transmission.

The PGs are a group of lipid-soluble unsaturated hydroxy-acids. The chemistry and nomenclature, the natural distribution and the pharmacological actions of compounds in this group have been summarized recently in two excellent reviews (Pickles, 1967; Bergstöm, Carlson & Weeks, 1968). The full identification of specific prostaglandins in tissues is particularly difficult, so that, in the words of Pickles (1967), 'an unknown PG-like substance tends to be identified with one or the other of the four authentic PGs of which pure samples are at present fairly readily available'. This tendency makes one cautious in accepting statements in the literature concerning the identity of PGs in tissues and perfusates, and makes the job of discussing the role of PGs more difficult. One should note that two well-established materials extracted from tissues have now been recognized as mixtures of PGs, namely Darmstoff (Suzuki & Vogt, 1965) and the irins (Änggård & Samuelsson, 1964; Ambache, Brummer, Rose & Whiting, 1966).

Prostaglandins have been postulated as transmitter substances in only two instances. One is the atropine-resistant, long-lasting miosis which is observed when the trigeminal nerve is stimulated antidromically, a response which is thought to be mediated by 'irins' (see Ambache, 1963). Since the response presumably does not involve autonomic fibres it will not be discussed further. On the other hand it has been postulated that PGs mediate the sustained non-adrenergic, non-cholinergic, non-histaminergic vasodilatation in the hind-limb caused by sympathetic nerve stimulation. This response has been observed in the hindlimb of dogs and monkeys but does not occur to any great extent, if at all, in cats (Beck, Pollard & Weiner, 1964; Beck, Pollard, Kayaalp & Weiner, 1966; Zimmerman, 1968). Beck & Pollard (1967) have shown that the sustained vasodilatation in dog hindlimbs is reduced by prostaglandin E_1, also a vaso-dilator, to a greater extent than are cholinergic dilatations. They suggest that the selective reduction of the sustained dilatation is due to PGE_1 occupying the same receptors as the endogenous vasodilator transmitter-substance, i.e. that the transmitter substance is PG-like. Dunham, Rolewicz & Zimmerman (1968) noted that the sustained dilation in dog hindlimb was observed in the perfused paw but not in the gracilis muscle, indicating that the nerve fibres responsible for the sustained dilatation are confined to the skin region. They related this observation to the fact that extractable PG-like material occurs in far greater amounts in skin than it does in muscle and suggested that the selective distribution could come about if the substance mediating sustained

16*

dilatation were a prostaglandin. Neither of these lines of argument is convincing and studies on this system, as on all other systems, are hampered by the lack of suitable pharmacological antagonists for the PGs.

In one region of the vasculature, the splenic vascular bed, there is clear evidence that sympathetic-nerve stimulation can cause the release of PGs. (Davies, Horton & Withrington, 1966; Ferreira & Vane, 1967). Davies, Horton & Withrington, (1966, 1968) found that the prostaglandin released into the splenic venous effluent after splenic nerve stimulation was mainly PGE_2, and that the amount released greatly exceeded the amount extractable from the spleen, indicating that the PG was formed *de novo* during nerve stimulation. However, it does not seem that the PG is released as a transmitter substance, for when the tissue is treated with the α-adrenergic-blocking drugs phenoxybenzamine or phentolamine, drugs which do not prevent the release of catecholamines but simply prevent them from activating post-synaptic receptors, nerve stimulation no longer causes the release of PGs. This indication that PGs are released as a result of the action of adrenergic transmitter substances on the effector organ is strengthened by the observation that injections of catecholamines into the spleen also cause PG release (Ferreira & Vane, 1967), an action which is prevented by α-adrenergic receptor-blocking drugs. A similar effect of adrenergic transmitter substances on vascular smooth muscle may well explain the observation that stimulation of the phrenic nerve causes the release of PGs, mainly PGE_1, from the isolated rat diaphragm. (Ramwell, Shaw & Kucharski, 1965; Ramwell & Shaw, 1967). The PG release is not a concomitant of transmission to striated muscle since it persists after treatment of the preparation with the neuro-muscular blocking drug tubocurarine. Furthermore, catecholamines cause a release of PGs from the diaphragm. In this case it seems likely that the PG is released from vascular smooth muscle in the diaphragm, following the excitatory action of catecholamines released from sympathetic adrenergic nerve fibres in the phrenic nerve. Alternatively the PGs may have been released by adrenergic stimulation of adipose tissue in the diaphragm, since these authors also noted a release of fatty acid on nerve stimulation. It has been found that both catecholamines and sympathetic nerve stimulation cause a release of mixed E and F prostaglandins from isolated rat epididymal fat pads (Shaw, 1966).

A link between excitatory autonomic nerve transmission and PG-release is not confined to adrenergic systems. Perfused adrenal glands of rats and cats release $PGF_{1\alpha}$ into the venous effluent spontaneously and the release is greatly increased by perfusion of the glands with acetylcholine solutions (Ramwell, Shaw, Douglas & Poisner, 1966; Shaw & Ramwell, 1967). Although it has not been tested, presumably stimulation of the sympathetic cholinergic innervation of the adrenal gland would cause a similar release of prostaglandins. In one other case a release of PG is associated with cholinergic excitatory transmission. There is a spontaneous release of PGs into the lumen of the gastro-intestinal tract (Coceani, Pace-Asciak, Volta & Wolfe, 1967; Bennett, Friedmann & Vane, 1967; Bartels, Vogt & Wille, 1968). The release from the frog intestine is accelerated by treatment with acetylcholine. Release into the lumen of the rat stomach is enhanced by stimulation of the vagus nerves, although there is disagreement about whether the PG is mainly E_1 (Bennett et al., 1967) or a mixture of E_2 and $F_{2\alpha}$ (Coceani et al., 1967). In addition, such stimuli as distending the stomach or stretching isolated strips of the gastric wall cause an accelerated PG release. Both of these procedures, like stimulation of the vagi, might be expected to cause a depolarization of the gastric smooth muscle. The PG release on stimulation of the vagi is prevented when the excitatory effects of acetylcholine have been blocked by hyoscine treatment. It is interesting that, in this preparation of rat stomach, stimulation of the perivascular sympathetic nerves does not cause a release of PG into the gastric lumen, for in this organ the sympathetic adrenergic innervation is inhibitory to the gastric smooth muscle. But the perivascular nerves should cause constriction in the gastric vascular bed and it would be interesting to know whether stimulation caused the appearance of PGs in a vascular perfusate.

In all the above examples of nerve-mediated prostaglandin release, it can be seen that PG is also released by the excitatory transmitter substance applied directly to the preparation. Further, when the action of the transmitter substance is prevented pharmacologically, nerve stimulation no longer causes PG release. It therefore seems clear that the PGs are released from the effector tissue itself as a result of its stimulation and not to any great extent from the nerves. Although it appears that the PGs are not likely to be transmitter substances in the autonomic nervous system, one must still ask whether their release on excitatory nerve stimulation is wholly coincidental or whether they have a true physiological role. No simple answer embracing all tissues can be given because of the diversity of action of the various prostaglandins. There is good reason to think that the PGs in some instances act as local hormones to moderate the excitatory influences of the innervation. For instance, the amount of PG released from rat epididymal fat pads following sympathetic nerve stimulation is adequate to inhibit completely the free fatty acid-mobilizing effects of nerve stimulation (Berti & Usardi, 1964), although clearly such mobilization does occur under normal conditions. Similarly, PGE_1 and PGE_2, both of which have been claimed to be released into the rat gastric lumen on vagus nerve stimulation, are potent inhibitors of gastric secretion in dogs (Robert, Nezamis & Phillips, 1967). In both instances, the prostaglandins could, and in fact should, act as local tissue hormones in a negative feedback mechanism, moderating the excitatory influence of the innervation. But the effects of PGs on intestinal and vascular smooth muscle are harder to rationalize. Although PGs cause contraction of most intestinal smooth muscle preparations, PGE_2 has been found to inhibit contractions of circular but not longitudinal muscle in human stomach (Bennett, Murray & Wyllie, 1968); guinea-pig ileam (Kottegoda, 1969). It is not clear that this difference has any functional significance, especially as most of the gastric PG is confined to the mucosa. Prostaglandins are known to antagonize specifically both inhibitory and excitatory actions of catecholamines on a number of smooth muscle preparations, often after an initial period of potentiation (Clegg, 1966). It might therefore seem that splenic prostaglandin could play a modulating role in sympathetic adrenergic transmission in that organ. However, PGE_1 and PGE_2 cause no change in splenic responses to catecholamines or nerve stimulation other than the change induced by the vasodilatation which the PGs produce (Davies & Withrington, 1968).

Non-cholinergic parasympathetic post-ganglionic neurones

There are a number of instances of parasympathetic nerves causing responses of cardiac and smooth muscle and of glands which are clearly not mediated by the release of acetylcholine. In some cases the transmitter substance involved appears to be a catecholamine. In other cases there is evidence that the transmitter substance is neither acetylcholine nor a catecholamine, but the nature of the transmitter substance is not yet known.

Effects mediated by catecholamines

The most intensely studied instance of a parasympathetic nerve causing an adrenergic response is the increase in rate and force of cardiac beats caused by stimulation of the vagus nerve. The response, first studied in detail by Dale, Laidlaw & Symons (1910), appears to be mediated by a catecholamine since it is inhibited by depletion of tissue catecholamines with reserpine (Greeff, Kasperat & Osswald, 1962; Leaders, 1963; Copen, Cirillo & Vassalle, 1968) and by blockade of adrenergic β-receptors (Misu & Kirpekar, 1968) and enhanced by treatment with cocaine (Greeff *et al.*, 1962), while it is associated with the release of a catecholamine from the heart (Middleton, Middleton & Toha, 1949). Although other examples of apparently adrenergic responses to stimulation of parasympathetic nerves are known, for instance in the vagal supply to the lung, only this one phenomenon will be discussed here. A number of explanations for such a response, no matter what tissue it occurs in, are possible and, fortunately or unfortunately, the large number of studies of vagal cardiostimulation has provided us with support for each of the explanations.

Before discussing the actual adrenergic mediation of responses to vagal stimulation, one point should be made. When stimulation of the vagi in a normal animal causes positive inotropic and chronotropic responses of the heart, the stimulant response is usually seen as a period of acceleration and enhanced beating at the end of the period of inhibition which is the 'normal' vagal response. On the other hand, when the vagi are stimulated in atropinized animals, or when only certain roots of the vagus nerves are stimulated in animals not treated with atropine, the cardio-inhibitory responses are obviated and cardiostimulation is observed during the period of nerve stimulation. A number of the papers concerned with vagal cardiostimulation deal only with the former type of response, the post-stimulatory enhancement of cardiac activity. But, at least under the experimental conditions of one group of workers (Misu & Kirpekar, 1968), the only post-stimulatory enhancement of cardiac activity observed was not a response mediated by catecholamines but appeared to be a passive over-compensatory recovery process of the heart muscle, set in motion only by the normal inhibitory action of released acetylcholine acting directly on the cardiac cells. A similar passive recovery from inhibitions caused by exogenous acetylcholine has also been reported (Hollenberg, Carriere & Barger, 1965). The fact that the inhibitory cholinergic nerve fibres in the vagi can cause cardiostimulation via a cholinergic action on the muscle should be borne in mind when considering post-stimulatory 'sympathomimetic' effects of vagus nerve stimulation.

The theories that will be discussed below are: that the adrenergic responses are mediated by admixed sympathetic adrenergic nerve fibres; that they are caused by adrenergic fibres of true parasympathetic nature; that the catecholamine involved is released from cardiac chromaffin cells innervated by parasympathetic fibres; and that the catecholamine is released from sympathetic adrenergic nerve fibres by the action on them of acetylcholine released from true parasympathetic cholinergic nerve fibres. These theories are outlined diagrammatically in Fig. 4.

Admixed sympathetic adrenergic fibres. While a number of authors consider that there are no vagal fibres *per se* which cause stimulation of the heart, there is complete agreement that sympathetic adrenergic fibres do join the vagi and innervate the heart (Fig. 4b). It has often been observed that the cardiac augmenting and accelerating effects of vagus nerve stimulation are greater when the stimulating electrodes are applied near the heart (e.g. Kabat, 1937, 1939; Pannier, 1946). The sympathomimetic responses caused by stimulating the vagi close to the heart in cats occur with the same optimal values of electrical pulse frequency and duration as the response to stimulation of the sympathetic accelerator nerve (Misu & Kirpekar, 1968). Dale *et al.* (1910) and Benítez, Holmgren & Middleton (1959) found that, after a high degenerative section of the vagus in cats, stimulation of the trunk in the cervical region was ineffective in causing cardiostimulation whereas stimulation of the thoracic vagus caused acceleration. Results of this type clearly indicate that sympathetic nerve fibres run from the paravertebral chains to join the vagi as they run to the heart. However, there is no agreement as to the highest point at which such fibres enter the vagi.

The main question for the purposes of this argument is how much of the observed cardiac acceleration and augmentation is mediated by the errant sympathetic fibres and how much by fibres which can properly be regarded as parasympathetic. At first sight it seems that this would be an easy question to solve, by carrying out sympathectomies and allowing the remains of the mixed

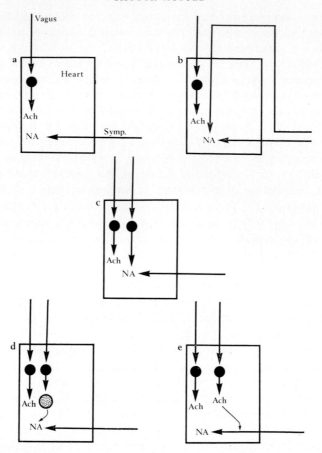

Fig. 4   Ways in which the vagus nerve could mediate cardiostimulation, as discussed in the text. In a, the classical innervation of the heart is shown. Cholinergic vagal preganglionic fibres form synapses (thick arrows) on post-ganglionic neurones (filled circles), which in turn release acetylcholine (Ach) in the heart; sympathetic post-ganglionic fibres, with their cell bodies in the paravertebral chains, release noradrenaline (NA). In b, some sympathetic adrenergic fibres are shown running with the vagal trunk. In c, adrenergic fibres of parasympathetic origin are provided to the heart. In d, some vagal cholinergic fibres innervate chromaffin cells (stippled circle) and cause them to release noradrenaline. In e, acetylcholine released from some vagal fibres stimulates sympathetic fibres and causes the release of noradrenaline from them.

sympathetic fibres to degenerate. In practice the results are difficult to interpret. There is general agreement that neither cervical sympathectomy (Middleton *et al.*, 1949) nor stellectomy or total thoracic sympathectomy (Samaan, 1935; Brouha, Cannon & Dill, 1936; Kabat, 1939; Okinaka, Nakao, Ikeda & Shizume, 1951; Misu & Kirpekar, 1968) abolishes vagal cardiostimulation. But there is considerable disagreement concerning the extent to which these treatments reduce the vagal responses, and presumably the actual amount of reduction depends on the site of vagal stimulation. To illustrate the prevailing confusion, one may refer to the results of Brown & Maycock (1942). These authors found that stimulation of the upper cervical vagi in normal cats did not give cardiostimulant responses, but did after a partial cervical sympathectomy had been carried out. They produced evidence that the appearance of responses after sympathectomy was due to a re-innervation of denervated cells in the middle cervical ganglion, which had been left *in situ*, by collaterals from vagal preganglionic fibres which had been damaged slightly during the denervation procedures.

Combinations of thoracic and cervical sympathectomy have not clarified the situation. In their original paper, Dale *et al.* (1910) found that vagal cardiostimulation in cats persisted after removal of the superior and inferior cervical and the stellate ganglia, as was also later noted by Burn & Rand (1958). Benítez *et al.* (1959) found, also in cats, that unilateral sympathectomy from the superior cervical ganglion to the third or fourth thoracic ganglia of the paravertebral chains abolished both vagal stimulation of atropinized hearts and stimulatory after-effects in untreated hearts; but, in the same species, Chiang & Leaders (1966) found that the stimulatory after-effects were only insignificantly reduced by sympathectomy from the superior cervical to the fifth thoracic ganglion. Presumably the results are inconsistent because of the enormous technical difficulties involved in achieving widespread and complete ablation of the sympathetic system, especially in view of the micro-anatomical variability which may be expected in this system. However, even assuming that a complete sympathetic denervation of the heart without interruption of the vagal pathways were possible, and that this procedure abolished vagal cardiostimulant responses, the result would not necessarily mean that the entire vagal sympathomimetic response is mediated by sympathetic fibres. Instead the result could mean simply that true parasympathetic fibres cause a release of catecholamines from structures which are loaded with amine from the sympathetic system or, as was originally suggested by Benítez *et al.* (1959), release from the sympathetic nerve terminations themselves. We must content ourselves at this stage by noting that sympathetic fibres do enter the vagus nerves and that the proportion of sympathetic fibres increases as the vagus approaches the heart.

Truly parasympathetic adrenergic fibres. Given that the sympathetic adrenergic fibres do enter into the vagal trunks, one must ask whether there are also truly vagal pathways which mediate positive chronotropic and inotropic responses of the heart. Clearly the results of sympathectomy do not, and probably can not, clarify the issue and another approach is needed. The most successful alternative appears to be to stimulate only certain of the intracranial roots of the vagus. All experiments of this type have been performed on dogs. Jourdan & Nowak (1934) found that stimulation of the vagus roots proper caused cardioacceleration in animals which had not been treated with atropine, whereas stimulation of the bulbar roots of the spinal accessory (XIth cranial) nerve, which in most mammals, including the dog, fuses immediately with the vagus and runs with it to the periphery, causes the normal vagal response of cardio-inhibition. Pannier (1946), in a careful and extensive study, reached the same conclusion, also on non-atropinized dogs. It is clearly difficult to place gross stimulating electrodes selectively on the fine intracranial rootlets without the stimulating current spreading to adjoining structures. This probably explains why Kabat (1939) needed to atropinize the animals and why Okinaka *et al.* (1951) had to carry out degenerative sections of the rostral roots of the spinal accessory nerve to reveal cardio-accelerator responses to intracranial vagus root stimulation. It is extremely unlikely that there are a considerable number of sympathetic fibres which loop up the vagus as far as the intracranial portions to mediate the cardio-accelerator responses recorded by all four groups of workers. So the results show that there are truly vagal pathways which can cause stimulation of the heart, at least in dogs.

It should be noted that, while there is clear proof that the cardiostimulation caused by electrical stimulation of the peripheral vagus trunk is mediated by a catecholamine, the evidence may refer only to the sympathetic component and there has as yet been no attempt to determine directly whether the vagal cardio-accelerator pathways stimulated at an intracranial site are also adrenergic. However, it seems unlikely that a considerable component of non-adrenergic stimulant pathways in the peripheral vagus would have escaped the attention of the numerous workers who have studied the system. Therefore, for the purposes of this argument, it will be assumed that vagal cardiostimulation mediated by parasympathetic pathways involves an adrenergic process.

The most obvious explanation for vagal cardiostimulant effects is that there are anatomically parasympathetic adrenergic neurones in vagal pathways (Fig. 4c). This explanation, at least in its simplest form, does not seem to be correct. Firstly, it appears that cardiostimulant responses to stimulation of the cervical vagus are inhibited by drugs which block nicotinic receptors for acetylcholine (Middleton *et al.*, 1949; Greeff *et al.*, 1962; Copen *et al.*, 1968), a result which must be contrasted with the lack of effect of nicotinic blocking drugs on stimu-lant responses to stimulation of the thoracic vagus (Benítez *et al.*, 1959; Misu & Kirpekar, 1968). The simplest explanation for the blocking action of antinicotinic drugs is that parasympathetic cholinergic preganglionic neurones make synaptic connections with post-ganglionic adrenergic neurones in the heart, whereas the sympathetic adrenergic fibres which join the vagi are post-ganglionic. If that is the case, one would expect to observe adrenergic neurone somata in the heart, using the fluorescent histochemical technique for visualizing catecholamines. But, in a careful study of the cardiac ganglia of several species, Jacobowitz (1967) was unable to find any catecholamine containing somata in normal tissue. Admittedly he observed that a few neurones developed specific fluorescence after inhibition of monoamine oxidase and subsequent loading of the tissue with the precursor of noradrenaline, 3,4-dihydroxyphenylalanine (DOPA), but this need not mean that the cells normally contain catecholamines. Secondly, Greeff *et al.* (1962) found that vagal positive chronotropic and inotropic effects on isolated atropinized guinea-pig atria were not affected by concentrations of the adrenergic neurone blocking drug guanethidine which completely prevented responses to stimulation of the sympathetic accelerator nerve. The result has not yet been verified, but seems too clear to doubt. At first sight, such a result indicates that the vagal cardiostimulant fibres are not adrenergic at all. However, since we do not have direct evidence concerning the mechanism of action of adrenergic neurone blocking drugs (but see p. 481), we cannot guarantee that all adrenergic nerves will be susceptible to their action. It would still be possible to suggest that the vagi provide the preganglionic input to post-ganglionic adrenergic neurones in the heart, provided that one postulates that the neurones contain lower levels of amine in the soma than is usual, to explain the negative results of fluorescence microscopy, and that they are in some unknown way resistant to the actions of adrenergic neurone blocking drugs. However, the same results fit at least two alternative theories which do not involve vagal adrenergic neurones.

Chromaffin cells innervated by parasympathetic fibres. It has been pointed out (see above) that the vagal cardiostimulation is prevented by nicotinic ganglion blocking drugs but is not affected by adrenergic neurone blockade. In these two respects the response is immediately reminiscent of the secretion of catecholamines from the adrenal medulla elicited by stimulation of the cholinergic sympathetic innervation. It could therefore be argued that vagal cardiostimulation occurs by essentially the same mechanism, that is that cholinergic vagal fibres mediate catecholamine secretion from chromaffin cells in the heart (Copen *et al.*, 1968; Fig. 4d). This theory has received clear support from the histochemical studies of Jacobowitz (1967). He found that in the heart of cats, rats, mice and guinea-pigs there are catecholamine-containing chromaffin-like cells. In fact only the cat tissue showed a positive chromaffin reaction, and even that was weak, but the fact remains that the more sensitive histochemical fluorescence technique for catecholamines revealed the amine content of the cells. The chromaffin cells were largely confined to the cardiac ganglia in the region of the sino-atrial and auriculo-ventricular nodes, but some of the chromaffin cells lay outside the ganglia. For some reason, Angelakos, Fuxe & Torchiana (1963) did not find cells of this type in the sino-atrial node or atria of guinea-pigs or rabbits using the same technique. By staining for cholinesterases, associated with what are presumably cholinergic nerve fibres, Jacobowitz was able to show that at least some of the chromaffin cells were innervated by fibres extending from neurones in the cardiac ganglia, which in turn probably receive their neural input from parasympathetic fibres in the vagi. It is clear that the chromaffin cells have the potential to mediate vagal cardio-acceleration, by virtue of their proximity to the pacemaker region of the sino-atrial node, but their ability to affect the force of atrial contractions is not obvious. As well as envisaging actions of amines released from the chromaffin cells on the heart muscle, Jacobowitz has also considered actions on coronary blood vessels and, as an inhibitory hormone, on transmission through the parasympathetic cardiac ganglia.

The theory that chromaffin cells are responsible for vagal cardiostimulation must take into account the observation that, after the most complete possible extrinsic denervations have been carried out, cardiac catecholamines are virtually entirely lost (see Cooper, 1966). This may mean that the chromaffin cells lack an appreciable ability to synthesize catecholamines and that they normally replete their amine stores by taking up catecholamines released from cardiac sympathetic adrenergic nerves. In this context, it is interesting that, after loading rat hearts with relatively large amounts of noradrenaline, specific catecholamine fluorescence appears microscopically in small non-neuronal cells in the atrial connective tissue (Farnebo, 1968). It is possible that the cells described by Farnebo are similar to the chromaffin cells discovered by Jacobowitz, but no full paper has yet been published on the subject.

Another explanation for the results of total denervation is possible. It may be that the amine content of the chromaffin cells represents only a very small proportion of the total cardiac content of noradrenaline. If that is the case, it is less likely that the release of noradrenaline from chromaffin cells can account for all of the stimulant effects on the heart mediated by parasympathetic pathways. One should therefore consider a further possible explanation of the phenomenon.

Sympathetic adrenergic fibres 'innervated' by parasympathetic fibres. It has been proposed that vagal cardiostimulation may be caused by acetylcholine, released from parasympathetic cholinergic nerves, acting on sympathetic adrenergic nerve fibres to cause a release of noradrenaline from them (Benítez *et al.*, 1959; Leaders, 1963; Fig. 4e). To support this proposal, it must be shown that cholinergic fibres are within 'striking distance' of adrenergic fibres, and that acetylcholine can cause the release of noradrenaline from the fibres. On the first point, we do not have detailed evidence for the heart. However it has been claimed on histochemical and microanatomical grounds that cholinergic and adrenergic nerve fibres may be enclosed by folds of the same Schwann cell in other tissues (Jacobowitz & Koelle, 1965; Robinson & Bell, 1967; Burnstock & Robinson, 1967; Graham, Lever & Spriggs, 1968). In other words, it is at least possible for cholinergic nerves in the heart to lie within a few microns of adrenergic fibres, and it is quite feasible that large enough amounts of acetylcholine are released to diffuse across this distance and arrive at adrenergic fibres in concentrations sufficient to elicit the release of noradrenaline.

On the second point, there is an abundance of evidence to show that acetylcholine can elicit a release of noradrenaline from sympathetic post-ganglionic fibres in the heart as well as in other tissues. The stimulant effects of acetylcholine on the atropinized heart were first shown by Hoffman, Hoffman, Middleton & Talesnik (1945) and have since been observed by many workers. The stimulant effect is clearly adrenergic, since it is associated with a release of noradrenaline from the heart (Hoffman et al., 1945; Richardson & Woods, 1959) and is prevented by reserpinization (Alvarado, Middleton & Beca, 1961; Cabrera, Cohen, Middleton, Utano & Viveros, 1966) or by treatment with the adrenergic β-receptor-blocking drug dichloroisopropyl-noradrenaline (Lee & Shideman, 1959; Blumenthal, 1964). It is established quite clearly that the stimulant effect of acetylcholine on the ventricle occurs exclusively via sympathetic nerves, since the ventricular effects are virtually abolished after extensive sympathectomy (Cabrera et al., 1966). On the other hand, the responses of the auricle are only partially reduced after sympathectomy (Cabrera et al., 1966) which perhaps indicates that the release of noradrenaline from chromaffin cells is more important in this region.

The receptors for the cardiostimulant actions of acetylcholine are of the nicotinic type, as was first shown by Hoffman et al. (1945). This would explain why vagal cardiostimulation is prevented by nicotinic blocking drugs. However, it appears that the receptors involved are situated on adrenergic nerve fibres and not necessarily on intracardiac ganglion cells, since the stimulant effects still occur in portions of tissue which do not contain neurone somas (Middleton, Oberti, Prager & Middleton, 1956; Lee & Shideman, 1959). But the manner in which nicotinic stimulation causes a release of noradrenaline from adrenergic nerves is still controversial. On the one hand, Burn and Rand argue that acetylcholine enters the fibres and combines with an intracellular nicotinic receptor to cause noradrenaline release, independent of any other effects that acetylcholine may produce (see Burn & Rand, 1965). On the other hand, it is postulated that acetylcholine combines with nicotinic receptors, presumably in the nerve fibre membrane, to cause depolarization of the membrane and the subsequent initiation of action potentials; the action potentials then cause the release of noradrenaline in the normal manner, whatever that may be. Direct evidence for such an action of acetylcholine on sympathetic adrenergic fibres was first shown on the spleen by Ferry (1963). He found that injections of acetylcholine into the spleen caused the appearance of antidromic action potentials emerging from the spleen in the splenic nerve. It has now been found that injection of acetylcholine into the left atrium of dogs, from where it should pass into the coronary circulation, causes antidromic firing in the sympathetic inferior cardiac nerve (Cabrera, Torrance & Viveros, 1966). Although the matter is not settled, the latter theory will be accepted in the following discussion, in view of the fact that action potential initiation by acetylcholine in sympathetic nerves has been shown directly, whereas the support for the theory of Burn and Rand is essentially indirect.

Assuming that acetylcholine released from parasympathetic neurones can reach and activate sympathetic nerve fibres, it still remains to explain why the vagal cardiostimulant response is not prevented by guanethidine (Greeff *et al.*, 1962). We should at least be able to show that acetylcholine still produces sympathomimetic responses of the heart after adrenergic neurone blockade has been established. To the contrary, it has been reported that the stimulation of isolated atropinized hearts by acetylcholine is abolished after treatment with the adrenergic neurone blocking drugs bretylium or xylocholine (Lee & Shideman, 1959; Huković, 1960). But one must treat these results with extreme caution, for in both studies very high concentrations of the blocking drugs were used, and since adrenergic neurone blocking drugs are also known to block nicotinic acetylcholine receptors (Rand & Wilson, 1967) it may well be that the reported blockade was not due to the 'specific' action of the drugs. In fact Huković (1960) observed that the sympathomimetic effect of acetylcholine was restored rapidly after removing bretylium from the organ bath in which the atria were suspended. Now it is well known that adrenergic neurone blockade is extremely persistent, even when the blocking drug is no longer in contact with the test organ. It therefore appears highly likely, although it was not tested by Huković, that in the period after washing out bretylium, acetylcholine was producing stimulation of sympathetic nerves when normal orthodromic action potentials were ineffective in producing noradrenaline release. Similarly, in the spleen, injections of acetylcholine can still cause the release of noradrenaline from the sympathetic nerves when electrical stimulation of the sympathetic nerves has been rendered ineffective by treatment with bretylium (Hertting & Widhalm, 1965; Fischer, Weise & Kopin, 1966). It is therefore likely that acetylcholine can cause noradrenaline release from sympathetic nerves in the heart under conditions of 'adrenergic neurone blockade', as is demanded by this theory of the origin of vagal cardiostimulation. The way in which acetylcholine action is resistant to adrenergic neurone blocking drugs is discussed in the following paragraphs.

Some further evidence concerning the selective blocking action of bretylium against noradrenaline release by nerve stimulation, as opposed to release by acetylcholine, has now been produced by Davey, Hayden & Scholfield (1968). These authors confirmed the earlier reports, finding that injected acetylcholine still caused noradrenaline release when the release elicited by stimulation of the splenic nerve with 30 pulses/sec had been abolished by treatment with bretylium. But they estimated that the injections of acetylcholine used were producing antidromic action potentials at the rate of only 5·4/sec. When they compared the blocking action of bretylium on acetylcholine-induced release with the blockade of the release elicited by lower frequencies of nerve stimulation (10 pulses/sec), they found that bretylium was about equally effective in preventing the release produced by either stimulus. It follows, of course, that at a time when bretylium had prevented the release caused by stimulation with 30 pulses/sec, stimulation with 10 pulses/sec was still capable of causing a release of noradrenaline. The fact that bretylium becomes more effective in preventing noradrenaline release as the frequency of nerve stimulation is increased gives a clue to the mode of action of bretylium and, presumably, of other adrenergic neurone blocking drugs.

There are currently two theories concerning the mode of action of adrenergic neurone blocking drugs. One, put forward by Burn and Rand (see Burn & Rand, 1965), is that bretylium enters the adrenergic fibre and occupies a nicotinic receptor for acetylcholine, thereby preventing a normal release of noradrenaline by intracellular acetylcholine. The other is that bretylium essentially acts as a local anaesthetic drug, preventing the conduction of action potentials (see Boura & Green, 1965). The latter theory proposes that adrenergic neurone blocking drugs are accumulated in adrenergic nerve fibres, by means of a membrane pump specific to adrenergic neurones, until the intracellular concentration is great enough to cause a conduction block in the fibre. Orthodromic action potentials cannot now penetrate to the nerve terminations and there is therefore a prevention of noradrenaline release. Since adrenergic neurone blocking drugs do not prevent conduction in the axons remote from the terminals (Exley, 1960), the theory demands either that greater concentrations of drug are accumulated in the terminal regions or that these regions are more sensitive to local anaesthesia, either of which is quite possible. Now the latter theory can be used to explain the results of Davey et al. (1968). If we assume that the adrenergic neurone blocking drug is gradually accumulated in the interior of the adrenergic nerve fibre and that the final result of this accumulation is a conduction block, it follows that during the onset of the blockade there will be a gradual reduction in the safety factor for action potential propagation along the fibres until finally conduction is prevented. The site at which conduction block first becomes evident will, all other things being equal, be the region which normally has the lowest safety factor for conduction. One can expect the point of lowest safety factor to be the region of preterminal branching which is found in adrenergic neurones (Malmfors & Sachs, 1965). Thus one can envisage bretylium and comparable drugs causing a conduction block in a preterminal region of the adrenergic neurone at a time when locally initiated action potentials can still be conducted in the terminal region, causing a normal release of transmitter substance (Fig. 5). In fact, Blinks (1966) found that the positive inotropic effects produced by field stimulation of nerve fibres within the atrial wall were not prevented by even very high concentrations of bretylium or xylocholine, a result which is consistent with a retention of excitability in terminal regions anatomically distal to the site of adrenergic neurone blockade. It can be seen that such a block at a branching site would similarly cause a blockade of responses to orthodromic stimulation of the axon remote from the terminals while acetylcholine, applied directly to the regions of the fibre peripheral to the site of conduction block, could still initiate action potential firing in the terminals and therefore cause the release of noradrenaline. Of course, it should be pointed out that as more blocking drug is taken up into the nerve fibre the conduction block may become more widespread.

The observations of Davey *et al*. (1968) can be interpreted in the light of, and in fact in support of, the 'local anaesthetic' theory, of adrenergic neurone blockade, as follows. It has been shown in the motor endplate that conduction block at regions of nerve fibre branching is more likely to occur if the action potential traffic passing through the region occurs at a higher rate (Krnjević & Miledi, 1959). It can therefore be argued that, while the preterminal branching region of the adrenergic neurones of the spleen can normally carry a traffic of action potentials at 30 cycles/sec, in a preparation partially narcotized under the influence of bretylium the safety factor for conduction across the branching region is reduced and branching block sets in at this frequency, although action

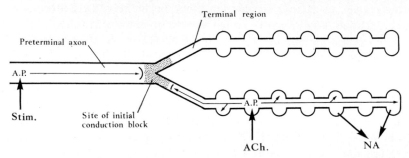

FIG. 5   Diagrammatic representation of the axonal portion of an adrenergic nerve fibre, to illustrate one theory of the mechanism of adrenergic neurone blockade. Electrical stimulation of the preterminal axon (Stim.) or application of acetylcholine (ACh) normally initiates action potentials (A.P.) which propagate into the varicose terminal regions, causing the release of noradrenaline (NA). As an adrenergic neurone blocking drug is accumulated in the nerve fibre, it exerts a local anaesthetic action which initially causes a conduction block in the preterminal region of branching (stippled). At this stage, orthodromically conducted action potentials elicited by stimulation of the pre-terminal axon cannot pass the branching region and therefore do not cause noradrenaline release. But acetylcholine, acting distally to the point of branching, still initiates action potentials conducted through the terminal region and therefore still causes transmitter release.

potentials arriving at lower frequencies, being less able to produce branching block, can still cross the region of branching and activate the transmitter release mechanisms in the terminal regions. A corollary of this interpretation of the nature of adrenergic neurone blockade is that the drugs should prevent anti-dromic firing elicited by acetylcholine before preventing the sympathomimetic actions of acetylcholine. This would, of course, hold true only if the acetylcholine acted solely peripheral to the region of preterminal branching and not proximal to it. In fact, Cabrera *et al*. (1966) found that both guanethidine and bretylium, in small doses, prevented the antidromic firing induced by acetylcholine in the sympathetic innervation of the heart, an observation which agrees well with the theory. On the other hand, Davey *et al*. (1968) found that antidromic firing persisted after bretylium had abolished the release of noradrenaline from the spleen by acetylcholine, an observation which neither agrees nor disagrees with the theory.

This theory of vagal cardiostimulation, in its full form, requires that acetylcholine released from parasympathetic neurones can reach the terminal regions of sympathetic post-ganglionic adrenergic neurones in the heart at a point distal to the site of the initial blockade produced by adrenergic neurone blocking drugs. By acting on the terminals, acetylcholine initiates action potentials, which cause a normal release of noradrenaline from the terminal regions.

One point which makes this theory attractive is, curiously enough, the inability of many workers to observe truly vagal stimulation of the heart, for it has now been found that acetylcholine applied to the heart, as well as stimulating nicotinic receptors which promote the release of noradrenaline from sympathetic fibres, also stimulates muscarinic receptors on the adrenergic neurones which antagonize the release process (Lindmar, Löffelholz & Muscholl, 1968). In fully atropinized hearts, acetylcholine can only activate the nicotinic receptors and therefore always causes the release of noradrenaline. But if the amount of atropine applied is too low, the muscarinic effects of acetylcholine start to predominate and release of noradrenaline is prevented. Such an effect of atropine on the sympathomimetic cardiac actions of acetylcholine has been reported by Barnett & Benforado (1966). It therefore follows that the presence or absence of a vagal cardiostimulant response by the mechanism proposed above would depend entirely on the amount of atropine with which the experimental animal or isolated organ was treated. In this respect, vagal cardiostimulation via the sympathetic adrenergic fibres in atropinized animals would seem to be entirely artifactual. However, there is still the observation that stimulation of certain roots of the vagi can cause cardio-acceleration even in animals not treated with atropine. The responses in non-atropinized animals might be due solely to the release of catecholamines from the chromaffin tissue source. Alternatively, it might be that the nicotinic and muscarinic receptors on the sympathetic adrenergic neurones are found in anatomically discrete regions of the fibres and that acetylcholine released from parasympathetic nerves has access only to the region containing the nicotinic, secretion-promoting receptors.

Summary. Four theories to explain the adrenergic positive inotropic and chronotropic actions of the vagi on the heart have been discussed. There is clear evidence that much of the effect observed when the thoracic vagi are stimulated is mediated by adrenergic fibres of sympathetic origin which run with the vagi. The remaining effect could be due to parasympathetic adrenergic nerve fibres, but is more likely to be due to a release of catecholamines from chromaffin tissue innervated by parasympathetic cholinergic neurones or to a release from the terminal regions of sympathetic adrenergic neurones activated by acetylcholine from parasympathetic fibres, or to release from both sites. The possibility is raised that parasympathetically mediated release from sympathetic adrenergic neurones can only occur when the preparation has been treated with muscarinic blocking drugs and is therefore completely artifactual.

Effects not mediated by catecholamines

Stimulation of parasympathetic nerves can cause responses of glands and smooth muscles which do not appear to be mediated by either acetylcholine or a catecholamine. The example to be considered here, the inhibition of gastro-intestinal muscle by parasympathetic nerves, is an example of a response going in a direction opposite to that of responses to acetylcholine and pharmacological evidence has been presented that the transmitter substance is not a catecholamine.

Parasympathetic inhibition of gut muscle. The classical response of the musculature of the gastro-intestinal tract to stimulation of the cranial or sacral parasympathetic nerve supply is a contraction. The sympathetic innervation of the gut is, classically, inhibitory. But it has been known for many years that stimulation of the vagus nerve can cause a relaxation of the stomach (e.g. Langley, 1898b). The response is best seen when the excitatory effect of stimulation, mediated by cholinergic nerve fibres, has been eliminated by treatment with atropine (May, 1904; McSwiney & Robson, 1929; Greeff et al., 1962), when the stomach muscle has a well-developed tonus (McCrea, McSwiney & Stopford, 1925; McSwiney & Wadge, 1928; Harrison & McSwiney, 1936), or when the vagi are stimulated with electrical pulses longer or stronger than those needed to cause excitatory responses (Veach, 1925; Martinson & Muren, 1963). The nerve fibres involved in the gastric inhibitory response are largely, if not completely, of parasympathetic origin and are not admixed sympathetic fibres, for stimulation of various structures in the brain of cats and dogs causes gastric inhibition mediated by the vagi (Eliasson, 1952; 1954; Hesser & Perret, 1960; Semba, Fujii & Kimura, 1964). Although vagal inhibitory effects on the small intestine and inhibition of the large intestine caused by stimulation of the pelvic nerves have also been reported, it is still not clear whether the responses are mediated by the same sort of nerve fibres as those causing gastric inhibition. Parasympathetic inhibitory effects on regions of the gut other than the stomach will therefore be considered on p. 492.

Until recently, the gastric inhibitory responses to stimulation of the vagus nerves were considered to be mediated by parasympathetic adrenergic nerve fibres. It has now become clear that the inhibitory nerve fibres are not adrenergic, but the nature of the transmitter substance which they release is still unknown. Much of the evidence for the non-adrenergic nature of the fibres has come from experiments in which the gut was stimulated transmurally, thus eliciting activity in nerve fibres running in the gut wall. These results will be considered firstly, before returning to the vagal responses. The first real indication that the gut wall contains inhibitory nerve fibres which are not adrenergic came from the work of Paton & Vane (1963) on the stomachs of cats, rabbits, rats and guinea-pigs. They found that the inhibitory response of the stomach to transmural stimulation, after excitatory responses had been abolished with hyoscine, was not prevented by treatment with adrenergic neurone blocking drugs. In contrast, these drugs completely inhibited the relaxant effects of stimulation of the perivascular sympathetic nerve supply to the stomach. Similar relaxations resistant to adrenergic neurone blockade have since been obtained from transmurally stimulated preparations of the small and large intestines of a number of mammals, including man (Burnstock, Campbell, Bennett & Holman, 1963a; 1964; Bucknell, 1964; Holman & Hughes, 1965a; Burnstock, Campbell & Rand, 1966; Day & Warren, 1967; 1968; Bianchi, Beani, Frigo & Crema, 1968). Paton & Vane (1963) originally ascribed this result to the stimulation of sympathetic adrenergic nerve fibres in the gut wall, at a site peripheral to the locus of action of the adrenergic neurone blocking drugs, as has been outlined above for the heart (see Fig. 5). But further evidence has shown that the nerve fibres involved are not the same as the fibres, the preterminal axons of which are stimulated in the perivascular sympathetic supply to the gut.

In an electrophysiological study of the innervation of the guinea-pig taeniae coli, the three bands of longitudinal muscle which are spaced around the circumference of the caecum, it was found that single electrical pulses passed across the taenia caused a hyperpolarization of the smooth muscle membrane (Burnstock et al., 1963a, b; 1964; Bennett, Burnstock & Holman, 1963; Suzuki & Inomata, 1964). The hyperpolarization occurred after a latent period of some 150msec and lasted in the order of 1 sec. The hyperpolarizing response, termed an inhibitory potential, interrupts spontaneous action potential firing and therefore causes a relaxation of the smooth muscle. When repeated stimuli are used, even at low frequencies of stimulation, the individual inhibitory potentials summate and the resulting hyperpolarization may be very large (see Fig. 6a). Hyperpolarizations which are almost certainly inhibitory potentials can be seen in the published records of an earlier investigation (Bülbring, Burnstock & Holman, 1958), and the responses have now been described in greater detail (Bennett, Burnstock & Holman, 1966b). Although the possibility was raised that the inhibitory potential is a direct response of the smooth muscle membrane to electrical stimulation (Bülbring, Kuriyama & Tomita, 1965), later studies showed that the response is neurogenic (Bülbring & Tomita, 1966, 1967). Inhibitory potentials have now been recorded from the smooth muscle of the guinea-pig small intestine (Kuriyama, Osa & Toida, 1967c).

It seems clear that the inhibitory potentials are caused by stimulation of the same nervous structures which mediate relaxant responses to transmural stimulation of the gut after treatment with adrenergic neurone blocking drugs. Bennett, Burnstock & Holman (1966b) found that the characteristics of the inhibitory potential were completely unaffected by treatment with guanethidine or bretylium.

The significance of the inhibitory potential for the present discussion is that it shows that the gut wall contains neural structures which are able to cause an intense inhibition of muscular activity when they are stimulated with low frequencies, or even with single pulses. In contrast, stimulation of the perivascular sympathetic nerve supply to the gut with single pulses causes no observable change in smooth muscle membrane activity; even when the nerves are stimulated repeatedly, the muscle membrane is only slowly and weakly hyperpolarized in both rabbit colon and guinea-pig taenia coli (Gillespie, 1962a; Bennett, Burnstock & Holman, 1966a; Fig. 6b). These differences between the inhibitory transmission processes are reflected in differences in the mechanical responses

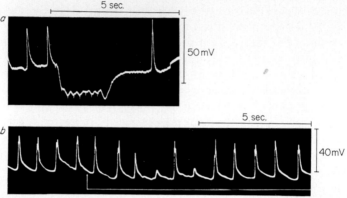

FIG. 6   Effects of transmural stimulation, in a, and of stimulation of the perivascular inhibitory nerves, in b, on the membrane potential and action potential firing recorded with micro-electrodes from smooth muscle cells of the guinea-pig taenia coli. Stimulation, in both panels, with 4 pulses/sec, marked by stimulus artefacts in a and by the arrow and line in b. Pulse duration 200 μsec. The micro-pipette was dislodged from the cell at the end of a. (From Bennett, Burnstock & Holman, 1966a, b)

of the atropinized taenia coli to stimulation of the extrinsic sympathetic innervation and of the intramural nerves (Burnstock et al., 1964; Burnstock et al., 1966). Relaxant responses to stimulation of the sympathetic perivascular nerves are rather slower in onset and develop less rapidly than responses to transmural stimulation. Stimulation of the perivascular nerves with low frequencies is far less effective in causing relaxation than stimulation of the intramural nerves; whereas the responses to perivascular stimulation do not become maximal until the stimulation frequency is raised to about 30 pulses/sec, the responses to transmural stimulation may reach their maximum at frequencies as low as 5 pulses/sec. The response to perivascular nerve stimulation at all frequencies is abolished by treatment with adrenergic neurone blocking drugs (Fig. 7a). However, these drugs cause a reduction of only the responses to higher frequencies of transmural stimulation while the responses to low frequencies of stimulation are virtually unaffected by adrenergic neurone blockade (Fig. 7b).

Now, Paton & Vane (1963) suggested that the inhibitory responses to transmural stimulation which are not prevented by adrenergic neurone blocking drugs might be caused by the stimulation of sympathetic adrenergic nerve fibres peripheral to the point of action of the blocking drugs. But it has been shown that the relaxations obtained by transmural stimulation, even during adrenergic neurone blockade, are associated with far more intense phenomena of inhibitory neuro-muscular transmission than are those obtained by stimulation of sympathetic adrenergic nerves at a point remote from their terminations. If the theory of Paton and Vane were correct, it would mean that the direct stimulation of adrenergic nerve terminals is more effective in releasing transmitter substance than is invasion of the terminal regions by orthodromically propagated action potentials. Specifically, in view of the ineffectiveness of single orthodromic volleys compared with the intense transmission phenomena noted on transmural

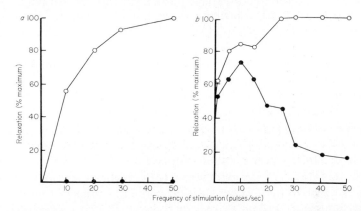

FIG. 7   Frequency-response curves for stimulation of the perivascular innervation, in *a*, and for transmural stimulation, in *b*, of the atropinized, isolated taenia coli of the guinea-pig. The curves obtained before (○) and after (●) treatment with guanethidine (10^{-6} g/ml) are shown. Note that low frequency stimulation applied transmurally is more effective in causing relaxation than stimulation of the perivascular adrenergic nerves. Guanethidine abolishes the effects of perivascular nerve stimulation, but has very little blocking action against responses to low frequency transmural stimulation.
(From Burnstock, Campbell & Rand, 1966)

stimulation with single pulses, the theory would imply that there is, in normal adrenergic nerve fibres, a region of almost complete preterminal conduction blockade, which is highly unlikely. It therefore appears that the inhibitory phenomena caused by transmural stimulation after adrenergic neurone blockade must be mediated by nervous elements other than the peripheral extensions of the sympathetic innervation. However, the evidence obtained with drugs like bretylium and guanethidine does not prove that the nerve elements, which will be called intramural inhibitory neurones in the following discussion, are non-adrenergic. For instance, there is a possibility that adrenergic neurone blocking drugs would be completely ineffective in preventing transmitter release from adrenergic fibres which do not branch in the preterminal region (see p. 482).

The argument that the intramural inhibitory neurones are not adrenergic relies on two lines of evidence. Firstly, the relaxations of intestinal preparations caused by stimulation of the perivascular innervation or by the application of catecholamines may be completely abolished by treatment with drugs which block α- and β-receptors for catecholamines, without preventing the inhibitory responses to transmural stimulation. (Bucknell, 1964; Day & Warren, 1967; 1968; Bianchi et al., 1968). Earlier reports that inhibitory responses of the stomach caused by transmural stimulation are reduced or abolished by a number of adrenaline blocking drugs (Greeff & Holtz, 1956; Greeff et al., 1962; Paton & Vane, 1963) could be explained by non-specific blocking actions of the drugs, a possibility which is often difficult to assess in pharmacological experiments. The evidence obtained with adrenaline blocking drugs is therefore consistent with the hypothesis that the intramural inhibitory neurones are not adrenergic. But the selective blocking action of the drugs does not in any way eliminate the possibility that the nerves are adrenergic. Experiments of this type have involved the use of competitive antagonists of adrenergic β-receptor stimulating drugs, and such drugs will cause less reduction of responses to high concentrations of catecholamines than of responses to low concentrations. We have already noted that the intramural inhibitory nerves produce a much more intense inhibition of the musculature than do the sympathetic nerves. If both types of nerve were adrenergic, the observed differences in transmission would demand that the intramural inhibitory nerves either release more transmitter per impulse or release the transmitter at a position nearer to the receptive surface of the muscle, resulting in the arrival of a larger concentration of catecholamines at the receptors. The adrenaline blocking drugs would therefore, under appropriate conditions, be expected to cause a selective reduction of the sympathetic responses without inhibiting the effects of transmural stimulation, even if the intramural inhibitory neurones were adrenergic.

The second, and more conclusive, line of evidence that the intramural inhibitory nerves are not adrenergic comes from studies made with the histochemical fluorescence technique for localizing catecholamines. It must be concluded that the cell bodies of the intramural inhibitory neurones lie within the gut wall; they cannot lie outside the gut, since fibres of this type have never been found in the many studies of the sympathetic innervation of the gut, and the only other pathways by which processes of the inhibitory nerve fibres might reach the intestine, the parasympathetic nerve trunks, appear to be entirely preganglionic (see p. 491). Therefore, if the intramural inhibitory nerves were adrenergic (Ambache, 1951), one would expect the histochemical technique to reveal fluorescent neurone somata in the enteric plexuses. But many histochemical studies of the adrenergic innervation of the gut wall have been made (Norberg, 1964; Tafuri & Raick, 1964; Hollands & Vanov, 1965; Jacobowitz, 1965; Åberg & Eränkö, 1967; Baumgarten, 1967; Bennett & Rogers, 1967; Gabella & Costa, 1967; Read & Burnstock, 1968) and there has been complete agreement that there are no adrenergic ganglion cells in the mammalian gut wall. The observations made also exclude the possibility, raised by Bucknell (1964), that the inhibitory effects of transmural stimulation during adrenergic neurone blockade are mediated by catecholamines released from chromaffin cells, for there are no chromaffin cells within the muscular layers. Only one possibility remains, that the intramural inhibitory neurones are not adrenergic and that therefore they cannot be detected by this histochemical technique.

The possibility may be raised that the inhibitory responses of the gut to transmural stimulation are mediated by sensory neurones in the enteric plexuses. The sensory neurones might conduct action potentials antidromically, causing a release of an inhibitory substance from the receptive processes, as is thought to occur in dorsal root antidromic vasodilator responses. There is no evidence to indicate whether the intramural inhibitory neurones do or do not also have a sensory function, but there is clear evidence that the neurones have a normal motor function. In studies of the taenia coli, it was found that stimulation of the nerves in a flap of caecal wall left attached to the longitudinal muscle strip causes a relaxation of the atropinized taenia (Burnstock *et al.*, 1966). Like the inhibitory responses to transmural stimulation of the taenia, these responses persisted after adrenergic neurone blockade had been established. But, unlike the responses to stimulation of the taenia, the relaxation caused by stimulation of the flap of caecal wall was reduced after treatment with pentolinium, a nicotinic receptor blocking drug. The results imply that some of the nerve fibres stimulated in the flap of caecal wall are cholinergic, making synaptic connections with intramural inhibitory neurones innervating the muscle of the taenia (Burnstock *et al.*, 1966). It is extremely unlikely that the cholinergic elements in the gut wall are transmitting their activity to the inhibitory neurones in the antidromic direction. It must therefore be concluded that transmission from the cholinergic neurones to the inhibitory intramural neurones is in fact the normal, orthodromic direction of conduction. The intramural inhibitory neurones must therefore be regarded as post-ganglionic neurones in autonomic nervous pathways, i.e. as motoneurones. Whether the inhibitory transmitter substance is released from axonal processes or from processes more properly regarded as dendritic does not affect this argument.

The nature of the transmitter substance released from the intramural inhibitory nerves is not yet known, but some speculation is possible. Most compounds which have been proposed as transmitter substances in one tissue or another may be eliminated, on the grounds that they excite or have no effect on muscle from most regions of the gastro-intestinal tract, even although some of them, e.g. some prostaglandins, do have inhibitory actions on specific regions of the gut of some animals. Other substances, such as 5-hydroxytryptamine, may be eliminated because their inhibitory effects on the gut appear to be initiated by stimulation of nerves in the gut wall and not to be due to a direct action on the smooth muscle. The list of substances ruled out in this way includes histamine, 5-hydroxytryptamine, γ-aminobutyric acid, the polypeptides bradykinin and substance P, and the prostaglandins. One group of compounds which would be well worth considering as the inhibitory transmitter substance is the group of nucleosides, nucleotides and their parent purines and pyrimidines. The nucleotide, adenosine triphosphate, has already been implicated in the transmission of dorsal root antidromic vasodilatation. The inhibitory action of the adenine ribonucleoside, adenosine, was noted forty years ago, in the first major study of the pharmacological actions of nucleic acid derivatives (Drury & Szent-Györgi, 1929). Other derivatives of this and other bases have since been shown to share the inhibitory action (for a review of the earlier work, see Drury, 1936). Some of the compounds cause inhibition of the gut when applied in quite low concentrations. For instance, adenosine causes relaxation of the isolated rabbit small intestine when applied in concentrations as low as 5×10^{-8} g/ml, a concentration well within the 'physiological' range accepted for other transmitter substances (Stafford, 1966). However, if the transmitter substance is a nucleotide, considerable difficulty may be found in proving the case. There are no all-purpose blocking or potentiating agents for the actions of the compounds, and nucleotides are known to be released from a variety of cells either as a specific or as a general concomitant of activation.

We can now return to the inhibitory effects of stimulation of the vagus nerve on the mammalian stomach. In view of the fact that the gut wall contains non-adrenergic inhibitory nerves, it might be expected that the vagal inhibitory pathways involve these neurones. This is in fact the case, as has been shown in three recent investigations of the innervation of the stomach in cats and guinea-pigs (Martinson, 1965a; Campbell, 1966; Bülbring & Gershon, 1967). It was found that treatment with adrenergic neurone blocking drugs could abolish the inhibitory effects of stimulation of the sympathetic adrenergic innervation of the stomach without causing any appreciable reduction of the gastric inhibition caused by vagal stimulation (Martinson, 1965a; Campbell, 1966; Fig. 8). Earlier reports that vagal inhibitory effects on the stomach could also be prevented by adrenergic neurone blocking drugs (Greeff et al., 1962; Paton & Vane, 1963) were confirmed, but it was shown that the blocking action of the drugs against the vagi in guinea-pigs was non-specific (Campbell, 1966). In addition, it was found that vagal inhibition of the stomach persisted after treatment with concentrations of adrenaline blocking drugs which abolished the response to stimulation of the sympathetic innervation (Martinson, 1965a; Bülbring & Gershon, 1967). Finally, it was shown that the vagal inhibition of the guinea-pig stomach, like the relaxations obtained by stimulation of the intramural inhibitory neurones in the gut wall, was better developed at low frequencies of stimulation than the sympathetic response (Campbell, 1966). The results make it obvious that the intramural inhibitory neurones in the stomach wall are connected with the neuraxis via the vagus nerve.

There is clear evidence that the fibres stimulated in the vagus nerve are preganglionic, making synaptic connections with the inhibitory neurones in the enteric plexuses. A number of studies have shown that vagal relaxation of the stomach is inhibited by treatment with nicotinic ganglion blocking drugs, indicating that the vagal pathways contain cholinergic preganglionic fibres. However, there has been some disagreement concerning the extent to which the inhibitory vagal responses are reduced. Greeff *et al.* (1962) and Paton & Vane (1963) found that the responses were abolished by hexamethonium, nicotine and a number of other antinicotinic drugs. But Bülbring & Gershon (1966; 1967) found that the competitive nicotinic blocking drug pentolinium reduced, but

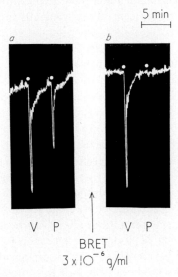

Fig. 8   Isolated, atropinized preparation of guinea-pig stomach with vagus and peri-vascular nerve supplies intact. Effect of the adrenergic neurone blocking drug bretylium. Stimulation of the vagi with 10 pulses/sec for 10 sec at V; stimulation of the perivascular nerves with 25 pulses/sec for 10 sec at P. Panel *a* shows control responses. In panel *b*, 46 min after the addition of bretylium (BRET, 3×10^{-6} g/ml), perivascular nerve effects are abolished but vagal effects are unaffected. Time marker, 5 min. (From Campbell, (1966)

never abolished, vagal relaxation of the guinea-pig stomach. Bülbring and Gershon also found that nicotine initially abolished the vagal effects but that, as the depolarizing phase of nictotine blockade was replaced by the competitive, non-depolarizing phase, there was a partial recovery of the inhibitory responses. They concluded that the vagal fibres were preganglionic, but that not all of the preganglionic fibres were cholinergic. The remaining synapses were operated by a transmitter substance other than acetylcholine.

In earlier studies (Gershon, Drakontides & Ross, 1965; Gershon & Ross, 1966*b*), mice had been injected with 5-hydroxytryptophane-^{3}H, the precursor of 5-hydroxytryptamine. The tritium label was then localized autoradiographic-ally and was found to lie in the region of preganglionic nerve terminal plexuses about ganglion cells in the enteric plexus. It was found that the label in the gut wall occurred as 5-hydroxytryptamine. In addition, a fluorescent histochemical study of the guinea-pig gut wall had been made by Tafuri & Raick (1964), who claimed that the fluorescence colour of the preganglionic terminals surrounding enteric ganglion cells was the typical yellow of 5-hydroxytryptamine reaction products. These lines of evidence therefore appear to indicate that 5-hydroxy-tryptamine occurs normally in some of the preganglionic nerve terminals in the enteric plexuses. However, it should be borne in mind that all of the other workers, listed above, who have made histochemical fluorescence studies of the gut wall claim that the fluorescent nerve terminals in Auerbach's plexus show the green colour of catecholamines, and not the yellow colour of 5-hydroxy-tryptamine. Also Taxi & Droz (1966) have presented electronmicroscopic autoradiographic evidence that the label from 5-hydroxytryptophane-^{3}H is taken up into apparently adrenergic axons in the rat vas deferens. But Bülbring & Gershon (1967) have tested, and have obtained evidence in support of, the theory that the non-cholinergic preganglionic nerve fibres in the vagal inhibitory pathway are tryptaminergic. They found that applications of 5-hydroxytrypt-amine caused relaxation of the stomach, mediated by nerves. They also found that a wide variety of blocking and desensitizing procedures eliminated both the inhibitory response to 5-hydroxytryptamine and the relaxation caused by stimulation of the vagi after establishing nicotinic blockade. In addition, they found that stimulation of the vagi caused the release of 5-hydroxytryptamine from the mouse stomach under conditions in which mucosal enterochromaffin cells would be moribund. Although the pharmacological evidence for vagal tryptaminergic nerve fibres supplying the intramural inhibitory neurones is very strong, a microspectrofluorometric study of the enteric nervous system should be undertaken to establish the true colour of the observed fluorescence in preganglionic nerve terminals.

It is more difficult to determine whether vagal inhibitory effects on the small intestine (Bayliss & Starling, 1899) are mediated by the intramural inhibitory neurones. One factor which is difficult to assess, and has not always been ade-quately controlled in experiments, is that stimulation of the vagi in non-atro-pinized animals will cause cardiac slowing, a fall in blood pressure and a com-pensatory secretion of pressor amines from the adrenal medulla; the secreted adrenaline alone might account for some of the observed inhibitory effects of vagal stimulation on the small intestine. However, intestinal inhibitory responses have been observed repeatedly in atropinized animals in which the vagi have no cardio-inhibitory effect, as in the original observations of Bayliss and Starling. One observation which makes it unlikely that the inhibitory responses are mediated by intramural inhibitory neurones is that vagal stimulation can still cause relaxation of intestinal segments transected at both oral and anal ends (Bunch, 1899). Under these experimental circumstances there is no pathway by which vagal fibres may reach the small intestine other than the route followed by vagal excitatory fibres, along the mesenteric blood vessels. If vagal inhibitory pathways follow the perivascular plexuses and involve non-adrenergic nerves, stimulation of the perivascular nerve supply in Finkleman preparations of the small intestine should cause relaxation even after treatment with both atropine and adrenergic neurone blocking drugs. In fact, after treatment with these drugs, the perivascular innervation is completely ineffective (Day & Rand, 1961; Bentley, 1962). The results might therefore be taken to indicate that there are typical adrenergic nerve fibres, whether sympathetic or parasympathetic, in the vagal supply to the small intestine. However, Kewenter (1965) has put forward a completely different explanation of the responses. He found, in contrast to the observations of other recent workers (e.g. Nakayama, 1965), that stimulation of the cervical vagi in cats did not cause an inhibition of activity in the small intestine. On the other hand, he found that stimulation of the intrathoracic vagus did inhibit intestinal activity, but that the response only occurred when the sympathetic innervation of the intestine was intact; the response was lost immediately after section of the splanchnic nerves. Kewenter proposed that the inhibitory responses are caused by the antidromic stimulation of afferent fibres in the vagi, which then, in some unexplained way, initiate activity on a central reflex arc, the efferent limb of which is the sympathetic adrenergic innervation of the gut. In view of the conflicting results, another investigation of the problem would be most welcome.

There is a similar confusion about whether the sacral parasympathetic outflow provides an inhibitory innervation to the large intestine via the pelvic nerves. Courtade & Guyon (1897) found that stimulation of the pelvic nerves in dogs caused contractions of the longitudinal muscle but relaxation of the circular muscle layer of the colon. Bayliss & Starling (1900) found that the initial effect of stimulation of the pelvic nerves on the colon in dogs was a relaxation, followed after some interval by a contraction. These results were not verified by Wells, Mercer, Gray & Ivy (1942), who found that both muscle coats showed only excitatory responses to stimulation. There is general agreement that the pelvic nerves do not mediate inhibition of the rabbit colon (Bayliss & Starling, 1900; Garry & Gillespie, 1955), but, to prove the point, the effects of the excitatory innervation should be abolished with atropine: unfortunately the cholinergic nerve fibres in the pelvic nerve supply to the rabbit colon are extremely resistant to blockade with atropine, and the enormous concentrations needed to achieve blockade could well prevent responses mediated by any type of nerve. In cats, Fülgraff & Schmidt (1963) were able to abolish the initial excitatory effects of pelvic nerve stimulation on the colon by treating the animals with atropine; after atropinization the initial response was a very slight fall of tone lasting throughout the period of stimulation. In guinea-pigs also it is easy to prevent the excitatory responses to pelvic nerve stimulation by atropinization, but inhibitory responses have not been observed (Rand & Ridehalgh, 1965), even in one study in which such inhibitory nerves were searched for specifically (Bianchi et al., 1968). There seems to be a good chance that there are inhibitory

17+

nerves in the parasympathetic supply to the colon, but that they occur only in certain species.

It will be extremely difficult to determine the extent of the distribution of neurones of the 'intramural inhibitory' type throughout the body. At the moment, we are able to recognize these neurones only because they react neither as adrenergic nor as cholinergic nerves. It seems reasonable to assume that all of the non-adrenergic, non-cholinergic inhibitory responses caused by stimulation of nerves in any region of the gut are mediated by one type of neurone only. But, if one found inhibitory neurones with the same two negative qualities in any other organ, it would be extremely rash to assume that the fibres release the same transmitter substance as the intramural inhibitory neurones. For instance, evidence has been presented that there are inhibitory nerve fibres in the rabbit portal vein (Hughes & Vane, 1967) and in the dog retractor penis (Luduena & Grigas, 1966) which are neither adrenergic nor cholinergic. In both cases the anatomical origin of the fibres involved is not known in detail, so we cannot even draw an anatomical parallel between these nerve fibres and the vagal fibres mediating gastric inhibition. Even in cases where the origin of the fibres is known, there may still be little that can be said about the nature of the fibres. For instance, Dorr & Brody (1967) have produced evidence that the parasympathetic fibres which cause the vascular changes of erection of the penis in dogs are neither cholinergic nor adrenergic. While we can draw the parallel between the parasympathetic nature of both the penial fibres and the gastric fibres, we are unable to say whether the parallel is valid since we have no evidence as yet that the pelvic parasympathetic outflow contains the non-adrenergic type of intestinal inhibitory fibres. In other organs again, it may be even more difficult to decide whether fibres are of the intramural inhibitory type or not for, since we do not know what the transmitter substance is, we cannot determine what intramural 'inhibitory' nerves should do to the tissue, i.e. they could be excitatory. For instance, it is well known that the pelvic nerves provide an excitatory innervation to the urinary bladder, but that excitatory responses are barely reduced by treatment with atropine. The normal modern view is that the excitatory innervation is cholinergic, but that transmission is for some reason resistant to the action of atropine (e.g. Huković, Rand & Vanov, 1965). However, it is possible, as was originally suggested by Henderson & Roepke (1934), that the pelvic nerve supply to the urinary bladder is non-cholinergic, and one may suspect that the nerves are of the 'intramural inhibitory' type.

On the other hand, some slight progress now appears to have been made in elucidating the distribution of 'intramural inhibitory' nerves in organs, other than the gastro-intestinal tract, innervated by the vagi. Recent studies on the composition of the vagus nerve in amphibians (personal unpublished observations) have shown, firstly, that the vagus nerve provides only inhibitory fibres to the stomach of a toad. The effects of stimulating the vagal fibres are not prevented by treatment with atropine or adrenergic neurone blocking drugs. It may therefore be reasonably argued that the vagal inhibitory fibres to the toad stomach are of the same type as the vagal inhibitory fibres to the mammalian stomach, i.e. non-adrenergic, non-cholinergic. Secondly, it has been found that the amphibian vagus nerve provides only inhibitory nerve fibres to the general musculature of the lung, and that the inhibitory effects of this nerve supply are also not affected by atropine or bretylium. It is highly likely that the inhibitory innervation of the lung is of the same type as the inhibitory innervation provided, in the same nerve trunk, to the stomach, i.e. 'intramural inhibitory'. It is reasonable to expect that a similar vagal innervation of the mammalian lung also exists. In fact there have been a number of reports of bronchodilator responses to stimulation of the vagi in mammals (see Widdicombe, 1963), but no decisive pharmacological studies of the responses have yet been made. It can be seen that acquiring information about the extent of the 'intramural inhibitory' system by this comparative method is a laborious and, in the final analysis, unsatisfying procedure.

CONCLUSION

The chapter has had two aims. Firstly it is hoped that workers entering the field of the autonomic nervous control of smooth and cardiac muscle and of glandular tissue will have seen that the rules governing the construction of the autonomic nervous system, as envisaged by Langley fifty years ago, at least with regard to the post-ganglionic neurones innervating the effector organs, are not rigid but are generalizations. Too often in the past, deviations from Langley's rules have been regarded as the result of accidental mingling of sympathetic and parasympathetic nerve fibres, but a study of deviant conditions may still be very rewarding in terms of the discovery of new types of neurones or of new levels of complexity in the interactions between nerve fibres. Secondly it is hoped that, by giving extended discussions of a few real or apparent deviations from Langley's rules, some indication has been given of the complexities envisaged by people currently working on the autonomic nervous system. While many of the arguments may eventually prove to be invalid, the existence of such arguments is evidence of the revival of interest in the autonomic nervous system which has occurred over the last few decades.

16

THE IDENTIFICATION OF NEUROTRANSMITTERS TO SMOOTH MUSCLE*

M. D. GERSHON

INTRODUCTION

For the neural control of smooth muscle, acetylcholine and noradrenaline are generally assumed to be the main transmitters involved. Most recent studies,

*Supported by National Institutes of Health Research Grant NB 07436.

therefore, have not questioned whether these or other substances are transmitters, but have focused on the details of such aspects as synthesis, storage, release and actions of noradrenaline or acetylcholine. With such general acceptance of these substances as neurotransmitters, it is time to re-examine the development of the hypothesis of chemical transmission and the evidence implicating noradrenaline and acetylcholine as the actual neurohumours.

DEVELOPMENT OF THE THEORY OF CHEMICAL TRANSMISSION

The neurone doctrine

The idea that a humoural agent might be responsible for transmission between autonomic nerves and smooth muscle effector cells evolved from the concept that the nervous system is not syncitial. This view, that neurones are individual cells whose cytoplasm is continuous, neither with that of other neurones nor with that of cells of effector organs such as muscle, developed together with, and partly as a result of, the related view that specialized points exist within nervous tissue at the junctions between neurones or where nerve and muscle meet. Later, morphological evidence strengthened the concept of the individuality of neurones. The anatomical non-continuity of neurones raised the question as to how information could be transferred from one cell to another, between neurones or between nerve and muscle. The problem was seen to involve at most synapses the transfer of information across a synaptic gap.

That a specialized area exists between neural tissue and skeletal muscle was first shown over a century ago by the classical experiments of Claude Bernard (1856) with curare. He showed that the effect of curare was to prevent passage of excitation from nerve to skeletal muscle although either nerve or muscle alone could still function. Thus the action of curare seemed to be directed at the junctional region indicating that this region must have properties different from that of the remainder of either the nerve or muscle cells. These experiments suggested discontinuity between nerve and muscle but did not prove it. In fact, the final establishment of the individuality of neurones has come about relatively recently.

The present view of the nervous system emerged from an, at times, acrimonious debate between holders of two opposing theories. The reticular theory, which viewed the nervous system as a complex anastomosing netlike structure, with cells at the internodes, but all in continuity, was proposed by von Gerlach (1872) and was long championed by Golgi (1885, 1890, 1891) and Held (1905, 1909). The reticular theory was opposed by His (1886, 1889), Forel (1887) and, most importantly, by Ramón y Cajal (1888, 1890a, 1890b, 1890c). The theory, so elegantly supported by Ramón y Cajal, of the independence of nerve cells, came to be known as the neurone theory or the neurone doctrine (Waldeyer, 1891). Synaptic contacts were recognized and a great deal of evidence accumulated in support of the neurone theory (Ramón y Cajal, 1909, 1911). By the second decade of the twentieth century, the neurone doctrine had achieved wide acceptance, and the question it raised about cell-to-cell transfer of information was under extensive investigation. However, some anatomists continued to believe in the reticular theory and support of the reticular theory was not finally dispelled until after the impressive weight of evidence for the neurone doctrine was assembled convincingly in the excellent reviews of Ramón y Cajal (1934), Nonidez (1944) and Bodian (1942, 1952). By the early 1950s the neuronal theory could be considered established. Yet because of the limited resolution of the light microscope, a true synaptic gap between neurones, or between neurones and skeletal or smooth muscles had not been conclusively demonstrated. In fact, there was generally thought to be no gap at all, but a single membrane between pre- and post-synaptic structures (Bodian, 1952). The advent of electron microscopy led to the determination that there was indeed a gap between neurones, of about 200 Å units in width (although material has been found in the gap between some neurones, and regions of very close apposition or actual fusion of membranes found between other neurones). Moreover the pre- and post-synaptic membranes were resolved, and were found to be separate membranes about 75 Å wide (Palade & Palay, 1954; Palay, 1956a). The characteristics of the vertebrate neuromuscular junction were found to be similar (Robertson, 1956). A still larger gap was found at nerve-smooth muscle junctions (Richardson, 1962, 1964). Thus the electron microscope not only provided the resolution necessary to confirm neuronal individuality, but revealed the existence of a gap across which the transmission process must operate.

Theories of neurotransmission

Since neurones are not in continuity with one another or with effector cells, some means must exist to transmit information across the gap between them. In general there have been two views as to how this is accomplished. According to one view the mechanism of transmission from cell to cell is similar to the mechanism of conduction along the fibre and depends on the spread of electrical current across the synapse. According to the other view a special chemical is released in response to the arrival of the nervous impulse at the axonal ending and this special chemical diffuses across the synaptic gap and excites (or inhibits) the post-synaptic membrane. Although both forms of transmission have been found to occur, the chemical process has been found to be the more general one in the vertebrate nervous system (see Eccles, 1964; Katz, 1966). Chemical mediation is compatible with the presence of a synaptic gap. The diffusion of a chemical from an axonal terminal across the gap to a sensitive post-junctional membrane can also account for the unidirectional polarity of synapses which was noted to be a synaptic property by Sherrington (see review by Liddell, 1960).

Chemical transmission at autonomic neuroeffector junctions

Although there had been an earlier suggestion by DuBois-Reymond in 1888 that neurotransmission might be chemical (see review by Dale, 1937a), the first clear evidence for the view came from the description by Lewandowsky (1899) of the effects of the administration of extracts of the adrenal glands. This work was followed by the work of Elliott (1904a, 1905) who not only studied the effects of adrenal extracts, but also worked out in detail the similarities between the stimulation of sympathetic nerves, the administration of adrenal extracts and the administration of adrenaline. On the basis of these observed similarities, Elliott suggested that sympathetic nerves might act through the liberation of adrenaline. Elliott's suggestion impressed Dixon, who had noted the resemblance between the actions of alkaloids like muscarine and the effects of vagal stimulation (Dixon & Brodie, 1903). Dixon (1906, 1907) tried to extend Elliott's concept to the parasympathetic nervous system, although, at the time, no specific transmitter was known. Dixon (1907) made an attempt to extract the hypothetical parasympathetic transmitter substance. He analysed extracts made from dog heart, removed from animals during intense vagal inhibition. His extract was said to be able to cause the inhibition of an isolated frog's heart, and was, like the vagus, susceptible to atropine blockade.

Thus, two of the critical points in establishing the identity of transmitter substances had already evolved by 1906, one, the observation of similarity (or identity) of effect between neural actions and those of application of the suspected transmitter substance, and the other, the attempt to extract the suspected substance and to demonstrate its release by nervous stimulation. The limitations of these observations should become clear when it is realized that adrenaline is not the sympathetic transmitter, as suggested by Elliott, and that Dixon probably extracted choline (see Dale, 1935) rather than acetylcholine from the dog's heart. Nevertheless, these experiments were important because they pointed in the direction that events were to follow. The use of atropine by Dixon also foreshadowed the use of 'specific blocking agents' in pharmacological analysis.

The implication of acetylcholine

The next significant insight into the nature of chemical transmission came in 1914 when Dale published his extensive studies of the action of a variety of esters and ethers of choline. This study established acetylcholine as the most likely substance to be the parasympathetic transmitter (rather than muscarine as proposed by Dixon, 1906). Actually, the intense cardio-inhibitory and vasodepressor action of acetylcholine had been shown years previously by Hunt & Taveau (1906) who found that acetylcholine was about 1000 times as potent at slowing the heart and lowering the blood pressure as choline itself. But it was Dale who noted how closely acetylcholine mimicked parasympathetic actions, just as adrenaline did those of sympathetic nerves. Dale also noted the evanescence of the actions of acetylcholine and adrenaline (less so), each reproducing, 'those effects of involuntary nerves which are absent from the action of the other, so that the two actions are in many directions complementary and antagonistic'. There was, however, no clear evidence that either of these substances were contained within nerve endings or were released on nerve stimulation. The acceptance of acetylcholine and adrenaline as mediators awaited this demonstration. For adrenaline (in mammals) the demonstration was not to be forthcoming but for acetylcholine the demonstration of its release came in 1921 when Otto Loewi published the first of his marvellously simple experiments with the vagus and sympathetic nerves to the frog's heart (1921a, b; 1924; Loewi & Navratil, 1926a, b). Loewi provided the first good evidence that nerves actually released a chemical able to exert effects at a distance. He found that the perfusion fluid obtained from isolated hearts of frogs or toads during stimulation of the vagus could inhibit the beat of the heart when the fluid was returned to the heart from which it was originally collected. When fluid was collected similarly, but the vagus was not stimulated, the fluid did not have the ability to slow the heart. The cardio-inhibitory properties of the fluid obtained during stimulation were blocked by atropine. Loewi (1921b) compared the material obtained from the perfused, stimulated heart, a material he called 'Vagusstoff', with acetylcholine and with choline and concluded that choline itself had too weak an action to be the inhibitory transmitter. Witanowsky (1925) described many of the chemical properties of 'Vagusstoff' and many investigators were able to further refine Loewi's techniques and confirm his findings (Kahn, 1926; Bain, 1933). Brinkman & van Dam (1922) extended Loewi's observations to show that the fluid collected from a frog's heart during vagal stimulation not only slowed heart rate but could also induce a contraction of a frog's stomach. This further observation was highly significant because it meant that 'Vagusstoff' not only possessed an activity reminiscent of vagal effects on the heart, but it also mimicked parasympathetic effects in general.

Further contributions were made by Loewi & Navratil (1926a, b) who found that the properties of 'Vagusstoff' were indistinguishable from those of acetylcholine. Activity could be destroyed by mild alkaline hydrolysis and restored (to even more than control activity) by acetylation. Moreover, although atropine abolished vagal effects on the heart, it did not antagonize the release of 'Vagusstoff'. Loewi was also able to find an esterase in heart tissue which inactivated 'Vagusstoff', which could then be re-activated by acetylation. Finally, Loewi and Navratil also showed that eserine potentiated both acetylcholine and 'Vagusstoff' and this action was probably due to inhibition of the esterase. The presence of the esterase in blood was shown by Engelhart & Loewi (1930) and by Matthes (1930) and thus the rapid inactivation of acetylcholine (Galehr & Plattner, 1928a, b) was accounted for. The esterase of blood could also be inhibited by eserine.

The demonstration of the action of eserine was of immense value because it permitted the extension of Loewi's kind of experimentation to mammals in which the blood and tissues usually contain a great deal of esterase activity. Thus the use of eserine made possible the study of stimulation of a variety of parasympathetic nerves (Dale & Gaddum, 1930; Feld-berg, 1933) and the study of the liberated material by parallel bioassay (see Dale, 1937a, b).

Evidence had accumulated therefore, which established the release of a parasympatho-mimetic substance on stimulation of parasympathetic nerves and there was evidence suggest-ing it was a choline ester with properties similar, if not identical, to acetylcholine. Until 1929, however, when acetycholine was found in the spleen of the horse and the ox by Dale and Dudley, an ester of choline had not been identified in tissues. Once found, there seemed to be no reason to object to acetylcholine as the mediator of the muscarine-like (or muscar-inic), parasympathetic effects. It remained to demonstrate the universality of acetylcholine as a mediator throughout the parasympathetic nervous system. Acetylcholine release was soon detected after stimulation of a variety of parasympathetic nerves and, impressed by the accumulating evidence, Dale proposed the division of autonomic nerves into those which released acetylcholine and those which released adrenaline (see Dale, 1937a).

THE CASE FOR ACETYLCHOLINE

These early studies by Loewi, Dale and others, have had a remarkably persuasive effect and there has been relatively little criticism of them. In retro-spect, however, it does seem that although elegant, these studies alone are not entirely adequate to establish acetylcholine as a transmitter. The case for acetylcholine, made in these studies, rests basically on four points:

1. The similarity or identity of action between acetylcholine and parasympathetic nerve stimulation.
2. The antagonism of both neural effects and those of exogenous acetylcholine by atropine (see Dale, 1914).
3. The potentiation of both neural effects and those of exogenous acetylcholine by eserine or other anticholinesterases.
4. The release of an acetylcholine-like substance upon parasympathetic nerve stimulation.

The great strength of the case lies in the association of all four points. Each of these points, considered individually can be challenged. They will be discussed in the following sections.

The effect of acetylcholine on smooth muscle

Both applied acetylcholine and parasympathetic nerve stimulation have similar effects on smooth muscle. In the consideration of acetylcholine as a neurotransmitter, it is important to determine whether acetylcholine, in mimicking neural effects, is doing so by acting *post-junctionally*. Acetylcholine, acts not only on post-synaptic surfaces but on pre-synaptic terminal axons as well. The demonstration of this action of acetylcholine has been made at motor nerve terminals in skeletal muscle (Hubbard & Yokota, 1964; Riker, 1966) but the effect seems to be a general one on most unmyelinated axons (Gray, 1959; Armett & Ritchie, 1960; Dettbarn & Davis, 1963; Paintal, 1964). It thus seems likely that acetylcholine would have a similar action at autonomic terminals. It has been found to activate splenic nerve terminals giving rise to sympathomimetic effects (Ferry, 1963). Since, in the application of acetylcholine to inner-vated smooth muscle preparations, the drug comes in contact with axon terminals as well as with the muscle, it is conceivable that part or even all of its action could be indirect and nerve mediated. This consideration, emphasized for the effect of acetylcholine on skeletal muscle by Riker (1966), makes it critical to establish the action of acetylcholine on nerve free preparations or on preparations in which neuronal elements have been inactivated. Chronically denervated preparations are not really adequate for this purpose because, after denervation, changes may occur in the muscle.

Recently, direct evidence has been provided which indicates that no part of the excitatory action of exogenous acetylcholine on smooth muscle actually is nerve mediated. In the guinea-pig ileum it is possible to separate the longitudinal muscle from the myenteric nerve plexus (Paton & Zar, 1965, 1968). When this is done, the resulting muscle strip does not release acetylcholine and is inexcitable by electrical stimulation with pulses of brief duration (Paton & Zar, 1965, 1968). The preparation appears nerve free, both by light and electron microscopy (see also Paton, 1964). The sensitivity of this preparation to acetylcholine has not been decreased by denervation relative to other drugs (Paton & Zar, 1965, 1968). Moreover, other studies have shown that the sensitivity of intestinal smooth muscle to acetylcholine is not reduced either by inactivation of neural elements by botulinum toxin (Ambache & Lessin, 1955) or by tetrodotoxin (Gershon, 1967). This recent evidence thus makes it possible to conclude that at least in some smooth muscle systems, such as the gut, acetylcholine does exert its parasympathomimetic action directly on the smooth muscle itself; the action not being influenced by concurrent stimulation of neural structures. This is probably not because neuronal elements, particularly intestinal ganglia, are insensitive to acetylcholine; but rather that intestinal smooth muscle itself is so exquisitely sensitive (Paton, 1957). For instance, Bülbring & Gershon (1967) found that concentrations of 10^{-6} to 10^{-5} g/ml were required to produce gang-lionic effects with acetylcholine in the isolated stomach of the guinea-pig. On the other hand, one tenth of this concentration is usually sufficient to produce a maximal contracture of the intestinal musculature.

The anticholinergic action of atropine

With respect to the ability of atropine and related drugs to antagonize the actions of acetylcholine and parasympathetic nerve stimulation, difficulties again arise. It is far easier to antagonize exogenous acetylcholine than it is to block the action of parasympathetic nerves and some parasympathetic effects are quite resistant to atropine. This latter point has posed a problem for a long time (Dale, 1935, 1937a, b; Henderson & Roepke, 1937; Ambache, 1955). The resistance of parasympathetic effects to atropine blockade seems not to be all or none, but some parasympathetic effects are more susceptible to atropine than others. Vagal cardio-inhibition is more sensitive to atropine than is vagal

17*

stimulation of the stomach, and vagal excitation of the stomach appears to be more sensitive to atropine (McSwiney & Robson, 1929) than does excitation of the small intestine (Bayliss & Starling, 1899; Cushny, 1910; Henderson, 1923). The parasympathetic effect most resistant to atropine appears to be excitation of the urinary bladder (Henderson & Roepke, 1934).

It is difficult to explain the failure of atropine to block some parasympathetic responses. All parasympathetic effects, even those resistant to atropine, are potentiated by eserine. Moreover, not only atropine sensitive parasympathetic effects such as the effect of stimulation of the vagus nerves to the heart (Loewi, 1921a, b), stimulation of the oculomotor nerve to the pupil (Engelhart, 1931) or of the chorda tympani to the salivary glands (Babkin, Stavraky & Alley, 1932; Gibbs & Szelöczey, 1932; Henderson & Roepke, 1933; Secker, 1934) are associated with the release of an acetylcholine-like substance. The effects of stimulation of vagus nerves to the stomach (Dale & Feldberg, 1934; Paton & Vane, 1963) and electrical stimulation of nerves in guinea-pig ileum myenteric plexus (Zar, 1966; Paton & Zar, 1968) which require more atropine for their abolition are also associated with acetylcholine release. In addition, even the highly atropine-insensitive effect of stimulation of parasympathetic nerves to the urinary bladder is associated with acetylcholine release (Carpenter & Rand, 1965). Furthermore, the acetylcholine released on electrical stimulation of the bladder is entirely derived from post-ganglionic axons since it is abolished after degeneration of nerves distal to bladder ganglia. When acetylcholine is no longer released on electrical stimulation, as after degenerative nerve section (Carpenter & Rand, 1965), contractile responses can no longer be elicited by electrical stimulation. Bladder contraction in response to electrical stimulation therefore seems dependent upon the neural release of acetylcholine. In vitro, bladder responses to transmural (post-ganglionic nerve) stimulation are greatly potentiated by eserine (Carpenter, 1963). Thus, acetylcholine does appear to be the neurotransmitter of excitation from pelvic post-ganglionic nerves to the bladder, despite the almost absolute resistance of this effect to atropine. The inability of atropine to block a neural response therefore, cannot be used as it has in the past (Henderson & Roepke, 1934), as evidence against the participation of acetylcholine.

The resistance of the bladder to atropine blockade of parasympathetic excitation is an example of atropine-resistance in an apparently cholinergic neuro-effector system (Ambache, 1955). Dale & Gaddum (1930) have suggested that this form of resistance may be due to the proximity of the nerve endings to the effector cells. Thus, the spatial relations of closer application of endogenous acetylcholine would be seen as an explanation for the greater ease of blocking the effects of exogenous than endogenous acetylcholine. Another explanation put forward by Carpenter & Rand (1965) suggests that endogenous and exogenous acetylcholine activate different receptor sites and the site available to exogenous acetylcholine is more exposed to atropine.

Recent electron microscopic observations of intestinal (Thaemert, 1966; Bennett & Rogers, 1967), and toad bladder innervation (Robinson & Bell, 1967) have not indicated a particularly close relationship between nerve and muscle and thus lend little support to the theory of Dale & Gaddum (1930). Nor do they reveal any specializations that would suggest a separate or unique receptor area for endogenous acetylcholine.

Besides this enigmatic problem there are other limitations to the use of atropine. One such, is the apparent resistance to atropine noted in rabbits, which is due to the occurrence in some (but not all) of these animals of an atropinesterase (Cloetta, 1908; Fleischmann, 1910; Bernheim & Bernheim, 1938; Glick & Glaubach, 1941; Lévy & Michel, 1945; Ammon & Savelsberg, 1949). In this case, atropine is split into relatively inert products of hydrolysis leading to either a transient anticholinergic effect (Ambache & Lippold, 1949) or none at all.

Still another difficulty involved in the use of atropine comes from the use of high concentrations of the drug. Like most other drugs, atropine loses specificity when excessive concentrations are used. For instance, atropine in high concentration inhibits hydrolysis of benzoylcholine by rabbit serum (Ellis, 1947); inhibits pseudocholinesterase in rat's intestinal mucosa (Todrick, 1954); releases histamine (Schachter, 1952), antagonizes histamine (Schild, 1947), and potentiates contractions of the nerve-free muscle of the chick amnion (Cuthbert, 1963a). High concentrations of atropine will also block transmission in ganglia (Bainbridge & Brown, 1960; Brown, 1967). In part, the ganglionic action of atropine appears to be directed against muscarinic receptors (Trendelenburg, 1954; Eccles & Libet, 1961). Atropine, acting specifically, would be expected to block muscarinic receptors for acetylcholine. These would, by definition, be activated by muscarine and blocked by atropine. Receptors stimulated by nicotine, or nicotinic receptors, should not be blocked by atropine. Atropine will, however, also block the nicotinic elevation of the blood pressure produced in dogs by neostigmine (Long & Eckstein, 1961) and the asynchronous post-ganglionic neural discharge evoked by neostigmine in both normal and denervated superior cervical ganglia of the cat (Takeshige & Volle, 1963). The ability of atropine in high concentrations to block ganglionic transmission appears to be a competitive antagonism directed non-specifically at nicotinic receptors and is overcome by higher concentrations of acetylcholine (Konzett & Rothlin, 1949). Another nicotinic effect which has been reported to be blocked by atropine is the release of adrenaline by the adrenal medulla upon splanchnic nerve stimulation (Feldberg, Minz & Tsudzimura, 1934; Lee & Trendelenburg, 1967).

The antagonism of a given parasympathetic or muscarinic effect by atropine may therefore be considered as significant evidence in favour of the mediation of that effect by acetylcholine only if the atropine has been used in a low concentration. The value of the observation would be considerably enhanced if it were accompanied by evidence that atropine was acting specifically, e.g. that nicotinic effects or histamine were not also blocked. On the other hand, a negative result, that is failure of atropine to block a response, should not, by itself, be taken as evidence against the participation of acetylcholine and does not itself require the invocation of additional neurotransmitters. In any case, atropine appears most useful when it is used as a part of a considerable body of independent evidence and is not the sole support of a cholinergic hypothesis.

Potentiation of cholinergic effects by anticholinesterases

Because of the problems, outlined above, associated with the use of atropine, great reliance has been placed on potentiation of responses by anticholinesterases. The primacy of this approach has been emphasized by Dale & Gaddum (1930). However here too problems arise, and these drugs should not be used uncritically.

A minor problem involves neural responses in which a ganglionic synapse participates. Potentiation of these responses, such as in gut, might be due to either a ganglionic or post-ganglionic site of action. In practice, it is usually possible to determine the effect of anticholinesterases on post-ganglionic stimulation alone, and, for parasympathetic responses, these are virtually all potentiated. For instance, the excitation of the intestine by ganglion stimulants such as nicotine is potentiated by anticholinesterases (Robertson, 1954) as is the twitch response to stimulation of post-ganglionic nerves (transmural stimulation; Paton, 1955). As mentioned, a similar potentiation of post-ganglionic responses has been observed *in vitro* for the bladder (Carpenter, 1963).

Another, more serious source of confusion in the use of anticholinesterase compounds involves other effects these compounds may have, unrelated to their ability to inhibit cholinesterase. Atropine-like side effects of BW284C51 (1,-5-bis(p-allyldimethylammoniumphenyl) pentan-3-one dibromide) have been noted by Tedeschi (1954) and by Ambache & Lessin (1955). Moreover, these compounds may themselves not be inert to neural tissue. Thus, Ambache & Lessin (1955) have suggested that BW284C51 activates neural structures in the gut wall because the spasm seen when this substance is added to a preparation is not seen after application of botulinum toxin, a substance known to prevent neural release of acetylcholine.

A similar problem has been encountered in investigations with tracheo-bronchiolar musculature. Eserine produces a contracture of this muscle (Dixon & Brodie, 1903). The contracture phenomenon is also noted after the application of a variety of organophosphorus anticholinesterases (King, Koppanyi & Karczmar, 1947; Douglas, 1951; DeCandole, Douglas, Lovatt Evans, Holmes, Spencer, Torrance & Wilson, 1953). The contracture is not a central phenomenon because it has been observed in isolated preparations (Trendelenburg, 1912; Douglas, 1951). The anticholinesterases also potentiate the tracheo-bronchiolar effects of parasympathetic nerve stimulation and exogenous acetylcholine (Douglas, 1951; Carlyle, 1963). The most common interpretation of the contracture is that it is due to an accumulation of acetylcholine released at postganglionic endings (Douglas, 1951). However, Carlyle (1963) did not observe a contracture after the addition of one of the organophosphorus anticholinesterases, diisopropyl phosphodiamidic fluoride (Mipafox), although this compound still potentiated the action of acetylcholine. Both eserine and neostigmine did produce contractures, however, in control preparations and in those treated with Mipafox. Since the contracture was prevented by lowering the Ca^{2+} concentration (although this would also affect the smooth muscle itself) or after hemicholinium, thus presumably depending upon acetylcholine release, Carlyle argued that eserine and neostigmine, but not Mipafox, actually release acetylcholine from parasympathetic terminals. Anticholinesterases also produce contractures of intestinal smooth muscle (Adrian, Feldberg & Kilby, 1947), which appears particularly susceptible to this effect (Krop & Kunkel, 1954). Most of these compounds also increase peristalsis and enhance spontaneous activity (Salerno & Coon, 1949; Shelley, 1955; Harry, 1963). As in the case of the bronchial musculature, Mipafox appears to be an exception to the rule, and does not produce these effects (Birmingham, 1961; Brownlee & Harry, 1963; Harry, 1963) although it does potentiate the action of acetylcholine. It seems clear that the contracture caused by the anticholinesterase compounds is related to the neural release of acetylcholine. It is not seen after the release of acetylcholine has been prevented by botulinum toxin (Ambache & Lessin, 1955) and is also blocked by inhibition of acetylcholine receptors with atropine (Hughes, 1955; Erdmann & Heye, 1958). Moreover, the intensity of contracture is directly proportional to the degree of inhibition of cholinesterase (Shelley, 1955) and is abolished by oximes, which also reverse organophosphorus inhibition of cholinesterase (Erdmann & Heye, 1958). Finally, and most importantly, the contracture is seen only in the innervated, but not the denervated, longitudinal muscle strip of guinea-pig ileum (Paton & Zar, 1965, 1968), showing that there is no direct stimulatory effect of anticholinesterases on the smooth muscle and that intact neural tissue is required for the response.

There are two possible explanations as to why the anticholinesterases produce contractures of intestinal muscle. One is that some of these compounds might actually themselves release acetylcholine (as postulated by Carlyle for bronchiolar musculature) and the other, that by inhibiting cholinesterase, they permit the accumulation of acetylcholine released as a consequence of normal neuronal activity or transmitter leakage. If the first of these possibilities were true, it would call into question the value of potentiation of neuronal effects by anticholinesterases as an argument in favour of acetylcholine as a mediator of transmission, because it would imply a primary pre-synaptic site of action of these drugs. That is, they might be acting by stimulating axonal terminals or by causing repetitive firing, etc. of neurones.

There is evidence that ganglionic activity is involved in the contractures because they are inhibited by ganglion blocking agents such as hexamethonium (Erdmann & Heye, 1958). Gershon (1967) found that the contracture which follows the administration of 10^{-7} g/ml of eserine is abolished by neuronal paralysis with tetrodotoxin, indicating that the contracture does result from neuronal activity. A contracture, sensitive to atropine, is noted however, when the eserine concentration is raised to 10^{-6} g/ml, despite the presence of a sufficiently high concentration of tetrodotoxin to inhibit conduction of neuronal action potentials. Thus, even when the neuronal elements are probably quiescent, there appears to be some leakage of acetylcholine, which will accumulate enough to contract intestinal smooth muscle when cholinesterase is completely inhibited. These experiments do not answer the question as to whether the neuronal activity was normally present or was, in fact, induced by the anticholinesterases. The failure of Mipafox to produce contractures could indicate that the other contracture-producing anticholinesterase compounds do release acetylcholine while Mipafox does not. This possibility has been suggested by Johnson (1963a). On the other hand, this view would not explain the correlation between the degree of cholinesterase inhibition and intensity of contracture noted by Shelley (1955). Moreover, Zar (1966) has found that Mipafox is a weak inhibitor of cholinesterase and may not completely inhibit the enzyme in the concentration (10^{-5} g/ml) in which it is commonly used (Birmingham, 1961; Brownlee & Harry, 1963; Harry, 1963; Johnson, 1963a, b). This might explain the failure of Mipafox to produce contractures. Furthermore, Zar (1966) has noted that treatment with Mipafox actually decreases the rate of acetylcholine release from guinea-pig ileum if it is given after eserine, thus indicating that Mipafox may actually be toxic to neuronal tissue. Since anticholinesterases are necessary for the measurement of acetylcholine released from mammalian tissues, electrical recordings from autonomic neuronal elements are required in order to determine whether anticholinesterases do act upon the neurones themselves.

Evidence from a variety of neuronal systems on this point is disquietingly conflicting. On the one hand, anticholinesterase compounds are poor blockers of conduction in axons (Crescitelli, Koelle & Gilman, 1946; Toman, Woodbury & Woodbury, 1947), even when they are applied directly to nodes of Ranvier (Dettbarn, 1960a, b) to desheathed vagal 'c' fibres (Armett & Ritchie, 1961) or when they are injected directly into axoplasm (Brady, Spyropoulos & Tasaki, 1958). In fact, Armett and Ritchie (1960, 1961, 1963), who studied the action of acetylcholine and anticholinesterases on mammalian 'c' fibres by means of the sucrose-gap technique, found the anticholinesterases had very little direct action on conduction; they did not potentiate the action of acetylcholine when it was effective, and some anticholinesterases actually antagonized acetylcholine (see also Dettbarn & Davis, 1963 and Rosenberg & Podleski, 1963).

On the other hand, there is a considerable body of evidence which indicates that the anticholinesterase compounds do act on the axonal terminations of nerves to skeletal muscle. Here, it appears that the drugs induce repetitive discharges in the terminals giving rise to antidromic impulses which can be recorded from the motor nerve or ventral root. This activity is apparently transmitted to the muscle (Riker, Roberts, Standaert & Fujimori, 1957; Riker, Werner, Roberts & Kuperman, 1959; Riker, 1960; Werner, 1960a, b; Barstad, 1962; Standaert 1963; Blaber & Bowman, 1963; Randić & Straughan, 1964; Kuperman & Okamoto, 1965; Riker & Standaert, 1966). The anticholinesterases are far more efficient than exogenous acetylcholine at inducing repetitive activity in the axonal terminals and exogenous acetylcholine appears to be more likely to induce depolarization and block of neuromuscular transmission than repetitive discharge or potentiation of transmission (Riker, 1966). The effectiveness of a variety of anticholinesterases at inducing neuromuscular facilitation is not necessarily correlated with their ability to inhibit cholinesterase. On the basis of this kind of evidence Riker (1966) has concluded that the facilitatory action of anticholinesterases at the neuromuscular junction is due to the effect these drugs have on motor nerve terminals and is not due to a build up of acetylcholine. Riker, pointing to the difference in effect between exogenous acetylcholine and the anticholinesterases, thus disagrees with the hypothesis of Koelle (1962) and Eccles (1964) (see also Randić & Straughan, 1964) who explain the effect of anticholinesterases on axonal terminals as being due to reflux of endogenously liberated acetylcholine. If the potentiation of responses of skeletal muscle to nerve stimulation by anticholinesterases is due to their direct action on the motor nerve terminal and not to a potentiation of the post-junctional action of acetylcholine, one might expect to encounter a similar phenomenon at smooth muscle neuroeffector junctions. In this case, the value of anticholinesterase potentiation of neural stimuli as an argument in favour of cholinergic transmission to smooth muscle would be seriously undermined.

In fact, with respect to autonomic junctions, the comparison of the effectiveness of anticholinesterases in potentiating responses at cholinergic junctions with their relative lack of potentiation at adrenergic junctions itself suggests the post-junctional nature of augmentation of acetylcholine at cholinergic sites. To be sure, potentiation of adrenergic effects by anticholinesterase compounds has been described (Burn & Rand, 1962, 1965; Burn, Rand & Wien, 1963; Burn & Huković, 1966; Ng, 1966; Birmingham, 1966; Bell, 1967a). This potentiation has been taken as a major support of the Burn-Rand hypothesis of a cholinergic link in the release of the adrenergic transmitter. (The evidence for and against this hypothesis has been reviewed by Ferry, 1966.) Potentiation of these effects, however, has been a much less striking phenomenon than potentiation at cholinergic junctions and often is observed only at low frequencies of nerve stimulation (Burn & Huković, 1966; Ng, 1966). Moreover, the nerves stimulated should be described as being anatomically sympathetic, rather than adrenergic, and in some cases the potentiation is probably due to the presence of cholinergic nerve fibres mixed with the adrenergic fibres (see Chapter 15). In other cases ganglionic synapses may exist all along the path of a nerve thought to be simply adrenergic or sympathetic post-ganglionic (Kuntz & Jacobs, 1955). In the guinea-pig vas deferens-hypogastric nerve preparation for instance, ganglia are found along the hypogastric nerve (Sjöstrand, 1965). The neural supply to this organ also contains fibres which are apparently cholinergic and the smooth muscle of the vas deferens has cholinergic receptors (Birmingham, 1966; Bell, 1967a). These may well account for anticholinesterase potentiation of responses to neuronal stimulation in vas deferens without necessitating the invocation of a cholinergic link in adrenergic transmission. For other instances of potentiation of adrenergic responses by anticholinesterases (Burn & Huković, 1966; Ng, 1966) a direct effect of the anticholinesterases on axon terminals may play a part. It seems unlikely, however, that the action on axon terminals is adequate to explain potentiation of cholinergic responses. This is particularly so since the anticholinesterase potentiation of the muscarinic neural effects is blocked by atropine or hyoscine. The effect of anticholinesterases on axon terminals appears to be nicotinic (Standaert, 1964; Standaert & Adams, 1965), is blocked by d-tubocurarine, and would probably not be blocked by atropine or hyoscine. The potentiation is thus very likely to be, as long believed, a post-junctional phenomenon.

The neural release of acetylcholine

The final link in what can be called the classical argument in favour of acetylcholine as the major parasympathetic post-ganglionic mediator is the detection of acetylcholine released by nerve stimulation. Although this form of experimentation began as early as 1921 (Loewi, 1921a, b), the identification of acetylcholine has relied on bioassay until very recently. Earlier chemical methods were inadequate to measure the small quantities of substance released. The use of a number of different, highly sensitive, bioassays in parallel as well as a variety of controls left little doubt as to the identity of the transmitter, but until chemical methods confirmed the results of bioassay the final definitive identification of the transmitter was lacking.

During the past few years a number of investigators have succeeded in chemically detecting acetylcholine release upon nerve stimulation. The use of choline-methyl-^{14}C and choline-methyl-^{3}H has made this experimentation possible. Acetylcholine release by pre-ganglionic stimulation of the nerve fibres to the superior cervical ganglion (Friesen, Kemp & Woodbury, 1964), and by stimulation of the phrenic nerve to the rat diaphragm (Saelens & Stoll, 1965) has been shown in this way. In the autonomic nervous system, choline-methyl-^{3}H has been used to label acetylcholine in the isolated cat heart and labelled acetylcholine has been shown to be released by stimulation of the vagus nerves (Wallach, Goldberg & Shideman, 1967). The authenticity of the labelled material as acetylcholine was shown by a variety of methods including chromatography and isotope dilution analysis. Thus the heart can synthesize labelled acetylcholine, and when the vagus nerves are stimulated, labelled acetylcholine is released in association with slowing of the heart. However, neither this experiment nor the others conclusively demonstrates that the source of the released acetylcholine was the vagal post-ganglionic axonal endings. In the experiments of Wallach *et al.* (1967) for instance, the source of acetylcholine could be either pre- or post-ganglionic nerves. These experiments have not established the neural release of acetylcholine, but they do verify that authentic acetylcholine is released upon nerve stimulation. Taken together with evidence from bioassay these chemical experiments tend to confirm that the material Loewi and his successors measured was actually acetylcholine. Further experiments are still necessary to verify that the measured acetylcholine was released from axon terminals in response to invasion by the nerve action potential.

A number of investigators have presented evidence for the non-neural synthesis of acetylcholine. In many tissues the evidence has been disputed, but the presence of acetylcholine in the nerve-free placenta and amnion (Comline, 1947; Cuthbert, 1963*b*) indicates that acetylcholine can be produced extra-neurally. The production of acetylcholine by a variety of tissues has been investigated by Dikshit (1938). It is often difficult, when measuring the release of acetylcholine from innervated organs, to determine whether the measured acetylcholine came from a neural source. For example, the data of Feldberg & Lin (1949*b*, 1950) and Chujyo (1953) support the extra-neural origin of much of the large amount of acetylcholine released by the intestine of the guinea-pig. On the other hand, Welsh & Hyde (1944) found that the concentration of acetylcholine in the myenteric plexus of the intestine is extremely high. Paton's (1957) and Schaumann's (1957) studies of the effect of morphine on acetylcholine release also support a neural origin of most of the released acetylcholine. Paton and Zar (Zar, 1966; Paton & Zar, 1968) have shown that strips of longitudinal muscle from guinea-pig ileum to which the myenteric plexus adheres synthesize acetylcholine whereas strips which are devoid of myenteric plexus do not.

Thus, from the external layers of the intestine, the origin of acetylcholine is clearly neuronal. However, since whole ileum releases more acetylcholine than does the innervated longitudinal muscle strip, there must be an additional source of acetylcholine in the whole ileum. Ogura, Mori & Watanabe (1964) and Gershon (1967) have found that paralysis of neuronal elements by tetrodotoxin reduces but does not abolish the release of acetylcholine from whole ileum. Gershon has postulated that the tetrodotoxin-resistant acetylcholine release represents transmitter leakage. Transmitter leakage and release at the skeletal neuromuscular junction is not affected by tetrodotoxin (Katz & Miledi, 1966). Ogura *et al.* have found, however, that the release of acetylcholine in the presence of tetrodotoxin is enhanced by stretch. They feel this represents release of extra-neural acetylcholine in agreement with Chujyo (1953). Ogura *et al.* thus believe in the existence of two pools of acetylcholine in intestine, one neural; one extra-neural.

The problem of the origin of the acetylcholine released from the whole gut has not yet been fully resolved. The studies of Zar (1966) and Paton & Zar (1968), however, have made it clear that there is a release of acetylcholine from neural structures, whether or not there may also be release from extra-neural sites. This release is enhanced by electrical stimulation of the gut and is causally related to the contraction of the smooth muscle which follows electrical stimulation.

Another autonomically innervated organ in which the origin of released acetylcholine has been carefully investigated is the bladder. Some of these studies have been referred to previously. Here, all of the store of acetylcholine appears to be related to the presence of cholinergic nerves (Carpenter & Rand, 1965; Carpenter & Rubin, 1967). After destruction of bladder ganglia and degeneration of post-ganglionic nerves, the tissue is depleted of acetylcholine.

The release of substances by pressure or stretch

The release of acetylcholine by stretch (Chujyo, 1953; Ogura et al., 1966) or, by raising intraluminal pressure (Kažić & Varagić, 1968) illustrates a difficulty common to many studies of release of substances suspected of being transmitters. It may be that stretch initiates reflex activity leading to acetylcholine release (Kažić & Varagić, 1968) but the possibility that release may be purely passive, perhaps even from an extra-neural source has not been excluded. Pressure and stretch lead to the release of other substances from the intestine in addition to acetylcholine. 5-Hydroxytryptamine (5-HT), for instance, is also released by pressure (or perhaps by mechanical distortion) from the intestinal mucosa (Bülbring & Crema, 1959; Bülbring, 1961a).

Studies which attempt to measure release of substances from the gut generally employ procedures which, while releasing the substance in question, also lead to contraction or spasm of the intestine. Under these conditions it may be difficult to determine whether the experimental procedure (neurone stimulation or administration of drugs) directly released the material, or whether release was secondary to the mechanical distortion or to the rise in intraluminal pressure related to the accompanying spasm.

This problem is more difficult in studies of the release of other substances than of acetylcholine. For example, Burks and Long have detected the release of 5-HT in response to parasympathetic nerve stimulation, acetylcholine, angiotensin, barium chloride and catecholamines (1966a, b), nicotine (1967a) and morphine and related agents (1967b). All of these, except catecholamines, have as a common feature a spasmogenic activity. Burks and Long have advanced evidence that 5-HT may itself be involved in producing the intestinal spasm that follows perfusion of loops of dog's intestine with agents related to morphine. For instance, the spasmogenic activity of the morphine group is reduced by tachyphylaxis to 5-HT and by reserpine-depletion of 5-HT (Burks & Long, 1967b). These authors also consider the 5-HT release by nicotine to be a secondary action attributed to ganglionic stimulation by nicotine. Post-ganglionic neurones are then thought to release acetylcholine which acts on

5-HT storing cells causing them, in turn, to release 5-HT. Since 5-HT release by nicotine is inhibited by atropine, the receptor for acetylcholine on the cells storing 5-HT (if such a receptor exists) must be muscarinic.

Burks and Long, however, have not really eliminated the possibility that 5-HT release in all or some of their experiments may not be a primary phenomenon but a secondary response to the mechanical activity of the gut. This would be consistent with the wide variety of spasmogenic agents which they found would release 5-HT. The inhibition of nicotine-induced 5-HT release by atropine might also be suggestive of mechanical release of 5-HT because, by blocking the action of acetylcholine on intestinal smooth muscle, atropine also prevents nicotine-induced spasm.

A similar problem arises in interpreting the work of a number of investigators who have detected 5-HT release following vagal or transmural stimulation of the stomach. Paton & Vane (1963) and Bennett, Bucknell & Dean (1966) concluded that the 5-HT was mainly released from mucosa and that the release was secondary to mechanical effects. On the other hand, Bülbring & Gershon (1967) found that 5-HT was released by electrical stimulation of the mouse stomach even after asphyxiation of the mucosa and in the presence of hyoscine to prevent contraction. The muscle of the stomach did relax after electrical stimulation in the presence of hyoscine, however, and the stomach wall therefore was not entirely free from movement. Since release of 5-HT was prevented by paralysis of nerves with tetrodotoxin, the release must not be simply a response to the passage of an electrical current, but must require the participation of neural tissue. Under these conditions 5-HT might be released actively from 'tryptaminergic' axonal endings, or passively in response to tissue movement. The experimental technique does not permit one to choose between these two possibilities. The observation that radioactive 5-HT is found in the myenteric plexus after the administration of its tritiated precursor, 5-hydroxytryptophan (Gershon, Drakontides & Ross, 1965; Hammarström, Ritzén & Ullberg, 1966; Gershon & Ross, 1966b), suggests that 5-HT might have been released in these experiments from the myenteric plexus (see also Chapters 15 and 17).

Old and new methods of study

The old methods of transmitter identification discussed so far, have been first and most successfully applied to the study of acetylcholine. The methods have been added to by more recent techniques of biochemistry, histochemistry and electrophysiology, and have also been applied to the study of other substances, such as 5-HT, suspected of being transmitters. Without supplementation by other methods, the pharmacological approach is limited. The limitations, as applied to acetylcholine, have been discussed, but these are more applicable in dealing with other substances. It is often difficult to establish the 'identity of action' between applied substance and actual neuronal stimulation. The specificity of 'specific antagonists' is frequently questionable. Mechanisms of transmitter inactivation may be unknown; inhibitors of inactivation may not be available and, if available, their specificity unreliable. Transmitter release is difficult to demonstrate, and when demonstrated, the source of the released material may be found to be a multiplicity of sites.

It may thus be hard to prove that the material has been released from axon terminals in response to the arrival of the action potential at the terminals. The value of the older methods lies in indicating the probability or possibility of a given substance acting as a transmitter. The danger arises when findings based on studies in which drugs were used as investigative tools are accepted as more than as support for a hypothesis.

In practice, mechanisms found to apply to acetylcholine as a neurotransmitter have often been used as a model for the investigation of other systems. The model can sometimes be less than helpful. For instance, the rapid inactivation of acetylcholine by hydrolysis, catalysed by a specific enzyme, acetylcholinesterase, for a time was taken as a good model for inactivation of transmitters in general. When applied to adrenergic systems, the model led only to complexity because it is probably not applicable here. The most important event in the termination of adrenergic neural effects appears to be not enzymatic destruction of noradrenaline, but re-uptake of released noradrenaline into the axon terminals or varicosities (see reviews by Kopin, 1967; Geffen, 1967).

In order to finally establish the identity of a neurotransmitter, modern supplements to the older methods are essential. First, the steps involved in synthesis and metabolism of the suspected transmitter should be determined. The machinery necessary for the production and storage of the transmitter should be shown to be present in neuronal tissue. If an effective neuronal storage mechanism exists, it may not be necessary to demonstrate transmitter synthesis in the axon terminals themselves. It is conceivable that transmitters could be synthesized primarily in neuronal perikarya and transported down the axons to concentrate at terminals in storage granules or vesicles.

Second, one would like to be able to show that neuronal elements not only are capable of synthesizing the suspected transmitter, but that they actually do contain the substance. This generally involves the use of histochemistry or radioautography. It is of even greater value if the subcellular storage unit can be identified and shown to be present in terminal axons.

Electrophysiological studies are of immense importance not only in determining the mechanics of transmitter action, but also in establishing the identity of action between the suspected transmitter and the actual transmitter released from nerves in the tissue. The contributions of electrophysiology to autonomic neurotransmission are discussed in Chapters 8, 12, 13 and 20 of this volume. This technique has been especially useful in the study of cholinergic transmission at the neuromuscular junction of skeletal muscles. Katz (1966) has written an excellent review of this subject.

ADRENERGIC TRANSMISSION

The limitations of the earlier methods

Although chemical mediation by an adrenergic substance (adrenaline; Elliott, 1904a) was suspected prior to the development of similar ideas about a muscarinic, parasympathomimetic substance, the final identification of noradrenaline as the adrenergic mediator, followed the identification of acetylcholine. In 1921 Loewi (1921a) reported that a substance similar to adrenaline, 'Acceleransstoff', was liberated when cardio-accelerator nerves were stimulated. Shortly thereafter Cannon & Uridil (1921–1922) found that stimulation of hepatic nerves would release a substance capable of accelerating the denervated cat heart and elevating the blood pressure. Most investigators believed the active substance to be adrenaline, especially following the elaborate studies of Bacq (1933).

The tentative identification of the sympathetic post-ganglionic transmitter as adrenaline was based first of all on the presumed identity of action of adrenaline with that of the actual transmitter, and secondly upon the supposed recovery of adrenaline upon neural stimulation. Difficulties were almost immediately apparent. The action of adrenaline was, in fact, not identical to the transmitter although often similar. The recovered transmitter was never chemically shown to be adrenaline. Studies of the adrenergic system were further handicapped by the absence of a 'specific antagonist' such as atropine and the lack of an effective agent to prevent transmitter inactivation such as eserine. The result was that studies of the adrenergic system such as those mentioned, and the elegant *in vitro* studies of Finkelman (1930), were able to provide convincing evidence in favour of chemical transmission but nothing certain about the true identity of the transmitter.

The lack of identity between the action of adrenaline and the transmitter or 'sympathin' was emphasized by Cannon & Rosenblueth (1935). These investigators introduced the theory that there were two sympathins, sympathin E and sympathin I, these agents being formed by the interaction of adrenaline with tissues (Cannon & Rosenblueth, 1933, 1937; Rosenblueth, 1950). A number of other investigators suggested that noradrenaline might function as a sympathetic transmitter, if not instead of adrenaline, then together with it (Greer, Pinkston, Baxter & Brannon, 1937, 1938; Stehle & Ellsworth, 1937; Melville, 1937). Actually, Barger & Dale (1910–11) had much earlier noted the potent sympathomimetic actions of noradrenaline. However, many investigators continued to find evidence in favour of the release of adrenaline (Gaddum, Jang & Kwiatkowski, 1939; Gaddum & Kwiatkowski, 1939) and for the presence of adrenaline in many sympathetically innervated tissues (Cannon & Lissák, 1939; Lissák, 1939).

Identification of noradrenaline

Starting with the work of von Euler in 1946, the evidence in favour of noradrenaline, rather than adrenaline as the actual sympathetic mediator, began to accumulate and the concept of the two sympathins and of adrenaline as the mediator are now (in mammals) largely of historical interest. von Euler (1946) used the method of parallel bioassay to demonstrate that the sympathetic transmitter behaved as noradrenaline not adrenaline. Following von Euler other investigators obtained evidence for the release of noradrenaline by splenic nerves (Peart, 1949; Mirkin & Bonnycastle, 1954) and vasoconstrictor nerves (Folkow & Uvnäs, 1948a). Further confirmation of noradrenaline as the mediator came from the experiments of Gaddum & Goodwin (1947) on liver 'sympathin'.

The earliest chemical study to indicate that adrenergically innervated organs contained at least one catechol, similar to, but different from adrenaline, was that of Raab (1943). Goodall, however, in 1951 finally found chemical evidence by column and paper chromatography for the presence of noradrenaline. Moreover, this study, coupled with denervation, strongly suggested that the noradrenaline was stored in association with sympathetic nerves. This suggestion was supported by the observation of Rexed & von Euler (1951) that the noradrenaline content of organs was proportional to their content of small unmyelinated nerves, an observation which also indicated that noradrenaline was confined to post-ganglionic nerves. The events leading to the identification of noradrenaline as 'sympathin' have been well reviewed by von Euler (1951, 1956).

The early development of knowledge of the adrenergic system was thus rather slow. In recent years the further development of knowledge of this system has been anything but slow and in some respects, more is now known about the adrenergic than the cholinergic system.

BIOSYNTHESIS OF NEUROTRANSMITTERS

Acetylcholine biosynthesis

For both transmitters, the biochemical routes of synthesis and metabolism have been worked out. For acetylcholine the first information on breaking down the synthetic mechanisms came from the laboratories of Feldberg and of Nachmansohn (Feldberg & Mann, 1945; 1946; Nachmansohn & Machado, 1943; Nachmansohn & Berman, 1946). These workers demonstrated the synthesis of acetylcholine by a system involving a soluble enzyme (from brain), a source of acetate, such as citrate or pyruvate, ATP, and an unknown co-factor. At about this time, during the 1940's, Lipmann was involved in his landmark studies of two carbon metabolism. Lipmann & Kaplan (1946) found that not only the acetylation of choline, but also that of sulphonamides, required the presence of a co-factor. They partially purified the co-factor and suggested that there was a common co-enzyme for acetylation, co-enzyme A (Lipmann & Kaplan, 1946; Lipmann, 1948). By 1952, this suggestion was substantiated and acetylcoenzyme A, choline and the enzyme, choline acetylase, were shown to be all that was required for the synthesis of acetylcholine (Korkes, Del Campillo, Kórey, Stern, Nachmansohn & Ochoa, 1952).

Choline acetylase

The enzyme, choline acetylase, is selectively distributed in the nervous system. The presence of the enzyme itself is strong evidence for the cholinergic nature of a given neurone. Cholinergic neurones contain a great deal of choline acetylase in their terminals while non-cholinergic neurones do not (see review by Hebb, 1957). When a cholinergic axon is severed, there is an extensive build-up of choline acetylase proximal to the cut a few days after nerve section (Hebb & Waites, 1956). This phenomenon indicates that choline acetylase is probably synthesized in neuronal perikarya, and moved to the terminals by axoplasmic flow. Flow of materials down the axon to neuronal terminals has been demonstrated by Weiss and co-workers (Weiss & Hiscoe, 1948; Weiss, 1961; Weiss, Taylor & Pillai, 1962). The proximal build up of enzyme which serves to concentrate it to detectable levels is useful in demonstrating cholinergic neurones.

Noradrenaline biosynthesis

The full route of synthesis of noradrenaline has been worked out only recently. Noradrenaline is synthesized from the amino acid tyrosine by three sequential enzymatic steps. Tyrosine is converted to dihydroxyphenylalanine (DOPA), a step catalysed by the enzyme tyrosine hydroxylase. This appears to be the rate-limiting step in the synthesis of norepinephrine. DOPA is then decarboxylated to 3,4-dihydroxyphenylethylamine (dopamine) by aromatic amino acid decarboxylase. Dopamine is finally hydroxylated in the β-position to form noradrenaline, a step again catalysed by an enzyme, dopamine β-hydroxylase (see review by Udenfriend, 1964; Kopin, 1967).

This pathway was first proposed by Blaschko (1939) and by Holtz (1939; see also Holtz, Heise & Lüdtke, 1938). The demonstration by Goodall & Kirshner (1958) of the formation of noradrenaline from labelled tyrosine supported the hypothesis and provided final evidence that sympathetic axons and terminals can manufacture their own neurotransmitter. This demonstration has been repeatedly confirmed (Austin, Livett & Chubb, 1967).

Tyrosine hydroxylase

The first step in the synthesis, the hydroxylation of tyrosine to DOPA, has been the most recently characterized and is the rate limiting step in the synthesis of noradrenaline (Nagatsu, Levitt & Udenfriend, 1964; Levitt, Spector, Sjoerdsma & Udenfriend, 1965; Udenfriend, 1966). Interestingly, the activity of the enzyme, tyrosine hydroxylase, appears to be proportional to the degree of activity in the sympathetic nerves (Weiner & Alousi, 1967; Sedvall & Kopin, 1967; Roth, Stjärne & von Euler, 1967). A reasonable explanation for this phenomenon is that of end-product inhibition of tyrosine hydroxylase by stored noradrenaline (Stjärne, 1966; Spector, Gordon, Sjoerdsma & Udenfriend, 1967). According to this view, increased activity in sympathetic neurones would, through the release of noradrenaline, lower the amount of stored intra-axonal noradrenaline and thus, by removing an inhibitor, increase enzyme activity. It would be somewhat hard to explain how this might occur if noradrenaline and tyrosine hydroxylase are stored in different intraneuronal compartments. A great deal of work has been done on the storage of noradrenaline and the enzymes involved in its synthesis.

Amine storage granules

Once formed, noradrenaline appears to be largely stored within sub-micro-scopic particles or amine storage granules (von Euler & Hillarp, 1956). The properties of these storage granules, which account for up to 80 to 90% of the noradrenaline stored in sympathetic nerves, have been reviewed by von Euler (1967). The three enzymes involved in the synthesis of noradrenaline have also been localized in fractions of bovine splenic nerve homogenates (Stjärne, 1966; Stjärne & Lishajko, 1967). Only one of the enzymes, dopamine β-hydroxylase (see also Stjärne, Roth & Lishajko, 1967) appears to be truly located within the amine storage particles. A similar localization of this enzyme has been found in the noradrenaline storing granules isolated from rat heart (Potter & Axelrod, 1963a, b) and in adrenal medullary granules (Laduron & Belpaire, 1968). On the other hand, tyrosine hydroxylase appears to be located outside these particles (Stjärne & Lishajko, 1967; Musacchio, 1968; Laduron & Belpaire, 1968) although a number of other investigators have found it to be particle bound (Nagatsu et al., 1964; Udenfriend, 1964; McGeer, Bagchi & McGeer, 1965). Aromatic amino acid decarboxylase, though usually described as cytoplasmic, shows some association with the amine-storage particles (Stjärne & Lishajko, 1967).

The last two enzymes in the synthesis of noradrenaline, aromatic amino acid decarboxylase and dopamine β-hydroxylase are not specific for noradrenaline precursors. The presentation of unusual substrates to sympathetic nerves can therefore lead to the synthesis and accumulation of false neurotransmitters (Day & Rand, 1963; Kopin, Fischer, Musacchio, Horst & Weise, 1965; see also reviews by Sourkes, 1966; Kaufman, 1966; Kopin, 1967).

The origin of the amine storage granules, like that of choline acetylase, appears to be neuronal perikarya. After compression or section of sympathetic nerves there is an accumulation of noradrenaline proximal to the point of constriction (Dahlström, 1965a). On the basis of data obtained from doubly constricted sciatic nerves of rat and cat, Dahlström & Häggendal (1966a) have estimated the rate of proximo-distal transport of the amine storage granules and their life-span. These are, respectively, 5 to 6 mm/hr and 35 days for the rat and 9 to 10 mm/hr and 70 days for the cat. Similar studies revealed a rate of 3 mm/hr and a life span of 50 days in the rabbit (Dahlström & Häggendal, 1967). The transport of granules will continue distal to a ligature and so appears to be independent of the perikaryon (Dahlström, 1967a). Experiments with labelled noradrenaline injected into the coeliac ganglion of the cat (Geffen & Livett, 1968; Geffen & Rush, 1968; Livett, Geffen & Austin, 1968) have also demonstrated a proximo-distal flow of noradrenaline and protein in splenic nerves. These experiments indicate that transport is of the storage granules and synthetic machinery, since the amount of actual norepinephrine transported from perikaryon to ending is quite small.

Further support for the proximo-distal transport of amine storage granules comes from studies with the amine depleting drug, reserpine. Reserpine has been shown to block the ATP-dependent uptake of noradrenaline by storage granules (von Euler & Lishajko, 1963; Kirshner, Halloway, Smith & Kirshner, 1966). This apparently leads gradually to the slow intraneuronal release of noradrenaline where it can be acted upon and depleted through metabolism by monoamine oxidase (Kopin & Gordon, 1962; Kopin, 1964; Carlsson, 1966). Reserpine does not interfere with the ability of the axon membrane to transport catecholamines, so that inhibition of monoamine oxidase prevents amine depletion following reserpine (see Carlsson, 1965, 1966). After a ligature has been placed on the sciatic nerve, and amines are depleted by reserpine, repletion begins in perikarya and reaches only to the ligature. Noradrenaline repletion does not occur distal to the ligature. This suggests that repletion is associated with the transport of newly-formed granules down axons to the axon terminals (Dahlström, 1967b). The noradrenaline storage granules have been identified by electron microscopy of adrenergic endings in many tissues including smooth muscle, as vesicles with an electron dense core, having a diameter in the range of 500 to 600 Å (Hager & Tafuri, 1959; Grillo & Palay, 1962; Richardson, 1962, 1964; Wolfe, Potter, Richardson & Axelrod, 1962; Wolfe & Potter, 1963; DeRobertis, 1964, 1966; Bondareff, 1965; Lever, Graham, Irvine & Chick, 1965; Hökfelt, 1966; Grillo, 1966; Bloom & Barrnett, 1966b; Van Orden, Bloom, Barrnett & Giarman, 1966; Potter, 1966; Kapeller & Mayor, 1967). These vesicles seem to be a basic unit of adrenergic structure. Other storage compartments have also been postulated for norepinephrine but these have not been visualized (see Trendelenburg, 1961; Kopin & Gordon, 1962; Carlsson, 1965; Kopin, 1966). Stjärne (1966) has found evidence from studies of splenic nerve that newly synthesized noradrenaline is stored separately from the major 'pool' of the amine, and also evidence for the possible existence of separate 'pools' of the amines within the storage granules. Noradrenaline can be induced to accumulate in axoplasm if granular storage is inhibited by reserpine and released noradrenaline is protected by inhibition of monoamine oxidase (Malmfors, 1965a; Van Orden, Bensch & Giarman, 1967). It is not clear whether this extra-granular noradrenaline is functional. Conditions which lead to its accumulation, such as addition of exogenous noradrenaline, actually block transmission when the hypogastric nerve to the guinea-pig vas deferens is stimulated (Van Orden *et al.*, 1967).

When a sympathetic post-ganglionic axon is constricted, there is an electron microscopically demonstrable build up of dense-core vesicles proximal to the ligature associated with the accumulation of noradrenaline (Kapeller & Mayor, 1966, 1967; Mayor & Kapeller, 1967). These observations substantiate the transport of amine storage granules. Thus, in both adrenergic and cholinergic neurones there is considerable evidence for the proximo-distal flow of material from perikaryon to axon terminals. Besides choline acetylase, there is also evidence for a similar flow of cholinesterase in cholinergic neurones (Lubinska, Niemierko & Oderfelt, 1961). The process therefore seems to be a general one, and appears to be a means of transporting essential protein-containing elements of the transmission mechanism to axon terminals. In view of the very low ribonucleic acid content of axoplasm (Koenig, 1965) it is unlikely that these elements could be synthesized anywhere but in the perikaryon. The speed of transport of amine storage granules (up to 3 to 10 mm/hr) greatly exceeds the value of 1 to 2 mm/day for axoplasmic flow given by Weiss (1961). It therefore appears that there are two separate systems for moving material from perikaryon to axon terminal. These two systems have been demonstrated by radioisotopic methods to be operative in splenic nerve (Geffen & Livett, 1968; Geffen & Rush, 1968; Livett, Geffen & Austin, 1968).

DIRECT DEMONSTRATION OF TRANSMITTER CONTENT

Monoamine histochemistry

One of the most striking modern developments for the identification of neurotransmitters has been the use of a sensitive histochemical procedure for demonstrating monoamines in tissue sections. The method evolved from the studies of Eränkö (1952, 1955), who used formaldehyde to induce fluorescence in the noradrenaline-storing cells of the adrenal medulla, and the studies of Barter & Pearse (1953, 1955) who used formaldehyde vapour to induce fluorescence in the 5-HT containing enterochromaffin cells. Falck and Hillarp and their associates developed the method for use with nervous tissues (Carlsson, Falck & Hillarp, 1962; Falck, 1962; Falck, Hillarp, Thieme & Torp, 1962). The procedure involves exposing tissues, either as dried whole mounts (Malmfors, 1965a) or as frozen-dried blocks, to formaldehyde vapour under controlled conditions of temperature and humidity. The methodological details have been discussed in a number of recent papers (Dahlström & Fuxe, 1964, 1965a; Falck & Owman, 1965; Hamberger, 1967; Eränkö, 1967). The method demonstrates catecholamines and 5-HT which fluoresce green and yellow respectively. The theoretical basis of the reactions which underlie the method has been discussed in several publications (Falck, 1962; Falck *et al.*, 1962; Corrodi & Hillarp, 1963, 1964; Corrodi, Hillarp & Jonsson, 1964; Corrodi & Jonsson, 1967). The specificity of the method has now been clearly established, but the distinction between the fluorophores of 5-HT and catecholamines can be difficult. This distinction is greatly enhanced by microspectrofluorometry (Caspersson, Hillarp & Ritzén, 1966; Ritzén, 1966), Although usually performed on dried or frozen-dried specimens, the reaction will proceed when tissues are fixed in appropriate aqueous solutions of formaldehyde (Eränkö & Räisänen, 1966) or by perfusion with phosphate buffered formaldehyde evolved from paraformaldehyde (Laties, Lund & Jacobowitz, 1967). It thus appears that it is not necessary to freeze-dry tissues in order to preserve catecholamines. If the osmolarity of a fixing solution is properly adjusted, diffusion can be eliminated even in aqueous solutions (Eränkö & Räisänen, 1966). Similarly, it has been shown that it is possible to fix 5-HT in tissues with aqueous fixatives without disturbing the physiological compartmentalization of the amine (Gershon & Ross, 1966a, b). The movement of 5-HT from tissues has been studied by analysis of curves of washout of labelled substances (Fischman & Gershon, 1964). A membrane appears to be important in the retention of 5-HT by tissue, and 5-HT binding is destroyed by hypotonic solutions or by freezing and thawing (Gershon & Ross, 1966a). It thus appears that drying of tissues to preserve amines may only be necessary if the tissues have been frozen.

Radioautography

The use of fixatives to preserve amines has made possible the use of radioautography for their demonstration in tissues. Radioautography after aqueous fixation has been used for the demonstration of both noradrenaline and 5-HT (Wolfe et al., 1962; Marks, Samorajski & Webster, 1962; Wolfe & Potter, 1963; Aghajanian & Bloom, 1966, 1967; Aghajanian, Bloom, Lovell, Sheard & Freedman, 1966; Gershon & Ross, 1966b). In the case of noradrenaline, the method has been particularly useful to directly demonstrate the uptake of exogenous noradrenaline by post-ganglionic sympathetic axon terminals (Wolfe et al., 1962; Gillespie & Kirpekar, 1966b).

NEURAL UPTAKE OF NORADRENALINE

Burn (1932) suggested that adrenaline added to the circulating blood in perfused preparations might replenish depleted stores of transmitter at sympathetic nerve endings. The uptake of exogenous noradrenaline by tissues was noted in studies of the disposition of infused radioactive noradrenaline (Axelrod, Weil-Malherbe & Tomchick, 1959; Whitby, Axelrod & Weil-Malherbe, 1961). Paton (1960) suggested that released adrenergic transmitter might also be reincorporated into axon terminals. The importance of intact axons in noradrenaline uptake is shown by observations that the uptake may be lost following denervation (Hertting, Axelrod, Kopin & Whitby, 1961; Strömblad & Nickerson, 1961; Gillespie & Kirpekar, 1965). Moreover, labelled noradrenaline and adrenaline, taken up by tissues, are released when sympathetic nerves are stimulated (Hertting & Axelrod, 1961; Rosell, Kopin & Axelrod, 1963; Rosell, Axelrod & Kopin, 1964; Gillespie & Kirpekar, 1966a; Boullin, Costa & Brodie, 1966, 1967). A number of substances interfere with the uptake of noradrenaline. These include cocaine (Whitby, Hertting & Axelrod, 1960; Muscholl, 1961; Burn & Burn, 1961; Dengler, Spiegel & Titus, 1961; Lindmar & Muscholl, 1963; Hillarp & Malmfors, 1964; Gillespie & Kirpekar, 1965), α-adrenergic blocking agents such as phenoxybenzamine (Hertting, Axelrod & Whitby, 1961; Kirpekar & Cervoni, 1963; Blakeley & Brown, 1963; Gillespie & Kirpekar, 1965), and psychoactive drugs such as imipramine, desmethylimipramine, amphetamine and chlorpromazine (Dengler *et al.*, 1961; Iversen, 1965a; Malmfors, 1965a; Carlsson, 1966; Berti & Shore, 1967). When sympathetic nerves are stimulated in the presence of this kind of drug there is a marked increase in the 'overflow' of noradrenaline reaching the circulation or perfusate (Brown & Gillespie, 1957; Brown, Davies & Ferry, 1961; Blakely, Brown & Ferry, 1963; Gillespie & Kirpekar, 1966a; Boullin *et al.*, 1966, 1967).

Inactivation of noradrenaline

This transmitter overflow is similar to that seen at high rates of nerve stimulation (Brown & Gillespie, 1957) and suggests that both are due to failure of the normal re-uptake mechanism of the sympathetic endings and that the uptake normally limits the movement of noradrenaline into the circulation. Furthermore, neither of the two enzymes concerned with the metabolism of noradrenaline, monoamine oxidase or catechol-o-methyl transferase, appear to have any pronounced ability to influence the actions of either exogenous or endogenously released noradrenaline, and probably are not concerned with transmitter inactivation (see Axelrod, 1966). On the other hand, cocaine, which blocks noradrenaline uptake, has long been known to potentiate sympathetic effects as well as those of exogenous noradrenaline (Trendelenburg, 1963, 1966). This potentiation appears to be due to the inhibition of noradrenaline uptake and the consequent failure of inactivation of the amine (Trendelenburg, 1965; Bhagat, Bovell & Robinson, 1967).

There is thus considerable evidence that re-uptake of noradrenaline occurs and that it is of physiological significance. Its function appears roughly to correspond to that of acetylcholinesterase for cholinergic neurones, that is, to terminate the action of released transmitter. It is also a means of conserving transmitter stores. The sympathetic axon terminal seems to be well equipped to maintain its concentration of noradrenaline. Besides the re-uptake mechanism, it will be recalled that loss of transmitter also may stimulate noradrenaline synthesis. Moreover, only a fraction of the noradrenaline stored in terminals appears to be available for release by nerve impulses (Boullin *et al.*, 1966; Costa, Boullin, Hammer, Vogel & Brodie, 1966; Gillespie & Kirpekar, 1966a).

There is yet another uptake mechanism for noradrenaline that functions in sympathetically innervated tissues (Iversen, 1965*b*). This mechanism operates at higher concentrations of noradrenaline than does the neural uptake mechanism, and has been shown histochemically to represent the uptake of noradrenaline by muscle (Avakian & Gillespie, 1968). Like the neural uptake of noradrenaline it too is blocked by phenoxybenzamine (Eisenfeld, Axelrod & Krakoff, 1967; Avakian & Gillespie, 1968).

CHOLINESTERASE HISTOCHEMISTRY

For cholinergic neurones, the use of histochemical and radioautographic methods have not been quite as useful as they have for adrenergic neurones. As yet, no method has been available for the localization of acetylcholine itself in tissues. Instead, histochemical procedures for acetylcholinesterase and microchemical methods for choline acetylase and acetylcholine have been used. The histochemical procedures for the localization of cholinesterase were introduced some years ago (Gomori, 1948; Koelle & Friedenwald, 1949) and have since been extensively modified to improve localization (see Eränkö, Koelle & Räisänen, 1967; El Badawi & Schenk, 1967; Davis & Koelle, 1967). As now refined, histochemical methods are useful not only for light, but also for electron microscopy (Barrnett, 1962; Karnovsky, 1964; Miledi, 1964; Bloom & Barrnett, 1966*a*; Brzin, Tennyson & Duffy, 1966; Schlaepfer & Torack, 1966; Tennyson, Brzin & Duffy, 1966; Davis & Koelle, 1967; Esterhuizen, Graham, Lever & Spriggs, 1968; Schlaepfer, 1968). It is tacitly assumed by most people who use the procedure that axons positive for acetylcholinesterase are cholinergic and the technique is thus often thought of as being an indirect method for the demonstration of cholinergic neurones (Koelle, 1963). The assumption is based on the presence of acetylcholinesterase activity in known cholinergic neurones and its apparent absence from other neurones. There also appears to be a correlation between acetylcholine content of fibres, choline acetylase activity and acetylcholinesterase activity (Hebb, 1957; Lewis, Shute & Silver, 1967) in both central and peripheral nervous systems. Some reservation should be made however before accepting acetylcholinesterase activity as a label for cholinergic neurones.

The studies of Schlaepfer & Torack (1966) clearly show that the enzyme is present not only at axon terminals or post-junctional membranes, but along the membrane of the entire length of most myelinated and all unmyelinated fibres in the sciatic nerve. The presence of the enzyme in all unmyelinated fibres implies that it is present in some afferent fibres as well as cholinergic efferent fibres and there is no convincing evidence that afferent fibres are cholinergic. Schlaepfer (1968) has even found acetylcholinesterase in unmyelinated dorsal root fibres and their ganglion cells. The enzyme is also found in cytoplasmic constituents such as the nuclear envelope and endoplasmic reticulum in neuronal perikarya (Brzin *et al.*, 1966; Tennyson *et al.*, 1966) and small vesicles in axoplasm (Schlaepfer & Torack, 1966). It is difficult to reconcile all the regions of enzyme localization with its presumed function.

There has been some question as to whether acetylcholinesterase might also be located in adrenergic fibres. According to the hypothesis of Burn and Rand, whereby acetylcholine is thought to participate in the release of noradrenaline, both acetylcholine and noradrenaline should be in the same fibre (see Ferry, 1966). If acetylcholinesterase could be demonstrated in the same fibres as noradrenaline by histochemistry, this would constitute evidence in favour of the Burn-Rand proposal. A number of workers have studied this point (Jacobowitz & Koelle, 1965; Åberg & Eränkö, 1967; Bell & McLean, 1967; Esterhuizen et al., 1968). Investigations with the light microscope are not adequate because the limited resolution of the instrument (0·2 microns) is not sufficient to differentiate whether noradrenaline and acetylcholinesterase activity are in the same fibre or in closely apposed fibres. However, Bell & McLean (1967) did find that noradrenaline and acetylcholinesterase were located in separate neurones supplying the guinea-pig vas deferens. Esterhuizen et al. (1968) combined the electron microscopic radioautographic localization of noradrenaline with the simultaneous demonstration of acetylcholinesterase activity, also on a fine structural level. These investigators found a sparse innervation of the cat's nictitating membrane with acetylcholinesterase-containing fibres, which, however, were always discrete from the adrenergic fibres. There is thus no evidence from these studies that adrenergic axon terminals contain a cholinergic mechanism. In fact the absence of acetylcholinesterase activity in adrenergic axon terminals might be taken as evidence in favour of using the reaction to locate cholinergic terminals.

SUMMARY AND CONCLUSIONS

For a long time after the confirmation of the chemical nature of the transmission process between autonomic nerves and effector cells, the identification of the mediators involved rested on pharmacological experiments. These experiments sought to verify the identity of action between substances suspected of being transmitters and the actual endogenously released material. Indirect confirmation was sought through the use of 'specific antagonists' and inhibitors of inactivation of the transmitters. Direct confirmation was sought by trying to obtain the release of the suspected substance upon nerve stimulation. These methods were, and still are, useful in obtaining ideas about the nature of the mediator involved. They are not sufficient by themselves to confirm the identity of a transmitter. Modern additions to these methods are necessary. These additions include the demonstration of the enzymatic machinery necessary to synthesize the suspected transmitter somewhere within the neurone thought to release the substance. Since neurones appear to be able to transport material from perikaryon to axon terminal, it is possible that this synthetic machinery may not be present in the terminals themselves. In this case an efficient storage compartment, such as is represented by amine storage granules, would have to be present. Some inactivating mechanism should be found at the synapse in question. This mechanism need not be a rapidly acting enzyme which destroys the transmitter like acetylcholinesterase, but may be an efficient uptake mechanism such as exists for noradrenaline. Obtaining release of substances upon nerve stimulation is useful if it can be shown that the release took place from axon terminals, but this may be difficult to show in the face of axonal re-uptake. The use of radioisotopes may be very helpful in this regard. The suspected transmitt? should also be shown to be stored within the axon terminals in question. Histochemistry and radioautography have exciting potential for this demonstration. Finally, electrophysiological methods are most revealing in establishing the identity of action between suspected and actual transmitters.

In retrospect, it seems that the current view of autonomic neurotransmission, accepting acetylcholine and noradrenaline as the major substances involved as mediators, is based on persuasive evidence and thus appears justified. These may not be the only neurotransmitter substances. Others have been suggested as possible mediators at a variety of synapses, including substance P and other polypeptides, prostaglandin E, histamine and 5-HT. None of these other substances could be said to have been established as a mediator as has acetylcholine or noradrenaline.

Certainly, however, there is good evidence that the smooth muscle of the gut receives a non-adrenergic inhibitory innervation (see Chapters 8, 15 and 17). This has been shown by analysis of the frequency-response characteristics and pharmacological sensitivity of responses of intrinsic neurones of the intestine to electrical stimulation (Burnstock, Campbell & Rand, 1966). The intrinsic inhibitory neurones are more effectively activated at lower frequencies of stimulation than sympathetic neurones. Unlike adrenergic neurones they are resistant to adrenergic neurone blocking agents such as bretylium. Similar inhibitory neurones have been found within the vagal pathway to the stomach (Martinson, 1965b; Campbell, 1966). The vagal inhibitory pathway involves a synapse in the wall of the gut. These ganglia have distinct receptors for acetylcholine and for 5-HT (Bülbring & Gershon, 1967). Since blockade of either of these receptors individually diminishes but does not block vagal relaxation of the stomach, and since both are released from the stomach on vagal stimulation, both may be involved in transmission to the intrinsic inhibitory ganglion cells. Intestinal inhibitory ganglia also have distinct receptors for acetylcholine and 5-HT (Drakontides & Gershon, 1968). The nature of the transmitter released by the post-ganglionic inhibitory neurone has not yet been determined.

There is also evidence that the circular smooth muscle of the gut receives a non-cholinergic excitatory innervation (see Chapter 17). For example, Kottegoda (1968) has found that contractions of the circular muscle in response to repetitive electrical stimulation (1 to 10 cps) were not blocked by antagonists of cholinergic transmission such as hyoscine, morphine, hemicholinium, or botulinum toxin, type D. Similar results have been obtained by Ambache & Freeman (1968) who find, in addition, that the unknown excitatory transmitter is not likely to be 5-HT, histamine or prostaglandin E. Ambache and Freeman conclude that the circular muscle receives both a cholinergic and a non-cholinergic excitatory innervation, although the possibility that the non-cholinergic component represented stimulation of afferent neurones normally involved in the peristaltic reflex arc could not be ruled out.

Histamine also has been implicated in neurotransmission to smooth muscle. In this case, the evidence has indicated that the vasodilatation that follows elevation of blood pressure may be due to a liberation of histamine from sympathetic nerves (Beck, Pollard, Kayaalp & Weiner, 1966; Brody, 1966; Tuttle, 1967; and Chapter 15). Histamine and enzymes involved in its biosynthesis and metabolism may be found in sympathetic post-ganglionic nerves (Torp, 1961; Enerbäck, Olsson & Sourander, 1965) although they may be present mainly in mast cells. Most of the evidence in favour of histamine as the vasodilator substance involves the ability of antihistamines to block the reflex vasodilatation. The evidence is therefore relatively weak in that so much depends on the specificity of these compounds. After administration of radioactive histamine, an increase of labelled histamine and methylhistamine in venous blood accompanies reflex vasodilatation. There is still no evidence, however, that post-ganglionic sympathetic axon terminals store histamine and release it from their varicosities to cause vasodilatation.

The history of neurotransmitter analysis suggests that it will be difficult to prove that one or another suspect substance actually is a neurohumoural agent at any given synapse. However, modern additions to the methods available hold out the promise that additional compounds will eventually be identified as transmitter substances. In any case, the pace of the advance in knowledge of the two established transmitters has been increasing rapidly and can be expected to continue to expand rapidly in the future.

17

PERISTALSIS OF THE SMALL INTESTINE

S. R. KOTTEGODA

INTRODUCTION

'The word "autonomic" does suggest a much greater degree of independence of the central nervous system than, in fact, exists except, perhaps, in that part which is in the wall of the alimentary canal'   *Langley* (1921)

In 1899 Bayliss and Starling observed that peristaltic contractions of the intestine are true co-ordinated reflexes which could be evoked by local stimuli and that these reflexes are carried out by local nervous mechanisms which are independent of the connections of the gut with the central nervous system. These and other observations formed the basis of the 'Law of the Intestine'. Since that time there have been several investigations into the nature of the peristaltic reflex. In most of these *in vitro* methods were used. Although the validity and importance of the 'Law of the Intestine' have been challenged, the original observations of Bayliss and Starling have not been questioned (Davenport, 1966).

THE DIFFERENT PHASES OF THE PERISTALTIC REFLEX

Trendelenburg (1917), having designed the preparation which was to remain classical in such studies, showed that when the lumen of the isolated small intestine was distended by fluid, the longitudinal muscle contracted first causing a shortening of the segment of intestine. This was followed by the contraction of the circular muscle with simultaneous relaxation of the longitudinal muscle. The contraction of the circular muscle took the form of a wave travelling in the aboral direction emptying the contents of the gut.

Bayliss and Starling had described inhibition of the peristaltic reflex by cocaine, and Feldberg & Lin (1949a) concluded that this action of cocaine, like the blocking of the emptying phase by tubocurarine or by large doses of nicotine, was due to inhibition of ganglionic transmission in the reflex pathway. Paton & Zaimis (1949) showed that the ganglion-blocking drug, hexamethonium, prevented the emptying phase of peristalsis. Later Bülbring, Lin & Schofield (1958) found that, when cocaine was applied to the mucosal surface of the gut, the peristaltic reflex was abolished while the response to nicotine remained at first intact. Thus, the sensory receptors sensitive to distension appeared to be located near the mucosal surface. When cocaine had been left in the lumen of the intestine for some time, the response to nicotine was also gradually abolished, presumably because of the diffusion of cocaine into the myenteric plexus. Bülbring, Lin & Schofield (1958) showed further that nerve fibres from the ganglion cells in the submucous plexus penetrate into the mucous membrane, and they drew attention to the importance of the mucous membrane and the structures in it for the initiation of the peristaltic reflex. Confirming the results of early investigations that the peristaltic reflex was unaffected by degenerative changes of all extrinsic nerves, they demonstrated that the reflex could be abolished by procedures which affected the integrity of the mucosal nerve fibres. Thus, asphyxiation of the mucosa, application of a local anaesthetic to it, or removal of the mucous membrane, all abolished the peristaltic reflex, while the response to nicotine remained intact.

Bülbring & Lin (1958) found that when 5-HT, which is a potent stimulant of sensory receptors in general, was introduced into the lumen of the intestine, there was a stimulation of peristalsis. They attributed this effect to stimulation of chemo-receptors and sensitization by 5-HT of mechano-receptors in the mucosa which normally trigger the peristaltic reflex. Substance P (Beleslin & Varagić, 1958) and phenyl diguanide (Bülbring & Lin, 1958; Bülbring & Crema, 1958) can have an action similar to that of 5-HT, when applied to the mucosa.

The stimulus which activates the receptors for the reflex contraction of the longitudinal muscle (Kosterlitz & Robinson, 1959) is qualitatively similar to that which elicits the reflex contractions of the circular muscle (Ginzel, 1959). This is: 'deformation of the sensory receptors by radial stretching' according to Kosterlitz & Lees (1964) who pointed out that 'the time lapse between the contractions of the two (muscle) layers is due to the fact that a greater degree of distension of the lumen is required for the contraction of the circular muscle than for the longitudinal muscle' (Kosterlitz & Lees, 1964). Schaumann, Jochum & Schmidt (1953) showed that slow or sudden stretch applied to the longitudinal muscle along its long axis does not provoke a shortening of this muscle.

It must be mentioned that the same or other sensory structures may initiate afferent impulses to the central nervous system via the vagus. In some animals, about 90% of the fibres of the abdominal vagus are afferent (Cragg & Evans, 1958) and distension of the gut (Iggo, 1957; Paintal, 1957) as well as substances such as 5-HT (Paintal, 1954) may stimulate such receptors. It has been suggested that the normal function of the mucosal mechanoreceptors with afferents in the vagus is to signal the passage of intestinal contents (Paintal, 1957). It is possible that mechanoreceptors sensitive to stretch may also exist in the myenteric plexus (Paton, personal communication).

It is unfortunate that a multiplicity of terms has been introduced to describe the various phases of the peristaltic reflex. This can lead to confusion especially when the effects of drugs on the reflex are discussed.

Trendelenburg (1917) in his systematic study, distinguished two phases in the reflex. The first or 'preparatory phase', in which the contraction of the longitudinal muscle shortens the gut, was followed by the 'emptying phase' when the wave-like contraction of the circular muscle emptied the contents of the gut. Kosterlitz and his co-workers (see Kosterlitz & Lees, 1964), on the premise that the preparatory phase was neither facilitatory nor necessary for the evacuation of the contents of the intestine, introduced other terminology. Thus, Kosterlitz & Robinson (1959) called that part of the reflex contraction of the longitudinal muscle which was resistant to ganglion-blocking drugs, the 'graded response' or 'type I contraction' or 'primary contraction' of the longitudinal muscle. They distinguished a second type of contraction of the longitudinal muscle which followed the first one, which they called the 'type II contraction'. The latter was associated with the contraction of the circular muscle in the 'emptying phase'. Thus the emptying phase, according to Kosterlitz and his colleagues, is composed of the type II contraction of the longitudinal muscle and the wave-like contraction of the circular muscle. Apparently, 'ejection phase' is synonymous with emptying phase (Kosterlitz & Robinson, 1957). Summarizing the phases of the reflex, Kosterlitz & Lees (1964) stated 'we therefore propose to call the first phase the *graded reflex of the longitudinal muscle* and the second phase the *peristaltic reflex proper*'; it would appear that 'peristaltic reflex proper' is synonymous with 'emptying phase'. 'The complete peristaltic reflex' according to these authors 'is made up of the graded reflex of the longitudinal muscle and the peristaltic reflex proper'. Thus, the type II contraction of the longitudinal muscle 'is a part of', or 'is synchronous with, the contraction of the circular coat' (Kosterlitz & Robinson, 1957); it occurs 'as soon as the emptying phase comes into play' (Kosterlitz & Robinson, 1959) or 'is observed during the emptying phase' (Kosterlitz & Lees, 1964).

Kosterlitz & Robinson (1957) believed that ganglion-blocking drugs inhibited the contraction of the circular muscle of the emptying phase but not the longitudinal muscle contraction of the preparatory phase. Later work from the same laboratory expressed the view that the type II contraction of the longitudinal muscle (part of the 'peristaltic reflex proper') was also inhibited by ganglion-blocking drugs (Kosterlitz & Robinson, 1959). In view of these findings the following extract from Kosterlitz & Lees (1964) is puzzling.

'Although the contractions of the longitudinal muscle and circular muscle layers follow each other in a fixed pattern; the contraction of one muscle layer does not trigger the contraction of the other, since ganglion-blocking agents block the contraction of the circular muscle *only* (italics mine) and high concentrations of acetylcholine cause a non-specific block of the longitudinal muscle without affecting the circular muscle.'

It is not clear which component of the longitudinal muscle contraction is referred to in the above statement since according to the same authors the 'graded response' and not the type II contraction is resistant to ganglion-blocking drugs. On the other hand, Schneider (1966) concluded that the longitudinal muscle contraction is sometimes sensitive and sometimes resistant to ganglion-blocking agents. From this it might appear that a subdivision of the longitudinal muscle contraction of the reflex on the basis of the sensitivity to ganglion-blocking agents is not satisfactory.

It is the opinion of the author that to associate the term 'emptying phase' with any contraction of the longitudinal muscle is liable to be misleading. Emptying the contents of the gut is entirely due to the co-ordinated contraction of the circular muscle, in response to the appropriate stimulus, travelling in an aboral direction. If any part of the peristaltic reflex is to be called the 'emptying phase' it should only be the contraction of the circular muscle. The early investigators in this field (Trendelenburg, 1917; Feldberg & Lin, 1949a) and recent authors as well (Harry, 1962; Bhattacharya & Sen, 1962) used the term in this sense. Further, the evacuating contraction of the circular muscle can occur without any immediately preceding activity of the longitudinal muscle. Therefore, in this paper the term 'emptying phase' will refer only to the contraction of the circular muscle, and no attempt will be made to subdivide the contraction of the longitudinal muscle.

The longitudinal muscle contraction

In his recent review, Kosterlitz (1967), again confining himself to the contribution of the longitudinal muscle to the peristaltic reflex, postulated 5 types of nervous pathways to this muscle:

(a) The pathway for the graded response with no synapse or a synapse with non-cholinergic transmission. The postganglionic fibre here would be cholinergic.

(b) A similar pathway where the excitatory transmitter is not acetylcholine.

(c) The classical cholinergic pathway to the longitudinal muscle with at least one cholinergic synapse and postganglionic cholinergic transmission.

(d) An adrenergic inhibitory pathway with a cholinergic synapse.

(e) A similar ('speculative') non-adrenergic inhibitory pathway with an unknown transmitter.

The observations discussed so far have not provided convincing evidence that the longitudinal muscle contraction can be divided into two strictly separate phases. However, there is evidence for the existence of an additional non-cholinergic excitatory transmitter which may be substance P (Paton & Zar, 1966).

There is no doubt that adrenergic inhibitory pathways exist (see Chapters 15 and 19) but there is no evidence that they come into play during normal peristalsis. On the other hand, evidence has accumulated recently indicating the existence of an intrinsic non-adrenergic inhibitory innervation which is very likely playing an important part in peristalsis (see Chapters 8, 15 and 16).

There is ample evidence that in the longitudinal muscle excitatory neuro-muscular transmission is mainly cholinergic. Paton (1955) using co-axial stimulation of the guinea-pig ileum, demonstrated that the contraction of the longitudinal muscle produced by single shocks of short duration (1 msec) was accompanied by an increase in the acetylcholine output from the gut. This contraction, while resisting ganglion-blocking drugs, could be enhanced by anti-cholinesterases and was blocked by atropine. Paton & Zar (1968) showed that stimulation of the innervated longitudinal muscle strip of the guinea-pig ileum released acetylcholine. Further, the longitudinal muscle, when innervated and, to a lesser extent, in the absence of nerves, contracted readily in response not only to minute doses of acetylcholine but also to low concentrations of numerous other substances. For example, like acetylcholine, the following substances caused contraction of the longitudinal muscle by a direct action: arecoline, muscarine, histamine, tremorine, oxytocin and substance P, while angiotensin, barium, potassium, bromophenylcholine ether and 5-HT had, in the order listed, a progressively increasing effect on the nerve plexus.

The circular muscle contraction

In strong contrast to the longitudinal muscle, the circular muscle is singularly insensitive to the stimulating substances listed above. Harry (1963) found that the circular muscle preparation of the guinea-pig ileum was insensitive to drugs known to act on autonomic effector tissues and even acetylcholine had to be used in doses between 10^{-5} to 10^{-4} to produce an effect; after inhibition of cholinesterase by mipafox 10^{-4} for 90 min, the tissue contracted in response to choline esters, 5-HT, histamine and nicotine. The contractions produced by 5-HT, histamine and nicotine were abolished by procaine, botulinum toxin, morphine and hemicholinium whilst the actions of the choline ester were unaffected. Brownlee & Harry (1963) drew attention to several differences between the responses of the longitudinal muscle and those of the circular muscle of the guinea-pig ileum to excitatory agents. They found that substance P and angiotensin were without effect on the circular muscle. Histamine, which acts directly on the longitudinal muscle, caused contraction of the circular muscle, probably by acting via the intrinsic nerve plexus since the action was antagonized by hyoscine or morphine. Brownlee & Johnson (1963), who studied the effects of 5-HT on the guinea-pig ileum, concluded that the main action of 5-HT in this tissue was located at the autonomic ganglion cells. Recently, Kottegoda (1969) found that bradykinin and gamma-aminobutyric acid (GABA) had no action on the circular muscle.

Thus, while there is extensive information concerning the response of the longitudinal muscle both to excitatory agents and to nerve stimulation, this is not so as far as the circular muscle is concerned. This is not surprising when one considers that a satisfactory innervated circular muscle preparation suitable for studying the circular muscle by itself has so far not been available because of the intimate connections between the two muscle coats and the myenteric plexus. Further, there seems to be a tacit acceptance, without much evidence, that the excitatory transmitter to the circular muscle, like that for the longitudinal muscle, is acetylcholine. Perhaps the main evidence which supports such an assumption lies in the observations of Harry (1962). He found that the contraction of the longitudinal muscle as well as the co-ordinated propagated contraction of the circular muscle and the concurrent release of acetylcholine produced by transmural stimulation of the guinea-pig ileum, were reduced or absent after previous exposure to botulinum toxin. This observation will have to be amended, as discussed below.

It is still puzzling, why the circular muscle is relatively resistant to the stimulant action of acetylcholine even after the inhibition of cholinesterase in view of the finding that there is more true cholinesterase associated with the circular muscle than with the longitudinal muscle (Koelle, Koelle & Friedenwald, 1950). However, the effects which anticholinesterases often produce on the innervated circular muscle do not appear to be due to actions at the myoneural junction. The initiation of the peristaltic type of contraction by organophosphate anticholinesterases in the rabbit intestine is believed to be due to action at autonomic ganglia as the effect is abolished by ganglion-blocking drugs (Erdmann & Heye, 1958). Similarly the spasm of the circular muscle of the isolated guinea-pig ileum observed when eserine is added to the bath, is blocked by pentolinium (Kottegoda, unpublished observation).

Temporal relationship between the two muscle contractions

When the longitudinal muscle contracts suddenly in response to the distension of the gut, at or near the peak of this contraction, the circular muscle begins to contract. This time lapse can be explained on the basis of different thresholds for the receptors subserving the two reflex contractions. The question remains whether there is any mutual relationship in the subsequent behaviour of the two muscle coats in the peristaltic reflex. While the contraction of the circular muscle proceeds, the longitudinal muscle relaxes. Thus, it is only for an extremely brief time, if any, that the two muscles are contracting together. The findings of Schaumann et al. (1953) indicate that the relaxation of the longitudinal muscle is an active process. They used the original Trendelenburg method and showed that, when the longitudinal muscle was first contracted by acetyl-choline, the reflex could still be initiated by distension of the lumen, i.e. the circular muscle contracted and the longitudinal muscle relaxed. If the relaxation of the longitudinal muscle in normal peristalsis is an active process, what is the mechanism responsible for its initiation and why does it remain relaxed as long as the circular muscle contraction continues? Is it possible that the nervous pathways for excitation and inhibition of the two muscle layers are arranged so that the muscles do not contract simultaneously? From a teleological point of view, it would seem essential to have an arrangement whereby during the contraction of one muscle coat the other relaxes or is prevented from contracting. Simultaneous contraction of both muscle layers would defeat the purpose of the peristaltic reflex, i.e. the propulsion of the contents of the gut.

RESPONSE TO ELECTRICAL STIMULATION OF INTRINSIC NERVES

In a recent investigation, Kottegoda (1968, 1969) used the isolated guinea-pig ileum set up for recording peristaltic contractions (Bülbring, Crema & Saxby, 1958) with electrodes suitably placed for co-axial electrical stimulation (Paton, 1955). Co-axial stimulation was applied at sub-threshold intraluminal pressure, to avoid interference by peristaltic movements. Pulses of 0·1 msec, at frequencies ranging from 1 to 10/sec, were used to stimulate the nerves. It was observed that, at a low frequency of 1 to 3/sec, the pattern of the responses of the two muscle layers were very similar to that in the peristaltic reflex: the longitudinal muscle contracted first, this was followed by a wave-like contraction of the circular muscle emptying the contents of the gut. As the contraction of the cir-cular muscle was developing, the longitudinal muscle relaxed although the stimulation was being continued (Fig. 1). On the basis of this observation and earlier work (e.g. Kosterlitz & Robinson, 1959; Ginzel, 1959) it was concluded that the contraction of the circular muscle seen under these conditions was reflex and not the response to stimulation of its excitatory nerves.

It is believed that the sensory receptors to the two muscles which are both activated by radial distension, are probably separate, those serving the circular muscle having a higher threshold than those of the longitudinal muscle (Koster-litz & Lees, 1964). Hence, during the reflex response to gradual distension of the lumen (Fig. 2a), the longitudinal muscle contracts first. It is also known that the excitatory nerves to the longitudinal muscle are activated by low frequencies of stimulation, even by single shocks (Paton, 1955). Therefore, the result shown in Fig. 1 and in Fig. 2b, which resembles that in Fig. 2a, can be interpreted as follows: When the longitudinal muscle contracted during stimulation at a frequency of 1/sec, the intraluminal pressure was increased by the shortening of the segment of ileum: this activated the receptors for the reflex contraction of the circular muscle. This interpretation is supported by the result in Fig. 2c: when the volume of intraluminal fluid was diminished, the response of the

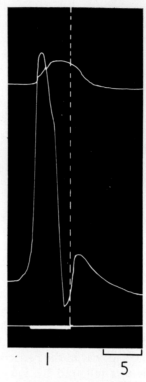

FIG. 1   Guinea-pig ileum. Co-axial electrical stimulation (0·1 msec pulses, 1/sec) for 5 sec (white bar) causes, first, a contraction of the longitudinal muscle (lower trace) followed by a contraction of the circular muscle (upper trace), during which the longitudinal muscle relaxes though stimulation continues. End of stimulation marked by broken line. Time marker: 5 sec.

longitudinal muscle remained unaltered, but it was unable to raise the intraluminal pressure sufficiently to reach the threshold necessary for activating the receptors for the circular muscle. Hence the circular muscle failed to contract. Moreover, in the absence of the circular muscle contraction the longitudinal muscle remained contracted until the stimulus was withdrawn.

The frequencies of stimulation necessary to activate the excitatory nerves to the circular muscle are higher than those required for the longitudinal muscle (Kottegoda, 1969). With frequencies between 3 and 10/sec, the contraction of the circular muscle was biphasic or, sometimes, merely prolonged (Fig. 3). The first part of the biphasic response was the reflex contraction described above, i.e. it corresponded to the single contraction seen when the frequency was 1/sec (Fig. 3a); the second component of the biphasic response was that due to activation of the excitatory fibres to the circular muscle by the electrical stimulus (Fig. 3b). Here too, with the contraction of the circular muscle, the longitudinal

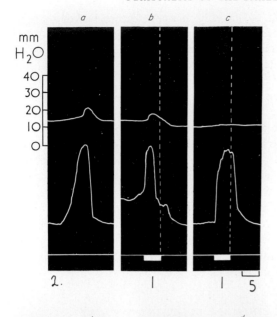

FIG. 2 (*a*) Peristaltic reflex in response to gradually rising intraluminal pressure (*top*: circular muscle, *bottom*: longitudinal muscle contraction). (*b*) Response to electrical stimulation at 1/sec. (*c*) Response to the same stimulus after the filling pressure had been reduced.

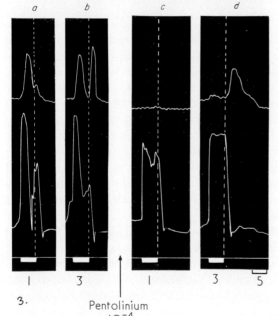

FIG. 3 The two components of the circular muscle contraction (top record), evoked by co-axial electrical stimulation, and the effect of ganglion block. (*a*) The response to low frequency stimulation (1/sec) is a reflex contraction (as in Fig. 1 and Fig. 2*b*). (*b*) The response to higher frequencies (3/sec) is a biphasic circular muscle contraction, the second component appearing after the end of stimulation. (*c*) Pentolinium abolishes the first component, but (*d*) not the second component. Note that, in the absence of a reflex circular muscle contraction, the longitudinal muscle (bottom record) remains contracted throughout the period of stimulation. Time marker: 5 sec.

muscle relaxed during stimulation. However, when the reflex component of the circular muscle response was suppressed by diminishing the volume of introluminal fluid, or by applying intraluminally a local anaesthetic, or in the presence

18*

of a ganglion blocking agent (Fig. 3c, d), only that contraction evoked by nerve stimulation was seen. Under these circumstances the longitudinal muscle remained contracted until the stimulus was withdrawn and the contraction of the circular muscle appeared only at the end of stimulation when the longitudinal muscle relaxed.

These observations were taken as indicative of a nerve-mediated reciprocal activity between the two muscle layers. During stimulation in the absence of drugs, the relaxation of the longitudinal muscle at a time when the circular muscle contracted was believed to have stemmed from the same nervous pathway which mediated the (reflex) contraction of the circular muscle. For, when the circular muscle contraction was absent the longitudinal muscle remained contracted until the stimulus was withdrawn. Similarly, at higher frequency of stimulation (which also excites the motor fibres to the circular muscle) the circular muscle contraction appeared only when the longitudinal muscle had relaxed, at the end of stimulation. This was interpreted as indicating that inhibitory fibres to the circular muscle were activated simultaneously with the excitatory fibres to the longitudinal muscle.

EFFECTS OF CHOLINERGIC BLOCKING DRUGS

These conclusions were further tested by the use of peripheral cholinergic blocking drugs which affected the responses of the two muscle coats to coaxial stimulation differently. On the basis of sensitivity or resistance to these substances it became clear that excitatory nervous transmission in the two muscle coats was different. Thus, hyoscine (Paton & Vane, 1963), morphine (Paton, 1957), hemicholinium (MacIntosh, Birks & Sastry, 1956; Wong & Long, 1961) and botulinum toxin (Ambache & Lessin, 1955) all greatly diminished or abolished the contractile response of the longitudinal muscle to co-axial stimulation at all frequencies of stimulation without abolishing the response of the circular muscle to the higher frequencies (Fig. 4A). In addition, it was observed that hyoscine inhibited the contraction of the longitudinal muscle during the peristaltic contractions of the circular muscle, though a much higher intraluminal pressure was required to trigger the reflex (Fig. 4B).

These observations are consistent with the view (Kottegoda, 1968, 1969) that the excitatory nerves to the circular muscle are not cholinergic. As stated earlier, Harry (1962) had found that botulinum toxin abolished both contractile responses of the circular muscle and of the longitudinal muscle to electrical stimulation (0·3 msec pulses). The discrepancy between these and Kottegoda's findings are explained by the differences in the frequencies of stimulation used in the two investigations. Higher frequencies of stimulation than 1/sec are usually necessary to activate the excitatory nerves to the circular muscle. Therefore, as explained above, the contractile response of the circular muscle to electrical stimulation at a frequency of 5/min, used by Harry, is a reflex response due to the rise in intraluminal pressure brought about by the contraction of the longitudinal muscle. Thus, in these circumstances, any agent (e.g. botulinum toxin) which merely blocked the contraction of the longitudinal muscle would also abolish that of the circular muscle since it was generated (reflexly) by the former.

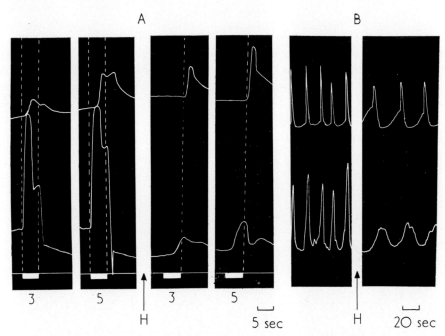

FIG. 4   The effect of hyoscine.   (A) The four panels show the response to co-axial electrical stimulation at two different frequencies (3/sec and 5/sec) before and after application of 2×10^{-7} g/ml hyoscine (H) at arrow. The longitudinal muscle contraction is almost abolished and, consequently, the reflex circular muscle contraction is also abolished, while the second component remains intact.   (B) Shows the peristaltic reflex contractions before and after hyoscine (H) at arrow. Note, that, in the presence of hyoscine, a much higher intraluminal pressure is required to elicit the reflex contraction of the circular muscle (*top record*). Note also that the longitudinal muscle contraction (*bottom record*) is much reduced and occurs later, i.e. when the circular muscle relaxes.

The contraction of the circular muscle evoked by higher frequencies of stimulation was not abolished by any cholinergic blocking agent, but it appeared only on discontinuing stimulation. This indicated that inhibitory fibres to the circular muscle were also activated and that this inhibition prevented the contraction. The delayed contraction of the circular muscle under these conditions is possibly due to a greater stability, and therefore persistence, of the excitatory transmitter whose effect breaks through as soon as stimulation stops.

Further evidence that the motor neurones which innervate the circular muscle are not cholinergic was obtained when it was observed (Kottegoda,

1969) that DMPP and 5-HT contracted the circular muscle, but not the longitudinal muscle, after the ileum had been exposed to botulinum toxin. The action of these two substances was presumably on intrinsic, non-cholinergic excitatory ganglia, since it was blocked by repeated application of DMPP.

RECIPROCAL INNERVATION

The diagram shown in Fig. 5 shows a possible arrangement of the reflex pathways to the two muscle layers (Kottegoda, 1969). It is possible to eliminate one portion (A) by exposure to high concentrations of acetylcholine. Schaumann *et al.* (1953) had shown that when acetylcholine (10^{-7} to 10^{-6}) was present in

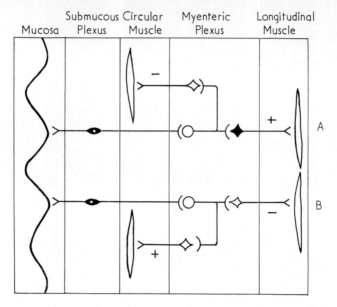

FIG. 5   Diagram of a possible arrangement of sensory, motor ($+$) and inhibitory ($-$) nerves in the intrinsic nerve plexus which may cause reciprocal activation of the two muscle coats. (From Kottegoda, 1969)

the bath, distension of the gut produced a propulsive contraction of the circular muscle with simultaneous relaxation of the longitudinal muscle. A similar reciprocal response of the two muscle layers is obtained with co-axial stimulation (Fig. 6) and also with 5-HT, or with dimethylphenylpiperazinium (DMPP) (Kottegoda, 1969) (Fig. 7). The action of 5-HT here was abolished by repeated doses of 5-HT (Fig. 8), or by DMPP. Since both substances are known to stimulate ganglia and then block them, and since their actions were not seen in the presence of tetrodotoxin, it was concluded that DMPP and 5-HT produced their effects by acting on neurones in a common pathway carrying

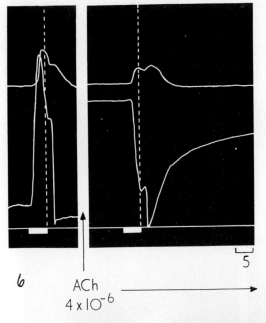

FIG. 6 The response to electrical stimulation (3/sec) before and after elimination of pathway A in Fig. 5, by exposure to a high concentration of acetylcholine (ACh 4 $\times 10^{-6}$ g/ml). Time marker: 5 sec. Note reciprocal response, i.e. simultaneous contraction of circular and relaxation of longitudinal muscle.

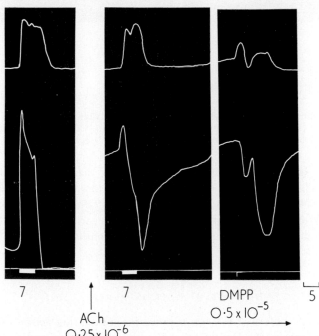

FIG. 7 The response to high frequency electrical stimulation (7/sec) before and during exposure to a high concentration of acetylcholine (ACh 0·25 $\times$ 10^{-6} g/ml). A reciprocal response is also elicited by DMPP (5 $\times$ 10^{-6} g/ml). Time marker: 5 sec.

excitatory fibres to the circular muscle and inhibitory fibres to the longitudinal muscle.

Some years ago, Ambache (1951) described a relaxant action of nicotine on the intestine after exposure of the gut to botulinum toxin. He suggested that the relaxant effect could be due to stimulation of peripheral adrenergic ganglia

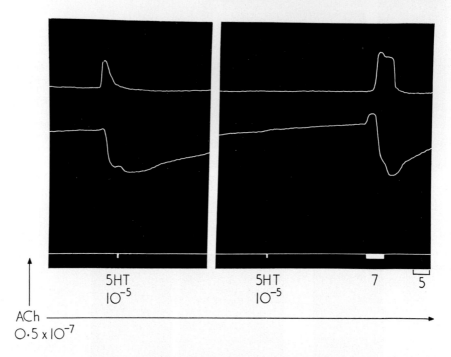

FIG. 8   In the presence of a high concentration of acetylcholine (ACh 5×10^{-8} g/ml) a reciprocal response is evoked by 5-HT (1×10^{-5} g/ml) A second application of 5-HT is ineffective, but the response to electrical stimulation (7/sec) is unimpaired.

by nicotine. In the experiments discussed above, the relaxation of the longitudinal muscle could not have been due to activation of adrenergic neurones since neither α- nor β-blockers abolished it (Kottegoda, 1969). Kosterlitz (1967) states that Watt has made a similar observation regarding the relaxation of the longitudinal muscle seen under the experimental conditions used by Schaumann *et al.* (1953).

THE ROLE OF 5-HT IN PERISTALSIS

As mentioned before, 5-HT has a powerful stimulating action on peristalsis when applied mucosally. On the other hand, it inhibits peristalsis when applied serosally (Kosterlitz & Robinson 1957; Ginzel 1957; Bülbring & Lin, 1958;

Bülbring & Crema 1958). It was observed (Kottegoda, 1969) that after repeated serosal application in the presence of acetylcholine, 5-HT blocked its own action as well as the reciprocal response of the two muscle layers to distension while the response to co-axial stimulation remained intact (Fig. 8). This inhibition of the *reflex* response by 5-HT in the presence of acetylcholine was similar to that produced by ganglion-blocking drugs, and this action may well be the basis for the inhibition of peristalsis which 5-HT causes on serosal application.

Small amounts of 5-HT are continuously released into the lumen of the in-testine; the amount released is directly proportional to the filling pressure and is greatly augmented during peristalsis (Bülbring & Lin, 1958; Bülbring & Crema, 1959a, b). In addition to the store in the mucosa (Feldberg & Toh, 1953) 5-HT is also located in the myenteric plexus (Gershon & Ross, 1966b). It has already been shown that 5-HT has no direct action on the circular muscle of the guinea-pig ileum (even its action on the longitudinal muscle here is mainly via the nerves), but there is evidence that it may have a role as a neural transmitter in the intramural plexus (see Chapters 18 and 20).

Bülbring & Gershon (1967) concluded that 5-HT, together with acetylcholine, participates as a preganglionic transmitter in inhibitory pathways in the stomach. Antagonists of 5-HT (e.g. LSD) are capable of inhibiting peristalsis (Ginzel, 1957; Bülbring & Lin, 1958). Parachlorophenylalanine, a specific inhibitor of 5-HT synthesis (Koe & Weissmann, 1966), when introduced into the fluid passing through the lumen of the ileum has been found to produce spasmodic arrest of peristalsis (Kottegoda, unpublished). Day & Warren (1968), who have put forward new evidence for the presence of inhibitory neurones in the intestinal wall, observed that reserpine (which depletes 5-HT stores) impaired the activation of these neurones by electrical stimulation.

If 5-HT is involved in some part of the nervous pathways mediating peristalsis the question of the relationship of such a role to the high concentrations of 5-HT found in the mucosa makes interesting speculation. It is known that nerves can take up 5-HT. Thus, it is possible that 5-HT is taken up by the sensory nerves from the mucosa and transferred along the axons to the preganglionic nerve endings in the myenteric plexus (see Masson, 1914; Masson & Berger, 1923). Williams (1967), in an electron microscope study, found evidence for the existence of interneurones between pre- and post-ganglion fibres in the rat; he observed small cells rich in 5-HT at the synaptic ends of the interneuronal bodies.

According to the observations of Kottegoda, neither 5-HT nor acetylcholine can be the post-ganglionic excitatory transmitter to the circular muscle in the guinea-pig ileum. Furthermore, the inhibitory transmitters to both muscle coats also remain to be identified. Thus, the only known transmitter in the nervous pathways postulated is the main excitatory transmitter to the longitud-inal muscle: acetylcholine. Recent observations indicate that prostaglandin may be the inhibitory transmitter in the circular muscle. Bennett, Friedmann & Vane (1967), who found that prostaglandin E_1 was released from the rat stomach during transmural stimulation, suggested that prostaglandins may have a role in the control of gastric mobility. Prostaglandin E_1 and E_2 inhibit selectively the contractile response of the circular muscle to co-axial stimulation; whereas they contract the longitudinal muscle (Kottegoda, 1969). PGE_2 has been identified in the mucosa of the human stomach and both PGE_1 and PGE_2 inhibit the circular muscle of the human stomach (Bennett, Murray & Wyllie, 1968).

It should be pointed out that the suggested nervous pathways from the mucosa to the two muscle layers (see Fig. 5) may not be applicable to all species. There is evidence that in the cat, strands of muscle passing between the longitudinal and circular muscle coats of the intestine may, without the intervention of nerves, initiate co-ordinated local movements in the two muscle coats (Kobayashi, Nagai & Prosser, 1966). In human embryos, wave-like, apparently myogenic, contractions of the circular muscle have been observed (Takita, personal communication) travelling in either direction along the gut at a stage when the myenteric plexus was not functionally developed. On the basis of these and other observations Takita proposed that 'basic peristalsis' was myogenic. However, it must be stressed that one cannot classify as peristalsis any kind of circular muscle contraction, whether associated with activity in the longitudinal muscle or not. In the true peristaltic reflex, the response of the circular muscle is a propagated wave-like contraction travelling in aboral direction emptying the contents of the gut.

SUMMARY AND CONCLUSIONS

Peristalsis, the co-ordinated reflex response of the gut to distension, bringing about the propulsion of the contents, can occur in the absence of all extrinsic nerves; it is mediated by nervous mechanisms located within the wall of the gastro-intestinal tract. The peristaltic reflex is abolished by tetrodotoxin which selectively blocks all nerve-mediated responses.

Radial distension is the mechanical stimulus which activates receptors, situated in or near the mucosa, initiating the reflex contractions of both the longitudinal and the circular muscle coat. Mucosal sensory receptors are also activated by substances which stimulate sensory receptors in general, including 5-HT. Local anaesthesia of the mucosa or any procedure which interferes with the integrity of the mucosa abolishes peristalsis. Ganglion blocking substances abolish the peristaltic reflex, indicating the presence of synapses in the reflex pathway.

The threshold of excitation by radial distension is lower for the longitudinal muscle than for the circular muscle. Therefore, when the intestinal lumen is distended slowly, the longitudinal muscle contracts first, shortening the intestine. The resulting increase in radial distension activates the sensory receptors of the excitatory pathway to the circular muscle. Its contraction is wave-like and travels in an aboral direction emptying the contents of the gut. During this circular muscle contraction the longitudinal muscle relaxes abruptly.

A similar sequence of events can be produced by co-axial stimulation (1/sec) of intrinsic nerves, indicating a nerve-mediated co-ordination between the activities of the two muscle layers. A hypothesis is put forward assuming that there are two separate motor pathways to the two muscle coats, each giving off collaterals which impinge on inhibitory neurones innervating the other muscle coat. This reciprocal innervation ensures that the two muscle layers never contract simultaneously and that, when one muscle contracts, the other relaxes or is prevented from contracting.

It is likely that 5-HT plays a role in interneuronal transmission in the inhibitory pathway. The nature of the final inhibitory transmitter to the muscles is, however, unknown. Moreover, the motor transmitter to the circular muscle, unlike that to the longitudinal muscle, does not appear to be acetylcholine.

The 'reciprocal inhibition' described in this chapter, i.e. the inhibitory mechanism which operates between the two muscle coats, which is very prominent in the small intestine, must be distinguished from the 'descending inhibition', which operates between two successive portions of the circular muscle coat and is very prominent in the large intestine. This is described in detail in Chapter 18. Unlike the inhibition of the circular muscle which takes place at the very beginning of the reflex, i.e. during the 'preparatory' longitudinal muscle contraction, the 'descending inhibition' is sensitive to atropine. This apparent contradiction might be a species difference. However, it is more likely that the two inhibitory mechanisms involve two different nervous pathways (e.g. as proposed in Fig. 5 of this chapter and in Fig. 5 of Chapter 18). The two nervous arrangements may be developed in a characteristic way for the small and for the large intestine respectively. They must necessarily be connected and the nervous mechanisms responsible for co-ordinating the two activities remain to be elucidated.

18

ON THE POLARITY OF THE PERISTALTIC REFLEX IN THE COLON

A. CREMA

INTRODUCTION

According to Bayliss & Starling (1900), the peristaltic reflex in the large intestine is due to a combination of 'ascending' excitatory and 'descending' inhibitory impulses. For the caudal propulsion of the solid content of the intestine, the relaxation below the bolus seems to play just as important a role as does the contraction above the bolus (Ritchie, 1968). Descending inhibition has been described by Bayliss & Starling (1900), Elliott & Barclay-Smith (1904), Langley & Magnus (1905), Cannon (1912); Auer & Krueger (1947), Hukuhara & Miyake (1959), Hukuhara, Nakayama & Nanba (1961), Williams (1967), and Hukuhara & Neya (1968).

The crucial problem is the mechanism which maintains the polarity of the peristalsis. Assuming that the propulsion is mainly due to ascending excitation, the polarity could be explained by the anatomical arrangement of either the smooth muscle (Nishio & Saito, 1965) or of the excitatory nerve fibres (Raiford & Mulinos, 1934; Bozler, 1949a, b). However, the existence of intrinsic inhibitory neurones must be assumed if one stresses the role of the descending inhibition (Ambache, 1951; Hukuhara, Yamagami & Nakayama, 1958). The presence in the intestine of an intrinsic inhibitory system has been postulated since 1922 by Langley.

PERISTALSIS ELICITED BY A LOCALIZED STIMULUS

In a recent investigation (Crema, Del Tacca, Frigo & Lecchini, unpublished) the question of polarity has been re-investigated, employing experimental procedures which simulate the physiological condition of the colon during peristalsis, i.e. by applying a localized stimulation with a solid bolus.

Isolated cat colon (12 to 15 cm in length) was mounted horizontally in an organ bath containing oxygenated Tyrode solution and was also perfused through its arterial supply. To elicit the reflex, the colonic wall was distended by filling with water a thin rubber balloon which had been introduced into the lumen through the fixed oral end of the colon. The rate of propulsion of the artificial bolus was measured by means of an isotonic transducer connected with the balloon. Simultaneously, the longitudinal movements were recorded by connecting the free aboral end of the colon to another isotonic transducer. Movements of the circular musculature were recorded, just above and below the bolus, by two strain-gauges which were applied 3·5 to 4 cm apart to the serosal surface of the colon.

The parasympathetic and sympathetic nerves were stimulated through electrodes placed around the pelvic nerves and the periarterial plexuses of the inferior mesenteric and common colonic arteries. The drugs were added to the fluid perfusing the organ.

THE 'DESCENDING INHIBITION'

The first response of the colonic musculature, subsequent to filling the balloon, is a longitudinal contraction which is followed by movement of the circular muscle consisting in a contraction above and a relaxation below the bolus. During the entire response period of the circular muscle, there is a shortening of the organ. The propulsion is dependent on the response of the circular muscle and occurs only in the presence of the descending inhibition (Fig. 1). There is a relationship between the degree of distension and the rate of propulsion.

Stimulation of the pelvic nerves improves the rate of propulsion (Fig. 3B), or initiates propulsive activity in the organ when spontaneous propulsion does not occur, or when the distension is subliminal (Fig. 2). Parasympathetic stimulation increases the contraction above, but only slightly affects the amplitude of the relaxation below the bolus (Fig. 3).

Hexamethonium (1×10^{-5} g/ml), added to the fluid perfusing the arteries, abolishes the response of both the longitudinal and circular musculature to distension, and also the effect of parasympathetic nerve stimulation (Fig. 3).

Stimulation of the periarterial plexuses prevents the peristaltic reflex in response to distension of the balloon, even during parasympathetic nerve stimulation. On the other hand, sympathetic denervation, achieved surgically by freezing the periarterial plexuses four days previously (Del Tacca, Lecchini, Frigo, Crema & Benzi, 1968), does not affect the descending inhibition and therefore does not impair the peristalsis.

Atropine and hyoscine, perfused into the arteries, interferes with the peristaltic reflex to a degree depending on their concentrations. At concentrations ranging from 1 to 3×10^{-8} g/ml, the longitudinal and circular contraction still occurs after distension but the propulsion is impaired. That is, the contraction of the

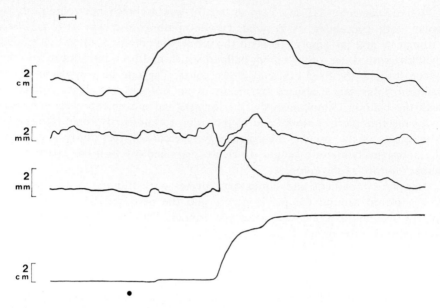

FIG. 1   Peristaltic reflex in the isolated colon. From top to bottom records of: longitudinal movements; movements of circular musculature below the bolus; movements of circular musculature above the bolus; longitudinal displacement of the balloon. Time marker: 10 sec. The mark (●) indicates maximal distension of the balloon.

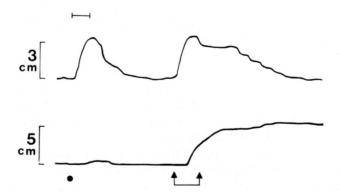

FIG. 2   Action of pelvic nerve stimulation on the propulsive activity in the isolated colon. Upper tracing: record of longitudinal movements. Lower tracing: record of the longitudinal displacement of the balloon. The mark (●) indicates subliminal distension of the balloon; the area between the arrows indicates pelvic nerve stimulation (for 15 sec, at 2/sec, 0·5 msec/pulse duration, supramaximal). Time marker: 10 sec. Note that parasympathetic stimulation is able to initiate colonic propulsion.

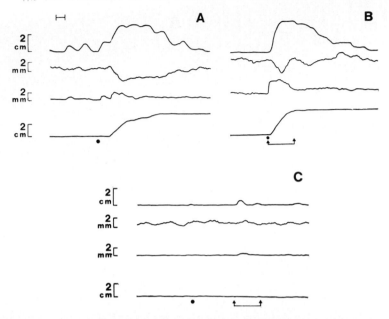

FIG. 3   Action of hexamethonium on the peristaltic reflex in the isolated cat colon. From top to bottom records of: longitudinal movements; movements of the circular musculature below the bolus; movements of the circular musculature above the bolus; longitudinal displacement of the balloon. Time marker: 10 sec. The mark (●) indicates maximal distension of the balloon. The period between the arrows indicates pelvic nerve stimulation (for 30 sec, 5/sec, 0·5 msec pulse duration, supramaximal). Between B and C, hexamethonium (1×10^{-5} g/ml) was added to the fluid perfused through the vascular supply. Note that pelvic nerve stimulation increases both the rate of propulsion and the contraction above the bolus, but does not affect the amplitude of the descending inhibition, which is of shorter duration due to the increased propulsion velocity. Note that hexamethonium abolishes all the components of the peristaltic reflex as well as the effects of pelvic nerve stimulation.

circular muscle above the bolus is maintained but the descending inhibition is abolished or even reversed into a contraction (Fig. 4). In the presence of such concentrations of atropine, propulsion is not initiated by pelvic nerve stimulation in spite of the fact that the musculature still contracts. The response of the longitudinal and circular musculature to distension is completely abolished by increasing the atropine concentration to 1×10^{-6} g/ml.

DISCUSSION AND CONCLUSIONS

The propulsion of the solid bolus in the colon is completely dependent upon the simultaneous development of contraction above and inhibition below the localized stimulus, in agreement with the myenteric reflex law (Bayliss & Starling, 1900; Cannon, 1912).

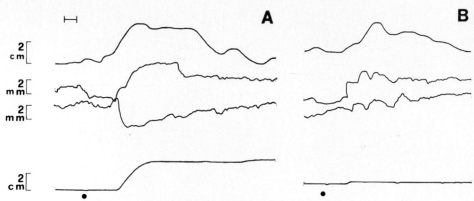

FIG. 4   Action of atropine on the peristaltic reflex in the isolated cat colon. From top to bottom records of: longitudinal movements; movements of circular musculature above the bolus; movements of circular musculature below the bolus; longitudinal displacement of the balloon. Time marker: 10 sec. The mark (●) indicates maximal distension of the balloon. Between A and B, atropine (1 × 10⁻⁸ g/ml) was added to the fluid perfused through the vascular supply. Note that atropine blocks the propulsion by reversing the descending inhibition.

The mechanism of propulsion seems to differ according to the nature of the intestinal content and, therefore, to the kind of stimulation (localized or widespread). If the content is fluid, which is the normal condition for the small intestine (Trendelenburg, 1917; Kosterlitz, Pirie & Robinson, 1956; Bülbring & Lin, 1958; Bülbring, Crema and Saxby, 1958; Lee, 1960; Kosterlitz & Lees, 1964), the descending inhibition can play only a secondary role. In contrast, when the content is solid, which is the normal condition in the colon, the descending inhibition is essential for the propulsion (Auer & Krueger, 1947; Truelove, 1966; Ritchie, 1968).

The intrinsic sympathetic nerves do not seem to be necessary for the initiation of peristalsis or for the descending inhibition in the colon, in agreement with results obtained by Bülbring, Lin & Schofield (1958), and Hukuhara, Yamagami & Nakayama (1958) in the ileum. Therefore, it seems reasonable to assume the presence in the intestinal wall of a non-adrenergic inhibitory system (see Chapters 8 and 15). Such a system has already been postulated both in animals (Burnstock, Campbell, Bennett & Holman, 1964; Holman & Hughes, 1965; Burnstock, Campbell & Rand, 1966; Bennett, Burnstock & Holman, 1966b; Bülbring & Tomita, 1967; Day & Warren, 1967; Kosterlitz, 1967; Day & Warren, 1968; Bianchi, Beani, Frigo & Crema, 1968) and in man (Crema, Del Tacca, Frigo & Lecchini, 1968).

Holman and Hughes (1965) have considered the possibility that the inhibitory neurons which innervate the longitudinal muscle of rat ileum are excited through cholinergic receptors. The fact that atropine or hyoscine, at concentrations ranging from 1 to 3 × 10⁻⁸ g/ml, block the descending inhibition without affecting the ascending contraction, leads us to the conclusion that muscarinic receptors are probably involved in the excitation of the inhibitory neurones to the circular musculature.

Since, in the presence of low concentrations of atropine, the distension of the balloon is able to produce a contraction both above and below the bolus, there apparently does not exist a polarity of the excitatory pathways. Therefore,

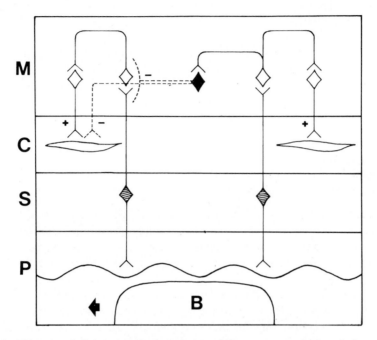

FIG. 5   Diagram of the proposed arrangement of the sensory, excitatory (+), and inhibitory (−) neurones involved in the circular muscle response during the peristaltic reflex of the colon as elicited by a solid bolus. B: balloon, M: myenteric plexus, C: circular muscle, S: submucosal plexus, P: mucosa, The arrow indicates the oral-aboral direction. The sensory neurone of the submucosal plexus excites a cholinergic interneurone which in turn impinges on the final motor neurone innervating the muscle, and through a collateral, on a caudal inhibitory neurone (black). This neurone is activated by muscarinic receptors. The broken lines indicate the two possible modes of inhibition. For further explanation see text.

we can assume an oral-aboral arrangement of the inhibitory system. Since hexamethonium abolishes all the reflex responses to distension, a nicotinic synapse must be involved in these pathways, including the one subserving the descending inhibition. The simplest explanation is to assume the existence of a caudally directed inhibitory pathway which originates (distally to the nicotinic synapse) from the excitatory neurones, so that the same sensory neurone can trigger an excitation above, and an inhibition below, the site of stimulation.

Figure 5 is a representation of the nervous mechanism proposed to explain the polarity. The scheme is oversimplified and assumes that the excitatory neurones of the myenteric plexus, excited by the sensory neurones of the sub-mucosal plexus through a nicotinic synapse, give off a collateral which impinges on a caudal inhibitory neurone (marked black in Fig. 5). The latter is excited through muscarinic receptors (Volle, 1966) and could cause the descending inhibition by two possible mechanisms: (1) by modulating the caudal motor neurones at the level of the ganglia, by means of an unknown mediator, or possibly by a presynaptic inhibition; (2) by acting directly on the smooth muscle.

Since atropine, in high concentration (1×10^{-6} g/ml), abolishes all the reflex responses of the circular muscle to distension, it is probable that the final trans-mitter from the excitatory nerve endings to the circular muscle is acetylcholine. This disagrees with the results obtained by Kottegoda (1968, 1969 and Chapter 17) in the guinea-pig ileum. This discrepancy could be due to:

i. the different parts of the intestine used in each investigation.
ii. the nature of the physiological stimulus used to elicit the peristaltic reflex, namely localized (bolus) in the colon, and widespread (fluid) in the ileum.
iii. the different animal species employed.

Apparently, the muscarinic receptors involved in the excitation of the inhibi-tory neurones are more sensitive to atropine than are the muscarinic receptors of the muscle, so that certain concentrations of atropine are able to block only the descending inhibition without affecting the strength of the muscular contrac-tion.

Parasympathetic stimulation augments the rate of propulsion, mainly by increasing the contraction above the bolus, while hardly affecting the descending inhibition. In our opinion, this indicates that the pelvic nerves impinge essen-tially on the excitatory neurones of Auerbach's plexus. On the other hand, the inhibitory system, when excited by the bolus, is not antagonized by parasym-pathetic nerve stimulation.

If we assume that the inhibitory neurones act directly on the smooth muscle by releasing a mediator which causes muscular relaxation, we can postulate a physiological role for the non-adrenergic inhibitory system described by the above-mentioned authors.

In conclusion, we believe that a particular anatomical arrangement of atropine-sensitive inhibitory neurones could explain the polarity of the propulsion of the solid contents of the colon.

19

ADRENERGIC RECEPTORS IN
THE INTESTINE

C. Y. Lee

INTRODUCTION

In his original description of the adrenergic receptive mechanism, Ahlquist (1948) designated the adrenergic receptors which subserve inhibition of the intestinal smooth muscle as the *alpha*-type. This conclusion was based primarily on the order of potency of a series of sympathomimetic amines in producing an inhibitory response in the intestinal muscle of several species. A strong argument against Ahlquist's classification of adrenergic inhibitory receptors of the intestine as the α-type was the inability of adrenergic blocking agents such as dibenamine and phentolamine to block effectively the relaxing effects of catecholamines on isolated segments of intestine (Nickerson, 1949; Furchgott, 1959). The findings that neither the α-blocker, dibenamine, nor the β-blocker, dichloroisoprenaline (DCI), effectively blocked the response of the rabbit small intestine to catecholamines, and that the potency ratio of three catecholamines in inhibiting intestinal smooth muscle was considerably different from that for the responses in other organs, led Furchgott (1959) to suggest that a third type of receptors, *delta*-type, mediated adrenergic relaxation of the intestine. Ahlquist & Levy (1959), however, found that a combination of DCI and an α-type blocking agent, dibozane, was able to antagonize the inhibitory effects of adrenaline on intestinal motility in the intact dog. In addition, they found that DCI alone was able to antagonize the inhibitory effect of isoprenaline, whereas dibozane alone was able to antagonize the inhibitory effect of phenylephrine. They, therefore, concluded that the dog intestine has both α- and β-receptors and that activation of either type leads to inhibition of motility. These results were confirmed subsequently by Furchgott (1960) in the *in vitro* experiments with isolated segments of the rabbit intestine and, more recently, also in the taenia of the guinea-pig caecum (Jenkinson & Morton, 1967a, c; Brody & Diamond, 1967; Bülbring &, Tomita, 1968a, 1969b) and in the taenia of the human colon (Bucknell & Whitney 1964).

Thus, it is now well-established that both α- and β-inhibitory receptors are present in the intestine. However, there is still no general agreement concerning the distribution of each type of receptors in the muscle layers and the neural elements of the intestinal wall, nor concerning the mechanisms of the inhibitory action of catecholamines. There is evidence to suggest that the receptors concerned with, and the mechanisms of, the inhibition of intestinal motility by catecholamines may vary, depending not only on experimental conditions but also on the type of motility to be inhibited.

ADRENERGIC INNERVATION OF THE INTESTINAL SMOOTH MUSCLE

The adrenergic nerves of the intestine generally have an inhibitory influence on the motor activity of the intestine, except in the regions of the sphincters (Elliot, 1904b; Dale, 1906). Until recently, it had generally been accepted that the adrenergic fibres terminate on the intestinal muscle layers (Hill, 1927) and that the adrenergic inhibitory function in the intestine is caused by a direct action of the liberated adrenergic transmitter on the smooth muscle cells (Youmans, 1949; Brown, Davies & Gillespie, 1958). There is now strong evidence to question the validity of this concept. Studies by Norberg (1964, 1965), Jacobowitz (1965), and Hollands & Vanov (1965) of the intestinal innervation with fluorescence microscopy show that most adrenergic nerves invest and terminate around the intramural cholinergic ganglion cells and not, as previously believed, in the muscle layers. The muscle layers receive very little adrenergic innervation, and what they do receive probably belongs to the blood vessels. A related observation was made by Marks, Samorajski & Webster (1962) who studied the localization of injected tritiated noradrenaline in the mouse small intestine. A considerable uptake of noradrenaline was localized in the neural elements contained in the myenteric plexus, whereas no uptake of noradrenaline was found in the nerve fibres in the muscle layers.

All these findings indicate that the muscle layers of the intestine are not directly innervated by adrenergic nerves and that neurogenic inhibition of intestinal motility may be exerted primarily by an indirect effect on the myenteric plexus. Nevertheless, a direct adrenergic inhibition of the smooth muscle could be produced either by diffusion of transmitter from sympathetic nerve endings in the myenteric plexus to the adjoining muscle (Gershon, 1967b), or by an overflow of transmitter from the vasoconstrictor nerve endings (Celander, 1959), or by the catecholamines released from the adrenal medulla.

LOCATION OF ADRENERGIC RECEPTORS IN THE
INTESTINAL WALL

Alpha-inhibitory receptors in the myenteric plexus

The presence of α-inhibitory receptors in the neural elements contained in the myenteric plexus was first advocated by McDougal & West (1954). They have shown that the inhibition of the peristaltic reflex of the isolated guinea-pig ileum by catecholamines can be specifically prevented by α-type adrenergic blocking agents. Experiments by Chiu & Lee (1961) and Lee & Tseng (1966) also indicate that α-receptors alone are concerned in the adrenergic inhibition of the peristaltic reflex of the isolated intestine of both guinea-pigs and rabbits, despite the presence of both α- and β-receptors in the intestinal smooth muscle. The nature of the mechanism of the inhibitory action of catecholamines on the peristaltic reflex is apparently different from that on the pendular movements. As shown in Table 1, the potency ratio of three catecholamines, adrenaline,

TABLE 1. Comparison between the potency ratio of adrenaline (A), noradrenaline (NA) and isoprenaline (ISO) in inhibiting peristalsis and that in inhibiting pendular movements

Response	Tissue	Potency ratio			Reference
		A	NA	ISO	
Inhibition of	Guinea-pig ileum	1	0·25	0·005	McDougal & West (1954)
peristalsis	Guinea-pig ileum	1	0·2	0·003	Chiu & Lee (1961)
	Rabbit ileum	1	0·4	0·003	Lee & Tseng (1966)
Inhibition of	Rabbit ileum	1	0·5	0·05	McDougal & West (1954)
pendular	Rabbit ileum	1	0·32	0·5	Lee & Tseng (1966)
movements	Rabbit duodenum	1	1–1·5	0·5–3	Furchgott (1960)

noradrenaline and isoprenaline, in inhibiting the peristaltic reflex of the isolated intestine consistently falls into the pattern characteristic of the α-receptors, as defined by Ahlquist (1948), whilst the relative activity of these catecholamines in inhibiting the pendular movements of the rabbit intestine does not fit the pattern characteristic of either α- or β-receptors. Moreover, while inhibition of the peristaltic reflex by these catecholamines can be completely prevented by α-type blocking agents but not by β-type blocking agents, (Chiu & Lee, 1961; Lee & Tseng, 1966) inhibition of the pendular movements of the rabbit intestine by adrenaline or noradrenaline can not be effectively antagonized by either kind of the adrenergic blocking agents alone, but is antagonized by a combination of the two (Furchgott, 1960). In addition, Lee & Tseng (unpublished) have also observed that inhibition of the peristaltic reflex of the guinea-pig ileum by periarterial sympathetic stimulation is completely blocked by tolazoline and not by propranolol (Fig. 1), while inhibition of the pendular movements of the rabbit ileum by the same sympathetic stimulation is blocked by neither one alone but blocked by the combination of the two (Fig. 2). All these findings indicate that adrenergic inhibition of the peristaltic reflex of the intestine, whether by applied catecholamines or by nerve stimulation, is effected through activation of α-receptors alone, whereas inhibition of the pendular movements involves

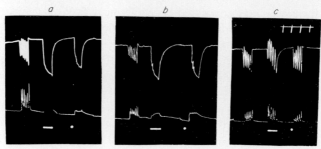

FIG. 1. Peristalsis in isolated guinea-pig ileum (Trendelenburg's method). Upper tracing, intestinal volume (downwards); lower tracing, contractions of longitudinal muscle. Time marker 30 sec. Effects of periarterial sympathetic nerve stimulation (indicated by the horizontal bars) and 1×10^{-7} g/ml adrenaline (indicated by dots) on the peristaltic reflex: a, control; b, in presence of 1×10^{-6} g/ml propranolol; c, in presence of 5×10^{-5} g/ml tolazoline. Note that inhibition of the peristaltic reflex either by sympathetic nerve stimulation or by adrenaline is antagonized by tolazoline but not by propranolol.

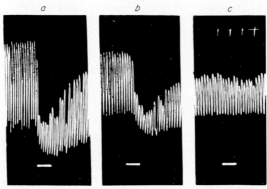

FIG. 2 Pendular movements in rabbit ileum (Finklemen's method). Horizontal bars indicate periarterial sympathetic nerve stimulation (0·2 msec, 60/sec, 30 V). Time marker 30 sec. a, control; b, in presence of 3×10^{-7} g/ml phenoxybenzamine; c, in presence of 3×10^{-7} g/ml phenoxybenzamine plus 1×10^{-6} g/ml propranolol.

both α- and β-receptors. As to the nature of the mechanism by which the inhibition of the peristaltic reflex is produced, McDougal & West (1954) have suggested a direct depressant action of catecholamines on transmission in the ganglionic synapses within the enteric plexus, largely based on these findings:

1. the contraction of the longitudinal muscle produced by nicotine is reduced by doses of the catecholamines which have little or no effect on the contractions produced by histamine and acetylcholine;

2. the inhibition of the nicotine contraction by the catecholamines is also antagonized by α-type blocking agents;

3. when the peristaltic reflex is inhibited by the catecholamines, eserine produces rhythmic contractions and an increase in tone of the circular muscle.

Although their conclusion was based on rather indirect evidence, recent histochemical studies with fluorescence microscopy (Norberg, 1964, 1965; Hollands & Vanov, 1965; Jacobowitz, 1965) have provided strong support for the possibility of depression of ganglionic transmission by the sympathetic nervous transmitter.

A related suggestion has been made by Kosterlitz & Watt (1965) who found that the effects of phenoxybenzamine and propranolol are additive in antagonizing the inhibitory actions of both adrenaline and noradrenaline on the responses of the longitudinal muscle of the guinea-pig ileum to co-axial stimulation in the presence of hexamethonium, whereas the inhibitory influence of the catecholamines on the contractions of the longitudinal muscle produced by acetylcholine or carbamylcholine was antagonized by propranolol but not by phenoxybenzamine. They interpreted these observations as showing that the α-receptors are situated on neurones (presumably cholinergic), innervating the longitudinal muscle, and the β-receptors are in the muscle itself.

Catecholamines acting on α-receptors situated on cholinergic neurones would appear to interfere with the conduction of the impulse in the nerve fibre and its terminals, or with the release of transmitter (Schaumann, 1958), and may play an important role as inhibitory modulators of ganglionic transmission (for references see McIsaac, 1966).

Recently, Paton & Vizi (1969) showed that adrenaline and noradrenaline reduce the release of acctylcholine from the longitudinal muscle strip of guinea-pig ileum by up to 80%, both at rest and during field stimulation. They concluded that the parasympathetic nervous activity in the longitudinal strip was under sympathetic control. On the other hand, Gershon (1967b), using guinea-pig stomach, rabbit jejunum and guinea-pig ileum (i.e. preparations including both muscle layers) found that sympathetic nerve stimulation caused no diminution of acetylcholine output. He came to the conclusion that the inhibition was due to direct action of the released noradrenaline on the muscle. He did not test whether the effects he was studying were on α- or β-receptors.

The two results are not necessarily contradictory since, as mentioned earlier, the possibility exists that the sympathetic nervous transmitter, though released in the nerve plexus, reaches the muscle.

Alpha- and beta-inhibitory receptors in the muscle layers

As already mentioned, Ahlquist & Levy (1959) have demonstrated that both α- and β-inhibitory receptors are present in the canine ileum, and their findings have been confirmed by many other investigators in the intestine of various species (Furchgott, 1960; Bucknell & Whitney, 1964; van Rossum & Mujić, 1965; Brody & Diamond, 1967; Jenkinson & Morton, 1967a, c; Bülbring & Tomita, 1968a, 1969b). However, it has been suggested that the two types of receptors may have different locations in the gut. As mentioned in the foregoing paragraph, Kosterlitz & Watt (1965) thought that the α-receptors are situated in the neurones innervating the longitudinal muscle and the β-receptors in the muscle itself in the guinea-pig ileum. Wilson (1964) also claimed that the adrenergic receptors located in the longitudinal muscle layer of the guinea-pig ileum are β-receptors, judging from the relative potencies of four sympathomimetic amines for the inhibition of histamine- and methacholine-induced contractions of the longitudinal muscle layer. Recent experiments by Lee & Cheng (unpublished) also failed to detect any α-inhibitory receptors in the atropine-treated longitudinal muscle layer of the guinea-pig ileum. In view of these findings together with those of Lum, Kermani & Heilman (1966) that cold storage produces a selective impairment or loss of α-receptor activity in the rabbit jejunum, it is tempting to speculate that all the α-inhibitory receptors in the intestine might be located on the intramural neural elements. However, in agreement with many other workers already mentioned, our attempts to eliminate the α-receptor activity from the isolated rabbit ileum and the taenia of the guinea-pig caecum by pre-treatment with atropine (10^{-6} g/ml) and/or tetrodotoxin (1 to 5×10^{-7} g/ml) have failed to substantiate such a generalized assumption. It was found that the relaxation produced by adrenaline or noradrenaline in these preparations could not be effectively antagonized by propranolol alone even after pre-treatment with atropine and/or tetrodotoxin.

On the other hand, a different distribution of adrenergic receptors between the circular and longitudinal muscle layers has been suggested by the experiments of Harry (1964) who distinguished only the α-inhibitory receptors in the circular muscle layer of the guinea-pig ileum. He found that noradrenaline and adrenaline were much more active than isoprenaline in inhibiting the contractions of the circular muscle strip produced by methacholine or carbachol and that the inhibitory action of adrenaline was specifically antagonized by piperoxane but not by dichloroisoprenaline (DCI). Based on these findings, Harry concluded that the site of action of the catecholamines on the circular muscle of the guinea-pig ileum is located at post-ganglionic neuroeffector junctions in the smooth muscle, rather than at the intramural nerve plexus as suggested by McDougal & West (1954). However, recent experiments by Lee & Chen (unpublished) indicate that both α- and β-receptors are present in the circular muscle layer of the guinea-pig ileum, although the α-receptors are more dominant in the circular muscle. It may be seen from the experiment illustrated in Fig. 3A, that the effect of adrenaline in inhibiting histamine-induced contractions of the circular muscle strip was antagonized not only by phentolamine but also by propranolol. It is also seen in Fig. 3A and B, that before atropine treatment, phentolamine (10^{-6} g/ml) was much more effective than propranolol (10^{-7} g/ml) in antagonizing the action of adrenaline, whereas the order of potency was reversed after atropine treatment. The difference between our results and those of Harry (1964) might be due to different experimental conditions employed. Harry tested the effects of the amines in inhibiting methacholine- or carbachol-induced contractions of the circular muscle strip, pre-treated with the anti-cholinesterase mipafox, whereas we used histamine as the stimulant without any

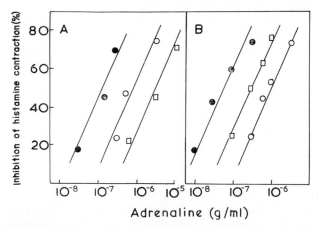

FIG. 3   Comparison of effectiveness of *alpha*- and *beta*-blocking agents in antagonizing the action of adrenaline in inhibiting histamine-induced contractions of the circular muscle strip of the guinea-pig ileum. A; in absence and B; in presence of atropine (10^{-6} g/ml). ●—●: control; □—□: after phentolamine (1×10^{-6} g/ml); ○—○: after propranolol (1×10^{-7} g/ml).

pretreatment. [Although Harry (1963) reported that, before inhibition of cholinesterase, the circular muscle strip did not respond to histamine, we obtained invariably good responses of the muscle strip with 10^{-6} g/ml histamine without any pre-treatment in our experiments.]

The finding that the β-receptors are also present in the circular muscle layer of the guinea-pig ileum was further supported by the following observations: As shown in Table 2, noradrenaline and adrenaline were much more active than isoprenaline on the non-treated circular muscle strip, whereas the order

TABLE 2. ED_{50} and relative potencies of catecholamines in inhibiting histamine-induced contractions of the circular muscle strip from the guinea-pig ileum

Catecholamine	No atropine		In presence of atropine (10^{-6} g/ml)	
	ED_{50} (g/ml)	Potency ratio	ED_{50} (g/ml)	Potency ratio
(−)-Noradrenaline	1.26×10^{-7}	1	1.99×10^{-7}	1
(−)-Adrenaline	1.59×10^{-7}	0.8	3.2×10^{-8}	6.2
(±)-Isoprenaline	1.26×10^{-6}	0.1	1.99×10^{-8}	10

of inhibitory potencies of these catecholamines was reversed in the atropine-treated circular muscle strip. The results described strongly suggest that, in guinea-pig ileum, at least part of the α-effects of catecholamines may be contributed by an action on receptors on the cholinergic neurones although both α- and β-receptors are present in the circular muscle layer itself.

Alpha-excitatory receptors in the intestine

The presence of the *α-excitatory* receptors in the intestinal sphincters was first demonstrated by the experiments of Dale (1906) who found that the motor response to stimulation of the sympathetic nerves or to adrenaline was abolished by ergot alkaloids. Apart from the regions of the sphincters, observations have also been reported from time to time suggesting the presence of α-excitatory receptors in other parts of the intestinal wall (Bunch, 1898; Magnus, 1903; Hoskins, 1912; Bernheim & Blocksom, 1932; Brunaud & Labouche, 1947; Lands, Luduena, Ananenko & Grant, 1950; Munro, 1951; Lands, 1952). Unlike in the oesophagus (Burnstock, 1960; Bailey, 1965) and stomach (Furchgott, 1967), however, the presence of α-excitatory receptors in the intestinal smooth muscle is usually difficult to demonstrate, except in the regions of the sphincters and the terminal ileum (Munro, 1951). It may be that, in these regions, α-excitatory receptors are more dominant than the *inhibitory* receptors so that the usual effect of adrenaline would be excitatory. No information has been available as to whether the α-excitatory receptors are present in the whole intestine. Only recently, Lee & Cheng (unpublished) found that α-excitatory receptors are present in the longitudinal muscle of the guinea-pig ileum but not in the duodenum. In strips of the guinea-pig ileum, taken at more than 20 cm distance from the ileo-caecal sphincter, adrenaline often produced a contraction after pre-treatment with atropine and propranolol. If phentolamine was added, however, no contraction was produced by adrenaline.

CONCLUDING REMARKS

It is evident that both α- and β-inhibitory adrenergic receptors are present in intestinal smooth muscle.

Apart from the regions of the sphincters and the terminal ileum, the presence of the α-excitatory receptors is usually difficult to demonstrate, presumably due to the predominance of the inhibitory receptors of both types. The mechanisms of inhibition induced by activation of the α- and β-receptors are essentially different (see Chapter 12).

The myenteric ganglia and post-ganglionic cholinergic fibres appear to possess α-inhibitory receptors only. There is no evidence to indicate the presence of β-receptors in the nerve elements. Recent fluorescent histochemical findings suggest that the myenteric ganglia are directly innervated by the adrenergic nerve fibres whereas the muscle cells are mostly, if not entirely, non-innervated.

The α-inhibitory receptors on myenteric ganglia are involved in the inhibition of the peristaltic reflex presumably due to the released catecholamine suppressing intrinsic ganglionic transmission. Both α- and β-receptors on the muscle are involved in the inhibition of the myogenic pendular movements, whether by sympathetic nerve stimulation or by administered catecholamines.

19+

20

EXCITATION OF VASCULAR SMOOTH MUSCLE

R. N. SPEDEN

INTRODUCTION

Vascular smooth muscle, by contracting or relaxing, changes the distensibility of blood vessel walls. The resulting changes in internal diameter of the vessels alter the distribution of the cardiac output to all tissues of the body. This review is concerned mainly with the initial stages of the sequence of events which leads to the contraction of the vascular smooth muscle cell, i.e. the access of vaso-constrictor agents to the cell and their action on the cell membrane. Emphasis has been placed on the ways in which the structure of the blood vessel wall influences the actions of vasoactive substances, including the adrenergic trans-mitter, on the muscle cells. At the cell membrane, attention has been focussed on the role of the membrane potential in the initiation of a contraction.

Aspects of the sequence of events involved in the contraction of vascular smooth muscle, i.e. excitation, excitation-contraction coupling and contraction of the contractile elements, have been reviewed recently by Bohr (1964a, b); Honig (1963) and Burnstock & Holman (1966a, b). Folkow (1964b) has evaluated the evidence favouring the myogenic hypothesis of vascular smooth muscle contraction and Bader (1963), Burton (1962) and Peterson (1966) have discussed the physical properties of contracted and relaxed blood vessel walls. Bohr (1965) has emphasized the extremely variable physiological and pharmacological characteristics of different blood vessels and, for this reason, care has been taken to specify the type of vessel used in all experiments.

THE INFLUENCE OF THE STRUCTURE OF THE BLOOD VESSEL WALL ON EXCITATION

There are now an impressive number of papers which show that frequently the adrenergic nerves are restricted to the medio-adventitial border in the majority of arteries. This restricted location of the nerves has important implications. Muscle cells other than the outermost cells must be activated either by diffusion of the transmitter over relatively long distances or by some form of electrical or mechanical coupling between cells. The larger the resistance vessel, the longer the diffusion distance to the innermost muscle cells and, possibly, the more inefficient the electrical coupling because of the increasing proportion of connective tissue in the wall. The restricted location of the adrenergic nerves thus places emphasis on the mechanism(s) by which excitation spreads throughout all muscle cells in the blood vessel wall. This has been the stimulus to the reviewer to examine the structural features underlying spread of excitation, and it seems appropriate to begin with the relationships between nerve and vascular smooth muscle.

The relationships between the nerve and vascular smooth muscle
Organization of the innervation

Since 1962, the electron microscope and the specific and highly sensitive fluorescence method of Falck and Hillarp (Falck, 1962; Falck, Hillarp, Thieme & Torp, 1962; Corrodi & Jonsson, 1967) have played an important role in studies of the innervation of blood vessels. The fluorescence pattern gives an overall view of the distribution and density of the innervation, whilst the greater resolution of the electron microscope allows a more detailed examination of the nerves and muscle cells and their relationships.

All adrenergic nerves undergo a characteristic alteration in their terminal region to form what Hillarp called 'the autonomic ground plexus' (Hillarp, 1946, 1959). It is from this ground plexus that the transmitter is released (Norberg, 1967; Van Orden, Bloom, Barrnett & Giarman, 1966). A typical autonomic plexus in the wall of a muscular blood vessel, as seen under the fluorescence microscope, consists of a two-dimensional network of fluorescing varicose nerve terminals lying just outside the media (Falck, 1962; Norberg & Hamberger, 1964; Fuxe & Sedvall, 1965; Ehinger, Falck & Sporrong, 1966). Several fine nerve terminals usually run together in each strand of the net which surrounds the entire length of the vessel. The mesh size of the network and therefore the density of the innervation is variable. Rat mesenteric arterioles (Norberg & Hamberger, 1964), arteries and arterioles of the nasal mucosa from a variety of mammals (Dahlström & Fuxe, 1965c) and the rabbit ear artery (Waterson & Smale, 1967) possess a particularly rich innervation as shown by the presence of a network of fine mesh. In contrast, pial arteries of the rabbit (Falck, Mehedlishvili & Owman, 1965) have a sparse innervation in the form of a wide mesh network. The innervation of veins shows similar variations in density of innervation, although their adrenergic nerve supply is usually scantier than that of the corresponding arteries (Norberg & Hamberger, 1964; Ehinger et al., 1966). Rabbit pial veins, for example, have an even sparser innervation than rabbit pial arteries (Falck et al., 1965). The wide veins of the erectile tissue in the inferior concha are unusual in possessing a rich adrenergic plexus equal in density to that of the mucosal arteries (Dahlström & Fuxe, 1965c). In general there is little difference in the density of innervation of arterioles and the larger resistance vessels. The number of adrenergic nerve fibres accompanying an artery decreases as the vessel diameter diminishes until just one or two remain at the level of the smallest arteriole (Fuxe & Sedvall, 1965; Ehinger et al., 1966). Large elastic arteries such as the cat aorta (Ehinger et al., 1966), the dog pulmonary artery (Fillenz, 1966) and the common carotid of sheep (Keatinge, 1966a) conform to the innervation pattern of the smaller arteries with the exception that nerve fibres may invade the outer half or third of the media. The invading nerves may be associated with the *vasa vasorum* (Ehinger et al., 1966; Norberg, 1967).

Although all blood vessels are involved in the distribution of the cardiac output, this common function does not impose a common location upon the nerve fibres in the blood vessel wall, or even require that they be present. Adrenergic nerve fibres are present throughout the media of some arteries in the iris (Ehinger, 1966) and pineal gland (Owman, 1965) as well as in the media of some thick-walled cutaneous veins (Ehinger et al., 1966). Blood vessels within the central nervous system, including the retina, are almost devoid of an adrenergic innervation (Carlsson, Falck & Hillarp, 1962; Malmfors, 1965b; Ehinger, 1966).

Conclusions based on fluorescence patterns have been derived from examinations of a wide variety of vessels from mice, rats, hamsters, guinea-pigs, rabbits, cats, dogs, sheep, cattle and monkeys. Adrenergic nerve fibres have also been found surrounding small blood vessels in the skin (Falck & Rorsman, 1963), testes and epididymis of man (Baumgarten & Holstein, 1967) and in human dental pulp (Waterson, 1967).

Electron microscopic observations have been made on a lesser variety of vessels from fewer species, but the existing results are in overall agreement with those of fluorescence studies. No nerve fibres have been detected in electron micrographs of the media of rat pyloric, pancreatic, mesenteric, cerebral and glomerular arterioles or in the media of auricular and cerebral arteries of the same species (Zelander, Ekholm & Edlund, 1962; Thaemert, 1963; Barajas, 1964; Appenzeller, 1964; Samarasinghe, 1965; Lever, Graham & Spriggs, 1966). Ultrastructure studies of the descending thoracic aorta, pulmonary artery and coronary arterioles of the rabbit have likewise failed to show nerves penetrating the media (Bierring & Kobayasi, 1963; Lever, Ahmed & Irvine, 1965; Verity & Bevan, 1966; Verity, Bevan & Ostrom, 1966). Similar findings have been reported for sheep kidney cortical arteries (Simpson & Devine, 1966), muscular pulmonary arteries of the dog (Fillenz, 1966), cat pial, pancreatic and splenic arteries (Pease & Molinari, 1960; Lever et al., 1966; Fillenz, 1966), monkey pial arteries and glomerular arterioles (Pease & Molinari, 1960; Barajas, 1964), guinea-pig pancreatic arterioles (Lever, Graham, Irvine & Chick, 1965) and human intracranial arteries (Dahl & Nelson, 1964). Bevan & Verity (1967) have provided further evidence for the presence of adrenergic nerves outside the media of the rabbit thoracic aorta by destroying the adventitio-medial border with a rapidly vibrating razor blade. After destruction of the adventitio-medial border, the aorta failed to respond to transmural electrical stimulation. The innervation of small arteries and arterioles in the fascia overlying rabbit medial thigh muscle may be unusual in that nerves penetrate the media (Rhodin, 1967). However, Rhodin emphasized the difficulty in distinguishing between nerves and endothelial protrusions and felt further investigation was required.

Electron microscope studies, like fluorescence studies, have failed to detect nerves in or on the wall of intracerebral arteries (Samarasinghe, 1965), retinal arteries and retinal arterioles (Hogan & Feeney, 1963a, b). Samarasinghe (1965) also noted the absence of nerves accompanying small extracerebral arterioles (diameter, 10 to 20 μ) in the sub-arachnoid space of rats. This observation is in accord with the inability of Forbes & Cobb (1938a, b) to see, through a cranial window, any consistent calibre changes in rat and monkey pial arteries of less than 50 μ in diameter. Constriction of cerebral arterioles would seem to be outside direct nervous control being instead produced by circulating and local vasoactive agents (Rosenblum, 1965).

Nerve-muscle separation

As the nerve terminals lie outside the media of most arteries, only the outermost muscle cells are in reasonably close proximity to nerve terminals. Even these muscle cells are separated from nerve terminals incompletely covered by Schwann cell by a gap which is rarely less than 1000 Å. The closest approach so far reported is 600 Å in the rat auricular artery (Appenzeller, 1964) if the observations of Rhodin (1967) are excluded. Rhodin obtained electron micrographs of rabbit fascia arteries and arterioles which show a separation varying from about 150 Å to about 350 Å between a muscle cell membrane and what he thought may be an axon profile. However, Rhodin felt further evidence was needed to confirm that these profiles were of axons because of the paucity of vesicles contained within the profiles. The minimum nerve-muscle separation reported for other arteries and arterioles varied from 700 Å for rat pancreatic arterioles to 4500 Å in rabbit pulmonary arteries (Verity & Bevan, 1966). Within these extremes was the innervation of resistance vessels from the rat pylorus (770 Å), pancreas (830 Å), mesentery (1000 Å) and kidney (1000 Å), the sheep (950 Å) and monkey kidney (1000 Å), dog pulmonary artery (1500 Å) and rabbit coronary artery (2000 Å). These minimum nerve-muscle separations are at least three times that of most other autonomically innervated tissues where the minimum gap is 70 to 200 Å wide (see Chapter 1). It seems generally agreed that most nerve-muscle separations in arteries are much greater than these minimum values, but quantitative information in the form of frequency distribution curves is lacking. The most quantitative estimate is that of Devine (1966) who reported that 50% of the 137 axons incompletely covered by Schwann cell in small mesenteric arteries of the rat were 1000 to 5000 Å away from the muscle cells. The remainder were between 5000 Å and 10,000 Å away. In intestinal arterioles 70% of the 124 axons were within 1000 to 5000 Å of the muscle cell membrane.

Diffusion of vasoactive substances

The adrenergic transmitter

The pattern of innervation of most blood vessels indicates that diffusion plays a more important role in access of the transmitter to the muscle cells than in most other smooth muscles. The combination of a wide gap between nerve fibres and muscle cells and the restriction of the nerves to outside the media means that diffusion distances are often long. The diffusion distances involved have led Fuxe & Sedvall (1964) and others (Thaemert, 1963; Folkow, 1964a, b; Simpson & Devine, 1966; Devine, 1966) to doubt if the transmitter reaches the innermost muscle cells of the larger arteries in sufficient concentration to activate the muscle. Folkow (1964a, b), on the basis of this doubt, has put forward an intriguing concept which allows the inner muscle cells to be more than an inert filling, which was one possibility considered by Fuxe & Sedvall. Folkow suggested that the smooth muscle cells may be organized into two functionally different layers; an inner spontaneously active layer largely unaffected by direct excitatory nerve control and an outer non-spontaneously active layer of a 'multi-unit' type of muscle (Bozler, 1948) dominated by the adrenergic nerves. The inner muscle layer was visualized as being responsible for basal vascular tone with the outer muscle layer providing a means of centralized nervous control of the vascular bed. Folkow has used his concept to explain how the vasodilatation produced by stimulating the vasodilator nerves to the vascular bed can be masked by the activity in the vasoconstrictor nerve supply to the vascular bed of cat skeletal muscle (Folkow, Öberg & Rubinstein, 1964). The tightening of the outer smooth muscle jacket produced by vasoconstrictor nerve stimulation would prevent, according to Folkow, Öberg and Rubenstein, any significant change in internal diameter of the resistance vessels arising from relaxation of the inner layer of muscle cells. Implicit in this explanation is the assumption that the distribution of the vasodilator nerves is restricted to the inner muscle layer. The recording of spontaneous electrical activity from the outermost cells of mammalian arteries *in vivo* suggests that some modification of Folkow's hypothesis is required (see p. 573).

The concentration of transmitter achieved within the vicinity of the muscle cells depends on a variety of factors of which the diffusion distance is only one. The concentration gradient, the diffusion coefficient of the transmitter in the extracellular space and the cross-sectional area must also influence diffusion of the transmitter. The concentration gradient is set up by release of transmitter from the nerve terminals with the quantity released being dependent on the density of the innervation (i.e. the number of release sites), the frequency of nerve stimulation and the amount released from each varicosity by a nerve impulse. It is probable that the quantity released by each nerve impulse is influenced by the past history of nerve stimulation. Excitatory junction potentials (EJP's) recorded from arterial muscle cells (Speden, 1964; 1967) showed facilitation during repetitive stimulation, a phenomenon also seen with EJPs recorded from other smooth muscle cells (see Holman, this volume). Burnstock, Holman & Kuriyama (1964) have provided evidence that the facilitation of EJPs seen in the guinea-pig vas deferens is a presynaptic event.

Not all of the released transmitter may be available for diffusion to the muscle cells; some may be taken up by the nerve terminals and some destroyed before reaching the cells (see Brown, 1965; Kopin, 1966; Iversen, 1967*b*; and Chapter 16). The ability of drugs which block the uptake of noradrenaline by nerve terminals to potentiate the response of blood vessels to adrenergic nerve stimulation indicates that such uptake competes effectively with the muscle receptors for the released transmitter (de la Lande, Cannell & Waterson, 1966; de la Lande, Frewin & Waterson, 1967; de la Lande & Waterson, 1967; Iversen, 1967*b*). The proportion of the released noradrenaline prevented from reaching the muscle by uptake into the nerve terminals is unknown, but it could well be appreciable. de la Lande and his group have compared the sensitization of the isolated rabbit ear artery to noradrenaline given intraluminally and extraluminally (addition to the organ bath) following cocaine and chronic denervation. The perfused innervated artery was much less sensitive to extraluminal noradrenaline (see also Gillespie, 1966) with the ratio of sensitivities for intraluminal noradrenaline and extraluminal noradrenaline varying from 2 to 40 (mean 17·9). Addition of cocaine or chronic denervation abolished or greatly diminished the difference in sensitivity mainly by increasing sensitivity to extraluminal noradrenaline. The ratio obtained in the presence of cocaine varied from 0·72 to 3·3. de la Lande *et al.* (1967) felt these results were best explained by the uptake of extraluminal noradrenaline into nerve terminals and its prevention by cocaine or by denervation. They proposed that chronic denervation or the presence of cocaine abolished the barrier to noradrenaline diffusion set up by the presence of the nerve terminals outside the media thereby allowing more extraluminal noradrenaline to reach and act on the muscle in the media. If the nerve terminals were as efficient in removing transmitter noradrenaline as in removing extraluminal noradrenaline, then only a small fraction of the released transmitter, perhaps less than 10%, would reach the muscle. However, de la Lande et al. (1967) noted that cocaine potentiated the constrictor responses to nerve stimulation much less than that to extraluminal noradrenaline. They suggested that a state of supersensitivity may prevail during nerve stimulation so that cocaine's ability to cause potentiation is diminished. Such supersensitivity may arise from immediate re-release of the re-absorbed noradrenaline, as Trendelenburg (1966b) has found with the cat nictitating membrane, or from blockade of the uptake of released transmitter during nerve stimulation. A further possibility is that the noradrenaline concentration is higher in the vicinity of the nerve terminals following release of noradrenaline from the nerves, than when noradrenaline is added to the organ bath, and a higher concentration of noradrenaline would be expected to compete more effectively with cocaine for a common uptake site in adrenergic nerves (Furchgott, Kirpekar, Rieker & Schwab, 1963; Iversen, 1967b). The relative contributions of these factors to the action of cocaine on the response of blood vessels to nerve stimulation is uncertain and consequently there is also uncertainty as to the proportion of released transmitter reaching the muscle. Cocaine itself does not appear to interfere with biosynthesis of noradrenaline or to inhibit catechol-O-methyl transferase and has only a weak inhibitory action on monoamine oxidase (see reviews by Iversen, 1967b; Kopin, 1964, 1966).

It seems generally agreed that uptake into sympathetic nerve terminals is the major route for removal of the released transmitter; inactivation by catechol-O-methyl transferase and monoamine oxidase are thought to play minor roles only (Kopin, 1964; Iversen, 1967b; Ferry, 1967b). Rosell, Kopin & Axelrod (1963) have loaded the sympathetic nerves supplying the vascular bed of the dog gracilis muscle with tritiated noradrenaline. The absorbed tritiated noradrenaline, when released by nerve stimulation, was not detectably metabolized during its diffusion into the venous blood; no increase in o-methylated or other metabolites of noradrenaline were seen. Similar experiments with the cat spleen have shown little (Hertting & Axelrod, 1961) or no increase (Gillespie & Kirpekar, 1966b) in the concentration of radioactive metabolites in the venous blood following adrenergic nerve stimulation despite a marked rise in the amount of tritiated noradrenaline present in the venous effluent.

In addition to the uncertainty about the quantity of noradrenaline available for diffusion, there is also uncertainty concerning the values of the physical parameters which influence diffusion. No estimates are available for the volume of diffusion, the diffusion coefficient of noradrenaline in arterial tissue, or the cross-sectional area of diffusion. The latter will be influenced by the number of elastic lamellae, which varies with the diameter of the vessels, their thickness and the size and number of fenestrations in the lamellae. Because of these unknowns, little reliance can be placed on calculations of the maximum distance a nerve need be from a muscle cell for transmitter release to be still effective. Bennett & Merrillees (1966) have attempted to evaluate the relative effectiveness of noradrenaline released 200 Å and 1000 Å away from a muscle cell, with the assumptions of instantaneous release of transmitter from a point source into a large fluid volume, and that the same quantity of transmitter is released from each varicosity. They calculated that some 200 varicosities (point sources) were required 1000 Å from the muscle if the effect was to be the same as one varicosity 200 Å away. A comparison of this result with the density of innervation of the guinea-pig vas deferens led them to conclude that only varicosities within about 1000 Å of the muscle membrane would have an appreciable effect during nerve-muscle transmission. If their calculations and conclusions are applicable to arteries, as they would be if the innervation of the arteries is no more dense than that of the vas deferens, then most arteries would be essentially uninnervated. Few nerve terminals have been detected within 1000 Å of vascular smooth muscle cells (see p. 561). It is undeniable, however, that most arteries and arterioles are under the influence of sympathetic nerves. Further, discrete depolarizations of the cell membrane have been recorded with an intracellular microelectrode from the outermost muscle cells in four types of artery following a single stimulus to the sympathetic nerves (Speden, 1964, 1967). If transmission is judged to be effective when a single nerve stimulus produces a discrete depolarization of a single cell membrane, then these cells are innervated effectively (see Chapter 8).

Attempts have been made to use the latency (or delay) from stimulation of the sympathetic nerves to onset of the EJP as a qualitative measure of the diffusion distances involved (see Burnstock & Holman, 1966a; and Chapter 8). For the isolated rabbit ear and mesenteric arteries, the minimum latency was 12 msec when the recording electrode was 1 to 2 mm from the nerve stimulating electrode (Speden, 1967). This delay is comparable to that reported for the guinea-pig vas deferens (Kuriyama, 1963b). Kuriyama found a minimum latency of 6 msec when the nerve stimulating electrode and recording electrode were separated by less than 1 mm. Increasing the distance between the two electrodes to 3 mm increased latency to 25 msec. This similarity in latencies for two preparations of differing patterns of innervation suggests that other factors such as delay in transmitter release (Katz & Miledi, 1965c; Bennett & Merrillees, 1966) and slow decremental conduction in the fine sympathetic nerve terminals (Kuriyama, 1963b; Bülbring & Tomita, 1967; Tomita, 1967a) overshadow the contribution of transmitter diffusion to the latency. The gap between the nerve terminals and the outermost muscle cells in the two rabbit arteries is unknown,

19*

although a nerve-muscle separation varying from about 1000 Å to about 10,000 Å would seem likely on the basis of information available for arteries and arterioles from twelve vascular beds from a variety of mammals (see p. 561). If this is the separation in rabbit ear and mesenteric arteries, then an increase in diffusion distance from about 200 Å in the guinea-pig vas deferens (Bennett & Merrillees, 1966) to 1000 Å or more in blood vessels fails to alter latency significantly. Nor does a separation of 1000 Å or greater prevent diffusion of transmitter in amounts sufficient to depolarize a muscle cell following a single nerve stimulus (Speden, 1964, 1967).

Whether the amount of transmitter released by one nerve impulse is sufficient to reach and excite muscle cells more distant from the nerve than the outermost muscle cells must depend in part on the sensitivity of the cells to noradrenaline. Other excitatory influences which are present *in vivo* may so raise sensitivity of the muscle cells to noradrenaline that the concentration reaching the innermost cells, although greatly diminished, is still effective. Repetitive stimulation of the vasomotor nerves would be expected to increase the amount of transmitter reaching the inner muscle cells provided a certain minimum rate of nerve stimulation is exceeded. There are indications that this minimum rate may be between one and two maximal nerve stimuli per second. The EJPs to successive maximal or near-maximal nerve stimuli recorded from the outermost muscle cells of guinea-pig mesenteric arteries with intact circulation and of isolated mesenteric, ear and caecal arteries of the rabbit summate at rates of stimulation of 2/sec or faster (Speden, 1964, 1967). No summation occurred during nerve stimulation at frequencies of less than 1 per sec.

The physiological discharge rate of vasoconstrictor nerves is such that summation of transmitter action seems likely to occur during maintenance of a normal blood pressure. Folkow (1952) has provided evidence that the normal physiological discharge rate of vasoconstrictor nerves to the skeletal muscle vascular bed of the cat hindlimb is equivalent to stimulating these nerves with maximal electrical pulses at frequencies of 1 to 2/sec. The upper limit for physiological excitation of the nerves was equal to 6 to 8 maximal electrical pulses per sec. A rate of electrical stimulation as slow as 1/sec is sufficient to maintain a significant peripheral resistance in a variety of vascular beds (Folkow, 1955) including the splanchnic bed (Folkow, Lewis, Lundgren, Mellander & Wallentin, 1964; Wallentin, 1966).

Blood borne vasoactive substances

Unlike the adrenergic transmitter, circulating vasoactive substances diffuse to the muscle cells from the intraluminal surface of the blood vessel. This different access to the muscle cells may modify the action of vasoactive agents both by the presence of different diffusion barriers and by the different accessibility and susceptibility to inactivation processes. Inactivation by nerve uptake seems of little importance in determining the sensitivity of the isolated ear artery to intraluminal noradrenaline since drugs and procedures which inhibit uptake of noradrenaline by nerve terminals had little effect on the sensitivity (de la Lande *et al.*, 1967). It was suggested (de la Lande *et al.*, 1967) that noradrenaline acted on the smooth muscle during its diffusion through the media before being taken up by the nerve terminals lying outside the media.

Adrenaline is less readily taken up by the nerve terminals and is more susceptible to inactivation by catechol-O-methyl transferase than noradrenaline (Iversen, 1967*b*). Catechol-O-methyl transferase inhibitors prolonged the pressor effect of intravenous adrenaline but not that of noradrenaline (Wylie, Archer & Arnold, 1960; Murnaghan & Mazurkiewicz, 1963). Enzymatic inactivation of the two catecholamines by monoamine oxidase seems to play little part in terminating the effect of injected adrenaline and noradrenaline in contrast to that of tyramine (Griesemer, Barsky, Dragstedt, Wells & Zeller, 1953). Unlike noradrenaline, tyramine is a more potent constrictor of the isolated rabbit ear artery when applied to the outside of the vessel (de la Lande & Waterson, 1968). These workers have speculated that monoamine oxidase, which is present in the media of the ear artery, may have degraded the tyramine during its diffusion from the lumen to its site of action, the nerve terminals. Frewin & Whelan (1968) have interpreted the slower onset of the constrictor effect of intra-arterial infusions of tyramine on human forearm vessels, compared to infusions of noradrenaline, to delays arising from the diffusion of tyramine through the blood vessel wall to the nerve terminals.

Variations in the density of innervation amongst different vascular beds (see p. 560) may lead to differences in sensitivity and duration of action of noradrenaline on the vascular beds of different tissues. The available evidence, which has been reviewed by Iversen (1967*b*), indicates that the rate of removal of circulating noradrenaline is directly related to the richness of innervation of the tissue. Any differences in the distribution of the inactivating enzymes may also modify the response to substances attacked by these enzymes. Khairallah, Page, Bumpus & Türker (1966) have provided some evidence that the ease with which isolated arteries develop tachyphylaxis to angiotensin reflects differences in the amount of angiotensinase present in the blood vessel wall.

Accessibility to adrenergic receptors may also be influenced by the route of administration of noradrenaline. Braunwald and his co-workers (Glick, Epstein, Wechsler & Braunwald, 1967) have shown that blood borne noradrenaline stimulates both α- and β-receptors in the splanchnic vessels and in the vascular bed of hindlimb skeletal muscle of dogs, whereas transmitter noradrenaline has little β-receptor stimulant effect. Pretreatment of the animals with phenoxybenzamine (an α-receptor blocker) converted the vasoconstriction to intra-arterial noradrenaline into a vasodilatation, whereas carotid sinus occlusion still caused a reflex vasoconstriction in the presence of phenoxybenzamine which was not potentiated by subsequent treatment with a β-receptor blocker, although the response to injected noradrenaline reverted to vasoconstriction.

Similar differences have been reported between the action of circulating and transmitter noradrenaline on human forearm blood vessels. Circulating noradrenaline stimulated both α- and β-adrenergic receptors of human forearm vessels (Brick, Hutchinson & Roddie, 1967a, whilst vasoconstrictor nerve stimulation activated α-receptors only (Brick, Hutchinson & Roddie, 1967b).

Spread of excitation

An alternative explanation to activation by diffusion of transmitter (or other vasoconstrictor substances) for the spread of excitation throughout the muscle cells in the blood vessel wall, is excitation by electrotonic spread from cell to cell once a threshold number of cells have been chemically excited. Spread of electrical current in the ureter, vas deferens and taenia coli of the guinea-pig is believed to involve a group of muscle cells interconnected in three dimensions by low resistance pathways (see Chapter 7). These low resistance pathways may be the regions where fusion of the outer leaflet of the plasma membranes of adjacent smooth muscle cells occurs — the nexus of Dewey & Barr (see Chapter 1).

Reports describing fusion of plasma membranes of adjacent arterial smooth muscle cells are rare. In one of the two types of arteries where membrane fusion has been seen, the rabbit pulmonary artery (Verity & Bevan, 1966), the incidence of nexuses was low. In the other, the aorta of young rats (Cliff, 1967), membrane fusion was less rare and close apposition of cell membranes was common.

Close apposition of cell membranes has been observed more frequently than has membrane fusion. Rhodin (1967) found that every muscle cell in any section of longitudinally cut terminal arterioles in the fascia of rabbit medial thigh muscles were, in places, close to neighbouring muscle cells. No basement membrane (ground substance) was present in the intervening gaps of about 150 Å. Such close contacts were less frequent in larger arterioles and were of shorter length. Separations of 200 Å or less with no intervening basement membrane have also been seen in human testicular arteries (Fawcett, 1959; Buck, 1963), rat pancreatic arterioles (Zelander et al., 1962), rabbit and rat aortae (Pease & Paule, 1960; Bierring & Kobayasi, 1963; Cliff, 1967), the mouse femoral artery (Rhodin, 1962) and a variety of small arteries with an internal elastic lamina (Movat & Fernando, 1963). The close contacts involved the lateral sides of muscle cell membranes apart from the femoral artery where close contact occurred at the interdigitated ends of the muscle cells. In most larger arteries, where layers of muscle cells are separated by elastic lamellae, close contacts are apparently restricted to muscle cells within a layer. The aorta of newborn rats seems unusual in that contact between separate muscle layers was made by pseudopod-like projections through fenestrations in the separating elastic lamina (Paule, 1963).

Although fusion of adjacent muscle cell membranes seems rare, the incidence of such fusion may be more frequent than the present observations indicate. Care needs to be taken during preparation of the section if fusion of the outer leaflets of adjacent cell membranes is to be retained. The methods of fixation used in most of the above studies would tend to cause separation of fused membranes (Dewey & Barr, 1964; Karlsson & Schultz, 1965). A careful study of the muscle cell relationships in those blood vessels where conduction of a contraction occurs may help in deciding the role of close apposition of muscle cell membranes in the spread of excitation

Contraction of a blood vessel wall may be highly localized or it may involve a considerable length of the wall. Numerous investigators have observed spontaneous contractions and relaxations of minute resistance vessels under a microscope — a phenomenon called vasomotion (see Furchgott, 1955; Wiedeman, 1963). The final muscle cells in terminal arterioles of the bat wing may show an extremely variable pattern of vasomotion not only at different times but also simultaneously with cells closely adjacent (Nicoll & Webb, 1955). This lack of synchrony between the activities of nearby cells indicates that the individual muscle cells tend to react independently (Nicoll, 1964). A greater co-ordination of activity was present with muscle cells further from the capillaries. Waves of contraction were frequently seen to move from the origins of the arcuate arterioles toward the terminal arterioles (Nicoll & Webb, 1955). The several muscle cells which make up the precapillary vascular sphincter of the rat meso-appendix may act independently of the muscle cells in the parent metarteriole, although their activity was usually synchronized (Chambers & Zweifach, 1944). There was no synchrony in the vasomotion of neighbouring metarterioles, but along the same metarteriole the contraction and relaxation tended to occur simultaneously. Some degree of varicosity of the wall was frequently observed during the constrictor phase. A contraction of the precapillary vascular sphincter in the vessels of the retrolingual membrane of the frog and the cheek pouch of the anaesthetized hamster may also take place in the absence of constriction of the parent arteriole (Lutz & Fulton, 1958).

The localized nature of the contractions of minute resistance vessels need not reflect the distribution pattern of the nerves. Vasomotion in the smaller arcuate and terminal arterioles in the bat wing was unaffected by stimulation of the nerve supply (Nicoll & Webb, 1955). The localized contraction of minute vessels in the retrolingual membrane of frogs was, however, restricted to muscle cells innervated by particular nerves, but the localization was not apparently determined by the nerve distribution (Lutz, Fulton & Akers, 1950). Direct electrical stimulation of vessels exposed to cocaine produced exactly the same vascular pattern of constriction as that produced by nerve stimulation before exposure to cocaine.

Localized contractions of larger blood vessels have been shown in a number of ways. Patel & Burton (1957) found that casts of rabbit small pulmonary arteries produced by injecting vinyl acetate were contorted, twisted and knotted.

The gnarled appearance of the arteries was believed to result from contraction of independent units of muscle cells as an infusion of noradrenaline increased the degree of contortion. Alvarez-Morujo (1966) observed marked variations in the shape of casts prepared from a variety of arteries. Constriction could follow a helical path, alternate with dilatations or form complex patterns of constrictions and dilatations. The well-known localized spasm of arteries which may follow manipulation is a sign of independent units of muscle in arteries.

The absence of conducted, spontaneous contractions in the majority of arteries under most circumstances led Bozler (1948) to classify vascular smooth muscle as multiunit smooth muscle. Prosser, Burnstock & Kahn (1960) have correlated the inability of pig carotid arteries to conduct a contraction (Burnstock & Prosser, 1960b) and to respond to stretch with a contraction (Burnstock & Prosser, 1960a), (which is another sign of a multiunit muscle) with the structure of the artery. Compared with visceral muscle, the muscle cells in pig carotid arteries were short, had narrow diameters and were separated by wide gaps from neighbouring muscle cells.

Although arteries may normally possess all the characteristics of multiunit muscle, there are many reports of isolated arteries and veins showing properties of single unit muscles (see Furchgott, 1955; Seller, Langhorst, Polster & Koepchen, 1967). These include large arteries like bovine carotid and mesenteric arteries which may develop conducted spontaneous activity (Peterson, 1936b; Bürgi, 1944; Monnier, 1944a, b). Another characteristic of single unit muscles, an active contraction following rapid stretch, has been observed in some but not all bovine mesenteric arteries. Burgi (1944) observed 23 active contractions to 101 rapid stretches. The carotid arteries of the dog (Bayliss, 1902) and horse (Wachholder, 1921), the human umbilical artery and dog mesenteric and cerebral arteries (Sparks, 1964) may also contract after being rapidly stretched.

The available evidence indicates that nervous mechanisms need not be essential for the initiation and propagation of contractions along blood vessels. One of the major factors which leads to development of spontaneous conducted contractions in bovine mesenteric arteries is storage in cold Ringer solution for several days before use, a procedure well known to cause deterioration of nerve fibres (Vaughan Williams, 1954). For example, the vessels of the isolated rabbit ear, after being stored for two days, responded poorly or not at all to electrical stimulation, which normally activates nerve fibres in fresh preparations, although responsiveness to injected noradrenaline was well maintained (de la Lande, personal communication). Conducted contractions of isolated bovine mesenteric arteries could be recorded from two to about seven days after removal of the artery from the animal and were not detectably affected by cold storage (Monnier, 1943). Chronic denervation, pretreatment with reserpine or guanethidine and exposure to an α-adrenergic blocking agent failed to modify the development and pattern of spontaneous activity in subcutaneous arteries of the dog's paw

(Johansson & Bohr, 1966). The human umbilical artery, which has no innervation, responds to stretch with an active contraction (Sparks, 1964) and may develop spontaneous activity (Somlyo, Woo & Somlyo, 1965).

Conduction of a contraction along arteries probably arises from spread of an electrical current along the artery in a way similar to that suggested for visceral muscle (see Chapter 7). The very success of external methods of electrical recording from strips of large arteries like the rabbit aorta (Shibata & Briggs, 1966), sheep, dog and bovine carotid arteries (Petersen, 1936b; Keatinge, 1964) indicates that electrical connections must exist between the muscle cells. These conducting pathways may be long. Keatinge (1964), using the sucrose-gap technique, was able to record a potential difference across a 10 mm-length of a 1 mm-wide, helically cut strip of sheep carotid artery. The conducting pathways seem less effective in blood vessels than in visceral smooth muscle for the longitudinal resistance of strips of rabbit anterior mesenteric veins (Cuthbert, 1966) and sheep carotid arteries (Keatinge, 1964), was high and the conduction velocities along arteries and veins were slow. Conduction of a contraction along bovine mesenteric arteries varied from 0·3 to 3·3 mm/sec (Monnier, 1944a; Burgi, 1944) which is slow compared to the velocities of usually 30 to 70mm/sec for conduction in visceral muscle (Prosser et al., 1960; Prosser 1962; Kuriyama, Osa & Toida, 1967b).

Many factors are known which favour the development of conducted spontaneous contractions in isolated arteries some of which are operative *in vivo*. Storage in cold Ringer solutions for several days before use, immersion in physiological salt solution at 37°C for several hours, and stretch, all predispose to the development of spontaneous activity (Ducret, 1931; Petersen, 1936a; Monnier, 1943; 1944a; Bürgi, 1944). Of these factors, stretch is operative *in vivo* and may be of particular importance. Brun (1947) noted that adrenaline applied topically to rat mesenteric arteries *in situ* failed to elicit rhythmic contractions of the vessels when the blood pressure had fallen to 40 mmHg or when the artery was compressed proximal to the point of observation. Instead, a non-rhythmic contraction developed. The presence of a local anaesthetic had no effect on the rhythmic contractions which were present at higher blood pressures or on their intensification by adrenaline. With isolated, perfused skin areas of dogs, perfusion pressures of 70 to 80 mmHg were needed before spontaneous activity of the perfused vessels was initiated (Seller et al., 1967).

Exposure of isolated arteries to vasoconstrictor substances like adrenaline may also initiate spontaneous contractions or increase their frequency if they were already present (Ducret, 1931; Bürgi, 1944, 1945; Johansson & Bohr, 1966), although mesenteric arteries needed to be stored for several days before such contractions were obtained. Certain isolated blood vessels like the small subcutaneous arteries of the dog's paw (Johansson & Bohr, 1966), the portal vein of the rat (Funaki & Bohr, 1964) and the anterior mesenteric vein of the rabbit (Cuthbert & Sutter, 1964) develop spontaneous activity particularly readily. Conducted contractions have been observed in arterial strips from arteries with outside diameters as small as 200 to 500 μ (Johansson & Bohr, 1966).

THE ROLE OF THE CELL MEMBRANE POTENTIAL IN THE INITIATION OF A CONTRACTION

Sufficient evidence had been obtained by the end of 1936 to indicate that electrical changes were associated with mechanical activity of blood vessels (Furchgott, 1955). Luisada (1933) was able to record small fluctuations in potential with external electrodes when constriction of isolated, perfused femoral arteries of dogs was initiated by sudden stretch, although in some arteries potential changes occurred in the absence of any mechanical activity. Potential changes were also seen following sudden stretch of sheep carotid arteries (Wybauw, 1935), human radial arteries and ox metatarsal arteries (Luisada, 1933). Evidence that the spontaneous rhythmic contractions of bovine carotid arteries may be initiated by an electrical event was provided by Petersen (1936b) who recorded transient potential changes of up to 3mV at the onset of each spontaneous contraction. Each of the three investigators presented evidence that the recorded potential changes were not artifacts arising from mechanical factors.

The success of Funaki (1958, 1960) in inserting a micro-electrode into a vascular smooth muscle cell marks the beginning of renewed interest in the electrophysiology of blood vessels. Subsequent investigators have recorded electrical activity from blood vessels using methods such as the sucrose-gap technique, external electrodes, pressure electrodes and intracellular electrodes. Each of these methods has its advantages and disadvantages (Burnstock, Holman & Prosser, 1963). Intracellular recording is the most informative, but it is the most difficult to apply to blood vessels. The individual muscle cells are small (Rhodin, 1962, 1967; see also Chapter 1), they are embedded in a mass of connective tissue and, in the larger vessels, protected on both the intimal and serosal surfaces by further sheets of connective tissue. As a result insertion of a micro-electrode into a cell without breaking or clogging its tip is difficult and there is uncertainty about the amount of damage caused to a small cell. External recording methods are also more difficult to apply to most blood vessels than to visceral smooth muscle. The electrical records represent the summed activity of a number of muscle cells and, because of their unco-ordinated and variable activity, interpretation is difficult. Pressure electrodes record electrical activity from a smaller number of cells than either the sucrose-gap method or external electrodes (Bortoff, 1961a; Gillespie, 1962b). The existing results should be considered as being qualitative rather than quantitative because of the above difficulties and uncertainties.

Electrical and mechanical activity of arteries and arterioles

Spontaneous activity

Some resistance vessels show spontaneous electrical activity, whilst others are electrically quiescent (Table 1). The phrase, spontaneous electrical activity, is used here to describe that activity which was present in the absence of applied external stimuli. Electrical quiescence tends to be associated with higher maximum resting membrane potentials, a correlation well illustrated by the effect of anaesthesia on the spontaneous electrical activity of small rat arteries (Steedman, 1966). During deep anaesthesia no action potentials were recorded with intracellular electrodes, slow fluctuations in membrane potential were either small or undetectable and the maximum membrane potential averaged 43 mV. With more lightly anaesthetized animals, however, the maximum resting potential was lower (mean, 35 mV) and both action potentials and slow potential fluctuations were seen.

The characteristics of the spontaneous electrical activity recorded from small mesenteric arteries of the rat and guinea-pig with intact circulation and innervation are similar (Speden, 1964; Steedman, 1966). The membrane potential undergoes slow rhythmic fluctuations of variable amplitude which, in the rat, were of reasonably constant periodicity (4·8 to 6·7 sec). Action potentials arose from these slow depolarizations provided a certain variable threshold was achieved. Such action potentials may occur singly or, more commonly, in groups of two or three, sometimes more, with each group being separated by a silent period. The action potential consisted of two components, an initial depolarization, which will be called a prepotential, followed by a more rapid component of variable amplitude — the spike. Repolarization was invariably more rapid than depolarization when the spike was present and the slow depolarization was transiently 'wiped out' when the amplitude of the spike was sufficient. The 'wiping out' of the slow depolarization gives rise to what looks like a positive after-potential, but is really a return to the maximum level of polarization achieved during the slow rhythmic potential fluctuations. These features of the spontaneous electrical activity of mesenteric arterial muscle cells are illustrated in Figs. 1 and 2. Figure 1a shows a typical slow depolarization which developed smoothly into a full action potential and abolished the slow depolarization on repolarization. A less common action potential (Fig. 1b) recorded from the same cell shows a clear separation between the prepotential and the slow depolarization. The ability of the prepotential to initiate a relatively large spike was extremely variable. The first prepotential superimposed on the slow depolarization in Fig. 2a failed to initiate a spike, the second gave rise to a spike of large amplitude which transiently abolished the slow depolarization, the third initiated a smaller spike and the fourth failed to do so. In the next burst of action potentials recorded 13 sec later from the same cell (Fig. 1c) the amplitude of successive spikes increased rather than decreased.

It seems unlikely that the spontaneous electrical activity is a consequence of damage following insertion of a micro-electrode into a small cell as Steedman (1966) was able to detect slow fluctuations in membrane potential and action potentials with both external electrodes and pressure electrodes. Moreover, reasonably stable membrane potentials could be recorded when guinea-pigs and rats were deeply anaesthetized (Speden, 1964; Steedman, 1966).

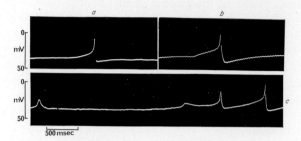

FIG. 1 Spontaneous action potentials recorded from smooth muscle cells in small mesenteric arteries of the anaesthetized guinea-pig; blood supply and innervation intact. The action potential shown in *b* was recorded from the same cell as *a*, but at twice the film speed. Those shown in *c* were from a different cell in the same artery.

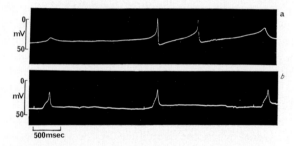

FIG. 2 Spontaneous action potentials (*a*) and action potentials (*b*) produced by splanchnic nerve stimulation. The action potentials in *a* were recorded 13 sec before those, from the same cell, shown in Fig. 1*c*. In *b*, the action potentials, recorded from a different cell, were produced by stimulating the splanchnic nerve with square pulses of 1 msec duration and 30 V at a frequency of 0·5/sec. (Speden, 1964)

The origin and significance of the three distinct components which together comprise the spontaneous activity of small mesenteric arteries is uncertain. The full action potential is similar in shape to that of a sinoatrial pacemaker cell (Hoffman & Cranefield, 1960; Burnstock *et al.*, 1963), a similarity which suggests that the arterial muscle cells may also be pacemaker cells or close to pacemaker cells. The prepotential recorded from both sinoatrial and arterial muscle cells develops smoothly into a more rapid depolarization which 'wipes out' the prepotential on repolarization. The electrical activity of mesenteric arterial smooth muscle cells does, however, differ from that of a cardiac pacemaker cell in showing slow waves of depolarization, a more irregular occurrence of prepotentials and a greater failure of the prepotential to initiate a spike. A further

TABLE 1. Intracellular membrane potentials of blood vessels

Blood vessel		Method	Spontaneous activity	Membrane potential (mV)*	Action potential (mV)*	Reference
Frog	Small vessels	*in situ*				
	Web and tongue		0	40·5	60	Funaki (1958, 1960)
	Tongue		0	64·7 (55–75)	60	Steedman (1966)
	Cutaneous		+	25	36	Funaki (1961)
	Cutaneous		0	43·6 (35–50)	—	Steedman (1966)
Turtle	Aorta and vena cava	*in vitro*	+	30–50	up to 50	Roddie (1962)
Rat	Mesenteric arteries	Intact circulation and innervation	+	39·4 (30–50)	22 (15–35)	Steedman (1966)
	Portal vein	*in vitro*	+	39 (30–65)	8–30	Funaki & Bohr (1964)
Guinea-pig	Mesenteric arteries	Intact circulation and innervation	+	38·7 (26–59)	26 (17–42)	Speden (1964)
	Ant. mesenteric vein	*in vitro*	+	51 (41–62)	45 (35–59)	Nakajima & Horn (1967)
Rabbit	Ear artery	*in vitro*	0	62·4	—	Speden (1967)
	Mesenteric artery	*in vitro*	0	54·1	—	
	Pulmonary artery	*in vitro*	0	51·5	—	Su, Bevan & Ursillo (1964)
	Ant. mesenteric vein	*in vitro*	+	33	19	Cuthbert, Matthews & Sutter (1965)

* Mean values with the maximum and minimum values in parenthesis.

difference is that the prepotential, unlike that of cardiac pacemaker cells, has yet to be shown to be an inherent (myogenic) property of the cell. The prepotential could also reflect electrotonic spread of activity from neighbouring cells or arise from an intermittent release of transmitter from the adrenergic nerves, although prepotentials recorded from other smooth muscle are not nervously mediated (Burnstock et al., 1963; Bülbring & Kuriyama, 1963a; Kuriyama et al., 1967b).

Whilst the electrical activity recorded from mesenteric arteries has features in common with other spontaneously active smooth muscles, there are significant differences. Both visceral and arterial smooth muscle show two forms of slow potentials; the slow fluctuating potential and the prepotential (see Kuriyama et al., 1967b; and Chapters 7 and 12). An important difference is the absence of typical conducted action potentials in the records so far obtained from mesenteric arteries. The full action potential invariably had the shape of a pacemaker potential. No large spike-like potential without relation to a prepotential (which is characteristic of a conducted action potential) has been seen. Another difference is the tendency for the spike to be graded rather than all or none (Fig. 1c), although more experiments are needed to substantiate this observation. This fluctuating amplitude of the spikes seems not to be an artifact arising from a dying cell for a small spike can preceed spikes of larger amplitude (Fig. 1c). In these respects, the spontaneous electrical activity of mesenteric arterial muscle cells resembles that of cells in strips of visceral muscle reduced by longitudinal cutting to a width of less than 100 μ and with fewer than 200 to 300 cells in cross section. Such strips were incapable of conducting a contraction more than 1 to 3 mm, although spontaneous activity persisted (Burnstock & Prosser, 1960a), and the spike to extracellular stimulation was graded (Tomita, 1966a). This similarity is not unexpected as the external diameters of the mesenteric arteries were 80 to 150 μ in the rat and 100 to 450 μ in the guinea-pig.

The number of vascular smooth muscle cells making up a functional unit in mesenteric arteries may exceed one. Steedman (1966) was able to record slow potential waves of surprisingly large amplitude, compared to intracellular recording, with external electrodes of 1 to 5 μ in tip diameter held in contact with the blood vessel wall. With intracellular recording the action potentials were some three times the amplitude of the slow waves, whereas the reverse held with extracellular recording. This observation, combined with evidence that the external records were not artifactual, led Steedman to conclude that adjacent cells were generating synchronous slow waves thereby allowing a given area of the vessel wall to become excitable more or less simultaneously. This conclusion is consistent with the absence of conducted action potentials in the electrical records. The number of muscle cells making up the functional unit is unknown, but there are reasons for expecting the number to be variable (see p. 571).

The action potentials developed spontaneously by muscle cells in frog abdominal arterioles *in situ* (Funaki, 1961) also have the appearance of pace-maker potentials. No slow fluctuations in membrane potential were noted, although these were prominent in muscle cells of the turtle aorta (Roddie, 1962). Roddie, by using strips of turtle aorta, was able to relate electrical to mechanical activity; every contraction was preceded by an action potential. The action potentials were unusual for vascular smooth muscle in showing a long plateau of depolarization lasting up to 4 sec following the rapid spike of depolarization (cf. Fig. 1).

Vasoconstrictor nerve stimulation

It is clear that the autonomic nervous system has a profound effect on the electrical activity of mesenteric arteries with an intact circulation and innervation. Activation of the autonomic nerves by asphyxia or by electrical stimulation of the greater splanchnic nerve increased the amplitude of the slow potential fluctuations and the frequency of the action potentials recorded extracellularly from rat mesenteric arteries (Steedman, 1966). Partial denervation by cutting the greater splanchnic nerves decreased both the frequency of action potentials and the amplitude of the slow potential waves and treatment of the rat with a ganglion blocking agent caused a temporary disappearance of all electrical activity, an effect which Steedman thought may arise from a number of causes including a direct action on the muscle cells. The fall in blood pressure following chemical or surgical interference with the nervous system may also remove any excitatory effect originating from stretch of the muscle cells. A distinct depolariz-ation to each nerve stimulus — (the excitatory junction potential (EJP)) — was recorded intracellularly from guinea-pig arterial muscle cells when the left splanchnic nerve was stimulated (Speden, 1964). These EJPs, like those recorded from other smooth muscle (see Chapter 8), were able to initiate a spike-like depolarization, although the ease with which they did so was greatly influenced by the depth of anaesthesia. Rates of stimulation of about 2/sec were needed when anaesthesia was deep, whereas a single stimulus was sufficient during light anaesthesia.

The electrical changes accompanying nerve stimulation of isolated arteries are much less dramatic and may be non-existent. EJPs of usually much smaller amplitude have been recorded from muscle cells in isolated and perfused mesen-teric, ear and caecal arteries of the rabbit, but none initiated a spike, although the rates of nerve stimulation used were sufficient to cause a marked constriction of the arteries (Speden, 1967). Instead, a sustained depolarization with a peak of only about 20 mV developed during nerve stimulation at frequencies of 10/sec. Su *et al.* (1964) were unable to detect any potential change with an intracellular micro-electrode during nerve stimulation of the isolated pulmonary artery of the rabbit. Evidence was cited which tends to exclude the possibility that the failure to detect spikes was due to cell damage produced by the micro-electrode. Speden (1967) reported resting potentials which were stable for 3 to 4 min and which were high for arterial muscle (Table 1); he could readily record action potentials from spontaneously active mesenteric veins of rabbits. Su & Bevan (1965), detected smooth sustained depolarizations in response to acetylcholine, histamine and serotonin, but not to noradrenaline. Sucrose-gap records of the electrical activity of another large artery, the common carotid artery of sheep, show an irregular electrical discharge of small amplitude (up to 3 mV) on stimulation of the sympathetic nerves with nicotine or acetylcholine (Keatinge, 1966a).

Su & Bevan (1966), Keatinge (1964) and Waugh (1962) have concluded from their results that the contraction of arterial muscle may be mediated both by electrical and non-electrical means. Speden (1967) was impressed by the high and stable membrane potentials of isolated arterial muscle compared with that of arteries in vivo (Table 1) and pointed out that removal of the arteries from their natural site must greatly reduce the excitatory influences normally acting on the muscle with the consequence that the muscle cells would be expected to have higher membrane potentials and to be less readily excited. A combination of the two views provides a working hypothesis which may explain most existing observations and can be verified. Contraction of arterial and arteriolar smooth muscle may be initiated both by electrical and non-electrical events with the relative contribution of each varying in a way dependent on environmental factors. When the physiological background excitatory influences are high electrical excitation is dominant; when low, non-electrical excitation becomes dominant. In the absence of excitatory influences as in vitro the high and stable membrane potential may make electrical excitation ineffectual, an ineffectiveness which can be removed by any factor that makes the membrane potential unstable and favours electrical interaction between neighbouring cells. Such instability may or may not be accompanied by depolarization, although excessive depolarization produced by high concentrations of depolarizing agents would again favour non-electrical excitation. A clear distinction needs to be made between two types of electrical activity both of which may conceivably initiate a contraction. One is a slow sustained depolarization, the other the more rapid and transient action potential. The very nature of the spontaneous electrical activity recorded from arterial muscle with intact circulation (see p. 573) suggests that it is the latter which is normally operative in vivo and, for this reason, instability rather than depolarization has been emphasized. Further evidence on which the above hypothesis is based is given in the following sections dealing with the membrane actions of vasoactive agents.

Vasoconstrictor substances

Vasoconstrictor amounts of adrenaline and noradrenaline have been reported to increase the spontaneous electrical activity of arterial muscle, to cause a slow sustained depolarization, to hyperpolarize or to have no effect on the membrane potential. Thus, adrenaline and noradrenaline increased the firing of action potentials by muscle cells in rat mesenteric arteries (Steedman, 1966) and the turtle aorta (Roddie, 1962). The catecholamines needed to be applied topically to rat mesenteric arteries in order to obtain a consistent effect; intravenous administration had variable effects on the electrical activity presumably because of varying autonomic reflex activity (Steedman, 1966). In the electrically and mechanically quiescent sheep carotid arterial strip, noradrenaline initiated a depolarization and a contraction (Keatinge, 1964). Usually the depolarization reached a plateau within seconds and remained relatively stable throughout the exposure to noradrenaline. Less often it took the form of a spike of up to 12 mV followed by a steady or fluctuating plateau of depolarization and, least often, that of a slow depolarization followed by one or more spikes. In some preparations, the contraction continued to develop at a time when repolarization was taking place despite the presence of noradrenaline which suggested to Keatinge (1964) either that adaptation was occurring or that the latter stages of the contraction were produced by non-electrical means. The spikes recorded with the sucrose gap were of long duration lasting some 15 sec, unlike those recorded in a similar way from the taenia coli (Burnstock & Straub, 1958) and the ureter (Bennett & Burnstock, 1966) which were over in less than a second. This long-lasting spike may reflect poor co-ordination of activity in the helically cut arterial strip (Keatinge, 1964), action potentials of long duration (Burnstock & Holman, 1966b) or a transient slow depolarization with absence of typical action potentials. The last possibility cannot be ignored as the depolarization recorded intracellularly from muscle cells in the isolated rabbit mesenteric artery during high frequency nerve stimulation was a sustained depolarization which showed a peak before falling to a lower level of depolarization (Speden, 1967). The absence of any increase in the amplitude of the more transient spike recorded in a divalent cation-free solution (Keatinge, 1965) suggests that poor co-ordination is not the cause of the long lasting spike. A sustained depolarization, but not spikes, has been recorded with the sucrose-gap technique from helically cut strips of canine carotid arteries following exposure to adrenaline (Barr, 1961b) and noradrenaline (Keatinge, 1966b).

The sustained depolarization of sheep carotid arterial strips following exposure to noradrenaline and the absence of spontaneous electrical activity led Keatinge (1964) to conclude that the 'cell membrane of arteries is relatively stable and adapts readily to depolarization'. An example of extreme stability of arterial cell membranes is the apparent absence of any potential change following noradrenaline stimulation of the pulmonary artery of the rabbit. Both noradrenaline and sympathetic nerve stimulation caused a constriction of the rabbit pulmonary artery which was unaccompanied by any intracellularly recorded potential change (Su et al., 1964). The remaining possible change, a hyperpolarization, was recorded extracellularly during an adrenaline-induced contraction of a rabbit aortic strip (Shibata & Briggs, 1966).

Angiotensin caused a partial contraction of the rabbit aortic strip which was unaccompanied by a potential change (Shibata & Briggs, 1966), whereas it produced both a slow depolarization and a contraction of sheep carotid arterial strips (Keatinge, 1966b). The depolarization and contraction, unlike those in response to adrenaline and noradrenaline, were transient despite the continued presence of the polypeptide, an effect which suggests that failure of receptor activation may be the cause of tachyphylaxis to angiotensin. Bradykinin had a similar effect, but there was no cross tachyphylaxis between bradykinin and angiotensin. With both substances repolarization started at a time when the contraction was still developing. Another polypeptide, vasopressin, caused a marked constriction of rat mesenteric arteries with intact circulation accompanied by an increased firing of action potentials (Steedman, 1966).

Acetylcholine and histamine, which usually constrict isolated, relaxed arteries, both depolarized and contracted strips of the sheep carotid (Keatinge, 1964, 1965, 1966a) and rabbit pulmonary artery (Su & Bevan, 1965). The irregular electrical discharge of sheep carotid arterial strips produced by acetylcholine reflected activation of sympathetic nerve fibres as the discharge was blocked by hexamethonium, nicotine and α-adrenergic blocking agents. In some preparations a slow smooth depolarization persisted in the presence of these agents and was attributed to a direct muscarinic action of acetylcholine (Keatinge, 1966a). Histamine, in contrast, produced a sustained depolarization both in the absence and in the presence of high concentrations (100 μg/ml) of the α-adrenergic blocking agent phentolamine. Keatinge (1966a) suggested that the numerous small spikes, seen on the sucrose-gap records following chemical activation of the sympathetic nerve, were muscle cell action potentials produced by an irregular release of transmitter. The records can equally well be interpreted as slow depolarizations which were transient because of rapid removal of the transmitter and irregular as a result of intermittent asynchronous nerve activity. Su & Bevan (1965) were able to record only slow depolarizations intracellularly from single cells in the rabbit pulmonary artery during the presence of histamine, acetylcholine and serotonin. The depolarizations reached a plateau in about a minute and maximum muscle tension was achieved several minutes later.

High concentrations of the vasoconstrictor substances (10 or 25 μg/ml) have been usually used to produce depolarization of isolated arteries, although Keatinge (1964) observed a small slow depolarization and a contraction with noradrenaline concentrations as low as 25 ng/ml. The necessity to use such high drug concentrations in order to produce vasoconstriction may be a further expression of the stability of the muscle cell membranes in the particular arteries examined. The mean resting potentials of both the sheep arterial strip and single cells in the rabbit pulmonary artery (61 mV and 51·5 mV respectively) were relatively high for vascular smooth muscle.

Vasodilator substances

As the effects of vasoconstrictor agents on isolated arteries is most commonly associated with a sustained depolarization, hyperpolarization would be expected to accompany relaxation. Keatinge (1966b) found that amyl nitrite and sodium nitrite relaxed and repolarized sheep carotid arterial strips which had been previously partially contracted and depolarized by either noradrenaline or histamine. Nitrites did not alter the membrane potential of relaxed arteries, a result which Keatinge thought was an indication of an electrically very stable membrane. Noradrenaline and adrenaline also relaxed strips contracted by histamine, but repolarization was slight hardly exceeding random baseline fluctuations (Keatinge, 1966b). High concentrations of acetylcholine needed to be applied topically to rat mesenteric arteries with intact circulation before spontaneous electrical activity was abolished (Steedman, 1966). The extracellular method of recording gave no indication of whether there was any change in membrane potential.

Action of vasoactive substances on depolarized arterial muscle

As with other smooth muscles, a polarized membrane is not essential for the initiation of a contraction. The potential difference between the recording and reference electrode of the sucrose-gap apparatus of about 60 mV for sheep carotid arterial strips approached zero ($+10$ mV to -10 mV) when both ends of the strip were exposed to physiological salt solutions in which the NaCl and NaHCO$_3$ had been replaced by K$_2$SO$_4$ and KHCO$_3$ (Keatinge, 1964). Such depolarized preparations responded to noradrenaline, adrenaline and histamine when the near-maximal sustained contraction produced by the high potassium solutions at 35°C was converted to a partial contraction by lowering the temperature to 15°C or 20°C. The contraction occurred in the absence of a detectable depolarization unlike that of preparations in normal physiological salt solutions at the same low temperatures.

Adequate information on the relative sensitivity of polarized and depolarized sheep carotid arteries to vasoactive substances is lacking, but specificity of action seems to be retained. Substances which contracted normal arteries also contracted depolarized arteries, whilst those which relaxed did so irrespective of the potential. Nitrites and adrenaline, for example, relaxed depolarized contracted arteries in the absence of any potential change provided, in the case of adrenaline, the preparations were treated with phentolamine (Keatinge, 1966b). Both bradykinin and angiotensin were able to contract strips which had been depolarized by high K solutions at 35°C and then partially relaxed by amyl nitrite without producing any detectable electrical change.

The rabbit pulmonary (Su & Bevan, 1966) and ear artery (de la Lande *et al.*, 1966) as well as intramesenteric arteries of dogs (Waugh, 1962) also retained mechanical responsiveness to catecholamines when the Na$^+$ ions in the physiological saline solutions were replaced by K$^+$ ions. The contractions to noradrenaline of the ear artery, were more prolonged for two of the three ear arteries and the noradrenaline dose-response curves were less steep than in normal Krebs solution. The sensitivity of the third artery to noradrenaline was identical in both normal and high K^+ solutions as was the sensitivity of dog intramesenteric arteries to submaximal doses of adrenaline. This ability of depolarized arterial smooth muscle to respond to vasoactive agents with little alteration in specificity and sensitivity, raises doubts over the role of slow depolarizations by themselves in mediating a contraction (see Chapter 9). The one indication that slow depolarizations may be of functional significance is that, in normal conditions, a depolarization invariably preceded the onset of a contraction. With the turtle aorta, however, slow depolarizations were unable by themselves to initiate a contraction (Roddie, 1962).

The role of calcium ions in the contraction of arterial muscle, both normal and depolarized, has been reviewed by Bohr (1964*a*) and evaluated more recently by a number of workers (Brecht, Estada & Götz, 1964; Hinke, 1965; Su & Bevan, 1966; Shibata & Briggs, 1966; Nash, Luchka & Jhamandas, 1966; Friedman & Friedman, 1967; Mokhme-Lundholm & Vamos, 1967).

Factors predisposing to electrical instability of the membrane in isolated arterial muscle

Vasoconstrictor substances have very different effects on the electrical properties of smooth muscle cells in most isolated arteries compared to those in arteries with intact circulation. A slow, well sustained depolarization is the most common response of isolated arteries to chemical stimulation, whereas electrical instability characterized by both rapid and slow potential changes is initiated or increased by stimulation of arteries with an intact circulation (see above). Such differences may reflect the absence of certain *in vivo* environmental factors under *in vitro* conditions with the additional possibility that large arteries may be less sensitive or even insensitive to these factors *in vivo*. The electrical activity of rat and guinea-pig mesenteric resistance vessels of up to 450 μ in external diameter is known to be greatly influenced by environmental factors (see p. 577). What is uncertain is the nature of those factors operative *in vivo* which permit the membranes of muscle cells in arteries to become electrically unstable. Keatinge (1964) found that drastic procedures such as inhibition of oxidative metabolism with cyanide or sodium fluoroacetate or oxygen deprivation were needed before isolated sheep arterial strips showed marked rhythmical electrical mechanical activity in the presence of noradrenaline. The same preparation developed rapid spontaneous electrical activity which, like that of visceral smooth muscle, was sensitive to both stretch and temperature when all divalent ions were removed from the physiological saline solution (Keatinge, 1965). A factor more likely to be operative *in vivo*, stretch, produced electrical instability of muscle cells in strips of bovine carotid arteries provided the strips were loaded with heavy weights (Petersen, 1936*b*). The effect of other factors on the electrical activity of isolated arterial smooth muscle which favour the development of spontaneous contractions (see p. 571) or have an important role in maintaining cardiovascular function such as the corticosteroids (Travis & Sayers, 1965) have yet to be studied.

Studies of these factors which predispose to instability of arterial smooth muscle cell membranes may help in determining the origins and nature of the slow waves of depolarization and prepotentials (see Chapter 12) which play an important role in the electrical and, in all probability, mechanical activity of arteries *in vivo*. Moreover, such information may be decisive in deciding whether there are fundamental differences in the properties of smooth muscle in large and small arteries and, if so, how these differences arise. The origin, nature and function of the electrical activity is one of the outstanding problems not only of arterial smooth muscle physiology but also of all smooth muscle physiology (Burnstock & Holman, 1966b).

Electrical and mechanical activity of veins

Electrophysiological investigations of veins have been confined almost exclusively to the longitudinal smooth muscle of mammalian portal and anterior mesenteric veins, the exceptions being Roddie's study (1962) on the turtle vena cava and those of Funaki (1958, 1960, 1961) on amphibian venules. The portal vein, which is formed by the junction of the anterior (superior) mesenteric and splenic vein, and the anterior mesenteric vein of a number of mammals are spontaneously active *in vitro* and, for the cat and rabbit portal vein (Funaki, 1966; Johansson & Ljung, 1967a), also *in vivo*.

The longitudinal smooth muscle in isolated, anterior mesenteric veins of such animals as cats, rabbits, rats, guinea-pigs and baboons and the portal vein of rabbits and rats has been reported to be spontaneously active (Maloff, 1934; Cuthbert *et al.*, 1965; Sutter, 1965; Funaki & Bohr, 1964; Axelsson, Johansson, Jonsson & Wahlström, 1966; Johansson & Ljung, 1967a, b; Nakajima & Horn, 1967). The spontaneous activity of these veins is remarkably similar and shows little species variation, but smooth muscle in other veins and venules may be expected to show different characteristics in view of the extreme variability in structure of vessels from different sites and in different species (Franklin, 1937). The relative proportions of circularly and longitudinally orientated smooth muscle vary considerably and may be present as a dense continuous layer or in widely spaced bundles of muscle fibres. The portal veins of the dog, rabbit, rat and man, for example, have longitudinal muscle fibres outside circular muscle fibres. The circular muscle is arranged as bundles in the rat or as a continuous layer or groups of bundles in the rabbit. (Suchard, 1901a, b). In the rat the longitudinal muscle is either in bundles or as a dense continuous layer (Suchard, 1901a, b) with adjacent smooth muscle cells in electron micrographs of the rat portal vein being separated by a small gap only (Funaki, 1966). The close packing of muscle cells in these portal veins might favour interaction between cells so that conduction occurs (see p. 571). In other veins, however, the separation between cells or groups of cells may be such that conduction cannot take place.

The circular smooth muscle in veins and venules, whose function is to alter calibre of the vessels, may have properties different from those of the longitudinal smooth muscle. Circular smooth muscle in the anterior mesenteric veins of rabbits, cats and guinea-pigs rarely showed spontaneous activity, unlike the longitudinal muscle in the same veins, and the posterior caval vein of the rabbit, which has little longitudinal muscle, was never spontaneously active (Sutter, 1965).

Spontaneous activity

The electrical activity of isolated portal and anterior mesenteric veins differs from that of arteries with an intact circulation in usually being a mixture of locally initiated and conducted action potentials. The variable and complex pattern of activity may take the form of a continuous discharge of action potentials, or bursts of variable duration, or there may be only a single action potential. Slow depolarizations, usually associated with action potentials, are of variable shape with durations of 1 to 2 sec to more than 10 sec and amplitudes which may exceed 20 mV. A good example of pacemaker type activity is shown in Fig. 1d published by Cuthbert et al. (1965). Signs of conducted electrical activity include action potentials which lack a prepotential, the occurrence of action potentials during any phase of the slow waves of depolarization, the presence of double spikes on the same prepotential and the failure of a large action potential to 'wipe out' slow depolarizations. Electrical activity which shows most of the above characteristics has been recorded intracellularly from the rat portal vein (Funaki & Bohr, 1964), the rabbit anterior mesenteric vein (Cuthbert et al., 1965; Matthews & Sutter, 1967) and the guinea-pig anterior mesenteric vein (Nakajima & Horn, 1967). Conduction of the action potential may be slow compared to visceral smooth muscle (see p. 571). An action potential produced by a threshold induction shock applied through an external micro-electrode in the rat portal vein propagated at a velocity of only about 17 mm/sec for a distance of 3·5 mm from the stimulating electrode (Funaki, 1966).

Conduction in the rat portal vein appears to be myogenic in nature. A local anaesthetic in concentrations sufficient to block nerve conduction failed to block conduction (Johansson & Ljung, 1967b) and neither phenoxybenzamine nor propranolol altered the electrical and mechanical activity (Johansson, Jonsson, Axelsson & Wahlström, 1967). Conduction does not seem to involve pull of muscle fibres upon adjacent muscle fibres, although the preparations are sensitive to stretch (Cuthbert et al. 1965; Holman, Kasby & Suthers, 1967). Conduction in the rat portal vein still occurred through a region of mechanical immobilization produced by anchoring the middle of a preparation and placing the distal and proximal halves of the vein at right angles to each other (Johansson

& Ljung, 1967*b*; Johansson *et al.*, 1967). The two parts of the vein contracted synchronously, although each was mechanically independent of the other part, which indicates that both were activated by the same pacemaker. Placing a tight ligature at the middle of the vein disrupted conduction and the two portions of the vein then contracted at different rhythms (Johansson *et al.*, 1967).

The presence of calcium ions was essential for the spontaneous electrical and mechanical activity of the rat portal vein (Axelsson *et al.*, 1966; Johansson *et al.*, 1967; Axelsson, Walhström, Johansson & Jonsson, 1967). Removal of calcium ions from the Krebs solution led to a marked reduction in the number of action potentials within a burst, until the electrical activity resembled that of cardiac pacemaker cells, each slow potential being accompanied by a single spike. Ultimately all electrical and mechanical activity ceased. The electrical and mechanical activity of the rabbit and rat portal vein was highly sensitive to both bath temperature and stretch (Cuthbert *et al.*, 1965; Funaki, 1966; Holman *et al.*, 1967).

The smooth muscle in venules of the frog tongue and web was electrically quiescent, although action potentials could be initiated by single induction shocks (Funaki, 1958, 1960), whereas muscle cells in venules in the skin of the frog abdomen were spontaneously active (Funaki, 1961). The action potentials, which arose spontaneously from a low membrane potential, were of a pace-maker type lasting some 200 msec. In contrast, the action potentials recorded intracellularly from cells in the turtle inferior vena cava consisted of a spike followed by a plateau of depolarization which persisted for several seconds (Roddie, 1962). An action potential propagated in the turtle vena cava at velocities of about 0·5 mm/sec.

Vasoconstrictor nerve stimulation

The nature of the innervation of the rabbit portal vein is uncertain. Trans-mural electrical stimulation of a longitudinally cut strip of the vein produced a relaxation or a contraction followed by an after-relaxation depending on the frequency of stimulation and the basal tone of the preparation (Hughes & Vane, 1967). A pharmacological analysis of the responses (Hughes & Vane, 1967) indicated that the contraction resulted from excitation of postganglionic sympathetic nerves with the released transmitter acting on α-receptors. Part of the relaxation seemed to arise from an action of the transmitter on beta receptors and part from excitation of non-adrenergic non-cholinergic nerve elements. Johansson & Ljung (1967*a*) have failed to confirm the results of Hughes & Vane. Stimulation of the splanchnic nerve increased the longitudinal tension in the wall of the rabbit or cat portal vein *in situ* with no sign of relaxation either during or after the end of stimulation. Following α-adrenergic blockade, splanchnic nerve stimulation neither contracted nor relaxed the portal vein *in situ* and no relaxation of the isolated rabbit portal vein was seen during stimulation of the post-ganglionic nerves either in the presence or absence of phenoxybenzamine (Johansson & Ljung, 1967a). Vagal nerve stimulation had weak and inconsistent excitatory effects on the longitudinal smooth muscle of rabbit and cat portal veins which were unaffected by atropine, phenoxybenzamine and neostigmine. Marked increases in activity of the circular smooth muscle in rat portal veins in situ were produced by stimulation of the vagus nerve (Funaki, 1966).

Holman et al. (1967) have used a modified sucrose-gap apparatus to record the electrical and mechanical activity of strips of rabbit portal vein during sympathetic nerve stimulation. Low frequency transmural electrical stimulation of the nerves increased the frequency of the bursts of action potentials, whilst high frequency stimulation produced a continuous discharge of action potentials.

Vasoactive substances

The effects of low and intermediate concentrations of noradrenaline on the electrical activity of the rabbit portal vein resembled those of low and high frequency nerve stimulation respectively (Holman et al., 1967). Cuthbert & Sutter (1965) noted that the correlation between the frequency of action potentials and tension held only during the initial phase of exposure of rabbit anterior mesenteric vein strips to noradrenaline. The discharge of action potentials declined to control levels at a time when the tension still exceeded control values. A similar dissocation between electrical and mechanical activity in the presence of noradrenaline has also been reported for the isolated rat portal vein (Johansson et al., 1967). The good correlation between frequency of action potentials and tension during exposure to angiotensin or a threefold increase in potassium ion concentration led Cuthbert & Sutter (1965) to conclude that the dissociation was not an artifact of the recording method — the sucrose-gap technique. Nor did the dissociation appear to be the result of depolarization which may cause cessation of spike activity at a time when tension is well maintained (see p. 587). The doses of noradrenaline used caused little or no depolarization and, further, depolarizations produced by potassium-rich solutions failed to dissociate electrical and mechanical activity. Other indications that smooth muscle in the rabbit anterior mesenteric vein may be activated both by electrical and non-electrical means have been provided by Matthews & Sutter (1967). These workers treated isolated preparations of the guinea-pig taenia coli and rabbit vein with the cardiac glycoside ouabain. Both preparations contracted immediately on exposure to high concentrations of the glycoside and the contraction was followed by a gradual relaxation over the next 30 to 60 min, but then the vein, and only the vein, contracted again. The initial contraction had been accompanied by an increased discharge of action potentials, recorded intracellularly, which subsequently decreased and then ceased after less than 35 min exposure to ouabain. There was a progressive fall in membrane potential to low values. Under these conditions of low membrane potential and absence of action potentials, noradrenaline and serotonin contracted the vein without altering the membrane potential. However, under the same conditions, acetylcholine, histamine and serotonin failed to stimulate the taenia coli. Noradrenaline also contracted rabbit anterior mesenteric and rat portal vein strips which had been depolarized by high potassium solutions (Cuthbert & Sutter, 1965; Axelsson et al., 1966; Johansson et al., 1967). This effect was blocked by dibenzyline. With this evidence, it is difficult to reach any conclusion other than that veins can be contracted by a mechanism which does not involve action potentials. Such a mechanism appears to be more highly developed in the anterior mesenteric vein of rabbits than in the guinea-pig taenia coli (Matthews & Sutter, 1967) and there is some evidence that calcium ions may play an important role in this mechanism (Cuthbert & Sutter, 1965; Axelsson et al., 1966; Johansson, et al., 1967; Axelsson et al., 1967; and Chapter 9).

High concentrations of noradrenaline depolarized the isolated rat portal vein and converted the bursts of action potentials into a continuous discharge of action potentials which declined in amplitude and then disappeared within about a minute (Johansson et al., 1967). Small spikes re-appeared a few seconds later. This reduction in amplitude of the action potentials may represent asynchrony of electrical activity, which would not be recorded by the sucrose-gap method, or depolarization to below the level required for action potential generation (Johansson et al., 1967). The effect of adrenaline on the intracellularly recorded electrical activity of the rat portal vein (Funaki & Bohr, 1964) and guinea-pig anterior mesenteric vein (Nakajima & Horn, 1967) suggest that the latter possibility is the more likely. High concentrations of adrenaline caused a continuous firing of action potentials which rapidly declined in amplitude to small oscillations as the depolarization developed. Adrenaline also accelerated the rate of discharge of action potentials recorded from smooth muscle cells in the turtle vena cava (Roddie, 1962).

Isoprenaline, in low concentrations (about 1×10^{-7}), increased the frequency of the bursts of action potentials recorded from the rat portal vein, but reduced the number of action potentials within a burst (Axelsson et al., 1966; Johansson et al., 1967). High concentrations (about 1×10^{-5}) first stimulated and then abolished all electrical and mechanical activity. Not all of the reduction in force of contraction caused by low concentrations could be attributed to the decreased number of spikes within a burst, for there was often a marked reduction in the amplitude of the contraction at a time when spike activity was well maintained. The ability of isoprenaline in low concentrations to relax portal veins depolarized by solutions in which all sodium had been replaced by potassium ions was taken as further evidence for the existence of non-electrical inhibition. This effect was abolished by propranolol. It was suggested by these workers that isoprenaline may act by reducing the availability of calcium ions because the effects of isoprenaline were similar to those produced by reduction in extracellular calcium ion concentration.

The effects of acetylcholine on the intracellularly recorded electrical activity of muscle cells in rat portal veins (Funaki & Bohr, 1964), guinea-pig anterior mesenteric veins (Nakajima & Horn, 1967) and the turtle vena cava (Roddie, 1962) are complex. There may be an initial transient hyperpolarization followed by increased firing of action potentials of large amplitude and depolarization (Funaki & Bohr, 1964; Nakajima & Horn, 1967). Hyperpolarization was particularly prominent in cells with low resting membrane potentials and, if they were electrically inactive, a discharge of action potentials was initiated. Low concentrations of acetylcholine inhibited the electrical and mechanical activity of the turtle vena cava or had no effect, whereas high concentrations invariably increased the firing of action potentials and caused summation of contractions (Roddie, 1962).

Histamine and angiotensin stimulated the anterior mesenteric vein of the rabbit in a way similar to noradrenaline, but the dissociation between electrical and mechanical activity was less marked particularly with angiotensin (Cuthbert & Sutter, 1965).

CONCLUSIONS

Emphasis has been placed on the role of the cell membrane potential in excitation of vascular smooth muscle and the way in which the structure of the blood vessel wall affects spread of excitation by influencing cell-to-cell interactions and diffusion of vasoactive agents. The information available is fragmentary, incomplete and mostly descriptive, but is sufficient to outline certain of the events involved in excitation of vascular smooth muscle. Alterations in membrane potential are usually, but not necessarily always, involved in excitation. The unknown nature of the events underlying the depolarizations, of the factors predisposing to electrical instability of the membrane, and of the link between depolarization and contraction, all act as incentives to further investigations.

ACKNOWLEDGEMENT — I should like to thank Professor R. F. Whelan and Dr. I. S. de la Lande for reading the manuscript critically and Mrs. A. Macaskill and Miss Pauline Boundy for secretarial assistance.

REFERENCES

Abe, Y. (1968). The effect of sodium and calcium on the action potential of pregnant rat myometrium. *J. Physiol.* **200**, 1–2*P*.

Abe, Y. & Tomita, T. (1968). Cable properties of smooth muscle. *J. Physiol.* **196**, 87–100.

Åberg, A. K. G. (1967). The series elasticity of active taenia coli *in vitro*. *Acta physiol. scand.* **69**, 348–354.

Åberg, A. K. G. & Axelsson, J. (1965). Some mechanical aspects of an intestinal smooth muscle. *Acta physiol. scand.* **64**, 15–27.

Åberg, G. & Eränkö, O. (1967). Localization of noradrenaline and acetylcholinesterase in the taenia of the guinea-pig caecum. *Acta physiol. scand.* **69**, 383–384.

Åberg, A. K. G., Mohme-Lundholm, E. & Vamos, N. (1967). The effect of H^+ and lactate ions on the electrical activity and content of high energy phosphate compounds of the taenia coli from the guinea-pig. *Acta physiol. scand.* **69**, 129–133.

Adrian, E. D., Feldberg, W. & Kilby, B. A. (1947). The cholinesterase inhibiting action of fluorophosphonates. *Br. J. Pharmac. Chemother.* **2**, 56–58.

Adrian, R. H. (1956). The effect of internal and external potassium concentration on the membrane potential of frog muscle. *J. Physiol.* **133**, 631–658.

—— (1960). Potassium chloride movement and the membrane potential of frog muscle. *J. Physiol.* **151**, 154–185.

Adrian, R. H. & Peachy. L. D. (1965). The membrane capacity of frog twitch and slow muscle fibres. *J. Physiol.* **181**, 324–336.

Adrian, R. H. & Slayman, C. L. (1966). Membrane potential and conductance during transport of sodium, potassium and rubidium in frog muscle. *J. Physiol.* **184**, 970–1014.

Aghajanian, G. K. & Bloom, F. E. (1966). Electron-microscopic autoradiography of rat hypothalamus after intraventricular H^3-norepinephrine. *Science, N.Y.* **153**, 308–310.

—— (1967). Localization of tritiated serotonin in rat brain by electron-microscopic autoradiography. *J. Pharmac. exp. Ther.* **156**, 23–30.

Aghajanian, G. K., Bloom, F. E., Lovell, R. A., Sheard, M. H. & Freedman, D. X. (1966). The uptake of 5-hydroxytryptamine ^3H from the cerebral ventricles: autoradiographic localization. *Biochem. Pharmac.* **15**, 1401–1403.

Ahlquist, R. P. (1948). A study of the adrenotropic receptors. *Am. J. Physiol.* **153**, 586–600.

—— (1962). The adrenotropic receptor-detector. *Arch. int. Pharmacodyn. Thér.* **139**, 38–41.

—— (1966). The adrenergic receptor. *J. pharm. Sci.* **55**, 359–367.

Ahlquist, R. P. & Levy, B. (1959). Adrenergic receptive mechanism of canine ileum. *J. Pharmac. exp. Ther.* **127**, 146–149.

Aiello, E. L. (1960). Factors affecting ciliary activity on the gill of the mussel *Mytilus edulis* *Physiol. Zoöl.* **33**, 120–135.

Akubue, P. I. (1966). A periarterial nerve-circular muscle preparation from the caecum of the guinea-pig. *J. Pharm. Pharmacol.* **18**, 390–395.

Alexander, R. S. (1967). Contractile mechanics of venous smooth muscle. *Am. J. Physiol.* **212**, 852–858.

Alpert, L. K. (1931). The innervation of the suprarenal glands. *Anat. Rec.* **50**, 221–234.

Alvarado, F., Middleton, S. & Beca, J. P. (1961). Efecto de la reserpina y de la adrenalectomia sobre la acción cardioestimulante de la acetilcolina. *Acta physiol. latinoam.* **11**, 236–237.

Alvarez, W. C. (1922). *The Mechanics of the Digestive Tract.* 2nd ed. pp. 66–81. New York: Hoeber.

Alvarez, W. C. & Mahoney, L. J. (1922). Action currents in the stomach and intestine. *Am. J. Physiol.* **58**, 476–493.

Alvarez-Morujo, A. (1966). The arterial segment. *Acta anat.* **64**, 107–116.

Ambache, N. (1951). Unmasking, after cholinergic paralysis by botulinum toxin, of a reversed action of nicotine on the mammalian intestine, revealing the probable presence of local inhibitory ganglion cells in the enteric plexuses. *Br. J. Pharmac. Chemother.* **6**, 51–67.

—— (1955). The use and limitations of atropine for pharmacological studies on autonomic effectors. *Pharmac. Rev.* **7**, 467–494.

—— (1963). Irin and a hydroxy-acid from brain. *Biochem. Pharmac.* **12**, 421–428.

Ambache, N., Brummer, H. C., Rose, J. G. & Whiting, J. (1966). Thin-layer chromatography of spasmogenic unsaturated hydroxy-acids from various tissues. *J. Physiol* **185**, 77–78P.

Ambache, N. & Freeman, M. A. (1968). Atropine resistant spasms due to excitation of non-cholinergic neurones in guinea-pig myenteric plexus. *J. Physiol.* **198**, 92–94P.

Ambache, N. & Lessin, A. W. (1955). Classification of intestinomotor drugs by means of type D botulinum toxin. *J. Physiol.* **127**, 449–478.

Ambache, N. & Lippold, O. C. J. (1949). Bradycardia of central origin produced by injections of tetanus toxin into the vagus nerve. *J. Physiol.* **108**, 186–196.

Amin, A. H., Crawford, T. B. B. & Gaddum, J. H. (1954). The distribution of substance P and 5-hydroxytryptamine in the central nervous system of the dog. *J. Physiol.* **126**, 596–618.

Ammon, R. & Savelsberg, W. (1949). Die enzymatische Spaltung von Atropin, Cocain und chemisch verwandten Estern. *Z. phys. Chem.* **284**, 135–156.

Anderson, N. & Moore, J. W. (1968). Voltage clamp of uterine smooth muscle. *Fedn Proc.* **27**, 2, 704P.

Angelakos, E. T., Fuxe, K. Torchiana, M. L. (1963). Chemical and histochemical evaluation of the distribution of catecholamines in the rabbit and guinea-pig hearts. *Acta physiol. scand.* **59**, 184–192.

Angelakos, E. T., Glassman, P. M., Millard, R. W. & King, M. (1965). Regional distribution and subcellular localization of catecholamines in the frog heart. *Comp. Biochem. Physiol.* **15**, 313–324.

Änggård, E. & Samuelsson, B. (1964). Smooth muscle stimulating lipids in sheep iris. The identification of prostaglandin $F_{2\alpha}$. *Biochem. Pharmac.* **13**, 281–283. *Pharmac.* **13**, 281–283.

Appenzeller, O. (1964). Electron microscopic study of the innervation of the auricular artery in the rat. *J. Anat. (Lond.)* **98**, 87–91.

Apter, J. T. & Koketsu, K. (1960). Temperature studies implicating calcium in regulation of muscle membrane potential. *J. cell. comp. Physiol.* **56**, 123–127.

Ariens, E. J. (1964). *Molecular Pharmacology.* New York: Academic Press.

Armett, C. J. & Ritchie, J. M. (1960). The action of acetylcholine on conduction in mammalian non-myelinated fibres and its prevention by an anticholinesterase. *J. Physiol.* **152**, 141–158.

—— (1961). The action of acetylcholine and some related substances on conduction in mammalian non-myelinated nerve fibres. *J. Physiol.* **155**, 372–384.

—— (1963). The ionic requirements for the action of acetylcholine on mammalian non-myelinated fibres. *J. Physiol.* **165**, 141–159.

Armin, J., Grant, R. T., Thompson, R. H. S. & Tickner, A. (1953). An explanation for the heightened vascular reactivity of the denervated rabbit's ear. *J. Physiol.* **121**, 603–622.

Armstrong, H. I. O., Milton, G. W. & Smith, A. W. M. (1956). Electropotential changes of the small intestine. *J. Physiol.* **131**, 147–153.

Aschoff, J. (1944). Mitteilung zur spontanen und reflektorischen Vasomotorik der Haut. *Pflügers Arch. ges. Physiol.* **248**, 171–177.

Ashley, C. C. (1967). The role of cell calcium in the contraction of single cannulated muscle fibres. *Am. Zool.* **7**, 647–659.

Ashley, C. C., Caldwell, P. C., Lowe, A. G., Richards, C. D. & Schirmer, H. (1965). The amount of injected EGTA needed to suppress the contractile responses of single *Maia* muscle fibres and its relation to the amount of calcium released during contraction. *J. Physiol.* **179**, 32–33*P*.

Astbury, W. T., Perry, S. V., Reed, R. & Spark, L. C. (1947). An electron microscope and X-ray study of actin. *Biochim. biophys. Acta.* **1**, 379–392.

Atzler, E. & Lehmann, G. (1927). Reaktionen der Gefässe auf direkte Reize. *Handbuch norm. u. pathol. Physiol.* VII, 2 pp. 963–997. Berlin: Springer.

Aubert, X. (1955). Intervention d'un élément élastique pur dans la contraction du muscle strié. *Archs int. Physiol.* **63**, (2), 197–202.

Auer, J. & Krueger, H. (1947). Experimental study of antiperistaltic and peristaltic motor and inhibitory phenomena. *Am. J. Physiol.* **148**, 350–357.

Auerbach L. (1864). Fernere vorläufige Mittheilung über den Nervenapparat des Darmes. *Virchows Arch. path. Anat. Physiol.* **30**, 457–460.

Aunap, E. (1936). Über die Form der glatten Muskelzellen und die Verbindungen zwischen ihnen. *Z. mikrosk. -anat. Forsch.* **40**, 587–598.

Austin, L., Livett, B. G. & Chubb, I. W. (1967). Biosynthesis of noradrenaline in sympathetic nervous tissue. *Circulation Res.* **21**, (Suppl. 3) 111–117.

Avakian, O. V. & Gillespie, J. S. (1968). Uptake of noradrenaline by adrenergic nerves, smooth muscle and connective tissue in isolated perfused arteries and its correlation with the vasoconstrictor response. *Br. J. Pharmac. Chemother.* **32**, 168–184.

Axelrod, J. (1966). Control of catecholamine and indoleamine metabolism by sympathetic nerves. In *Mechanisms of Release of Biogenic Amines*, Ed. von Euler, U. S., Rosell, S. & Uvnäs, B. 189–209. Oxford: Pergamon Press.

Axelrod, J., Weil-Malherbe, H. & Tomchick, R. (1959). The physiological disposition of ^{3}H-epinephrine and its metabolite metanephrine. *J. Pharmac. exp. Ther.* **127**, 251–256.

Axelsson, J. (1961a). The effect of nitrate on electrical and mechanical activity of smooth muscle. *J. Physiol.* **155**, 9*P*.

—— (1961b). Dissociation of electrical and mechanical activity in smooth muscle. *J. Physiol.* **158**, 381–398.

—— (1961c). Effects of nitrate on the electrical and mechanical activity of intestinal smooth muscle. *Biochem. Pharmac.* **8**, 155.

—— (1962). Studies of electrical and mechanical activity in intestinal smooth muscle. Phil. D. Thesis, Gleerups, Lund.

—— (1964). A study of the relationship of mechanical to electrical activity in smooth muscle. D. Phil. Thesis, Oxford.

Axelsson, J., Bueding, E., & Bülbring, E. (1959). The action of adrenaline on phosphorylase activity and membrane potential of smooth muscle. *J. Physiol.* **148**, 62–63*P*.

—— (1961). The inhibitory action of adrenaline on intestinal smooth muscle in relation to its action on phosphorylase activity. *J. Physiol.* **156**, 357–374.

Axelsson, J. & Bülbring, E. (1959). Some means of abolishing the tension response in smooth muscle during continued electrical activity at the cell membrane. *J. Physiol.* **149**, 50–51*P*.

—— (1961). Metabolic factors affecting the electrical activity of intestinal smooth muscle. *J. Physiol.* **156**, 344–356.

Axelsson, J., Gudmundsson, G. & Wahlström, B. (1968). Quantitative analysis of the correlation between electrical and mechanical activity in smooth muscle. *Acta physiol. scand.* **73**, *c* 23.

Axelsson, J. & Högberg, S. G. R. (1967). Effects of caffeine on electrical and mechanical activity of smooth muscle. *Acta pharmac. tox.* **25**, suppl. 4, 53.

Axelsson, J., Högberg, S. G. R. & Timms, A. R. (1965). The effect of removing and readmitting glucose on the electrical and mechanical activity and glucose and glycogen content of the intestinal smooth muscle from the taenia coli of the guinea-pig. *Acta physiol. scand.* **64**, 28–42.

Axelsson, J. & Holmberg, B. (1969). The effects of extracellularly applied ATP and related compounds on the electrical and mechanical activity of the smooth muscle of taenia coli of the guinea-pig. *Acta physiol. scand.* (In press).

Axelsson, J., Holmberg, B. & Högberg, G. (1965). Some effects of ATP and adrenaline on intestinal smooth muscle. *Life Sci.*, **4**, 817–821.

—— (1966). ATP and intestinal smooth muscle. *Acta physiol. scand.* **68**, suppl. 277, 19.

Axelsson, J., Holmberg, B., Högberg, S. G. R. & Wahlström, B. (1969). Internal potassium and tension development in an intestinal smooth muscle. *Acta physiol. scand. Suppl.* **330**, 177.

Axelsson, J., Johansson, B., Jonsson, O. & Wahlström, B. (1966). The effects of adrenergic drugs on electrical and mechanical activity of the portal vein. *Symp. Electr. Activ. Innerv. Blood Vessels*, Cambridge, 1966. *Biblphie anat.* **8**, 16–20.

Axelsson, J. & Thesleff, S. (1958). Activation of the contractile mechanism in striated muscle. *Acta physiol. scand.* **44**, 55–66.

Axelsson, J. & Wahlström, B. (1966). Communication to the XII Scand. Physiol. Congress.

Axelsson, J., Wahlström, B., Johansson, B. & Jonsson, O. (1967). Influence of the ionic environment on spontaneous electrical and mechanical activity of the rat portal vein. *Circulation Res.* **21**, 609–619.

Babkin, B. P., Stavraky, G. W. & Alley, A. (1932). Humoral transmission of chorda tympani hormone. *Am. J. Physiol.* **101**, 2–3.

Bacq, Z. M. (1933). Recherches sur la physiologie du système nerveux autonome. III. Les propriétés biologiques et physicochimiques de la sympathine comparées à celles de l'adrénaline. *Archs int. Physiol.* **36**, 167–246.

—— (1935). Recherches sur la physiologie et la pharmacologie du système nerveux autonome. XII. Nature cholinergique et adrénergique des diverses innervations vasomotrices du pénis chez le chien. *Archs int. Physiol.* **40**, 311–321.

Bacq, Z. M. & Frédéricq, H. (1934). L'innervation sympathique post-ganglionnaire de la membrane nictitante du chat peut-être en partie de nature cholinergique. *C. r. Séanc. Soc. Biol.* **117**, 482–485.

Bader, H. (1963). The anatomy and physiology of the vascular wall. In *Handbook of Physiology*, Section 2: Circulation, Vol. II. Ed. Hamilton, W. F. & Dow, P. Washington: Amer. Physiol. Soc.

Bailey, D. M. (1965). The action of sympathomimetic amines on circular and longitudinal smooth muscle from the isolated oesophagus of the guinea-pig. *J. Pharm. Pharmacol.* **17**, 782–787.

Bain, W. A. (1933). The mode of action of vasodilator and vasoconstrictor nerves. *Q. Jl. exp. Physiol.* **23**, 381–389.

Bainbridge, J. G. & Brown, D. M. (1960). Ganglion-blocking properties of atropine-like drugs. *Br. J. Pharmac. Chemother.* **15**, 147–151.

Baker, P. F., Blaustein, M. P., Hodgkin, A. L. & Steinhardt, R. A. (1969). The influence of calcium on sodium efflux in squid axons. *J. Physiol.* **200**, 431–458.

Baker, R. V., Blaschko, H. & Born, G. V. R. (1959). The isolation from blood platelets of particles containing 5-hydroxytryptamine and adenosine triophosphate. *J. Physiol.* **149**, 55–56P.

Balassa, G. & Gurd, M. R. (1941). Action of adrenaline and potential changes in the cat uterus. *J. Pharmacol. exp. Ther.* **72**, 63–73.

Banerjee, A. K. & Lewis, J. J. (1963). The effects of smooth muscle stimulants on the movements of calcium-47 in the guinea-pig ileum *in vitro*. *J. Pharm. Pharmacol.* **15**, 409–410.

—— (1964) Effects of smooth muscle stimulants and their antagonists upon potassium ion uptake and release in strips of guinea-pig ileum. *J. Pharm. Pharmacol.* **16**, 134–136.

Barajas, L. (1964). The innervation of the juxtaglomerular apparatus. An electron microscopic study of the innervation of the glomerular arterioles. *Lab. Invest.* **13**, 916–929.

Barcroft, J., Khanna, L. C. & Nisimaru, Y. (1932). Rhythmical contraction of the spleen. *J. Physiol.* **74**, 294–298.

Barcroft, J. & Nisimaru, Y. (1932a). Cause of rhythmical contraction of the spleen. *J. Physiol.* **74**, 299–310.

—— (1932b). Undulatory changes of blood pressure. *J. Physiol.* **74**, 311–320.

Barfurth, D. (1891). Über Zellbrücken glatter Muskelfasern. *Arch. mikrosk. Anat. Entw-Mech*, **38**, 38–51.

Barger, G. & Dale, H. H. (1910–11). Chemical structure and sympathomimetic action of amines. *J. Physiol.* **41**, 19–59.

Barlow, R. B. (1961). Effects on amine oxidase of substances which antagonise 5-hydroxytryptamine more than tryptamine on the rat fundus strip. *Br. J. Pharmac. Chemother.* **16**, 153–162.

Barlow, R. B. & Khan, I. (1959). Actions of some analogues of 5-hydroxytryptamine on the isolated rat uterus and the rat fundus strip preparations. *Br. J. Pharmac. Chemother.* **14**, 265–272.

Barnett, A. & Benforado, J. M. (1966). The nicotinic effects of choline esters and of nicotine in guinea-pig atria. *J. Pharmac. exp. Ther.* **152**, 29–36.

Barr, L. (1961a). Transmembrane resistance of smooth muscle cells. *Am. J. Physiol.* **200**, 1251–1255.

—— (1961b). The responsiveness of arterial smooth muscle. In *Biophysics of Physiological and Pharmacological Actions*. Ed. Shanes, A. M. Washington: Amer. Assoc. Adv. Sci. Publ. 69.

—— (1963). Propagation in vertebrate visceral smooth muscle. *J. theor. Biol.* **4**, 73–85.

—— (1965). Membrane potential profiles and the Goldman equation. *J. theor. Biol.* **9**, 351–356.

Barr, L. & Berger, W. (1964). The role of current flow in the propagation of cardiac muscle action potentials. *Pflügers Archiv. ges. Physiol.* **279**, 192–194.

Barr, L., Berger, W. & Dewey, M. M. (1968). Electrical transmission at the nexus between smooth muscle cells. *J. gen. Physiol.* **51**, 347–369.

Barr, L., Dewey, M. M. & Berger, W. (1965). Propagation of action potentials and the structure of the nexus in cardiac muscle. *J. gen. Physiol.* **48**, 797–823.

Barr, L., Dewey, M. M. & Evans, H. (1965). The role of the nexus in the propagation of action potentials of cardiac and smooth muscle. *Fedn Proc.* **24**, 142.

Barrnett, R. J. (1962). The fine structural localization of acetylcholinesterase at the myoneural junction. *J. cell Biol.* **12**, 247–262.

—— (1966). Ultrastructural histochemistry of normal neuromuscular junctions. *Ann. N.Y. Acad. Sci.* **135**, 27–34.

Barstad, J. A. B. (1962). Presynaptic effect of neuromuscular transmitter. *Experientia* **18**, 579–580.

Bartels, J., Vogt, W. & Wille, G. (1968). Prostaglandin release from and formation in perfused frog intestine. *Arch. exp. Path. Pharmak.* **259**, 153–154.

Bartelstone, H. J., Nasmyth, P. A. & Telford, J. M. (1967). The significance of adenosine cylic 3′5′-monophosphate for the contraction of smooth muscle. *J. Physiol.* **188**, 159–176.

Barter, R. & Pearse, A. G. E. (1953). Detection of 5-hydroxytryptamine in mammalian enterochromaffin cells. *Nature, Lond.* **172**, 810.

—— (1955). Mammalian enterochromaffin cells as the source of serotonin (5-hydroxytryptamine.) *J. Path. Bact.* **69**, 25–31.

Bartholini, G., Pletscher, A. & Bruderer, H. (1964). Formation of 5-hydroxytryptophol from endogenous 5-hydroxytryptamine by isolated blood platelets. *Nature, Lond.* **203**, 1281–1283.

Bass, A. D. Hurwitz, L. & Smith, B. (1964). Smooth muscle contraction in presence of an inhibitor of K efflux. *Am. J. Physiol.* **206**, 1021–1024.

Bass, P., Code, C. F. & Lambert, E. H. (1961). Motor and electric activity of the duodenum. *Am. J. Physiol.* **201**, 287–291.

Bass, P. & Wiley, J. N. (1965). Effects of ligation and morphine on electric and motor activity of dog duodenum. *Am. J. Physiol.* **208**, 908–913.

Bauer, H., Goodford, P. J. & Hüter, J. (1965). The calcium content and ^{45}calcium uptake of the smooth muscle of the guinea-pig taenia coli. *J. Physiol.* **176**, 163–179.

Baumgarten, H. G. (1967). Über die Verteilung von Catecholaminen im Darm des Menschen, *Z. Zellforsch. mikrosk. Anat.* **83**, 133–146.

Baumgarten, H. G. & Holstein, A. F. (1967). Catecholaminhaltige Nervenfasern im Hoden des Menschen. *Z. Zellforsch. mikrosk. Anat.* **79**, 389–395.

Baumgartner, H. R. & Born, G. V. R. (1967). 5-hydroxytryptamine in rabbit platelets and their aggregation. *J. Physiol.* **194**, 92–93P.

—— (1968). Effects of 5-hydroxytryptamine on platelet aggregation. *Nature, Lond.* **218**, 137–141.

Baumgartner, H. R. & Born, G. V. R. (1969). The relation between the 5-hydroxytryptamine content and aggregation of rabbit platelets. *J. Physiol.* **201**, 397–408.

Baur, M. (1928). Versuche am Amnion von Huhn und Gans. *Naunyn-Schmiedeberg's Arch. exp. Path. Pharmak.* **134**, 49–65.

Bayliss, W. M. (1902). On the local reactions of the arterial wall to changes of internal pressure. *J. Physiol.* **28**, 220–231.

Bayliss, W. M. & Starling, E. H. (1899). The movements and innervation of the small intestine. *J. Physiol.* **24**, 99–143.

—— (1900). The movements and the innervation of the large intestine. *J. Physiol.* **26**, 107–118.

Bear, R. S. (1945). Small angle X-ray diffraction studies on muscle. *J. Am. chem. Soc.* **67**, No. 2. 1625–1626.

Beck, L. (1961a). Active reflex dilatation in the innervated perfused hind leg of the dog. *Am. J. Physiol.* **201**, 123–128.

—— (1961b). Blockade of reflex dilatation by ergot alkaloids. *Pharmacologist*, **3**, 50.

—— (1964). A new concept of autonomic interaction in the peripheral sympathetic nervous system. *Texas Rept. Biol. Med.* **22**, 375–409.

—— (1965). Histamine as the potential mediator of active reflex vasodilatation. *Fedn. Proc.*, **24**, 1298–1310.

Beck, L. & Brody, M. J. (1961). The physiology of vasodilatation. *Angiology* **12**, 202–222.

Beck, L. & Pollard, A. A. (1967). Selective reduction of sustained dilatation by prostaglandin E_1 and phenoxybenzamine. *Proc. Can. Fed. Biol. Sci.* **10**, 23.

Beck, L., Pollard, A. A., Kayaalp, S. O. & Weiner, L. M. (1966). Sustained dilatation elicited by sympathetic nerve stimulation. *Fedn Proc.* **25**, 1596–1606.

Beck, L., Pollard, A. & Weiner, L. M. (1964). Transient and sustained components of vaso-dilatation associated with sympathetic stimulation. *Fedn. Proc.* **23**, 540.

Beleslin, D. & Varagić, V. (1958). Effect of substance P on the peristaltic reflex of the isolated guinea-pig ileum. *Br. J. Pharmac. Chemother.* **13**, 321–325.

Bell, C. (1967a). An electrophysiological study of the effects of atropine and physostigmine on transmission to the guinea-pig vas deferens. *J. Physiol.* **189**, 31–42.

—— (1967b). Effects of cocaine and of monoamine oxidase and catechol-o-methyltransferase inhibitors on transmission to the guinea-pig vas deferens. *Br. J. Pharmac. Chemother.* **31**, 276–289.

—— (1969). Fine structural localisation of acetylcholinesterase at a cholinergic vasodilator nerve-arterial smooth muscle synapse. *Circulation Res.* **24**, 61–70.

Bell, C. & McLean, J. R. (1967). Localization of norepinephrine and acetylcholinesterase in separate neurons supplying the guinea-pig vas deferens. *J. Pharmac. exp. Ther.* **157**, 69–73.

Belleau, B. (1960). Relationships between agonists, antagonists and receptor sites. In *Adrenergic Mechanisms.* Ed. Vane, J. R., Wolstenholme, G. E. W. & O'Connor, M. London: Churchill.

Belleau, B. & Triggle, D. J. (1962). Blockade of adrenergic α-receptors by a carbonium ion. *J. med. pharm. Chem.* **5**, 636–639.

Bencosme, S. A. (1959). Studies on the terminal autonomic nervous system with special reference to the pancreatic islets. *Lab. Invest.* **8**, 629–646.

Benítez, D., Holmgren, B. & Middleton, S. (1959). Sympathetic cardiac stimulating fibres in the vagi. *Am. J. Physiol.* **197**, 739–742.

Bennett, A., Bucknell, A. & Dean, A. C. B. (1966). The release of 5-hydroxytryptamine from the rat stomach *in vitro*. *J. Physiol.* **182**, 57–65.

Bennett, A., Friedmann, C. A. & Vane, J. R. (1967). Release of prostaglandin E_1 from the rat stomach. *Nature, Lond.* **216**, 873–876.

Bennett, A., Murray, J. G. & Wyllie, J. H. (1968). Occurrence of prostaglandin E_2 in the human stomach, and a study of its effects on human isolated gastric muscle. *Br. J. Pharmac. Chemother.* **32**, 339–349.

Bennett, H. S. (1956). The concept of membrane flow and membrane vesiculation as mechanisms for active transport and ion pumping. *J. biophys. biochem. Cytol.* **2**, no. 4 suppl. pp. 99–103.

Bennett, M. R. (1966a). Rebound excitation of the smooth muscle cells of the guinea-pig taenia coli after stimulation of intramural inhibitory nerves. *J. Physiol.* **185**, 124–131.

—— (1966b). Transmission from intramural excitatory nerves to the smooth muscle cells of the guinea-pig taenia coli. *J. Physiol.* **185**, 132–147.

—— (1966c). Model of the membrane of smooth muscle cells of the guinea-pig taenia coli muscle during transmission from inhibitory and excitatory nerves. *Nature, Lond.* **211**, 1149–1152.

—— (1967a). The effect of cations on the electrical properties of the smooth muscle cells of the guinea-pig vas deferens. *J. Physiol.* **190**, 465–479.

—— (1967b). The effect of intracellular current pulses in smooth muscle cells of the guinea-pig vas deferens at rest and during transmission. *J. gen. Physiol.* **50**, 2459–2475.

Bennett, M. R. & Burnstock, G. (1966). Application of the sucrose gap method to determine the ionic basis of the membrane potential of the smooth muscle. *J. Physiol.* **183**, 637–648.

—— (1968). Electrophysiology of transmission in the intestine. In: *The Handbook of Physiology 1968*. Alimentary Canal IV. Amer. Physiol. Soc.: Washington, D.C.

Bennett, M. R., Burnstock, G. & Holman, M. E. (1963). The effect of potassium and chloride ions on the inhibitory potential recorded in the guinea-pig taenia coli. *J. Physiol.* **169**, 33–34P.

—— (1966a). Transmission from perivascular inhibitory nerves to the smooth muscle of the guinea-pig taenia coli. *J. Physiol.* **182**, 527–540.

—— (1966b). Transmission from intramural inhibitory nerves to the smooth muscle of the guinea-pig taenia coli. *J. Physiol.* **182**, 541–558.

Bennett, M. R., Burnstock, G., Holman, M. E. & Walker, J. W. (1962). The effect of Ca^{2+} on plateau-type action potentials in smooth muscle. *J. Physiol.* **161**, 47P.

Bennett, M. R. & Merrillees, N. C. R. (1966). An analysis of the transmission of excitation from autonomic nerves to smooth muscle. *J. Physiol.* **185**, 520–535.

Bennett, M. R. & Rogers, D. C. (1967). A study of the innervation of the taenia coli *J. cell Biol.* **33**, 573–596.

Bennett, M. V. L. (1966). Physiology of electrotonic junctions. *Ann. N.Y. Acad. Sci.* **137**, 509–539.

Bennett, M. V. L., Aljure, E., Nakajima, Y. & Pappas, G. D. (1963). Electrotonic junctions between teleost spinal neurons; electrophysiology and ultrastructure. *Science, N.Y* **141**, 262–264.

Bentley, G. A. (1962). Studies on sympathetic mechanisms in isolated intestinal and vas deferens preparations. *Br. J. Pharmac. Chemother.* **19**, 85–98.

Berger, E. & Marshall, M. (1961). Interactions of oxytocin, potassium, and calcium in the rat uterus. *Am. J. Physiol.* **201**, 931–934.

Berger, W. (1963). Die Doppelsaccharosetrennwandtechnik; Eine Methode zur Untersuchung des Membranpotentials und der Membraneigenschaften glatter Muskelzellen. *Pflügers. Arch. ges. Physiol.* **277**, 570–576.

Bergman, R. A. (1958). Intercellular bridges in ureteral smooth muscle. *Johns Hopkins Hosp. Bull.* **102**, 195–202.

—— (1968). Uterine smooth muscle fibres in castrate and estrogen-treated rats. *J. cell Biol.* **36**, 639–648.

Bergström, S., Carlson, L. A. & Weeks, J. R. (1968). The prostaglandins: a family of biologically active lipids. *Pharmac. Rev.* **20**, 1–48.

Berkson, J., (1933). Electromyographic studies of the gastrointestinal tract, IV: An inquiry into the origin of the potential variations of the rhythmic contractions in the intestine; evidence in disfavour of muscle action currents. *Am. J. Physiol.* **104**, 67–72.

—— (1934). Electromyographic studies of the gastrointestinal tract. V: Further inquiries into the origin of potential variations of the small intestine by means of certain drugs. *Am. J. Physiol.* **105**, 450–453.

Berkson, J., Baldes, E. J. & Alvarez, W. C. (1932). Electromyographic studies of the gastrointestinal tract. I. The correlation between mechanical movement and changes in electrical potential during rhythmic contraction of the intestine. *Am. J. Physiol.* **102**, 683–692.

Bernard, C. (1856). Physiological analysis of the properties of the muscular and nervous systems by means of curare. *C. r. hebd. Séanc. Acad. Sci., Paris.* **43**, 825–829. Reprinted in Shuster, L., *Readings in Pharmacology.* pp. 75–81 (1962). Boston: Little, Brown & Co.

Bernheim, F. & Bernheim, M. L. C. (1938). The hydrolysis of homatropine and atropine by various tissues. *J. Pharmac. exp. Ther.* **64**, 209–216.

Bernheim, F. & Blocksom, B. H. (1932). Action of epinephrine on the intestine following stimulation by parasympathetic drugs. *Am. J. Physiol.* **100**, 313–316.

Bernstein, J. (1912). *Elektrobiologie.* Braunschweig: Vieweg.

Berti, F. & Shore, P. A. (1967). A kinetic analysis of drugs that inhibit the adrenergic neuronal membrane amine pump. *Biochem. Pharmac.* **16**, 2091–2094.

Berti, F. & Usardi, M. M. (1964). Investigations on a new inhibitor of free fatty acid mobilization. *Gior. Arterioscl.* **2**, 261–265.

Bertler, Å., Falck, B. & Owman, C. (1964). Studies on 5-hydroxytryptamine stores in the pineal gland of rat. *Acta physiol. scand.* **63**, Suppl. 239, 1–18.

Bevan, J. A. (1960). The use of the rabbit aorta strip in the analysis of the mode of action of 1-epinephrine on vascular smooth muscle. *J. Pharmac. exp. Ther.* **129**, 417–427.

Bevan, J. A. & Verity, M. A. (1967). Sympathetic nerve-free vascular muscle. *J. Pharmac. exp. Ther.* **157**, 117–124.

Bevergård, S. Holmgren, A. & Jonsson, B. (1963). Circulatory studies in well trained athletes at rest and during heavy excercise, with special reference to stroke volume and the influence of body position. *Acta physiol. scand,* **57**, 26–50.

Beviz, A. & Mohme-Lundholm, E. (1969). The energy consumption of isometric contraction of vascular smooth muscle by K-ions. *Acta physiol. scand.* (In press).

Beviz, A., Mohme-Lundholm, E. & Vamos, N. (1969). Energetics of isometric contraction of taenia coli from the guinea-pig. *Life Sci.* (In press).

Bhagat, B., Bovell, G. & Robinson, I. M. (1967). Influence of cocaine on the uptake of H^3-norepinephrine and on the responses of isolated guinea-pig atria to sympathomimetic amines. *J. Pharmac. exp. Ther.* **155**, 472–478.

Bhattacharya, B. K. & Sen, A. B. (1962). The action of BQ 20, a polymethylene bis-quinolinium compound on the isolated guinea-pig intestine. *Archs int. Pharmacodyn. Thér.* **139**, 109–119.

Biamino, G., Vorthaler, H. & Thron, H. L. (1967). Das Verhalten der Spontanrhythmik am isolierten Aortenstreifenpräparat der Ratte bei Veränderungen im Ionenmilieu. *Pflügers Arch. ges. Physiol.* **297**, R 37–R 38.

Bianchi, C., Beani, L., Frigo, G. M. & Crema, A. (1968). Further evidence for the presence of non-adrenergic inhibitory structures in the guinea-pig colon. *Europ. J. Pharmacol.* **4**, 51–61.

Bianchi, C. P. & Shanes, A. M. (1959). Calcium influx in skeletal muscle at rest, during activity and during potassium contracture. *J. gen. Physiol.* **42**, 803–815.

Bierring, F. & Kobayasi, T. (1963). Electron microscopy of the normal rabbit aorta. *Acta path. microbiol. scand.* **57**, 154–168.

Birks, R. Huxley, H. E. & Katz, B. (1960). The fine structure of the neuromuscular junction of the frog. *J. Physiol.* **150**, 134–144.

Birmingham, A. T. (1961). The absence of spasm in a sensitive assay for acetylcholine. *J. Pharm. Pharmacol.* **13**, 510.

—— (1966). The potentiation of anticholinesterase drugs of the responses of the guinea-pig isolated vas deferens to alternate preganglionic and postganglionic stimulation. *Br. J. Pharmac. Chemother.* **27**, 145–156.

Birmingham, A. T. & Wilson, A. B. (1963). Preganglionic and postganglionic stimulation of the guinea-pig isolated vas deferens preparation. *Br. J. Pharmac. Chemother.* **21**, 569–580.

Bitman, J., Cecil, H. C., Hawk, H. W. & Sykes, J. F. (1959). Effect of estrogen and progesterone on water and electrolyte content of rabbit uteri. *Am. J. Physiol.* **197**, 93–98.

Blaber, L. C. & Bowman, W. C. (1963). Studies on the repetitive discharges evoked in motor nerve and skeletal muscle after injection of anticholinesterase drugs. *Br. J. Pharmac. Chemother.* **20**, 326–344.

Blakeley, A. G. H. & Brown, G. L. (1963). Uptake of noradrenaline by the isolated perfused spleen. *J. Physiol.* **169**, 98P.

Blakeley, A. G. H., Brown, G. L. & Ferry, C. B. (1963). Pharmacological experiments on the release of the sympathetic transmitter. *J. Physiol.* **167**, 505–514.

Blaschko, H. (1939). The specific action of L-Dopa decarboxylase. *J. Physiol.* **96**, 50*P*.

—— (1952). Amine oxidase and amine metabolism. *Pharmac. Rev.* **4**, 415–458.

Blaschko, H., Born, G. V. R., D'Iorio, A. & Eade, N. R. (1956). Observations on the distribution of catecholamines and adenosine triphosphate in the bovine adrenal medulla. *J. Physiol.* **133**, 548–557.

Blaschko, H. & Hope, D. B. (1957). Observations on the distribution of amine oxidase in invertebrates. *Archs. Biochem.* **69**, 10–15.

Blaschko, H. & Milton, A. S. (1960). Oxidation of 5-hydroxytryptamine and related compounds by *Mytilus* gill plates. *Br. J. Pharmac. Chemother.* **15**, 42–46.

Blaschko, H. & Welch, A. D. (1953). Localization of adrenaline in cytoplasmic particles of the bovine adrenal medulla. *Arch. exp. Path. Pharmak.* **219**, 17–22.

Blaustein, M. P. & Hodgkin, A. L. (1969). The effect of cyanide on the efflux of calcium from Squid axons. *J. Physiol.* **200**, 497–528.

Bleichmar, H. & De Robertis, E. (1962). Submicroscopic morphology of the infrared receptor of the pit viper. *Z. Zellforsch. mikrosk. Anat.* **56**, 748–761.

Blinks, J. R. (1965). Influence of osmotic strength on cross-section and volume of isolated single muscle fibres. *J. Physiol.* **177**, 42–57.

—— (1966). Field stimulation as a means of effecting the graded release of autonomic transmitters in isolated heart muscle. *J. Pharmac. exp. Ther.* **151**, 221–235.

Bloom, F. E. & Barrnett, R. J. (1966a). Fine structural localization of acetylcholinesterase in electroplaque of the electric eel. *J. cell Biol.* **29**, 475–495.

—— (1966b). Fine structural localization of noradrenaline in vesicles of autonomic nerve endings. *Nature, Lond.* **210**. 599–601.

Blumenthal, M. R. (1964). Effects of acetylcholine on the heart. *Fedn Proc.* **23**, 123.

Bodian, D. (1942). Cytological aspects of synaptic function. *Physiol. Rev.* **22**, 146–169.

—— (1952). Introductory survey of neurons. *Cold Spring Harb. Symp. quant. Biol.* **17**, 1–13.

Boeke, J. (1933). Innervationsstudien. V. Der sympathische Grundplexus und seine Beziehungen zu den quergestreiften Muskelfasern und zu den Herzmuskelfasern. *Z. mikrosk.-anat. Forsch.* **34**, 330–378.

—— (1949). The sympathetic endformation, its synaptology, the interstitial cell, the periterminal network and its bearing on the neurone theory. Discussion and critique. *Acta Anat.* **8**, 18–61.

Bohr, D. F. (1964a). Electrolytes and smooth muscle contraction. *Pharmac. Rev.* **16**, 85–111.

—— (1964b). The vascular smooth muscle cell. In: *The Heart and Circulation*, Second National Conference on Cardiovascular Diseases, Vol. 2, Research. Ed. E. C. Andrus.

—— (1965). Individualities among vascular smooth muscles. In: *Electrolytes and Cardiovascular Diseases.* Vol. I. Ed. Bajusz, E. Basel: Karger.

Boldireff. (1904). *Zbl Physiol.* **18**, 457. (Quoted from Klee, 1927).

Bolme, P. & Novotny, J. (1968). Vasodilatation and oxygen uptake in skeletal muscle of the dog. *Br. J. Pharmac.* **34**, 234–235*P*.

Bondareff, W. (1965). Submicroscopic morphology of granular vesicles in sympathetic nerves of rat pineal body. *Z. Zellforsch. mikrosk. Anat.* **67**, 211–218.

Bondareff, W. & Gordon, B. (1966). Submicroscopic localization of norepinephrine in sympathetic nerves of rat pineal. *J. Pharmac. exp. Ther.* **153**, 42–47.

Bonney, W. R. & Ferguson, J. K. W. (1950). Reactions of isolated uterine muscle of rats and guinea pigs. *Arch. Int. Pharmacodyn. Thér.* **83**, 566–572.

Borgstedt, H. H. (1959). The effect of serotonin and LSD on the isolated dog ureter. *Fedn Proc.* **18**, 370.

Born, G. V. R. (1956a). Adenosine triphosphate (ATP) in blood platelets. *Biochem. J.* **62**, 33*P*.

—— (1956b). The relation between the tension and the high-energy phosphate content of smooth muscle. *J. Physiol.* **131**, 704–711.

—— (1962). The fate of 5-hydroxytryptamine in a smooth muscle and in connective tissue. *J. Physiol.* **161**, 160–174.

20*

Born, G. V. R. (1967). The effect of 5-hydroxytryptamine on the potassium exchange of human platelets. *J. Physiol.* **190**, 273–280.

Born, G. V. R. & Bricknell, J. (1959). The uptake of 5-hydroxytryptamine by blood platelets in the cold. *J. Physiol.* **147**, 153–161.

Born, G. V. R. & Bülbring, E. (1956). The movement of potassium between smooth muscle and the surrounding fluid. *J. Physiol.* **131**, 690-703.

Born, G. V. R., Day, M. & Stockbridge, A. (1967). The uptake of amines by human erythrocytes *in vitro*. *J. Physiol.* **193**, 405–418.

Born, G. V. R. & Gillson, R. E. (1959). Studies on the uptake of 5-hydroxytryptamine by blood platelets. *J. Physiol.* **146**, 472–491.

Born, G. V. R., Hornykiewicz, O. & Stafford, A. (1958). The uptake of adrenaline and noradrenaline by blood platelets of the pig. *Br. J. Pharmac. Chemother.* **13**, 411–414.

Born, G. V. R., Ingram, G. I. C. & Stacey, R. S. (1958). The relation between 5-hydroxytryptamine and adenosine triphosphate in blood platelets. *Br. J. Pharmac Chemother.* **13**, 62–64.

Born, G. V. R. & Vane, J. R. (1961). The penetration of labelled tryptamine and 5-hydroxytryptamine into smooth muscles. Communication to the British Pharmacological Society.

Bortoff, A. (1961a). Slow potential variations of small intestine. *Am. J. Physiol.* **201**, 203–208.

—— (1961b). Electrical activity of intestine recorded with pressure electrode. *Am. J. Physiol.* **201**, 209–212.

—— (1965). Electrical transmission of slow waves from longitudinal to circular intestinal muscle. *Am. J. Physiol.* **209**, 1254–1260.

Boschman, Th. A. C. (1968). Dissertation K. Universiteit, Nijmegen.

Botár, J. (1966). *The Autonomic Nervous System.* Budapest: Akademiai Kiadó.

Boucek, R. J., Takashita, R. & Fojaco, R. (1963). Relation between microanatomy and functional properties of the coronary arteries. (Dog). *Anat. Rec.* **147**, 199–207.

Boullin, D. J., Costa, E. & Brodie, B. B. (1966). Apparent depletion of NE stores after repetitive stimulation of cat colon in presence of phenoxybenzamine. *Int. J. Neuropharmac.* **5**, 293–298.

—— (1967). Evidence that blockade of adrenergic receptors causes overflow of norepinephrine in cat's colon after nerve stimulation. *J. Pharmac. exp. Ther.* **157**, 125–134.

Boulpaep. E. (1963). Permeability of heart muscle to choline. *Archs int. Physiol.* **71**, 623–625.

Boura, A. L. A. & Green, A. F. (1965). Adrenergic neurone blocking agents. *Ann. Rev. Pharmacol.* **5**, 183–212.

Bowman, W. C. & Everett, S. D. (1964). An isolated parasympathetically-innervated oesophagus preparation from the chick. *J. Pharm. Pharmacol.* **16**, 72–79T.

Boyd, G., Gillespie, J. S. & Mackenna, B. R. (1962). Origin of the cholinergic response of the rabbit intestine to stimulation of its extrinsic sympathetic nerves after exposure to sympathetic blocking agents. *Br. J. Pharmac. Chemother.* **19**, 258–270.

Boyd, H., Burnstock, G. & Rogers, D. (1964). Innervation of the large intestine of the toad (*Bufo marinus.*) *Br. J. Pharmac. Chemother.* **23**, 151–163.

Boyle, P. J. & Conway, E. J. (1941). Potassium accumulation in muscle and associated changes. *J. Physiol.* **100**, 1–63.

Boyle, P. J., Conway, E. J., Kane, F. & O'Reilly, H. L. (1941). Volume of interfibre spaces in frog muscle and the calculation of concentrations in the fibre water. *J. Physiol.* **99**, 401–414.

Bozler, E. (1938a). Electric stimulation and conduction of excitation in smooth muscle. *Am. J. Physiol.* **122**, 614–623.

—— (1938b). The action potentials of visceral smooth muscle. *Am. J. Physiol.* **124**, 502–510.

—— (1941). Action potentials and conduction of excitation in muscle. *Biol. Symp.* **3**, 95–110.

—— (1942). The action potentials accompanying conducted responses in visceral smooth muscles. *Am. J. Physiol.* **136**, 553–560.

—— (1948). Conduction, automaticity and tonus of visceral muscles. *Experientia* **4**, 213–218.

—— (1949a). Myenteric reflex. *Am. J. Physiol.* **157**, 329–337.

—— (1949b). Reflex peristalsis of the intestine. *Am. J. Physiol.* **157**, 338–342.

Bozler, E. (1959). Osmotic effects and diffusion of non electrolytes in muscle. *Am. J. Physiol.* **197** 505 510.

—— (1961a). Distribution of non electrolytes in muscle. *Am. J. Physiol.* **200**, 651–655.

—— (1961b). Electrolytes and osmotic balance of muscle in solutions of non electrolytes. *Am. J. Physiol.* **200**, 656–657.

—— (1962). Osmotic phenomena in smooth muscle. *Am. J. Physiol.* **203**, 201–205.

—— (1963). Distribution and exchange of calcium in connective tissue and smooth muscle. *Am. J. Physiol.* **205**, 686–692.

—— (1965). Osmotic properties of amphibian muscles. *J. gen. Physiol.* **49**, 37–45.

Bozler, E., Calvin, M. E. & Watson, D. W. (1958). Exchange of electrolytes in smooth muscle. *Am. J. Physiol.* **195**, 38–44.

Bozler, E. & Lavine, D. (1958). Permeability of smooth muscle. *Am. J. Physiol.* **195**, 45–49.

Brading, A. F. (1967). A constant flow apparatus for measuring radio-active ion effluxes from guinea-pig taenia coli. *J. Physiol.* **192**, 15–16P.

Brading, A. F., Bülbring, E. & Tomita, T. (1969a). The effect of temperature on the membrane conductance of the smooth muscle of the guinea-pig taenia coli. *J. Physiol.* **200**, 621–635.

—— (1969b). The effect of sodium and calcium on the action potential of the smooth muscle of the guinea-pig taenia coli. *J. Physiol.* **200**, 637–654.

Brading, A. F. & Jones, A. W. (1969). Distribution and kinetics of CoEDTA in smooth muscle, and its use as an extracellular marker. *J. Physiol.* **200**, 387–402.

Brading, A. F., & Setekleiv, J., (1968). The effect of hypo- and hypertonic solutions on volume and ion distribution of smooth muscle of guinea-pig taenia coli. *J. Physiol.* **195**, 107–118.

Brading, A. F. & Tomita, T. (1968a). Effect of anions on the volume of smooth muscle. *Nature, Lond.* **218**, 276–277.

—— (1968b). The action potential of the guinea-pig taenia coli in low sodium solution *J. Physiol.* **197**, 30–31P.

—— (1968c). Volume changes of the smooth muscle of the guinea-pig taenia coli, and the influence of calcium. *J. Physiol.* **197**, 68–69P.

Brady, R. O., Spyropoulos, C. S. & Tasaki, I. (1958). Intra-axonal injection of biologically active materials. *Am. J. Physiol.* **194**, 207–213.

Brandon, K. W. & Rand, M. J. (1961). Acetylcholine and the sympathetic innervation of the spleen. *J. Physiol.* **157**, 18–32.

Brandt, P. W., Reuben, J. P., Girardier, L. & Grundfest, H. (1965). Correlated morphological and physiological studies on isolated single muscle fibres. *J. cell Biol.* **25**, Part 2, 233–260.

Bratton, C. B., Hopkins, A. L. & Weinberg, J. W. (1965). Nuclear magnetic resonance studies of living muscle. *Science, N.Y.* **147**, 738–739.

Brecht, K., Estada, J. A. & Götz, A. (1964). Zur Beeinflussung der Gefässmotorik durch Ca-$^{++}$ und K^{+}-Ionen. *Pflügers Arch. ges. Physiol.* **279**, 330–340.

Brecht, K. & D. Jeschke. (1960). Über die Wirkung des Serotonins auf die Froschlunge. *Naturwissenschaften.* **47**, 20–21.

Breeman van, C. & Daniel, E. E. (1966). The influence of high potassium depolarization and acetylcholine on calcium exchange in the rat uterus. *J. gen. Physiol.* **49**, 1299–1317.

Breeman van, C., Daniel, E. E. & Breeman van, D. (1966). Calcium distribution and exchange in the rat uterus. *J. gen. Physiol.* **49**, 1265–1298.

Brettschneider, H. (1962). Elektronenmikroskopische Untersuchungen über die Innervation der glatten Muskulatur des Darmes. *Z. mikrosk.-anat. Forsch.* **68**, 333–360.

—— (1963). Elektronenmikroskopische Beobachtungen über die Innervation der Schilddrüse. *Z. mikrosk.-anat. Forsch.* **69**, 630–649.

—— (1964). Elektronenmikroskopische Studien zur vegetativen Gefässinnervation. *Anat. Anz.* **58**, 54–69.

Brick, I., Hutchinson, K. J. & Roddie, I. C. (1967a). The vasodilator properties of noradrenaline in the human forearm. *Br. J. Pharmac. Chemother.* **30**, 561–567.

Brick, I., Hutchinson, K. J. & Roddie, I. C. (1967b). A comparison of the effects of circulating noradrenaline and vasoconstrictor nerve stimulation on forearm blood vessels. *J. Physiol.* **189**, 27–28P.

Briggs, A. H. (1962). Calcium movements during potassium contracture in isolated rabbit aortic strips. *Am. J. Physiol.* **203**, 849–852.

Brinkman, R. & Vandam, E. (1922). Die chemische Übertragbarkeit der Nervenreizwirkung. *Pflügers Arch. ges. Physiol.* **196**, 66–82.

Brocklehurst, W. E. (1957). The action of 5-hydroxytryptamine on smooth muscle. In *5-Hydroxytryptamine.* Ed. Lewis, G. P. pp. 172–176. London: Pergamon Press.

Brodie, T. G. & Dixon, W. E. (1904). Contributions to the physiology of the lungs. Part II. On the innervation of the pulmonary blood vessels; and some observations on the action of suprarenal extract. *J. Physiol.* **30**, 476–502.

Brody, M. J. (1966). Neurohumoral mediation of active reflex vasodilatation. *Fedn Proc.* **25**, 1583–1592.

—— (1968). Mechanisms of pharmacologic blockade of reflex vasodilatation. *Fedn Proc.* **27**, 756.

Brody, M. J., Du Charme, D. W. & Beck, L. (1967). Active reflex vasodilatation induced by veratrine and dopamine. *J. Pharmac. exp. Ther.* **155**, 84–90.

Brody, S. & Wiqvist, N. (1961). Ovarian hormones and uterine growth: Effects of estradiol, progesterone and relaxin on cell growth and cell division in the rat uterus. *Endocrinology.* **68**, 971–977.

Brody, T. M. & Diamond, J. (1967). Blockade of the biochemical correlates of contraction and relaxation in uterine and intestinal smooth muscle. *Ann. N.Y. Acad. Sci.* **139**, 772–780.

Brooks, J. R., Schaeppi, U. & Pincus, G. (1965). Evidence for the presence of alpha adrenergic excitatory receptors in the rat uterus. *Life Sci.* **4**, 1817–1821.

Brouha, L., Cannon, W. B. & Dill, D. B. (1936). The heart rate of the sympathectomized dog in rest and exercise. *J. Physiol.* **87**, 345–359.

Brown, A. M. (1967). Cardiac sympathetic adrenergic pathways in which synaptic transmission is blocked by atropine sulphate. *J. Physiol.* **191**, 271–288.

Brown, G. L. (1965). The release and fate of the transmitter liberated by adrenergic nerves. *Proc. R. Soc. B.* **162**, 1–19.

Brown, G. L., Davies, B. N. & Ferry, C. B. (1961). The effect of neuronal rest on the output of sympathetic transmitter from the spleen. *J. Physiol.* **159**, 365–380.

Brown, G. L., Davies, B. N. & Gillespie, J. S. (1958). The release of chemical transmitter from the sympathetic nerves of the intestine of the cat. *J. Physiol.* **143**, 41–54.

Brown, G. L. & Gillespie, J. S. (1957). The output of sympathetic transmitter from the spleen of the cat. *J. Physiol.* **138**, 81–102.

Brown, G. L. & Maycock, W. D'A. (1942). Acceleration of the heart by the vagus in cats after complete sympathectomy. *J. Physiol.* **101**, 369–374.

Brownlee, G. & Harry, J. (1963). Some pharmacological properties of the circular and longitudinal muscle strips from the guinea-pig isolated ileum. *Br. J. Pharmac. Chemother.* **21**, 544–554.

Brownlee, G. & Johnson, E. S. (1963). Site of 5-hydroxytryptamine receptors in the intramural nerve plexus of the guinea-pig ileum. *Br. J. Pharmac. Chemother.* **21**, 306–322.

—— (1965). The release of acetylcholine from the isolated ileum of the guinea-pig induced by 5-hydroxytryptamine and dimethylphenylpiperazinium. *Br. J. Pharmac. Chemother.* **24**, 689–700.

Brücke, F., Kaindl, F. & Mayer, H. (1952). Über die Veränderung in der Zusammensetzung des Nebennierenmarkinkretes bei elektrischer Reizung des Hypothalamus. *Archs. int. Pharmacodyn. Thér.* **88**, 407–412.

Brun, G. C. (1947). Mechanism of the vasoconstrictor action of ephedrine. I. Arterial contraction before and after local anaesthesia. *Acta pharmacol. (Kbh.)* **3**, 225–238.

Brunaud, M. & Labouche, C. (1947). L'adrénaline, agent contracturant des fibres longitudinales du duodénum du cheval. *C. r. Séanc. Soc. Biol.* **141**, 167–169.

Bruns, R. R. & Palade, G. E. (1968). Studies on blood capillaries. *J. cell Biol.* **37**, 244–276.

Brzin, M., Tennyson, V. M. & Duffy, P. E. (1966). Acetylcholinesterase in frog sympathetic and dorsal root ganglia. A study by electron microscope cytochemistry and micro-gasometric analysis with the magnetic diver. *J. cell. Biol.* **31**, 215–242.

Buchtal, F. & Rosenfalck, P. (1957). Elastic properties of striated muscle. In *Tissue Elasticity.* Ed. Remington, J. W. Amer. Physiol. Soc.: Washington, D.C. pp. 73–93.

Buck, B. & Goodford, P. J. (1966). The distribution of ions in the smooth muscle of the guinea-pig taenia coli. *J. Physiol.* **183**, 551–569.

Buck, R. C. (1963). Histogenesis and morphology of arterial tissue. In *Atherosclerosis and its Origin.* Eds. Sandler, M. & Bourne, G. H. London: Academic Press.

Bucknell, A. (1964). Effects of direct and indirect stimulation on isolated colon. *J. Physiol.* **177**, 58–59P.

Bucknell, A. & Whitney, B. (1964). A preliminary investigation of the pharmacology of the human isolated taenia coli preparation. *Br. J. Pharmac. Chemother,* **23**, 164–175.

Bueding, E. & Bülbring, E. (1964). The inhibitory action of adrenaline. Biochemical and biophysical observations. In: *Pharmacology of Smooth Muscle.* Ed. Bülbring, E. Oxford: Pergamon Press.

—— (1967). Relationship between energy metabolism of intestinal smooth muscle and the physiological action of epinephrine. *Ann. N.Y. Acad. Sci.* **139**, 758–761.

Bueding, E. Bülbring, E., Gercken, G., Hawkins, J. T. & Kuriyama, H. (1967). The effect of adrenaline on the adenosine triphosphate and creatine phosphate content of intestinal smooth muscle. *J. Physiol.* **193**, 187–212.

Bülbring, E. (1944). The action of adrenaline on transmission in the superior cervical ganglion. *J. Physiol.* **103**, 55–67.

—— (1954). Membrane potentials of smooth muscle fibres of the taenia coli of the guinea-pig. *J. Physiol.* **125**, 302–315.

—— (1955). Correlation between membrane potential, spike discharge and tension in smooth muscle. *J. Physiol.* **128**, 200–221.

—— (1957a). Changes in configuration of spontaneously discharged spike potentials from smooth muscle of the guinea-pig's taenia coli. The effect of electrotonic currents and of adrenaline, acetylcholine and histamine. *J. Physiol.* **135**, 412–425.

—— (1957b). The actions of humoral transmitters on smooth muscle. *Br. med. Bull.* **13**, 172–175.

—— (1957c). Physiology and pharmacology of intestinal smooth muscle. *Lect. sci. Basis Med.* **7**, 374–397.

—— (1960). Biophysical changes produced by adrenaline and noradrenaline. Ciba Foundation Symp. *Adrenergic Mechanisms.* pp. 275–287. London: Churchill.

—— (1961a). The intrinsic nervous system of the intestine and local effects of 5-hydroxytryptamine. In *Regional Neurochemistry.* Ed. Kety, S. & Elkes, J. pp. 437–441. New York: Pergamon Press.

—— (1961b). Die Physiologie des glatten Muskels. *Pflügers Arch. ges. Physiol.* **273**, 1–17.

—— (1962). Electrical activity in intestinal smooth muscle. *Physiol. Rev.* **42**, Suppl. 5, 160-178.

—— (1964). Pharmacology of smooth muscle. Second International Pharmacological Meeting. Vol. 6. Oxford: Pergamon Press.

—— (1967). Recent observations on smooth muscle. *Acta Biologica Jugoslavica,* **3** No. 3 239–248.

Bülbring, E. & Burn, J. H. (1935). The sympathetic dilator fibres in the muscles of the cat and dog. *J. Physiol.* **83**, 483–501.

—— (1936). Sympathetic vaso-dilatation in the skin and intestine of the dog. *J. Physiol.* **87**, 254–274.

Bülbring, E., Burn, J. H. & Shelley, H. J. (1953). Acetylcholine and ciliary movement in the gill plates of *Mytilus edulis. Proc. R. Soc. B.* **141**, 445–466.

Bülbring, E. & Burnstock, G. (1960). Membrane potential changes associated with tachyphylaxis and potentiation of the response to stimulating drugs in smooth muscle. *Br. J. Pharmac. Chemother.* **15**, 611–624.

Bülbring, E., Burnstock, G. & Holman, M. E. (1958). Excitation and conduction in the smooth muscle of the isolated taenia coli of the guinea-pig. *J. Physiol.* **142**, 420–437.

Bülbring, E., Casteels, R. & Kuriyama, H. (1968). Membrane potential and ion content in cat and guinea-pig myometrium and the response to adrenaline and noradrenaline. *Br. J. Pharmac.* **34**, 388–407.

Bülbring, E. & Crema, A. (1958). Observations concerning the action of 5-hydroxytryptamine on the peristaltic reflex. *Br. J. Pharmac. Chemother.* **13**, 444–457.

—— (1959a). The release of 5-hydroxytryptamine in relation to pressure exerted on the intestinal mucosa. *J. Physiol.* **146**, 18–28.

—— (1959b). The action of 5-hydroxytryptamine, 5-hydroxytryptophan and reserpine on intestinal peristalsis in anaesthetized guinea-pigs. *J. Physiol.* **146**, 29–53.

Bülbring, E., Crema, A. & Saxby, O. B. (1958). A method for recording peristalsis in isolated intestine. *Br. J. Pharmac. Chemother.* **13**, 440–443.

Bülbring, E. & Gershon, M. D. (1966). 5-Hydroxytryptamine participation in vagal relaxation of the stomach. *J. Physiol.* **186**, 95–96P.

—— (1967). 5-hydroxytryptamine participation in the vagal inhibitory innervation of the stomach. *J. Physiol.* **192**, 823–846.

—— (1968). Serotonin participation in the vagal inhibitory pathway to the stomach. In *Adv. Pharmacol.* Ed. Garattini, S. & Shore, P. A. Vol. 6A, pp. 323–333. London: Academic Press.

Bülbring, E. & Golenhofen, K. (1967). Oxygen consumption by the isolated smooth muscle of guinea-pig taenia coli. *J. Physiol.* **193**, 213–224.

Bülbring, E. Goodford, P. J. & Setekleiv, J. (1966). The action of adrenaline on the ionic content and on sodium and potassium movements in the smooth muscle of the guinea-pig taenia coli. *Br. J. Pharmac. Chemother.* **28**, 296–307.

Bülbring, E. & Hooton, I. N. (1954). Membrane potential of smooth muscle fibres in the rabbit's sphincter pupillae. *J. Physiol.* **125**, 292–301.

Bülbring, E. & Kuriyama, H. (1963a). Effects of changes in the external sodium and calcium concentrations on spontaneous electrical activity in smooth muscle of guinea-pig taenia coli. *J. Physiol.* **166**, 29–58.

—— (1963b). Effects of changes in ionic environment on the action of acetylcholine and adrenaline on the smooth muscle cells of guinea-pig taenia coli. *J. Physiol.* **166**, 59–74.

—— (1963c). The effect of adrenaline on the smooth muscle of guinea-pig taenia coli in relation to the degree of stretch. *J. Physiol.* **169**, 198–212.

Bülbring, E., Kuriyama, H. & Tomita, T. (1965). Discrimination between the 'inhibitory potential' and electrotonic hyperpolarisation in smooth muscle. *J. Physiol.* **181**, 8–10P.

Bülbring, E. & Lin, R. C. Y. (1958). The effect of intraluminal application of 5-hydroxytryptamine and 5-hydroxytryptophan on peristalsis; the local production of 5-HT and its release in relation to intraluminal pressure and propulsive activity. *J. Physiol.* **140**, 381–407.

Bülbring, E., Lin, R. C. Y. & Schofield, G. (1958). An investigation of the peristaltic reflex in relation to anatomical observations. *Q. Jl. exp. Physiol.* **43**, 26–37.

Bülbring, E. & Lüllmann, H. (1957). The effect of metabolic inhibitors on the electrical and mechanical activity of the smooth muscle of the guinea-pig's taenia coli. *J. Physiol.* **136**, 310–323.

Bülbring, E. & Tomita, T. (1966). Evidence supporting the assumption that the 'inhibitory potential' in the taenia coli of the guinea-pig is a post synaptic potential due to nerve stimulation. *J. Physiol.* **185**, 24–25P.

—— (1967). Properties of the inhibitory potential of smooth muscle as observed in the response to field stimulation of the guinea-pig taenia coli. *J. Physiol.* **189**, 299–315.

—— (1968a). The effect of catecholamines on the membrane resistance and spike generation, in the smooth muscle of the guinea-pig taenia coli. *J. Physiol.* **194**, 74–76P.

—— (1968b). The effects of Ba^{2+} and Mn^{2+} on the smooth muscle of guinea-pig taenia coli. *J. Physiol.* **196**, 137–139P.

—— (1969a). Increase of membrane conductance by adrenaline in the smooth muscle of guinea-pig taenia coli. *Proc. R. Soc. B.* **172**. 89–102.

—— (1969b). Suppression of spontaneous spike generation by catecholamines in the smooth muscle of the guinea-pig taenia coli. *Proc. R. Soc. B.* **172**, 103–119.

—— (1969c). Effect of calcium, barium and manganese on the action of adrenaline in the smooth muscle of the guinea-pig taenia coli. *Proc. R. Soc. B.* **172**, 121–136.

Bülbring, E. & Tomita T. (1969d). Calcium and the action potential in smooth muscle. In *Calcium and Cellular Function*. Ed. Cuthbert, A. W. London: Macmillan (in press).

Bullon, A. & Stiefel, E. (1955). Ueber die efferente Innervation der glatten Muskulatur. *Acta neuroveg (Wien)*. **12**, 375–388.

Bunch, J. L. (1898). On the origin, course and cell-connections of the viscero-motor nerves of the small intestine. *J. Physiol.* **22**, 357–379.

—— (1899). On the vaso-motor nerves of the small intestine. *J. Physiol.* **24**, 72–98.

Burch, G. E., Cohn, A. E. & Neumann, C. (1942). A study by quantitative methods of the spontaneous variations in volume of the finger tip, toe tip, and postero-superior portion of the pinna of resting normal white adults. *Am. J. Physiol.* **136**, 433–447.

Burch, G. E. & DePasquale, N. (1960). Relation of arterial pressure to spontaneous variations in digital volume. *J. appl. Physiol.* **15**, 23–24.

Burgen, A. S. V. & Spero, L. (1968). The action of acetylcholine and other drugs on the efflux of potassium and rubidium from smooth muscle of the guinea-pig intestine. *Br. J. Pharmac.* **34**, 99–115.

Bürgi, S. (1944). Zur Physiologie und Pharmakologie der überlebenden Arterie. *Helv. physiol. pharmacol. Acta.* **2**, 345–365.

—— (1945). Förderung, Hemmung und Desorganisation der arteriellen Eigenrhythmen. *Helv. physiol. pharmacol. Acta.* **3**, 215–229.

Burke, W. & Ginsborg, B. L. (1956a). The electrical properties of the slow muscle fibre membrane. *J. Physiol.* **132**, 586–598.

—— (1956b). The action of the neuromuscular transmitter on the slow fibre membrane. *J. Physiol.*, **132**, 599–610.

Burks, T. F. & Long, J. P. (1966a). Catecholamine-induced release of 5-hydroxytryptamine (5-HT) from perfused vasculature of isolated dog intestine. *J. pharm. Sci.* **55**, 1383–1386.

—— (1966b). 5-Hydroxytryptamine release into dog intestinal vasculature. *Am. J. Physiol.* **211**, 619–625.

—— (1967a). Release of 5-hydroxytryptamine from isolated dog intestine by nicotine. *Br. J. Pharmac. Chemother.* **30**, 229–239.

—— (1967b). Release of intestinal 5-hydroxytryptamine by morphine and related agents. *J. Pharmac. exp. Ther.* **156**, 267–276.

Burn, G. P. & Burn, J. H. (1961). Uptake of labelled noradrenaline by isolated atria. *Br. J. Pharmac. Chemother.* **16**, 344–351.

Burn, J. H. (1932). On vasodilator fibres in the sympathetic, and on the effect of circulating adrenaline in augmenting the vascular response to sympathetic stimulation. *J. Physiol.* **75**, 144–159.

—— (1950). A discussion on the action of local hormones. *Proc. Roy. Soc.* B. **137**, 281–320.

—— (1956). *Functions of Autonomic Transmitters.* Baltimore: Williams & Wilkins.

—— (1967). Release of noradrenaline from the sympathetic postganglionic fibre. *Br. med. J.* (1967) No. 2. 197–201.

—— (1968a). The development of the adrenergic fibre. *Br. J. Pharmac. Chemother.* **32**, 575–582.

—— (1968b). p. 39 in Discussion of the mechanism of the release of noradrenaline. In: *Adrenergic Neurotransmission Ciba Foundation Study Group No. 33.* Ed. Wolstenholme & O'Connor. London: Churchill.

Burn, J. H., Dromey, J. J. & Large, B. J. (1963). The release of acetylcholine by sympathetic nerve stimulation at different frequencies. *Br. J. Pharmac. Chemother.* **21**, 97–103.

Burn, J. H. & Huković, S. (1966). The effect of anticholinesterases on the increase in rate of the isolated rabbit heart in response to stimulation of the accelerator fibres. *J. Physiol.* **186**, 33–34*P*.

Burn, J. H. & Rand, M. J. (1958). Excitatory action of the vagus in the isolated atria in relation to adrenaline. *J. Physiol.* **142**, 173–186.

—— (1959). Sympathetic postganglionic mechanism. *Nature, Lond.* **184**, 163–165.

—— (1960). Sympathetic postganglionic cholinergic fibres. *Br. J. Pharmac. Chemother.* **15**, 56–66.

—— (1962). A new interpretation of the adrenergic nerve fibre. In: *Advances in Pharmacology*

Vol. 1. Eds. Garattini, S. & Shore, P. A. pp. 1–30. New York & London: Academic Press.

Burn J. H. & Rand, M. J. (1965). Acetylcholine in adrenergic transmission. *Ann. Rev. Pharmacol.*, **5** 163–182.

Burn, J. H., Rand, M. J. & Wien, R. (1963). The adrenergic mechanism in the nictitating membrane. *Br. J. Pharmac. Chemother.* **20**, 83–94.

Burn, J. H. & Weetman, D. F. (1963). The effect of eserine on responses of the vas deferens to hypogastric nerve stimulation. *Br. J. Pharmac. Chemother.* **20**, 74–82.

Burn, R. R. & Palade, G. E. (1968). Studies on blood capillaries. *J. Cell Biol.* **37**, 244–276.

Burnstock, G. (1958a). The effects of acetylcholine on membrane potential, spike frequency, conduction, velocity and excitability in the taenia coli of the guinea-pig. *J. Physiol.* **143**, 165–182.

—— (1958b). The action of adrenaline on excitability and membrane potential in the taenia coli of the guinea-pig and the effect of DNP on this action and on the action of acetyl-choline. *J. Physiol.* **143**, 183–194.

—— (1960). Membrane potential changes associated with stimulation of smooth muscle by adrenaline. *Nature, Lond.*, **186**, 727–728.

—— (1968). The Autonomic Neuromuscular Junction. Proc. 24th Int. Congr. Physiol. Sci. (Washington) **6**, 7–8.

Burnstock, G. & Campbell, G. (1963). Comparative physiology of the vertebrate autonomic nervous system. II. Innervation of the urinary bladder of the ring-tail possum. (*Pseudocheirus peregrinus*.) *J. exp. Biol.* **40**, 421–436.

Burnstock, G., Campbell, G., Bennett, M. & Holman, M. E. (1963a). The effect of drugs on the transmission of inhibition from autonomic nerves to the smooth muscle of the guinea-pig taenia coli. *Biochem. Pharmac.* **12**, (Suppl.) 134.

—— (1963b). Inhibition of the smooth muscle of the taenia coli. *Nature, Lond.* **200**, 581–582.

—— (1964). Innervation of the guinea-pig taenia coli: are there intrinsic inhibitory nerves which are distinct from sympathetic nerves? *Int. J. Neuropharmac.* **3**, 163–166.

Burnstock, G., Campbell, G. & Rand, M. J. (1966). The inhibitory innervation of the taenia of the guinea-pig caecum. *J. Physiol.* **182**, 504–526.

Burnstock, G., Dewhurst, D. J. & Simon, S. E. (1963). Sodium exchange in smooth muscle. *J. Physiol.* **167**, 210–228.

Burnstock, G. & Holman, M. E. (1961). The transmission of excitation from autonomic nerve to smooth muscle. *J. Physiol.* **155**, 115–133.

—— (1962a). Spontaneous potentials at sympathetic nerve endings in smooth muscle. *J. Physiol.* **160**, 446–460.

—— (1962b). Effect of denervation and of reserpine treatment on transmission at sympathetic nerve endings. *J. Physiol.* **160**, 461–469.

—— (1963). Smooth muscle: autonomic nerve transmission. *Ann. Rev. Physiol.* **25**, 61–90.

—— (1964). An electrophysiological investigation of the actions of some autonomic blocking drugs on transmission in the guinea-pig vas deferens. *Br. J. Pharmac. Chemother.* **23**, 600–612.

—— (1966a). Junction potentials at adrenergic synapses. *Pharmac. Rev.* **18**, 481–493.

—— (1966b). Effects of drugs on smooth muscle. *Ann. Rev. Pharmacol.* **6**, 129–156.

Burnstock, G., Holman, M. E. & Kuriyama, H. (1964). Facilitation of transmission from autonomic nerve to smooth muscle of guinea-pig vas deferens. *J. Physiol.* **172**, 31–49.

Burnstock, G., Holman, M. E. & Prosser, C. L. (1963). Electrophysiology of smooth muscle. *Physiol. Rev.* **43**, 482–527.

Burnstock, G. & Merrillees, N. C. R. (1964). Structural and experimental studies on autonomic nerve endings in smooth muscle. In: *Pharmacology of Smooth Muscle.* pp. 1–17. Proc. 2nd. Int. Pharmacol. Meeting (Prague). Oxford: Pergamon Press.

Burnstock, G., O'Shea, J. & Wood, M. (1963). Comparative physiology of the vertebrate autonomic nervous system. I. Innervation of the urinary bladder of the toad. (*Bufo marinus*.) *J. exp. Biol.* **40**, 403–419.

Burnstock, G. & Prosser, C. L. (1960a). Responses of smooth muscles to quick stretch; relation of stretch to conduction. *Am. J. Physiol.* **198**, 921–925.

Burnstock, G. & Prosser, C. L. (1960b). Conduction in smooth muscles: comparative electrical properties. *Am. J. Physiol.* **199**, 553–559.

—— (1960c). Delayed repolarization in smooth muscles. *Proc. Soc. exp. Biol. Med.* **103**, 269–270.

Burnstock, G. & Robinson, P. M. (1967). Localization of catecholamines and acetylcholinesterase in autonomic nerves. *Circulation Res.* **21**, Suppl. 3, 43–55.

Burnstock, G. & Straub, R. W. (1958). A method for studying the effects of ions and drugs on the resting and action potentials in smooth muscle with external electrodes. *J. Physiol.* **140**, 156–167.

Burton, A. C. (1939). The range and variability of the blood flow in the human fingers and the vasomotor regulation of body temperature. *Am. J. Physiol.* **127**, 437–453.

—— (1962). Physical principles of circulatory phenomena: the physical equilibria of the heart and blood vessels. In: *Handbook of Physiology, Section 2: Circulation, Vol. I.* Ed. Hamilton, W. F. & Dow, P. Washington: Amer. Physiol. Soc.

Burton, A. C. & Taylor, R. M. (1940). A study of the adjustment of peripheral vascular tone to the requirements of the regulation of body temperature. *Am. J. Physiol.* **129**, 565–577.

Cabrera, R., Cohen, A., Middleton, S., Utano, L. & Viveros, H. (1966). The immediate source of noradrenaline released in the heart by acetylcholine. *Br. J. Pharmac. Chemother.* **27**, 46–50.

Cabrera, R., Torrance, R. W. & Viveros, H. (1966). The action of acetylcholine and other drugs upon the terminal parts of the postganglionic sympathetic fibre. *Br. J. Pharmac. Chemother.* **27**, 51–63.

Caesar, R., Edwards, G. A. & Ruska, H. (1957). Architecture and nerve supply of mammalian smooth muscle tissue. *J. biophys. biochem. Cytol.* **3**, 867–878.

Cajal, S. R. (1893). Sur les ganglions et plexus nerveux d'intestin. *C. r. Soc. Biol. (Paris)* **5**, 217–223.

Caldeyro-Barcia, R. (1965). Regulation of myometrial activity in pregnancy. In *Muscle.* Ed. W. M. Paul, E. E. Daniel, C. M. Kay & G. Monckton. pp. 317–347. Pergamon Press: Oxford.

Caldeyro-Barcia, R. & Poseiro, J. J. (1959). Oxytocin and contractility of the pregnant human uterus. *Ann. N.Y. Acad. Aci.* **75**, 813–830.

Caldeyro-Barcia & Sico-Blanco. (1962–63). In *Greenhill's Yearbook of Obstetrics and Gynecology.* Pp. 151.

Caldwell, P. C. (1958). Studies on the internal pH of large muscle and nerve fibres. *J. Physiol.* **142**, 22–62.

—— (1968). Factors governing movement and distribution of inorganic ions in nerve and muscle. *Physiol. Rev.* **48**, 1–64.

Caldwell, P. C. & Walster, G. (1963). Studies on the micro-injection of various substances into crab muscle fibres. *J. Physiol.* **169**, 353–372.

Cambridge, G. W. & Holgate, J. A. (1955a). Superfusion as a method for the study of drug antagonism. *Br. J. Pharmac. Chemother.* **10**, 326–335.

—— (1955b). A method for identification of 5-hydroxytryptamine. *J. Physiol.* **130**, 22P.

Campbell, G. (1966). The inhibitory nerve fibres in the vagal supply to the guinea-pig stomach. *J. Physiol.* **185**, 600–612.

Campbell, G. & Burnstock, G. (1968). Comparative physiology of gastro-intestinal motility. In *Handbook of Physiology.* Alimentary Canal IV. Amer. Physiol. Soc.: Washington, D.C.

Cannon, W. B. (1912). Peristalsis, segmentation and the myenteric reflex. *Am. J. Physiol.* **30**, 114–128.

Cannon, W. B. & Lissák, K. (1939). Evidence for adrenaline in adrenergic neurons. *Am. J. Physiol.* **125**, 765–777.

Cannon, W. B. & Rosenblueth, A. (1933). Studies on conditions of activity in endocrine organs. XXIX. Sympathin E & Sympathin I. *Am. J. Physiol.* **104**, 557–574.

—— (1935). A comparison of the effects of sympathin and adrenine on the iris, *Am. J. Physiol.* **113**, 251–258.

—— (1937). *Autonomic Neuroeffector Systems.* New York: Macmillan.

Cannon, W. B. & Uridil, J. E. (1921-1922). Studies on the conditions of activity in endocrine glands. VIII. Some effects on the denervated heart of stimulating the nerves of the liver. *Am. J. Physiol.* **58**, 353–364.

Caputo, C. (1966). Caffeine- and potassium-induced contractures of frog striated muscle fibres in hypertonic solutions. *J. gen. Physiol.* **50**, 129–140.

Carlsson, A. (1965). Drugs which block the storage of 5-hydroxytryptamine and related amines. In *Handbuch der Exp. Pharmacol.* XIX. Ed. Erspamer, V., pp. 529–592. Berlin: Springer Verlag.

—— (1966). Physiological and pharmacological release of monoamines in the central nervous system. In *Mechanisms of Release of Biogenic Amines.* Ed. von Euler, U. S. Rosell, S. & Uvnäs, B., pp. 331–346. Oxford: Pergamon Press.

Carlsson, A., Falck, B. & Hillarp, N. Å. (1962). Cellular localisation of brain monoamines. *Acta physiol. scand.* **56**, Suppl. 196, 1–28.

Carlyle, R. F. (1963). The mode of action of neostigmine and physostigmine on the guinea-pig trachealis muscle. *Br. J. Pharmac. Chemother,* **21**, 137–149.

Carmeliet, E. E. (1961a). *Chloride and Potassium Permeability in Cardiac Purkinje fibres.* Arscia Brussels.

—— (1961b). Chloride ions and the membrane potential of Purkinje fibres. *J. Physiol.* **156**, 375–388.

Carpenter, F. G. (1963). Excitation of rat urinary bladder by coaxial electrodes and by chemical agents. *Am. J. Physiol.* **204**, 727–731.

Carpenter, F. G. & Rand, S. A. (1965). Relation of acetylcholine release to responses of the rat urinary bladder. *J. Physiol.* **180**, 371–382.

Carpenter, F. G. & Rubin, R. M. (1967). The motor innervation of the rat urinary bladder. *J. Physiol.* **192**, 609–618.

Carroll, P. M. & Sereda, D. D. (1968a). Cell membrane of uterine smooth muscle. *Nature, Lond.* **217**, 666–667.

—— (1968b). Sialic acid of uterine smooth muscle: reappraisal of the serotonin receptor on smooth muscle. *Nature, Lond.* **217**, 667–668.

Carsten, M. E. (1965). A study of uterine actin. *Biochemistry, N. Y.* **4**, (6) 1049–1054.

Caspersson, T., Hillarp, N-Å. & Ritzen, M. (1966). Fluorescence microspectrophotometry of cellular catecholamines and 5-hydroxytryptamine. *Expl. Cell Res.* **42**, 415–428.

Casteels, R. (1964). The relation between the membrane potential and the ion content of smooth muscle cells. Ph.D. Thesis, Oxford.

—— (1965). The chloride distribution in the smooth muscle of the guinea-pig taenia coli. *J. Physiol.* **178**, 10–11P.

—— (1966). The action of ouabain on the smooth muscle cells of the guinea-pig's taenia coli. *J. Physiol.* **184**, 131–142.

Casteels, R. & Kuriyama, H. (1965). Membrane potential and ionic content in pregnant and non-pregnant rat myometrium. *J. Physiol.* **177**, 263–287.

—— (1966). Membrane potential and ion content in the smooth muscle of the guinea pig's taenia coli at different external potassium concentrations. *J. Physiol.* **184**, 120–130.

Casteels, R. & Meuwissen, H. (1968). Interaction between potassium and chloride in smooth muscle cells of the guinea-pig taenia coli. *J. Physiol.* **194**, 76–77P.

Cavallito, C. J., Arrowood, J. G. & O'Dell, T. B. (1956). Influence of anesthesia on the neuromuscular blocking activity of mylaxen. *Anesthesiology,* **17**, 547–558.

Cauna, N. & Ross, L. L. (1960). The fine structure of Meissner's touch corpuscles of human fingers. *J. biophys. biochem. Cytol.* **8**, 467–482.

Celander, O. (1959). Are there any centrally controlled sympathetic inhibitory fibres to the musculature of the intestine? *Acta physiol. scand.* **47**, 299–309.

Celander, O. & Folkow, B. (1953). The nature and distribution of afferent fibres provided with the axon reflex arrangement. *Acta physiol. scand.* **29**, 359–370.

Cervoni, P., West, T. C. & Fink, L. D. (1956). Autonomic postganglionic innervation of the nictitating membrane of the cat. *J. Pharmacol. exp. Ther.* **116**, 90–97.

Chambers, R. & Zweifach, B. W. (1944). Topography and function of the mesenteric capillary circulation. *Am. J. Anat.* **75**, 173–205.

Chapman, G. & McLauchlan, K. A. (1967). Oriented water in the sciatic nerve of rabbit. *Nature, Lond.* **215**, 391–392.

Charles, A. (1960). Electronmicroscopic observations of the arrector pili muscle of the human scalp. J, invest. Derm. 35, 27–30.

Chiang, T S. & Leaders, F. E. (1966). Cardiostimulatory responses to vagal stimulation, nicotine and tyramine. Am. J. Physiol. 211, 1443–1446.

Chiu, C. Y. & Lee, C. Y. (1961). The effects of some adrenergic blocking agents on the inhibition of peristalsis caused by sympathominetic amines. J. Formosan M.A. 60, 1128.

Choi, J. K. (1962). Fine structure of the smooth muscle of the chicken's gizzard. In Electron Microscopy. 5th Int. Congress for Electron Microscopy. Vol. 2, M–9. Ed. Breese, S. S. New York: Academic Press.

—— (1963). The fine structure of the urinary bladder of the toad Bufo marinus. J. cell Biol. 16, 53–72.

Chujyo, N. (1953). Site of acetylcholine production in the wall of intestine. Am. J. Physiol. 174, 196–198.

Chujyo, N. & Holland, W. C. (1962). Na, K, Ca, and Cl exchange in guinea-pig ileum. Am. J. Physiol. 202, 909–912.

—— (1963). Potassium-induced contracture and calcium exchange in the guinea-pig's taenia coli. Am. J. Physiol. 205, 94–100.

Clark, S. L. (1937). Innervation of the intrinsic muscles of the eye of the cat. J. comp. Neurol. 66, 307–325.

—— (1959). The ingestion of proteins and colloidal materials by columnar absorptive cells of the small intestine in suckling rats and mice. J. biophys. biochem. Cytol. 5, 41–49.

Clegg, P. C. (1966). Antagonism by prostaglandins of the responses of various smooth muscle preparations to sympathomimetics. Nature, Lond. 209, 1137–1139.

Clementi, F. (1962). Electron microscope observations on the action of vitamin E on the uterine smooth muscle cells. Experientia. 18, 406–407.

—— (1965). Modifications ultrastructurelles provoqués par quelques médicaments sur les terminaisons nerveuses adrénergiques et sur la médullaire surrénale. Experientia, 21, 171–188.

Clementi, F. & Garbagnati, E. (1965). Different action of granules depleting drugs on sympathetic nerve endings of the pineal and of the vas deferens. In Electron Microscopy, Ed. Titlbach, M. Czech Acad. Sci. (Prague) Vol. B, 305–306.

Clementi, F., Mantegazza, P. & Botturi, M. (1966). A pharmacologic and morphologic study of the nature of the dense-core granules present in the presynaptic endings of sympathetic ganglia. Int. J. Neoropharmac. 5, 281–285.

Cliff, W. J. (1966). Electron microscopic study of the structure of the aortic wall. Proc. Aust. Physiol. Soc. August, 1966.

—— (1967). The aortic tunica media in growing rats studied with the electronmicroscope. Lab. Invest. 17, 599–615.

Cloetta, M. (1908). Über das Verhalten des Atropins bei verschieden empfindlichen Tierarten. Arch. exp. Path. Pharmak. 1908, 119–125.

Cobb, J. L. S. & Bennett, T. (1969). A study of nexuses in visceral smooth muscle. J. Cell. Biol. 41, 287–297.

Coceani, F., Pace-Asciak, C., Volta, F. & Wolfe, L. S. (1967). Effect of nerve stimulation on prostaglandin formation and release from the rat stomach. Am. J. Physiol. 213, 1056–1064.

Code, C. F., Hightower, N. C. jr. & Morlock, C. G. (1952). Motility of the alimentary canal in man. Review of recent studies. Am. J. Med. 13, 328–351.

Cohen, C. & Holmes, K. C. (1963). X-ray diffraction evidence for α-helical coiled-coils in native muscle. J. molec. Biol. 6, 423–431.

Colburn, R. W. & Maas, J. W. (1965). Adenosine triphosphate-metal-norepinephrine ternary complexes and catecholamine binding. Nature, Lond. 208, 37–41.

Cole, D. F. (1950). The effects of oestradiol on the rat uterus. J. Endocr. 7, 12–23.

Cole, K. S. & Curtis, H. J. (1941). Membrane potential of the squid giant axon during current flow. J. gen. Physiol. 24, 551–563.

Comline, R. S. (1947). Synthesis of acetylcholine by non-nervous tissue. J. Physiol. 105, 6–7P.

Connell, A. M. (1968). Motor action of the large bowel. In Handbook of Physiology, Alimentary Canal IV. pp. 2075–2091. Amer. Physiol. Soc., Washington, D.C.

Connelly, C. M. (1959). Recovery processes and metabolism of nerve. In *Biophysical Sciences—A study programme*. Ed. J. L. Oncley, 475–484. New York: John Wiley.

Conti, G., Haenni, B., Laszt, L. & Rouiller, C. H. (1964). Structure et ultrastructure de la cellule musculaire lisse de la paroi carotidienne à l'état de repos et à l'état de contraction. *Angiologica*, **1**, 119–140.

Cooper, T. (1966). Surgical sympathectomy and adrenergic function. *Pharmac. Rev.* **18**, 611–618.

Cope, F. W. (1967). NMR evidence for complexing of Na$^+$ in muscle, kidney and brain, and by actomyosin. The relation of cellular complexing of Na$^+$ to water structure and to transport kinetics. *J. gen. Physiol.* **50**, 1353–1375.

Copen, D. L., Cirillo, D. P. & Vassalle, M. (1968). Tachycardia following vagal stimulation. *Am. J. Physiol.* **215**, 696–703.

Coraboeuf, E. & Weidmann, S. (1954). Temperature effects on the electrical activity of Purkinje fibres. *Helv. physiol. pharmac. Acta* **12**, 32–41.

Corrodi, H. & Hillarp, N.-Å. (1963). Fluoreszenzmethoden zur histochemischen Sichtbarmachung von Monoaminen. (1) Identifizierung der fluoreszierenden Produkte aus Modellversuchen mit 6,7-Dimethoxyisochinolinderivaten und Formaldehyd. *Helv. chim. Acta*. **46**, 2425–2430.

—— (1964). Fluoreszenzmethoden zur histochemischen Sichtbarmachung von Monoaminen. (2) Identifizierung des fluoreszierenden Produktes aus Dopamin und Formaldehyd. *Helv. chim. Acta*. **47**, 911–918.

Corrodi, H., Hillarp, N.-Å. & Jonsson, G. (1964). Fluorescence methods for the histochemical demonstration of monoamines. 3. Sodium borohydride reduction of the fluorescent compounds as a specificity test. *J. Histochem. Cytochem.* **11**, 582–586.

Corrodi, H. & Jonsson, G. (1967). The formaldehyde fluorescence method for the histochemical demonstration of biogenic monoamines. A review on the methodology. *J. Histochem. Cytochem.* **15**, 65–78.

Costa, E., Boullin, D. J., Hammer, W., Vogel, W. & Brodie, B. B. (1966). Interactions of drugs with adrenergic neurons. *Pharmac. Rev.* **18**, 577–597.

Coupland, R. E. (1962). Nerve-endings on chromaffin cells in rat adrenal medulla. *Nature, Lond.* **194**, 310–312.

—— (1965). Electronmicroscopic observations on the structure of the rat adrenal medulla. 11. Normal Innervation. *J. Anat.* **99**, 255–272.

Coupland, R. E. & Hopwood, D. (1966). The mechanism of the differential staining reaction for adrenaline- and noradrenaline-storing granules in tissue fixed in gluteraldehyde *J. Anat.* **100**, 227–244.

Coupland, R. E., Pyper, A. S. & Hopwood, D. (1964). A method of differentiating between noradrenaline- and adrenaline-storing cells in the light and electron microscope. *Nature, Lond.* **201**, 1240–1242.

Courtade, D. & Guyon, J. F. (1897). Influence motrice du grand sympathique et du nerf érecteur sacré sur le gros intestin. *Archs. Physiol.* **9**, 880–890.

Coutinho, E. M. (1966). The effect of magnesium on the staircase phenomenon of the rat uterus. *Acta. physiol. Latinoam.* **16**, 318–323.

Coutinho, E. M. & Csapo, A. (1959). Effect of oxytocics on the "Ca-deficient" uterus. *J. gen. Physiol.* **43**, 13–27.

Cragg, B. G. & Evans, D. H. L. (1958). Some reflexes mediated by the afferent abdominal vagus in the rabbit and the cat. *J. Physiol.* **142**, 6–7*P*.

Crank, J. (1956). *The Mathematics of Diffusion*. Oxford: Clarendon Press.

Cravioto, H. (1962). Electronenmikroskopische Untersuchungen am sympathischen Nervensystem des Menschen. I. Nervenzellen. *Z. Zellforsch. mikrosk. Anat.* **58**, 312–330.

Creed, K. E. & Wilson, J. A. F. (1969). The latency of responses of secretory acinar cells to nerve stimulation in submaxillary gland of the cat. *Aust. J. exp. Biol. med. Sci.* **47**, 135–144.

Creese, R., Jenden, D. J. & Steinborn, J. (1969). Diffusion and exponential processes in flat muscles: The interpretation of curves obtained with radio-active sodium. *J. Physiol.* (In press).

Creese, R., Neil, M. W. & Stephenson, G. (1956). Effect of cell variation on potassium exchange of muscle. *Trans. Faraday Soc.* **52**, 1022–1032.

Crema, A., Del Tacca, M., Frigo, G. M. & Lecchini, S. (1968). Presence of a non-adrenergic inhibitory system in the human colon. *Gut* **9**, 638–640.

Crescitelli, F., Koelle, G. B. & Gilman, A. (1946). Transmission of impulses in peripheral nerve treated with di-isopropyl fluorophosphate (DFP). *J. Neurophysiol.* **9**, 241–252.

Cross, S. B., Keynes, R. D. & Rybová, R. (1965). The coupling of sodium efflux and potassium influx in frog muscle. *J. Physiol.* **181**, 865–880.

Csapo, A. (1954). Dependence of isometric tension and isotonic shortening of uterine muscle on temperature and on strength of stimulation. *Am. J. Physiol.* **177**, 348–354.

—— (1955). In: *Modern Trends in Obstetrics and Gynaecology.* 2nd Ser., Ch. 2. Ed. Bowes, K. London: Butterworth.

—— (1956a). The mechanism of effect of the ovarian steroids. *Recent Prog. Horm. Res.* **12**, 404–431.

—— (1956b). Progesterone "block". *Am. J. Anat.* **98**, 273–291.

—— (1959). Function and regulation of the myometrium. *Ann. N. Y. Acad. Sci.* **75**, 790–808.

—— (1960). Molecular structure and function of smooth muscle. In *Structure and Function of Muscle.* Ed. G. Bourne. **1**, 229–264. New York: Academic Press.

—— (1961a). The *in vivo* and *in vitro* effects of estrogen and progesterone on the myometrium. In *Mechanisms of Action of Steroid Hormones.* Ed. Villee, C. A. & Engel, L. L. pp. 126–147. Oxford: Pergamon Press.

—— (1961b). Defence mechanism of pregnancy. *Progesterone and the Defence Mechanism of Pregnancy.* Ciba Foundation Study Group **9** pp. 3–27. London: Churchill.

—— (1962). Smooth muscle as a contractile unit. *Physiol. Rev.* **42** (suppl. 5), 7–33.

Csapo, A. & Corner, G. W. (1952). The antagonistic effects of estrogen and progesterone on the staircase phenomenon in uterine muscle. *Endocrinology* **51**, 378–385.

Csapo, A., Erdös, T., De Mattos, C. R., Gramss, E. & Moscowitz, C. (1965). Stretch-induced uterine growth, protein synthesis and function. *Nature, Lond.* **207**, 1378–1379.

Csapo, A. & Kuriyama, H. (1963). Effects of ions and drugs on cell membrane activity and tension in the postpartum rat myometrium. *J. Physiol.* **165**, 575–592.

Csapo, A. & Pinto-Dantas, C. A. (1965). The effect of progesterone on the human uterus. *Proc. natn. Acad. Sci. U.S.A.* **54**, 1069–1076.

Cushny, A. R. (1910). The action of atropine, pilocarpine and physostigmine. *J. Physiol.* **41**, 233–245.

Cuthbert, A. W. (1962). Electrical and mechanical activity of the chick amnion. *Nature, Lond.* **193**, 488–489.

—— (1963a). Some effects of atropine on smooth muscle. *Br. J. Pharmac. Chemother.* **21**, 285–294.

—— (1963b). An acetylcholine-like substance and cholinesterase in the smooth muscle of the chick amnion. *J. Physiol.* **166**, 284–295.

—— (1966). Electrical activity in mammalian veins. *Biblphie anat.* **8**, 11–15.

Cuthbert, A. W., Matthews, E. K. & Sutter, M. C. (1965). Spontaneous electrical activity in a mammalian vein. *J. Physiol.* **176**, 22–23P.

Cuthbert, A. W. & Sutter, M. C. (1964). Electrical activity of a mammalian vein. *Nature, Lond.* **202**, 95.

—— (1965). The effects of drugs on the relation between the action potential discharge and tension in a mammalian vein. *Br. J. Pharmac. Chemother,* **25**, 592–601.

Dagliesh, C. E. Toh, C. C. & Work, T. S. (1953). Fractionation of the smooth muscle stimulants present in extracts of the gastrointestinal tract. Identification of 5-hydroxytryptamine and its distinction from substance P. *J. Physiol.* **120**, 298–310.

Dahl, E. & Nelson, E. (1964). Electron microscopic observations on human intracranial arteries, II. Innervation. *Archs. Neurol. Chicago* **10**, 158–164.

Dahlström, A. (1965). Observations on the accumulation of noradrenaline in the proximal and distal parts of the peripheral adrenergic nerves after compression. *J. Anat.* **99**, 677–689.

—— (1967a). The transport of noradrenaline between two simultaneously performed ligations of the sciatic nerves of rat and cat. *Acta physiol. scand.* **69**, 158–166.

Dahlström, A. (1967b). The effect of reserpine and tetrabenazine on the accumulation of noradrenaline in the rat sciatic nerve after ligation. *Acta. physiol. scand.* **69**, 167–179.

Dahlström, A. & Fuxe, K. (1964a). A method for the demonstration of monoamine-containing nerve fibres in the central nervous system. *Acta physiol. scand.* **60**, 293–294.

—— (1964b). Evidence for the existence of monoamine-containing neurons in the central nervous system: 1. Demonstration of monoamines in the cell bodies of brain stem neurons. *Acta physiol. scand.* **62**, Suppl. 232, 1–55.

—— (1965a). Evidence for the existence of monoamine neurons in the central nervous system: II. Experimentally induced changes in the intraneuronal amine levels of the bulbospinal neuron systems. *Acta physiol. scand,* **64**, Suppl. 247, 1–36.

—— (1965b). Evidence for the existence of an outflow of noradrenaline nerve fibres in the ventral roots of the rat spinal cord. *Experientia,* **21**, 409–410.

—— (1965c). The adrenergic innervation of the nasal mucosa of certain mammals. *Acta oto-laryng. (Stockh.)* **59**, 65–72.

Dahlström, A. & Häggendal, J. (1966a). Some quantatative studies of the noradrenaline content in the cell bodies and terminals of a sympathetic adrenergic neuron system. *Acta physiol. scand,* **67**, 271–277.

—— (1966b). Studies on the transport and life span of amine storage granules in a peripheral adrenergic neuron system. *Acta physiol. scand.* **67**, 278–288.

—— (1967). Studies on the transport and life span of amine storage granules in the adrenergic neuron system of the rabbit sciatic nerve. *Acta physiol. scand.* **69**, 153–157.

Dahlström, A., Häggendal, J. & Hökfelt, T. (1966). The noradrenaline content of the varicosities of sympathetic adrenergic nerve terminals in the rat. *Acta. physiol. scand.* **67**, 289–294.

Dale, H. H. (1906). On some physiological actions of ergot. *J. Physiol.* **34**, 163–206.

—— (1914). The action of certain esters and ethers of choline, and their relation to muscarine. *J. Pharmac. exp. Ther.* **6**, 147–190.

—— (1933). Nomenclature of fibres in the autonomic system and their effects. *J. Physiol.* **80**, 10–11P.

—— (1934). Chemical transmission of the effects of nerve impulses. *Br. med. J.* **No. 1**, 835–841.

—— (1935). Pharmacology and nerve-endings. *Proc. R. Soc. Med.* **28**, 319–332.

—— (1937a). Acetylcholine as a chemical transmitter of the effects of nerve impulses. I. History of ideas and evidence. Peripheral autonomic actions. Functional nomenclature of nerve fibres. *J. Mt Sinai Hosp.* **4**, 401–415.

——— (1937b). Acetylcholine as a chemical transmitter of the effects of nerve impulses. II. Chemical transmission at ganglionic synapses and voluntary motor nerve endings, some general considerations. *J. Mt Sinai Hosp.* **4**, 416–429.

—— (1952). *Transmission of Effects from Nerve Endings.* London: Oxford University Press.

Dale, H. H. & Dudley, H. W. (1929). The presence of histamine and acetylcholine in the spleen of the ox and the horse. *J. Physiol.* **68**, 97–123.

Dale, H. H. & Feldberg, W. (1934). The chemical transmission of secretory impulses to the sweat glands of the cat. *J. Physiol.* **82**, 121–128.

Dale, H. H. & Gaddum, J. H. (1930). Reactions of denervated voluntary muscle, and their bearing on the mode of action of parasympathetic and related nerves. *J. Physiol.* **70**, 109–144.

Dale, H. H., Laidlaw, P. P. & Symons, C. T. (1910). A reversed action of the vagus on the mammalian heart *J. Physiol.* **41**, 1–18.

Dalton, A. J. & Felix, M. D. (1956). A comparative study of the golgi complex. *J. biophys. biochem. Cytol.* **2**, no. 4, suppl. pp. 79–83.

Daly, I. de B. & Hebb, C. (1952). Pulmonary vasomotor fibres in the cervical vagosympathetic nerve of the dog. *Q. Jl exp. Physiol.* **37**, 19–44.

Daniel, E. E. (1958). Smooth muscle electrolytes. *Can. J. Biochem. Physiol.* **36**, 805–818.

—— (1963a). On roles of calcium, strontium and barium in contraction and excitability of rat uterine muscle. *Archs int. Pharmocodyn. Thér.* **146**, 298–349.

—— (1963b). Potassium movements in rat uterus studied *in vitro.* I. Effects of temperature. *Can. J. Biochem. Physiol.* **41**, 2065–2084.

Daniel, E. E. (1964). The interconnection between active transport and contracture in uterine tissues. *Canad. J. Physiol. Pharmac.* **42**, 453–495.

— (1965a). Attempted synthesis of data regarding divalent ions in muscle function. In *Muscle*. Ed. Paul, W. M., Daniel, E. E., Kay, C. M. & Monckton, G. pp. 295–313. Oxford: Pergamon Press.

— (1965b). Electrical and contractile activity of pyloric region in dogs and effect of drugs administered intra-arterially. *Gastroenterology*, **49**, 403–418.

Daniel, E. E. & Chapman, K. M. (1963). Electrical activity of the gastrointestinal tract as an indication of mechanical activity. *Am. J. dig. Dis.* **8**, 54–102.

Daniel, E. E. & Daniel, B. N. (1957). Effects of ovarian hormones on the content and distribution of cation in intact and extracted rabbit and cat uterus. *Can. J. Biochem. Physiol.* **35**, 1205–1223.

Daniel, E. E., Honour, A. J. & Bogoch, A. (1960). Electrical activity of the longitudinal muscle of dog small intestine studied *in vivo* using microelectrodes. *Am. J. Physiol.* **198**, 113–118.

Daniel, E. E. & Robinson, K. (1960). Relation of sodium secretion to metabolism in isolated sodium-rich uterine segments. *J. Physiol.* **154**, 445–460.

Daniel, E. E., Sehdev, H. & Robinson, K. (1962). Mechanisms for activation of smooth muscle. *Physiol. Rev.* **42**, suppl. 5, 228–260.

Daniel, E. E. & Singh, H. (1958). Electrical properties of the smooth muscle cell membrane. *Can. J. Biochem. Physiol.* **36**, 959–975.

Daniel, E. E. Wachter, B. T., Honour, A. J. & Bogoch, A. (1960). The relationship between electrical and mechanical activity of the small intestine of dog and man. *Can. J. Biochem.* **38**, 777–801.

Davenport, H. W. (1966). *Physiology of the Digestive Tract.* 2nd Ed., Chicago: Year Book Medical Publishers Inc.

Davey, M. J., Hayden, M. L. & Scholfield, P. C. (1968). The effects of bretylium on C fibre excitation and noradrenaline release by acetylcholine and electrical stimulation. *Br. J. Pharmac.* **34**, 377–387.

Davies, B. N., Horton, E. W. & Withrington, P. G. (1966). The occurrence of prostaglandin E_2 in splenic venous blood of the dog following nerve stimulation. *J. Physiol.* **188**, 38–39*P*.

— (1968). The occurrence of prostaglandin E_2 in splenic venous blood of the dog following splenic nerve stimulation. *Br. J. Pharmac. Chemother.* **32**, 127–135.

Davies, B. N. & Withrington, P. G. (1968). The effects of prostaglandin E_1 and E_2 on the smooth muscle of the dog spleen and on its responses to catecholamines, angiotensin and nerve stimulation. *Br. J. Pharmac. Chemother.* **32**, 136–144.

Davis, R. B. & Kay, D. (1965). Demonstration of 5-hydroxytryptamine in blood platelets by electronmicroscope autoradiography. *Nature, Lond.* **207**, 650–651.

Davis, R. B. & Koelle, G. B. (1965). Electron microscopic localization of acetylcholinesterase (AchE) at the motor endplate by the gold thiolacetic acid (THAC) and gold-thiocholine (ThCh) methods. *J. Histochem. Cytochem.* **13**, 703–704.

— (1967). Electronmicroscopic localization of acetylcholinesterase and non specific cholinesterase at the neuromuscular junction by the gold-thiocholine and gold-thiolacetic acid methods. *J. cell Biol.* **34**, 157–171.

Davis, R. B. & White, A. G. (1968). Localization of 5-hydroxytryptamine in blood platelets: an autoradiographic and ultra structural study. *Br. J. Haemat.* **15**, 93–99.

Day, M. D. & Rand, M. J. (1961). Effect of guanethidine in revealing cholinergic sympathetic fibres. *Br. J. Pharmac. Chemother.* **17**, 245–260.

— (1963). A hypothesis for the mode of action of α-methyldopa in relieving hypertension. *J. Pharm. Pharmac.* **15**, 221–224.

Day, M. D. & Warren, P. R. (1967). Inhibitory responses to transmural stimulation in isolated intestinal preparations. *J. Pharm. Pharmac.* **19**, 408–410.

— (1968). A pharmacological analysis of the responses to transmural stimulation in isolated intestinal preparations. *Br. J. Pharmac. Chemother,* **32**, 227–240.

Dean, R. B. (1941). Theories of electrolyte equilibrium in muscle. *Biol. Symp.* **3**, 331–348.

De Candole, C. A., Douglas, W. W., Lovatt Evans, C., Holmes, R., Spencer, K. E. V.,

Torrance, R. W. & Wilson, K. M. (1953). The failure of respiration in death by acetylcholinesterase poisoning. *Br. J. Pharmac. Chemother.* **8**, 466–475.

De Duve, C. (1959). Lysosomes, a new group of cytoplasmic particles. In *Subcellular Particles*. Ed. Hayashi, T. pp. 128–158. New York: Ronald Press Co.

De Harven, E. & Bernhard, W. (1956). Etude au microscope électronique de l'ultrastructure du centriole chez les vertébrés. *Z. Zellforsch. mikrosk. Anat.* **45**, 378–398.

De Kleijn, A. & Socin, Ch. (1915). Zur näheren Kenntnis des Verlaufs der postganglionären Sympathicusbahnen für Pupillenerweiterung, Lidspaltenöffnung und Nickhautretraktion bei der Katze. *Pflügers Arch. ges. Physiol.* **160**, 407–415.

de la Lande, I. S., Cannell, V. A. & Waterson, J. G. (1966). The interaction of serotonin and noradrenaline on the perfused artery. *Br. J. Pharmac. Chemother.* **28**, 255–272.

de la Lande, I. S., Frewin, D. & Waterson, J. G. (1967). The influence of sympathetic innervation on vascular sensitivity to noradrenaline. *Br. J. Pharmac. Chemother.* **31**, 82–93.

de la Lande, I. S. & Waterson, J. G. (1967). Site of action of cocaine on the perfused artery. *Nature, Lond.* **214**, 313–314.

—— (1968). The action of tyramine on the rabbit ear artery. *Br. J. Pharmac.* **34**, 8–18.

De Lorenzo, A. J. (1958). Electron microscopic observations on the taste buds of the rabbit. *J. biophys. biochem. Cytol.* **4**, 143–150.

—— (1959). The fine structure of synapses. *Biol. Bull. mar. biol. Lab. Woods Hole.* **117**, 390.

De Robertis, E., (1958). Submicroscopic morphology and function of the synapse. *Expl. Cell Res.* **5**, 347–369.

—— (1963). Contribution of electronmicroscopy to some neuropharmacological problems. In *Proc. First. int. Pharmac. Meeting, Stockholm, 1961*, **5**, p. 49. Oxford: Pergamon Press.

—— (1964). *Histophysiology of Synapses and Neurosecretion*. Oxford: Pergamon Press.

—— (1966). Adrenergic endings and vesicles isolated from brain. *Pharmac. Rev.* **18**, 413–424.

De Robertis, E., & Bennett, H. S. (1955). Some features of the submicroscopic morphology of synapses in frog and earthworm. *J. biophys. biochem. Cytol.* **1**, 47–65.

De Robertis, E., & Ferreira, A. V. (1957). Submicroscopic changes of the nerve endings in the adrenal medulla after stimulation of the splanchnic nerve. *J. biophys. biochem. Cytol.* **3**, 611–614.

De Robertis, E. & Pellegrino De Iraldi, A. (1961). Plurivesicular secretory processes and nerve endings in the pineal gland of the rat. *J. biophys. biochem. Cytol.* **10**, 361–372.

De Robertis, E., Pellegrino De Iraldi, A., Rodriguez De Lores Arnaiz, G. & Zieher, L. M. (1965). Synaptic vesicles from the rat hypothalamus. Isolation and norepinephrine content. *Life. Sci. Oxford.* **4**, 193–201.

Del Castillo, J. & Katz, B. (1954). Statistical factors involved in neuromuscular facilitation and depression. *J. Physiol.* **124**, 574–585.

—— (1955a). On the localization of acetylcholine receptors. *J. Physiol.* **128**, 157–181.

—— (1955b). Local activity at a depolarized nerve-muscle junction. *J. Physiol.* **128**, 396–411.

—— (1955c). Production of membrane potential changes in the frog's heart by inhibitory nerve impulses. *Nature, Lond.* **175**, 1035.

Del Castillo, J., De Mello, W. C. & Morales, T. (1964). Influence of some ions on the membrane potential of Ascaris muscle. *J. gen. Physiol.* **48**, 129–140.

Del Tacca, M., Lecchini, S., Frigo, G. M., Crema, A. & Benzi, G. (1968). Antagonism of atropine towards endogenous and exogenous acetylcholine before and after sympathetic system blockade in the isolated distal guinea-pig colon. *Europ. J. Pharmacol.* **4**, 188–197.

Della Bella, D., Benelli, G. & Gandini, A. (1964). Eserine and autonomic nervous control of the guinea-pig vas deferens. *J. Pharm. Pharmac.* **16**, 779–787.

Dengler, H. J., Spiegel, H. E. & Titus, E. D. (1961). Effects of drugs on uptake of isotopic norepinephrine by cat tissue. *Nature, Lond.* **191**, 816–817.

D'Ermo, F. & Bonomi. L. (1961). Ricerche farmacologische sull'azione della serotonina (enteramina) sulla muscolatura iridea isolata. *Boll. Oculist.* **40**, 483–491.

Dettbarn, W. D. (1960a). New evidence for the role of acetylcholine in conduction. *Biochim. biophys. Acta* **41**, 377–386.

—— (1960b). The effect of curare on conduction in myelinated isolated nerve fibres of the frog. *Nature, Lond.* **186**, 891–892.

Dettbarn, W. D. & Davies, F. A. (1963). Effects of acetylcholine on axonal conduction of lobster nerve. *Biochim. biophys. Acta* **66**, 397–405.

Devine, C. E. (1966). Neuromuscular relationships in rat intestinal and mesenteric blood vessels. *Proc. Univ. Otago med. Sch.* **44**, 9–11.

Devine, C. E., Robertson, A. A. & Simpson, F. O. (1967). Effect of sympatholytic drugs on sympathetic axonal fine structure and tissue catecholamine levels. *N.Z. med. J.* **66**, 390–391.

Devine, C. E. & Simpson, F. O. (1967). The fine structure of vascular sympathetic neuromuscular contacts in the rat. *Am. J. Anat.* **121**, 153–174.

—— (1968). Localisation of tritiated norepinephrine in vascular sympathetic axons of the rat intestine and mesentery by electron microscope radioautography. *J. Cell. Biol.* **38**, 184–192.

Dewey, M. M. & Barr, L. (1962). Intercellular connection between smooth muscle cells: The nexus. *Science, N.Y.* **137**, 670–672.

—— (1964). A study of the structure and distribution of the nexus. *J. cell. Biol.* **23**, 553–585.

Diamond, J. & Brody, T. M. (1966). Hormonal alteration of the response of the rat uterus to catecholamines. *Life Sci.* **5**, 2187–2193.

Dick, D. A. T. (1959). Osmotic properties of living cells. *Int. Rev. Cytol.* **8**, 387–448.

—— (1966). *Cell Water.* London: Butterworths.

Dick, D. A. T. & Lea, E. J. A. (1967). The partition of sodium fluxes in isolated toad oocytes. *J. Physiol.* **191**, 289–308.

Dietlen, H. (1913). Ergebnisse des medizinischen Röntgenverfahrens für die Physiologie. *Ergebn. Physiol.* **13**, 47–95.

Dikshit, B. B. (1938). Acetylcholine formation by tissues. *Q. Jl. exp. Physiol.* **28**, 243–251.

Dixon, W. E. (1906). Vagus inhibition. *Br. med. J.* **2**, 180–181.

—— (1907). On the mode of action of drugs. *Med. Mag. (Lond.)* **16**, 454–457.

Dixon, W. E. & Brodie, T. G. (1903). Contributions to the physiology of the lungs. Part I. The bronchial muscles, their innervation, and the action of drugs upon them. *J. Physiol.* **29**, 97–173.

Dogiel, A. S. (1894). Die Nervenendigungen in den Nebennieren der Säugethiere. *Arch. Anat. Physiol.* (Anat. Abt.) **1894**, 90–104.

—— (1898). Die sensiblen Nervenendigungen im Herzen und in den Blutgefässen der Säugethiere. *Arch. mikrosk. Anat. EntwMech.* **52**, 44–70.

Dorr, L. D. & Brody, M. J. (1965). Functional separation of adrenergic and cholinergic fibres to skeletal muscle vessels. *Am. J. Physiol.* **208**, 417–424.

—— (1967). Hemodynamic mechanisms of erection in the canine penis. *Am. J. Physiol.* **213**, 1526–1531.

Dorward, P. & Holman, M. E. (1967). Excitatory action of noradrenaline on smooth muscle. *Aust. J. exp. Biol. med. Sci.* **45**, 48P.

Douglas, W. W. (1951). The effect of some anticholinesterase drugs on the isolated tracheal muscle of the guinea-pig. *J. Physiol.* **112**, 20P.

Drakontides, A. B. & Gershon, M. D. (1968). 5-Hydroxytryptamine receptors in the mouse duodenum. *Br. J. Pharmac. Chemother.* **33**, 480–492.

Draper, M. H. & Weidmann, S. (1951). Cardiac resting and action potentials recorded with an intracellular electrode. *J. Physiol.* **115**, 74–94.

Dreifuss, J. J., Girardier, L. & Forssman, W. G. (1966). Etude de la propagation de l'excitation dans le ventricule de rat au moyen de solutions hypertoniques. *Pflügers Arch. ges. Physiol.* **292**, 13–33.

Drury, A. N. (1936). The physiological activity of nucleic acid and its derivatives. *Physiol. Rev.* **16**, 292–325.

Drury, A. N. & Szent-Györgyi, A. (1929). The physiological activity of adenine compounds with especial reference to their action upon the mammalian heart. *J. Physiol.* **68**, 213–237.

Ducret, S. (1931). Rhythmische Tonusschwankungen und Adrenalinerregbarkeit der Mesenterialgefässe. *Pflügers Arch. ges. Physiol.* **227**, 753–758.

Dudel, J. & Kuffler, S. W. (1961a). The quantal nature of transmission and spontaneous miniature potentials at the crayfish neuromuscular junction. *J. Physiol.* **155**, 514–529.

Dunham, E., Rolewicz, T. & Zimmerman, B. (1968). Prostaglandin as the possible mediator of cutaneous sympathetic vasodilatation. *Fedn. Proc.* **27**, 536.

Dupont, J. R. & Sprinz, H. (1964). The neurovegatative periphery of the gut. A revaluation with conventional technics in the light of modern knowledge. *Am. J. Anat.* **114**, 393–402.

Durbin, R. P. & Jenkinson, D. H. (1961a). The effect of carbachol on the permeability of depolarised smooth muscle to inorganic ions. *J. Physiol.* **157**, 74–89.

—— (1961b). The calcium dependence of tension development in depolarised smooth muscle. *J. Physiol.* **157**, 90–96.

Durbin, R. P. & Monson, R. R. (1961). Ionic composition and permeability of smooth muscle. *Fedn Proc.* **20**, 134.

Dydyńska, M. & Wilkie, D. R. (1963). The osmotic properties of striated muscle fibres in hypertonic solutions. *J. Physiol.* **169**, 312–329.

Eakins, K. E. & Katz, R. L. (1967). The effects of sympathetic stimulation and epinephrine on the superior rectus muscle of the cat. *J. Pharmac. exp. Ther.* **157**, 524–531.

Ebashi, S. (1961). Calcium binding activity of vesicular relaxing factor. *J. Biochem., Tokyo* **50**, 236–244.

—— (1969). Comparative aspect of regulatory structural proteins of muscle, with particular reference to troponin. In *Symposium on Vascular Neuroeffector Systems*, Interlaken, August, 1969. (In press).

Eccles, J. C. (1936). Synaptic and neuro-muscular transmission. *Ergebn. Physiol.* **38**, 339–444.

—— (1961). The mechanism of synaptic transmission. *Ergebn. Physiol.* **51**, 299–430.

—— (1964). *The Physiology of Synapses.* Berlin: Springer Verlag.

Eccles, J. C. & Magladery, J. W. (1937a). The excitation and response of smooth muscle. *J. Physiol.* **90**, 31–67.

—— (1937b). Rhythmic responses of smooth muscle. *J. Physiol.* **90**, 68–99.

Eccles, R. M. & Libet, B. (1961). Origin and blockade of the synaptic response of curarized sympathetic ganglia. *J. Physiol.* **157**, 484–503.

Eckert, R. (1963). Electrical interaction of paired ganglion cells in the leech. *J. gen. Physiol.* **46**, 573–587.

Edman, K. A. P., & Schild, H. O. (1961). Interactions of acetylcholine, adrenaline and magnesium with calcium in the contraction of depolarized rat uterus. *J. Physiol.* **155**, 10–11P.

—— (1962). The need for calcium in the contractile responses induced by acetylcholine and potassium in the rat uterus. *J. Physiol.* **161**, 424-441.

—— (1963). Calcium and the stimulant and inhibitory effects of adrenaline in depolarized smooth muscle. *J. Physiol.* **169**, 404–411.

Edwards, C. & Harris, E. J. (1957). Factors influencing the sodium movement in frog muscle with a discussion of the mechanism of sodium movement. *J. Physiol.* **135**, 567–580.

Ehinger, B. (1966). Ocular and orbital vegetative nerves. *Acta physiol. scand.* **67**, Suppl. 268, 1–35.

Ehinger, B., Falck, B. & Sporrong, B. (1966). Adrenergic fibres to the heart and to peripheral vessels. *Biblphie. anat.* **8**, 35–45.

Ehinger, B., Håkanson, R., Owman, C. & Sporrong, B. (1968). Histochemical demonstration of histamine in paraffin sections by a fluorescence method. *Biochem. Pharmac.* **17**, 1997–1998.

Eisenfeld, A. J., Axelrod, J. & Krakoff, L. (1967). Inhibition of the extraneuronal accumulation and metabolism of norepinephrine by adrenergic blocking agents. *J. Pharmac. exp. Ther.* **156**, 107–113.

Eisenfield, A. J., Landsberg, L. & Axelrod, J. (1967). Effect of drugs on the accumulation and metabolism of extraneuronal norepinephrine in the rat heart. *J. Pharmac. exp. Ther.* **158**, 378–385.

El-Badawi, A. & Schenk, E. A. (1967). Histochemical methods for separate, consecutive and simultaneous demonstration of acetylcholinesterase and norepinephrine in cryostat sections. *J. Histochem. Cytochem.* **15**, 580–588.

Elford, B. C. (1967). Deuterium oxide exchange in isolated guinea-pig taenia-coli. *J. Physiol.* **188**, 29–30P.

Elfvin, L.-G. (1958). The ultrastructure of unmyelinated fibres in the splenic nerve of the cat. *J. Ultrastruct. Res.* **1**, 428–454.

—— (1961). The electron-microscopic investigation of filament structures in unmyelinated fibres of cat splenic nerve. *J. Ultrastruct. Res.* **5**, 51–64.

—— (1963). The ultrastructure of the superior cervical sympathetic ganglion of the cat. II. The structure of the preganglionic end fibres and the synapses as studied by serial sections. *J. Ultrastruct. Res.* **8**, 441–476.

—— (1965). The fine structure of the cell surface of chromaffin cells in the rat adrenal medulla. *J. Ultrastruct. Res.* **12**, 263–286.

Eliasson, S. (1952). Cerebral influence on gastric motility in the cat. *Acta physiol. scand.* **26**, suppl. 95.

—— (1954). Activation of gastric motility from the brainstem of the cat. *Acta. physiol. scand.* **30**, 199–214.

Eliasson, S., Folkow, B., Lindgren, P. & Uvnäs, B. (1951). Activation of sympathetic vasodilator nerves to the skeletal muscles in the cat by hypothalamic stimulation. *Acta physiol. scand.* **23**, 333–351.

Eliasson, S., Lindgren, P. & Uvnäs, B. (1952). Representation in the hypothalamus and the motor cortex in the dog of the sympathetic vasodilator outflow to the skeletal muscles. *Acta physiol. scand.* **27**, 18–37.

—— (1954). The hypothalamus, a relay station of the sympathetic vasodilator tract. *Acta physiol. scand,* **31**, 290–300.

Elliot, G. F. (1964). X-ray diffraction studies on striated and smooth muscles. *Proc. R. Soc.* B. **160**, 467–472.

—— (1967). Variations of the contractile apparatus in smooth and striated muscles. X-ray diffraction studies at rest and in contraction. *J. gen. Physiol.* **50**, No. 6, 171–184.

Elliott, T. R. (1904a). On the action of adrenalin. *J. Physiol.* **31**, 20–21.

—— (1904b). On the innervation of the ileo-colic sphincter. *J. Physiol.* **31**, 157–168.

—— (1905). The action of adrenalin. *J. Physiol.* **32**, 401–467.

—— (1913). The innervation of the adrenal glands. *J. Physiol.* **46**, 285–290.

Elliott, T. R. & Barclay-Smith, E. (1904). Antiperistalsis and other muscular activities of the colon. *J. Physiol.* **31**, 272–304.

Ellis, S. (1947). Benzoylcholine and atropine esterase, *J. Pharmac. exp. Ther.* **91**, 370–378.

—— (1959). Relation of biochemical effects of epinephrine to its muscular effects. *Pharmac. Rev.* **11**, 469–479.

Enerbäck, L., Olsson, Y. & Sourander, P. C. (1965). Mast cells in normal and sectional peripheral nerve. *Z. Zellforsch. mikrosk. Anat.* **66**, 596–608.

Engel, P., Hildebrandt, G. & Scholz, H. G. (1968). Die Messung der Phasenkoppelung zwischen Herzschlag und Atmung beim Menschen mit einem neuen Koinzidenzmessgerät. *Pflügers. Arch. ges. Physiol.* **298**, 258–270.

Engelhart, E. (1931). Der humorale Wirkungsmechanismus der Oculomotoriusreizung. *Pflügers Arch. ges. Physiol.* **227**, 220–234.

Engelhart, E. & Loewi, O. (1930). Fermentative Azetylcholinspaltung im Blut und ihre Hemmung durch Physostigmin. *Arch. exp. Path. Pharmak.* **150**, 1–13.

Engström, H. (1958). On the double innervation of the sensory epithelia of the inner ear. *Acta Oto-lar.* **49**, 109–118.

Eränkö, O. (1952). On the histochemistry of the adrenal medulla of the rat, with special reference to acid phosphatase. *Acta anat.* **16**, Suppl. 17, 308.

—— (1955). Distribution of fluorescing islets, adrenaline and noradrenaline in the adrenal medulla of the hamster. *Acta endocr., Copnh.* **18**, 174–179.

—— (1956). Histochemical demonstration of noradrenaline in the adrenal medulla of the hamster. *J. Histochem. Cytochem.* **4**, 11–13.

—— (1960). Cell types of the adrenal medulla. In *Adrenergic Mechanisms*, Ed. Vane, J. R., Wolstenholme, G. & O'Connor, M. pp. 103–110. London: Churchill.

—— (1967a). The practical histochemical demonstration of catecholamines by formaldehyde-induced fluorescence. *Jl. R. microsc. Soc.* **87**, 259–276.

—— (1967b). Histochemistry of nervous tissues, catecholamines and cholinesterases. *Ann. Rev. Pharmacol.* **7**, 203–222.

Eränkö, O, & Hänninen, L. (1960). Electron microscopic observations on the adrenal medulla of the rat. *Acta. path. microbiol. scand.* **50**, 126–132.

Eränkö, O., Koelle, G. B. & Räisänen, L. (1967). A thiocholine-lead ferrocyanide method for acetylcholinesterase. *J. Histochem. Cytochem.* **15**, 674–679.

Eränkö, O. & Räisänen, L. (1966). Demonstration of catecholamines in adrenergic nerve fibers by fixation in aqueous formaldehyde solution and fluorescence microscopy. *J. Histochem. Cytochem.* **14**, 690–691.

Eränkö, O., Rechardt, L. & Hänninen, L. (1967). Electron microscopic demonstration of cholinesterases in nervous tissue. *Histochemie*, **8**, 369–376.

Erdmann, W. D. & Heye, D. (1958). Analyse der erregenden und lähmenden Wirkung von Alkylphosphaten (Parathion, Paraoxon, Systox) am isolierten Kaninchendarm. *Arch. exp. Path. Pharmak.* **232**, 507–521.

Erspamer, V. (1952). Biological activity of some enteramine-related substances. *Nature, Lond.* **170**, 281–282.

—— (1954a). Pharmacology of indolealkylamines. *Pharmacol. Rev.* **6**, 425–487.

—— (1954b). Il sistema cellulase enterocromaffine e l'enteramina (5-idrossitriptamina). *Rend. sci. Farmitalia.* **1**, 1–193.

—— (1966). Occurrence of indolealkylamines in nature. In *5-Hydroxytryptamine and related Indolealkylamines*, Handbook of exp. pharmacology. Ed. Erspamer, V., Vol. 19, Ch. 4. pp. 132–181. Berlin: Springer.

Esterhuizen, A. C., Graham, J. D. P., Lever, J. D. & Spriggs, T. L. B. (1967). The innervation of the smooth muscle of the nictitating membrane of the cat. *J. Physiol.* **192**, 41P.

—— (1968). Catecholamine and acetylcholinesterase distribution in relation to noradrenaline release. An enzyme histochemical and autoradiographic study on the innervation of the cat nictitating muscle. *Br. J. Pharmac. Chemother.* **32**, 46–56.

Euler, U. S. von, (1946). A specific sympathomimetic ergone in adrenergic nerve fibres (sympathin) and its relations to adrenaline and noradrenaline. *Acta. Physiol. scand.* **12**,73–97.

—— (1949). Histamine as a specific constituent of certain autonomic nerve fibres. *Acta physiol. scand.* **19**, 85–93.

—— (1951). The nature of adrenergic nerve mediators. *Pharmacol. Rev.* **3**, 247–277.

—— (1956). *Noradrenaline.* Springfield: Thomas.

—— (1958). The presence of the adrenergic neurotransmitter in intraaxonal structures. *Acta physiol. scand.* **43**, 155–166.

—— (1963). *Comparative Endocrinology.* Vol. 2, Eds. Euler U. S. von & Heller, H. pp. 209–233. New York: Academic Press.

—— (1967). Some factors affecting catecholamine uptake, storage and release in adrenergic nerve granules. *Circulation Res.* **21**, III-5–III-12.

Euler, U. S. von & Folkow, B. (1953). Einfluss verschiedener afferenter Nervenreize auf die Zusammensetzung des Nebennierenmarkinkretes bei der Katze. *Arch. exp. Path. Pharmak.* **219**, 242–247.

Euler, U. S. von & Gaddum, J. H. (1931). Pseudomotor contractures after degeneration of the facial nerve. *J. Physiol.* **73**, 54–66.

Euler, U. S. von & Hillarp, N.-Å. (1956). Evidence for the presence of noradrenaline in submicroscopic structures of adrenergic axons. *Nature, Lond.* **177**, 44–45.

Euler, U. S. von & Lishajko, F. (1963). Effect of reserpine on the uptake of catecholamines in isolated nerve storage granules. *Int. J. Neuropharmac,* **2**, 127–134.

Euler, U. S. von & Purkhold, A. (1951). Histamine in organs and its relation to the sympathetic nerve supply. *Acta physiol. scand.* **24**, 218–224.

Euler, U. S. von & Swanbeck, G. (1964). Some morphological features of catecholamine storing nerve vesicles. *Acta physiol. scand.* **62**, 487–488.

Evans, C. Lovatt (1926). The physiology of plain muscle. *Physiol. Rev.* **6**, 358–398.

Evans, D. H. L. & Evans, E. M. (1964). The membrane relationships of smooth muscles: an electronmicroscope study. *J. Anat.* **98**, 37–46.

Evans, D. H. L. & Schild, H. O. (1953). Reactions of nerve-free and chronically denervated plain muscle to drugs. *J. Physiol.* **122**, 63P.

—— (1957a). Reactions of the isolated amnion of the chick suspended in isotonic KCl to acetylcholine and to electrical stimulation. *J. Physiol.* **136**, 36–37P.

Evans, D. H. L. & Schild H. O. (1957b). Mechanism of contraction of smooth muscle by drugs. *Nature, Lond.* **180**, 341–342.

Evans, D. H. L., Schild, H. O. & Thesleff, S. (1958). Effects of drugs on depolarized plain muscle. *J. Physiol.* **143**, 474–485.

Exley, K. A. (1960). The persistence of adrenergic nerve conduction after TM10 or bretylium in the cat. In *Adrenergic Mechanisms*. Eds. Vane, J. R., Wolstenholme, G. E. W. & O'Connor, M. pp. 158–161. London: Churchill.

Fajans, K. (1923). Struktur und Deformation der Elektronenhüllen in ihrer Bedeutung für die chemischen und optischen Eigenschaften anorganischer Verbindungen. *Naturwissenschaften* **11**, 165–172.

Falck, B. (1962). Observations on the possibilities of the cellular localization of monoamines by a fluorescence method. *Acta physiol. scand*, **56**, Suppl. 197, 1–25.

Falck, B. Häggendal, J. & Owman, Ch. (1963). The localization of adrenaline in adrenergic nerves in the frog. *Q. Jl exp. Physiol.* **48**, 253–257.

Falck, B., Hillarp, N.-Å. & Högberg, B. (1956). Content and intracellular distribution of adenosine triophosphate in cow adrenal medulla. *Acta physiol. scand.* **36**, 360–376.

Falck, B., Hillarp, N.-Å., Thieme, G. & Torp, A. (1962). Fluorescence of catecholamines and related compounds condensed with formaldehyde. *J. Histochem. Cytochem.* **10**, 348–354.

Falck, B., Mehedlishvili, G. I. & Owman, Ch. (1965). Histochemical demonstration of adrenergic nerves in cortex-pia of rabbit. *Acta pharmacol. (Kbh.)* **23**, 133–142.

Falck, B. & Owman, Ch. (1965). A detailed methodological description of the fluorescence method for the cellular demonstration of biogenic monoamines. *Acta Univ. Lund.* (*Sect. II*) 7, 1–23.

—— (1968). 5-hydroxytryptamine and related amines in endocrine cell systems. *Adv. Pharmacol.* Eds. Garattini, S. & Shore, P. A. Vol. 6A, pp. 211–231. London: Academic Press.

Falck, B. & Rorsman, H. (1963). Observation on the adrenergic innervation of the skin. *Experientia*, **19**, 205–206.

Falk, G. & Fatt, P. (1964). Linear electrical properties of striated muscle fibres observed with intracellular electrodes. *Proc. R. Soc. B.* **160**, 69–123.

Fänge, R. (1955). Use of the isolated heart of a freshwater mussel (*Anodonta cygnea* L.) for biological estimation of 5-hydroxytryptamine. *Experientia.* **11**, 156.

Farber, S. J. & Schubert, M. (1957). The binding of cations by chondroitin sulfate. *J. clin. Invest*, **36**, 1715–1722.

Farnebo, L. O. (1968). Histochemical studies on the uptake of noradrenaline in the perfused rat heart. *Br. J. Pharmac.* **34**, 227–228P.

Farquhar, M. G. & Palade, G. E. (1963). Junctional complexes in various epithelia. *J. cell Biol.* **17**, 375–412.

Farrar, J. T. (1963). Gastrointestinal smooth muscle function. *Am. J. dig. Dis.* **8**, 103–104.

Fatt, P. & Ginsborg, B. L. (1958). The ionic requirements for the production of action potentials in crustacean muscle fibres. *J. Physiol.* **142**, 516–543.

Fatt, P. & Katz, B. (1951). An analysis of the end-plate potential recorded with an intracellular electrode. *J. Physiol.* **115**, 320–370.

—— (1952). Spontaneous subthreshold activity at motor nerve endings. *J. Physiol.* **117**, 109–128.

—— (1953). The electrical properties of crustacean muscle fibres. *J. Physiol.* **120**, 171–204.

Fawcett, D. W. (1959). The fine structure of capillaries, arterioles and small arteries. In *The Microcirculation.* Eds. Reynolds, S. R. M. & Zweifach, B. W. pp. 1–27. Urbana: University of Illinois Press.

Feeney, L. & Hogan, M. J. (1961). Electron microscopy of the human choroid. II. The choroidal nerves. *Am. J. Opthal.* **51**, 1072–1083.

Feldberg, W. (1933). Der Nachweis eines acetylcholinähnlichen Stoffes im Zungenvenenblut des Hundes bei Reizung des Nervus lingualis. *Pflügers Arch. ges. Physiol.* **232**, 88–104.

Feldberg, W. & Lin, R. C. Y. (1949a). The action of local anaesthetics and d-tubocurarine on the isolated intestine of the rabbit and guinea-pig. *Br. J. Pharmac. Chemother.* **4**, 33–44.

Feldberg, W. & Lin, R. C. Y. (1949b). The effect of cocaine on the acetylcholine output of the intestinal wall. *J. Physiol.* **109**, 475–487.

—— (1950). Synthesis of acetylcholine in the wall of the digestive tract. *J. Physiol.* **111**, 96–118.

Feldberg, W. & Mann, T. (1945). Formation of acetylcholine in cell-free extracts from the brain. *J. Physiol.* **104**, 8–20.

—— (1946). Properties and distribution of the enzyme system which synthesizes acetylcholine in nervous tissue. *J. Physiol.* **104**, 411–425.

Feldberg, W., Minz, B. & Tsudzimura, H. (1934). The mechanism of the nervous discharge of adrenaline. *J. Physiol.* **81**, 286–304.

Feldberg, W. & Toh, C. C. (1953). Distribution of 5-hydroxytryptamine (serotonin, enteramine) in the wall of the digestive tract. *J. Physiol.* **119**, 352–362.

Feltham, P. (1961). Creep and stress-relaxation in crystalline solids at low temperature. Svenska nationalkommittén för mekanik. Reologisektionen, Stockholm, 2–14.

Fenn, W. O. (1924). The relation between the work performed and the energy liberated in muscular contraction. *J. Physiol.* **58**, 373–395.

—— (1957). Some elasticity problems in the human body. In *Tissue Elasticity*. Ed. by J. W. Remington pp. 98–101. Amer. Physiol. Soc.: Washington, D.C.

Fernandez-Moran, H. (1958). Fine structure of the light receptors in the compound eye of insects. *Expl. Cell. Res. Suppl.* **5**, 586–643.

Ferreira, S. H. & Vane, J. R. (1967). Prostaglandins: their disappearance from and release into the circulation. *Nature, Lond.* **216**, 868–873.

Ferry, C. B. (1963). The sympathomimetic effect of acetylcholine on the spleen of the cat. *J. Physiol.* **167**, 487–504.

—— (1966). Cholinergic link hypothesis in adrenergic neuroeffector transmission. *Physiol. Rev.* **46**, 420–456.

—— (1967a). The innervation of the vas deferens of the guinea-pig. *J. Physiol.* **192**, 463–478.

—— (1967b). The autonomic nervous system. *Ann. Rev. Pharmacol.* **7**, 185–202.

Fillenz, M. (1966). Innervation of blood vessels of lung and spleen. *Biblphie. anat.* **8**, 56–59.

Filo, R. S., Rüegg, J. C. & Bohr, D. F. (1963). Acto-myosin-like protein of arterial wall. *Am. J. Physiol.* **205**, 1247–1252.

Finkleman, B. (1930). On the nature of inhibition in the intestine. *J. Physiol.* **70**, 145–157.

Fischer, J. E., Weise, V. K. & Kopin, I. J. (1966). Interactions of bretylium and acetylcholine at sympathetic nerve endings. *J. Pharmac. exp. Ther.* **153**, 523–529.

Fischer, G. M. & Llaurado, J. G. (1966). Collagen and elastin content in canine arteries selected from functionally different vascular beds. *Circulation Res.* **19**, 394–399.

Fischman, D. A. & Gershon, M. D. (1964). A method of studying intracellular movement of water-soluble isotopes prior to radioautography. *J. cell Biol.* **21**, 139–143.

Fisher, R. B. & Lindsay, D. B. (1956). The action of insulin on the penetration of sugars into the perfused rat heart. *J. Physiol.* **131**, 526–541.

Fisher, R. B. & Young, D. A. B. (1961). Direct determination of extracellular fluid in the rat heart. *J. Physiol.* **158**, 50–58.

Fleischmann, P. (1910). Atropinentgiftung durch Blut. *Arch. exp. Path. Pharmak.* **62**, 518–526.

Florey, E. & Florey, E. (1954). Über die mögliche Bedeutung von Enteramin (5-Oxy-Tryptamin) als nervöse Aktionssubstanz bei Cephalopoden und decapoden Crustaceen. *Z. Naturf.* **9b**, 58–68.

Folkow, B. (1952). Impulse frequency in sympathetic vasomotor fibres correlated to the release and elimination of the transmitter. *Acta physiol. scand.* **25**, 49–76.

—— (1955). Nervous control of the blood vessels. *Physiol. Rev.* **35**, 629–663.

—— (1964a). Autoregulation in muscle and skin. *Circulation Res.* **15**, Suppl. 1, 19–29.

—— (1964b). Description of the myogenic hypothesis *Circulation Res.* **15**, Suppl. 1, 279–285.

Folkow, B., Frost, J., Haeger, K. & Uvnäs, B. (1949). The sympathetic vasomotor innervation of the skin of the dog. *Acta. physiol. scand.* **17**, 195–200.

Folkow, B., Haeger, K. & Uvnäs, B. (1948). Cholinergic vasodilator nerves in the sympathetic outflow to the muscles of the hind limbs of the cat. *Acta physiol. scand.* **15**, 401–411.

Folkow, B., Häggendal, J. & Lisander, B. (1967). Extent of release and elimination of nor-adrenaline at peripheral adrenergic nerve terminals. *Acta physiol. scand.* Suppl. 307.

Folkow, B., Lewis, D. H., Lundgren, O., Mellander, S. & Wallentin, I. (1964). The effect of graded vasoconstrictor fibre stimulation on the intestinal resistance and capacitance vessels. *Acta physiol. scand.* **61**, 445–457.

Folkow, B. Öberg, B. & Rubinstein, E. H. (1964). A proposed differentiated neuro-effector organization in muscle resistance vessels. *Angiologica*, **1**, 197–208.

Folkow, B. & Uvnäs, B. (1948a). The chemical transmission of vasoconstrictor impulses to the hind limbs and the splanchnic region of the cat. *Acta physiol. scand.* **15**, 365–388.

—— (1948b). The distribution and functional significance of sympathetic vasodilators to the hind limbs of the cat. *Acta physiol. scand.* **15**, 389–400.

—— (1949). The chemical transmission of vasoconstrictor impulses to the hind limbs of the dog. *Acta physiol. scand.* **17**, 191–194.

Forbes, H. S. & Cobb, S. (1938a). Vasomotor control of cerebral vessels. *Res. Publs Ass. Res. nerv. ment. Dis.* **18**, 201–217.

—— (1938b). Vasomotor control of cerebral vessels. *Brain*, **61**, 221–233.

Forel, A. (1887). Einige hirnanatomische Betrachtungen und Ergebnisse. *Arch. Psychiat. Nervenkr.* **18**, 162–198.

Fozzard, H. (1964). Membrane capacitance of the cardiac Purkinje fibre. *J. Physiol.* **175**, 47–48P.

—— (1966). Membrane capacity of the cardiac Purkinje fibre. *J. Physiol.* **182**, 255–267.

Fraenkel, A. (1926). *Arch. VerdauKrankh.* **37**, 408.

Frank, G. B. (1960). Effects of changes in extracellular calcium concentration on the potassium induced contracture of frog's skeletal muscle. *J. Physiol.* **151**, 518–538.

Frankenhaeuser, B. & Hodgkin, A. L. (1957). The action of calcium on the electrical properties of squid axons. *J. Physiol.* **137**, 218–244.

Franklin, K. J. (1937). *A Monograph on Veins.* London: Thomas.

Freeman-Narrod, M. & Goodford, P. J. (1962). Sodium and potassium content of the smooth muscle of the guinea-pig taenia coli at different temperatures and tensions. *J. Physiol.* **163**, 399–410.

Frewin, D. B. & Whelan, R. F. (1968). The mechanism of action of tyramine on the blood vessels of the forearm in man. *Br. J. Pharmac Chemother.* **33**, 105–116.

Freygang, W. H., Rapoport, S. I. & Peachey, L. D. (1967). Some relations between changes in the linear electrical properties of striated muscle fibers and changes in ultrastructure. *J. gen. Physiol.* **50**, 2437–2458.

Friedman, S. M. & Friedman, C. L. (1967). The ionic matrix of vasoconstriction. *Circulation Res.* **21**, Suppl. 2, 147–155.

Friedman, S. M., Gustafson, B. & Friedman, C. L. (1968). Characteristics of temperature-dependent sodium exchanges in a small artery. *Canad. J. Physiol. Pharmac.* **46**, 681–685.

Friedman, S. M., Gustafson, B., Hamilton, D. & Friedman, C. L. (1968). Compartments of sodium in a small artery. *Canad. J. Physiol. Pharmac.* **46**, 673–679.

Friesen, A. J. D., Kemp, J. W. & Woodbury, D. M. (1964). Identification of acetylcholine in sympathetic ganglia by chemical and physical methods. *Science, N.Y.* **145**, 157–159.

Fritz, O. G. Jr. & Swift, T. J. (1967). The state of water in polarized and depolarized frog nerves, a proton magnetic resonance study. *Biophys. J.* **7**, 675–687.

Frumento, A. S. (1965). The electrical effects of an ionic pump. *J. theor. Biol.* **9**, 253–262.

Fujino, S. & Fujino, M. (1964). Removal of the inhibitory effect of hypertonic solutions on the contractility in muscle cells and the excitation-contraction link. *Nature, Lond.* **201**, 1331–1333.

Fujita, H., Machino, M., Nakagami, K., Imai, Y. & Yamamoto, Y. (1964). Electronmicro-scopic studies on the submandibular glands of normal and parasympathicus-stimulated dogs. *Acta anat. Nippon.* **39**, 269–293.

Fülgraff, G. & Schmidt, L. (1963). Die Wirkung elektrischer Reizung sympathischer und parasympathischer Nerven auf das proximale und distale Colon und ihre pharmako-logische Beeinflussbarkeit. *Arch. exp. Path. Pharmak.* **245**, 106–107.

Funaki, S. (1958). Studies on membrane potentials of vascular smooth muscle with intra-cellular micro-electrodes. *Proc. Japan Acad.* **34**, 534–536.

Funaki, S. (1960). Electrical activity of single vascular smooth muscle fibers. In *Electrical Activity of Single Cells*. Ed. Y. Katsuki pp. 233–241. Igaku Shoin, Tokyo.

—— (1961). Spontaneous spike-discharges of vascular smooth muscle. *Nature, Lond.***191**, 1102–1103.

—— (1966). Electrical and mechanical activity of isolated smooth muscle from the portal vein of the rat. *Biblphie. anat.* **8**, 5–10.

Funaki, S. & Bohr, D. F. (1964). Electrical and mechanical activity of isolated vascular smooth muscle of the rat. *Nature, Lond.* **203**, 192–194.

Furchgott, R. F. (1954). Dibenamine blockade in strips of rabbit aorta and its use in differentiating receptors. *J. Pharmac. exp. Ther.* **111**, 265–284.

—— (1955). The pharmacology of vascular smooth muscle. *Pharmac. Rev.* **7**, 183–265.

—— (1959). The receptors for epinephrine and norepinephrine (Adenergic receptors.) *Pharmac. Rev.* **11**, 429–442.

—— (1960). Receptors for sympathomimetic amines. In *Ciba Foundation Symposium on Adrenergic Mechanisms*. Ed. J. R. Vane, Wolstenholme, G. E. W. & O'Connor, M. pp. 246–252. London: Churchill.

—— (1967). The pharmacological differentiation of adrenergic receptors. *Ann. N.Y. Acad. Sci.* **139**, 553–570.

Furchgott, R. F., Kirpekar, S. M., Rieker, M. & Schwab, A. (1963). Actions and interactions of norepinephrine, tyramine and cocaine on aortic strips of rabbit and left atria of guinea-pig and cat. *J. Pharmac. exp. Ther.* **142**, 39–58.

Furness, J. and Burnstock, G. (1969). A comparative study of spike potentials in response to nerve stimulation of the vas deferens of the mouse, rat and guinea-pig. *Comp. Biochem. Physiol.* (in press).

Furness, J. B., McLean, J. R. and Burnstock, G. (1969). Distribution of adrenergic nerves and neuro-muscular transmission in the mouse vas deferens during post-natal development. *Develop. Biol.* (in press).

Furshpan, E. J. (1964). Electrical transmission at an excitatory synapse in a vertebrate brain. *Science, N.Y.* **144**, 878–880.

Furshpan, E. J. & Potter, D. D. (1957). Mechanism of nerve-impulse transmission at a crayfish synapse. *Nature, Lond.* **180**, 342–343.

—— (1959). Transmission at the giant motor synapses of the crayfish. *J. Physiol.* **145**, 289–325.

Fusari, R. (1891). De la terminaison des fibres nerveuses dans les capsules surrénales des mammifères. *Archs ital. Biol.* **16**, 262–275.

Fuxe, K., Höckfelt, T. & Ungerstedt, U. (1968). Localization of indolealkylamines in central nervous System. *Adv. Pharmacol.* Ed. Garattini, S. & Shore, P. A. Vol. 6A, pp. 235–251. London: Academic Press.

Fuxe, K. & Sedvall, G. (1964). Histochemical and biochemical observations on the effect of reserpine on noradrenaline storage in vasoconstrictor nerves. *Acta physiol. scand.* **61**, 121–129.

—— (1965). The distribution of adrenergic nerve fibres to the blood vessels in skeletal muscle. *Acta physiol. scand.* **64**, 75–86.

Gabella, G. (1967). Fibre nervose adrenergiche nello strato musculare circolare dell intestino tenue di ratto. *Bull. Soc. ital. Biol. Sper.* **XLIII**, fasc. 20.

Gabella, G. & Costa, M. (1967). Le fibre adrenergiche nel canale alimentare. *G. Accad. Med. Torino* **CXXX**, fasc. 1–6.

Gaddum, J. H. (1940). *Pharmacology*. Oxford University Press.

—— (1953). Tryptamine receptors. *J. Physiol.* **119**, 363–368.

Gaddum, J. H. & Goodwin, L. G. (1947). Experiments on liver sympathin. *J. Physiol.* **105**, 357–369.

Gaddum, J. H. & Hameed, K. A. (1954). Drugs which antagonise 5-hydroxytryptamine. *Br. J. Pharmac. Chemother.* **9**, 240–248.

Gaddum, J. H., Hameed, K. A., Hathway, D. E. & Stephens, F. F. (1955). Quantitative studies of antagonists for 5-hydroxytryptamine. *Q. Jl exp. Physiol.* **40**, 49–74.

Gaddum, J. H., Hebb, C. O., Silver, A. & Swann, A. A. B. (1953). 5-Hydroxytryptamine.

Pharmacological action and destruction in perfused lungs. *Q. Jl exp. Physiol.* **38**, 255–262.

Gaddum, J. H. Jang, C. S. & Kwiatkowski, H. (1939). The effect on the intestine of the substance liberated by adrenergic nerves in a rabbit's ear. *J. Physiol.* **96**, 104–108.

Gaddum, J. H. & Kwiatkowski, H. (1939). Properties of the substance liberated by adrenergic nerves in the rabbit's ear. *J. Physiol.* **96**, 385–391.

Gaddum, J. H. & Paasonen, M. K. (1955). The use of some molluscan hearts for the estimation of 5-hydroxytryptamine. *Br. J. Pharmac. Chemother.* **10**, 474–483.

Gaddum, J. H. & Picarelli, Z. P. (1957). Two kinds of tryptamine receptor. *Br. J. Pharmac. Chemother.* **12**, 323–328.

Galehr, O. & Plattner, F. (1928a). Über das Schicksal des Acetylcholins im Blute. I. Mitteilung. *Pflügers Arch. ges. Physiol.* **218**, 488–505.

—— (1928b). Über das Schicksal des Acetylcholins im Blute. II. Mitteilung. Seine Zerstörung im Blute verschiedener Säugetiere. *Pflügers Arch. ges. Physiol.* **218**, 506–513.

Gansler, H. (1956). Electronenmikroskopische Untersuchungen am Uterusmuskel der Ratte unter Follikelhormonwirkung. *Virchows Arch. path. Anat. Physiol.* **329**, 235–244.

—— (1960). Phasenkontrast-und electronenmikroskopische Untersuchungen zur Morphologie und Funktion der glatten Muskulatur. *Z. Zellforsch. mikrosk. Anat.* **52**, 60–92.

—— (1961). Struktur und Funktion der glatten Muskulatur. II. Licht-und electronenmikroskopische Befunde an Hohlorganen von Ratte, Meerschweinchen und Mensch. *Z. Zellforsch. mikrosk. Anat.* **55**, 724–762.

Garrahan, P., Villamil, M. F. & Zadunaisky, J. A. (1965). Sodium exchange and distribution in the arterial wall. *Am. J. Physiol.* **209**, 955–960.

Garrett, W. J. (1954). The effects of adrenaline and noradrenaline on the intact human uterus in late pregnancy and labour. *J. Obst. Gynaec.* **61**, 586–589.

—— (1955). The effects of adrenaline, noradrenaline and dihydroergotamine on excised human myometrium. *Br. J. Pharmac. Chemother,* **10**, 39–44.

Garry, R. C. & Gillespie, J. S. (1955). The responses of the musculature of the colon of the rabbit to stimulation, *in vitro*, of the parasympathetic and of the sympathetic outflows. *J. Physiol.* **128**, 557–576.

Gaskell, W. H. (1886). On the structure, distribution and function of the nerves which innervate the visceral and vascular systems. *J. Physiol* **7**, 1–80.

—— (1889). On the relation between the structure, function, distribution and origin of the cranial nerves; together with a theory of the origin of the nervous system of Vertebrata. *J. Physiol.* **10**, 153–211.

Gasser, H. (1952). Discussion of saltatory conduction hypothesis. *Cold. Spring Harb. Symp. quant. Biol.* **17**, 32–36.

—— (1958). Comparison of the structure, as revealed with the electronmicroscope, and the physiology of the unmedullated fibres in the skin nerves and in the olfactory nerves. *Expl. Cell Res. Suppl.* **5**, 3–17.

Gauer, O. H. & Henry, J. P. (1963). Circulatory basis of fluid volume control. *Physiol. Rev.* **43**, 423–481.

Geffen, L. B. (1967). Noradrenaline storage, release and inactivation in sympathetic nerves. *Circulation Res.* **21**, Suppl. 3, III-57–III-62.

Geffen, L. B. & Livett, B. G. (1968). Axoplasmic transport of C^{14} noradrenaline (NA) and protein and their release by nerve impulses. *Proc. XXIV. Int. Cong. Physiol. Sci.* **VII**, 454.

Geffen, L. B. & Rush, R. A. (1968). Transport of noradrenaline in sympathetic nerves and the effect of nerve impulses on its contribution to transmitter stores. *J. Neurochem.* **15**, 925–930.

Génis-Galvez, J. M. & Clements, C. O. (1957). Analisis comparativo de la inervacion del musculo liso de la membrana nictitante y del musculo cilar. Estudio experimental tras la extirpacion del ganglio sympático cervical superior. *Arch. esp. Morfol.* **13**, 159–179.

George, E. P. (1961). Resistance values in a syncytium. *Aust. J. exp. Biol. med. Sci.* **39**, 267–274.

Gerlach, J. von (1872). The spinal cord. In *A Manual of Human and Comparative Histology.*

21+

(English translation by Henry Power). Vol. *II*, Ed. Stricker, S. pp. 327–366. London: New Sydenham Society.

Gerova, M., Gero, J. & Dolezel, S. (1967). Mechanisms of sympathetic regulation of arterial smooth muscle. *Experientia.* **23**, 639–640.

Gerschenfeld, H. M. & Stefani, E. (1966). An electrophysiological study of 5-hydroxytryptamine receptors of neurones in the molluscan nervous system. *J. Physiol.* **185**, 684–700.

Gershon, M. D. (1967a). Effects of tetrodotoxin on innervated smooth muscle preparations. *Br. J. Pharmac. Chemother.* **29**, 259–279.

—— (1967b). Inhibition of gastrointestinal movement by sympathetic nerve stimulation: The site of action. *J. Physiol.* **189**, 317–327.

Gershon, M. D., Drakontides, A. B. & Ross, L. L. (1965). Serotonin: synthesis and release from the myenteric plexus of the mouse intestine. *Science.* **149**, 197–199.

Gershon, M. D. & Ross, L. L. (1966a). Radioisotopic studies of the binding, exchange and distribution of 5-hydroxytryptamine synthesised from its radioactive precursor. *J. Physiol.* **186**, 451–476.

—— (1966b). Location of sites of 5-hydroxytryptamine storage and metabolism by radioautography. *J. Physiol.* **186**, 477–492.

Gibbs, O. S. & Szelöczey, J. (1932). Die humorale Übertragung der Chorda tympani-Reizung. II. Mitteilung. *Arch. exp. Path. Pharmak.* **168**, 64–88.

Gielen, W. (1966). Über die Funktion von Gangliosiden, ein Serotonin-und Ca^{2+}-Receptor. *Z. Naturf.* **21b**, 1007–1008.

Gillespie, J. S. (1962a). Spontaneous mechanical and electrical activity of stretched and unstretched intestinal smooth muscle cells and their response to sympathetic nerve stimulation. *J. Physiol.* **162**, 54–75.

—— (1962b). The electrical and mechanical responses of intestinal smooth muscle cells to stimulation of their extrinsic parasympathetic nerves. *J. Physiol.* **162**, 76–92.

—— (1964). Cholinergic junction potentials in intestinal smooth muscle. In *Pharmacology of Smooth Muscle.* Ed. Bülbring, E. pp. 81–85. Oxford: Pergamon Press.

—— (1966). Review Lecture. Tissue binding of noradrenaline. *Proc. R. Soc. B.* **166**, 1–10.

—— (1968). Electrical activity in the colon. In *Handbook of Physiology, Alimentary Canal IV.* pp. 2093–2120, Amer. Physiol. Soc., Washington, D.C.

Gillespie, J. S. & Kirpekar, S. M. (1965). The inactivation of infused noradrenaline by the cat spleen. *J. Physiol.* **176**, 205–227.

—— (1966a). The histological localization of noradrenaline in the cat spleen. *J. Physiol.* **187**, 69–79.

—— (1966b). The uptake and release of radioactive noradrenaline by the splenic nerves of cats. *J. Physiol.* **187**, 51–68.

Gillespie, J. S. & Mackenna, B. R. (1961). The inhibitory action of the sympathetic nerves on the smooth muscle of the rabbit gut, its reversal by reserpine, and restoration by catecholamines and by dopa. *J. Physiol.* **156**, 17–34.

Gillis, C. N. (1964). The retention of exogenous norepinephrine by rabbit tissues. *Biochem. Pharmac.* **13**, 1–12.

Gillis, C. N., Schneider, F. H., Van Orden, L. S. & Giarman, N. J. (1966). Biochemical and microfluorometric studies of norepinephrine redistribution accompanying sympathetic nerve stimulation. *J. Pharmac. exp. Ther.* **151**, 46–54.

Ginsborg, B. L. (1967). Ion movements in functional transmission. *Pharmac. Rev.* **19**, 289–316.

Ginzel, K. H. (1957). The action of lysergic acid diethylamide (LSD 25), its 2-brom derivative (BOL 148) and of 5-hydroxytryptamine (5-HT) on the peristaltic reflex of the guinea-pig ileum. *J. Physiol.* **137**, 62–63P.

—— (1959). Investigations concerning the initiation of the peristaltic reflex in the guinea-pig ileum. *J. Physiol.* **148**, 75P.

Giradier, L., Reuben, J. P., Brandt, P. W. & Grundfest, H. (1963). Evidence for anion-permselective membrane in crayfish muscle fibres and its possible role in excitation-contraction coupling. *J. gen. Physiol.* **47**, 189–214.

Glick, D. & Glaubach, S. (1941). The occurrence and distribution of atropinesterase and the specificity of atropinesterases. *J. gen. Physiol.* **25**, 197–205.

Glick, G., Epstein, S. E., Wechsler, A. S. & Braunwald, E. (1967). Physiological differences between the effects of neuronally released and blood borne norepinephrine on beta adrenergic receptors in the arterial bed of the dog. *Circulation Res.* **21**, 217–227.

Glick, G., Wechsler, A. S. & Epstein, S. E. (1968). Mechanism of reflex vasodilatation: Assessment of the role of neural reuptake of norepinephrine and release of histamine. *J. clin. Invest.* **47**, 511–520.

Goldman, D. E. (1943). Potential, impedance and rectification in membrane. *J. gen. Physiol.* **27**, 37–60.

Golenhofen, K. (1962a). Physiologie des menschlichen Muskelkreislaufes. *Marb. Sitzungsber.* **83/84**, 167–254.

—— (1962b). Zur Reaktionsdynamik der menschlichen Muskelstrohmbahn. *Arch. Kreisl.-Forsch.* **38**, 202–223.

—— (1964). 'Resonance' in the tension response of smooth muscle of guinea-pig's taenia coli to rhythmic stretch. *J. Physiol.* **173**, 13–15P.

—— (1965). Rhythmische Dehnung der glatten Muskulatur vom Blinddarm des Meerschweinchens. *Pflügers Arch. ges. Physiol.* **284**, 327–346.

—— (1966a). Untersuchungen zur Minuten-Rhythmik der glatten Muskulatur an der isolierten Taenia coli des Meerschweinchens. *Pflügers Arch. ges. Physiol.* **292**, 34–45.

—— (1966b). Physiologische Aspekte zur Soziosomatik des Kreislaufs. *Verh. dtsch. Ges. Kreisl.-Forsch.* **32**, 23–37.

Golenhofen, K., Blair, D. A. & Seidel, W. (1961). Zur Natur affektiver Muskeldurchblutungssteigerungen beim Menschen. *Pflügers Arch. ges. Physiol.* **272**, 223–236.

Golenhofen, K. & Hildebrandt, G. (1957a). Über spontan-rhythmische Schwankungen der Muskeldurchblutung des Menschen. *Z. Kreislaufforsch.* **46**, 257–270.

—— (1957b). Zur Ursache spontaner Muskeldurchblutungsschwankungen im I-Minuten-Rhythmus. *Verh. dtsch. Ges. Kreisl.-Forsch.* **23**, 380–385.

—— (1957c). Psychische Einflüsse auf die Muskeldurchblutung. *Pflügers Arch. ges. Physiol.* 637–646.

—— (1958). Die Beziehungen des Blutdruckrhythmus zu Atmung und peripherer Durchblutung. *Pflügers. Arch. ges. Physiol.* **267**, 27–45.

Golenhofen, K. & v. Loh, D. (1966). Temperatureinflüsse auf die Spontanaktivität von isolierter glatter Muskulatur (Taenia coli des Meerschweinchens). *Pflügers Arch. ges. Physiol.* **291**, R64-R65.

Golgi, C. (1885). *Sulla fina Anatomia dégli Organi Centrali del Sistema Nervoso.* Milano: V. Hoepli.

—— (1890). Über den feineren Bau des Rückenmarkes. *Anat. Anz.* **5**, 372–396.

—— (1891). Le réseau nerveux diffus des centres du système nerveux. Ses attributs physiologiques. Méthode suivie dans les recherches histologiques. *Arch ital. Biol.* **15**, 434–463.

Gomori, G. (1948). Histochemical demonstration of sites of choline esterase activity. *Proc. Soc. exp. Biol. Med.* **68**, 354–358.

Gonella, T. (1965). Variation de l'activité électrique spontanée du duodénum de Lapin avec le lieu de dérivation. *C. r. Acad. Sci.* (*Paris*) **260**, (May) 5362–5365.

Goodall, F. R. (1965). Degradative enzymes in the uterine myometrium of rabbits under different homonal conditions. *Arch. Biochem. Biophys.* **112**, 403–410.

—— (1966). Progesterone retards postpartum involution of the rabbit myometrium. *Science*, **152**, 356-358.

Goodall, M. C. (1951). Studies of adrenaline and noradrenaline in mammalian heart and suprarenals. *Acta physiol. scand.* **24**, Suppl. 85, 1–51.

Goodall, M. C. & Kirshner, N. (1958). Biosynthesis of epinephrine and norepinephrine by sympathetic nerves and ganglia. *Circulation* **17**, 366–371.

Goodford, P. J. (1962). The sodium content of the smooth muscle of the guinea-pig taenia coli. *J. Physiol.* **163**, 411–422.

—— (1964). Chloride content and ^{36}Cl uptake in the smooth muscle of the guinea-pig taenia coli. *J. Physiol.* **170**, 227–237.

—— (1965). The loss of radioactive ^{45}calcium from the smooth muscle of the guinea-pig taenia coli. *J. Physiol.* **176**, 180–190.

Goodford, P. J. (1966). An interaction between potassium and sodium in the smooth muscle of the guinea-pig taenia coli. *J. Physiol.* **186**, 11–26.

—— (1967). The calcium content of the smooth muscle of the guinea-pig taenia coli. *J. Physiol.* **192**, 145–157.

—— (1968). The distribution and exchange of electrolytes in intestinal smooth muscle. *Handbook of Physiology—Alimentary canal IV.* pp. 1743–1766. Am. Physiol. Soc., Washington, D.C.

Goodford, P. J. & Hermansen, K. (1961). Sodium and potassium movements in the unstriated muscle of the guinea-pig taenia coli. *J. Physiol.* **158**, 426–448.

Goodford, P. J., Johnson, F. R., Krasucki, Z. & Daniel, V. (1967). The transport of sodium in smooth muscle cells. *J. Physiol.* **194**, 77–78P.

Goodford, P. J. & Leach, E. H. (1966). The extracellular space of the smooth muscle of the guinea-pig taenia coli. *J. Physiol.* **186**, 1–10.

Goto, M. & Csapo, A. (1959). The effect of the ovarian steroids on the membrane potential of uterine muscle. *J. gen. Physiol.* **43**, 455–466.

Goto, M., Kuriyama, H. & Abe, Y. (1961). Refractory period and conduction of excitation in the uterine muscle cells of the mouse. *Jap. J. Physiol.* **11**, 369–377.

Goto, M. & Woodbury, J. W. (1958). Effects of stretch and NaCl on transmembrane potentials and tension of pregnant rat uterus. *Fedn. Proc.* **17**, 58.

Graf, K., Graf, W. & Rosell, S. (1958). Spontan-rhythmische und unregelmässige Schwankungen der Leberdurchblutung des Menschen. *Acta physiol. scand.* **43**, 233–253.

—— (1959). Zusammenhänge der Durchblutungsrhythmik in Haut-, Muskel- und Intestinalstrombahn des Menschen. *Pflügers Arch. ges. Physiol.* **270**, 43.

Graham, J. D. P. & Gurd, M. R. (1960). Effects of adrenaline on the isolated uterus of the cat. *J. Physiol.* **152**, 243–249.

Graham, J. D. P., Lever, J. D. & Spriggs, T. L. B. (1968). An examination of adrenergic axons around pancreatic arterioles of the cat for the presence of acetylcholinesterase by high resolution autoradiographic and histochemical methods. *Br. J. Pharmac. Chemother.* **33**, 15–20.

Gray, E. G. & Whittaker, V. P. (1962). The isolation of nerve endings from brain: an electron-microscope study of cell fragments derived by homogenization and centrifugation. *J. Anat.* **96**, 79–88.

Gray, J. A. B. (1959). Initiation of impulses at receptors. In *Handbook of Physiology*, Sect. I, Vol. I, Neurophysiology, Ed. Field, J. & Magoun, H. W. pp. 123–145. Washington, D.C.: American Physiological Society.

Greeff, K., & Holtz, P. (1951). Über die Uteruswirkung des Adrenalins und Arterenols. Ein Beitrag zum Problem der Uterusinnervation. *Arch. int. Pharmacodyn. Thér.* **88**, 228–252.

—— (1956). Untersuchungen am isolierten Vagus-Magenpräparat. *Naunyn Schmeidebergs Ach. exp. Path. Pharmak.* **227**, 427–435.

Greeff, K., Kasperat, H. & Osswald, W. (1962). Paradoxe Wirkungen der elektrischen Vagusreizung am isolierten Magen- und Herzvorhofpräparat des Meerschweinchens sowie deren Beeinflussung durch Ganglienblocker, Sympathicolytica, Reserpin und Cocain. *Arch. exp. Path. Pharmak*, **243**, 528–545.

Greenawalt, J. W. & Carafoli, E. (1966). Electron microscope studies on the active accumulation of Sr^{++} by rat-liver mitochondria. *J. cell. Biol.* **29**, 37–62.

Greenawalt, J. W., Rossi, C. S. & Lehninger, A. L. (1964). Effect of active accumulation of calcium and phosphate ions on the structure of rat liver mitochondria. *J. cell Biol.* **23**, 21–38.

Greenberg, M. J. (1960). Structure-activity relationships of tryptamine analogues on the heart of *Venus Mercenaria*. *Br. J. Pharmac. Chemother.* **15**, 375–388.

Greer, C. M., Pinkston, J. O., Baxter, J. H. & Brannon, E. S. (1937). Comparison of responses of smooth muscle to dl-β-(3,4-dihydroxyphenyl)-β-hydroxyethylamine (arterenol) 1-epinephrine and "liver sympathin". *J. Pharmac. exp. Ther.* **60**, 108–109.

—— (1938). Norepinephrine (β-(3,4-dihydroxyphenyl)-β-hydroxyethylamine) as a possible mediator in the sympathetic division of the autonomic nervous system. *J. Pharmac. exp. Ther.* **62**, 189–227.

Greven, K. (1954). Über den Mechanismus der Regulierung der Kontraktionsstärke beim glatten Muskel durch tetanische und quantitative (räumliche) Summation. *Z. Biol.* **100**, 377–385.

Grette, K. (1957). Determination of 5-hydroxytryptamine with the isolated, perfused pig ear. *Acta pharmac. tox.* **13**, 177–183.

Griesemer, E. C., Barsky, J., Dragstedt, C. A., Wells, J. A. & Zeller, E. A. (1953). Potentiating effect of iproniazid on the pharmacological action of sympathomimetic amines. *Proc. Soc. exp. Biol. (N. Y.)* **84**, 699–701.

Grigor'eva, T. A. (1962). *The Innervation of Blood Vessels.* New York: Pergamon Press.

Grillo, M. A. (1966). Electron microscopy of sympathetic tissues. *Pharmac. Rev.* **18**, 387–399.

Grillo, M. A. & Palay, S. L. (1962). Granule-containing vesicles in the autonomic nervous system. In Proc. 5th Int. Congr. *Electron Microscopy.* Vol. 2, U-1. Ed. Breese, S. S. New York: Academic Press.

Gryglewski, R. & Supniewski, J. (1963). Influence of 5-hydroxytryptamine and other biologically active substances on the movements of the isolated stomach of *Helix pomatia. Bull. Acad. pol. Sci. Cl. 2* **11**, 53–56.

Gudmundsson, G. (1966). Interpretation of one-dimensional magnetic anomalies by use of the Fourier-transform. *Geophys. J. R. astr. Soc.* **12**, 87–97.

—— (1968). Fourier analysis of the electrical and mechanical activity in the rat portal vein. Staff paper, Dep. of Maths., Univ. of Manchester, Inst. of Sc. and Techn.

Gunn, J. A. (1944). The action of adrenaline and of choline esters on the uterus of the sheep. *J. Physiol.* **103**, 290–296.

Gunn, M. (1951). A study of enteric plexuses in some amphibians. *Q. Jl microsc. Sci.* **92**, 65–78.

Hagen, P. & Barrnett, R. J. (1960). The storage of amines in the chromaffin cell. In *Adrenergic Mechanisms.* Ed. Vane, J. R., Wolstenholme, G. & O'Connor, M., pp. 83–99. Ciba. Found. Symp. London: Churchill.

Hagemeijer, F., Rorive, G. & Schoffeniels, E. (1965). The ionic composition of rat aortic smooth muscle fibres. *Arch. int. Physiol. Biochem.* **73**, 453–475.

Hager, H. & Tafuri, W. L. (1959). Electronenoptischer Nachweis sog. neurosekretorischer Elementargranula in marklosen Nervenfasern des Plexus myentericus (Auerbach) des Meerschweinchens. *Naturwissenschaften,* **46**, 332–333.

Hägqvist, G. (1956). Gewebe und Systeme der Muskulatur. In *Handbuch der Mikroskopschen Anatomie des Menschen.* II. Ed. Möllendorff, W. & Bargmann, W. Berlin: Springer-Verlag.

Hagiwara, S. & Morita, H. (1962). Electrotonic transmission between two nerve cells in leech ganglion. *J. Neurophysiol.* **25**, 721–731.

Hagiwara, S. & Naka, K. (1964). The initiation of spike potentials in barnacle muscle fibres under low intracellular Ca^{++}. *J. gen. Physiol.* **48**, 141–162,

Hagiwara, S. & Nakajima, S. (1966). Differences in Na and Ca spikes as examined by application of tetrodotoxin, procaine and manganese ions. *J. gen. Physiol.* **49**, 793–806.

Hagiwara, S. & Takahashi, K. (1967). Resting and spike potentials of skeletal muscle fibres of salt-water elasmobranch and teleost fish. *J. Physiol.* **190**, 499–518.

Hamberger, B. (1967). Reserpine-resistant uptake of catecholamines in isolated tissues of the rat. *Acta physiol. scand.* **70**, Suppl. 295, 1–56.

Hamberger, B. & Norberg, K.-A. (1965). Studies on some systems of adrenergic synaptic terminals in the abdominal ganglia of the cat. *Acta physiol. scand.* **65**, 235–242.

Hammarström, L., Ritzén, M. & Ullberg, S. (1966). Combined autoradiography and fluorescence microscopy. Localization of labelled 5-hydroxytryptophan in relation to endogenous 5-hydroxytryptamine in the gastrointestinal tract. *Experientia* **22**, 213–215.

Hampton, J. C. (1960). An electronmicroscopic study of mouse colon. *Diseases of Colon and Rectum,* **3**, 423–440.

Handschumacher, R. E. & Vane, J. R. (1967). The relationship between the penetration of tryptamine and 5-hydroxytryptamine into smooth muscle and the associated contractions. *Br. J. Pharmac. Chemother.* **29**, 105–118.

Hanson, J. & Lowy, J. (1963). The structure of F. Actin and of Actin filaments isolated from muscle. *J. molec. Biol.* **6**, 46–60.

21*

Hanson, J. & Lowy, J. (1964a). Comparative studies on the structure of contractile systems. *Circulation Res.* **15** (suppl. 2), 4–13.

—— (1964b). The structure of actin filaments and the origin of the axial periodicity in the I-substance of vertebrate striated muscle. *Proc. R. Soc. B.* **160**, 449–460.

Harman, J. W., O'Hegarty, M. T. & Byrnes, C. K. (1962). Ultrastructure of human smooth muscle. 1. Studies of cell surface and connections on normal and achalasic oesophageal smooth muscle. *Exp. molec. Path.* **1**, 204–228.

Harris, E. J. (1960). *Transport and Accumulation in Biological systems.* London: Butterworths Scientific Publications.

Harris, E. J. & Burn, G. P. (1949). The transfer of sodium and potassium ions between muscle and the surrounding medium. *Trans. Faraday Soc.* **45**, 508–528.

Harris, E. J. & Steinbach, H. B. (1956). The extraction of ions from muscle by water and sugar solution with a study of the degree of exchange with tracer of sodium and potassium in the extracts. *J. Physiol.* **133**, 385–401.

Harrison, J. S. & McSwiney, B. A. (1936). The chemical transmitter of motor impulses to the stomach. *J. Physiol.* **87**, 79–86.

Harry, J. (1962). Effect of cooling, local anaesthetic compounds and Botulinum toxin on the responses of and the acetylcholine output from the electrically transmurally stimulated isolated guinea-pig ileum. *Br. J. Pharmac. Chemother.* **19**, 42–55.

—— (1963). The action of drugs on the circular muscle strip from the guinea-pig isolated ileum. *Br. J. Pharmac. Chemother.* **20**, 399–417.

—— (1964). The site of action of sympathomimetic amines on the circular muscle strip from the guinea-pig isolated ileum. *J. Pharm. Pharmac.* **16**, 332–336.

Hart, L. G. & Long, J. P. (1965). Influence of hemicholinium No. 3 on mammalian tissue levels of acetylcholine. *Proc. Soc. exp. Biol. Med.* **119**, 1037–1040.

Hasama, B. (1931). Pharmakologische Studien: Über den bioelektrischen Strom am isolierten Harnleiter. *Arch. exp. Path. Pharmak.* **160**, 107–116.

—— (1933). Die elektrischen Vorgänge am isolierten Eileiter bei Ablauf der peristaltischen Kontraktion. *Pflügers Arch. ges. Physiol.* **231**, 311–331.

—— (1936). Über die Aktionsströme am Kaninchenmagen in situ. *Pflügers Arch. ges. Physiol.* **236**, 545–553.

Hashimoto, Y. & Holman, M. E. (1967). Effect of manganese ions on the electrical activity of mouse vas deferens. *Aust. J. exp. Biol. med. Sci.* **45**, 533–539.

Hashimoto, Y., Holman, M. E. & McLean, A. J. (1967). Effect of tetrodotoxin on the electrical activity of the smooth muscle of the vas deferens. *Nature, Lond.* **215**, 430–432.

Hashimoto, Y., Holman, M. E. & Tille, J. (1966). Electrical properties of the smooth muscle membrane of the guinea-pig vas deferens. *J. Physiol.* **186**, 27–41.

Hasselbach, W. (1965a). Relaxation and the sarcotubular calcium pump. *Fedn Proc.* **23**, 909–912.

—— (1965b). Relaxing factor and the relaxation of muscle. *Progr. Biophys.* **14**, 167–222.

—— (1966). Structural and enzymatic properties of the calcium transporting membranes of the sarcoplasmic reticulum. *Ann. N.Y. Acad. Sci.* **137**, 1041–1048.

Hasselbach, W. & Ledermair, O. (1958). Der Kontraktions-zyklus der isolierten kontraktilen Strukturen der Uterusmuskulatur und seine Besonderheiten. *Pflügers Arch. ges. Physiol.* **267**, 532–542.

Hastings, A. B. (1940). The electrolytes of tissues and body fluids. *Harvey Lect.* **36**, 91–125.

Hattingberg, M. von, Kuschinsky, G. & Rahn, K. H. (1966). Der Einfluss von Pharmaka auf Calcium-Gehalt und ^{45}Calciumaustausch der glatten Muskulatur der Taenia coli von Meerschweinchen. *Naunym-Schmiedeberg's Arch. exp. Path. Pharmak.* **253**, 438–443.

Haugaard, N. & Hess, M. E. (1965). Actions of autonomic drugs on phosphorylase activity and function. *Pharmac. Rev.* **17**, 27–69.

Hazlewood, C. F., Nichols, B. L. & Chamberlain, N. F. (1969). Evidence for the existence of a minimum of two phases of ordered water in skeletal muscle. *Nature, London.* **222**, 747–750.

Headings, V. E., Rondell, P. A. & Bohr, D. F. (1960). Bound sodium in artery wall. *Am. J. Physiol.* **199**, 783–787.

Hebb, C. O. (1957). Biochemical evidence for the neural function of acetylcholine. *Physiol. Rev.* **37**, 196–220.

Hebb, C. O. & Waites, G. M. H. (1956). Choline acetylase in antero- and retro-grade degeneration of a cholinergic nerve. *J. Physiol.* **132**, 667–671.

Held, H. (1905). Zur Kenntniss einer neurofibrillären Continuität im Centralnervensystem der Wirbelthiere. *Arch. für Anat. u. Physiol. Anat. Abt.* (Leipzig) 55–78.

—— (1909). *Die Entwicklung des Nervengewebes bei den Wirbeltieren.* p. 378. Leipzig Barth.

Henderson, V. E. (1923). On the action of atropine on intestine and urinary bladder. *Arch. int. Pharmacodyn. Thér.* **27**, 205–211.

Henderson, V. E. & Roepke, M. H. (1933). On the mechanism of salivary secretion. *J. Pharmac. exp. Ther.* **47**, 193–207.

—— (1934). The role of acetylcholine in bladder contractile mechanisms and in parasympathetic ganglia. *J. Pharmac. exp. Ther.* **51**, 97–111.

—— (1937). Drugs affecting parasympathetic nerves. *Physiol. Rev.* **17**, 373–407.

Henle, J. (1841). *Allgemeine Anatomie.* Lehre von den Mischungs-und Vorbestandtheilen des menschlichen Körpers. Pp. 573–577. Leipzig: Leopold Voss.

Hermansen, K. (1961). The effect of adrenaline, noradrenaline and isoprenaline on the guinea-pig uterus. *Br. J. Pharmac.* Chemother **16**, 116–128.

Herrlinger, J. D., Lüllmann, H. & Schuh, F. (1967). Über die nach maximaler Stimulierung auftretende Empfindlichkeitsverminderung glatter Muskulatur gegenüber Agonisten. *Naunyn-Schmiedeberg's Arch. exp. Path. Pharmak.* **256**, 348–359.

Hertting, G. & Axelrod, J. (1961). Fate of tritiated noradrenaline at the sympathetic nerve endings. *Nature, Lond.* **192**, 172–173.

Hertting, G., Axelrod, J., Kopin, I. J. & Whitby, L. G. (1961). Lack of uptake of catecholamines after chronic denervation of sympathetic nerves. *Nature, Lond.* **189**, 66.

Hertting, G., Axelrod, J. & Whitby, G. L. (1961). Effect of drugs on the uptake and metabolism of H^3-norepinephrine. *J. Pharmac. exp. Ther.* **134**, 146–153.

Hertting, G. & Widhalm, S. (1965). Über den Mechanismus der Noradrenalin-Freisetzung aus sympathischen Nervenendigungen. *Arch. exp. Path. Pharmak.* **250**, 257–258.

Herz, R. & Weber, A. (1965). Caffeine inhibition of Ca uptake by muscle reticulum. *Fedn Proc.* **24**, 208.

Hess, A. (1956). The fine structure and morphological organization of non-myelinated nerve fibres. *Proc. R. Soc. B.* **144**, 496–506.

Hess, B. (1968). Biochemical regulations. In *Systems Theory in Biology.* Ed. Mesarovic, M. D. New York: Springer.

Hess, B., Brand, K. & Pye, K. (1966). Continuous oscillations in a cell-free extract of *S. carlsbergensis. Biochem. biophys. Res. Commun.* **23**, 102–108.

Hesser, F. H. & Perret, G. E. (1960). Studies on gastric motility in the cat. II. Cerebral and infracerebral influence in control, vagectomy and cervical cord preparations. *Gastroenterology* **38**, 231–246.

Hidaka, T. & Kuriyama, H. (1969). Responses of the smooth muscle cell membrane of the guinea-pig jejunum to the field stimulation. *J. gen. Physiol.* **53**, 471-486.

Hidaka, T., Kuriyama, H. & Tasaki, H. (1969). Electrophysiological study of guinea-pig stomach. *J. gen. Physiol.* In press.

Hildebrandt, G. (1961). Rhythmus und Regulation. *Med. Welt.* 73–81.

—— (1963). Störungen der rhythmischen Koordination und ihre balneotherapeutische Beeinflussung. *Z. angew. Bäder-u Klimaheilk.* **10**, 402–420.

—— (1967). Die Koordination rhythmischer Funktionen beim Menschen. *Verh. dtsch. Ges. inn. Med.* **73**, 921–941.

Hildebrandt, G. & Golenhofen, K. (1958). Zur Physiologie der Muskelruhedurchblutung des Menschen. *Arch phys. Ther.* **10**, 217–223.

Hill, A. V. (1928). The diffusion of oxygen and lactic acid through tissues. *Proc. R. soc. B.* **104**, 39–96.

—— (1938). The heat of shortening and the dynamic constants of muscle. *Proc. R. Soc. B.* **126**, 136–195.

—— (1958a). The priority of the heat production in a muscle twitch. *Proc. R. Soc. B.* **148**, 397–402.

Hill, A. V. (1958b). The relation between force developed and energy liberated in an isometric twitch. *Proc. R. Soc. B.* **149**, 58–62.

—— (1965). *Trails and Trials in Physiology.* London: Edward Arnold.

Hill, C. J. (1927). A contribution to our knowledge of the enteric plexuses. *Phil. Trans. R. Soc.* **215**, 355–387.

Hill, D. K. (1965). The organization of the inter-fibre space in the striated muscle of the toad, and the alignment of striations of neighbouring fibres. *J. Physiol.* **179**, 368–384.

Hillarp, N.-Å. (1946). Structure of the synapse and the peripheral innervation apparatus of the autonomic nervous system. *Acta anat.* Suppl. **4**, 1–153.

—— (1959). The construction and functional organisation of the autonomic innervation apparatus. *Acta physiol. scand.* **46**, Suppl. 175, 1–38.

—— (1960). Peripheral autonomic mechanisms. In *Handbook of Physiology*, Sect. 1, Vol. II (Neurophysiology). Ed. Field, J. pp. 979–1006. Washington: Amer. Physiol. Soc.

Hillarp, N.-Å. & Hökfelt, B. (1953). Evidence of adrenaline and noradrenaline in separate adrenal medullary cells. *Acta physiol. scand.* **30**, 55–68.

Hillarp, N.-Å., Lagerstedt, S. & Nilson, B. (1953). The isolation of a granular fraction from the suprarenal medulla, containing the sympathomimetic catecholamines. *Acta physiol. scand.* **29** 251–263.

Hillarp, N.-Å. & Malmfors, T. (1964). Reserpine and cocaine blocking of the uptake and storage mechanisms in adrenergic nerves. *Life Sci. Oxford* **3**, 703–708.

Hinke, J. A. M. (1961). The measurement of sodium and potassium activities in the squid axon by means of cation-selective glass micro-electrodes. *J. Physiol.* **156**, 314–335.

—— (1965). Calcium requirements for noradrenaline and high potassium ion contraction in arterial smooth muscle. In *Muscle.* Ed. Paul, W. M., Daniel, E. E., May, C. M. & Monckton, G., pp. 269–285. New York: Pergamon Press.

Hinke, J. A. M. & McLaughlin, S. G. A. (1967). Release of bound sodium in single muscle fibres. *Canad. J. Physiol. Pharmac.* **45**, 655–667.

His, W. (1886). *Zur Geschichte des menschlichen Rückenmarkes und der Nervenwurzeln.* Leipzig: S. Hirzel.

—— (1889). Die Neuroblasten und deren Entstehung im embryonalen Marke. *Abhandlung d. math.-physik. Klasse d. Königl. Sächs. Gesellsch. d. Wissensch.* (Leipzig) **15**, 311–372.

Hodgkin, A. L. (1939). The relation between conduction velocity and the electrical resistance outside the nerve fibre. *J. Physiol.* **94**, 560–570.

—— (1951). The ionic basis of electrical activity in nerve and muscle. *Biol. Rev.* **26**, 339–409.

—— (1958). Ionic movements and electrical activity in giant nerve fibres. *Proc. R. Soc. B.* **148**, 1–37.

Hodgkin, A. L. & Horowicz, P. (1957). The differential action of hypertonic solutions on the twitch and action potential of a muscle fibre. *J. Physiol.* **136**, 17–18P.

—— (1959a). Movements of sodium and potassium in single muscle cells. *J. Physiol.* **145**, 405–432.

—— (1959b). The influence of potassium and chloride ions on the membrane potential of single muscle fibres. *J. Physiol.* **148**, 127–160.

Hodgkin, A. L. & Huxley, A. F. (1939). Action potentials recorded from inside a nerve fibre. *Nature, Lond.* **144**, 710–711.

—— (1945). Resting and action potentials in single nerve fibres. *J. Physiol.* **104**, 176–195.

—— (1952a). Currents carried by sodium and potassium ions through the membrane of the giant axon of *Loligo. J. Physiol.* **116**, 449–472.

—— (1952b). The components of membrane conductance in the giant axon of *Loligo. J. Physiol.* **116**, 473–496.

—— (1952c). The dual effect of membrane potential on sodium conductance in the giant axon of *Loligo. J. Physiol.* **116**, 497–506.

—— (1952d). A quantitative description of membrane current and its application to conduction and excitation in nerve. *J. Physiol.* **117**, 500–544.

Hodgkin, A. L. & Katz, B. (1949). The effect of sodium ions on the electrical activity of the giant axon of the squid. *J. Physiol.* **108**, 37–77.

Hodgkin, A. L. & Keynes, R. D. (1953). The mobility and diffusion coefficient of potassium in giant axons from *Sepia. J. Physiol.* **119**, 513–528.

Hodgkin, A. L. & Keynes, R. D. (1955a). Active transport of cations in giant axons from *Sepia* and *Loligo*. *J. Physiol.* **128**, 28–60.

—— (1955b). The potassium permeability of a giant nerve fibre. *J. Physiol.* **128**, 61–88.

—— (1957). Movements of labelled calcium in squid giant axons. *J. Physiol.* **138**, 253–281.

Hodgkin, A. L. & Rushton, W. A. H. (1946). The electrical constants of a crustacean nerve fibre. *Proc. R. Soc. B.* **133**, 444–479.

Hoditz, H. & Lüllmann, H. (1964). Die Calcium-Umsatzgeschwindigkeit ruhender und kontrahierender Vorhofsmuskulatur *in vitro*. *Pflügers Arch. ges. Physiol.* **280**, 22–29.

Hoffman, B. F. & Cranefield, P. F. (1960). *Electrophysiology of the Heart*. New York: McGraw-Hill.

Hoffmann, F., Hoffmann, E. J., Middleton, S. & Talesnik, J. (1945). The stimulating effect of acetylcholine on the mammalian heart, and the liberation of an epinephrine-like substance by the isolated heart. *Am. J. Physiol.* **144**, 189–198.

Hogan, M. J. & Feeney, L. (1963a). The ultrastructure of the retinal blood vessels. I. The large vessels. *J. Ultrastruct. Res.* **9**, 10–28.

—— (1963b). The ultrastructure of the retinal vessels II. The small vessels. *J. Ultrastruct. Res.* **9**, 29–46.

Hökfelt, T. (1966a). Electronmicroscopic observations on nerve terminals in the intrinsic muscles of the albino rat iris. *Acta physiol. scand.* **67**, 255–256.

—— (1966b). The effect of reserpine on the intraneuronal vesicles of the rat vas deferens. *Experientia*, **22**, 56.

—— (1968). *In vitro* studies on central and peripheral monoamine neurons at the ultrastructural level. *Z. Zellforsch.* **91**, 1–74.

Hökfelt, T. & Nilsson, O. (1965). Electron microscopy of the adrenergic and cholinergic innervation of the iris muscle. *J. Ultrastruct. Res.* **12**, 237.

Hollands, B. C. S. & Vanov, S. (1965). Localization of catechol amines in visceral organs and ganglia of the rat, guinea-pig and rabbit. *Br. J. Pharmac. Chemother.* **25**, 307–316.

Hollenberg, M., Carriere, S. & Barger, A. C. (1965). Biphasic action of acetylcholine on ventricular myocardium. *Circulation Res.* **16**, 527–536.

Holman, M. E. (1957). The effect of changes in sodium chloride concentration on the smooth muscle of the guinea-pig's taenia coli. *J. Physiol.* **136**, 569–584.

—— (1958). Membrane potentials recorded with high-resistance microelectrodes and the effects of changes in ionic environment on the electrical and mechanical activity of the smooth muscle of the taenia coli of the guinea-pig. *J. Physiol.* **141**, 464–488.

—— (1964). Electrophysiological effects of adrenergic nerve stimulation. In *Pharmacology of Smooth Muscle*. Ed. Bülbring, E. pp. 19–35. Oxford: Pergamon Press.

—— (1967). Some electrophysiological aspects of transmission from noradrenergic nerves to smooth muscle. *Circulation Res.* **21**: Suppl. 3. 71–82.

—— (1968). An introduction to the electrophysiology of smooth muscle. In *Handbook of Physiology—Alimentary Canal IV*, pp. 1165–1708. Am. Physiol. Soc. Washington, D.C.

Holman, M. E. & Hughes, J. R. (1965a). Inhibition of intestinal smooth muscle. *Aust. J. exp. Biol. med. Sci.* **43**, 277–290.

—— (1965b). An inhibitory component of the response to distension of rat ileum. *Nature, Lond.* **207**, 641–642.

Holman, M. E. & Jowett, A. (1964). Some actions of catecholamines on the smooth muscle of the guinea-pig vas deferens. *Aust. J. exp. Biol. med. Sci.* **42**, 40–53.

Holman, M. E., Kasby, C. B. & Suthers, M. B. (1967). Electrical activity of rabbit portal vein. *Aust. J. exp. Biol. Sci.* **45**, 50P.

Holman, M. E., Kasby, C. B. Suthers, M. B. & Wilson, J. A. F. (1968). Some properties of the smooth muscle of rabbit portal vein. *J. Physiol.* **196**, 111–132.

Holman, M. E. & McLean, A. (1967). The innervation of sheep mesenteric veins. *J. Physiol.* **190**, 55–69.

Holmgren, A., Jonsson, B. & Sjöstrand, T. (1960). Circulatory data in normal subjects at rest and during exercise in recumbent position, with special reference to the stroke volume at different work intensities. *Acta physiol. scand.* **49**, 343–364.

Holst, E. v. (1939). Die relative Koordination als Phänomen und als Methode zentralnervöser Funktionsanalyse. *Ergebn. Physiol.* **42**, 228–306.

Holton, P. (1959). The liberation of adenosine triphosphate on antidromic stimulation of sensory nerves. *J. Physiol.* **145**, 494–504.

Holton, F. A. & Holton, P. (1954). The capillary dilator substances in dry powders of spinal roots; a possible role of adenosine triphosphate in chemical transmission from nerve endings. *J. Physiol.* **126**, 124–140.

Holton, P. & Rand, M. J. (1962). Sympathetic vaso-dilatation in the rabbit ear. *Br. J. Pharmac. Chemother.* **19**, 513–526.

Holtz, P. (1939). Dopadecarboxylase. *Naturwissenschaften*, **27**, 724–725.

Holtz, P., Heise, R. & Lüdtke, K. (1938). Fermentativer Abbau von 1-Dioxy-phenylalanin (DOPA) durch Niere. *Arch. exp. Path. Pharmak.* **191**, 87–118.

Holtz, P. & Wölpert, K. (1937). Die Reaktion des Katzen- und Meerschweinchen-Uterus auf Adrenalin während der verschiedenen Stadien des Sexualzyklus und ihre hormonale Beeinflussung. *Arch. exp. Path. Pharm.* **185**, 20–41.

Honig, C. R. (1963). The intrinsic properties and functional organization of vascular smooth muscle. In *The Peripheral Blood Vessels*. Ed. Orbison, J. L. & Smith, D. E. Baltimore: Williams and Wilkins.

Honjin, R., Takahashi, A., Shimasaki, S. & Maruyama, H. (1965). Two types of synaptic nerve processes in the ganglia of Auerbach's plexus of mice, as revealed by electron microscopy. *J. Electronmic.* **14**, 43–49.

Horvath, B. (1954). Ovarian hormones and the ionic balance of uterine muscle. *Proc. natn. Acad. Sci. U.S.A.* **40**, 515–521.

Hoskins, R. G. (1912). The sthenic effect of epinephrine upon intestine. *Am. J. Physiol.* **29**, 363–366.

Hotta, Y. & Tsukui, R. (1968). Effect on the guinea-pig taenia coli of the substitution of strontium or barium ions for calcium ions. *Nature, Lond.* **217**, 867–869.

Howarth, J. V. (1958). The behaviour of frog muscle in hypertonic solutions. *J. Physiol.* **144**, 167–175.

Hubbard, J. I. & Yokota, T. (1964). Direct evidence for an action of acetylcholine on motor-nerve terminals. *Nature, Lond.* **203**, 1072–1073.

Hughes, F. B. (1955). The muscularis mucosae of the oesophagus of the cat, rabbit and rat. *J. Physiol.* **130**, 123–130.

Hughes, J. & Vane, J. R. (1967). An analysis of the responses of the isolated portal vein of the rabbit to electrical stimulation and to drugs. *Br. J. Pharmac. Chemother.* **30**, 46–66.

Huković, S. (1960). The action of sympathetic blocking agents on isolated and innervated atria and vessels. *Br. J. Pharmac. Chemother.* **15**, 117–121.

Huković, S., Rand, M. J. & Vanov, S. (1965). Observations on an isolated innervated preparation of rat urinary bladder. *Br. J. Pharmac. Chemother.* **24**, 178–188.

Hukuhara, T., & Fukuda, H. (1968). The electrical activity of guinea-pig small intestine with special reference to the slow wave. *Jap. J. Physiol.* **18**, 71–86.

Hukuhara, T. & Miyake, T. (1959). The intrinsic reflexes in the colon. *Jap. J. Physiol.* **9**, 49–55.

Hukuhara, T., Nakayama, S. & Nanba, R. (1961). The role of the intrinsic mucosal reflex in the fluid transport through the denervated colonic loop. *Jap. J. Physiol.* **11**, 71–79.

Hukuhara, T., Nanba, R. & Fukuda, H. (1964). The problem whether the intramural ganglion cells play any role in the ureteral motility. *Jap. J. Physiol.* **14**, 188–196.

Hukuhara, T. & Neya, T. (1968). The movements of the colon of rats and guinea-pigs. *Jap. J. Physiol.* **18**, 551–562.

Hukuhara, T., Yamagami, M. & Nakayama, S. (1958). On the intestinal intrinsic reflexes. *Jap. J. Physiol.* **8**, 9–20.

Hunt, R. (1901). Further observations on the blood-pressure-lowering bodies in extracts of suprarenal gland. *Am. J. Physiol.* **5**, VI-VII.

Hunt, R. & Taveau, R. de M. (1906). On the physiological action of certain cholin derivatives and new methods for detecting cholin. *Br. med. J.* **2**, 1788–1791.

Hurwitz, L. (1960). Potassium transport in isolated guinea-pig ileum. *Am. J. Physiol.* **198**, 94–98.

—— (1965). Calcium and its interrelations with cocaine and other drugs in contraction of intestinal smooth muscle. In *Muscle*. Eds. Paul, W. M., Daniel, E. E., Kay, C. M. & Monckton, G. pp. 239–251. Oxford: Pergamon Press.

Hurwitz, L., Battle, F. & Weiss, G. B. (1962). Action of the calcium antagonists cocaine and ethanol on contraction and potassium efflux of smooth muscle. *J. gen. Physiol.* **46**, 315–332.

Hurwitz, L., Joiner, P. D. & von Hagen, S. (1967). Calcium pools utilized for contraction in smooth muscle. *Am. J. Physiol.* **213**, 1299–1304.

Hurwitz, L., Tinsley, B. & Battle, F. (1960). Dissociation of contraction and potassium efflux in smooth muscle. *Am. J. Physiol.* **199**, 107–111.

Hüter, J., Bauer, H. & Goodford, P. J. (1963). Die Wirkung von Adrenalin auf Kalium-Austausch und Kalium-Konzentration im glatten Muskel (Taenia coli des Meerschweinchens.) *Arch. exp. Path. Pharmak.* **246**, 75–76.

Hutter, O. F. & Noble, D. (1960). The chloride conductance of frog skeletal muscle. *J. Physiol.* **151**, 89–102.

Hutter, O. F. & Warner, A. E. (1967a). The pH sensitivity of the chloride conductance of frog skeletal muscle. *J. Physiol.* **189**, 403–425.

—— (1967b). The effect of pH on the ^{36}Cl efflux from frog skeletal muscle. *J. Physiol.* **189**, 427–443.

Huxley, A. F. (1957). Muscle structure and theories of contraction. *Progr. Biophys.* **7**, 257–318.

—— (1959). Ion movements during nerve activity. *Ann. N.Y. Acad. Sci.* **81**, 221–246.

—— (1960). *Mineral Metabolism.* **1**, 163–167. New York: Academic Press.

—— (1964). Muscle. *Ann. Rev. Physiol.* **26**, 131–152.

Huxley, A. F. & Stämpfli, R. (1951). Effect of potassium and sodium on resting and action potentials of single myelinated nerve fibres. *J. Physiol.* **112**, 496–508.

Huxley, A. F. & Taylor, R. E. (1958). Local activation of striated muscle fibres. *J. Physiol.* **144**, 426–441.

Huxley, H. E. (1963). Electron microscope studies on the structure of natural and synthetic protein filaments from striated muscle. *J. molec. Biol.* **7**, 281–308.

—— (1965). Structural evidence concerning the mechanism of contraction in striated muscle. In *Muscle*. Eds. Paul, W. M., Daniel, E. E., Kay, C. M. & Monckton, G. pp. 3–28. New York: Pergamon Press.

—— (1966). Personal communication, cited in Shoenberg, C. F., Ruegg, J. C., Needham, D. M., Schirmer, R. H. and Nemetchek-Gansler, H. *Biochem. Z.* **345**, 255–266.

Iggo, A. (1957). Gastro-intestinal tension receptors with unmyelinated afferent fibres in the vagus of the cat. *Q. Jl. exp. Physiol.* **42**, 130–143.

Imai, S. & Takeda, K. (1967a). Effects of vasodilators on the isolated taenia coli of the guinea-pig. *Nature, Lond.* **213**, 509–511.

—— (1967b). Actions of calcium and certain multivalent cations on potassium contracture of guinea-pig's taenia coli. *J. Physiol.* **190**, 155–169.

Irisawa, H. & Kobayashi, M. (1963). Effects of repetitive stimuli and temperature on ureter action potentials. *Jap. J. Physiol.* **13**, 421–430.

Irvine, G., Lever, J. D. & Ahmed, J. (1965). Muscle surface specializations and the neuro-muscular relationship in coronary and splanchnic arterioles. *J. Anat. Lond.* **99**, 409.

Ishikawa, T. (1962). Fine structure of the human ciliary muscle. *Invest. Ophthal.* **1**, 587–608.

Ishiko, N. & Sato, M. (1960). The effect of stretch on the electrical constants of muscle fibre membrane. *Jap. J. Physiol.* **10**, 194–203.

Isojima, C., & Bozler, E., (1963). Role of calcium in initiation of contraction in smooth muscle. *Am. J. Physiol.* **205**, 681–685.

Iversen, L. L. (1965a). The inhibition of noradrenaline uptake by drugs. *Adv. Drug Res.* **2**, 1–46. Ed. Harper, N. J. & Simmonds, A. B.

—— (1965b). The uptake of catechol amines at high perfusion concentrations in the rat isolated heart: a novel catechol amine uptake process. *Br. J. Pharmac. Chemother.* **25**, 18–33.

—— (1967a). The catecholamines. *Nature, Lond.* **214**, 8–14.

—— (1967b). *The Uptake and Storage of Noradrenaline in Sympathetic Nerves.* Cambridge: University Press.

Jabonero, V. (1954). Der anatomische Aufban der peripheren Neurosekretion. *Acta neuroveg. (Wien)* Suppl. 5–6, 159–211.

―― (1959). Die plexiforme Synapse auf Distanz und die Bedeutung der sogenannten interkalären Zellen. *Acta neuroveg. (Wien)* **19**, 276–302.

―― (1960). El problema de las "Neuronas simpaticas intersticiales de Cajal" y el modo de terminar las vias vegetativas eferentes. *Trab. Inst. Cajal Invest. biol.* **52**, 1–79.

―― (1962). Ueber die Brauchbarkeit der Osmiumtetroxyd-zinkjodid Methode für Analyse der vegetativen Peripherie. *Acta neuroveg. (Wien)* **26**, 184–210.

―― (1965). Studien über die Synapsen des peripheren vegetativen Nervensystems. *Acta neuroveg. (Wien)* **27**, 101–120.

Jacobowitz, D. (1965). Histochemical studies of the autonomic innervation of the gut. *J. Pharmac. exp. Ther.* **149**, 358–364.

―― (1967). Histochemical studies of the relationship of chromaffin cells and adrenergic nerve fibers to the cardiac ganglia of several species. *J. Pharmac. exp. Ther.* **158**, 227–240.

Jacobowitz, D. & Koelle, G. B. (1963). Demonstration of both acetylcholinesterase and catecholamines in the same nerve trunk. *Pharmacologist* **5**, 270.

―― (1965). Histochemical correlations of acetylcholinesterase and catecholamines in postganglionic autonomic nerves of the cat, rabbit and guinea-pig. *J. Pharmac. exp. Ther.* **148**, 225–237.

Jaeger, J. (1962). Elektronenoptische Untersuchungen an der glatten Muskulatur des menschlichen graviden Uterus. *Gynaecologia (Basel)* **154**, 193–205.

―― (1963). Zur Ultrastruktur der menschlichen Uterusmuskelzelle unter der Geburt. *Arch. Gynaek.* **199**, 173–181.

Jaeger, J. & Pohlmann, G. (1962). Zur Ultrastruktur der menschlichen Uterusmuskelzelle. *Beitr. path. Anat.* **126**, 113–126.

Jaisle, F. (1960). Tropomyosin in der menschlichen Uterusmuskulatur. *Arch. Gynaek.* **194**, 277–286.

Jansson, G. & Martinson, J. (1966). Studies on the ganglionic site of action of the sympathetic outflow to the stomach. *Acta physiol. scand.* **68**, 184–192.

Jenkinson, D. H. & Morton, I. K. M. (1965). Effects of noradrenaline and isoprenaline on the permeability of depolarized intestinal smooth muscle to inorganic ions *Nature, Lond.* **205**, 505–506.

―― (1967a). Adrenergic blocking drugs as tools in the study of the actions of catecholamines on the smooth muscle membrane. *Ann. N.Y. Acad. Sci.* **139**, 762–771.

―― (1967b). The effect of noradrenaline on the permeability of depolarized intestinal smooth muscle to inorganic ions. *J. Physiol.* **188**, 373–386.

―― (1967c). The role of α and β-adrenergic receptors in some actions of catecholamines on intestinal smooth muscle. *J. Physiol.* **188**, 387–402.

Jensen, K. B. & Vennerød, A. M. (1961). Reversal of the inhibitory action of adrenaline and histamine on rat uterus. *Acta pharmacol. (Kbh.).* **18**, 298–306.

Johansson, B. & Bohr, D. F. (1966). Rhythmic activity in smooth muscle from small subcutaneous arteries. *Am. J. Physiol.* **210**, 801–806.

Johansson, B. & Jonsson, O. (1968). Cell volume as a factor influencing electrical and mechanical activity of vascular smooth muscle. *Acta physiol. scand.* **72**, 456–468.

Johansson, B., Jonsson, O., Axelsson, J. & Wahlstrom, B. (1967). Electrical and mechanical characteristics of vascular smooth muscle response to norepinephrine and isoproterenol, *Circulation Res.* **21**, 619–633.

Johansson, B. & Ljung, B. (1967a). Sympathetic control of rhythmically active vascular smooth muscle as studied by a nerve-muscle preparation of portal vein. *Acta physiol. scand.* **70**, 299–311.

―― (1967b). Spread of excitation in the smooth muscle of the rat portal vein. *Acta physiol. scand.* **70**, 312–322.

Johnson, E. S. (1963a). A note on the relation between the resting release of acetylcholine and increase in tone of the isolated guinea-pig ileum. *J. Pharm. Pharmac.* **15**, 69–72.

Johnson, E. S. (1963b). The origin of the acetylcholine released spontaneously from the guinea-pig isolated ileum. *Br. J. Pharmac. Chemother.* 21, 555–568.

Jones, A. W. (1968). Influence of oestrogen and progesterone on the electrolyte accumulation in the rabbit myometrium. *J. Physiol.* 197, 19–20P.

—— (1969). Factors affecting sodium exchange and distribution in rabbit myometrium. In press.

Jones, A. W. & Karreman, G. (1969a). Ion exchange properties of canine carotid artery. *Biophys. J.* 9, 884–909.

—— (1969b). Potassium accumulation and permeability in the canine carotid artery. *Biophys. J.* 9, 910–924.

Jones, A. W. & Tomita, T. (1967). The longitudinal tissue resistance of the guinea-pig taenia coli. *J. Physiol.* 191, 109–110P.

Jonsson, G. (1967). The formaldehyde fluorescence method for the histochemical demonstration of biogenic monoamines. Thesis, Stockholm.

Jourdan, F. & Nowak, S. J. G. (1934). Les fibres cardio-accélératrices dans le nerf pneumogastrique du chien; leur origine et leur trajet. *C. r. Séanc. Soc. Biol.* 117, 234–238.

Jung, H. (1960). Erregungsphysiologische Regelwirkungen von 17 β-oestradiol am Myometrium. *Acta endocr., Copenh.* 35, 49–58.

Kabat, H. (1937). An analysis of cardioaccelerator fibres in the vago-sympathetic trunk of the dog. *Am. J. Physiol.* 119, 345–346.

—— (1939). The cardio-accelerator fibres in the vagus nerve of the dog. *Am. J. Physiol.* 128, 246–257.

Kaestle (1913). Röntgenuntersuchung des Magens. In *Röntgenkunde.* Ed. Rieder-Rosenthal. Leipzig: Barth.

Kahn, R. H. (1926). Über humorale Übertragbarkeit der Herznervenwirkung. *Pflügers Arch. ges. Physiol.* 214, 482–498.

Kaiser, I. H. & Harris, J. S. (1950). Effect of adrenaline on pregnant human uterus. *Am. J. Obst. Gynec.* 59, 775–784.

Kameya, Y. (1964). An electronmicroscope study on the uterine smooth muscle. *J. Jap. Obstet. & Gynaec. Soc.* 11, 33–47.

Kao, C. Y. (1961). Contents and distribution of potassium, sodium and chloride in uterine smooth muscle. *Am. J. Physiol.* 201, 717–722.

—— (1966). Tetrodotoxin, saxitoxin and their significance in the study of excitation phenomena. *Pharmac. Rev.* 18, 997–1050.

—— (1967). Ionic basis of electrical activity in uterine smooth muscle. In *Cellular Biology of the Uterus.* 386–448. Ed. Wynn, R. M. Amsterdam: North Holland publishing Co.

Kao, C. Y. & Nishiyama, A. (1964). Ovarian hormones and resting potential of rabbit uterine smooth muscle. *Am. J. Physiol.* 207, 793–799.

Kao, C. Y. & Siegman, M. J. (1963). Nature of electrolyte exchange in isolated uterine smooth muscle. *Am. J. Physiol.* 205, 674–680.

Kapeller, K. & Mayor, D. (1966). Ultrastructural changes proximal to a constriction in sympathetic axons during the first 24 hours after operation. *J. Anat.* 100, 439–441.

—— (1967). The accumulation of noradrenaline in constricted sympathetic nerves as studied by fluorescence and electron microscopy. *Proc. R. Soc. B.* 167, 282–292.

Karlsson, U. & Schultz, R. L. (1965). Fixation of the central nervous system for electron microscopy by aldehyde perfusion. 1. Preservation with aldehyde perfusates versus direct perfusion with osmium tetroxide with special reference to membranes and the extracellular space. *J. Ultrastruct. Res.* 12, 160–186.

Karnovsky, M. J. (1964). The localization of cholinesterase activity in rat cardiac muscle by electron microscopy. *J. cell Biol.* 23, 217–232.

Karnovsky, M. J. & Roots, L. (1964). A 'direct-colouring' thiocholine method for cholinesterases. *J. Histochem. Cytochem.* 12, 219–221.

Karreman, G. (1964). Adsorption of ions at charged sites and phase boundary potentials. *Bull. math. Biophys.* 26, 275–290.

—— (1965). Cooperative specific absorption of ions at charged sites in an electric field. *Bull. math. Biophys.* 27, 91–104.

Karreman, G. & Jones, A. W. (1965). Potassium and sodium uptake in vascular smooth muscle. Proc. 18th Ann. Conf. Eng'g Med. Biol.

Karrer, H. E. (1959). The striated musculature of blood vessels. I. General cell morphology. *J. biophys. biochem. Cytol.* **6**, 383–392.

—— (1960). Cell interconnections in normal human cervical epithelium. *J. biophys. biochem. Cytol.* **7**, 181–184.

—— (1961). An electron microscope study of the aorta in young and ageing mice. *J. Ultrastruct. Res.* **5**, 1–27.

Karrer, H. E. & Cox, J. (1960). The striated musculature of blood vessels. II. Cell interconnections and cell surface. *J. biophys. biochem. Cytol.* **8**, 135–150.

Katz, B. (1939). The relation between force and speed in muscular contraction. *J. Physiol.* **96**, 45–64.

—— (1948). The electrical properties of the muscle fibre membrane. *Proc. Roy. Soc. B.* **135**, 506–534.

—— (1950). Depolarization of sensory terminals and the initiation of impulses in the muscle spindle. *J. Physiol.* **111**, 261–282.

—— (1962). The transmission of impulses from nerve to muscle, and the subcellular unit of synaptic action. *Proc. R. Soc. B.* **155**, 455–477.

—— (1966). *Nerve, Muscle & Synapse.* New York: Mcgraw-Hill.

Katz, B. & Miledi, R. (1965a). The quantal release of transmitter substances. In: *Studies in Physiology.* Ed. Curtis, D. R. & McIntyre, A. K. pp. 118–125. Berlin: Springer-Verlag.

—— (1965b). The effect of temperature on the synaptic delay at the neuromuscular junction. *J. Physiol.* **181**, 656–670.

—— (1965c). The measurement of synaptic delay, and the time course of acetylcholine release at the neuromuscular junction. *Proc. Roy. Soc. B.* **161**, 483–495.

—— (1966). The production of endplate potentials in muscles paralysed by tetrodotoxin. *J. Physiol.* **185**, 5–6P.

Katz, B. & Schmitt, O. H. (1940). Electric interaction between two adjacent nerve fibres. *J. Physiol.* **97**, 471–488.

Katz, B. & Thesleff, S. (1957). On the factors which determine the amplitude of the 'miniature end-plate potential.' *J. Physiol.* **137**, 267–278.

Kaufman, S. (1966). Coenzymes and hydroxylases; ascorbate and dopamine-β-hydroxylase; tetrahydropteridines and phenylalanine and tyrosine hydroxylases. *Pharmac. Rev.* **18**, 61–69.

Kaufmann, R. & Kienböck, R. (1911). Über den Rhythmus der Antrumperistaltik des Magens. *Münch. med. Wschr.* **58**, No. 1. 1237–1238.

Kaufmann, R. & Fleckenstein, A. (1965). Die Bedeutung der Aktionspotential-Dauer und der Ca^{++}-Ionen beim Zustandekommen der positiv-inotropen Kältewirkungen am Warmblüter-Myokard. *Pflügers Arch. ges. Physiol.* **285**, 1–18.

Kažić, T. & Varagić, V. M. (1968). Effect of increased intraluminal pressure on the release of acetylcholine from the isolated guinea-pig ileum. *Br. J. Pharmac. Chemother.* **32**, 185–192.

Keatinge, W. R. (1964). Mechanism of adrenergic stimulation of mammalian arteries and its failure at low temperatures. *J. Physiol.* **174**, 184–205.

—— (1965). Electrical activity of arterial smooth muscle in calcium-free solution. *J. Physiol.* **177**, 32–33P.

—— (1966a). Electrical and mechanical response of arteries to stimulation of sympathetic nerves. *J. Physiol.* **185**, 701–715.

—— (1966b). Electrical and mechanical responses of vascular smooth muscle to vasodilator agents and vasoactive polypeptides. *Circulation Res.* **18**, 641–649.

Keech, M. K. (1960). Electronmicroscope study of the normal rat aorta. *J. biophys. biochem. Cytol.* **7**, 533–538.

Kelly, R. E. & Rice, R. V. (1968). Localization of myosin filaments in smooth muscle. *J. cell Biol.* **37**, 105–116.

Kerkut, G. A. & Cottrell, G. A. (1963). Acetylcholine and 5-hydroxytryptamine in the snail brain. *Comp. Biochem. Physiol.* **8**, 53–63.

Kernan, R. P. (1962). Membrane potential changes during sodium transport in frog sartorius muscle. Nature, Lond. **193**, 986–987.

Kewenter, J. (1965). The vagal control of the jejunal and ileal motility and blood flow. Acta physiol. scand. **65**, Suppl. 251, 5–68.

Keynes, R. D. (1954). The ionic fluxes in frog muscle. Proc. R. Soc. B. **142**, 359–382.

—— (1963). Chloride in the squid giant axon. J. Physiol. **169**, 690–705.

Keynes, R. D. & Lewis, P. R. (1951a) The resting exchange of radioactive potassium in crab nerve. J. Physiol. **113**, 73–98.

—— (1951b). The sodium and potassium content of cephalopod nerve fibres. J. Physiol. **114**, 151–182.

Khairallah, P. A., Page, I. H., Bumpus, F. M. & Turker, R. K. (1966). Angiotensin tachyphylaxis and its reversal. Circulation Res. **19**, 247–254.

Kimizuka, H. & Koketsu, K. (1963). Changes in the membrane permeability of frog's sartorious muscle fibres in Ca-free EDTA solution. J. gen. Physiol. **47**, 379–392.

King, T. O., Koppanyi, T. & Karczmar, A. G. (1947). Pharmacology of a new bronchoconstrictor drug, hexaethyltetraphosphate. Ann. Allergy **5**, 570–571.

Kirk, J. E. & Dyrbye, M. (1956). Hexosamine and acid-hydrolyzable sulfate concentrations of the aorta and pulmonary artery in individuals of various ages. J. Geront. **11**, 273–281.

Kirpekar, S. M. & Cervoni, P. (1963). Effect of cocaine, phenoxybenzamine and phentolamine on the catecholamine output from spleen and adrenal medulla. J. Pharmac. exp. Ther. **142**, 59–70.

Kirpekar, S. M., Cervoni, P. & Furchgott, R. F. (1962). Catecholamine content of the cat nictitating membrane following procedures sensitizing it to norepinephrine. J. Pharmac. exp. Ther. **135**, 180–190.

Kirshner, N., Holloway, C., Smith, W. J. & Kirschner, A. G. (1966). Uptake and storage of catecholamines. In Mechanisms of Release of Biogenic Amines, Ed. von Euler, U. S., Rosell, S. & Uvnäs, B., pp. 109–123. Oxford: Pergamon Press.

Klaus, W. & Lüllmann, H. (1964). Calcium als intracelluläre Überträgersubstanz und die mögliche Bedeutung dieses Mechanismus für pharmakologische Wirkungen. Klin. Wschr. **42**, 253–259.

Klee, P. (1927). Die Magenbewegungen. Handb. norm. u. pathol. Physiol. Bd. **3**, 398–440. Berlin: Julius Springer.

Kobayashi, M. (1965). Effects of Na and Ca on the generation and conduction of excitation in the ureter. Am. J. Physiol. **208**, 715–719.

Kobayashi, M. & Irisawa, H. (1964). Effect of sodium deficiency on the action potential of the smooth muscle cell of ureter. Am. J. Physiol. **206**, 205–210.

Kobayashi, M., Nagai, T. & Prosser, C. L. (1966). Electrical interaction between muscle layers of cat intestine. Am. J. Physiol. **211**, 1281–1291.

Kobayashi, M., Prosser, C. L. & Nagai, T. (1967). Electrical properties of intestinal muscle as measured intracellularly and extracellularly. Am. J. Physiol. **213**, 275–286.

Kochmann, M. & Seel, H. (1929). Über die Abhängigkeit der Adrenalinwirkung auf den isolierten Meerschweinchenuterus vom Zyklushormon. Z. ges. exp. Med. **68**, 238–244.

Koe, B. K. & Weissmann, A. (1966). p-Chlorophenylalanine: a specific depletor of brain serotonin. J. Pharmac. exp. Ther. **154**, 499–516.

Koella, W. P. & Schaeppi, U. (1962). The reaction of the isolated cat iris to serotonin. J. Pharmac. exp. Ther. **138**, 154–158.

Koelle, G. B. (1962). A new general concept of the neurohumoral functions of acetylcholine and acetylcholinesterase. J. Pharm. Pharmac. **14**, 65–90.

—— (1963). Cytological distributions and physiological functions of cholinesterases. In Cholinesterases and Anticholinesterase Agents; Handbuch der Experimentellen Pharmakologie, Suppl. 15, pp. 187–298. Ed. Koelle, G. B. Berlin: Springer Verlag.

Koelle, G. B. & Foroglou-Kerameos, C. (1965). Electronmicroscopic localization of cholinesterases in a sympathetic ganglion by a gold-thiolacetic acid method. Life Sci. Oxford. **4**, 417–424.

Koelle, G. B. & Friedenwald, J. S. (1949). A histochemical method for localizing cholinesterase activity. Proc. Soc. exp. Biol. Med. **70**, 617–622.

Koelle, G. B., Koelle, E. S. & Friedenwald, J. S. (1950). The effect of inhibition of specific and non-specific cholinesterase on motility of the isolated ileum. *J. Pharmac. exp. Ther.* **100**, 180–191.

Koenig, E. (1965). Synthesis mechanisms in the axon. II. RNA in myelin-free axons of the cat. *J. Neurochem.* **12**, 357–361.

Koepchen, H. P. (1962). *Die Blutdruckrhythmik.* Darmstadt: Dr. D. Steinkopff.

Kölliker, A. (1849). Beiträge zur Kenntnis der glatten Muskeln. *Z. wiss. Zool.* **1**, 48–87.

Konzett, H. & Rothlin, E. (1949). Beeinflussung der Nikotin-artigen Wirkung von Acetylcholin durch Atropin. *Helv. physiol. pharmac. Acta.* **7**, C46–C47.

Kopin, I. J. (1964). Storage and metabolism of catecholamines: The role of monoamine oxidase. *Pharmac. Rev.* **16**, 179–191.

—— (1966). Biochemical aspects of release of norepinephrine and other amines from sympathetic nerve endings. *Pharmac. Rev.* **18**, 513–523.

—— (1967). The adrenergic synapse. In *The Neurosciences. A study Program.* Ed. Quarton, G. C., Melnenchuk, T. & Schmitt, F. O., pp. 427–432. New York: Rockefeller University Press.

Kopin, I. J., Fischer, J. E., Musacchio, J. M., Horst, W. D. & Weise, V. K. (1965). 'False neurochemical transmitters' and the mechanism of sympathetic blockade by monoamine oxidase inhibitors. *J. Pharmac. exp. Ther.* **147**, 186–193.

Kopin, I. J. & Gordon, E. K. (1962). Metabolism of norepinephrine-H^3 released by tyramine and reserpine. *J. Pharmac. exp. Ther.* **138**, 351–359.

Korkes, S., Del Campillo, A., Korey, S. R. Stern, J. R., Nachmansohn, D. & Ochoa, S. (1952). Coupling of acetyl donor systems with choline acetylase. *J. biol. Chem.* **198**, 215–220.

Koshtoyants, K. S., Buznikov, G. A. & Manukhin, B. N. (1961). The possible role of 5-HT in the motor activity of embryos of some marine gastropods. *Comp. Biochem. Physiol.* **3**, 20–26.

Kosterlitz, H. W. (1967). Intrinsic intestinal reflexes. *Am. J. dig. Dis.* **12**, 245–254.

Kosterlitz, H. W. & Lees, G. M. (1964). Pharmacological analysis of intrinsic intestinal reflexes. *Pharmac. Rev.* **16**, 301–339.

Kosterlitz, H. W., Pirie, V. W. & Robinson, J. A. (1956). The mechanism of the peristaltic reflex in the isolated guinea-pig ileum. *J. Physiol.* **133**, 681–694.

Kosterlitz, H. W. & Robinson, J. A. (1955). Mechanism of the contraction of the longitudinal muscle of the isolated guinea-pig ileum, caused by raising the pressure in the lumen. *J. Physiol.* **129**, 18P.

—— (1957). Inhibition of the peristaltic reflex of the isolated guinea-pig ileum. *J. Physiol.* **136**, 249–262.

—— (1959). Reflex contractions of the longitudinal muscle coat of the isolated guinea-pig ileum. *J. Physiol.* **146**, 369–379.

Kosterlitz, H. W. & Watt, A. J. (1965). Adrenergic receptors in the guinea-pig ileum. *J. Physiol.* **177**, 11P.

Kottegoda, S. R. (1968). Are the excitatory nerves to the circular muscle of the guinea-pig ileum cholinergic? *J. Physiol.* **197**, 17–18P.

—— (1969). An analysis of possible nervous mechanisms involved in the peristaltic reflex. *J. Physiol.* **200**, 687–712.

Krapp, J. (1962). Elektronenmikroskopische Untersuchungen über die Innervation von Iris und Corpus ciliare der Hauskatze unter besonderer Berücksichtigung der Muskulatur. *Z. mikrosk.-anat. Forsch.* **68**, 418–447.

Krnjević, K. & Miledi, R. (1959). Presynaptic failure of neuromuscular propagation in rats. *J. Physiol.* **149**, 1–22.

Krnjević, K. & Phyllis, J. W. (1963). Actions of certain amines on cerebral cortical neurones. *Br. J. Pharmac. Chemother.* **20**, 471–490.

Krop, S. & Kunkel, A. M. (1954). Observations on pharmacology of the anticholinesterases sarin and tabun. *Proc. Soc. exp. Biol. Med.* **86**, 530–533.

Kubát, J. (1953). Zur statistischen Behandlung von Relaxationsprozessen. *Koll.-Z.* **134**, 197–206.

—— (1954). Über Relaxation mechanischer Spannungen. Proc. of the second int. congress on rheology. 178–180P. London: Butterworths.

Kubát, J. (1965). A similarity in the stress relaxation behaviour of high polymers and metals. Stockholm. Thesis.

Kuffler, S. W. & Vaughan Williams, E. M. (1953a). Small nerve junctional potentials. The distribution of small motor nerves to frog skeletal muscle, and the membrane characteristics of the fibres they innervate. *J. Physiol.* **121**, 289–317.

—— (1953b). Properties of the 'slow' skeletal muscle fibres of the frog. *J. Physiol.* **121**, 318–340.

Kuhn, R. & Wiegandt, H. (1964). Weitere Ganglioside aus Menschenhirn. *Z. Naturf.* **19b**, 256–257.

Kumar, D., Wagatsuma, T. & Barnes, A. C. (1965). In vitro hyperpolarizing effect of adrenaline on human myometrial cell. *Am. J. Obstet. Gynec.* **91**, 575–576.

Kuntz, A. (1953). *The Autonomic Nervous System*, 4th Edition. Philadelphia: Lea & Febiger.

Kuntz, A. & Jacobs, M. W. (1955). Components of periarterial extensions of celiac and mesenteric plexuses. *Anat. Rec.* **123**, 509–520.

Kuperman, A. S. & Okamoto, M. (1965). Comparison of the effects of some ethonium ions and their structural analogues on neuromuscular transmission in the cat. *Br. J. Pharmac. Chemother.* **24**, 223–239.

Kuriyama, H. (1961a). The effect of progesterone and oxytocin on the mouse myometrium. *J. Physiol.* **159**, 26–39.

—— (1961b). Recent studies on the electrophysiology of the uterus. In Ciba Foundation Study Group, no. 9, *Progesterone and the Defence Mechanism of Pregnancy*. p. 51. London: Churchill.

—— (1963a). The influence of potassium, sodium and chloride on the membrane potential of the smooth muscle of taenia coli. *J. Physiol.* **166**, 15–28.

—— (1963b). Electrophysiological observations on the motor innervation of the smooth muscle cells in the guinea-pig vas deferens. *J. Physiol.* **169**, 213–228.

—— (1964a). Effect of calcium and magnesium on neuromuscular transmission in the hypogastric nerve-vas deferens preparation of the guinea-pig. *J. Physiol.* **175**, 211–230.

—— (1964b). Effect of electrolytes on the membrane activity of the uterus. In *Pharmacology of Smooth Muscle*. Ed. Bülbring, E. p. 127–140. Oxford: Pergamon Press.

Kuriyama, H. & Csapo, A. (1959). The 'evolution' of membrane and myoplastic activity of uterine muscle. *Biol. Bull. mar. biol. Lab. Woods Hole* **117**, 417–418.

—— (1961a). A study of the parturient uterus with the micro-electrode technique. *Endocrinology* **68**, 1010–1025.

—— (1961b). Placenta and myometrial block. *Am. J. Obstet. Gynec.* **82**, 592–599.

Kuriyama, H., Osa, T. & Toida, N. (1966). Effects of tetrodotoxin on smooth muscle cells of the guinea pig taenia coli. *Br. J. Pharmac. Chemother.* **27**, 366–376.

—— (1967a). Membrane properties of the smooth muscle of the guinea-pig ureter. *J. Physiol.* **191**, 225–238.

—— (1967b). Electrophysiological study of the intestinal smooth muscle of the guinea-pig. *J. Physiol.* **191**, 239–255.

—— (1967c). Nervous factors influencing the membrane activity of intestinal smooth muscle. *J. Physiol.* **191**, 257–270.

Kuriyama, H. & Tomita, T. (1965). The responses of single smooth muscle cells of guinea-pig taenia coli to intracellularly applied currents, and their effects on the spontaneous electrical activity. *J. Physiol.* **178**, 270–289.

Kuschinsky, G., Lüllmann, H. & Muscholl, E. (1954). Untersuchungen über die Einwirkung von verschiedenen Pharmaka auf die Spontanrhythmik des isolierten Hühneramnion. *Arch. exp. Path. Pharmak.* **223**, 369–374.

Laduron, P. & Belpaire, F. (1968). Tissue fractionation and catecholamines. II. Intracellular distribution patterns of tyrosine hydroxylase, DOPA decarboxylase, dopamine-β-hydroxylase, phenylethanolamine, N-methyltransferase and monoamine oxidase in adrenal medulla. *Biochem. Pharmac.* **17**, 1127–1140.

Laguens, R. & Lagrutta, J. (1964). Fine structure of human uterine muscle in pregnancy. *Am. J. Obstet. Gynaec.* **89**, 1040.

Lahrtz, Hg., Lüllmann, H. & Reis, H. E. (1967). Über den Einfluss von Kalium und Carbachol auf die Calcium-Abgabe normaler und chronisch denervierter Rattenzwerchfelle. *Pflügers Arch. ges. Physiol.* **297**, 10–18.

Lamb, J. F. & McGuigan, J. A. S. (1968). The effect of potassium, sodium, chloride, calcium and sulphate ions and of sorbitol and glycerol during the cardiac cycle in frog's ventricle. *J. Physiol.* **195**, 283–315.

Lands, A. M. (1952). Sympathetic receptor action. *Am. J. Physiol.* **169**, 11–21.

Lands, A. M., Luduena, F. P., Ananenko, E. & Grant, J. I. (1950). A comparison of the sympathetic inhibitory action of 1-(3,4-dihydroxyphenyl)-2-aminoethanol with that of several of its analogs. *Arch. int. Pharmacodyn. Thér.* **83**, 602–616.

Lane, B. P. (1965). Alterations in the cytologic detail of intestinal smooth muscle cells in various stages of contraction. *J. cell Biol.* **27**, 199–213.

—— (1967). Localization of products of ATP hydrolysis in mammalian smooth muscle cells. *J. cell Biol.* **34**, 713–720.

Lane, B. P. & Rhodin, J. A. G. (1964a). Cellular interrelationships and electrical activity in two types of smooth muscle. *J. Ultrastruct. Res.* **10**, 470–488.

—— (1964b). Fine structure of the lamina muscularis mucosae. *J. Ultrastruct. Res.* **10**, 489–497.

Langendorff, O. (1908). Untersuchungen über die Natur des periodischaussetzenden Rhythmus, insbesondere des Herzens. *Pflügers Arch ges. Physiol.* **121**, 54–74.

Langer, S. Z., Draskóczy, P. R. & Trendelenburg, U. (1967). Time course of the development of supersensitivity to various amines in the nictitating membrane of the pithed cat after denervation or decentralization. *J. Pharmac. exp. Ther.* **157**, 255–273.

Langley, J. N. (1896). Observations on the medullated fibres of the sympathetic system and chiefly on those of the grey rami communicantes. *J. Physiol.* **20**, 55–76.

—— (1898a). On the union of cranial autonomic (visceral) fibres with the nerve cells of the superior cervical ganglion. *J. Physiol.* **23**, 240–270.

—— (1898b). On inhibitory fibres in the vagus for the end of the oesophagus and the stomach. *J. Physiol.* **23**, 407–414.

— — (1900). The sympathetic and other related systems of nerves. In *Textbook of Physiology*. Vol. **2**, Ed. Schäfer, E. A. pp. 616–696. Edinburgh: Pentland.

—— (1901). Observations on the physiological action of extracts of the supra-renal bodies. *J. Physiol.* **27**, 237–256.

—— (1905). On the reaction of cells and of nerve-endings to certain poisons; chiefly as regards the reaction of striated muscle to nicotine and to curari. *J. Physiol.* **33**, 374–413.

—— (1921). *The Autonomic Nervous System.* Part I. Cambridge: Heffer.

—— (1922a). The secretion of sweat. *Part 1.* Supposed inhibitory nerve fibres in the posterior nerve roots. Secretion after denervation. *J. Physiol.* **56**, 110–119.

—— (1922b). Connexions of the enteric nerve cells. *J. Physiol.* **56**, 39P.

Langley, J. N. & Anderson, H. K. (1895). The innervation of the pelvic and adjoining viscera, Part IV. The internal generative organs. *J. Physiol.* **19**, 122–130.

Langley, J. N. & Magnus, R. (1905). Some observations of the movements of the intestine before and after degenerative section of the mesenteric nerves. *J. Physiol.* **33**, 34–51.

Langley, J. N. & Orbeli, L. A. (1911). Some observations on the degeneration in the sympathetic and sacral autonomic nervous system of Amphibia following nerve section. *J. Physiol.* **42**, 113–124.

Laszt, L. (1960). Correlation between the electrolyte and water content of the organs and hypertension after administration of corticosteroids. *Nature, Lond.* **185**, 695.

Laszt, L. & Hamoir, G. (1961). Etude par electrophorèse et ultracentrifugation de la composition protéinique de la couche musculaire des carotides de bovidé. *Biochim. biophys. Acta* **50**, 430–449.

Laties, A. M., Lund, R. & Jacobowitz, D. (1967). A simplified method for the histochemical localization of cardiac catecholamine-containing nerve fibers. *J. Histochem. Cytochem.* **15**, 535–541.

Lauricella, E., D'Alessandro, P. & Fiumara, D. (1960). Considerazioni sulla struttura della fibrocellula muscolare uterina umana al microscopio elettronico. *Minerva ginec.* XII, **21**, 1047–1053.

Lawrentjew, B. J. (1926). Über die Verbreitung der nervösen Elemente (einschliesslich der 'interstitiellen Zellen' Cajals) in der glatten Muskulatur; ihre Endigungsweise in den glatten Muskelzellen. *Z. mikrosk.-anat. Forsch.* **6**, 467–488.

Lawrentjew, B. J. & Borowskaja, A. J. (1936). Die Degeneration der post-ganglionären Fasern des autonomen Nervensystems und deren Endigungen. *Z. Zellforsch. mikrosk. Anat.* **23**, 761–778.

Leaders, F. E. (1963). Local cholinergic-adrenergic interaction: mechanism for the biphasic chronotropic response to nerve stimulation. *J. Pharmac. exp. Ther.* **142**, 31–38.

—— (1965). Separation of adrenergic and cholinergic fibres in sympathetic nerves in the hind limbs of the dog by hemicholinium. (HC-3). *J. Pharmac. exp. Ther.* **148**, 238–246.

Leaders, F. E. & Dayrit, C. (1965). The cholinergic component in the sympathetic innervation to the spleen. *J. Pharmac. exp. Ther.* **147**, 145–152.

Leaming, D. B. & Cauna, N. (1961). A qualitative and quantitative study of the myenteric plexus of the small intestine of the cat. *J. Anat.* **95**, 160–169.

Lee, C. Y. (1960). The effect of stimulation of extrinsic nerves on peristalsis and on the release of 5-hydroxytryptamine in the large intestine of the guinea-pig and of the rabbit. *J. Physiol.* **152**, 405–418.

Lee, C. Y. & Tseng, L. F. (1966). A further study on the adrenergic inhibition of the peristaltic reflex of the gut. Abstracts, III. Intern. Pharmacol. Congress; São Paulo, Brazil. P. 117.

Lee, F.-L. & Trendelenburg, U. (1967). Muscarinic transmission of preganglionic impulses to the adrenal medulla of the cat. *J. Pharmac. exp. Ther.* **158**, 73–79.

Lee, W. C. & Shideman, F. E. (1959). Mechanism of the positive inotropic response to certain ganglionic stimulants. *J. Pharmac. exp. Ther.* **126**, 239–249.

Leeson, C. R. & Leeson, T. S. (1965a). The fine structure of the rat umbilical cord at various times of gestation. *Anat. Rec.* **151**, 183–198.

—— (1965b). The rat ureter. Fine structural changes during its development. *Acta anat.* **62**, 60–79.

Leeuwe, M. (1937). Over de interstitieele cel (Cajal). Een onderzoek van de periphere sympathicus met behulp van de vitale methylen-blau-kleuring. *Diss. Utrecht.*

Lembeck, F. & Strobach, R. (1956). Kaliumabgabe aus glatter Muskulatur. *Arch. exp. Path. Pharmak.* **228**, 130–131.

Le Fevre, P. G. & Marshall, J. K. (1958). Conformational specificity in a biological sugar transport system. *Am. J. Physiol.* **194**, 333–337.

Lenn, N. J. (1967). Localization of uptake of tritiated norepinephrine by rat brain *in vivo* and *in vitro* using electron microscopic autoradiography. *Am. J. Anat.* **120**, 377–390.

Lev, A. A. (1964). Determination of activity and activity coefficients of potassium and sodium ions in frog muscle fibres. *Nature, Lond.* **201**, 1132–1134.

Lever, J. D., Ahmed, M. & Irvine, G. (1965). Neuromuscular and intercellular relationships in the coronary arterioles. A morphological and quantitative study by light and electron-microscopy. *J. Anat.* **99**, 829–840.

Lever, J. D. & Esterhuizen, A. C. (1961). Fine structure of the arteriolar nerves in the guinea-pig pancreas. *Nature, Lond.* **192**, 566–567.

Lever, J. D., Graham, J. D. P., Irvine, G. & Chick, W. J. (1965). The vesiculated axons in relation to arteriolar smooth muscle in the pancreas. A fine structural and quantitative study. *J. Anat.* **99**, 299–313.

Lever, J. D., Graham, J. D. P. & Spriggs, T. L. B. (1966). Electron microscopy of nerves in relation to the arteriolar wall. *Biblphie. anat.* **8**, 51–55.

Levi, H. & Ussing, H. H. (1948). The exchange of sodium and chloride ions across the fibre membrane of the isolated frog sartorius. *Acta physiol. scand.* **16**, 232–249.

Levitt, M., Spector, S., Sjoerdsma, A. & Udenfriend, S. (1965). Elucidation of the rate limiting step in norepinephrine biosynthesis in the perfused guinea-pig heart. *J. Pharmac, exp. Ther.* **148**, 1–8.

Levy, B. & Tozzi, S. (1963). The adrenergic receptive mechanism of the rat uterus. *J. Pharm. Pharmacol.* **142**, 178–184.

Lévy, J. & Michel, E. (1945). Sur l'hydrolyse enzymatique de l'atropine. *Bull. Soc. Chim. biol. (Paris)* **27**, 570–577.

Lewandowsky, M. (1899). Ueber die Wirkung des Nebennierenextraktes auf die glatten Muskeln, im Besonderen des Auges. *Arch. Physiol.* Suppl. 1899, 360–366.

Lewartowski, B. & Bielecki, K. (1963). The influence of hemicholinium No. 3. and vagal stimulation on acetylcholine content of rabbit atria. *J. Pharmac. exp. Ther.* **142**, 24–30

Lewis, P. R. & Shute, C. C. D. (1964). Demonstration of cholinesterase activity with the electron microscope. *J. Physiol.* **175**, 5–7P.

—— (1966). The distribution of cholinesterase in cholinergic neurons demonstrated with the electron microscope. *J. Cell Sci.* **1**, 381–390.

—— (1967). The simultaneous demonstration of catecholamines and cholinesterase with the electronmicroscope. *J. Physiol.* **186**, 53–55P.

Lewis, P. R., Shute, C. C. D. & Silver, A. (1967). Confirmation from choline acetylase analyses of a massive cholinergic innervation to the rat hippocampus. *J. Physiol.* **191**, 215–224.

Liddell, E. G. T. (1960). *The Discovery of Reflexes.* Oxford: Clarendon Press.

Lindgren, P. (1955). The mesencephalon and the vasomotor system: an experimental study on the central control of peripheral blood flow in the cat. *Acta physiol. scand.* **35**, suppl. 121.

Lindgren, P., Rosen, A., Strandberg, P. & Uvnäs, B. (1956). The sympathetic vasodilator outflow—a cortico-spinal autonomic pathway. *J. comp. Neurol.* **105**, 95–109.

Lindgren, P. & Uvnäs, B. (1953). Vasodilator responses in the skeletal muscles of the dog to electrical stimulation in the oblongate medulla. *Acta physiol. scand.* **29**, 137–144.

—— (1954). Postulated vasodilator centre in the medulla oblongata. *Am. J. Physiol.* **176**, 68–76.

—— (1955). Vasoconstrictor inhibition and vasodilator activation—two functionally separate vasodilator mechanisms in the skeletal muscles. *Acta physiol. scand.* **33**, 108–119.

Lindmar, R., Löffelholz, K. & Muscholl, E. (1968). A muscarinic mechanism inhibiting the release of noradrenaline from peripheral adrenergic nerve fibres by nicotinic agents. *Br. J. Pharmac. Chemother.* **32**, 280–294.

Lindmar, R. & Muscholl, E. (1963). Bilanzversuche mit Noradrenalin-Infusionen am perfundierten Rattenherzen. *Arch. exp. Path. Pharmak.* **245**, 99–100.

Ling, G. N. (1962). *A Physical Theory of the Living State: The Association-Induction Hypothesis.* New York: Blaisdell.

—— (1965a). The membrane theory and other views for solute permeability, distribution, and transport in living cells. *Perspect Biol. Med.* **9**, 87–106.

—— (1965b). The physical state of water in living cell and model systems. *Ann. N.Y. Acad. Sci.* **125**, 401–417.

—— (1966a). All-or-none adsorption by living cells and model protein-water systems: discussion of the problem of 'permease-induction' and determination of secondary and tertiary structures of proteins. *Fedn Proc.* **25**, 958–970.

—— (1966b). Cell membrane and cell permeability *Ann. N.Y. Acad. Sci.* **137**, 837–859.

—— (1966c). Elektrische Potentiale lebender Zellen. In *Die Zelle.* Ed. Metzner, H. Stuttgart: Wissenschaftliche Verlagsgesellschaft. m. b. h.

Ling, G. N. & Cope, F. W. (1969). Is the bulk of intracellular K^+-ion adsorbed? *Science, N.Y.* **163**, 1335–1336.

Ling, G. N. & Gerard, R. W. (1949). The normal membrane potential of frog sartorius fibres. *J. cell. comp. Physiol.* **34**, 383–396.

Ling, G. N. & Ochsenfeld, M. M. (1965). Studies on the ionic permeability of muscle cells and their models. *Biophys. J.* **5**, 777–807.

—— (1966). Studies on ion accumulation in muscle cells. *J. gen. Physiol.* **49**, 819–843.

Ling, G. N., Ochsenfeld, M. M. & Karreman, G. (1967). Is the cell membrane a universal rate-limiting barrier to the movement of water between the living cell and its surrounding medium? *J. gen. Physiol.* **50**, 1807–1820.

Lipmann, F. (1948). Biosynthetic mechanisms. *Harvey lect.* Ser. **44**, pp. 99–123.

Lipmann, F. & Kaplan, N. O. (1946). A common factor in the enzymatic acetylation of sulphanilamide and of choline. *J. biol. Chem.* **162**, 743–744.

Lissák, K. (1939). Effects of extracts of adrenergic fibers on the frog heart. *Am. J. Physiol.* **125**, 778–785.

Livett, B. G. & Geffen, L. B. (1901). Transport of C^{14}-noradrenaline down sympathetic nerves. Presented to *Aust. Soc. Clin. & Expl. Pharmac.* Nov. 29th–30th.

Livett, B. G., Geffen, L. B. & Austin, L. (1968). Proximo-distal transport of (^{14}C) noradrenaline and protein in sympathetic nerves. *J. Neurochem.* **15**, 931–939.

Loewenstein, W. R. (1966). Permeability of membrane junctions. *Ann. N.Y. Acad. Sci.* **137**, 441–472.

—— (1967a). On the genesis of cellular communication. *Develop. Biol.* **15**, 503–520.

Loewenstein, W. R., Nakas, M. & Socolar, S. J. (1967). Junctional membrane uncoupling: Permeability transformations at a cell membrane junction. *J. gen. Physiol.* **50**, 1865–1891.

Loewi, O. (1921a). Über humorale Übertragbarkeit der Herznervenwirkung. I. Mitteilung. *Pflügers Arch. ges. Physiol.* **189**, 239–242.

—— (1921b). Über humorale Übertragbarkeit der Herznervenwirkung. II. Mitteilung. *Pflügers Arch. ges. Physiol.* **193**, 201–203.

—— (1924). Über humorale Übertragbarkeit der Herznervenwirkung. III. Mitteilung. *Pflügers Arch. ges. Physiol.* **203**, 408–412.

Loewi, O. & Hellauer, H. (1938). Über das Azetylcholin in peripheren Nerven. *Pflügers Arch. ges. Physiol.* **240**, 769–775.

Loewi, O. & Navratil, E. (1926a). Über humorale Übertragbarkeit der Herznervenwirkung. X. Mitteilung: Über das Schicksal des Vagusstoffs. *Pflügers Arch. ges. Physiol.* **214**, 678–688.

—— (1926b). Über humorale Übertragbarkeit der Herznervenwirkung. XI. Mitteilung. Über den Mechanismus der Vaguswirkung von Physostigmin und Ergotamin. *Pflügers Arch. ges. Physiol.* **214**, 689–696.

Long, J. P. & Eckstein, J. W. (1961). Ganglionic actions of neostigmine methylsulphate. *J. Pharmac. exp. Ther.* **133**, 216–222.

Lorento de Nó, R. (1947). *A Study of Nerve Physiology.* New York: The Rockefeller Inst. for Medical Research.

Lowenstein, O. & Lowenfeld, I. E. (1962). The Pupil. In *The Eye*, vol. 3. Ed. Davson, H. 231–267. New York: Academic Press.

Lubínska, L., Niemierko, S. & Oberfeld, B. (1961). Gradient of cholinesterase activity and of choline acetylase activity in nerve fibres. *Nature, Lond.* **189**, 122–123.

Luciani, A. (1873). Eine periodische Funktion des isolierten Frosch-herzens. *Ber. d. sächs. Ges. d. Wissensch. math.-physik. Klasse.*

Lucké, B. & McCutcheon, M. (1932). The living cell as an osmotic system and its permeability to water. *Physiol. Rev.* **12**, 68–139.

Luduena, F. P. & Grigas, E. O. (1966). Pharmacological study of autonomic innervation of dog retractor penis. *Am. J. Physiol.* **210**, 435–444.

Luisada, A. (1933). Beitrag zum Studium der Gefässtätigkeit. I Mitt. Die electrischen Phänomene an isolierten Gefässen. *Z. ges. exp. Med.* **91**, 440–449.

Lüllmann, H. & Mohns, P. (1969). The Ca metabolism of intestinal smooth muscles during forced electrical stimulation. *Europ. J. Physiol.* (Pflügers Arch.) **308**, 214–224.

Lüllmann, H. & Siegfriedt, A. (1968). Über den Calcium-Gehalt und den ^{45}Calcium-Austausch in Längsmuskulatur des Meerschweinchendünndarms. *Pflügers Arch. ges. Physiol.* **300**, 108–119.

Lum, B. K. B., Kermani, M. H. & Heilman, R. D. (1966). Intestinal relaxation produced by sympathomimetic amines in the isolated rabbit jejunum: selective inhibition by adrenergic blocking agents and by cold storage. *J. Pharmac. exp. Ther.* **154**, 463–471.

Lundberg, A. (1958). Electrophysiology of salivary glands. *Physiol. Rev.* **38**, 21–40.

Lundholm, L. & Mohme-Lundholm, E. (1965). Energetics of isometric and isotonic contraction in isolated vascular smooth muscle under anaerobic conditions. *Acta physiol. scand.* **64**, 275–282.

Lüttgau, H. C. & Niedergerke, R. (1958). The antagonism between Ca and Na ions on the frog's heart. *J. Physiol.* **143**, 486–505.

Lutz, B. R. & Fulton, G. P. (1958). Smooth muscle and blood flow in small blood vessels. In *Factors Regulating Blood Flow*. Ed. Fulton, G. P. & Zweifach, B. Washington: Amer. Physiol. Soc.

Lutz, B. R., Fulton, G. P. & Akers, R. P. (1950). The neuromotor mechanism of the small blood vessels in membranes of the frog (*Rana pipiens*) and the hamster (*Mesocricetus auratus*) with reference to the normal and pathological conditions of blood flow. *Expl. Med. Surg.* **8**, 258–287.

MacIntosh, F. C., Birks, R. I. & Sastry, P. B. (1956). Pharmacological inhibition of acetylcholine synthesis. *Nature, Lond.* **178**, 1181.

McCrea, E. D., McSwiney, B. A. & Stopford, J. S. B. (1925). The effect on the stomach of stimulation of the peripheral end of the vagus nerve. *Quart. J. exp. Physiol.* **15**, 201–233.

McDougal, M. D. & West, G. B. (1954). The inhibition of the peristaltic reflex by sympathomimetic amines. *Br. J. Pharmac. Chemother.* **9**, 131–137.

McGeer, P. L., Bagchi, S. P. & McGeer, E. G. (1965). Subcellular localization of tyrosine hydroxylase in beef caudate nucleus. *Life Sci. Oxford* **4**, 1859–1867.

McGill, C. (1909). The structure of smooth muscle in the resting and in the contracted condition. *Am. J. Anat.* **9**, 493–545.

McGovern, J. P., Ozkaragoz, K., Hensel, A. E. & Burdon, K. L. (1961). Qualitative and quantitative studies with 5-hydroxytryptamine (serotonin) in the Schultz-Dale apparatus. *J. Allergy* **32**, 321–326.

McIsaac, R. J. (1966). Ganglionic blocking properties of epinephrine and related amines. *Int. J. Neuropharmac.* **5**, 15–26.

McLaughlin, S. G. A. & Hinke, J. A. M. (1966). Sodium and water binding in single striated muscle fibers of the giant barnacle. *Canad. J. Physiol. Pharmacol.* **44**, 837–848.

—— (1968). Optical density changes of single muscle fibers in sodium-free solutions. *Canad. J. Physiol. Pharmacol.* **46**, 247–260.

McLean, J. R. & Burnstock, G, (1966). Histochemical localization of catecholamines in the urinary bladder of the toad. (*Bufo marinus*). *J. Histochem. Cytochem.* **14**, 538–548.

—— (1967a). Innervation of the urinary bladder of the sleepy lizard (*Trachysaurus rugosus*) 1. Fluorescent histochemical localization of catecholamines. *Comp. Biochem. Physiol.* **20**, 667–673.

—— (1967b). Innervation of the lungs of the toad (*Bufo marinus*) II. Fluorescent histochemistry of catecholamines. *Comp. Biochem. Physiol.* **22**, 767–773.

—— (1967c). Innervation of the lungs of the sleepy lizard (*Trachysaurus rugosus*) I. Fluorescent histochemistry of catecholamines. *Comp. Biochem. Physiol.* **22**, 809–813.

McSwiney, B. A. & Robson, J. M. (1929). The response of smooth muscle to stimulation of the vagus nerve. *J. Physiol.* **68**, 124–131.

McSwiney, B. A. & Wadge, W. J. (1928). Effects of variations in intensity and frequency on the contractions of the stomach obtained by stimulation of the vagus nerve. *J. Physiol.* **65**, 350–356.

Magnus, R. (1903). Pharmakologie der Magen-und Darmbewegungen. *Ergebn. Physiol.* **2**, 637–672.

Mahler, H. R. (1961). The use of amine buffers in studies with enzymes. *Ann. N.Y. Acad. Sci.* **92**, 426–439.

Mallin, M. L. (1965). Actomyosin and myosin of vascular muscle. *Nature, Lond.* **207**, 1297–1298.

Malmfors, T. (1965a). Studies on adrenergic nerves. The use of rat and mouse iris for direct observations on their physiology and pharmacology at cellular and subcellular levels. *Acta physiol. scand.* **64**, Suppl. 248, 1–93.

—— (1965b). The adrenergic innervation of the eye as demonstrated by fluorescence microscopy. *Acta physiol. scand.* **65**, 259–267.

Malmfors, T. & Sachs, C. (1965). Direct demonstration of the systems of terminals belonging to an individual adrenergic neuron and their distribution in rat iris. *Acta physiol. scand.* **64**, 377–382.

Maloff, G. A. (1934). Zur Pharmakologie der Venen. Über selbständige Venenkontraktionen. *Arch. int. Pharmacodyn. Thér.* **48**, 333–353.

Mann, M. (1949). Sympathin and the rat uterus. *J. Physiol.* **110**, 11P.

Mann, M. & West, G. B. (1950). The nature of hepatic and splenic sympathin. *Br. J. Pharmac. Chemother.* **5**, 173–177.

Mansour, T. E. (1959). The effect of serotonin and related compounds on the carbohydrate metabolism of the liver fluke, *Fasciola hepatica*. *J. Pharmac. exp. Ther.* **126**, 212–216.

Mark, J. S. T. (1956). An electron microscope study of uterine smooth muscle. *Anat. Rec.* **125**, 473–493.

Marks, B. H., Samorajski, T. & Webster, E. J. (1962). Radioautographic localization of norepinephrine-H^3 in the tissues of mice. *J. Pharmac. exp. Ther.* **138**, 376–381.

Marrazzi, A. S. (1939). Electrical studies on the pharmacology of autonomic synapses. II. The action of a sympathomimetic drug (epinephrine) on sympathetic ganglia. *J. Pharmac. exp. Ther.* **65**, 395–404.

Marshall, J. M. (1959). Effects of estrogen and progesterone on single uterine muscle fibers in the rat. *Am. J. Physiol.* **197**, 935–942.

—— (1962). Regulation of activity in uterine smooth muscle. *Physiol. Rev.* **42**, (suppl. 5), 213–227.

—— (1963). Behaviour of uterine muscle in Na-deficient solutions; effects of oxytocin. *Am. J. Physiol.* **204**, 732–738.

—— (1964). The action of oxytocin on uterine smooth muscle. In *Pharmacology of Smooth Muscle*. Ed. Bülbring, E. pp. 143–153. Pergamon Press.

—— (1965). Calcium and uterine smooth muscle membrane potentials In *Muscle*. Ed. Paul, W. M., Daniel, E. E., Kay, C. M. and Monckton. G. pp. 229–238. Oxford: Pergamon Press.

—— (1967). Comparative aspects of the pharmacology of smooth muscle. *Fedn Proc.* **26**, 1104–1110.

—— (1968). Relation between the ionic environment and the action of drugs on the myometrium. *Fedn Proc.* **27**, 115–119.

Marshall, J. M. & Csapo, A. I. (1961). Hormonal and ionic influences on the membrane activity of uterine smooth muscle cells. *Endocrinology* **68**, 1026–1035.

Marshall, J. M. & Miller, M. D. (1964). Effects of metabolic inhibitors on the rat uterus and on its response to oxytocin. *Am. J. Physiol.* **206**, 437–442.

Martin, A. R. (1966). Quantal nature of synaptic transmission. *Physiol. Rev.* **46**, 51–66.

Martin, A. R. & Pilar, G. (1963a). Dual mode of synaptic transmission in the avian ciliary. ganglion. *J. Physiol.* **168**, 443–463.

—— (1963b). Transmission through the ciliary ganglion of the chick. *J. Physiol.* **168**, 464–475.

—— (1964a). An analysis of electrical coupling at synapses in the avian ciliary ganglion. *J. Physiol.* **171**, 454–475.

—— (1964b). Quantal components of the synaptic potential in the ciliary ganglion of the chick. *J. Physiol.* **175**, 1–16.

Martinson, J. (1965a). Vagal relaxation of the stomach. Experimental re-investigation of the concept of the transmission mechanism. *Acta. physiol. scand.* **64**, 453–462.

—— (1965b). Studies on the efferent vagal control of the stomach. *Acta physiol. scand.* **65**, Suppl. 255, 1–23.

Martinson, J. & Muren, A. (1963). Excitatory and inhibitory effects of vagus stimulation on gastric motility in the cat. *Acta physiol. scand.* **57**, 309–316.

Mashima, H. & Yoshida, T. (1965). Effect of length on the development of tension in guinea-pig's taenia coli. *Jap. J. Physiol.* **15**, 463–477.

Mashima, H., Yoshida, T. & Handa, M. (1966). Contraction and relaxation of guinea-pig's taenia coli in relation to spike discharges. *Jap. J. Physiol.* **16**, 304–315.

Masson, P. (1914). La glande endocrine de l'intestin chez l'homme. *C. R. Acad. Sci.* (*Paris*) **158**, 59–61.

Masson, P. & Berger, L. (1923). Sur un nouveau mode de sécrétion interne: La neurocrinie. *C. R. Acad. Sci.* (*Paris*) **176**, 1748–1750.

Masuda, T. (1927). The action of the vagus on the spleen. *J. Physiol.* **62**, 289–300.

Matsuda, K. (1960). Some electrophysiological properties of terminal Purkinje fibers of heart. In *Electrical Activity of Single Cells*. Ed. Katsuki, Y. pp. 283–294. Tokyo: Igakushoin.

Matthes, K. (1930). The action of blood on acetylcholine. *J. Physiol.* **70**, 338–348.

Matthes, K. (1951). Kreislaufuntersuchungen am Menschen mit fortlaufend registrierenden Methoden. Stuttgart: Georg Thieme Verlag.

Matthews, E. K. & Sutter, M. C. (1967). Ouabain-induced changes in the contractile and electrical activity, potassium content, and response to drugs of smooth muscle cells. *Can. J. Physiol. Pharmacol.* **45**, 509–520.

May, W. P. (1904). The innervation of the sphincters and musculature of the stomach. *J. Physiol.* **31**, 260–271.

Mayor, D. & Kapeller, K. (1967). Fluorescence microscopy and electron microscopy of adrenergic nerves after constriction at two points. *Jl R. microsc. Soc.* **87**, 277–294.

Meigs, E. B. (1912). Contributions to the general physiology of smooth and striated muscle. *J. exp. Zool.* **13**, 497–571.

Meigs, E. B. & Ryan, L. A. (1912). The chemical analysis of the ash of smooth muscle. *J. biol. Chem.* **11**, 401–414.

Meissner, G. (1857). Über die Nerven der Darmwand. *Z. rat. Med.* **3**, 364–366.

Mellander, S., Johansson, B., Gray, S., Jonsson, O., Lundvall, J. & Ljung, B. (1967). The effects of hyperosmolarity on intact and isolated vascular smooth muscle. Possible role in exercise hyperemia. *Angiologica*, **4**, 310–322.

Melton, C. E. (1956). Electrical activity in the uterus of the rat. *Endocrinology* **58**, 139–149.

—— (1962). Conduction of impulses in rat myometrium. Proc. 22nd Int. Congr. Physiol. **2**, 532.

Melville, K. I. (1937). The antisympathomimetic action of dioxane compounds (F883 + F933) with special reference to the vascular responses to dihydroxyphenyl ethanolamine (arterenol) and nerve stimulation. *J. Pharmac. exp. Ther.* **59**, 317–327.

Merrillees, N. C. R. (1960). The fine structure of muscle spindles in the lumbrical muscles of the rat. *J. biophys. biochem. Cytol.* **7**, 725–742.

—— (1968). The nervous environment of individual smooth muscle cells of the guinea-pig vas deferens. *J. cell Biol.* **37**, 794–817.

Merrillees, N. C. R., Burnstock, G. & Holman, M. E. (1963). Correlation of fine structure and physiology of the innervation of smooth muscle in the guinea pig vas deferens. *J. cell Biol.* **19**, 529–550.

Meyling, H. A. (1953). Structure and significance of the peripheral extension of the autonomic nervous system. *J. comp. Neurol.* **99**, 495–535.

Michaelis, L. (1926). Die Permeabilität von Membranen. *Naturwissenschaften* **3**, 33–42.

Michaelson, I. A., Richardson, K. C., Snyder, S. N. & Titus, E. O. (1964). The separation of catecholamine storage vesicles from rat heart. *Life Sci. Oxford* **3** (2), 971–978.

Middleton, S., Middleton, H. H. & Toha, J. (1949). Adrenergic mechanism of vagal cardio-stimulation. *Am. J. Physiol.* **158**, 31–37.

Middleton, S., Oberti, C., Prager, R. & Middleton, H. H. (1956). Stimulating effect of acetylcholine on the papillary myocardium. *Acta physiol. latinoam.* **6**, 82–89.

Miledi, R. (1964). Electron microscopical localization of products from histochemical reactions used to detect cholinesterase in muscle. *Nature, Lond.* **204**, 293–295.

Miller, J. W. (1967). Adrenergic receptors in the myometrium. *Ann. N. Y. Acad. Sci.* **139**, 788–798.

Miller, M. D. & Marshall, J. M. (1965). Uterine response to nerve stimulation; relation to hormonal status and catecholamines. *Am. J. Physiol.* **209**, 859–865.

Milofsky, A. (1957). The fine structure of the pineal in the rat with special reference to parenchyma. *Anat. Rec.* **127**, 435–436.

Milton, G. W. & Smith, A. W. M. (1956). The pacemaking area of the duodenum. *J. Physiol.* **132**, 100–114.

Milton, G. W., Smith, A. W. M. & Armstrong, H. I. O. (1955). The origin of the rhythmic electropotential changes in the duodenum. *Q. Jl exp. Physiol.* **40**, 79–88.

Mirkin, B. L. & Cervoni, P. (1962). The adrenergic nature of neuro-humoral transmission in the cat nictitating membrane following treatment with reserpine. *J. Pharmac. exp. Ther.* **138**, 301–308.

Mirkin, R. I. & Bonnycastle, D. D. (1954). A pharmacological and chemical study of humoral mediators in the sympathetic nervous system. *Am. J. Physiol.* **178**, 529–534.

Misu, Y. & Kirpekar, S. M. (1968). Effects of vagal and sympathetic nerve stimulation on the isolated atria of the cat. *J. Pharmac. exp. Ther.* **163**, 330–342.

Mitchell, G. A. G. (1953). *Anatomy of the Autonomic Nervous System.* Edinburgh: Livingstone.

—— (1956). *Cardiovascular Innervation.* Edinburgh: Livingstone.

Miyamoto, Y. (1963). *Fac. Sci. Hokkaido, Univ. Ser. VI,* **15**, 235–247.

Mohme-Lundholm, E. & Vamos, N. (1967). Influence of Ca^{++} deficiency and pronethalol, on metabolic and contractile effects of catecholamines and K-ions in vascular muscle. *Acta pharmac. (Kbh.)* **25**, 87–96.

Monnier, M. (1943). Erregungsleitung in der Arterienwand. *Helv. physiol. pharmac. Acta* **1**, 249–264.

—— (1944a). Reizbildung in der Arterienwand. *Helv. physiol. pharmac. Acta.* **2**, 279–303.

—— (1944b). Die funktionellen Potenzen der isolierten Arterie (Erregbarkeit, Reizbildung, Erregungsleitung, autonome Anpassung). *Helv. physiol. pharmac. Acta.* **2**, 533–539.

Moore, D. H. & Ruska, H. (1957). Electron microscope study of mammalian cardiac muscle cells. *J. biophys. biochem. Cytol.* **3**, 261–268.

Moore, K. E., Milton, A. S. & Gosselin, R. E. (1961). Effect of 5-hydroxytryptamine on the respiration of excised lamellibranch gill. *Br. J. Pharmac. Chemother.* **17**, 278–285.

Mosher, H. S. Fuhrman, F. A., Buchwald, H. D. & Fischer, H. G. (1964). Tarichatoxin-tetrodotoxin: a potent neurotoxin. *Science, N.Y.* **144**, 1100–1110.

Mori, S., Maeda, T. & Shimizu, M. (1964). Electron microscopic histochemistry of cholinesterases in the rat brain. *Histochemie* **4**, 65–72.

Movat, H. Z. & Fernando, N. V. P. (1963). The fine structure of the terminal vascular bed. I. Small arteries with an internal elastic lamina. *Exp. molec. Path.* **2**, 549–563.

Mullins, L. J. & Noda, K. (1963). The influence of sodium-free solutions on the membrane potential of frog muscle fibres. *J. gen. Physiol.* **47**, 117–132.

Munger, B. L. (1961). The ultrastructure and histophysiology of human eccrine sweat glands. *J. biophys, biochem. Cytol.* **11**, 385–402.

Munro, A. F. (1951). The effect of adrenaline on the guinea-pig intestine. *J. Physiol.* **112**, 84–94.

Murnaghan, M. F. & Mazurkiewicz, I. M. (1963). Some pharmacological properties of 4-methyltropolone. *Rev. canad. Biol.* **22**, 99–102.

Musacchio, J. M. (1968). Subcellular distribution of adrenal tyrosine hydroxylase. *Biochem. Pharmacol.* **17**, 1470–1473.

Muscholl, E. (1961). Effect of cocaine and related drugs on the uptake of noradrenaline by heart and spleen. *Br. J. Pharmac. Chemother,* **16**, 352–359.

Nachmansohn, D. & Berman, M. (1946). Studies on choline acetylase. III. On the preparation of the coenzyme and its effect on the enzyme. *J. biol. Chem.* **165**, 551–563.

Nachmansohn, D. & Machado, A. L. (1943). The formation of acetylcholine. A new enzyme: 'choline acetylase.' *J. Neurophysiol.* **6**, 397–403.

Nagai, T. & Prosser, C. L. (1963a). Patterns of conduction in smooth muscle. *Am. J. Physiol.* **204**, 910–914.

—— (1963b). Electrical parameters of smooth muscle cells. *Am. J. Physiol.* **204**, 915–924.

Nagasawa, J. (1963). The effects of temperature and some drugs on the ionic movements in the smooth muscle of guinea-pig taenia coli. *Tohoku J. exp. Med.* **81**, 222–237.

Nagasawa, J. & Mito, S. (1967). Electron microscopic observations on the innervation of the smooth muscle. *Tohoku J. exp. Med.* **91**, 277–293.

Nagasawa, J. & Suzuki, T. (1967). Electron microscopic study on the cellular interrelationships in the smooth muscle. *Tohoku J. exp. Med.* **91**, 299–313.

Nagatsu, T., Levitt, M. & Udenfriend, S. (1964). Tyrosine hydroxylase. The initial step in norepinephrine biosynthesis. *J. biol. Chem.* **239** (3), 2910–2917.

Nakajima, A. & Horn, L. (1967). Electrical activity of single vascular smooth muscle fibres. *Am. J. Physiol.* **213**, 25–30.

Nakanishi, C., McLean, J. R., Wood, C. & Burnstock, G. (1969). The role of sympathetic nerves in control of the nonpregnant and pregnant human uterus. *J. Reprod. Med.* **2**, 20–23.

22+

Nakayama, S. (1965). Effects of stimulation of the vagus nerve on the movements of the small intestine. *Jap. J. Physiol.* **15**, 243–252.

Narahashi, T., Moore, J. W. & Scott, W. R. (1964). Tetrodotoxin blockage of sodium conductance increase in lobster giant axons. *J. gen. Physiol.* **47**, 965–974.

Nash, C. W., Luchka, E. V. & Jhamandas, K. H. (1966). An investigation of sodium and calcium competition in vascular smooth muscle. *Canad. J. Physiol. Pharmacol.* **44**, 147–156.

Nasonov, D. N. & Aizenberg, E. I. (1937). The effect of non-electrolytes on the water content of live and dead muscles. *Biol. Zbl.* **6**, 165–183.

Nastuk, W. L. & Hodgkin, A. L. (1950). The electrical activity of single muscle fibres. *J. cell. comp. Physiol.* **35**, 39–74.

Nauss, K. M. & Davies, R. E. (1966). Changes in inorganic phosphate and arginine during the development, maintenance and loss of tension in the anterior byssus retractor muscle of *Mytilus edulis*. *Biochem. Z.* **345**, 173–187.

Needham, D. M. (1961). Molecular aspects of the contractile mechanism of the uterus and its changes during pregnancy. In *Progesterone and the Defence Mechanism of Pregnancy*. Ciba Foundation Study Group. No. 9. pp. 40–50. London: Churchill.

Needham, D. M. & Shoenberg, C. F. (1964). Proteins of the contractile mechanism of mammalian smooth muscle and their possible location in the cell. *Proc. R. Soc. B.* **160**, 517–522.

— — (1967). The biochemistry of the myometrium. In *Cellular Biology of The Uterus*. Ed. Wynn, R. M. pp. 291–352. New York: Appleton-Century-Crofts.

Needham, D. M. & Williams, J. M. (1959). Some properties of uterus actomyosin and myofilaments. *Biochem. J.* **73**, 171–181.

—— (1963a). The protein of the dilution precipitate obtained from salt extracts of pregnant and non-pregnant uterus. *Biochem. J.* **89**, 534–545.

—— (1963b). Salt soluble collagen in extracts of uterus muscle and in foetal metamyosin. *Biochem. J.* **89**, 546–552.

—— (1963c). Proteins of the uterine contractile mechanism. *Biochem. J.* **89**, 552–561.

Nemetschek-Gansler, H. (1967). Ultrastructure of the myometrium. In *Cellular Biology of the Uterus*. Ed. Wynn, R. M. pp. 353-385. New York: Appleton-Century-Crofts.

Ng, K. K. F. (1966). The effect of some anticholinesterases on the response of the taenia to sympathetic nerve stimulation. *J. Physiol.* **182**, 233–243.

Nickerson, M. (1949). The pharmacology of adrenergic blockade. *Pharmac. Rev.* **1**, 27–101.

Nicol, J. A. C. (1952). Autonomic nervous systems in lower chordates. *Biol. Rev.* **27**, 1–49.

Nicoll, P. A. (1964). Structure and function of minute vessels in autoregulation. *Circulation Res.* **15**, Suppl. 1, 245–253.

Nicoll, P. A. & Webb, R. L. (1955). Vascular patterns and active vasomotion as determiners of flow through minute vessels. *Angiology*, **6**, 291–308.

Niedergerke, R. & Orkand, R. K. (1966a). The dual effect of calcium on the action potential of the frog's heart. *J. Physiol.* **184**, 291–311.

—— (1966b). The dependence of the action potential of the frog's heart on the external and intracellular sodium concentration. *J. Physiol.* **184**, 312–334.

Nilsson, O. (1964). The relationship between nerves and smooth muscle cells in the rat iris. I. The dilatator muscle. *Z. Zellforsch. mikrosk. Anat.* **64**, 166–171.

Nishihara, H. (1969). Some observations on the fine structure of the guinea-pig taenia coli in hypertonic solution, with special reference to the nexus. *J. Anat* (in press)

Nishio, T. & Saito, K. (1965). Muscular architecture of large intestine. Proc. II Int. Congr. Study Dis. Colon and Rectum, Hedrologicum, Tokyo. 151–155.

Noble, D. (1962a). A modification of the Hodgkin-Huxley equations applicable to Purkinje fibre action and pacemaker potentials. *J. Physiol.* **160**, 317–352.

—— (1962b). The voltage dependence of the cardiac membrane conductance. *Biophys. J.* **2**, 381–393.

—— (1966). Applications of Hodgkin-Huxley equations to excitable tissues. *Physiol. Rev.* **46**, 1–50.

Noble, D. & Stein, R. B. (1966). The threshold conditions for initiation of action potentials by excitable cells. *J. Physiol.* **187**, 129–162.

Nonidez, J. F. (1944). The present status of the neurone theory. *Biol. Rev.* **19**, 30–40.

Nonomura, Y., Hotta, Y. & Ohashi, H. (1966). Tetrodotoxin and manganese ions; effects on electrical activity and tension in taenia coli of guinea-pig. *Science* (*N. Y.*) **152**, 97–99.

Norberg, K.-A. (1964). Adrenergic innervation of the intestinal wall studied by fluorescence microscopy. *Int. J. Neuropharmac.* **3**, 379–382.

—— (1965). The sympathetic adrenergic neuron and certain adrenergic mechanisms. A histochemical study. M.D. Thesis. Stockholm.

—— (1967). Transmitter histochemistry of the sympathetic adrenergic nervous system. *Brain Res.* **5**, 125–170.

Norberg, K.-A. & Hamberger, B. (1964). The sympathetic adrenergic neuron. Some characteristics revealed by histochemical studies on the intraneuronal distribution of the transmitter. *Acta physiol. scand.* **63**, Suppl. 238, 1–42.

Norberg, K.-A. & Sjöqvist, F. (1966). New possibilities for adrenergic modulation of ganglionic transmission. *Pharmac. Rev.* **18**, 743–751.

Ogston, A. G. (1955). Removal of acetylcholine from a limited volume by diffusion. *J. Physiol.* **128**, 222–223.

Ogston, A. G. & Phelps, C. F. (1961). The partition of solutes between buffer solutions and solutions containing hyaluronic acid. *Biochem. J.* **78**, 827–833.

Ogura, Y., Mori, Y. & Watanabe, Y. (1964). Inhibition of the release of acetylcholine from isolated guinea-pig ileum by crystalline tetrodotoxin. *J. Pharmac. exp. Ther.* **154**, 456–462.

Ohashi, H. & Ohga, A. (1967). Transmission of excitation from the parasympathetic nerve to the smooth muscle. *Nature, Lond.* **216**, 291–292.

Okinaka, S., Nakao, K., Ikeda, M. & Shizume, K. (1951). The cardio-accelerator fibres in the vagus nerve. *Tohoku J. exp. Med.* **54**, 393–398.

Olmstead, E. G. (1966). *Mammalian Cell Water.* London: Henry Kimpton.

Oosaki, T., & Ishii, S. (1964). Junctional structure of smooth muscle cells. The ultrastructure of the regions of junction between smooth muscle cells in the rat small intestine. *J. Ultrastruct. Res.* **10**, 567–577.

Orlov, R. S. (1962). On impulse transmission from motor sympathetic nerve to smooth muscle. *Fiziol. Zh. SSSR.* **48**, 342–348.

—— (1963a). Spontaneous electrical activity of smooth muscle cells before and after denervation. *Fiziol. Zh. SSSR.* **49**, 115–121.

—— (1963b). Transmission of inhibitory impulses from nerve to smooth muscle. *Fiziol. Zh. SSSR.* **49**, 575–582.

Overton, E. (1902a). Beiträge zur allgemeinen Muskel und Nervenphysiologie. I. Ueber die osmotischen Eigenschaften der Muskeln. *Pflügers Arch. ges. Physiol.* **92**, 115–280.

—— (1902b). Beiträge zur allgemeinen Muskel und Nervenphysiologie. II. Mittheilung ueber die Unentbehrlichkeit von Natrium (oder Lithium) Ionen für den Contractionsact des Muskels. *Pflügers Arch. ges. Physiol.* **92**, 346–386.

—— (1904). Beiträge zur allgemeinen Muskel und Nervenphysiologie. III. Mittheilung. Studien über die Wirkung der Alkali-und Erdalkalisatze auf Skelettmuskeln und Nerven. *Pflugers Arch. ges. Physiol.* **105**, 176–290.

Owman, C. (1964). Sympathetic nerves probably storing two types of monoamines in the rat pineal gland. *Int. J. Neuropharmac.* **3**, 105–112.

—— (1965). Localization of neuronal and parenchymal monoamines under normal and experimental conditions in the mammalian pineal gland. In *Progr. Brain Res.* Vol. 10. Ed. Ariëns Kappers, J. & Schadé, J. P. pp. 423–453. London: Elsevier.

Paintal, A. S. (1954). The response of the gastric stretch receptors and certain other abdominal and thoracic vagal receptors to some drugs. *J. Physiol.* **126**, 271–285.

—— (1957). Responses from mucosal mechanoreceptors in the small intestine of the cat. *J. Physiol.* **139**, 353–368.

—— (1964). Effects of drugs on vertebrate mechanoreceptors. *Pharmac. Rev.* **16**, 341–380.

Pak, M. J., Walker, Jr. J. L., Greene, E. A., Loh, C. K. & Lorber, V. (1966). Measurement of tracer efflux during cardiac cycle. *Am. J. Physiol.* **211**, 1455–1460.

Palade, G. E. & Palay, S. L. (1954). Electron microscope observations of interneuronal and neuromuscular synapses. *Anat. Rec.* **118**, 335–336.

Palay, S. L. (1956a). Synapses in the central nervous system. *J. biophys. biochem. Cytol.* **2**, Suppl., 193–202.

—— (1956b). Structure and function in the neuron. *Progress in Neurobiology.* (1) *Neurochemistry.* Ed. Korey & Nurnberger. pp. 64–82. New York: Hoeber Inc.

Palay, S. L. & Karlin, L. J. (1959). An electron microscopic study of the intestinal villus I. The fasting animal. *J. biophys. biochem. Cytol.* **5**, 363–371.

Pallie, W. & Pease, D. C. (1958). Prefixation use of hyaluronidase to improve in situ preservation for electron microscopy. *J. Ultrastruct. Res.* **2**, 1–7.

Panagiotis, N. M. & Hungerford, G. F. (1966). Response of pineal sympathetic nerve processes and endings to angiotension. *Nature, Lond.* **211**, 374–376.

Panner, B. J. & Honig, C. R. (1967). Filament ultrastructure and organization in vertebrate smooth muscle. *J. cell Biol.* **35**, 303–321.

Pannier, R. (1946). Contribution à l'innervation sympathique du coeur.—Les nerfs cardio-accélérateurs. *Archs int. Pharmacodyn. Thér.* **73**, 193–259.

Papasova, M. P., Nagai, T. & Prosser, C. L. (1968). Two component slow waves in smooth muscle of cat stomach. *Am. J. Physiol.* **214**, 695–702.

Parker, F. (1958). An electron microscope study of coronary arteries. *Am. J. Anat.* **103**, 247–259.

Parmley, W. W. & Sonnenblick, E. H. (1967). Series elasticity in heart muscle. Its relation to contractile element velocity and proposed muscle models. *Circulation Res.* **20**, 112–123.

Passow, H. (1961). Zusammenwirken von Membranstruktur und Zellstoffwechsel bei der Regulierung der Ionenpermeabilität roter Blutkörperchen. In *Biochemie des Aktiven Transports* (Colloq. Ges. Chem. Mosbach), Berlin: Springer.

Patel, D. J. & Burton, A. C. (1957). Active constriction of small pulmonary arteries in rabbit. *Circulation Res.* **5**, 620–628.

Patlak, C. S. (1960). Derivation of an equation for the diffusion potential. *Nature, Lond.* **188**, 944–945.

Paton, W. D. M. (1955). The response of the guinea-pig ileum to electrical stimulation by coaxial electrodes. *J. Physiol.* **127**, 40–41P.

—— (1957). The action of morphine and related substances on contraction and on acetylcholine output of coaxially stimulated guinea-pig ileum. *Br. J. Pharmac. Chemother*, **12**, 119–127.

—— (1960). In *Adrenergic Mechanisms.* Ciba Foundation Symposium, Eds. Vane, J. R., Wolstenholme, G. E. W. & O'Connor, M. pp. 124–127. London: Churchill.

—— (1961). A theory of drug action based on the rate of drug-receptor combination. *Proc. Roy. Soc. B.* **154**, 21–69.

—— (1964). Electron microscopy of the smooth muscle and nerve networks of guinea-pig ileum. *J. Physiol.* **173**, 20P.

Paton, W. D. M. & Rang, H. P. (1965). The uptake of atropine and related drugs by intestinal smooth muscle of the guinea-pig in relation to acetylcholine receptors. *Proc. Roy, Soc. B.* **163**, 1–44.

Paton, W. D. M. & Rothschild, A. M. (1965). The changes in response and in ionic content of smooth muscle produced by acetylcholine action and by calcium deficiency. *Br. J. Pharmac. Chemother.* **24**, 437–448.

Paton, W. D. M. & Vane, J. R. (1963). An analysis of the responses of the isolated stomach to electrical stimulation and to drugs. *J. Physiol.* **165**, 10–46.

Paton, W. D. M. & Vizi, E. S. (1969). The inhibitory action of noradrenaline and adrenaline on acetylcholine output by guinea-pig ileum longitudinal muscle strip. *Br. J. Pharmac.* **35**, 10–28.

Paton, W. D. M. & Zaimis, E. J. (1949). The pharmacological action of polymethylene bistrimethylammonium salts. *Br. J. Pharmac. Chemother.* **4**, 381–400.

Paton, W. D. M. & Zar, A. M. (1965). A denervated preparation of the longitudinal muscle of the guinea-pig ileum. *J. Physiol.* **179**, 85–86P.

Paton W. D. M. & Zar A. M. (1966). Evidence for transmission of nerve effects by substance P in guinea-pig longitudinal muscle strip. *Abst. III Int. Pharmac. Congr.* São Paulo, Brazil, p. 9.

—— (1968). The origin of acetylcholine released from guinea-pig intestine and longitudinal muscle strips. *J. Physiol.* **194**, 13–34.

Paule, W. J. (1963). Electron microscopy of the newborn rat aorta. *J. Ultrastruct Res.* **8**, 219–235.

Payton, B. W. & Loewenstein, W. R. (1968). Stability of electrical coupling in leech giant nerve cells: Divalent cations, propionate ions, tonicity and pH. *Biochim. biophys. Acta* **150**, 156–158.

Peachey, L. D. (1964). Electron-microscopic observations on the accumulation of divalent cations in intramitochondrial granules. *J. cell. Biol.* **20**, 95–112,

Peachey, L. D. & Huxley, A. F. (1962). Structural identification of twitch and slow striated muscle fibres of the frog. *J. cell. Biol.* **13**, 177–180.

Peachey, L. D. & Porter, K. R. (1959). Intracellular impulse conduction in smooth muscle. *Science, N.Y.* **129**, 721–722.

Peart, W. S. (1949). The nature of splenic sympathin. *J. Physiol.* **108**, 491–501.

Pease, D. C. & Molinari, S. (1960). Electron microscopy of muscular arteries; pial vessels of the cat and monkey. *J. Ultrastruct. Res.* **3**, 447–468.

Pease, D. C. & Paule, W. J. (1960). Electron microscopy of elastic arteries; the thoracic aorta of the rat. *J. Ultrastruct. Res.* **3**, 469–483.

Pease, D. C. & Quilliam, T. A. (1957). Electron microscopy of the Pacinian corpuscle. *J. biophys. biochem. Cytol.* **3**, 331–342.

Pellegrino de Iraldi, A. & De Robertis, E. (1961). Action of reserpine on the submicroscopic morphology of the pineal gland. *Experientia*, **17**, 122–124.

—— (1963). Action of reserpine, iproniazid and pyrogallol on nerve endings of the pineal gland. *Int. J. Neuropharmac.* **2**, 231–239.

Pellegrino de Iraldi, A., Duggan, H. F. & De Robertis, E. (1963). Adrenergic synaptic vesicles in the anterior hypothalamus of the rat. *Anat. Rec.* **145**, 521–531.

Pellegrino de Iraldi, A., Zieher, L. M. & De Robertis, E. (1965). Ultrastructure and pharmacological studies of nerve endings in pineal organs. *Prog. Brian Res.* **10**, 389–421.

Perry, S. V. & Corsi, A. (1958). Extraction of proteins other than myosin from the isolated rabbit myofibril. *Biochem. J.* **68**, 5–12.

Persoff, D. A. (1960). A comparison of methods for measuring efflux of labelled potassium from contracting rabbit atria. *J. Physiol.* **152**, 354–366.

Peterson, H. (1936a). Rhythmische Spontankontraktionen an Gefässen. *Z. Biol.* **97**, 378–392.

—— (1936b). Die elektrischen Erscheinungen an Arterienstreifen von Warmblütern. *Z. Biol.* **97**, 393–398.

Peterson, L. H. (1966). Physical factors which influence vascular caliber and blood flow. *Circulation Res.* **18**, Suppl. 1, 3–11.

Phelps, C. F. (1965). The physical properties of inulin solutions. *Biochem. J.* **95**, 41–47.

Picarelli, Z. P., Hyppolito, N. & Valle, J. R. (1962). Synergistic effect of 5-hydroxytryptamine on the response of rats' seminal vesicles to adrenaline and noradrenaline. *Archs. int. Pharmacodyn. Thér*, **138**, 354–363.

Pick, J. (1963). The submicroscopic organisation of the sympathetic ganglion in the frog (*Rana pipiens*). *J. comp. Neurol.* **120**, 409–462.

Pickles, V. R. (1967). The prostaglandins. *Biol. Rev.* **42**, 614–652.

Policard, A., Collet, A. & Prégermain, S. (1960). Observations au microscope électronique sur quelques vaisseaux pulmonaires. *Bull. micr. appl.* **10**, 17–27.

Politoff, A., Socolar, S. J. & Loewenstein, W. R. (1967). Metabolism and the permeability of cell membrane junctions. *Biochim. Biophys. Acta* **135**, 791–793.

Poloni, A. (1955). Il muscolo dorsale di sanguisuga quale test biologico per l'evidenziamento dell'attivita serotoninica nei liquidi organici. *Cervello*, **31**, 472–476.

Ponder, E. (1948). *Haemolysis and related phenomena*. London: Churchill.

Porter, K. R. & Franzini-Armstrong, C. (1965). The sarcoplasmic reticulum. *Scientific Am.* **212** (3) 73–80.

Portzehl, H. (1957). Die Bindung des Erschlaffungsfaktors von Marsh an die Muskelgrana. *Biochim. biophys. Acta* **26**, 373–377.

Potter, J. M. & Sparrow, M. P. (1968). The relationship between calcium content of depolarized mammalian smooth muscle and its contractility in response to acetylcholine. *Aust. J. exp. Biol. med Sci.* **46**, 435–446.

Potter, L. T. (1966). Storage of norepinephrine in sympathetic nerves. *Pharmac. Rev.* **18**, 439–451.

—— (1967). Role of intraneuronal vesicles in the synthesis, storage and release of catecholamines. *Circulation Res.* **21**, (Suppl. 3), 13–24.

Potter, L. T. & Axelrod, J. (1962). Intracellular localization of catecholamines in tissues of the rat. *Nature, Lond.* **194**, 581–582.

—— (1963a). Studies on the storage of norepinephrine and the effect of drugs. *J. Pharmac. exp. Ther.* **140**, 199–206.

—— (1963b). Properties of norepinephrine storage particles of the rat heart. *J. Pharmac. exp. Ther.* **142**, 299–305.

Potter, L. T., Axelrod, J. & Wolfe, D. (1962). Norepinephrine storage vesicles. *Pharmacologist* **4**, 168.

Potter, L. T., Cooper, T., Willman, V. L. & Wolfe, D. E. (1965). Synthesis, binding, release and metabolism of norepinephrine in normal and transplanted dog hearts. *Circulation Res.* **16**, 468–481.

Prosser, C. L. (1962). Conduction in nonstriated muscles. *Physiol. Rev.* **42**, Suppl. 5, 193–206.

Prosser, C. L., Burnstock, G. & Kahn, J. (1960). Conduction in smooth muscle: Comparative structural properties. *Am. J. Physiol.* **199**, 545–552.

Prosser, C. L. & Sperelakis, N. (1956). Transmission in ganglion-free circular muscle from the cat intestine. *Am. J. Physiol.* **187**, 536–545.

Prosser, C. L., Smith, C. E. & Melton, C. E. (1955). Conduction of action potentials in the ureter of the rat. *Am. J. Physiol.* **181**, 651–660.

Prusoff, W. H. (1960). The distribution of 5-hydroxytryptamine and adenosine triphosphate in cytoplasmic particles of the dog's small intestine. *Br. J. Pharmac. Chemother.* **15**, 520–524.

Puestow, C. B. (1932). The activity of isolated intestinal segments. *Archs Surg., Chicago* **24**, 565–573.

Pye, K. & Chance, B. (1966). Sustained sinusoidal oscillations of reduced pyridine nucleotide in a cell-free extract of *Saccharomyces carlsbergensis*. *Proc. natn. Acad. Sci. U.S.A.* **55**, 888–894.

Raab, W. (1943). Adrenaline and related substances in blood and tissues. *Biochem. J.* **37**, 470–473.

Raiford, T. & Mulinos, M. G. (1934). The myenteric reflex as exhibited by the exteriorized colon of the dog. *Am. J. Physiol.* **110**, 129–136.

Ramón Y Cajal, S. (1888). Estructura de los centros nerviosos de las aves. *Revist. trimestral de Histologia normal y patologica.* **1**, Moya: Madrid.

—— (1890a). Sur les fibres nerveuses de la couche granuleuse du cervelet et sur l'évolution des éléments cerebelleux. *Int. Mschr. Anat. Physiol.* **7**, 12–31.

—— (1890b). Sur l'origine et les ramifications des fibres nerveuses de la moelle embryonnaire. *Anat. Anz.* **5**, 85–95.

—— (1890c). Réponse à Mr. Golgi à propos des fibrilles collatérales de la moèlle épinère et de la structure générale de la substance grise. *Anat. Anz.* **5**, 579–587.

—— (1909). *Histologie du système nerveux de l'homme et des vertébrés, Vol.* 1, p. 986. Paris: Maloine.

—— (1911). *Histologie du système nerveux de l'homme et des vertébrés.* Vol. 2, p. 933. Paris: Maloine.

—— (1934). Les preuves objectives de l'unité anatomique des cellules nerveuses. *Trab. Lab. Invest. biol. Univ. Madr.* **29**, 1–137.

Ramwell, P. W. & Shaw, J. E. (1967). Prostaglandin release from tissues by drug, nerve and hormone stimulation. *Prostaglandins.* Proc. 2nd Nobel Symp. Stockholm. Ed. Bergström, S. & Samuelsson, B. pp. 283–292. New York: Interscience.

Ramwell, P. W., Shaw, J. E. & Kucharski, J. (1965). Prostaglandin release from the rat phrenic nerve-diaphragm preparation. *Science, N.Y.* **149**, 1390–1391.

Ramwell, P. W., Shaw, J. E., Douglas, W. W. & Poisner, A. M. (1966). Efflux of prostaglandin from adrenal glands stimulated with acetylcholine. *Nature, Lond.* **210**, 273–274.

Rand, M. J. & Ridehalgh, A. (1965). Actions of hemicholinium and triethylcholine on responses of guinea-pig colon to stimulation of autonomic nerves. *J. Pharm. Pharmac.* **17**, 144–156.

Rand, M. J. & Stafford, A. (1964). Responses of oesophageal preparations from various species to stimulation of the vagus nerves. *Br. J. Pharmac. Chemother.* **22**, 4.

Rand, M. J. & Wilson, J. (1967). The actions of some adrenergic neurone blocking drugs at cholinergic junctions. *Europ. J. Pharmacol.* **1**, 210–221.

Randić, M. & Straughan, D. W. (1964). Antidromic activity in the rat phrenic nerve-diaphragm preparation. *J. Physiol.* **173**, 130–148.

Rapport, M. M. (1968). Discussion of the possible mechanism of action of serotonin on molluscan muscle. *Adv. Pharmacol.* Ed. Garattini, S. & Shore, P. A. Vol. 6B, 16–17.

Read, J. B. & Burnstock, G. (1968). Comparative histochemical studies of adrenergic nerves in the enteric plexuses of vertebrate large intestine. *Comp. Biochem. Physiol.* **27** 505–517.

—— (1969). Adrenergic innervation of the gut musculature in vertebrates. *Histochemie.* **17**, 263–272.

Reas, H. W. & Tsai, T. H. (1966). The antagonism by atropine of the response of the nictitating membrane to sympathetic nerve stimulation. *J. Pharmac. exp. Ther.* **152**, 186–196.

Reid, G. & Rand, M. (1952). Pharmacological actions of synthetic 5-hydroxytryptamine (serotonin, thrombocytin). *Nature, Lond.* **169**, 801–802.

Reinke, D. A., Rosenbaum, A. H. & Bennett, D. R. (1967). Patterns of dog gastrointestinal contractile activity monitored *in vivo* with extraluminal force transducers. *Am. J. dig. Dis.* **12**, 113–141.

Remington, J. W. Ed. (1957). *Tissue Elasticity.* Baltimore: Waverley Press.

Renkin, E. M. (1961). Permeability of frog skeletal muscle cells to choline. *J. gen. Physiol.* **44**, 1159–1164.

Renkin, E. M. & Rosell, S. (1962). Effects of different types of vasodilator mechanisms on vascular tonus and on transcapillary exchange of diffusible material in skeletal muscle. *Acta physiol. scand.* **54**, 241–251.

Reuben, J. P., Brandt, P. W., Garcia, H. & Grundfest, H. (1967). Excitation-contraction coupling in crayfish. *Am. Zool.* **7**, 623–645.

Revel, J. P., Olson, W. & Karnovsky, M. J. (1967). A twenty-Angström gap junction with hexagonal array of subunits in smooth muscle. *J. Cell. Biol.* **35**, 112A.

Rexed, B. & Von Euler, U. S. (1951). The presence of histamine and noradrenaline in nerves as related to their content of myelinated and unmyelinated fibres. *Acta psychiat. neurol. scand.* **26**, 61–65.

Reynolds, S. R. M. (1965). *Physiology of the Uterus.* 2nd Ed. 2nd printing. New York: Hafner.

Reynolds, S. R. M., Harris, J. S. & Kaiser, I. H. (1954). *Clinical Measurements of the Uterine Forces in Pregnancy and Labor.* 1st ed. Springfield: Thomas.

Rhodin, J. A. G. (1962). Fine structure of vascular walls in mammals, with special reference to smooth muscle component. *Physiol Rev.* **42**, Suppl. 5., 48–81.

—— (1963). *An Atlas of Ultrastructure.* Philadelphia: Saunders.

—— (1967). The ultrastructure of mammalian arterioles and precapillary sphincters. *J. Ultrastruct. Res.* **18**, 181–223.

Richardson, J. A. & Woods, E. F. (1959). Release of norepinephrine from the isolated heart. *Proc. Soc. exp. Biol. Med.* **100**, 149–151.

Richardson, K. C. (1958). Electron microscopic observations on Auerbach's plexus in the rabbit with special reference to the problem of smooth muscle innervation. *Am. J. Anat.* **103**, 99–136.

—— (1960). Studies on the structure of autonomic nerves in the small intestine correlating the silver-impregnated image in the light microscopy with the permanganate-fixed ultrastructure in electronmicroscopy. *J. Anat.* **94**, 457–472.

Richardson, K. C. (1962). The fine structure of autonomic nerve endings in smooth muscle of the rat vas deferens. *J. Anat.* **96**, 427–442.

—— (1963). Structural and experimental studies on autonomic nerve endings in smooth muscle. *Biochem. Pharmac.* **12**, Suppl., p. 10.

—— (1964). The fine structure of the albino rabbit iris with special reference to the identification of adrenergic and cholinergic nerves and nerve endings in its intrinsic muscles. *Am. J. Anat.* **114**, 173–205.

—— (1966). Electronmicroscopic identification of autonomic nerve endings. *Nature, Lond.* **210**, 756.

Riker, W. F. (1960). Pharmacologic considerations in a reevaluation of the neuromuscular synapse. *Archs. Neurol. Psychiat. (Chicago).* **3**, 488–499.

—— (1966). Actions of acetylcholine on mammalian motor nerve terminal. *J. Pharmac. exp. Ther.* **152**, 397–416.

Riker, W. F., Roberts, J., Standaert, F. G. & Fujimori, H. (1957). The motor nerve terminal as the primary focus for drug-induced facilitation of neuromuscular transmission. *J. Pharmac. exp. Ther.* **121**, 286–312.

Riker, W. F. & Standaert, F. G. (1966). The action of facilitatory drugs and acetylcholine on neuromuscular transmission. *Ann. N.Y. Acad. Sci.* **135**, 163–176.

Riker, W. F., Werner, G., Roberts, J. & Kuperman, A. (1959). Pharmacological evidence for the existence of a presynaptic event in neuromuscular transmission. *J. Pharmac. exp. Ther.* **125**, 150–158.

Ringer, S. (1883). A further contribution regarding the influence of the different constituents of the blood on the contraction of the heart. *J. Physiol.* **4**, 29–42.

Ritchie, J. A. (1968). Colonic motor activity and bowel function. *Gut* **9**, 442-456.

Ritchie, J. A., Ardran, G. M. & Truelove, S. C. (1962). Motor activity of the sigmoid colon of humans. A combined study by intraluminal pressure recording and cineradiography. *Gastroenterology* **43**, 642–668.

Ritzén, M. (1966). Quantitative fluorescence microspectrophotometry of catecholamine-formaldehyde products. *Expl Cell Res.* **44**, 505–520.

—— (1967). Cytochemical identification and quantitation of biogenic monoamines—a microspectrophotometric and autoradiographic study. M.D. Thesis, Stockholm.

Robert, A., Nezamis, J. E. & Phillips, J. P. (1967). Inhibition of gastric secretion by prostaglandins. *Am. J. dig. Dis.* **12**, 1073–1076.

Roberts, G. C. K. (1967). Studies on mechanisms of action and storage of biologically active amines. Ph.D. Thesis. University of London.

Roberts, M. H. T. & Straughan, D. W. (1967). An excitatory effect of 5-hydroxytryptamine on single cerebral cortical neurones. *J. Physiol.* **188**, 27P.

Robertson, J. D. (1955). Recent electron microscope observations on the ultrastructure of the crayfish median-to-motor giant synapse. *Expl. Cell Res.* **8**, 226–229.

—— (1956). The ultrastructure of a reptilian myoneural junction. *J. biophys. biochem. Cytol.* **2**, 381–394.

—— (1961). Ultrastructure of excitable membranes and the crayfish median-giant synapse. *Ann. N.Y. Acad. Sci.* **94**, 339–387.

—— (1963). The occurrence of a subunit pattern in the unit membranes of club endings in Mauthner cell synapses in goldfish brain. *J. cell Biol.* **19**, 201–221.

Robertson, J. D., Bodenheimer, T. S. & Stage, D. E. (1963). The ultrastructure of Mauthner cell synapses and nodes in goldfish brains. *J. cell Biol.* **19**, 159–199.

Robertson, P. A. (1954). Potentiation of 5-hydroxytryptamine by the true-cholinesterase inhibitor 284C51. *J. Physiol.* **125**, 37–38P.

—— (1960). Calcium and contractility in depolarized smooth muscle. *Nature, Lond.* **186**, 316–317.

Robinson, G. A., Butcher, R. W. & Sutherland, E. W. (1967). Adenyl cyclase as an adrenergic receptor. *Ann. N.Y. Acad. Sci.* **139**, 703–723.

Robinson, P. M. (1969a). The fine structure of the innervation of the seminal vesicle of the rat (unpublished observation).

—— (1969b). A cholinergic component of the innervation of the guinea-pig vas deferens: The fine structural localization of cholinesterase. *J. Cell Biol.* **41**, 462–476.

Robinson, P. M. & Bell, C. (1967). The localization of acetylcholinesterase at the autonomic neuromuscular junction. *J. cell Biol.* **33**, 93–102.

Robinson, R. A. & Stokes, R. H. (1955). *Electrolyte solutions*. London: Butterworths.

Robson, J. M. & Schild, H. D. (1938). Response of the cat's uterus to the hormones of the posterior pituitary lobe. *J. Physiol.* **92**, 1–8.

Rocha E Silva, M., Valle, J. R. & Picarelli, Z. P. (1953). A pharmacological analysis of the mode of action of serotonin (5-hydroxytryptamine) upon the guinea-pig ileum. *Br. J. Pharmac. Chemother.* **8**, 378–388.

Roddie, I. C. (1962). The transmembrane potential changes associated with smooth muscle activity in turtle arteries and veins. *J. Physiol.* **163**, 138–150.

Rogers, D. C. (1964). Comparative electromicroscopy of smooth muscle and its innervation. Ph.D. Thesis, Zoology Dept., University of Melbourne.

Rogers, D. C. & Burnstock, G. (1966a). The interstitial cell and its place in the concept of the autonomic ground plexus. *J. comp. Neurol.* **126**, 255–284.

—— (1966b). Multiaxonal autonomic functions in intestinal smooth muscle of the toad. (*Bufo marinus.*) *J. comp. Neurol.* **126**, 625–652.

Rosell, S., Kopin, I. J. & Axelrod, J. (1964). Release of tritiated epinephrine following sympathetic nerve stimulation. *Nature, Lond.* **201**, 301.

—— (1963). Fate of H³ noradrenaline in skeletal muscle before and following sympathetic stimulation. *Am. J. Physiol.* **205**, 317–321.

Rosell, S. & Uvnäs, B. (1962). Vasomotor nerve activity and oxygen uptake in skeletal muscle of the anaesthetized cat. *Acta physiol. scand.* **54**, 209–222.

Rosenbaum, A. H., Reinke, D. A. & Bennett, D. R. (1967). *In vivo* force, frequency and velocity of dog gastro-intestinal contractile activity. *Amer. J. dig. Dis.* **12**, No. 2. 142–153.

Rosenberg, P. & Podleski, T. R. (1963). Ability of venoms to render squid axons sensitive to curare and acetylcholine. *Biochim. biophys. Acta* **75**, 104–115.

Rosenblueth, A. (1950). *The Transmission of Nerve Impulses at Neuroeffector Junctions and Peripheral Synapses*. Technol. Press, Mass. Inst. Technol. New York: J. Wiley.

Rosenblueth, A. & Bard, P. (1932). The innervation and functions of the nictitating membrane in the cat. *Am. J. Physiol.* **100**, 537–544.

Rosenblueth, A. & Cannon, W. B. (1935). The chemical mediation of sympathetic vasodilator nerve impulses. *Am. J. Physiol.* **112**, 33–40.

Rosenblum, W. I. (1965). Cerebral microcirculation: a review emphasizing the interrelationship of local blood flow and neuronal function. *Angiology* **16**, 485–507.

Rosenbluth, J. (1965a). Ultrastructure of somatic muscle cells in *Ascaris lumricoides*. *J. cell Biol.* **26**, 579–592.

—— (1965b). Smooth muscle: an ultrastructural basis for the dynamics of its contraction. *Science, N.Y.* **148**, 1337–1339.

Ross, L. L. (1959). Electron microscopic observations of the carotid body of the cat. *J. biophys. biochem. Cytol.* **6**, 253–262.

Ross, R. & Klebanoff, S. J. (1967). Fine structural changes in uterine smooth muscle and fibroblasts in response to oestrogen. *J. cell Biol.* **32**, 155–168.

Rostgaard, J. & Barrnett, R. J. (1964). Fine structure localization of nucleoside phosphates in relation to smooth muscle cells and unmyelinated nerves in the small intestine of the rat. *J. Ultrastruct. Res.* **11**, 193–207.

Roth, R. H., Stjärne, L. & Von Euler, U. S. (1967). Factors influencing the rate of norepinephrine biosynthesis in nerve tissue. *J. Pharmac. exp. Ther.* **158**, 373–377.

Rothberger, C. J. (1926). Allgemeine Physiologie des Herzens. *Handb. norm. u. pathol. Physiol.* **7**, 523–662. Berlin: Springer.

Rothlin, Von E. & Brügger, J. (1945). Quantitative Untersuchungen der sympathikolytischen Wirkung genuiner Mutterkornalkaloide und derer Dihydroderivate am isolierten Uterus des Kaninchens. *Helv. physiol. pharmac. Acta.* **3**, 519–535.

Roy, C. S. (1881). The physiology and pathology of the spleen. *J. Physiol.* **3**, 203–228.

Rudolph, L. & Ivy, A. C. (1930). The physiology of the uterus in labor, an experimental study of the dog and rabbit. *Am. J. Obstet. Gynec.* **19**, 317–335.

Rudzik, A. D. & Miller, J. W. (1962). The effect of altering the catecholamine content of

22*

the uterus on the rate of contractions and the sensitivity of the myometrium to relaxin. *J. Pharmacol.* **138**, 88–95.

Rugh, R. (1962). *Experimental Embryology.* Burgess: Minneapolis.

Rushton, W. A. H. (1927). The effect upon the threshold for nervous excitation of the length of nerve exposed, and the angle between current and nerve. *J. Physiol.* **63**, 357–377.

—— (1937). The initiation of the nervous impulse. *J. Physiol.* **90**, 5–6P.

—— (1938). Initiation of the propagated disturbance. *Proc. R. Soc. B.* **124**, 210–243.

Saelens, J. K. & Stoll, W. R. (1965). Radiochemical determination of choline and acetylcholine flux from isolated tissue. *J. Pharmac. exp. Ther.* **147**, 336–343.

Sakuma, A. & Beck, L. (1961). Pharmacological evidence for active reflex dilatation. *Am. J. Physiol.* **201**, 129–133.

Salerno, P. R. & Coon, J. M. (1949). A pharmacologic comparison of hexaethyltetraphosphate (HETP) tetraethylpyrophosphate (TEPP), with physostigmine, neostigmine and DFP. *J. Pharmac. exp. Ther.* **95**, 240–255.

Samaan, A. (1935). The antagonistic cardiac nerves and heart rate. *J. Physiol.* **83**, 332–340.

Samarasinghe, D. D. (1965). The innervation of the cerebral arteries in the rat: an electron microscope study. *J. Anat.* **99**, 815–828.

Sato, A. (1960). Electrophysiological studies of the working mechanism of muscle walls of the stomach. *Jap. J. Physiol.* **10**, 359–373.

Sato, S. (1966). An electronmicroscopic study on the innervation of the intra-cranial artery of the rat. *Am. J. Anat.* **118**, 873–890.

Sauer, M. E. & Rumble, C. T. (1946). The number of nerve cells in the myenteric and submucous plexuses of the small intestine of the cat. *Anat. Rec.* **96**, 373–381.

Schabadasch, A. (1934). Studien zur Architektonik des vegetativen Nervensystems. I. Neue intramurale Nervengeflechte der Harnblase und des Harnleiters. *Z. Zellforsch. mikrosk. Anat.* **21**, 657–732.

Schachter, M. (1952). The release of histamine by pethidine, atropine, quinine and other drugs. *Br. J. Pharmac. Chemother.* **7**, 646–654.

Schanker, L. S., Nafpliotis, P. A. & Johnson, J. M. (1961). Passage of organic bases into human red cells. *J. Pharmac. exp. Ther.* **133**, 325–331.

Schaeppi, U. & Koella, W. P. (1964). Reaction of isolated pig iris sphincter to electrical stimulation and acetylcholine. *Am. J. Physiol.* **206**, 255–261.

Schanne, O., Kawata, H., Schäfer, B. & Lavallée, M. (1966). A study on the electrical resistance of the frog sartorius muscle. *J. gen. Physiol.* **49**, 897–912.

Schatzmann, H. J. (1961). Calciumaufnahme und -abgabe am Darmmuskel des Meerschweinchens. *Pflügers Arch. ges. Physiol.* **274**, 295–310.

—— (1964a). Erregung und Kontraktion glatter Vertebratenmuskeln. *Ergebn. Physiol.* **55**, 28–130.

—— (1964b). Excitation, contraction and calcium in smooth muscle. In *Pharmacology of Smooth Muscle.* Ed. Bülbring, E. pp. 57–69. Pergamon Press.

Schaumann, O., Jochum, K. & Schmidt, H. (1953). Analgetika und Darmmotorik. III Zum Mechanismus der Peristaltik. *Naunyn-Schmiedebergs. Arch. exp. Path. Pharmak.* **219**, 302–309.

Schaumann, W. (1957a). Inhibition by morphine of the release of acetylcholine from the intestine of the guinea-pig. *Br. J. Pharmac. Chemother* **12**, 115–118.

—— (1957b). Resistenz einiger Wirkungen peripherer Vagusreizung gegenuber Ganglienblockern. *Arch. exp. Path. Pharmak.* **231**, 378–387.

—— (1958). Zusammenhänge zwischen der Wirkung der Analgetica und Sympathicomimetica auf den Meerschweinchen Dünndarm. *Arch. exp. Path. Pharmak.* **233**, 112–124.

Schild, H. O. (1947). pA, a new scale for the measurement of drug anatogism. *Br. J. Pharmac. Chemother.* **2**, 189–206.

—— (1964). Calcium and the effects of drugs on depolarized smooth muscle. In *Pharmacology of Smooth Muscle.* Ed. E. Bülbring pp. 95–104. Oxford: Pergamon Press.

—— (1966). Calcium and the relaxant effect of isoproterenol in the depolarized rat uterus. *Pharmac. Rev.* **18**, 495–501.

Schlaepfer, W. W. (1968). Acetylcholinesterase activity of motor and sensory nerve fibres in spinal nerve roots of rat. *Z. Zellforsch. mikrosk. Anat.* **88**, 441–456.

Schlaepfer, W. W. & Torack, R. M. (1966). The ultrastructural localization of cholinesterase activity in the sciatic nerve of the rat. *J. Histochem. Cytochem.* **14**, 369–378.

Schofield, B. M. (1952). The innervation of the cervix and cornu uteri in the rabbit. *J. Physiol.* **117**, 317–328.

—— (1954). The influence of estrogen and progesterone on the isometric tension of the uterus in the intact rabbit. *Endocrinology.* **55**, 142–147.

—— (1955). The influence of the ovarian hormones on myometrial behaviour in the intact rabbit. *J. Physiol.* **129**, 289–304.

—— (1964). Hormonal control of myometrial contractions. In *Pharmacology of Smooth Muscle.* Ed. Bülbring, E. pp. 105–111. Oxford: Pergamon Press.

Schofield, G. C. (1962). Experimental studies on the myenteric plexus in mammals. *J. comp. Neurol.* **119**, 159–186.

Schneider, R. (1966). The longitudinal muscle component of the peristaltic reflex in the guinea-pig isolated ileum. *Br. J. Pharmac. Chemother.* **27**, 387–397.

Schümann, H. J. (1958b). Über die Verteilung von Noradrenalin und Hydroxytyramin in symathischen Nerven (Milznerven.) *Arch. exp. Path. Pharmak.* **234**, 17–25.

Schwann, Th. (1847). *Microscopical Researches into the Accordance in Structure and Growth of Animals and Plants.* Translated from German, by Henry Smith, London, Sydenham Society.

Scott, B. L. & Pease, D. C. (1959). Electronmicroscopy of the salivary and lachrimal glands of the rat. *Am. J. Anat.* **104**, 115–161.

Secker, J. (1934). The humoral control of the secretion by the submaxillary gland of the cat following chorda stimulation. *J. Physiol.* **81**, 81–92.

Sedvall, G. C. & Kopin, I. J. (1967). Acceleration of norepinephrine synthesis in the rat submaxillary gland *in vivo* during sympathetic nerve stimulation. *Life Sci. Oxford.* **6**, 45–51.

Seller, H., Langhorst, P., Polster, T. & Koepchen, H. P. (1967). Zeitliche Eigenschaften der Vasomotorik II. Erscheinungsformen und Entstehung spontaner und nervös induzierter Gefässrhythmen. *Pflügers Arch. ges. Physiol.* **296**, 110–132.

Selverston, A. (1967). Structure and function of the transverse tubular system in crustacean muscle fibers. *Am. Zool.* **7**, 515–525.

Semba, T., Fujii, K. & Kimura, N. (1964). The vagal inhibitory responses of the stomach to stimulation of the dog's medulla oblongata. *Jap. J. Physiol.* **14**, 319–327.

Setekleiv, J. (1964). Uterine motility of the estrogenized rabbit. I. Isotonic and isometric recording *in vivo*. Influence of anaesthesia and temperature. *Acta physiol. scand.* **62**, 68–78.

—— (1967). Factors influencing the ^{42}K efflux from the smooth muscle of guinea-pig taenia coli. *J. Physiol.* **188**, 39–40P.

Shanes, A. M. (1958). Electrochemical aspects of physiological and pharmacological action in excitable cells. *Pharmac. Rev.* **10**, 59–273.

Shaw, J. E. (1966). Prostaglandin release from adipose tissue *in vitro* evoked by nerve stimulation or catecholamines. *Fedn Proc.* **25**, 770.

Shaw, J. E. & Ramwell, P. W. (1967). Prostaglandin release from the adrenal gland. *Prostaglandins*, Proc. 2nd Nobel Symp. Stockholm. Ed. Bergström, S. & Samuelsson, B. pp. 293–299. New York: Interscience.

Shelley, H. (1955). A correlation between cholinesterase inhibition and increase in muscle tone in rabbit duodenum. *Br. J. Pharmac. Chemother.* **10**, 26–35.

Sheng, P.-K. & Tsao, T.-C. (1955). A comparative study of nucleotropomyosins from different sources. *Sci. Sinica* **4**, 157–176.

Shestopalova, N. M. (1965). The fine structure of smooth muscle cells in the small intestine of cotton rats. In *Electron Microscopy*, Ed. Titlbach, M. Czech Acad. Sci. (Prague) Vol. B, 91–92.

Shibata, S. & Briggs, A. H. (1966). The relationships between electrical and mechanical events in rabbit aortic strips. *J. Pharmac. exp. Ther.* **153**, 466–470.

Shimizu, N. & Ishii, S. (1964). Electronmicroscopic observation of catecholamine-containing granules in the hypothalamus and area postrema and their changes following reserpine injection. *Arch. histol. jap.* **24**, 489–497.

Shimizu, N. & Ishii, S. (1966). Electron microscopic histochemistry of acetylcholinesterase of rat brain by Karnovsky's method. *Histochemie* **6**, 24–33.

Shoenberg, C. F. (1958). An electron microscope study of smooth muscle in pregnant uterus of rabbit. *J. biophys. biochem. Cytol.* **4**, 609–614.

—— (1962). Some electron microscope observations on the contraction mechanisms in vertebrate smooth muscle. In *Proc. 5th Int. Congr. Electron Microscopy* Vol. 2, M-8. Ed. Breese, S. S. New York: Academic Press.

—— (1965). Contractile proteins of vertebrate smooth muscle. *Nature, Lond.* **206**, 526–527.

—— (1969). An electron microscopic study of the influence of divalent ions on myosin filament formation in chicken gizzard extracts and homogenates. *Tissue & Cell.* **1**, 83–96.

Shoenberg, C. F., Rüegg, J. C., Needham, D. M., Schirmer, R. H. & Nemetchek-Gansler, H. (1966). A biochemical and electron microscope study of the contractile proteins in vertebrate smooth muscle. *Biochem. Z.* **345**, 255–266.

Shuba, M. F. (1961). Electrotonus in smooth muscle. *Biofizika* **6**, 56–64.

—— (1965). Electrical properties of smooth muscle. *Biofizika* **10**, 67–76.

Shute, C. C. D. & Lewis, P. R. (1966). Electron microscopy of cholinergic terminals and acetylcholinesterase-containing neurones in the hippocampal formation of the rat. *Z. Zellforsch. mikrosk. Anat.* **69**, 334–343.

Siliotti, I. & Grebella, P. (1959). Isolamento di mitocondri da omogento di miometrio umano. *Attual Ostet. Ginec.* **4**, 821–826.

Simpson, F. O. & Devine, C. E. (1966). The fine structure of autonomic neuromuscular contacts in arterioles of sheep renal cortex. *J. Anat.* **100**, 127–137.

Sjöstrand, F. S. (1958). Ultrastructure of retinal rod synapses of the guinea-pig eye as revealed by three-dimensional reconstructions from serial sections. *J. Ultrastruct. Res.* **2**, 122–170.

Sjöstrand, F. S., Andersson-Cedergren, E. & Dewey, M. M. (1958). The ultrastructure of the intercalated discs of frog, mouse and guinea-pig cardiac muscle. *J. Ultrastruct. Res.* **1**, 271–287.

Sjöstrand, F. S. & Elfvin, L.-G. (1962). The layered asymmetric structure of the plasma membrane in the exocrine pancreas cells of the cat. *J. Ultrastruct. Res.* **7**, 504–534.

Sjöstrand, F. S. & Wetzstein, R. (1956). Elektronenmikroskopische Untersuchung der phäochromen (chromaffinen) Granula in den Markzellen der Nebenniere. *Experientia* **12**, 196–199.

Sjöstrand, N. O. (1965). The adrenergic innervation of the vas deferens and the accessory male genital glands. An experimental and comparative study of its anatomical and functional organization in some mammals, including the presence of adrenaline and chromaffin cells in these organs. *Acta physiol. scand.* **65**, Supp.. 257, 1–82.

Smith, A. N. (1953a). The effect of compound 48/80 on acid gastric secretion in the cat. *J. Physiol.* **119**, 233–243.

—— (1953b). Release of histamine by the histamine liberator compound 48/80 in cats. *J. Physiol.* **121**, 517–538.

Smith, C. A. & Sjöstrand, F. S. (1961). Structure of the nerve endings on the external hair cells of the guinea-pig cochlea as studied by serial sections. *J. Ultrastruct. Res.* **5**, 523–555.

Smith, D. S. & Treherne, J. E. (1965). The electron microscopic localization of cholinesterase activity in the central nervous system of an insect (*Periplaneta americanal.*) *J. cell. Biol.* **26**, 445–459.

Sola, O. M. & Martin, A. W. (1953). Denervation hypertrophy and atrophy of hemidiaphragm of the rat. *Am. J. Physiol.* **172**, 324–332.

Solomon, A. K. (1960). Compartmental methods of kinetic analysis. In *Mineral Metabolism*. Vol. 1A. Comar, C. L. & Bronner, F. pp. 119–167. New York: Academic Press.

Somlyo, A. P. & Somlyo, A. V. (1968a). Vascular smooth muscle. I. Normal structure, pathology, biochemistry and biophysics. *Pharmacol. Rev.* **20**, 197–272.

—— (1968b). Electromechanical and pharmacomechanical coupling in vascular smooth muscle. *J. Pharmac. exp. Ther.* **159**, 129–145.

Somlyo, A. V., Woo, C.-Y. & Somlyo, A. P. (1965). Responses of nerve-free vessels to vasoactive amines and polypeptides. *Am. J. Physiol.* **208**, 748–753.

Sourkes, T. L. (1966). Dopa decarboxylase: substrates, coenzyme, inhibitors. *Pharmac. Rev.* **18**, 53–60.

Sparks, H. V. (1964). Effect of quick stretch on isolated vascular smooth muscle. *Circulation Res.* **15**, Suppl. 1, 254–260.

Sparrow, M. P., Mayrhofer, G. & Simmons, W. J. (1967). Uptake and increased binding by smooth muscle in half isotonic sucrose and its relationship to contractility. *Aust. J. exp. Biol. med. Sci.* **45**, 469–484.

Sparrow, M. P. & Simmonds, W. J. (1965). The relationship of the calcium content of smooth muscle to its contractility in response to different modes of stimulation. *Biochem. biophys. Acta* **109**, 503–511.

Spector, S., Gordon, R., Sjoerdsma, A. & Udenfriend, S. (1967). End-product inhibition of tyrosine hydroxylase as a possible mechanism for regulation of norepinephrine synthesis. *Molec. Pharmacol.* **3**, 549–555.

Speden, R. N. (1964). Electrical activity of single smooth muscle cells of the mesenteric artery produced by splanchnic nerve stimulation of the guinea-pig. *Nature, Lond.* **202** 193–194.

—— (1967). Adrenergic transmission in small arteries. *Nature, Lond.* **216**, 289–290.

Sperelakis, N. (1962). Ca^{45} and Sr^{89} movements with contraction of depolarized smooth muscle. *Am. J. Physiol.* **203**, 860–866.

Sperelakis, N. & Hoshiko, T. (1961). Electrical impedance of cardiac muscle. *Circulation Res.* **9**, 1280–1283.

Sperelakis, N. & Prosser, C. L. (1959). Mechanical and electrical activity in intestinal smooth muscle. *Am. J. Physiol.* **196**, 850–856.

Sperelakis, N. & Tarr, M. (1965). Weak electrotonic interaction between neighbouring visceral smooth muscle cells. *Am. J. Physiol.* **208**, 737–747.

Spero, L. (1967). The action of drugs on ionic fluxes in smooth muscle. Thesis. Cambridge University.

Stahl, M. (1963). Electronenmikroskopische Untersuchungen über die vegetative Innervation der Bauchspeicheldrüse. *Z. mikrosk.-anat. Forsch.* **70**, 63–102.

Stafford, A. (1966). Potentiation of adenosine and the adenine nucleotides by dipyridamole. *Br. J. Pharmac. Chemother.* **28**, 218–227.

Stämpfli, R. & Nishie, K. (1956). Effects of calcium-free solutions on membrane-potential of myelinated fibers of the Brazilian frog, *Leptodactylus ocellatus. Helv. physiol. pharmac. Acta* **14**, 93–104.

Standaert, F. G. (1963). Post-tetanic repetitive activity in the cat soleus nerve. *J. gen. Physiol.* **47**, 53–70.

—— (1964). The action of d-tubocurarine on the motor nerve terminal. *J. Pharmac. exp. Ther.* **143**, 181–186.

Standaert, F. G. & Adams, J. E. (1965). The actions of succinylcholine on the mammalian motor nerve terminal. *J. Pharmac. exp. Ther.* **149**, 113–123.

Steedman, W. M. (1966). Micro-electrode studies on mammalian vascular muscle. *J. Physiol.* **186**, 382–400.

Stehle, R. L. & Ellsworth, H. C. (1937). Dihydroxyphenyl ethanolamine (arterenol) as a possible sympathetic hormone. *J. Pharmac. exp. Ther.* **59**, 114–121.

Stephenson, E. W. (1967). Cation regulation in the smooth muscle of frog stomach. *J. gen. Physiol.* **50**, 1517–1546.

Stjärne, L. (1964). Studies of catecholamine uptake, storage and release mechanisms. *Acta physiol. scand.* **62**, Suppl. 228, 5–97.

—— (1966). Studies of noradrenaline biosynthesis in nerve tissue. *Acta physiol. scand.* **67**, 441–454.

Stjärne, L. & Lishajko, F. (1967). Localization of different steps in noradrenaline synthesis to different fractions of a bovine splenic nerve homogenate. *Biochem. Pharmac.* **16**, 1719–1728

Stjärne, L, Roth, R. H. & Lishajko, F. (1967). Noradrenaline formation from dopamine in isolated subcellular particles from bovine splenic nerve. *Biochem. Pharmac.* **16**, 1729–1739.

Stöhr, Ph. Jr. (1954). Zusammenfassende Ergebnisse über die Endigungsweise des vegetativen Nervensystems. II. Das Verhalten des nervösen Terminalretikulums zum Muskelgewebe. *Acta neuroveg.* (*Wien*) **10**, 62–109.

Strömblad, B. C. R. & Nickerson, M. (1961). Accumulation of epinephrine and norepinephrine by some rat tissues. *J. Pharmac. exp. Ther.* **134**, 154–159.

Strong, K. C. (1938). A study of the structure of the media of the distributing arteries by the method of microdissection. *Anat. Rec.* **72**, 151–167.

Su, C. & Bevan, J. A. (1965). The electrical response of pulmonary artery muscle to acetylcholine, histamine and serotonin. *Life Sci.* **4**, 1025–1029.

—— (1966). Electrical and mechanical responses of pulmonary artery muscle to neural and chemical stimulation. *Biblphie. anat.* **8**, 30–34.

Su, C., Bevan, J. A. & Ursillo, R. C. (1964). Electrical quiescence of pulmonary artery smooth muscle during sympathomimetic stimulation. *Circulation Res.* **15**, 20–27.

Suchard, E. (1901a). Observations nouvelles sur la structure du tronc de la veine porte du rat, du lapin, du chein, de l'homme et du poulet. *C. R. Soc. Biol.* (*Paris*) **53**, 192–194.

—— (1901b). De la disposition et de la forme des cellules endothéliales du tronc de la veine porte. *C. R. Soc. Biol.* (*Paris*) **53**, 300–302.

Šumbera, J., Bravený, P. & Kruta, V. (1967). Effects of temperature on the duration and velocity of myocardial contraction in normal and low calcium media. *Archs int. Physiol. Biochim*, **75**, 261–276.

Sunano, S. & Miyazaki, E. (1968). The initiation of contraction by extracellular calcium in smooth muscle of the guinea-pig taenia coli. *Experientia* **24**, 364–365.

Sutherland, E. W. & Rall, T. W. (1960). The relation of adenosine-3′,5′-phosphate and phosphorylase to the actions of catecholamines and other hormones. *Pharmac. Rev.* **12**, 265–299.

Sutherland, E. W. Rall, T. W. & Menon, T. (1962). Adenylcyclase. I. Distribution, preparation and properties. *J. biol. Chem.* **237**, 1220–1227.

Sutter, M. C. (1965). The pharmacology of isolated veins. *Br. J. Pharmac. Chemother.* **24**, 742–751.

Suwa, K. (1962). An electronmicroscope study on the aortic media in human with special reference to the innervation of the tunica media. *Acta Med. Okayama*, **16**, Suppl. 1–13.

Suzuki, T. & Inomata, H. (1964). The inhibitory post-synaptic potential in intestinal smooth muscle investigated with intracellular microelectrodes. *Tohoku J. exp. Med.* **82**, 48–51.

Suzuki, T., Nishiyama, A. & Inomata, H. (1963). Effect of tetraethyl ammonium ion on the electrical activity of smooth muscle cell. *Nature, Lond.* **197**, pp. 908–909.

Suzuki, T. & Vogt, W. (1965). Prostaglandine in einem Darmstoffpräparat aus Froschdarm. *Arch exp. Path. Pharmak.* **252**, 68–78.

Swift, T. J. & Fritz, O. G., Jr. (1969). A proton spin-echo study of the state of water in frog nerves. *Biophys. J.* **9**, 54–59,

Tafuri, W. L. (1964). Ultrastructure of the vesicular component in the intramural nervous system of the guinea-pig's intestines. *Z. Naturf.* **19B** 622–625.

Tafuri, W. L. & Raick, A. (1964). Presence of 5-hydroxytryptamine in the intramural nervous system of guinea-pig's intestine. *Z. Naturf.* **19B**, 1126–1128.

Takeda, H. & Csapo, A. I. (1961). Uterine function in experimental missed abortion. *Biol. Bull.* **121**, 410.

Takeda, H. & Nakanishi, H. (1965). Electrical and mechanical activity of the guinea-pig vas deferens *in situ*. *Ann. Report Shionogi Res. Lab.* No. **15**, 163–166.

Takeshige, C. & Volle, R. L. (1963). Asynchronous post-ganglionic firing from the cat superior cervical ganglion treated with neostigmine. *Br. J. Pharmac. Chemother.* **20**, 214–220.

Takeuchi, A. & Takeuchi, N. (1959). Active phase of frog's end-plate potential. *J. Neurophysiol.* **22**, 395–411.

—— (1960). On the permeability of end-plate membrane during the action of transmitter. *J. Physiol.* **154**, 52–67.

Tamai, T. & Prosser, C. L. (1966). Differentiation of slow potentials and spikes in longitudinal muscle of cat intestine. *Am. J. Physiol.* **210**, 452–458.

Tanaka, I. & Sasaki, Y. (1966). On the electrotonic spread in cardiac muscle of the mouse. *J. gen. Physiol.* **49**, 1089–1110.

Tasaki, I. & Hagiwara, S. (1957). Capacity of muscle fibre membrane. *Am. J. Physiol.* **188**, 423–429.

Taxi, J. (1961). Sur l'innervation des fibres musculaires lisses de l'intestin de souris. *C. r. hebd. seanc. Acad. Sci., Paris* **252**, 331–333.

—— (1965). Contribution à l'étude des connexions des neurones moteurs du systeme nerveux autonome. *Naturelles Zoologie.* 12ᵉ série. **VII**, 413–674.

Taxi, J. & Droz, B. (1966). Etude de l'incorporation de noradrénaline-³H (NA-³H) et de 5-hydroxytryptophane-³H (5-HTP-³H) dans les fibres nerveuses du canal déférent et de l'intestin. *C. r. hebd. Séance. Acad. Sci., Paris D.* **263**, 1237–1240.

Tedeschi, R. E. (1954). Atropine-like activity of some anticholinesterases on the rabbit atria. *Br. J. Pharmac. Chemother.* **9**, 367–369.

Templeton, R. D. & Lawson, H. (1931). Studies in the motor activity of the large intestine. I. Normal motility in the dog, recorded by the tandem balloon method. *Am. J. Physiol.* **96**, 667–676.

Tennyson, V. M., Brzin, M. & Duffy, P. (1966). Cholinesterase localization in the developing nervous system of the rabbit embryo and human fetus by electron microscopic histochemistry. *Anat. Rec.* **154**, 432.

Theobald, G. W. (1965). The part played by nerve impulses in the regulation of uterine activity in man. In *Muscle.* Ed. Paul, W. M., Daniel, E. E., Kay, C. M. & Monckton, G. pp. 363–373. Oxford: Pergamon Press.

Teplov, S. I. & Vasil'eva, L. I. (1968). Cholinergic mechanism of coronary vasodilatation. *Bull. Biol. Méd. exp. U.S.S.R.* **65**, 119–122. (English Translation.)

Thaemert, J. C. (1959). Intercellular bridges as protoplasmic anastomoses between smooth muscle cells. *J. biophys. biochem. Cytol.* **6**, 67–70.

—— (1963). The ultrastructure and disposition of vesiculated nerve processes in smooth muscle. *J. cell Biol.* **16**, 361–377.

—— (1966). Ultrastructural interrelationships of nerve processes and smooth muscle cells in three dimensions. *J. cell. Biol.* **28**, 37–49.

Thiersch, J. B., Landa, J. F. & West, T. C. (1959). Transmembrane potentials in the rat myometrium during pregnancy. *Am. J. Physiol.* **196**, 901–904.

Thoenen, H., Tranzer, J. P., Hürlimann, A. & Haefely, W. (1966). Untersuchungen zur Frage eines cholinergischen Gliedes in der postganglionären sympathischen Transmission. *Helv. physiol. pharmac. Acta* **24**, 229–246.

Thompson, J. W. (1961). The nerve supply to the nictitating membrane of the cat. *J. Anat.* **95**, 371–385.

Tichowa, W. A. (1961). Die Innervation des Ziliarmuskels. *Z. mikrosk.-anat. Forsch.* **67**, 452–468.

Todrick, A. (1954). The inhibition of cholinesterases by antagonists of acetylcholine and histamine. *Br. J. Pharmac. Chemother.* **9**, 76–84.

Toman, J. E. P., Woodbury, J. W. & Woodbury, L. A. (1947). Mechanism of nerve conduction block produced by anticholinesterases. *J. Neurophysiol.* **10**, 429–441.

Tomita, T. (1966a). Electrical responses of smooth muscle to external stimulation in hypertonic solution. *J. Physiol.* **183**, 450–468.

—— (1966b). Membrane capacity and resistance of mammalian smooth muscle. *J. theor. Biol.* **12**, 216–227.

—— (1966c). Electrical properties of the smooth muscle of the guinea-pig vas deferens. *J. Physiol.* **186**, 9–10P.

—— (1967a). Current spread in the smooth muscle of the guinea-pig vas deferens. *J. Physiol.* **189**, 163–176.

—— (1967b). Spike propagation in the smooth muscle of the guinea-pig taenia coli. *J. Physiol.* **191**, 517–527.

—— (1969). The longitudinal tissue impedance of the smooth muscle of guinea-pig taenia coli. *J. Physiol.* **201**, 145–159.

Torack, R. M. & Barrnett, R. J. (1962). Fine structural localization of cholinesterase activity in the rat brain stem. *Expl Neurol.* **6**, 224–244.

Torp, A. (1961). Histamine and mast cells in nerves. *Medicina Experimentalis,* **4**, 180–182.

Tousimis, A. J. & Fine, B. S. (1959). Ultrastructure of the iris: an electron microscopic study. *Proceedings Amer. J. Opthal.* **48**, 397–417.
—— (1961). Electron microscopy of the pigment epithelium of the iris. In *The Structure of the Eye.* Ed. Smelser, G. K. pp. 441–452. New York: Academic Press.
Tramezzani, J. H., Chiocchio, S. & Wasserman, G. F. (1964). A technique for light and electronmicroscopic identification of adrenalin and noradrenalin storing cells. *J. Histochem. Cytochem.* **12**, 890–899.
Tranzer, J. P., Da Prada, M. & Pletscher, A. (1966). Ultrastructural localisation of 5-hydroxytryptamine in blood platelets. *Nature, Lond.* **212**, 1574–1575.
Tranzer, J. P. & Thoenen, H. (1967a). Significance of 'empty vesicles' in post-ganglionic sympathetic nerve terminals. *Experientia* **23**, 123–124.
—— (1967b). Electronmicroscopic localisation of 5-hydroxytryptamine (3,4,5-trihydroxy-phenylethylamine) a new 'false' sympathetic transmitter. *Experientia* **23**, 743–745.
—— (1968). Various types of amine-storing vesicles in peripheral adrenergic nerve terminals. *Experientia,* **24**, 484–486.
Travis, R. H. & Sayers, G. (1965). Adrenocorticotropic hormone: adrenocortical steroids and their synthetic analogs. In: *The Pharmacological Basis of Therapeutics,* 3rd. ed. Ed. Goodman, L. S. & Gilman, A. London, Macmillan.
Trelstad, R. L., Revel, J. P. & Hay, E. D. (1966). Tight junctions between cells in the early chick embryo as visualised with the electronmicroscope. *J. cell. Biol.* **31**, C6.
Trendelenburg, P. (1912). Physiologische und pharmakologische Untersuchungen an der isolierten Bronchialmuskulatur. *Arch. exp. Path. Pharmak.* **69**, 79–107.
—— (1917). Physiologische und pharmakologische Versuche über die Dünndarmperistaltik. *Arch. exp. Path. Pharmak.* **81**, 55–129.
—— (1927). Bewegungen des Darmes. In *Handb. norm. u. pathol. Physiol.* **3**, pp. 452–471. Berlin: Julius Springer.
Trendelenburg, U. (1954). The action of histamine and pilocarpine on the superior cervical ganglion and the adrenal glands of the cat. *Br. J. Pharmac. Chemother.* **9**, 481–487.
—— (1961). Modification of the effect of tyramine by various agents and procedures. *J. Pharmac. exp. Ther.* **134**, 8–17.
—— (1963). Supersensitivity and subsensitivity to sympathomimetic amines. *Pharmac. Rev.* **15**, 225–276.
—— (1965). Supersensitivity by cocaine to dextrorotary isomers of norepinephrine. *J. Pharmac. exp. Ther.* **148**, 329–338.
—— (1966a). Mechanisms of supersensitivity and subsensitivity to sympathomimetic amines. *Pharmac. Rev.* **18**, 629–640.
—— (1966b). Supersensitivity to norepinephrine induced by continuous nerve stimulation. *J. Pharmac. exp. Ther.* **151**, 95–102.
Trial, W. M. (1963). Intracellular studies on vascular smooth muscle. *J. Physiol.* **167**, 17–18P.
Troshin, A. S. (1966). *Problems of Cell Permeability.* Oxford: Pergamon Press.
Trouton, F. T. & Rankine, A. C. (1904). On the stretching and torsion of lead wire beyond elastic limit. *Phil. Mag.* (6), **8**, 538–556.
Truelove, S. C. (1966). Movements of the large intestine. *Physiol. Rev.* **46**, 457–512.
Tsai, T. H. & Fleming, W. W. (1964). The adrenotropic receptors of the cat uterus. *J. Pharm. Pharmacol.* **143**, 268–272.
Türker, R. K., Page, I. H. & Khairallah, P. A. (1967). Angiotensin alteration of sodium fluxes in smooth muscle. *Arch. int. Pharmacodyn Thér.* **165**, 394–404.
Tuttle, R. S. (1965). Relationship between blood histamine and centrally evoked hypotensive response. *Am. J. Physiol.* **209**, 745–750.
—— (1966). Histaminergic component in the baroreceptor reflex of the pyramidal cat. *Fedn Proc.* **25**, 1593–1595.
—— (1967). Physiological release of histamine-^{14}C in the pyramidal cat. *Am. J. Physiol.* **213**, 620–624.
Tuttle, R. S. & Connor, R. S. (1965). An intensive hypotensive response evoked by medullary stimulation in the pyramidal cat. *Am. J. Physiol.* **208**, 693–697.
Twarog, B. M. (1954). Responses of a molluscan smooth muscle to acetylcholine and 5-hydroxytryptamine. *J. cell. comp. Physiol.* **44**, 141–164.

Twarog, B. M. (1968). Possible mechanism of action of serotonin on molluscan muscle. *Adv. Pharmacol.* Ed. Garattini, S. & Shore, P. A. **6B**, London: Academic Press.

Uchizono, K. (1964). On different types of synaptic vesicles in the sympathetic ganglia of amphibia. *Jap. J. Physiol.* **14**, 210–220.

Udenfriend, S. (1964). Biosynthesis of the sympathetic neurotransmitter, norepinephrine. *Harvey Lect.* **Ser. 60**, p. 57–83.

—— (1966). Tyrosine hydroxylase. *Pharmac. Rev.* **18**, 43–51.

Uehara, Y. & Burnstock, G. (1969). Demonstration of 'gap junctions' between smooth muscle cells of the sheep ureter. *J. Cell. Biol.* (in press).

Urakawa, N. & Holland, W. C. (1964). Ca^{45} uptake and tissue calcium in K-induced phasic and tonic contraction in taenia coli. *Am. J. Physiol.* **207**, 873–876.

Ursillo, R. C. (1961). Electrical activity of the isolated nerve-urinary bladder strip preparation of the rabbit. *Am. J. Physiol.* **201**, 408–412.

Ursillo, R. C. & Clark, P. B. (1956). The action of atropine on the urinary bladder of the dog and on isolated nerve-bladder strip preparation of the rabbit. *J. Pharmac. exp. Ther.* **118**, 338–347.

Ussing, H. H. (1949). Transport of ions across cellular membrane. *Physiol. Rev.* **29**, 127–155.

Utterback, R. A. (1944). The innervation of the spleen. *J. comp. Neurol.* **81**, 55–68.

Uvnäs, B. (1967). Cholinergic vasodilator innervation to skeletal muscles. *Circulation Res.* **20–21**, Suppl. 1, 83–90.

Van der Kloot, W. G. & Dane, B. (1964). Conduction of the action potential in the frog ventricle. *Science, N. Y.* **146**, 74–75.

Van der Pol. M. C. (1956). The effect of some sympathomimetics in relation to the two receptor theory. *Acta physiol. pharmacol. neerl.* **4**, 524–531.

Van Orden, L. S. III., Bensch, K. G. & Giarman, N. J. (1967). Histochemical and functional relationships of catecholamines in adrenergic nerve endings. II. Extra vesicular norepinephrine. *J. Pharmac. exp. Ther.* **155**, 428–439.

Van Orden, L. S. III., Bloom, F. E., Barrnett, R. J. & Giarman, N. J. (1966). Histochemical and functional relationships of catecholamines in adrenergic nerve endings. I. Participation of granular vesicles. *J. Pharmac. exp. Ther.* **154**, 185–199.

Van Orden, L. S., Vugman, I. & Giarman, N. J. (1965). 5-hydroxytryptamine in single neoplastic mast cells: A microscopic spectrofluorometric study. *Science, N. Y.* **148**, 642–644.

Van Rossum, J. M. & Mujić, M. (1965). Classification of sympathomimetic drugs on the rabbit intestine. *Arch. Int. Pharmacodyn. Thér.* **155**, 418–431.

Vane, J. R. (1957). A sensitive method for the assay of 5-hydroxytryptamine. *Br. J. Pharmac. Chemother.* **12**, 344–349.

—— (1959). The relative activities of some tryptamine analogues on the isolated rat stomach strip preparation. *Br. J. Pharmac. Chemother.*, **14**, 87–98.

Van Liew, H. D. (1967). Graphic analysis of aggregates of linear and exponential processes. *J. theor. Biol.* **16**, 43–53.

Vaughan Williams, E. M. (1954). The mode of action of drugs upon intestinal motility. *Pharmacol. Rev.* **6**, 159–190.

—— (1958a). Some observations concerning the mode of action of acetylcholine in isolated rabbit atria. *J. Physiol.* **140**, 327–346.

—— (1958b). The mode of action of quinidine on isolated rabbit atria interpreted from intracellular potential records. *Br. J. Pharmac. Chemother.* **13**, 276–287.

Vayo, H. W. (1965). Determination of the electrical parameters of vertebrate visceral smooth muscle. *J. theor. Biol.* **9**, 263–277.

Veach, H. O. (1925). Studies on the innervation of smooth muscle. I. Vagus effects on the lower end of the oesophagus, cardia and stomach of the cat, and the stomach and lung of the turtle in relation to Wedensky inhibition. *Am. J. Physiol.* **71**, 229–264.

Veldstra, H. (1956). Synergism and potentiation with special reference to the combination of structural analogues. *Pharmacol. Rev.* **8**, 339–388.

Verity, M. A. & Bevan, J. A. (1966). A morphopharmacologic study of vascular smooth muscle innervation. *Biblphie. anat.* **8**, 60–65.

Verity, M. A. & Bevan, J. A. (1968). Fine structural study of the terminal effector plexus, neuromuscular and intermuscular relationships in the pulmonary artery. *J. Anat. Lond.* **103**, 49–63.

Verity, M. A., Bevan, J. A. & Ostrom, R. J. (1966). Plurivesicular nerve endings in the pulmonary artery. *Nature, Lond.* **211**, 537–538.

Vijai, K. K. & Foster, J. F. (1967). The amphoteric behaviour of bovine plasma albumin. Evidence for masked carboxylate groups in the native protein. *Biochemistry, N.Y.* **6**, No. 4. 1152–1159.

Villamil, M. F., Rettori, V., Barajas, L. & Kleeman, C. R. (1968). Extracellular space and the ionic distribution in the isolated arterial wall. *Am. J. Physiol.* **214**, 1104–1112.

Villamil, M. F., Rettori, V., Yeyati, N. & Kleeman, C. R. (1968). Chloride exchange and distribution in the isolated arterial wall. *Am. J. Physiol.* **215**, 833–839.

Vogt, M. (1965). Transmitter released in cat uterus by stimulation of the hypogastric nerves. *J. Physiol.* **179**, 163–171.

Volle, R. L. (1966). Muscarinic and nicotinic stimulant actions at autonomic ganglia. *International Encyclopedia of Pharmacology and Therapeutics.* Ed. Karczmar, A. G. Section 12, Ganglionic blocking and stimulating agents. Oxford: Pergamon.

Wachholder, K. (1921). Haben die rhythmischen Spontankontraktionen der Gefässe einen nachweisbaren Einfluss auf den Blutstrom? *Pflügers Arch. ges. Physiol.* **190**, 222–229.

Waldeyer, W. (1891). Ueber einige neuere Forschungen im Gebiete der Anatomie des Centralnervensystems. *Dt. med. Wschr.* **17**, 1213–1218.

Wallach, M. B., Goldberg, A. M. & Shideman, F. E. (1967). The synthesis of labelled acetylcholine by the isolated cat heart and its release by vagal stimulation. *Int. J. Neuropharmac.* **6**, 317–323.

Wallentin, I. (1966). Studies on intestinal circulation. *Acta physiol. scand.* **69**, Suppl. 279, 1–38.

Wardell, W. M. & Tomita, T. (1967). Coupling resistance of double barrelled microelectrodes. *Nature, Lond.* **216**, 1007–1008.

Waser, P. G. (1966). Autoradiographic investigations of cholinergic and other receptors in the motor endplate. *Adv. Drug Res.* **3**, 81–120.

Washizu, Y. (1966). Grouped discharges in ureter muscle. *Comp. Biochem. Physiol.* **19**, 713–728.

—— (1968). Conversion of ureteral action potential by metal ions. *Comp. Biochem. Physiol.* **24**, 301–305.

Watanabe, A & Grundfest, H. (1961). Impulse propagation at the septal and commissural junctions of crayfish lateral giant axons. *J. gen. Physiol.* **45**, 267–308.

Waterson, J. G. (1967). Fluorescent structures in the rabbit dental pulp. *Aust. J. exp. Biol. med. Sci.* **45**, 309–311.

Waterson, J. G. & Smale, D. E. (1967). Location of noradrenergic structures in the central artery of the rabbit ear. *Aust. J. exp. Biol. med. Sci.* **45**, 301–308.

Waud, D. R. (1962). 'Cross-protection' method as a means of differentiating receptors. *Nature, Lond.* **196**, 1107–1108.

—— (1968). Pharmacological receptors. *Pharmacol. Rev.* **20**, 49–88.

Waugh, W. H. (1962). Adrenergic stimulation of depolarized arterial muscle. *Circulation Res.* **11**, 264–276.

Weatherall, M. (1962a). Quantitative analysis of movements of potassium in rabbit auricles. *Proc. R. Soc. B.* **156**, 57–82.

—— (1962b). Location of fractions of potassium in rabbit auricles. *Proc. R. Soc. B.* **156**, 83–95.

Weber, A. & Herz, R. (1963). The binding of calcium to actomyosin systems in relation to their biological activity. *J. biol. Chem.* **238**, 599–605.

Weber, A., Herz, R. & Reiss, I. (1964). The regulation of myofibrillar activity by calcium. *Proc. R. Soc. B.* **160**, 489–501.

Wegmann, A. & Kako, K. (1961). Particle bound and free catecholamines in dog hearts and the uptake of injected norepinephrine. *Nature, Lond.* **192**, 978.

Weidmann, S. (1951). Electrical characteristics of *Sepia* axons. *J. Physiol.* **114**, 372–381.

—— (1952). The electrical constants of Purkinje fibres. *J. Physiol.* **118**, 348–360.

Weidmann, S. (1955a). The effect of the cardiac membrane potential on the rapid availability of the sodium-carrying system. *J. Physiol.* **127**, 213–224.

— (1955b). Effects of calcium ions and local anaesthetics on electrical properties of purkinje fibres. *J. Physiol.* **129**, 568–582.

Weiner, N. & Alousi, A. (1967). Influence of nerve stimulation on rate of synthesis of norepinephrine. *Fedn Proc.* **25**, 259.

Weiner, N. & Jardetzky, O. (1964). A study of catecholamine nucleotide complexes by nuclear magnetic resonance spectroscopy. *Arch. exp. Path. Pharmak.* **248**, 308–318.

Weinstein, H. J. & Ralph, P. H. (1951). Myofilaments from smooth muscle. *Proc. Soc. exp. Biol. Med. N.Y.* **78**, 614–615.

Weiss, G. B., Coalson, R. E. & Hurwitz, L. (1961). K transport and mechanical responses of isolated longitudinal smooth muscle from guinea-pig ileum. *Am. J. Physiol.* **200**, 789–793.

Weiss, G. B. & Hurwitz, L. (1963). Physiological evidence for multiple calcium sites in smooth muscle. *J. Gen. Physiol.* **47**, 173–187.

Weiss, P. (1961). The concept of perpetual neuronal growth and proximo-distal substance convection. In *Regional Neurochemistry.* Ed. Kety, S. S. & Elkes, J. pp. 220–242. New York: Pergamon Press.

Weiss, P. & Hiscoe, H. B. (1948). Experiments on the mechanism of nerve growth. *J. exp. Zool.* **107**, 315–395.

Weiss, P., Taylor, A. C. & Pillai, P. A. (1962). The nerve fiber as a system in continuous flow: microcinematographic and electronmicroscopic demonstrations. *Science, N.Y.* **136**, 330.

Weitz, W. & Vollers, W. (1925). Studien über Magenbewegungen. *Z. ges. exp. Med.* **47**, 42–69.

—— (1926). Über rhythmische Kontraktionen der glatten Muskulatur an verschiedenen Organen (Magen, Darm, Harnblase, Scrotum, Penis, Uterus, Milz und Gefässe). *Z. ges. exp. Med.* **52**, 723–746.

Wellens, D. (1964). Inhibition of norepinephrine-induced reflex vasodilatation in dog hind limb. *Archs int. Pharmocodyn. Thér.* **151**, 281–285.

Wells, J. A., Mercer, T. H., Gray, J. S. & Ivy, A. C. (1942). The motor innervation of the colon. *Am. J. Physiol.* **138**, 83–93.

Welsh, J. H. (1957). Serotonin as a possible neurohumoral agent: evidence obtained in lower animals. *Ann. N.Y. Acad. Sci.* **66**, 618–630.

—— (1968). Distribution of serotonin in the nervous system of various animal species. *Adv. Pharmacol.* Ed. Garattini, S. & Shore, P. A. Vol. 6A, pp. 171–188. London: Academic Press.

Welsh, J. H. & Hyde, J. E. (1944). Acetylcholine content of the myenteric plexus and resistance to anoxia. *Proc. Soc. exp. Biol. Med.* **55**, 256–257.

Werle, E. & Schievelbein, H. (1964). Über eine Hemmung der 5-Hydroxytryptamin- und Nikotinwirkung am isolierten Darm durch Adenosin-derivate. *Biochem. Pharmac.* **13** 855–859.

Werman, R., McCann, F. V. & Grundfest, H. (1961). Graded and all-or-none electrogenesis in arthropod muscle. I. Effects of alkali-earth cations on the neuromuscular system of *Romalea microptera. J. gen. Physiol.* **44**, 979–995.

Werner, G. (1960a). Neuromuscular facilitation and antidromic discharges in motor nerves; their relation to activity in motor nerve terminals. *J. Neurophysiol.* **23**, 171–187.

—— (1960b). Generation of antidromic activity in motor nerves. *J. Neurophysiol.* **23**, 453–461.

Wesemann, Von W. & Zilliken, F. (1967). Receptors of neurotransmitters 11. Sialic acid metabolism and the serotonin induced contraction of smooth muscle. *Biochem. Pharmac.* **16**, 1773–1779.

—— (1968). Serotoninrezeptor und Neuraminsäurestoffwechsel der glatten Muskulatur. *Hoppe-Seylers Z. Physiol. Chem.* **349**, 823–830.

Whitby, L. G., Axelrod, J. & Weil-Malherbe, H. (1961). The fate of H³-norepinephrine in animals. *J. Pharmac. exp. Ther.* **132**, 193–201.

Whitby, L. G., Hertting, G. & Axelrod, J. (1960). Effect of cocaine on the disposition of noradrenaline labelled with tritium. *Nature, Lond.* **187**, 604–605.

Whittaker, V. P. (1966). Catecholamine storage particles in central nervous system. *Pharmac. Rev.* **18**, 401–412.

Widdicombe, J. G. (1963). Regulation of tracheobronchial smooth muscle. *Physiol. Rev.* **43**, 1–37.

Wiedeman, M. P. (1963). Patterns of the arteriovenous pathways. In *Handbook of Physiology*. Section 2: Circulation **II**. p. 27. Ed. Hamilton, W. F. & Daw, P. Washington: American Physiological Society.

Wienbeck, M., Golenhofen, K. & Lammel, E. (1968). Der Effekt von CO_2 auf die Spontanaktivität von isolierter glatter Muskulatur (Taenia coli des Meerschweinchens). *Pflügers Arch. ges. Physiol.* **300**, R 78.

Wilbrandt, W. & Rosenberg, T. (1961). The concept of carrier transport and its corollaries in pharmacology. *Pharmac. Rev.* **13**, 109–183.

Wilkie, D. R. (1954). Facts and theories about muscle. *Progr. Biophys.* **4**, 288–324.

—— (1956). The mechanical properties of muscle. *Br. med. Bull.* **12**, No. 3. 177–182.

—— (1966). Muscle. *Ann. Rev. Physiol.* **28**, 17–38.

Willems, J. L., Bernard, P. J., Delaunois, A. L. & De Schaepdryver, A. F. (1965). Adrenergic receptors in the progesterone dominated rabbit uterus. *Arch. Int. Pharmacodyn. Thér.* **157**, 243–250.

Williams, I. (1967). Mass movements (mass peristalsis) and diverticular disease of the colon. *Br. J. Radiol.* **40**, 2–14.

Williams, T. H. W. (1967). Electron microscopic evidence for an autonomic interneuron. *Nature, Lond.* **214**, 309–310.

Wilson, A. B. (1964). *Beta* sympathetic inhibitory receptors in the small intestine of the guinea-pig. *J. Pharm. Pharmac.* **16**, 834–835.

Winter, J. C. & Gessner, P. K. (1968). Phenoxybenzamine antagonism of tryptamines, their indene isosteres and 5-hydroxytryptamine in the rat stomach fundus preparation. *J. Pharmac. exp. Ther.* **162**, 286–293.

Winter, J. C., Gessner, P. K. & Godse, D. D. (1967). Synthesis of some 3-indenealkylamines. Comparison of the biological activity of 3-indenealkylamines and 3-benzo(b)thiophenealkylamines with their tryptamine isosteres. *J. med. Chem.* **10**, 856–859.

Witanowsky, W. R. (1925). Über humorale Übertragbarkeit der Herznervenwirkung. VIII, Mitteilung. *Pflügers Arch. ges. Physiol.* **208**, 694–704.

Wolfe, D. E. & Potter, L. T. (1963). Localization of norepinephrine in the atrial myocardium. *Anat. Rec.* **145**, 301.

Wolfe, D. E., Potter, L. T., Richardson, K. C. & Axelrod, J. (1962). Localizing tritiated norepinephrine in sympathetic axons by electron microscopic autoradiography. *Science, N.Y.* **138**, 440–444.

Wong, K. C. & Long, J. P. (1961). Autonomic blocking properties of hemicholinium (HC-3). *J. Pharmac. exp. Ther.* **133**, 211–215.

Wood, J. G. (1966). Electronmicroscopic localization of amines in central nervous tissue. *Nature, Lond.* **209**, 1131–1133.

Wood, J. G. & Barrnett, R. J. (1963). Histochemical differentiation of epinephrine and norepinephrine granules in the adrenal medulla with the electronmicroscope. *Anat. Rec.* **145**, 301–302.

—— (1964). Histochemical demonstration of norepinephrine at a fine structural level. *J. Histochem. Cytochem.* **12**, 197–209.

Woodbury, R. A. & Abreu, B. E. (1944). Influence of epinephrine upon human gravid uterus. *Am. J. Obstet. Gynec.* **48**, 706–708.

Woodbury, J. W. & Crill, W. E. (1961). On the problem of impulse conduction in the atrium. In *International Symposium on Nervous Inhibition*. pp. 124–135. New York: Pergamon Press.

Woodbury, J. W. & McIntyre, D. M. (1954). Electrical activity of single muscle cells of pregnant uteri studied with intracellular ultramicroelectrodes. *Am. J. Physiol.* **177**, 355–360.

Woolley, D. W. (1958). A probable mechanism of action of serotonin. *Proc. natn. Acad. Sci. U.S.A.* **44**, 197–201.

Woolley, D. W. & Shaw, E. (1953). Yohimbine and ergotoxin as naturally-occurring antimetabolites of serotonin. *Fedn Proc.* **12**, 293.

Woolley, D. W. & Gommi, B. W. (1964). Serotonin receptors: V, Selective destruction by neuraminidase plus EDTA and reactivation with tissue lipids. *Nature, Lond.* **202**, 1074–1075.

—— (1966). Serotonin receptors VI. Methods for the direct measurements of the isolated receptors. *Archs int. Pharmacodyn. Thér,* **159**, 8–17.

Wright, A. M., Moorhead, M. & Welsh, J. H. (1962). Actions of derivatives of lysergic acid on the heart of *Venus mercenaria. Br. J. Pharmac. Chemother.* **18**, 440–450.

Wybauw, R. (1935). Les phénomènes électriques artériels. *Bull. Acad. roy. Méd. Belg.* **15**, 604–626.

Wylie, D. W., Archer, S. & Arnold, A. (1960). Augmentation of pharmacological properties of catecholamines by O-methyl transferase inhibitors. *J. Pharmac. exp. Ther.* **130**, 239–244.

Yamada, H. & Miyake, S. (1960). Electronenmikroskopische Untersuchungen an Nervenfasern in menschlichen Schweissdrüsen. *Z. Zellforsch. mikrosk. Anat.* **52**, 129–139.

Yamaguchi, T. Matsushima, T., Fujino, M. & Nagai, T. (1962). The excitation-contraction coupling of the skeletal muscle and the 'glycerol effect'. *Jap. J. Physiol.* **12**, 129–142.

Yamamoto, T., (1960). Electron microscope investigation on the relationship between the smooth muscle cell of the proc. vermiformis and the autonomic peripheral nerves. *Acta neuroveg. (Wien)* **21**, 406–425.

Yamamoto, I. (1961). An electronmicroscope study on development of uterine smooth muscle. *J. Electronmic.* **10**, 145–160.

—— (1962). An electronmicroscope study of the rabbit arterial wall. Smooth muscle and the intercellular components with special reference to the elastic fibre. *J. Electronmic.* 212–225.

Yamauchi, A. (1964). Electron microscopic studies on the autonomic neuromuscular junction in the taenia coli of the guinea-pig. *Acta anat. Nippon.* **39**, 22–38.

—— (1965). Electronmicroscopic observations on the development of S-A and A-V nodal tissues in the human embryonic heart. *Z. Anat. Entwickl.-Gesch.* **124**, 562–587.

—— (1968). An electron microscope study on the innervation of the trout heart. *J. comp. Neurol.* **32**, 567–588.

Yamauchi, A. & Burnstock, G. (1967). Nerve myoepithelium and nerve-glandular epithelium contacts in the lacrimal gland of the sheep. *J. cell Biol.* **34**, 917–919.

—— (1969a). Post-natal development of smooth muscle cells in the mouse vas deferens. A fine structural study. *J. Anat.* **104**, 1–15.

—— (1969b). Post-natal development of the innervation of the mouse vas deferens. A fine structural study. *J. Anat.* **104**, 17–32.

Yamauchi, A., Orlov, R., McLean, J. R. & Burnstock, G. (1969). Correlation of fine structure, fluorescent histochemistry and electrophysiology of the innervation of the dog retractor penis. In preparation.

Youmans, W. B. (1949). *Nervous and Neurohumoral Regulations of Intestinal Motility.* New York: Interscience publishers.

Zacks, S. & Blumberg, J. M. (1961). The histochemical localization of acetylcholinesterase in the fine structure of neuromuscular junctions of mouse and human intercostal muscle. *J. Histochem. Cytochem.* **9**, 317–324.

Zar M. A. (1966). Factors influencing the output of acetylcholine. D.Phil Thesis. University of Oxford.

Zatzman, M., Stacey, R. W., Randall, J. & Eberstein, A. (1954). Time course of stress relaxation in isolated arterial segments. *Am. J. Physiol.* **177**, 299–302.

Zelander, T., Ekholm, R. & Edlund, Y. (1962). The ultrastructural organisation of the rat exocrine pancreas. III. Intralobular vessels and nerves. *J. Ultrastruct. Res.* **7**, 84–101.

Zepf, S. (1966). Eine serienmässige Calcium-Bestimmung in kleinen Gewebeproben mit dem Spektralfluorometer. *Zeiss-Mitteilungen.* **4**, 43–57.

Zimmerman, B. G. (1966a). Influence of sympathetic stimulation on segmental vascular resistance before and after adrenergic neuronal blockade. *Arch. int. Pharmocodyn. Thér.* **160**, 66–82.

—— (1966b). Sympathetic vasodilatation in the dog's paw. *J. Pharmac. exp. Ther.* **152**, 81–87.

—— (1968). Comparison of sympathetic vasodilator innervation of hindlimb of the dog and cat. *Am. J. Physiol.* **214**, 62–66.

Zuspan, F. P., Cibils, L. A. & Pose, S. V. (1962). Myometrial and cardiovascular responses to alterations in plasma epinephrine and norepinephrine. *Am. J. Obstet. Gynec.* **84**, 841–851.

INDEX